风雨一百年

栉沐

The Remarkable Progress of
Life Science in China:
A Century History

——中国生命科学历史回眸

主编 康乐 饶子和

中国教育出版传媒集团
高等教育出版社·北京

内容简介

生命科学的发展对于生命规律的探索和生物技术的变革具有重要意义。粮食安全、清洁能源、环境保护、健康和老龄化等问题为生命科学提出新的挑战。许多国家都将生命科学确定为优先发展领域，以推动生物经济的发展为抓手，实现社会经济的可持续发展。近年来，我国在生命科学领域取得了令人瞩目的成绩。回顾中国百年来近现代生命科学史，前辈科学家为我国生命科学发展铺垫的基石至今仍然焕发着光彩。本书从*Protein & Cell*的"回忆"（Recollection）专栏选取100篇文章，重点介绍我国近现代生命科学的先驱和奠基人、取得的突出成就以及一些重要历史事件，为我们展开了一幅前辈科学家栉风沐雨、砥砺前行的画卷。

本书采用中英文对照形式，文章作者包括故事人物的亲友、弟子、研究人员等。本书适合对我国生命科学百年发展史感兴趣的相关专业师生和大众阅读。

图书在版编目（CIP）数据

栉风沐雨一百年 ： 中国生命科学历史回眸 ： 汉、英 / 康乐，饶子和主编. -- 北京 ： 高等教育出版社，2023.3

ISBN 978-7-04-059412-6

Ⅰ. ①栉… Ⅱ. ①康… ②饶… Ⅲ. ①生命科学－科学史－中国－汉、英 Ⅳ. ①Q1-0

中国版本图书馆CIP数据核字(2022)第172042号

策划编辑 李冰祥 殷 鸽　责任编辑 殷 鸽 贾祖冰　封面设计 赵 阳　版式设计 杜微言
责任绘图 邓 超　责任校对 吕红颖　责任印制 存 怡

出版发行 高等教育出版社
社 址 北京市西城区德外大街4号
邮政编码 100120
印 刷 鸿博昊天科技有限公司
开 本 787mm×1092mm 1/16
印 张 48.75
字 数 1120千字
购书热线 010-58581118
咨询电话 400-810-0598
网 址 http://www.hep.edu.cn
http://www.hep.com.cn
网上订购 http://www.hepmall.com.cn
http://www.hepmall.com
http://www.hepmall.cn
版 次 2023年3月第1版
印 次 2023年3月第1次印刷
定 价 148.00元

物 料 号 59412-00

ZHIFENGMUYU YIBAINIAN——ZHONGGUO SHENGMING KEXUE LISHI HUIMOU

前言

生命科学的发展对于生命规律的探索和生物技术的变革具有重要意义。粮食安全、清洁能源、环境保护、健康和老龄化等问题为生命科学提出新的挑战。世界许多国家都将生命科学确定为优先发展领域，以推动生物经济的发展为抓手，实现社会经济的可持续发展。近年来，伴随着我国经济的快速发展和国家研究与试验发展经费投入的不断增长，我国在生命科学领域取得了令人瞩目的成绩。根据国际上公开的报告，我国发表在国际生命科学领域期刊的论文数量全球排名第一，引用排名第一；科睿唯安公布的2021年全球高被引科学家名单中，中国有935人次入选（港、澳、台数据除外），全球排名第二。在体现技术创新的专利技术方面，中国生物科技领域的国际专利数量全球排名第二。多个生物高科技公司已进入世界500强企业。近5年来，每年毕业的博士、硕士研究生数量全球第一。这样巨大的变化实际上是用了仅仅100年的时间实现的，或者说最大的变化发生在最近的40年。

Protein & Cell，一个创建于2010年、立足于中国的国际生命科学期刊，专门开辟了一个“回忆”（Recollection）专栏，介绍近100年来对我国生命科学作出重要贡献的人和事。至今该专栏已经发表120多篇相关文章，吸引了许多读者，为我们展开了一幅我国前辈科学家栉风沐雨、砥砺前行的画卷。40多年的改革开放和经济发展，为生命科学与技术的发展提供了强大的动力。当我们回顾百年来的近现代生命科学史，前辈科学家为我国生命科学发展铺垫的基石至今仍然焕发着光彩。为了铭记前辈科学家对我国生命科学发展所作的贡献，我们整理出版了这本书。书中的文章来源于Recollection专栏。为了让青年一代科学家对我国生命科学的历史有一个回顾性的了解，本书重点介绍了20世纪70年代之前的人和事件。由于受到专栏字数的限制，每篇文章只能介绍一个人或事件的一个侧面。因为原文是以英文发表的，为了扩大读者面，我们特邀请了一些科技工作者将其翻译成中文，结集出版。由于涉及的历史比较久远，为了使年轻一代的科学家能全面地了解这段历史，了解中国生命科学发展的历程，深入理解特定时期里特定人和事件的历史背景，我们在本书的前言部分略费笔墨，将中国生命科学史做一个简单的回顾。

在鸦片战争之前，我国经历了2000多年的封建社会，整个社会在旧制度下缓慢发展，对兴起于西方的近代科学和技术缺乏认识。对大众而言，科举制度成为遴选人才和社会等级晋升的唯一渠道。19世纪中后期，如果说有现代意义的科学知识的话，在社会上流行传播的主要是对一些西方生物学著作的翻译和介绍。1859年，李善兰（早期的数学家）与韦廉臣（A. Williamson）合译我国第一本介绍西方近代植物学的著作。1886年，英国传教士艾约瑟（J. Edkins）基于相关著作，用中文比较系统地介绍了动物学、植物学和生理学的初步知识。1889年，颜永京翻译出版了美国心理学家海文（J. Haven）的《心灵学》（*Mental Philosophy*），该书是我国第一部心理学译著。1894年中日甲午战争爆发，中华民族受到空前震动。社会各阶层在深感耻辱和愤恨之余，开始意识到变革社会、救国图存的紧迫性，并且进行了救国救民的各种道路的探索。1897年，著名启蒙思想家严复翻译了赫胥黎的《天演论》，首次将进化论介绍到中国，“物竞天择、适者生存”的理论推动了当时的社会变革和维新变法思想的形成。知识分子阶层开始走出国门，向西方学习科学技术，并将西方科学引进中国。1905年，科举制度的废除标志着封建教育制度的终结。众多知识分子放弃了获取功名利禄的科举之路，将引进西方科学技术的通道打开。随后，我国开始建立新式学堂，向西方学习，自然科学教育兴起。随着西方科学知识的传播，人们逐渐意识到生物学与人民生活和社会需求密切相关，是发展农业和医药卫生、开发资源、提升民众健康的基础学科，因此生物学受到了越来越多青年学者的青睐，他们怀抱着“科学救国”的理念，积极投身于这一领域。

出国留学成为当时向西方学习自然科学的主要途径之一。自1875年开始，我国到欧美和日本留学的学生逐渐增多。受到资金和语言的限制，早期是以留学日本居多。1909年起，我国利用美国退回的庚子赔款资助了一批优秀的年轻人到美国留学。根据双方商定，前四年里，中国每年选派不少于100名的留学生赴美，由此拉开了我国大批留学生到欧美高校和研究机构留学的序幕。1916年，邹秉文在美国康奈尔大学获得植物病理学学士学位后，回国在金陵大学任教。1920年，作为中国第一位生物学博士，秉志从美国回国。1921年，秉志等建立了第一个国人自建的现代意义的大学生物系。1922年，中国第一个生物学的研究机构“中国科学社生物研究所”成立，秉志任所长。1928年，秉志和胡先骕又牵头建立了北平静生生物调查所。成立于1928年6月的中央研究院和成立于1929年9月的国立北平研究院，是当时我国两个最重要的大型综合科研机构，标志着我国科学研究的体制化进程基本完成。成立于1917年的北京协和医学院，是我国当时生物医学研究的重要机构，集中了一大批优秀的科学家，是我国实验生物学的重镇。20世纪30年代，我国生物科学的大学院系、研究机构和主要分支学科学会相继成立，期刊和书籍也得以出版。1937年，社会经济和科技的发展被日本全面侵华战争打断。由于战争，北京协和医学院人才流失非常严重。抗日战争前后返回美国的有药理学家陈克恢、谢和平，1948—1949年赴美国的有吴宪和林可胜等。许多大学和研究机构迁移到大西南，办学和研究条件极为艰苦。第二次世界大战结束后，中国的教育和科研工作得到初步的恢复，1948年，中央研究院评选出生命科学领域的25位生物组院士和1位数理化组院士（吴宪），他们几乎都是抗日战争全面爆发前从欧美国家留学归来的学者（仅罗宗洛1人留学日本）。抗战胜利后，出国留学的人数达到了相当的规模。当时，我国共有5541位海外留学生，大部分在1946—1948年出国，其中以留学欧美国家的人数最多，仅留美学生即占留学生总数的63%。留美学生中，学习生命科学的主要集中在哈佛大学和康奈尔大学等著名高校。早期回国的留学生，在经费奇缺的环境中克服艰难险阻，将西方的科学知识和研究方法引入我国并使之本土化，在我国建立起基本的科研和教学体系，努力培养本土人才，并通过有效的国际合作发展科学，他们的工作为我国生命科学的发展奠定了重要的基础。

中华人民共和国成立之后，前期主要是向苏联和东欧国家派出留学生，同时自主培养自然科学各类专门人才。1950—1960年代，我国派出了1万多名留学生到苏联学习，派出了1000多名留学生到东欧国家学习。1976年之后，高等教育和研究生培养得到恢复，高等院校毕业的本科生和研究生逐年增加。1977年恢复高考，在570万考生中，录取学生仅有25万人。从1977年到2017年，录取的大学本科生大约有5760万人。自1995年以来，我国共招收研究生771万人。1982年，我国只有6人获得博士学位；2002年，当年获得博士学位的人数达到14368人。从1987年到2017年，博士毕业生总数达到70万左右。从1978年开始，我国再次开启选派出国留学生计划，当年有52名青年学生被派往美国的大学和研究机构，这是新中国成立之后第一批赴美留学生。2007年，我国在美国各大学和研究机构的留学生超过35万人。1979年，美籍物理学家、诺贝尔奖获得者李政道教授发起中美物理学联合招生项目（China-United States Physics Examination and Application Program，简称CUSPEA项目），每年选派100多名中国学生赴美国一流大学和科研机构学习物理。之后，美籍分子生物学家、康奈尔大学教授吴瑞发起了中美生物化学联合招生项目（China-United States Biochemistry Examination and Application Program，简称CUSBEA项目）。这一项目从1982年开始到1989年结束，共派遣8届422名留学生。与此同时，国内的教育与科研事业也得到了飞速发展，培养了一大批本土科学家。特别是在20世纪的最后20年中，本土培养的科学家，在科研条件简陋和科研经费缺乏的情况下，承担起振兴祖国科技的重任。他们的贡献很大程度上填补了当时科研和教育队伍代际转移的空隙，奠定了我国生命科学研究赶超世界先进水平的基础。进入21世纪，大批留学生回到祖国，加入教育和科技事业中。投入科学研究和开发的经费在20世纪90年代前有了一定的增长，但总量仍然偏低。即使到了1990年，全国的研发经费也只有125亿元。但是，到了2018年，中国的研究与试验发展投入达到2万亿元，居世界第二。准确地说，我国生命科学的翻天覆地的变化更多应该归于最近的40年。但是，这40年的飞速发展，是过去近100年来前辈科学家的艰苦奋斗的结果。

本书介绍了一批我国近现代生命科学的先驱和奠基人，对今天许多年轻人来说可能非常陌生，例如近代生物学和动物学奠基人秉志，动物学奠基人伍献文，植物学奠基人胡先骕和钱崇澍，微生物学奠基人戴芳澜，细胞生物学奠基人贝时璋，医学奠基人伍连德，生理学奠基人林可胜，生物化学奠基人吴宪，病理学先驱谢和平，药理学奠基人陈克恢，遗传学奠基人陈桢以及农学奠基人邹秉文等著名科学家。他们大多都是早年留学海外，学成回国后在各自的研究领域中作出奠基性的重要贡献，成为我国近代生命科学和各分支学科的创始人。他们培养的许多学生成为20世纪后半叶我国生命科学的领军人物。

读者也可以从本书中了解到我国近现代生命科学取得的突出成就，这些成就都具有重要的国际影响力。例如，伍连德发现并用科学的方法战胜了肺鼠疫，在100多年后的今天依然对公共卫生事业具有借鉴意义。汤飞凡率先发现引起沙眼的病原体——沙眼衣原体，解决了沙眼这一严重威胁世人眼睛健康的流行性疾病。童第周建立了核移植技术并培育出世界上第一群克隆鱼。顾方舟研发生产脊髓灰质炎疫苗，在我国消灭了小儿麻痹症，成为世界公共卫生史上继消灭天花之后的又一创举。王应睐等来自三个单位的科研人员精诚合作，在1965年首次实现了结晶牛胰岛素的全合成。王德宝等科学家历时13年完成酵母丙氨酸转移核糖核酸（酵母丙氨酸tRNA）人工全合成，其组成、序列和生物功能与天然的酵母丙氨酸tRNA完全相同。1971年，屠呦呦发现了可以治疗疟疾的青蒿素，2015年获得诺贝尔生理学或医学奖。1999年，中国科学家参加人类基因组计划，后来6个国家的科学家一起发布项目的完成图。

本书还介绍了我国近现代生命科学发展史的一些重要事件，例如20世纪30年代描述生物学和实验生物学之间的论战、遗传学的重要转折点——1956年青岛遗传学座谈会、中日联合实验室的设立等。同时，读者还可以了解到哈佛大学、康奈尔大学等著名大学与我国现代生命科学的渊源，以及一些国际友人对我国生命科学事业的热心帮助。例如美国生物学家祁天锡在我国工作生活了几十年，在生物学教育、动植物调查、生物学研究建制化等方面推动了我国生物学的发展。“DNA之父”詹姆斯·沃森邀请中国学者到冷泉港实验室访学，并对冷泉港亚洲的设立作出了巨大努力和贡献。

回顾历史，是为了开拓未来。“科学成就离不开精神支撑”，这在早期科学家身上得到了充分的印证。在社会动荡、战火不断的极端困境里，他们为发展祖国的生命科学事业付出了巨大的努力，取得了卓越的成就。支撑他们的是“科学救国”的殷殷爱国情和坚定信念。回顾我国生命科学的发展历程，对我们追寻人才成长的道路、探索科技发展的规律以及国际合作与交流都具有重要的启示。因此，我们新时代的科研工作者要将老一辈科学家的科学精神传承下去，为实现中华民族的伟大复兴而努力奋斗。

感谢*Protein & Cell*编辑部，感谢陈宜瑜院士和罗桂环研究员在成书过程中提出的宝贵意见。中国科学院北京生命科学研究院的同事和高等教育出版社的李冰祥编辑等为本书的编辑和出版付出了辛勤劳动。希望本书能够成为了解生命科学发展史、向世界展示我国学术研究的重要窗口。由于部分英文文章发表时间较早，内容有待更新，因此译者在翻译时会新增或删减部分内容，或对文章进行重新创作，可能存在中、英文文章并不完全一致的情况。同时，近现代生命科学发展历程中群星璀璨，还有诸多作出卓越贡献的科学家没有收录。例如，我国著名的医学教育家和公共卫生学家颜福庆，他创立了湘雅医学院和上海医学院，为我国培养了大批医学人才，是我国近代医学的奠基人。此外，未被收录的科学家还有心理学家潘菽，植物学家张景钺，植物生理学家殷宏章，人类学家吴定良，热带病学家李宗恩，医学家袁贻瑾等。我们热切期盼关心生命科学发展史的朋友撰写相关文章，投稿*Protein & Cell*的Recollection专栏，后续我们也将继续结集出版。

康乐

2022年9月20日于北京

Preface

The development of the life science is of great significance for the exploration of the law of life and the evolution of biotechnology. Importantly, there are urgent challenges for the life science to tackle, such as food security, clean energy, environmental protection, health, and aging. Many countries around the world have identified the life science as a priority area of development, leveraging the promotion of bioeconomics to achieve sustainable social and economic development. With the rapid development of China's economy and a continuous increase in national R&D funding, China has made remarkable achievements in the field of life science. In fact, according to internationally published reports, China ranks first in the world in terms of the number of journal publications in the field of life sciencc, and the highest number of citations. Among all Highly Cited Researchers 2021 reported by Clarivate Analytics, 935 are from China (excluding the data of Hong Kong, Macao and Taiwan), the number of which ranks second in the world. The number of international patents of China, an indicator of technological innovation, ranks second worldwide in the field of biotechnology. Furthermore, several Chinese biotechnology companies have been listed among the world's top 500 enterprises. And in the past five years, China has the largest number of doctoral and master's students graduating annually in the world. Notably, such a great progress took only about 100 years to achieve, especially with the biggest changes occurring in the last 40 years.

Protein & Cell, an international life science journal launched in China in 2010, has dedicated a Recollection column to introduce people and events that have made important contributions to China's life science over the past 100 years. To date, this column has published more than 120 related articles, attracting many readers. These stories unfold a scroll of portraits of our predecessors, who were combed by the wind and washed by the rain while forging ahead to advance the science and technology. It is the reform and opening-up policy combined with rapid economic development in China that has provided such powerful impetus for the boost of the life science and biotechnology over the past 40 years. When we reflect on the history of modern life science of China over the past century, our predecessors' scientists still shine as the cornerstone of the development of life science in China. In keeping with their contributions to life science, we have compiled and published this book based on articles from the Recollection column.

Aiming at providing young scientists a retrospective understanding of the history of China's life science, this book focuses on people and events that had an important influence on China's life science before the 1970s. However, due to word limit in the Recollection column, each article can only introduce one person or a single facet of an event once at a time. To expand readership, we invited researchers to translate these articles into Chinese and publish them as a collection, which were originally published in English. Since the history involved is relatively long, we will review the history of China's life science briefly in the preface. Thus, we can help the young generation of scientists with in-depth understanding of the course of life science development as well as the contexts of specific people and events.

Prior to the Opium War, China had experienced more than 2000 years of a feudal society. Under this system, China developed slowly and lacked understanding of science and technology. For the general public, the imperial examination system was the only channel for talent selection and social rank promotion. And in mid- and late-19th century, the spread of scientific knowledge of modern significance was primarily disseminated through the translation of publications in biology of western countries. For example, in 1859, Shanlan Li (an early Chinese mathematician) and A. Williamson jointly translated China's first book on modern Western botany. In 1886, J. Edkins, a British missionary, systematically introduced preliminary knowledge of zoology, botany, and physiology in Chinese based on relevant works. In 1889, Yongjing Yan translated and published *Mental Philosophy* by American psychologist J. Haven. Notably, this book was the first work of psychology to be translated into Chinese. However, in the Sino–Japanese War of 1894, China was unprecedentedly shocked. Apart from shame and resentment, the Chinese population began to realize the urgency of changing their society, which led them to explore various ways to save their homeland. Regarding such changes, in 1897, Fu Yan, a famous enlightenment thinker, translated Huxley's *Evolution and Ethics* into Chinese, and introduced the theory of evolutionism to China for the very first time. He creatively modified the concept of "Natural selection, survival of the fittest" to justify the reformation of social systems and politics. The theory inspired many insightful people of intellectual class to act and save the nation through Constitutional Reform and Modernization. These intellectuals began to travel abroad to learn science and technology from the West, and introduced these to China in return. With abolition of the imperial examination system in 1905 marking the end of the feudal education system, many Chinese scholars turned to Western science and technology. Subsequently, China began to establish new schools based on Western philosophy. Modern education in the natural sciences emerged. Along with the spread of Western scientific knowledge, the Chinese people gradually realized that biological sciences are what closely relate to people's lives and social needs. The disciplines were fundamental to agriculture, medicine, natural resource development, as well as public health. Therefore, it was favored by an increasing number of young scholars who actively devoted themselves to the field based on the concept of "saving the country through science".

At that time, studying abroad became one of the main ways to learn the natural sciences from the West. The number of Chinese students went to Europe, America, and Japan gradually increased since 1875. In the beginning, most of the Chinese students chose to study in Japan due to funding and language constraints. When it came to 1909, Chinese government began to support outstanding young scholars to study in the United States through the Boxer Indemnity Scholarship program. According to this bilateral agreement, China should send no less than 100 students to the United States every year for the first four years. And it opened the prelude to numerous Chinese students studying in European and American universities and research institutions. In 1916, after receiving the bachelor's degree in plant pathology from Cornell University in the United States, Ping-Wen Tsou (Bingwen Zou) assumed a teaching position at Jinling University in China. As the first Chinese student with PhD degree in biology from the United States, Chih Ping (Zhi Bing) also returned China in 1920, and then established the first modern biology department in a university in China along with his fellow Chinese colleagues in 1921. He also acted as the director of China's first research institution of biology, the Institute of Biology of the Chinese Society of Science, which was officially established one year later. Chih Ping and Hsen-Hsu Hu (Xiansu Hu) went on to establish the Fan Memorial Institute of Biology in 1928. In the same year, the Academia Sinica was established, followed by the National Academy of Peiping in September 1929. These two institutions represented integrated research institutions of national significance, marking the institutionalization of scientific research in China. Another important institution for biomedical research in China was the Peking Union Medical College. Founded in 1917, the institute has brought together many outstanding scientists as the basecamp for experimental biology in China. In the 1930s, university departments, research institutes, and various scholar societies successively launched, with academic journals and books published at an unprecedented rate. But the pace was broken by the Japanese army's full-scale invasion in 1937. It has brought destructive effect to the development of China's society, economy, and technology. It has also taxed Peking Union Medical College with significant loss of talents. For example, the pharmacologists Ko-Kuei Chen (Kehui Chen) and Richard Ho-Ping Sia (Heping Xie) left for the United States during the war, and Hsien Wu (Xian Wu) and Robert Kho-Seng Lim (Kesheng Lin) left in 1948—1949. Many universities and research institutions moved to the southwest part of China, with poor teaching and research conditions. China's education and scientific research were not restored until the end of World War II. In 1948, the Academia Sinica selected 26 academicians in the field of life science, including 25 academicians in the biology group and one in the math/physics/chemistry group (Hsien Wu) . Almost all academicians had studied in Europe and America before full outbreak of the War of Resistance against Japanese Aggression, except for Tsung-Lo Lo (Zongluo Luo) in Japan. Ever since the end of war, the number of students studying abroad had reached a considerable scale. In fact, among all 5541 overseas students, most went abroad between 1946 and 1948. And many studied in Europe and the United States. It is worth mentioning that about 63% of these international students were in the United States, majoring in life science at famous universities such as Harvard University and Cornell University. These returned scholars had overcome arduousness of funding scarcity but managed to introduce and adapt scientific knowledge as well as research methods to China.

They also established the fundamental research and education system to cultivate local talents. Embracing extensive international collaborations, the striving of our pioneers had landed China with the foundation of life science.

After the founding of the People's Republic of China, China started to send students to the Soviet Union and Eastern European countries while also cultivate local talents in various natural sciences disciplines. During the 1950s and 1960s, China sent more than 10000 students to the Soviet Union and more than 1000 students in Eastern European countries. Upon the end of 1976, China had immediately restored higher education and postgraduate training, with gradual increase in the number of undergraduates and postgraduates. In 1977, when college entrance examination was resumed, 5.7 million candidates applied for the tests with 250 thousand were admitted. This number has grown into 57.6 million undergraduate students during 1977—2017. As for the graduate students, 7.71 million postgraduate students were admitted since 1995. While there were only six PhD graduates in 1982, a total of 14368 students had obtained doctoral degree within 2002. And the total number of doctoral graduates was approximately 700 thousand between 1987 and 2017. In 1978, China re-launched the program sending students to study abroad. And a group of 52 young students were sent to American universities and research institutions by the end of that year. The delegation was the first batch to study in the United States after the founding of the People's Republic of China. In 2007, more than 350000 Chinese students were studying at American universities and research institutions. In 1979, Professor Tsung-Dao Lee (Zhengdao Li) , an American Chinese physicist and Nobel Prize laureate, initiated the China–United States Physics Examination and Application (CUSPEA) Program. More than 100 Chinese students were selected to study physics in top American universities and research institutions on annual basis. Later, Professor Ray Wu (Rui Wu) , an American Chinese molecular biologist at Cornell University, initiated the China–United States Biochemistry Examination and Application (CUSBEA) Program. This project started in 1982 and ended in 1989, admitted a total of 422 Chinese students to study biology in the USA. Meanwhile, China's domestic education and scientific research have developed rapidly. We have trained many local scientists, especially over the last two decades of the 20th century. In regard to the difficulties of poor working conditions and funding shortages, these domestically-trained scientists took on the important tasks of revitalizing sciences and technologies in China. Their contributions have largely filled the gaps in the intergenerational transfer of research and educational achievements. It also laid the foundation for China's life science research to catch up with that of the world's leading countries. Entering the 21st century, many overseas students returned homeland and join the venture of Chinese education and technology. While China's investment in scientific research had moderate increases, it was still significantly lower than those in the leading countries in 1990s. Nevertheless, the national R&D funding had reached 2 trillion RMB in 2018, ranking second in the world, as opposed to 12.5 billion RMB in 1990. We can recognize the rapid advances and tremendous progress in China's life science over the last 40 years. And we should acknowledge that it is based on the perseverance and striving of our predecessors throughout the past 100 years.

This book introduces a group of pioneers and founders of the modern biological sciences in China. Many young people nowadays may not be familiar with these predecessors, such as Chih Ping, the founder of modern biology and zoology; Xianwen Wu, the founder of zoology; Hsen-Hsu Hu and Chong-Su Chien (Chongshu Qian) , the founders of botany; Fan-Lan Tai (Fanglan Dai) , the founder of microbiology; Shi-Tsan Pai (Shizhang Bei) , the founder of cell biology; Lien-Teh Wu (Liande Wu) , the founder of medicine; Robert Kho-Seng Lim (Kesheng Lin) , the founder of physiology; Hsien Wu, the founder of biochemistry; Richard Ho-Ping Sia (Heping Xie) , a pioneer of pathology; Ko-Kuei Chen, the founder of pharmacology; Shisan C. Chen (Zhen Chen) , the founder of genetics; Ping-Wen Tsou (Bingwen Zou) , the founder of agronomy, and so on so forth. Most of them have background of oversea education, and they have made important contributions in respective research fields. They were also deemed as founders of modern Chinese life science and its various branches. Furthermore, many of their students have become leaders of China's life science in the second half of the last century.

Readers can also learn about the outstanding achievements of modern Chinese life science with international influence. For example, Lien-Teh Wu discovered pneumonic plague and defeated the plague taking scientific research approaches, making cases for public health for more than a hundred years later. Another important pioneer, Feifan Tang, was the first to identify the pathogen that causes trachoma, *Chlamydia trachomatis*. His research findings had led to control of trachoma endemics, a serious threat to the eye health of the world. Ti-Chow Tung (Dizhou Tong) established nuclear transfer technology and cultivated the world's first colony of cloned fish. Fangzhou Gu developed a polio vaccine and eliminated polio in China, following the glorious history of public health of elimination of smallpox in the world. Yinglai Wang and collaborating researchers from three different institutes achieved full synthesis of crystalline bovine insulin for the first time in 1965. Debao Wang and colleges collaborated and artificially synthetized yeast alanine transfer ribonucleic acid (yeast alanine tRNA) , whose compositions, sequences, and biological functions were exactly the same of their natural counterparts. With her discovery of artemisinin in 1971 that can treat malaria, Youyou Tu was the laureate of the Nobel Prize in Physiology or Medicine in 2015. In 1999, Chinese scientists participated in the international Human Genome Project. Later, the scientists from 6 countries including China released a complete map of human genome.

This book will also introduce important events over the course of modern life science development in China. These events include: The debates between descriptive biology and experimental biology in the 1930s; an important turning point in genetics, the Qingdao Genetics Symposium in 1956; and the establishment of the Sino-Japanese Joint Laboratory. Readers will also learn about the connections between China and famous American universities such as Harvard University and Cornell University in modern life science, as well as the enthusiastic assistances from international research communities. Among these friendly collaborators, Nathaniel Gist Gee, a biologist from the United States, lived in China for decades and promoted the development of Chinese biology in terms of education, investigation in animals and plants, and the institutionalization of biological research. Another researcher, James Watson, known as the "father of DNA", has invited Chinese scholars to visit the Cold Spring Harbor Laboratory and made great efforts and contributions to the establishment of Cold Spring Harbor Asia.

Knowing the past opens a door to the future. "Spiritual support stimulates scientific achievements", a motto that has been fully confirmed by early scientists. Even in the extreme plight of social unrest and constant wars, our predecessors have made great efforts to develop Chinese life science. And in doing so, they have made outstanding achievements. What supported them was the ardent patriotism and firm belief of "saving the country through science". Reviewing the history of the development of China's life science will inspire us to train talented people, to explore the laws of science and technology, and to pursue international cooperation. Therefore, scientific researchers of our time should pass on the scientific spirit of the older generation of scientists and strive hard to realize the great rejuvenation of the Chinese nation in the new era.

I would like to thank the editorial department of *Protein & Cell*, Academician Yiyu Chen, and Professor Guihuan Luo for their invaluable suggestions and comments during the preparation of this book. I would also like to thank my colleagues from the Beijing Institutes of Life Science of the Chinese Academy of Sciences, and editor Bingxiang Li and her colleagues from the Higher Education Press, who worked hard with us to edit and publish this book. This book is dedicated to help readers understand the history of life science development as well as current research achievements of life science in China. We have a huge hall of fames. It is worth mentioning that our book may not be able to cover all scientists and their outstanding contributions. In particular, Prof. Fu-Ching Yen (Fuqing Yan) , a famous medical educator and public health scientist, founded Hsiang-Ya School of Medicine as well as Shanghai Medical College and trained many medical talents and experts. He is recognized as one of the founders of modern medical science in China. In addition, those unlisted scientists include: psychologist Shu Pan, botanist Ching-Yue Chang (Jingyue Zhang) , plant physiologist Hung-Chang Yin (Hongzhang Yin) , anthropologist Ting-Liang Woo (Dingliang Wu) , tropical pathologist Zong'en Li, and medical scientist Yijin Yuan. We sincerely invite friends concerned about the history of life science development to submit relevant articles to the Recollection column of *Protein & Cell* as we will continue to collect and publish these data in the future.

Le Kang

September 20, 2022 · Beijing

中文目录

三、微生物学

四、细胞生物学和遗传学

五、生物化学

六、病毒学和免疫学

七、生理学和医学

八、心理学

九、生命科学领域重要成就与重要事件

十、中外学术交流

英文目录

一、Zoology

二、Botany

六、Virology and Immunology

七、Physiology and Medicine

八、Psychology

一、动物学

1. 秉志：中国现代生物学先驱

图1 秉志（1886—1965）

秉志先生的铜像坐落在中国科学院动物研究所大厅。每当人们经过这里时，尤其是年轻人，都会想要了解他是谁？他作出了怎样的贡献？根据铜像下的介绍，我们了解到秉志先生（图1）是中国现代生物学领域的奠基人、动物学家和教育家。作为中国现代生物学的先驱，他赋予了“第一”多个含义。作为生物学家，他是第一个非政府科学学会的联合创始人，是第一个中国科学杂志的联合创办者，创办了中国第一个国立大学生物系，创建了中国第一个生物学研究所，还是中国动物学会的创始人。而作为教育家，他可谓桃李满天下，培养的学生日后都成了各自领域的著名学术大家。

秉志先生生于1886年，在家乡河南接受了传统文化教育，在17岁时通过科举考试成为秀才。然后到北京京师大学堂经过四年的学习，获得了庚子赔款资助赴美国接受高等教育。秉志先生在北京学习期间，对自然科学表现出极大的兴趣，尤其是达尔文提出的物种起源激发他后来选择跟随康奈尔大学的J. G. Needham教授学习昆虫学。秉志先生1913年获得了学士学位，然后继续在康奈尔大学攻读研究生并成为第一位获得昆虫学博士学位的中国人。他的导师Needham教授称他为“一个学者、令人尊敬的同事、最亲爱的朋友”。后来，秉志先生来到费城威斯特解剖学和生物学研究所，在著名的神经生物学家H. H. Donaldson教授指导下开展脊椎动物神经学研究。在此期间，他还在宾夕法尼亚大学进行课程学习，为他将来广阔的研究范围奠定了扎实的基础。

秉志先生还在康奈尔大学学习的时候，他已经开始积极推动国内现代科研的开展。1915年，他联合中国留学生在康奈尔大学创立了第一个中国非政府学术组织——中国科学社。他也是第一个中国科学类杂志《科学》的联合创办人，这个杂志的主要目的

译者：苏瑞凤
中国科学院北京生命科学研究院，北京100101，中国
邮箱：surf@biols.ac.cn

是将现代科学介绍给中国大众。杂志创立之初的几年，秉志先生先后发表了15篇论文，包括第一篇关于医学昆虫——疟蚊的论文。

他于1920年回国后受聘于南京高等师范学校（1921年更名为东南大学，1928年更名为国立中央大学）。秉志先生在农业系教授动物学，期间他摒弃了按照门类讲解动物学的传统教学方式，而是采用了将进化理论和各门类动物结合进行授课的创新教学方式。这种全新的教学使得动物学这门课程广受学生欢迎。受秉志先生的影响，超过一半农学专业的学生转为学习动物学专业。那时候，中国大学只设自然科学系但没有生物系。在秉志先生不断的努力和影响下，东南大学决定设立生物系，这标志着中国第一个国立大学生物系的诞生。秉志先生任生物系主任，该系分为动物学和植物学两个分支，秉志先生和钱崇澍先生分别担任学术委员会主席。由于新建实验室的仪器和标本购置资金不足，秉志先生和他的教职工就使用自己的积蓄购置诸如切片机等实验仪器。生物系大多数的动物标本都是秉志和他的学生利用暑期从浙江和山东采集的。

秉志先生回国后，中国科学社同期迁回中国并且落址南京。秉志先生作为中国科学社的创始人，在1922年发起创建了中国科学社生物研究所。相对于东南大学生物系的定位主要是教学，中国科学社生物研究所则主要聚焦于科学研究，尤其是分类学和形态学。开始的阶段，研究所与东南大学生物系一样，财政经费非常困难。然而幸运的是，研究所的刊物《中国科学社生物研究所报告》得到了广泛的关注和好评，也为研究所带来了捐助。其中最大的捐助者是1924年成立的负责使用美国退还的庚子赔款的中华教育文化基金会。在基金会的资助下，研究所建成了两层楼的实验室，研究范围也从分类学扩大到了生理学、生物化学和遗传学。

随着时代发展，秉志和他的同事日益感受到在中国北方建立生物研究所的紧迫性，以便研究北方的动物和植物。这个想法再一次得到中华教育文化基金会和时任基金会主席范源濂的大力支持。但是在新研究所创建不到一年主席不幸离世。在范主席去世后，他的家庭继续支持研究所的建设，甚至将老房子捐出作为研究所办公地点。1928年，新的生物学研究所正式在北平建立。为纪念范先生作出的贡献，研究所以他的字“静生”为名，命名为“北平静生生物调查所”，秉志先生任该所所长。同时管理南北两个研究所，秉志先生比以往更为勤奋。因为研究所的经费难以满足实验仪器和标本的购置，秉志先生使用自己的薪水支付来往两个城市间的差旅费。在秉志先生和同事的努力下，随着经费资助的增加，北平静生生物调查所的规模逐步扩大，原来的实验楼已经变得十分拥挤。在教育部和中华教育文化基金会的资助下，一栋新的实验楼于1931年建成并投入使用（图2）。到了1935年，北平静生生物调查所已经成为中国最好的动物学研究所，许多来自大学和其他研究所的学者都慕名而来开展他们的研究工作。

秉志先生除了在东南大学生物系和两个研究所工作外，他还在厦门大学、复旦大学以及中国科学院水生生物研究所和中国科学院动物研究所工作过。直至去世前一天，秉志先生仍然在实验室工作。在他50多年的学术生涯里，秉志先生在形态学、生理学、分类学、昆虫学、古生物学，尤其是解剖学和神经学方面取得了瞩目的成就。他特别强调生物学研究中，形态结构和生理功能结合研究的重要性。他从博士阶段开始发表的富有创造性的研究成果得到了世界认可。

图2 北平静生生物调查所新实验楼

秉志先生教育和培养了许多年轻学生，这些人大多数成为著名的生物学家和教育家，他们沿着秉志先生的步伐为我国的生命科学事业作出了巨大贡献。在老年时期，秉志先生将自己的房子和超过3000本藏书的图书馆捐献给了国家。他是将自己整个生命献给祖国科研事业的中国老一辈科学家中的榜样。作为将现代生物学带到中国的第一人，秉志先生是公认的中国现代生物学先驱。

1. Bing Zhi: Pioneer of modern biology in China

Located in the lobby of the Institute of Zoology at Chinese Academy of Sciences, is a bronze statue of Dr. Bing Zhi. Many passers-by, especially the young ones, would ask who this person is and what he has accomplished. The short answer is, according to the plate of the statue, that he is a pioneer zoologist and educator and the founder of modern biology in China. The more detailed answer involves the word "first" for quite a few significances: As a biologist, Bing Zhi is the co-founder of the first non-governmental scientific society of China, the co-founder of the first Chinese scientific journal, the founder of the first biology department in China, the founder of the first biological research institute in China, and the founder of China Zoological Society; as an educator, he trained many young men who later became renowned specialists in their own areas.

Born in 1886, Bing Zhi had received traditional education in his home province Henan, and passed the imperial exams to become a first-degree scholar at age 17. After four-year preparatory study at the Imperial Capital University in Peking, he was sent to the United States for higher education, supported by the Boxer Indemnity Fund. During his study in Beijing, he had shown great interests in natural sciences and was particularly inspired by Charles Darwin's Origin of Species, which explained why he later chose to study entomology under Prof. J. G. Needham when he came to Cornell University. After getting B.S. degree in 1913, he continued for graduate study in Cornell University and became the first Chinese to get Ph.D. degree in entomology. Prof. Needham called him "a scholar, a respectable colleague and a dearest friend" . Later, he went to the Wistar Institute of Anatomy and Biology and studied vertebrate neurology under Prof. H. H. Donaldson, who is a famous neurologist. During this period, he also registered classes in University of Pennsylvania, which provided solid knowledge foundation for his wide range research work in future.

Even when he was a student at Cornell University, Bing Zhi had always been active in promoting modern science in China. In 1915, Bing Zhi and other Chinese students at Cornell founded the first non-governmental academic organization of China—the Science Society of

Ming Li, Le Kang

Beijing Institutes of Life Science, Chinese Academy of Sciences, Beijing 100101, China

Correspondence: lkang@ioz.ac.cn (L. Kang)

China. He was also the co-founder of the first Chinese scientific journal—*Ke Xue* (meaning "Science" in Chinese). The main purpose of the journal was to introduce modern science to ordinary people in China. During the first couple of years, Bing Zhi published 15 articles on *Ke Xue*, including the first Chinese article on medical insects—*Malaria Mosquito*.

When he returned from United States in 1920, Bing Zhi was hired by the Nanjing Normal College, which became Southeast University in 1921, then National Central University in 1928. At the Nanjing Normal College, Bing Zhi taught Zoology in the department of Agriculture. Instead of the old-fashioned teaching style where selected animals from each phylum were introduced to the students one by one, Bing Zhi used the evolution theory to connect animals from different phyla. His coherent teaching style made zoology very popular in his class. In fact, more than half of the students changed their major from Agriculture to Zoology, largely due to Bing Zhi's influence. Back then, some of the Chinese universities and colleges had a department of natural sciences, but none had a department of biology. Because of Bing Zhi's continuous efforts and influence, the Southeast University decided to establish the Department of Biology, which would be the first biology department in China. The department, with Bing Zhi as the Dean, comprised of two divisions—Zoology and Botany, of which Bing Zhi and botanist Qian Chongshu were the chairs, respectively. The newly-established department was underfunded for lab equipments and specimens, so Bing Zhi and other faculties used their own savings to buy equipments, such as section machines. Most of the zoological specimens were collected by Bing Zhi and his students in Zhejiang and Shandong provinces during summer breaks.

Meanwhile, the Science Society of China moved back to China and had its physical location in Nanjing. As a founding member of the society, Bing Zhi initiated the establishment of the Biological Laboratory of the Science Society of China in 1922. While the Department of Biology at Southeast University was more teaching-oriented, the Laboratory was getting more focused on research, especially Taxonomy and Morphology. The Laboratory went through a financially difficult time at the beginning, just like the Department of Biology at Southeast. Fortunately, the Laboratory's publication, *Contributions from the Biological Laboratory of the Science Society of China*, received quite a lot of attention and praises, thereby brought donations to the Laboratory. The biggest sponsor was the China Foundation for the Promotion of Education and Culture, an organization set up in 1924 to manage the Boxer Indemnity Fund. With the yearly donation from the China Foundation, the Laboratory was able to build a two-story building for the labs, and expanded its research focuses from taxonomy to physiology, biochemistry and genetics.

As time went by, it appeared increasingly urgent for Bing Zhi and his colleagues to set up another institute in the northern China so they could study the animals and plants in the north. Again, the idea was applaused by the China Foundation and its director Fan Yuanlian, who passed away less than one year before the launch of the new institute. Even after his death, Fan's family continued to support the institute and even donated his old house to accommodate the institute. When the new institute was finally launched in Beijing in 1928, it was named "Fan Memorial Institute of Biology" in memory of Fan Yuanlian. Again, Bing Zhi was made Director of the institute. Running two institutes at the same time, one in northern China and the other in the south, Bing Zhi had to be more diligent than ever. Financially, since the funds donated to the institute were never sufficient for the equipments and specimens, Bing Zhi voluntarily used his own salary to cover transportation costs between the two cities. With the devotion of Bing Zhi and his colleagues, along with relatively abundant funding, the Fan Memorial Institute of Biology soon expanded to an extent that the original building (Fan's old house) could not accommodate. With the help of the Education Ministry and the China Foundation, a new laboratory building for the institute was constructed and put into use in 1931. By 1935, the institute became the number one zoological research institution in the country and many scholars from universities and other institutes kept coming in to carry out their research projects.

Other than the one department and two institutes mentioned above, Bing Zhi also worked in Xiamen University, Fudan University, Institute of Hydrobiology, and Institute of Zoology of Chinese Academy of Sciences. He worked in laboratory till the day before he passed away. During over 50 years of his academic life, Bing Zhi accomplished significant achievements in morphology, physiology, taxonomy, entomology, paleontology, etc., especially in anatomy and neurology. He emphasized the significance to correlate morphological structure to physiological functions, and vice versa. His productive research publications since the Ph.D. study won worldwide recognition.

Bing Zhi had taught and trained hundreds of young people. Most of them later became biologists and educators themselves, and followed his steps to further contribute to the life science in China. At the old age, he also donated his house and private library with over three thousand books to the country. He is an exact example of the old-generation Chinese scientists who devoted his whole life to science and the country. Together with the fact that he first brought modern biology to China, it would be appropriate to deem Bing Zhi as the pioneer of modern biology in China.

Figures

Fig.1 Dr. Bing Zhi (1886—1965)

Fig.2 The new laboratory building of the Fan Memorial Institute of Biology

References

Hu ZG(2005) History of Fan Memorial Institute of Biology. Jinan: Shandong Education Press. (胡宗刚. 2005. 静生生物调查所史稿. 济南: 山东教育出版社.)

Hu ZG(2005) Unforgettable Xiansu Hu. Wuhan: Changjiang Literature and Art Press.(胡宗刚. 2005. 不该遗忘的胡先骕. 武汉: 长江文艺出版社.)

Schneider L(2005) Biology and Revolution in Twentieth-century China. Lanham: Rowman & Littlefield Publishers.

Zhai QH, Hu ZG (2006) Bing Zhi Selection. Beijing: Peking University Press. (翟启慧, 胡宗刚. 2006. 秉志文存. 北京: 北京大学出版社.)

2. 朱洗：杰出的实验生物学开拓者

朱洗是我国细胞生物学与实验生物学的创始人和奠基者之一（图1）。他1900年出生于浙江临海，1931年在法国取得博士学位。学成归国后，他致力于研究单性蟾蜍的培育等问题，以其勇攀高峰的创新精神以及潜心研究的奉献精神为生物学领域作出了巨大贡献。同时，他也是一名伟大的爱国者，以其忧国奉公的爱国精神促进了农业等多个产业的蓬勃发展。

图1　朱洗（1900—1962）（童第周，1962）

朱洗生平

1900年10月14日，朱洗出生于浙江临海店前村。自幼聪慧勤奋的他，在读书道路上颇为顺遂，早年入私塾读书，15岁便以优异成绩考入县城回浦学校，3年后又考入浙江省立第六中学。然而，1919年，五四运动以势不可挡之势席卷中华，朱洗怀揣着满腔热血加入罢课队伍，却遭到学校开除。这一坎坷经历并未消磨他求学的热忱之心。正在家乡当排字工人的朱洗听闻蔡元培与李石曾等人在上海招收赴欧留学生时，他当机立断，从临海赶赴上海。在上海，他凭借过人的才智与大志得到赏识。1920年，他踏上了远赴法国求学的漫漫长路。

“天将降大任于斯人也，必先苦其心志，劳其筋骨。”初至法国，朱洗几乎身无分文，吃穿住行尚且困难，遑论上学读书。别无他法，他只得先后辗转多所工厂，白天辛苦工作，挣得微薄工资补贴生计，夜间不知疲倦勤学法语，徜徉在生物学的海洋。就这样缩衣减食、勤工俭学了整整五年，他终于攒够学费，并成功考入蒙彼利埃大学，成为著名胚胎学家J. E. Bataillon教授的得意门生。他与教授合作研究，发表论文多达

译者：钱雨菁，谢群

浙江师范大学教师教育学院，金华 321000，中国

邮箱：xiequn@zjnu.cn（谢群）

14篇。支撑朱洗在逆境中坚定前行的是他坚韧、向学与专注的精神。

那时的朱洗初露锋芒。毋庸置疑，法国先进的实验条件将为他的研究工作提供诸多便利，无须多时他便可成为闻名遐迩的生物学家。然而，九一八事变后，坚信“科学救国”的朱洗不仅婉辞了Bataillon教授的盛情邀请，还耗费历年积蓄购置书籍和实验仪器。这是他“淡泊名利”的态度，更是他“天下兴亡，匹夫有责”的信念。

就这样，一颗赤子心，两袖鼓清风，朱洗回到了他日思夜想的祖国。之后，朱洗先后任职于多所大学，还曾在上海筹建生物研究所。当时的中国内忧外患，朱洗不仅面临实验材料短缺等困难，还需躲避炮火的攻击。上海沦陷后，朱洗回到临海，虽然暂时无法从事精细的生物研究，但他仍旧兴医办学，为家乡贡献自己的力量。直至新中国成立，朱洗任中国科学院北京实验生物研究所研究员，才获得了较为安定的研究环境。在此之前，哪怕抗日战争结束后，他也常需频繁奔波，往返多地做指导与研究。永不言弃的精神以及对生物学和祖国的热爱使万千险阻皆如汤沃雪。

朱洗始终践行“搞科学工作需要人的全部生命，八小时工作制是行不通的”。在培育出世界上第一批“没有外祖父的癞蛤蟆”，解决蓖麻蚕养育以及人工养殖鱼类繁殖问题的同时，他还著书25本，翻译国际名著4本。即使晚年因积劳成疾而缠绵病榻，他仍忘乎所以地写文章，直到生命最后一刻（杨春晓，1992）。朱洗的一生，是为生物学与祖国无私奉献的一生。

朱洗对实验生物学的贡献

朱洗与生物学的缘分始于幼年时其祖父经营的中药铺与饲养的小动物，在师从Bataillon教授前，他也对进化论与新拉马克主义颇感兴趣。他与教授共同完成的青蛙单性生殖研究为研究者津津乐道，但他在生物学领域最著名的贡献当属对蟾蜍的单性繁殖研究。

单性蟾蜍的培育

1951年，朱洗即与他的助手王幽兰等人开始了这项艰难的研究。此前，虽也有研究者以蟾蜍的卵为人工单性发育材料，但效果不好。朱洗也亲自对广东与上海的两种蟾蜍的卵做过针刺实验，只得到与前人相差无几的结果。但朱洗并未因此灰心丧气，反而将目光聚焦于针刺方法的改良，缘因他发现蟾蜍卵带的胶质过厚，常用细针难以刺入血球；而使用无膜的裸卵，卵又质地娇嫩不耐针刺。经过再三考量，因裸卵具有绝无受精可能等诸多优点，朱洗最终选择使用裸卵，改良细针。于是，在经历了最初用白金丝针刺破6840个蟾蜍卵，仅得13个未能进一步发育的畸形原肠胚，更无出膜蝌蚪的失败后，朱洗改用直径更细的玻璃丝针。这种丝针尖端直径10～13 μm，初时缺点显著，勉强可用。对其多次改良后，方得到差强人意的实验结果。第2年，朱洗团队使用玻璃丝针刺破3650个裸卵，几乎每次都获得了出膜蝌蚪。大胆创新与坚持不懈终于换来振奋人心的结果。

然而，之前的辛劳工作对精益求精的朱洗而言都只是试探性的，1958—1959年才是他开展正式工作的时期。在此期间，朱洗团队主要探究了三个问题：卵球激动的表现、单性蟾蜍的获得及其发育、卵球成熟程度与单性发育的关系。通过刺破40140个卵，在仔细观察与比较的基础上，这三个难题都得到了确切的答案。尤为引人注目的是，朱洗团队成功培育了25只单性蟾蜍，即“没有父亲的蟾蜍”（朱洗和王幽兰，1959）。这一实验结果一举打破了过去常用蛙做单性生殖研究，认为蟾蜍不能成为相关研究材料的误解。

1961年3月，无父蟾蜍中仅存的一只长成成体的雌性蟾蜍产下3000多枚卵球，繁殖了世界上第一批“没有外祖父的癞蛤蟆”（朱洗等，1961）。朱洗的这一研究工作是站在前人肩膀上想前人之未想，思维广阔如雄鹰自由翱翔于天地，行动却如屈居于一小方土地的树，扎根实验室一隅，潜心钻研。

蓖麻蚕的引种驯化与推广

实际上，朱洗并不是“两耳不闻窗外事，一心闭门做实验”的人。他所具备的百折不挠、细致入微、通达开阔等科学家品质，既是天性使然，也是时代淬炼的结晶，更是他走入现实生活拾起的平凡而珍贵的宝物。

他曾呼吁同道结合我国实际，研究养殖印度蓖麻蚕以促进农业经济发展并改善生态环境。从这一目的出发，他率先对蓖麻蚕的饲育法与保种法进行了研究，对蓖麻蚕蚕丝的用途也提出了卓有成效的建议（朱洗等，1953）。此外，他还利用我国樗蚕等本土蚕种，将其与蓖麻蚕杂交，定向选育集合多种优良性状的新蚕种。在此理论与实践基础上，他指导蚕农养育蓖麻蚕，编写蓖麻蚕养殖小册子，在多个省市开设蓖麻蚕养殖培训课程……最终成功地在全国范围内完成了蓖麻蚕的引种驯化与推广。

对家蚕养育的研究

朱洗在潜心研究与大力推广印度蓖麻蚕的同时，对本土家蚕仍保持着密切关注。为响应“科学为人民服务”的号召，他率领研究团队共采集50多种植物叶子，一一养育家蚕。对于每种植物叶子的育蚕效果，朱洗都设计了严密且巧妙的对照实验，使其与桑叶做比较。以蒲公英为例，他将实验所用家蚕分为六组，第一组和第六组分别全食蒲公英和桑叶，第二组自五龄起改食桑叶，第三组自四龄起改食桑叶，以此类推（朱洗等，1951）。设置有梯度的对照组不仅能够比较两种叶子养育家蚕的效果，还能大致确立最优养育方式。

同样的对照实验法也被朱洗应用于对家蚕节食疾病的研究中。他注意到江浙一带的家蚕常发肠胃病，且发病与季候相关。为寻找这种疾病的预防方法，朱洗及其团队设计并制造了“定温定湿箱”，研究者可通过设置不同的温度与湿度，调节箱内气候状态。朱洗将家蚕分成不同组别，对照组给予足够饱食的桑叶量，实验组则依次递减桑叶量。他将这些家蚕放入同一高温高湿箱内，实验结果证明，在湿热条件下家蚕饱食易发病，适当节食则能有效避免肠胃病（朱洗等，1952）。这些关于家蚕养育的对照研究，体现了他作为实验生物学开拓者的智慧与细致入微。

人工养殖鱼类的繁殖

即使身处和平时期，朱洗的研究仍与祖国的经济发展和人民的生养根基息息相关。他对家鱼人工繁殖的研究也深刻地体现了其信奉的“研究须与实际结合”的理念。为解决家鱼的生殖腺在池塘中难以发育这一现实问题，朱洗与其研究团队的足迹近到鱼塘渔场，远至江河湖泊，先后对金鱼、鲤、鳊的卵球受精、成熟与胚胎发育进行了一系列研究，成果颇丰（朱洗等，1960）。至1958年，他使用绒毛膜促性腺激素对几种家鱼催产，使鲢与鳙成功产卵，实现了家鱼的人工繁殖。之后，他又建立了一整套人工养殖鱼类的技术方法，并进行大规模推广，使我国的淡水养殖业走上了崭新的发展道路。

朱洗从事生物学研究工作数十年，著有专著《现代生物学丛书》等，翻译著作《脊椎动物发生学》等，还出版了《动物学》等科普读物。他是一位杰出的生物学家，他令人敬仰的科学家精神激励了一代又一代青年人投身科学事业，他的研究工作推动了我国细胞生物学与实验生物学的发展，他的科普工作更是惠及万家，加深了大众对动物学的认识，促进了我国多项产业的发展。同时，他更是一位对祖国怀有深厚情感的爱国者。这种爱，隐于内心，见于行动。不论是乱世中毅然决然地离法救国，还是和平年代所从事的每一项生物研究，朱洗都将自己的命运与热爱的生物学和祖国紧密地结合在了一起。

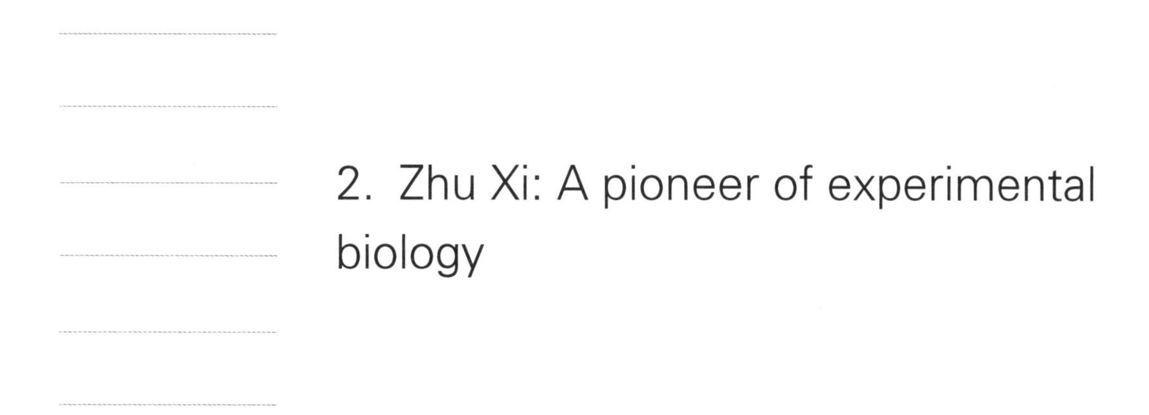

2. Zhu Xi: A pioneer of experimental biology

Zhu Xi (朱洗, 1900—1962), one of the founders of cell biology and experimental biology in China, was born in Linhai, Zhejiang Province in 1900. In 1931, he obtained his doctorate in France. After returning China, he devoted himself to toad reproduction research. With the spirit of innovation and dedication to research, he had made marvelous contributions to China, especially in biology. At the same time, he put his patriotism into action, promoting the vigorous development of agriculture and other industries.

The life of Zhu Xi

Zhu Xi was born on October 14, 1900 in Dianqian village of Linhai, Zhejiang. For those who enjoy learning, in fact, learning has become part of their meaningful lives. Zhu Xi is exactly that kind of person. In his early years, influenced by traditional Chinese culture, he studied at a private school and started to study Confucian classics. At the age of 15, he enrolled to the Hui Pu School, a local county school, owing to excellent grades. Three years later, he was admitted to Zhejiang No. 6 Middle School. In 1919, the May 4th Movement took place in China. Zhu Xi joined the strike but was expelled from the school. However, his passion for learning never faded away no matter what he suffered. At the time, Zhu Xi, as a typesetter in Linhai, responded to call of Cai Yuanpei (蔡元培), Li Shizeng (李石曾) and others and rushed to Shanghai to apply for going to Europe without hesitation. In 1920, Zhu Xi was sent to study abroad by virtue of his great intelligence and capacity.

All things are difficult before they are easy. After his arriving in France, financial hardship impelled him to make a living. He took many jobs, worked hard during the day, and studied French and biology at night. Five years later, he managed to save enough money to go to Montpellier University, where he met Professor Jean Eugène Bataillon, a famous embryologist. Before long, Zhu Xi displayed his talent for the first time in biology, publishing about 14 papers

Yujing Qian, Xiazhao Yu, Yangqing Sun, Qun Xie, Ge Liang, Yanping Cheng
Teacher Education College, Zhejiang Normal University, Jinhua 321000, China
Correspondence: xiequn@zjnu.cn (Q. Xie)

with Professor Bataillon. His perseverance, diligence, concentration enabled him to move forward firmly in adversity.

Undoubtedly, the advanced experimental facilities in France paved the way for his research, and he soon became famous as a biologist. The September 18th Incident shocked the world. Zhu Xi firmly believed that saving the country as important as science, and declined Professor Bataillon's detainment. He spent years of savings on books and laboratory equipment, which he brought back to China.

After returning to China, Zhu Xi worked at Sun Yat-sen University in Guangzhou, Institute of Zoology, Peking Research Institute, and Institute Franco-Chinois. He also set up a Biological Research Institute in Shanghai. At that time, China was suffering from internal and external troubles. Zhu Xi not only faced difficulties such as the shortage of experimental materials, but also had to avoid the attack of artillery fire. After the fall of Shanghai, Zhu Xi returned to Linhai, where although he was temporarily unable to engage in biological research, he continued to devote himself to running schools for his hometown. After the founding of the People's Republic of China, Zhu Xi was appointed as a researcher at the Beijing Institute of Experimental Biology, Chinese Academy of Sciences, which provided him a relatively stable research environment. Before that, he had to frequently travel to many places to do instruction and research.

Zhu Xi held that scientific work requires all one's life, and the eight-hour working day is not enough. In addition to cultivating the world's first "toads without grandfather" , solving the problems of rearing Eri silkworms and breeding artificially farmed fish, he had written 25 books and translated 4 international classics. Even though he was plagued by illness in his later years, he continued to write until the end (Yang, 1992). Zhu Xi had devoted his whole life to biology and homeland.

Zhu Xi's contributions to experimental biology

Zhu Xi had a deep connection with biology. He was close to biology from an early age, when his family kept a variety of animals. Before learning from Professor Bataillon, he already studied evolutionism and neo-Lamarckism. People usually talk about Zhu Xi's research on parthenogenesis in frogs, which he completed with Professor Bataillon, but in fact his most famous contribution in biology is the research on parthenogenesis in toads.

The breeding of parthenogenetic toad

In 1951, Zhu Xi, along with his assistant Wang Youlan (王幽兰) et al., carried out the difficult research. Previously, although some researchers had used toad eggs as artificial parthenogenetic materials, the effect was disappointing. Zhu Xi also experimented with acupuncture on the eggs of two kinds of toads from Guangdong and Shanghai, and got similar results. But Zhu Xi was not discouraged, instead focused on improving acupuncture methods. He found that the gelatin layer surrounding toad eggs bands were too thick to penetrate with the usual fine needle, and naked eggs without membrane were delicate and not resistant to acupuncture. After careful consideration, Zhu Xi finally chose to use naked eggs and improved the fine needles because of its many advantages, such as the possibility of no fertilization. Therefore, after the initial failure of needling 6840 toad eggs with platinum needles, resulting in only 13 malformed gastrulas that failed to develop further, let alone hatched tadpoles, Zhu Xi switched to glass needles with a smaller diameter. The needle, with a tip diameter of 10–13 μm, did not work well at first as an experimental tool, but after several refinements, it worked well in experiments. The next year, Zhu Xi's team punctured 3650 naked eggs with glass needles, and almost every time they obtained hatched tadpoles.

The period, 1958—1959, was actually the time when he started his formal work. During this period, Zhu Xi's team mainly explored three questions: The performance of ovum activation, the acquisition and development of parthenogenetic toads, and the relationship between ovary maturation and parthenogenetic development. After puncturing 40140 eggs, on the basis of careful observation and comparison, three difficult questions had been precisely solved. The most remarkable thing was that they managed to breed 25 parthenogenetic toads (Zhu and Wang, 1959). And the results suggested that toads can be used as research material in the same way that frogs had been used for parthenogenesis studies in the past.

Not only have a dream, but try to achieve it by working hard. In March 1961, the only remaining parthenogenetic toad, an adult female, laid more than 3000 eggs, giving birth to the world's first toads without a maternal grandfather (Zhu et al., 1961). Such satisfactory research results could be attributed not only to his innovation, but also to his dedication to research on the basis of predecessors.

Introduction, domestication and popularization of the Eri silkworm

Zhu Xi didn't divorce himself from the masses and from reality and act blindly. On the contrary, he concerned about the real life and spared no effort to make contributions to improving people's lives. He once called on his colleagues to study and breed Indian Eri silkworms according to the reality of China to promote agricultural economic development and improve the ecological environment. Then he took the lead in studying the methods of rearing and conservation of Eri silkworms, and put forward valid suggestions on the use of Eri silk (Zhu et al., 1953). Furthermore, he also crossed the Eri silkworm with native silkworms, and selected new silkworms of good traits.

Based on this theory and practice, he guided the silkworm farmers to raise Eri silkworms, composed brochures on its breeding, and set up training courses for its breeding in many cities and provinces. Finally, the introduction, domestication and popularization of Eri silkworm were successfully completed in the whole country.

Studies on breeding of silkworms

Zhu Xi was devoting himself to the research and promotion of the Eri Silkworm, while still keeping a watchful eye on the native silkworm. In response to the call of "Science serves the people" , he led the research team to collect more than 50 kinds of plant leaves and raised silkworms one by one. In elaborate experiments, control experiments were conducted to compare the results of silkworm rearing of each plant leaves with that of mulberry leaves. In the case of the dandelion, he divided the silkworms which used in experiments into 6 groups. The 1st group and the 6th group ate dandelion and mulberry leaves respectively, the 2nd group ate mulberry leaves from its fifth instar, and the 3rd group ate mulberry leaves from its fourth instar, and so on (Zhu et al., 1951). The gradient control group experiment could not only compare the effects of two kinds of leaves on rearing silkworm, but also roughly establish the optimal rearing mode.

The same controlled experiment method was also used in the study of silkworm dieting disease. Zhu Xi noticed that the silkworm in Jiangsu and Zhejiang Provinces often suffered from gastrointestinal diseases and the incidence was related to the season. To find an effective way to prevent the disease, he led his team designed and manufactured the "constant temperature and humidity chamber" , which allowed researchers to adjust the climate conditions by setting different temperatures and humidity. Then Zhu Xi divided silkworms which used in experiments into experimental groups and control groups, and the control group was given enough mulberry leaves, while the experimental group was given successively reduced mulberry leaves. Subsequently, he put these silkworms into the same chamber with high temperature and humidity, and the experimental results showed that silkworms were apt to disease when they were fed well in hot and humid conditions, and a proper diet could effectively avoid gastrointestinal diseases (Zhu et al., 1952). These comparative studies on the rearing of silkworms showed the wisdom and carefulness of this great pioneer of biology.

Propagation of farmed fish

Even in peacetime, Zhu Xi's research was still closely related to the economic development of the motherland and the livelihood of the people. His research on artificial reproduction of domestic fish also reflected his belief that research should be combined with practice. Zhu Xi and his research team visited nearby fish ponds and even distant rivers and lakes, and carried out a series of studies on the fertilization, maturation and embryonic development of goldfish, carp, and roach in order to solve the practical problem that the gonads of domestic fish were difficult to develop in ponds, and finally obtained satisfactory results (Zhu et al., 1960). In 1958, he used chorionic gonadotropin to induce the spawning of several kinds of domestic fish, which made silver carp and bighead carp spawn, and realized the artificial reproduction of domestic fish. Afterwards he established a whole set of artificial fish farming techniques and popularized them on a large scale, which brought the country's freshwater farming industry to a new level.

Zhu Xi had been engaged in biological research for decades. During his lifetime, he wrote monographs like *Modern Biology Series*, translated famous works like *Embryology of Vertebrates*, and published popular science books like *Zoology*. He is an outstanding biologist whose spirit inspires generations of young people. He promoted the development of cell biology and experimental biology in China, and contributed to the popularization of science. Zhu Xi is also a patriot who has a deep affection for his country. No matter when and where, he closely combined his destiny with biology and his motherland.

Figures

Fig.1 Zhu Xi (1900—1962) (Tong, 1962)

References

Tong DZ (1962) The life and academic achievements of Mr. Zhu Xi. Science Bulletin 10: 27–34. (童第周. 1962. 朱洗先生的生平及其学术成就. 科学通报, 10:27–34.)

Yang CX (1992) Zhu Xi. Zhejiang Archives 3:49. (杨春晓. 1992. 朱洗. 浙江档案, 3:49.)

Zhu X, Wang GS, Jiang TJ, et al (1951) Test report on silkworm breeding of mulberry in dandelion generation. Science Bulletin 8:819–822. (朱洗, 王高顺, 蒋天骥, 等. 1951. 蒲公英代桑育蚕的试验报告. 科学通报, 8:819–822.)

Zhu X, Wang YL (1959) Artificial parthenogenesis and the relationship between different maturation degree and parthenogenesis of Bufo toad maturation oosphere in vitro. Chinese Journal of Experimental Biology 4:276–298. (朱洗, 王幽兰. 1959. 蟾蜍体外成熟卵球的人工单性发育及卵球成熟程度之不同与单性发育的关系. 实验生物学报, 4:276–298.)

Zhu X, Wang YL, Lin ZC (1960) The relationship between fertilization and embryonic development of oosphere of different maturation levels in goldfish, carp, and roach. Chinese Journal of Experimental Biology Z1:47–58+190–192. (朱洗, 王幽兰, 林志春. 1960. 金鱼、鲤、鳊的不同成熟程度卵球的受精和胚胎发育的关系. 实验生物学报, Z1:47–58+190–192.)

Zhu X, Wang YL, Xu GJ (1961) World’s first fatherless female toad to lay eggs. Science Bulletin 4:50. (朱洗,王幽兰,徐国江. 1961. 世界第一只无父的母蟾蜍产卵传种. 科学通报, 4:50.)

Zhu X, Zhang G, Jiang TJ, et al (1953) Please pay attention to raising Eri silkworm. Science Bulletin 1:28–41. (朱洗, 张果, 蒋天骥, 等. 1953. 请大家注意养蓖麻蚕. 科学通报, 1:28–41.)

Zhu X, Zhang G, Wang GS, et al (1952) Study on dieting disease of silkworm. Science Bulletin 4:72. (朱洗,张果,王高顺,等. 1952. 家蚕节食却病研究的报导. 科学通报, 4:72.)

3. 王家楫及其原生动物学研究

王家楫是著名的原生动物学家，我国原生动物学的奠基人，轮虫生物学的开创者（图1）。

图1　王家楫（1898—1976）

王家楫出生于江苏奉贤的一个书香之家，六岁开始在肇文书院接受教育。在国势衰微之际，他的父亲提倡读书人应学习先进的技术以报效祖国。父亲的开拓精神及对新事物的开明态度，对王家楫的一生产生了深远的影响。

1920年，王家楫从南京高等师范学校毕业，此后师从动物学家秉志教授，从事原生动物的研究。1925年他凭借优异的成绩考取官费留学生，赴美国宾夕法尼亚大学深造，深入研究海洋原生动物。1928年，他以博士学位论文《淡水池塘原生动物季节性分布的生态学研究》（Wang，1928）获得优秀生物工作者金质奖章，并获博士学位。在美期间，他多次在*Science*、*Morphology and Physiology*和*Animal Physiology*等杂志上发表有关原生动物生理学和分类学的论文，引起了美国学者的广泛关注，随后被耶鲁大学特聘为斯特林研究员。

1929年，王家楫获悉外国科学考察团将前往中国采集动植物标本，他认为中国的生物资源是属于中国人的，我们应当让外国人知道，中国人自己的事应由中国人自己解决，中国的生物资源应该由中国人自己来开采。于是，他毅然放弃耶鲁大学优越的工作机会和生活条件，怀着一颗赤子之心回国潜心研究原生动物学。回国后，王家楫被聘为中国科学社生物研究所动物学部教授兼国立中央大学动物学教授。1934年担任中央研究院动植物研究所所长，后来该所以王家楫为首的生物学家开创了我国鱼类学、原生动物学和藻类学等领域的研究。1948年当选中央研究院院士，1949年被任命为中央研究院常务委员、动物研究所所长。同年，中国科学院水生生物研究所成立，王家

译者：付雷
浙江师范大学，金华321004，中国
邮箱：ful527@163.com

楫担任所长，1955年当选中国科学院学部委员（院士）。

王家楫是我国原生动物学研究的第一人。在秉志的影响下，王家楫对达尔文的进化论产生了极大的兴趣，当他发现我国在原生动物研究领域还处于一片空白时，便立志为此奉献一生。1925年王家楫发表的处女作《南京原生动物之研究》（Wang，1925）轰动一时，这是我国首篇关于原生动物的文章，标志着我国原生动物学研究领域的开端。他在回国后延续了他的博士论文的相关研究。在之后的四年中，他的足迹遍布山东、福建、广东、四川和其他沿海地区，发现了许多海洋与淡水原生动物的新种属，为开展我国原生动物学调查奠定了基础。后来，他用蛋白银染色法发现缘毛类纤毛虫虽然虫体纤毛退化，但膜下纤维系统仍然存在，并发表《壁累枝虫之纤维系统》论文，对原生动物系统发育的研究起到了重要的推动作用。晚年时王家楫患有严重的白内障和1200度近视，但是他仍然没有停止对原生动物学的研究，发表了《珠穆朗玛峰地区的原生动物》（王家楫，1974）和《西藏高原部分地区的原生动物》（王家楫，1977），为西藏地区原生动物学的考察作出了巨大贡献。此外，他还发现了原生动物3个新属、58个新种、4个新变种、8个新亚种，为后人研究原生动物学奠定了基础。

王家楫是我国淡水轮虫学的开创者，他出版的《中国淡水轮虫志》（王家楫，1961）（图2）首次对分布在我国沼泽、池塘、湖泊及水库内的常见轮虫种类做了详细的分类和描述，拓展了我国无脊椎动物学的研究领域。该著作涉及已观察到的淡水轮虫种类252种，隶属于79属，15科。其中有4个新种及2个新族，且自目到属的特征都扼要地加以叙述，并附有检索表。该书对种的描述特别详细，所有的252个种都附有一个或多个图像，共计533幅图，可以归纳成27个版块。该著作对轮虫的亲缘关系以及形态、生理、生态等问题，都有深入的探讨。该项成果获得1978年全国科学大会奖和湖北省科学大会奖。

王家楫的毕生贡献不仅限于自己的工作，他还为祖国培养了朱树屏、沈韫芬等一大批优秀的生物学家。1927年，王家楫在中国科学社生物研究所和国立中央大学任教授，教授普通生物学、无脊椎动物学、组织学及胚胎学等课程。受其父影响，他在论述大学教授的任务时说，如果大学教授"议论所及，不出教本"，则"无异蕴匾而藏也"，所以应该"从大处落墨，视实时之急需，用我所长，尽力于教课以外之义务。"（王家楫，1934，1936）这表明在当时的形势下，他认为当务之急是培养经世致用的人才（图3）。

王家楫还是一名极具影响力的社会活动家。1934年他参与发起和创立了中国动物学会并任理事，担任《动物学杂志》总编辑。1952年加入九三学社，1954年任九三学社武汉直属小组组长、武汉分社主任委员。1960年12月至1961年2月，他代表中国为越南规划设立中央水产研究所。1962年代表我国赴苏联列宁格勒参加太平洋西部渔业研究委员会第二次会议。

王家楫是我国伟大的生物学家、教育家和爱国主义学者，是我国动物学家的典范，为我国原生动物学和轮虫学作出了不可磨灭的历史性贡献。

图2 《中国淡水轮虫志》

图3 王家楫和他的学生

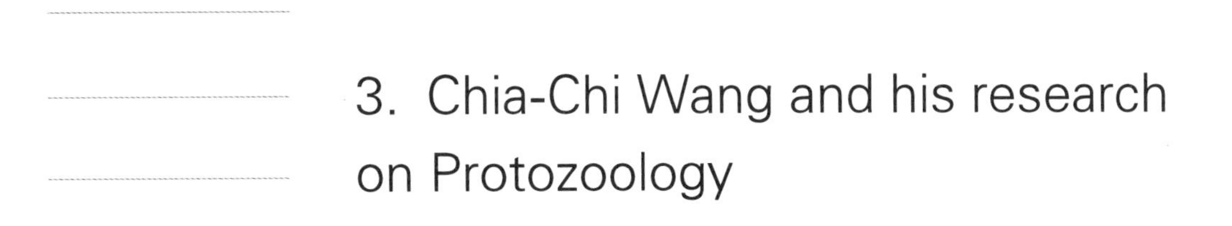

3. Chia-Chi Wang and his research on Protozoology

Chia-Chi Wang (王家楫, 1898—1976) was a famous protozoologist, the founder of protozoology and pioneer of rotiferology in China.

Chia-Chi Wang was born in a scholarly family in Fengxian, Jiangsu. At the age of six, he began to study at the Zhaowen Academy. When the country was declining, his father advocated that scholars should learn advanced technology to change the status quo. His father's pioneering spirit and enlightened attitude towards new things had a profound impact on Chia-Chi Wang's life.

In 1920, Chia-Chi Wang graduated from the Nanking Ecole Normal School (which was later incorporated into Southeast University), where he later engaged in protozoology-related research under the guidance of Professor Chih Ping (秉志). In 1925, he studied at the University of Pennsylvania as a state-financed student because of his excellent performance. While in the United States, he delved into the study of marine protozoa. In 1928, his doctoral thesis Ecological studies of the seasonal distribution of protozoa in a fresh-water pond (Wang, 1928) won the Gold Medal, and earned him a doctorate. While studying in the United States, he published a number of papers on the physiology and taxonomy of protozoa in *Science*, *Morphology and Physiology*, as well as *Animal Physiology*, which attracted much attention from American scholars. Later, he was hired as a Sterling Fellow at Yale University.

In 1929, Chia-Chi Wang heard that foreign scientific groups would go to China to collect plant and animal samples. He thought that Chinese biological resources belonged to China, and claimed foreigners should know that Chinese affairs could be solved by Chinese people, and that biological resources in China could be exploited by Chinese people only. Therefore, he gave up the superior job opportunities and living conditions provided by Yale University, and returned to China to continue his research on protozoa. Then, Chia-Chi Wang was hired as a professor of zoology at the Institute of Biology of the Science Society of China, and a professor of biology at the National Central University at the same time. In 1934, he served as a director of the Institute of Zoology and Botany of the National Academia Sinica. Later, an advanced research program on

Ting Shi, Lei Fu
Zhejiang Normal University, Jinhua 321004, China
Correspondence: ful527@163.com (L. Fu)

the fish, protozoa and algae of China was initiated at this institute, which was headed by Chia-Chi Wang. In 1948, he was elected as an academician of the National Academia Sinica. In 1949, he was appointed as a standing committee member of the National Academia Sinica and the director of the Institute of Zoology. In the same year, the Institute of Hydrobiology of the Chinese Academy of Sciences was established and Chia-Chi Wang was appointed as the director. In 1955, he was elected as an academician of the Chinese Academy of Sciences.

Chia-Chi Wang was a great biologist, who was the first to study protozoa in China, and the founder of rotiferology. Under the influence of Chih Ping, Chia-Chi Wang developed a great interest in Darwin's theory of evolution. He devoted himself to this theory, and personally went into the field to study protozoa. When he found that the research on protozoa in China was still underdeveloped, he decided to dedicate to this field. In 1925, Chia-Chi Wang's first paper Study of the protozoa of Nanking (Wang, 1925) was published, and this was the first article on protozoology in China, which meant the beginning of Chinese protozoology research. In Pennsylvania, his PhD thesis Ecological studies of the seasonal distribution of protozoa in a fresh-water pond was a continuation of his research on protozoa, which he later continued after returning to China. In the following four years, he travelled around Shandong, Fujian, Guangdong, Sichuan and other coastal areas and territorial seas of China to study protozoa. As a result, many marine and freshwater protozoa were discovered, which laid the foundation for future studies. Later, he found that ciliophora still retain a fiber membrane although the cilia are degenerated. Subsequently, he published the paper The fiber system of the wall worm. His research played an important role in accelerating the development of protozoan phylogeny. Chia-Chi Wang suffered from severe cataract and 1200-degree myopia in his later years, but he never stopped his research, and then addressed Protozoa in the Mount Qomolangma Region (Wang, 1974) and Protozoa in parts of the Tibetan Plateau (Wang, 1977), which made great contributions to the investigation of protozoa in Tibet. Moreover, he discovered three new genera, fifty-eight new species, four new varieties, and eight new subspecies of protozoa, illustrating the extent of his groundbreaking work.

Chia-Chi Wang was the pioneer of Chinese freshwater rotiferology. His book *Chinese Freshwater Rotifers* (Wang, 1961), which was the first to classify and describe the common rotifer species in Chinese swamps, ponds, lakes and reservoirs in detail, expanded the area of Chinese invertebrate zoology. This book included 252 species divided into 79 genera and 15 families. There were 4 new species and 2 new races, which were summed up in detail from class to genus with a search index attached. The description of the species was particularly detailed. One or more images were attached to all of the 252 species, with 533 pictures in total, which were classified into 27 plates. The kinship, morphology, physiology and ecology of rotifers was discussed in depth, and this work won him the National Science Conference Award in 1978, as well as the Hubei Science Conference Award.

Chia-Chi Wang's contribution was not limited to his own research, since he also instructed several famous biologists, including Shuping Zhu (朱树屏) and Yunfen Shen (沈韫芬). In 1927, Chia-Chi Wang taught general biology, invertebrate zoology, histology and embryology at the

Institute of Biology of the Science Society of China and the National Central University. Under the influence of his father, he believed that only teaching knowledge through textbooks rather than in person didn't have any effect. Thus, he emphasized that the goal of university education was not only imparting textbook knowledge, but also life skills, to improve the students' ability to solve the urgent problems of the nation (Wang, 1934, 1936), which was especially important at the birth of modern China.

Accordingly, Chia-Chi Wang was also an influential social activist. In 1934, the China Zoological Society was founded, and he served as a board council member as well as the editor-in-chief of the *Journal of Zoology*. In 1952, he became a member of the Jiu San Society (九三学社). In 1954, he was the team leader of the group in Wuhan, and the chairman of the Wuhan Branch. From December 1960 to February 1961, he helped set up the Central Fisheries Research Institute for Vietnam. In 1962, he participated in the second meeting of the Western Pacific Fisheries Commission in Leningrad, Soviet Union.

Chia-Chi Wang is clearly one of the great biologists of China, an educator and a patriotic scholar. He was a role model for Chinese zoologists. He made an indelible historical contribution to Chinese protozoology and rotiferology.

Figures

References

Wang CC (1925) Study of the protozoa of Nanking. Part I. Cont Biol Lab Sci Soc China 1:92–177.

Wang CC (1928) Ecological studies of the seasonal distribution of protozoa in a fresh-water pond. J Morphol 46:431–478.

Wang CC (1934) The value of science in education. Sci Pict 2(6):201.

Wang CC (1936) Mission of university professors. Guo Feng Mon 8 (5): 170–173.

Wang JJ (1961) Chinese Freshwater Rotifer. Beijing: Science Press. (王家楫. 1961. 中国淡水轮虫志. 北京：科学出版社.)

Wang JJ (1974) Protozoa in the Mount Qomolangma Region. In: Scientific Investigation Report in the Mount Qomolangma Region (1966—1968) Biological and Alpine Physiology. Beijing: Science Press. (王家楫. 1974. 珠穆朗玛峰地区的原生动物. 见：珠穆朗玛峰地区科学考察报告(1966—1968) 生物与高山生理. 北京：科学出版社.)

Wang JJ (1977) Protozoa in parts of Tibetan Plateau. Acta Zool Sin 23(2):131–160. (王家楫. 1977. 西藏高原部分地区的原生动物. 动物学报, 23(2): 131–160.)

4. 童第周：中国生物克隆技术之父

图1　童第周（1902—1979）

在我国，童第周（T. C. Tung）（图1）是个家喻户晓的名字，在小学课本中就有一则关于他的故事。故事讲述了他由于家庭贫困直到17岁才进入中学，以及他如何依靠毅力和勤奋最终成为一位成功的生物学家。但是在科学界，他之所以被人们铭记和尊敬，则是由于他在种间克隆技术上所取得的杰出成就。

谈到克隆技术，可能大多数人脑海中马上会闪现出多利羊的身影，这个动物克隆技术上的突破性进展是由Ian Wilmut与Keith Campbell以及其他苏格兰罗斯林学院的合作者在1997年共同实现的。然而，多利并不是第一个克隆动物，早在半个多世纪以前，脊椎动物克隆技术就已经得到了发展。第一个被成功克隆的脊椎动物是一只北方豹蛙（*Rana pipiens*），在1952年，Robert Briggs与Thomas King通过将囊胚细胞的细胞核注入同种生物的去核卵细胞中，完成了这项工作。另外，人们还希望了解分化完全的体细胞是否会像胚胎细胞那样含有全部发育所需的遗传信息，或者说，体细胞是否也可以用于动物克隆。为了解释这个问题，1962年，John Gurdon与他的同事利用小肠上皮细胞成功克隆了南非蛙（*Xynopus*），由此证明即使分化后的细胞也能够保持完整的遗传信息以维持发育的全能性。然而，来自不同种属的细胞核与去核卵细胞的相容性一直困扰着众多实验生物学家。

作为一名实验胚胎学家，在数十年的研究经历中，童第周开展的研究涉及诸多领域，包括胚胎学、细胞生物学及遗传学，利用了不同的生物模型，例如海胆、文昌鱼及硬骨鱼类（图2）。无论童第周使用何种技术手段进行何种课题的研究，他始终对胚胎产生与发育过程中细胞核与细胞质的相容性问题抱有极大的兴趣——是否仅仅是细胞核包含了个体发育所必需的全部遗传物质？如果一个物种的卵细胞核被其他物种的细胞核所代替的话，将会发生什么？他一遍又一遍地反复思考这些问题。但在那个年

译者：郝宁

中国科学院生物物理研究所，北京100101，中国

邮箱：haoning@ibp.ac.cn

图2 童第周与他的妻子、胚胎学家叶毓芬

代，技术手段的缺乏使得他不能将其付诸实践。

当童第周读到Briggs和King在1952年发表的关于在两栖动物身上进行细胞核移植的文章时，他认识到卵细胞核的显微操作可能是找到答案的一种有效手段。在20世纪50年代的中国，与研究工作相关的一切资源都极度匮乏，想要进行核移植等高精度研究几乎是不可能的。在经历了数年的不懈努力和无数次的失败后，童第周与他的同事成功地创立了他们自己的显微操作体系（图3）。通过这个外形不佳但方便易用的装置，他们将雄性亚洲鲤鱼的细胞核转移到雌性亚洲鲤鱼的去核卵细胞中，从而产生了世界上第一群克隆鱼。这些克隆鱼非常健康，它们完成了完整的个体发育期，在水中畅游，并最终产生了后代。这些研究发生在1963年，比多利羊的诞生早了34年。遗憾的是，由于实验结果只用中文发表，甚至连英文摘要都没有，这项研究的重大科学意义并没有被国际科学界所关注和认识。

图3 童第周与他的同事成功地创立了他们自己的显微操作体系
（中国科学院遗传与发育生物学研究所，1989）

在取得克隆鱼的成功后，童第周并没有止步不前，他翻开了研究工作的下一页——种间克隆。在10年后，也就是1973年，他和同事从鲤鱼的胚胎细胞中分离出了细胞核，并将其转移到鲫鱼的去核卵细胞中，一部分核移植体完全发育成熟。这是世界上第一例成功的种间动物克隆工作！然而由于当时我国正处于艰难的政治经济环境，这项研究并没有得到应有的关注。实际上，这项研究结果在童第周去世一年后，即1980年才发表。童第周曾非常高兴地看到，细胞核–细胞质杂交鱼具有某些与其细胞核供体不一样的形态特性，他梦想着这些新特征，例如生长速度快以及营养含量高等，能在未来水产养殖业中发挥积极作用。

4. Father of biological cloning in China

Dizhou Tong (T. C. Tung) is a well-familiar name in China, as there is a story about him in one of the elementary school textbooks. The story describes how he came from a poor family and could not go to middle school until the age of 17, and how his resolution and diligence made him a successful biologist. But in the academic community, he is remembered and respected for another reason—the outstanding achievement of cross-species cloning in fish.

When talking about the issue of cloning, the first thing that come to the minds of most people is probably the sheep Dolly, the breakthrough in animal cloning by Ian Wilmut, Keith Campbell and the collaborators at the Roslin Institute, Scotland in 1997. However, Dolly is not the first cloned animal because the art of animal cloning in vertebrates has been developing for more than half a century. The first successfully cloned vertebrate is a northern leopard frog *Rana pipiens*, which was done by Robert Briggs and Thomas King through introducing nuclei from cells of the hemisphere blastula into enucleated eggs of the same species in 1952. Nevertheless, people wondered whether a fully differentiated somatic cell would contain the same genetic information as an embryonic cell did, i.e., whether cloning with somatic cells was possible. To answer this question, John Gurdon and his colleagues used the nucleus of an adult intestinal epithelial cell to clone South African frogs *Xynopus* in 1962, proving that even a well-specialized cell still maintains the full genetic information of developmental totipotency. On the other hand, the incompatibility of nuclei and enucleated eggs among different species had been puzzling experimental biologists.

As an experimental embryologist, during his decades of research experience, Dizhou Tong laid out his research in a wide spectrum including embryology, cell biology and genetics with different models such like sea urchins, amphioxux, and teleosts. No matter what subject he was studying and what techniques he was using, Dizhou Tong was very much curious about the

Zuoyan Zhu[1,2], Ming Li[3], Le Kang[3]

1 Institute of Hydrobiology, Chinese Academy of Sciences, Wuhan 430072, China

2 College of Life Science, Peking University, Beijing 100871, China

3 Beijing Institutes of Life Science, Chinese Academy of Sciences, Beijing 100101, China Contact: zyzhu@ihb.ac.cn (Z.-Y. Zhu)

compatibility of nucleus and cytoplasm in embryogenesis and development—did the nucleus alone have all the genetic materials necessary to support the development of an organism? What would happen if the egg nucleus was replaced by another one from different species? He thought of this issue over and over again, but the lack of techniques in those years did not allow him to test his curiosity.

When he read the paper about nuclear transplantation performed on amphibians by Briggs and King in 1952, Dizhou Tong realized that the nuclear micro-manipulation of eggs could be an appropriate way to approach his question. In the 1950s when all kinds of resources for research were extremely deficient in China, it was almost impossible to perform highly precise research like nuclear transplantation. After a few years of enduring efforts and countless failures, Dizhou Tong and his colleagues successfully made their own set of micro-manipulator. With this not-so-pretty but very handy equipment, they managed to transfer the nuclei from cells of a male Asian carp to enucleated eggs of a female Asian carp, and therefore, generated the first cloned fish in the world. The cloned fish were perfectly healthy—they developed to full term, swam around and finally produced offspring. This happened in 1963, thirty-four years before Dolly the sheep was born. Sadly, the significance of this study was not recognized by the international scientific community, as the result was only published in Chinese even without English abstract.

Dizhou Tong did not stop at the success of fish cloning. He moved onto the next page—cross-species cloning. Ten years later, in 1973, he and his coworkers isolated nuclei from embryonic cells of a common carp *Cyprinus carpio* and transferred it into enucleated eggs of crucian carp *Carassius auratus*. Some of the nuclear-transplants later fully developed into adulthood. It was the first case of cross-species animal cloning in the world! However, the study did not attract much attention as it should have been, due to the difficult circumstances in China in the 1970s. In fact, the result of this study did not get published until 1980, one year after Dizhou Tong passed away. Dizhou Tong was very happy to see the nucleo-cytoplasmic hybrid fish with some distinct characteristics from the nuclear donor species, and dreamed that the novel traits, such as faster growth rate and nutritional contents, would be beneficial to aquaculture in the future.

Figures

Fig.1 Dizhou Tong(1902—1979)

Fig.2 Dizhou Tong and his wife, embryologist Yufen Ye

Fig.3 The micro-injection machine designed by Dizhou Tong and his colleagues (Institute of Genetics and Developmental Biology, Chinese Academy of Sciences, 1989)

References

Institute of Genetics and Developmental Biology, Chinese Academy of Sciences (1989) Dizhou Tong Collection. Beijing: Academic Journal Press.

5. 伍献文：中国鱼类学研究的奠基者

2021年，历时近半个世纪，三卷本的《中国动物志·硬骨鱼纲·鲤形目》经过四代鱼类学工作者的努力终于全部完成，最后一卷正式交科学出版社付印，本书的全部完成和出版是我国著名鱼类学家伍献文院士（图1）一生的夙愿。

图1　伍献文（1900—1985）

出生于1900年的伍献文是浙江瑞安人，自1932年从巴黎国家自然历史博物馆获得博士学位以来，一直在中央研究院和中国科学院工作。他的研究工作除鱼类以外，还包括线虫、两栖类和爬行类。1964—1977年，他主编的两卷本《中国鲤科鱼类志》（包括113属，412种），系统性地对我国淡水鱼类进行全面描述，结束了我国淡水鱼类研究需要查阅过去几十年零乱出版的鱼类学研究论文的现状，为后来的《中国动物志·硬骨鱼纲·鲤形目》的编著打下了良好的基础（伍献文等，1964，1977）。

自卡尔·林奈1758年出版《自然系统》一书以来，双名法在分类学中得到了广泛应用。双名法虽然以模式标本解决了物种的命名问题，但是，物种间的相互关系和分类的等级在那时还是人为的经验系统，由最大相似性来决定，而通过祖先性状、衍生性状来决定演化极向的系统发育研究刚刚开始。1977年起，伍献文院士指导陈宜瑜、陈湘粦等对鲤形目及其以下的鲤科、鳅科、平鳍鳅科、亚口鱼科开展了广泛的骨骼学研究，使用当时最前沿的分支系统学（cladistic systematics）方法，以祖离性状特征分析了鲤亚目的进化过程（图2）。终于在1981年的《中国科学》和1984年的《动物分类学报》上，分别发表了重要论文（伍献文等，1981；陈湘粦等，1984），阐述了鲤科鱼类系统发育关系，成功地建立了鲤形目鱼类的自然分类系统，论文发表后得到了国际动物学界鱼类学家群体的广泛关注。英国鱼类学家Greenwood邀请陈宜瑜去大英博物馆开展合作研究；美国密歇根大学教授S. V. Fink和W. L. Fink高度评价伍献文团队的工作，写来长信探讨研究内

作者：何舜平，陈宜瑜

淡水生态与生物技术国家重点实验室，中国科学院水生生物研究所，武汉430072，中国

邮箱：clad@ihb.ac.cn（何舜平）；chenyy@nsfc.gov.cn（陈宜瑜）

图2 1981年，伍献文院士（中）与团队成员陈宜瑜（右）和陈湘粦（左）

容。伍献文团队的研究工作，为我国鱼类学研究后来的快速发展奠定了坚实的基础。这一阶段的工作被《世界鱼类》第二版收录并运用在新的鱼类分类系统中（Nelson，1984）。

伍献文教授高度重视鱼类标本的收藏、采集，强调野外工作的重要性、长期性。团队的陈宜瑜、曹文宣长期致力于青藏高原野外科考，不断补充标本，使得我国的鲤形目鱼类标本收藏世界第一，这些标本收藏和相关研究使得中国科学院水生生物研究所（简称水生所）成为世界鱼类学研究中心之一，与法国、瑞典、英国和美国的鱼类研究中心一起引领世界鱼类学的发展。伍献文院士创建的鱼类标本馆目前收集了鱼类、原生动物等无脊椎动物标本共40余万号，鱼类1200余种，是世界鱼类学研究重要的基地。

伍献文先生对科学前沿研究的重视和长远的眼光使得水生所鱼类学研究团队飞速发展，他们对鲤科鲤形目（何舜平等，2004；He et al.，2008b；Wang et al.，2006；Jian et al.，2020）和高原鱼类（曹文宣等，1981；武云飞，1984；He et al.，2004；He and Chen，2007；Yang et al.，2021）的进化以及深渊鱼类的适应研究（Shen et al.，2017；Wang et al.，2019）都源自这种精神的传承、发展和创新。正因为伍献文院士打下了鲤形目、鲤科鱼类系统分类学的坚实基础，以及老一辈学者的承先启后，我国的鱼类学研究工作逐步跟上了时代的脚步。随着分子系统学的逐步深入应用，我们从形态学趋同、最大相似性的分类学泥潭中走了出来，纠正了过去在鲤形目、鲤科分类学中的错误，提出了东亚特有鲤科鱼类在东亚季风气候条件下独立起源的观点，重建的鲤形目鱼类系统发育得到了广泛承认。大量有关论文发表在《分子系统发育与进化》（Wang et al.，2006；He et al.，2008a，2008b；Li et al.，2008a；Duan et al.，2009；Zhao et al.，2011；Tao et al.，2013；Yang et al.，2015）、《中国科学》（何舜平等，2004；Zhao et al.，2005；Wang et al.，2016；Tao et al.，2019）、《中国科学通报》（He et al.，2001，2004；Li et al.，2008b；He and Chen，2007）、《自然科学进展》（Peng et al.，2002；Wang et al.，2004；Li et al.，2005）等杂志，研究成果被知名的《世界鱼类》第五版大量引用（Nelson et al.，2016）。

伍献文在近代鱼类学研究上所作的贡献主要体现在集中描述了鲤形目、鲤科鱼类，出版了《中国鲤科鱼类志》《中国动物志·硬骨鱼纲·鲤形目》，领导开展了鲤形目、鲤科鱼类的系统发育研究，客观反映了它的系统关系。伍献文的贡献还包括建立了亚洲最大的鱼类标本馆，以鲤形目鱼类收藏为特色。

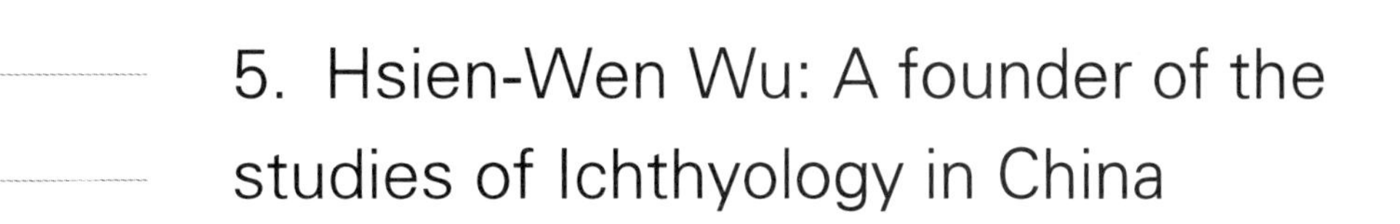

5. Hsien-Wen Wu: A founder of the studies of Ichthyology in China

The three-volume *Fauna Sinica, Osteichthyes, Cypriniformes* was finally completed in 2021. After the efforts of four generations of ichthyologists, the last volume has been officially submitted to the Science Press for printing. The publication of this book is a lifelong dream of the Academician Hsien-Wen Wu (伍献文, H. W. Wu) , a famous ichthyologist in China.

Hsien-Wen Wu was born in the city of Ruian, Zhejiang Province in 1900. After graduating from Xiamen (Amoy) University in 1927, he joined the teaching staff in the Central University and began research work on fishes and their parasitic nemathelminths. His first ichthyological report, entitled "Study of fishes of Amoy" , was published in 1929. In 1932, he graduated from the National Museum of Natural History in Paris, obtained a Doctor of Science degree with his dissertation "Contribution a l' Etude, morphologique, biologique et systematique des poisons heterosomata (Pisces Heterosomata) de la Chine" and was recommended to preside over the research work of the Institute of Zoology and Botany in Academia Sinica. In the same time, he held a position concurrently as professor in zoology, anatomy, embryology and parasitology in the National Central University, Fudan University and Jiangsu Medical College, respectively. His research objects were not only fish, but also included helminths, amphibians and reptiles. During World War Ⅱ , the Institute of Zoology and Botany moved from Nanjing to Chonqing, where Professor Wu devoted himself to studies on experimental ichthyology, including physiology and functional morphology such as the mechanism of aerial respiration in fish. From 1940—1947, Professor Wu guided his early students, Dr. Jiankang Liu (刘建康) and Dr. Xiaowei Zhang (张孝威), the famous ichthyologists in China, serially published many scientific reports on the respiration mechanism of rice eel, relative in detailly exploring the structure and function of respiration organ of rice eel based on the morphology, histology and physiology. In 1950, the Institute of Zoology and Botany was renamed as the Institute of Hydrobiology (IHB), Chinese Academy of Sciences (CAS) and moved from Shanghai to Wuhan. From 1964 to 1977, Dr. Wu

---

Shunping He, Yiyu Chen

State Key Laboratory of Freshwater Ecology and Biotechnology, Institute of Hydrobiology, Chinese Academy of Sciences, Wuhan 430072, China

Correspondence: clad@ihb.ac.cn (S.-P. He) ; chenyy@nsfc.gov.cn (Y.-Y. Chen)

devoted himself to the editing of the two-volume monography *The Cyprinid Fishes of China* (including 113 genera, 412 species), and systematically gave a comprehensive description of freshwater fishes in China. The publication of this book ended an era of consulting many ichthyological articles scattered in scientifical journals in the past decades, and laid a good foundation for the later compilation of *Fauna Sinica, Osteichthyes, Cypriniformes* (Wu et al., 1964, 1977). Also in 1953, one of the excellent students of Professor Wu, Xiaowei Zhang organized a 5-year duration survey of resource of mackerel fish in fishery outside Yangtai and nearby, this survey fill the gap of marine fish ecology in China.

Since *Systema Natura* was published by Carolus Linnaeus in 1758, binomial nomenclature has been widely applied in taxonomy. The binomial nomenclature has solved the name issue of species with the help of type specimens. However, the relationships among species and the level of classification were still an artificial empirical system, determined by maximum similarity at that time. The phylogenetic studies with ancestral and derived traits (the cladistic systematics) were not developed until 1970s. Since 1977, encouraged and promoted by Hsien-Wen Wu, Yiyu Chen (陈宜瑜), Xianglin Chen (陈湘粦) and the members of team have conducted an extensive skeletal studies on the representative species in the order Cypriniformes, as well as the descending families such as Cyprinidae, Cobitidae, Homalopteridae, and Catostomidae. They applied the cutting edge cladistic systematics to analyze the evolutionary process of the Cyprinoidei, and finally published two important papers (Wu et al., 1981; Chen et al., 1984). These papers have expounded the phylogenetic relationship of Cyprinid fish and successfully established the natural classification system of cypriniform fishes. The publication of these papers attracted extensive attention from the international zoological and ichthyological communities. British ichthyologist invited Dr. Yiyu Chen to conduct cooperative research in the British Museum. Professor S. V. Fink and W. L. Fink from the University of Michigan spoke highly of the work of Hsien-Wen Wu's team and wrote a long letter to discuss the scientific topics. The work of Hsien-Wen Wu's team has laid a solid foundation for the rapid development of ichthyology research in China, and these works have been included in the second edition of *Fishes of the World* and have been applied in the fish classification system at that time (Nelson, 1984).

Professor Hsien-Wen Wu paid high attention to the collection of fish specimens and emphasized the importance of long-term fieldwork. The team members Yiyu Chen and Wenxuan Cao (曹文宣) have long devoted themselves to field research on the Qinghai-Tibetan Plateau to collect specimens. Due to their efforts, China currently has the largest collection of Cypriniformes specimens in the world, and IHB has been rewarded to be one of the world's leading centers for ichthyology, along with fish research centers in France, Sweden, the United Kingdom, and the United States. Until now, the fish museum founded by Hsien-Wen Wu has collected more than 400000 specimens, accounting for more than 1200 species of fish.

Dr. Hsien-Wen Wu emphasized the importance of frontier scientific research, which promoted the rapid development of the ichthyology research team in IHB. Following his heritage, the team has conducted a series of cutting-edge studies, including the evolution of Cyprinidae, Cypriniformes (He et al., 2004b, 2008b; Wang et al., 2006; Jian et al., 2020) and plateau fishes (Cao et al., 1981; Wu, 1984; He et al., 2004a; He and Chen, 2007; Yang et al., 2021), as well as the adaptation of abyssal fish (Shen et al., 2017; Wang et al., 2019). Due to the solid foundation built by Hsien-Wen Wu and the good inheritance by Dr. Yiyu Chen et al., the research work of ichthyology in China has gradually kept pace with the times. With the further application of molecular systematics, we have stepped out of the mire of morphological convergence and maximum similarity, and corrected many mistakes in the taxonomy of Cypriniformes and Cyprinidae. Now we have suggested an independent origination of the East Asian Cyprinids under the East Asian monsoon climate and reconstructed the phylogeny of Cypriniformes. A large number of papers related to molecular systematics of Cyprinidae and Cypriniformes has been published in *Molecular Phylogenetics and Evolution* (Wang et al., 2006; He et al., 2008a, 2008b; Li et al., 2008a; Duan et al., 2009; Zhao et al., 2011; Tao et al., 2013; Yang et al., 2015), *Science China* (He et al., 2004b; Zhao et al., 2005; Wang et al., 2016; Tao et al., 2019), *Chinese Science Bulletin* (He et al., 2001, 2004a; Li et al., 2008b; He and Chen, 2007), *Progress in Natural Science* (Wang et al., 2004; Li et al., 2005) and so on, which has been widely accepted by the international ichthyologist community and has been cited in the famous fifth edition of *Fishes of the World* (Nelson et al., 2016).

In summary, Hsien-Wen Wu's great contribution to ichthyological research in China is mainly manifested in the concentrated description of Cyprinidae and Cypriniformes, the publication of *The Cyprinid Fishes of China* and *Fauna Sinica, Osteichthyes, Cypriniformes*, and the launch of the phylogenetic studies on Cyprinidae and Cypriniformes.

Hsien-Wen Wu's great contribution also includes the establishment of the largest fish museum in Asia featuring Cypriniformes.

Figures

Fig.1 Hsien-Wen Wu (1900—1985)

Fig.2 Hsien-Wen Wu (middle) with the team members Dr. Yiyu Chen (right) and Dr. Xianglin Chen (left) in 1981

References

Cao WX, Chen YY, Wu YF, et al (1981) Origin and evolution of Schizothoracine fishes in relation to the upheaval of the Qinghai-Tibetan Plateau. In: Qinghai-Tibetan Plateau Comprehensive Scientific Expedition, Chinese Academy of Sciences. Studies on the Period, Amplitude and Type of the Uplift of the Qinghai-Tibetan Plateau. Beijing: Science Press, 118–130. (曹文宣，陈宜瑜，武云飞，等. 1981. 裂腹鱼类的起源和演化及其与青藏高原隆起的关系. 见：中国科学院青藏高原综合科学考察队. 青藏高原隆起的时代、幅度和形式问题. 北京：科学出版社，118–130.)

Chen XL, Yue PQ, Lin RD (1984) Major groups within the family Cyprinidae and their phylogenetic relationships. Acta Zootaxonomica Sinica 9(4): 424–440. (陈湘粦，乐佩琦，林人端. 1984. 鲤科的科下类群及其宗系发生关系. 动物分类学报，9(4):424–440.)

Chen YY (1986) Wu Hsien-Wen, 1900—1985. Copeia 2:557–558.

Duan ZY, Zhao K, Peng ZG, et al (2009) Comparative phylogeography of the Yellow River schizothoracine fishes (Cyprinidae): Vicariance, expansion, and recent coalescence in response to the Quaternary environmental upheaval in the Tibetan Plateau. Molecular Phylogenetics and Evolution 53(3):1025–1031.

He DK, Chen YF (2007) Molecular phylogeny and biogeography of the highly specialized grade schizothoracine fishes (Teleostei: Cyprinidae) inferred from cytochrome b sequences. Chinese Science Bulletin 52(6):777–788.

He DK, Chen YF, Chen YY, et al (2004a) Molecular phylogeny of the specialized schizothoracine fishes (Teleostei: Cyprinidae), with their implications for the uplift of the Qinghai-Tibetan Plateau. Chinese Science Bulletin 49(1):39–48.

He SP, Chen YY, Tsuneo N (2001) Sequences of cytochrome b gene for primitive cyprinid fishes in East Asia and their phylogenetic concerning. Chinese Science Bulletin 46(8):661–665.

He SP, Gu X, Mayden RL, et al (2008a) Phylogenetic position of the enigmatic genus *Psilorhynchus* (Ostariophysi: Cypriniformes): Evidence from the mitochondrial genome. Molecular Phylogenetics and Evolution 47(1):419–425.

He SP, Liu HZ, Chen YY, et al (2004b) Molecular phylogenetic relationships of Eastern Asian Cyprinidae (Pisces: Cypriniformes) inferred from cytochrome b sequences. Science China 34(1):96–104. (何舜平，刘焕章，陈宜瑜，等. 2004. 基于细胞色素b基因序列的鲤科鱼类系统发育研究(鱼纲：鲤形目). 中国科学，34(1):96–104.)

He SP, Mayden RL, Wang X, et al (2008b) Molecular phylogenetics of the family Cyprinidae (Actinopterygii: Cypriniformes) as evidenced by sequence variation in the first intron of S7 ribosomal protein-coding gene: Further evidence from a nuclear gene of the systematic chaos in the family. Molecular Phylogenetics and Evolution 46(3):818–829.

Jian JB, Yang LD, Gan XN, et al (2020) Whole genome sequencing of silver carp (*Hypophthalmichthys molitrix*) and bighead carp (*Hypophthalmichthys nobilis*) provide novel insights into their evolution and speciation. Molecular Ecology Resources 21(3):912–923.

Li JB, Wang XZ, He SP, et al (2005) Phylogenetic studies of Chinese labeonine fishes (Teleostei: Cyprinidae) based on the mitochondrial 16S rRNA gene. Progress in Natural Science 15(3):213–219.

Li JB, Wang XZ, Kong XH, et al (2008a) Variation patterns of the mitochondrial 16S rRNA gene with secondary structure constraints and their application to phylogeny of cyprinine fishes (Teleostei: Cypriniformes). Molecular Phylogenetics and Evolution 47(2):472–487.

Li ZQ, Guo BC, Li JB, et al (2008b) Bayesian mixed models and divergence time estimation of Chinese cavefishes (Cyprinidae: *Sinocyclocheilus*). Chinese Science Bulletin 53(15):2342–2352.

Nelson JS, Grande TC, Wilson MVH (2016) Fishes of the World, 5th edition. New Jersey: John Wiley & Sons.

Nelson JS (1984) Fishes of the World, 2nd edition. New York: John Wiley & Sons.

Shen YJ, Dai W, Gao ZM, et al (2017) Molecular phylogeny and divergence time estimates using the mitochondrial genome for the hadal snailfish from the Mariana trench. Science Bulletin 16:1106–1108.

Tao WJ, Mayden RL, He SP (2013) Remarkable phylogenetic resolution of the most complex clade of Cyprinidae (Teleostei: Cypriniformes): A proof of concept of homology assessment and partitioning sequence data integrated with mixed model Bayesian analyses. Molecular Phylogenetics and Evolution 66(3):603–616.

Tao WJ, Yang L, Mayden RL, et al (2019) Phylogenetic relationships of Cypriniformes and plasticity of pharyngeal teeth in the adaptive radiation of cyprinids. Science China-Life Sciences 4:553–565.

Wang K, Shen Y, Yang Y, et al (2019) Morphology and genome of a snailfish from the Mariana Trench provide insights into deep-sea adaptation. Nature Ecology and Evolution 5:823–833.

Wang XZ, Gan XN, Li JB, et al (2016) Cyprininae phylogeny revealed independent origins of the Tibetan Plateau endemic polyploid cyprinids and their diversifications related to the Neogene uplift of the plateau. Science China-Life Sciences 59(11):1149–1165.

Wang XZ, Li JB, He SP (2006) Molecular evidence for the monophyly of East Asian groups of Cyprinidae (Teleostei: Cypriniformes) derived from the nuclear recombination activating gene 2 sequences. Molecular Phylogenetics and Evolution 42(1):157–170.

Wang XZ, Liu HZ, He SP, et al (2004) Sequence analysis of cytochrome b gene indicates that East Asian group of cyprinid subfamily Leuciscinae (Teleostei: Cyprinidae) evolved independently. Progress in Natural Science 14(2):132–137.

Wu HW, et al (1964) The Cyprinid Fishes of China I. Shanghai: Shanghai Science and Technology Press. (伍献文，等. 1964. 中国鲤科鱼类志(上卷). 上海: 上海科学技术出版社.)

Wu HW, et al (1977) The Cyprinid Fishes of China Ⅱ. Shanghai: Shanghai Science and Technology Press. (伍献文，等. 1977. 中国鲤科鱼类志(下卷). 上海: 上海科学技术出版社.)

Wu HW, Chen YY, Chen XL, et al (1981) Families division of the Cyprinidei and their systematic relationships. Science China 3:369–376. (伍献文，陈宜瑜，陈湘粦，等. 1981. 鲤亚目鱼类分科的系统和科间系统发育的相互关系. 中国科学，3: 369–376.)

Wu YF (1984) Study on systematic classification of Chinese subfamily Schizothoracinae. Bulletin of Plateau Biology 3:119–140. (武云飞. 1984. 中国裂腹鱼亚科鱼类的系统分类研究. 高原生物学集刊，3: 119–140.)

Yang L, Sado T, Hirt MV, et al (2015) Phylogeny and polyploidy: Resolving the classification of cyprinine fishes (Teleostei: Cypriniformes). Molecular Phylogenetics and Evolution 85: 97–116.

Yang LD, Wang Y, Sun N, et al (2021) Genomic and functional evidence reveals convergent evolution in fishes on the Tibetan Plateau. Molecular Ecology 30(22):5752–5764.

Zhao K, Duan ZY, Peng ZG, et al (2011) Phylogeography of the endemic *Gymnocypris chilianensis* (Cyprinidae): Sequential westward colonization followed by allopatric evolution in response to cyclical Pleistocene glaciations on the Tibetan Plateau. Molecular Phylogenetics and Evolution 59(2):303–310.

Zhao K, Li JB, Yang GS, et al (2005) Molecular phylogenetics of *Gymnocypris* (Teleostei: Cyprinidae) in Lake Qinghai and adjacent drainages. Chinese Science Bulletin 50(13):1325–1333.

6. 马世骏：为满足人类和社会发展需求不断开拓新领域的科学家

马世骏（图1）先生生于1915年，是我国著名的生态学家，中国科学院院士。在我国的传统文化中，"世骏"意指宝马良驹，这个美好的名字也恰好描述了他一生辉煌的学术生涯——50年马不停蹄地拓展新的研究领域，从昆虫到人类，从实验科学到系统科学，从生态学到环境科学，以及从自然科学到社会科学。

图1 马世骏（1915—1991）

马世骏先生1937年毕业于北京大学，获得生物学学士学位。进行了几年番茄害虫防治方面的工作之后，他赴美国继续深造。最初他在犹他大学学习昆虫生态学并获得硕士学位，然后他在明尼苏达大学开展玉米螟幼虫防控研究并获得博士学位。1951年，马世骏先生先后在荷兰、比利时、奥地利、法国和英国参加一个多月的学术会议，在那里遇见了著名的生态学家和昆虫学家Charles Elton博士和B. P. Uvarov博士。

1952年，马世骏先生结束了他的国外留学生活回到祖国。他就职于中国科学院昆虫学研究所（之后成为中国科学院动物研究所的重要组成部分之一），并建立了我国第一个昆虫生态学实验室。他的第一要务就是控制我国的蝗虫灾害。蝗虫作为世界上臭名昭著的害虫，是中国农业经济的心腹大患。马世骏先生花费了大量的时间对偏远地区进行了细致的野外考察和调研，从高原到平原，从湖边到河岸地区。经过了细致全面的调查研究，马世骏先生制定了蝗虫治理对策。他没有单纯地去研发高效的农药，而是提出了蝗害的综合防控方案，不仅包括改善农药使用，而且兼顾环境自身的调控作用。在项目执行过程中，他组织了来自昆虫分类学、生态学、生理学和毒理学等多个领域的科学家研究蝗虫群体动力学、调控途径和环境变化等。在他的不懈努力下，我国的蝗害到20世纪70年代就成功降到了最低值。

在20世纪70年代以前，马世骏先生的研究工作多集中在我国主要害虫的控制上，

译者：苏瑞凤
中国科学院北京生命科学研究院，北京100101，中国
邮箱：surf@biols.ac.cn

包括具有迁徙能力的蝗虫、黏虫、棉铃虫和松毛虫等。但是，他并没有把自己局限在传统的害虫生态学和防控研究的方法上。马世骏先生是一位具有远见卓识的科学家，他高瞻远瞩地开拓多领域学科交叉融合来研究昆虫生态学，在中国科学院动物研究所昆虫生态学实验室建立了数学生态学、物理生态学、化学生态学、地理生态学以及经济生态学等多个方向的研究组。他邀请了包括数学生态学研究组的丁岩钦、李典谟和兰仲雄，生理生态学研究组陈宁生和侯无危，化学生态学研究组的陈德明和吴坤君，地理生态学研究组的陈永林和李鸿昌，经济生态学研究组的盛成发等科学家加入团队。这些团队成员都成长为我国昆虫害虫防控领域的专家。他具备超前科学视野，在20世纪60年代就组织计算机科学家和生物学家召开系列跨学科研讨会。他的所有研究都领先于时代，极大地拓展了我国同时代生态学家的研究范畴。

在昆虫生态学研究发展的同时，马世骏先生意识到了解整个生态系统的重要性。为此，他逐渐将研究重心从昆虫生态学转移到了生态系统生态学，这也是当时我国最前沿科学之一。在20世纪70年代，他创立了中国生态学学会，1981年创办了《生态学报》。为了推动人类活动对生态系统影响的研究，他极力推动中国加入“人与生物圈”（MAB）计划和“国际地圈-生物圈”（IGBP）计划。

随着我国工业的快速发展，环境问题日益凸显。作为一个有强烈社会责任感的科学家，马世骏先生也最先意识到这些问题的严重性，他可能是我国最早严肃考虑全球变化、环境污染和核冬天的科学家。他还认识到，迫切需要将生态学研究和环境科学研究结合起来才能解决环境问题。1975年，他推动建立了中国科学院生态环境研究中心，将土壤污染、水污染和大气污染等生态问题列为该中心的研究重点。目前该中心已经成为我国生态环境研究领域的领跑者。

在20世纪80年代，马世骏先生开始意识到人类在生态和环境问题中的重要作用。他认为，“解决环境危机不仅仅需要科学技术，更重要的是需要人们具备生态环境保护的自觉意识。”在60多岁的时候，马世骏先生为了努力提升公众的生态和环境问题意识，工作足迹遍布全国。他应用生态学知识做了大量的工作以影响决策者，最终形成了最著名的“可持续发展”国家政策。

基于马世骏先生的学术成就和社会影响力，他当选为多个国际生态环境组织学会和专家委员会的主席或委员，包括G. H. Brundtland博士的世界环境发展委员会。作为唯一的中国成员，他在1987年7月向联合国提交的著名的《布伦特兰报告：我们共同的未来》的起草工作中作出了巨大贡献，他让世界听到了中国作为发展中国家在环境和发展中的立场。

1991年5月30日，马世骏先生在河北省参加“全国生态农业林业县建设经验交流会”并做大会报告，在返回的途中遇到车祸不幸离世。这对于我国整个生态学界和环境科学界都是巨大的损失。他的一生都在不断探索新领域，如果他没有突然离世，也许还将带领我国的生态学家和环境科学家走入另一个全新的领域。

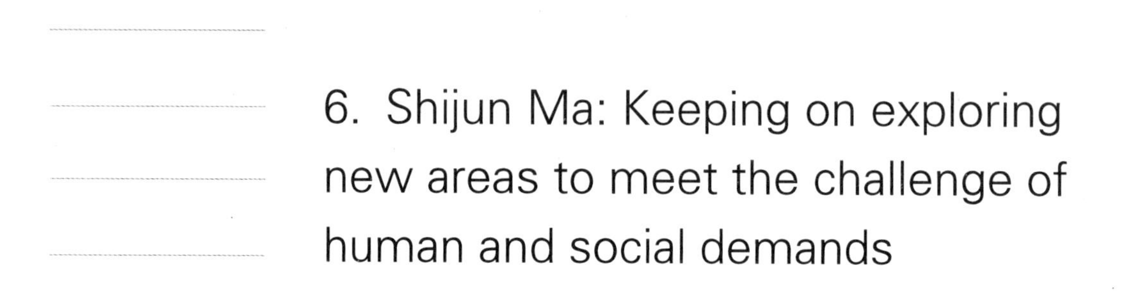

6. Shijun Ma: Keeping on exploring new areas to meet the challenge of human and social demands

Professor Shijun Ma (1915—1991) is a renowned Chinese ecologist. Shijun, literally in Chinese, means the finest horse, which also perfectly describes his academic life—50 years of non-stop traveling and exploring new areas, from insects to human being, from experimental science to systematic science, from ecology to environmental science, and from nature to society.

After getting his BS degree from the Department of Biology at the Peking University in 1937, followed by working on the pest control of tobacco field for a few years, Shijun Ma went to study in the United States. He first studied insect ecology at the Utah State University (where he got his MS degree) and then at the University of Minnesota (where he was awarded PhD degree), doing research on the control of the maizesnout moth's larva. In 1951, Shijun Ma spent over a month traveling in Holland, Belgium, Austria, France and Britain, attending academic conferences and meeting with famous ecologists and entomologists such as Dr. Charles Elton and Dr. Boris P. Uvarov.

In 1952, Shijun Ma finished his long travel and returned to China. He joined the Institute of Entomology (which later became a major part of the Institute of Zoology, Chinese Academy of Sciences, Beijing), Academia Sinica and established the first insect ecology laboratory in China. His first major project was to control the migratory locust plague in China. This notorious pest worldwide was also a severe problem for a country heavily relying on agricultural economy. Before deciding on the locust-controlling strategy, he spent a great amount of time on traveling and researching in the remote regions, from village to village, from plateau to plain, and from lakeside to river basin. Rather than simply looking for more effective pesticides, he proposed an integrated approach to control locust plagues, which not only included improved pesticide application, but also focused on regulation of the environment itself. He organized many scientists from different disciplines such as insect taxonomy, ecology, physiology, and toxicology and they worked together to study the locust population dynamics, control approaches and environmental changes. With his continuous efforts, the locust plague in China was successfully

Le Kang, Ming Li

Beijing Institutes of Life Science, Chinese Academy of Sciences, Beijing 100101, China

Correspondence: lkang@ioz.ac.cn (L. Kang); liming@ioz.ac.cn (M. Li)

reduced to a minimal extent by the 1970s.

Before the 1970s, Shijun Ma's research work mainly focused on controlling major insect pests in China, including migratory locusts, armyworms, cotton boll-worms and pine caterpillars. However, he did not limit himself to the traditional approaches commonly seen in pest ecology and control study. As a foreseeing scientist, he started to explore in new disciplines to study insect ecology, and established several research groups covering areas such as mathematical ecology, physical ecology, chemical ecology, geographical ecology, and even economic ecology at the Insect Ecology Laboratory in the Institute of Zoology. The scientists he recruited to these groups, including Yanqin Ding, Dianmo Li and Zhongxiong Lan in the mathematical ecology group, Ningsheng Chen and Wuwei Hou in the physical ecology group, Deming Chen and Kunjun Wu in the chemical ecology group, Yonglin Chen and Hongchang Li (plus himself) in the geographical ecology group, and Chengfa Sheng in the economic ecology group, all became the backbones of Chinese insect pest fighting experts later on. His vision was so ahead of his time that he even organized a series of workshops of interdisciplinary research between computer scientists and biologists as early as in the 1960s. All his studies in these disciplines were brand new at that time and greatly broadened the horizons of his peer ecologists in China.

While his insect ecological study moved forward, on the other hand, Shijun Ma began to realize the importance of understanding the whole ecosystem no matter what organism he was working on. Thus, he gradually shifted his research focus from insect ecology to ecosystem ecology, which also was a cutting edge area in China at that time. In the 1970s, he founded the Chinese Society of Ecology, and then established the scientific journal *Acta Ecologica Sinica* in 1981. To push the studies on the effects of human activities on ecosystem, he made big efforts to promote the Man and the Biosphere (MAB) programme and the International Geosphere-Biosphere Programme (IGBP) in China.

As the development of Chinese industry picked up its pace, the environmental issues also started to emerge. As a scientist with strong social consciousness, Shijun Ma was among the first to see the severeness of the issues, and was probably the first Chinese to seriously worry about global change, environmental pollution, and even nuclear winter. He also realized that it was urgently needed to combine the power of ecological research and environmental science to fight the environmental issues. In 1975, he promoted the establishment of the Research Center for Eco-Environmental Sciences, Chinese Academy of Sciences, in which ecological questions linked with pollutions of soil, water, and atmosphere are study emphasis. The Center has now become the leading institute of ecoenvironmental research in China.

Since the 1980s, Shijun Ma started to notice the important role of human being in ecological and environmental issues. He believed that "it is not only the science and technology, but also the consciousness of ecology by the public, which can resolve the environmental crisis in the country." In his late 60s, Shijun Ma put tremendous efforts into raising public awareness of ecological and environmental problems, leaving his footprints all over the country. He also did a great deal of work on influencing the decision-makers with his ecological knowledge, which greatly contributed to the well known state policy "sustainable development" .

Because of his academic achievements and social influences, Shijun Ma also served as chairs or board members of many international eco-environmental societies/committees, including Dr. Gro Harlem Brundtland's World Commission on Environment & Development. As the only member from China, he contributed a great deal in drafting the famous Brundtland report: Our common future in 1987, and letting the world hear China's standpoint on environment and development as a developing country.

Unfortunately, Shijun Ma died in a tragic car accident on his way back from giving a plenary speech in the Rural Eco-environmental Construction Meeting in Hebei Province on May 30, 1991. It is a great loss for the whole Chinese ecology society as well as the environmental science society. With his tireless quest of new areas in his whole life, we wonder where his new exploration would lead us if not for his sudden death.

Figures

Fig.1 Shijun Ma (1915—1991)

References

Jiang G, Li H, Polunin N (1991) Shijun Ma (1915—1991). Environmental Conservation 18: 365.

State Key Laboratory of Urban and Regional Ecology, Research Center for Eco-Environmental Sciences, Chinese Academy of Sciences, et al (1995) Shijun Ma Collection. Beijing: China Environmental Science Press. (中国科学院生态环境研究中心城市与区域生态国家重点实验室, 等. 1995. 马世骏文集. 北京: 中国环境科学出版社.)

7. 迂回与回归——记胡经甫的交叉科学研究之旅

图1 胡经甫（1896—1972）

胡经甫教授（图1）是我国著名的现代昆虫分类学创始人和奠基人，《中国昆虫名录》是他的经典著作。但他对我国生物学发展的贡献绝不仅限于昆虫分类学方面。实际上很少有人知道胡经甫教授还获得过医学博士学位并且从事了一段时间临床医学工作，这段经历也为他以后开展医学昆虫学领域的交叉研究工作奠定了基础。

胡经甫先生1896年出生于一个书香门第家庭，15岁前在家跟随自己父亲学习。在东吴大学附中学习两年后，他考取了东吴大学的生物系。东吴大学是一所教会学校，大多数的任教老师都来自美国。胡经甫先生在大学期间学习刻苦，成绩优秀，在大四时被生物学的任课老师、美国生物学家N. G. Gee（祁天锡）选为助手（图2）。在祁天锡先生身边的工作经验让他对分类学的兴趣日益加深。他一直追随祁天锡先生学习和工作，并分别在1917年和1919年获得学士学位和硕士学位，他也成为我国历史上获得生物学硕士学位的第一人。胡经甫先生随后赴美国康奈尔大学，在昆虫学家J. G. Needham的指导下继续学习，并于1922年获得博士学位，他的博士论文《叉襀形态解剖及生活史研究》发表后获得广泛称赞。

和大多数在美国接受教育的年轻中国生物学家一样，胡经甫先生毕业后就回到中国。他最初在东南大学从事了一年的教学工作，之后在东吴大学工作两年，然后来到北平开启了他在燕京大学长达23年的教学工作（图3）。他博学多才并且兴趣广泛，除了昆虫学以外，他还在燕京大学和东吴大学讲授医学预科课程。他的学生后来很多都成为我国生物学和医学界的翘楚，包括昆虫学家徐荫祺、陆近仁和林昌善，还有医学家邓家栋、吴光和祝海如。

译者：苏瑞凤
中国科学院北京生命科学研究院，北京100101，中国
邮箱：surf@biols.ac.cn

图2　胡经甫（后排左一）和祁天锡（前）（1915年）

图3　胡经甫和他的妻女（1936年）

在教学之外，胡经甫先生的大部分时间都在进行研究工作，于1916年至1949年共撰写出版了80多篇论文和论著。他最著名的论著是《中国昆虫名录》Ⅰ—Ⅵ卷，记录了我国两万余种昆虫。他从1929年开始撰写初稿，历经12年编纂和校订，直至1941年完成了本书的出版。这本论著初稿于1933年完成，当时他正作为访问学者受邀到康奈尔大学工作。在那里他不仅结识了许多从事昆虫学研究的同行，并且有机会参观访问美国、英国、法国、比利时、德国、瑞士和意大利的博物馆，在这些地方进一步核实了他研究的中国昆虫标本和记录，他非常细致地对论著的初稿进行了校订，并最终完成出版。胡经甫先生细致全面、精益求精的工作使得这本论著成为我国昆虫学研究的里程碑。

1941年，胡经甫先生再次受邀作为访问教授前往美国大学工作。由于太平洋战争爆发，他在赴美途中滞留菲律宾的马尼拉。然而幸运的是，在当地华人的资助下，他在当地医学院进修了一直都非常感兴趣的医学专业。就这样，在45岁的时候，他阴差阳错地成了菲律宾大学医学院的一名医学生。在这里他完成了为期4年的医学院课程学习并继续做了一些实习医师工作，最后在战争结束前他回到祖国。回国后，他在长沙湘雅医学院继续完成医学培训并且获得了医学博士学位。随后他返回北平，在燕京大学继续他的教授生涯，同时兼任燕京大学和清华大学的校医工作。

1949年胡经甫先生辞去燕京大学的职务，在北京开业行医。行医期间，他依然从事他热爱的昆虫学研究工作，而且他昆虫学和医学的双重背景也使得他备受重视。这一期间众多研究所都向他发出邀请，1953年他接受了军事医学科学院寄生虫学系的教授职位，专门从事医学昆虫学工作。从那时起，医学昆虫，尤其是蠓科和虻科领域，成为他第二段学术生涯的研究重点。他招收的很多学生后来都成为医学研究和预防医学领域的专家。回顾他在菲律宾曲折的经历，看似让他的职业生涯走了回头路，但反而成为一条通往成功的新路。

7. Detour and return—Chenfu Wu's interdisciplinary research journey

Chenfu Wu (胡经甫) is best known as the founder of modern entomotaxonomy of China, and famous for his monumental work—*Catalogus Insectorum Sinensium* (*Catalogue of Chinese Insects*). However, his contribution to biology in China is beyond entomotaxonomy. Only few know that Chenfu Wu also held an M.D. degree and had been practicing medicine for years, and this experience led to his later interdisciplinary work on medical entomology.

Born in 1896 in a teacher's family, Chenfu Wu was home-schooled until 15 years old. After spending two years in high school, he enrolled in the Department of Biology in Soochow University, a missionary school where most teachers came from the United States. He did so well at school that he was chosen as a teaching assistant in his senior year by his biology teacher, American biologist N. Gist Gee. The TA experience gave him more opportunities to work side by side with Gist Gee and deepened his interests in taxonomy. He continued his study (and the job as a TA) with Gist Gee after getting his B.S. degree in 1917, and received his master degree in 1919. As a matter of fact, he was the first one to receive a M.S. degree in Biology in China. Chenfu Wu later went to Cornell University to study with entomologist J. G. Needham and got his Ph.D. degree in 1922. His thesis work "Morphology, anatomy and ethology of *Nemoura*" was published later and received wide acclaim.

As most of the young American-trained Chinese biologists, Chenfu Wu returned to China after graduation. He took on a teaching position in National Southeast University for a year, then Soochow University for two years, before he moved to Beijing and started his 23-years of teaching at Yanjing University. He had always had broad interests and a wide spectrum of knowledge, thus in addition to entomology, he also taught pre-med courses in Soochow and Yanjing universities. Many of his students later became the backbone of Chinese biology and medicine, including entomologists Yinqi Xu (徐荫祺), Jinren Lu (陆近仁), Changshan Lin (林昌善) and medical scientists Jiadong Deng (邓家栋), Guang Wu (吴光), Hairu Zhu (祝海如).

Teaching aside, Chenfu Wu spent most of his time on research and had published more than 80 articles/books from 1916 to 1949. His most famous work is "*Catalogus Insectorum Sinensium*

Ming Li
Beijing Institutes of Life Science, Chinese Academy of Sciences, Beijing 100101, China
Contact: liming@ioz.ac.cn

(*Catalogue of Chinese Insects*)" Vol. I—VI, which has recorded over twenty thousands of species of insects in China. It took him twelve years (1929—1941) to write, revise and publish the book. The manuscript was finished by 1933, when he was invited to work in Cornell University as a visiting scholar. This gave him opportunities to meet peer entomologists and visit museums in US, then in UK, France, Belgium, Germany, Switzerland and Italy, to double-check their specimens and records about Chinese insects. During and after his long travel, he meticulously revised the book before sending it for print. The book later became a milestone in China's entomological research.

In 1941, Chenfu Wu was once again invited by American universities to work there as a visiting professor. Unfortunately the Pacific War broke out while he was enroute to the US, and he ended up stranded in Manila, Philippines. Luckily enough, he was always interested in medicine and a local Chinese offered to sponsor him for medical school. Thus, at the age of 45, he became a medical student at the Medical School of University of Philippines. It took him four years to finish the medical courses and even some internship, before the war ended and he was able to go home. Back to China, he completed his medical training and obtained the M.D. degree in Hsiang-Ya Medical College in Changsha City. When he eventually returned to Beijing and resumed his professor position in Yanjing University, he also served as school doctor in both Yanjing University and Tsinghua University.

After 1949, he resigned from Yanjing University and opened his own clinic and started to practice medicine. Even as a physician, he was still known for his entomological work, and his dual background in both entomology and medicine made him even more valuable. He had been offered jobs from various institutions, and in 1953, he accepted the offer from the Academy of Military Medical Sciences (AMMS) and became a professor at the Department of Parasitology of AMMS, working specifically on medical entomology. From then on, medical insects, especially Ceratopogonidae and Tabanidae, became the focus of the second half of his academic career. Many of his students later became the experts of medical research and epidemic prevention. Looking back to his experience in Philippines, what started out as a series of detours in his career has turned out to be a new main route toward success.

Figures

Fig.1 Chenfu Wu (1896—1972)

Fig.2 Chenfu Wu (first on the left back) and Gist Gee (front) in 1915

Fig.3 Chenfu Wu with his wife and daughter in 1936

References

Tan JZ (1988) Chinese Modern Biologist. Changsha: Hunan Science & Technology Press. (谈家桢. 1988. 中国现代生物学家传. 长沙: 湖南科学技术出版社.)

William JH (1996) China Voyager: Gist Gee's Life in Science. New York: Sharpe, M.E.

8. 陈世骧：中国昆虫学的先驱和奠基人之一

陈世骧，中国科学院院士，杰出的生物学家、昆虫学家和进化分类学家（图1）。作为一名严谨勤奋的科学家，他一生发表180余篇文章、专著和会议论文等，发现76个昆虫新属、949个昆虫新种，前后数次被列入多个国家的《世界名人录》，至今仍受到国际科学界的高度评价。

图1　陈世骧（1905—1988）

陈世骧出生于浙江嘉兴郊区的一个书香门第家庭，那里既是鱼米之乡，也是稻螟猖獗之地。为了科学治理虫害，其父陈志巩发起并参与了我国第一个民间治虫组织——治螟委员会。五四运动爆发后，正在嘉兴第二高等小学读书的小世骧跟随老师在街上表演爱国主义隐喻戏剧（章有为，1988），自此在心底埋下了科学救国的种子。在该信念的支持下，他于1924年考入复旦大学生物系并在学习期间表现出极致勤勉与热忱（图2）。巧合的是，这期间法国昆虫学家、昆虫学会会员松梁材（Auguste Savio）正在此任教。1928年毕业后，陈世骧毅然决定继续未竟的昆虫学事业。出色的表现使他顺利申请成为法国国家自然历史博物馆昆虫学实验室的博士研究生。1933年1月15日，陈世骧与谢蕴贞在法国订婚。随后完成博士论文《中国和越南北部叶甲亚科的系

作者：葛业静

中国科学技术大学科技史与科技考古系，合肥230026，中国

邮箱：geyejing@mail.ustc.edu.cn

统研究》并于一年后获得博士学位（Chen，1934）。该文堪称昆虫学史上的经典之作，曾获1935年法国昆虫学会巴赛奖金。陈世骧曾在大英博物馆、法国国家自然历史博物馆和德国昆虫研究所等著名机构工作，亦有多次机会长聘留任，但作为一个真正的爱国者，他最终决定回到祖国，投身于昆虫学研究。1934年8月，任中央研究院动植物研究所研究员；1944年5月，任中央研究院动物研究所研究员。中华人民共和国成立之后，先后担任中国科学院昆虫研究室主任（1950—1953年），昆虫研究所所长（1953—1962年），动物研究所所长（1962—1982年）、名誉所长（1982—1988年），1955年被选聘为中国科学院第一批学部委员（院士）。

1952年春，我国东北地区涌现一批带有细菌的昆虫，这无疑对整个国家的社会和经济相当不利。得知该情况后，陈世骧迅速进行昆虫研究，并在反细菌战的科学调查中鉴定了这些昆虫的各类标本。几乎与此同时，严重的蝗虫灾害暴发。1952年初，陈世骧在上海带头成立蝗虫研究小组，工作内容是饲养飞蝗和研究其繁殖力（图3）。次

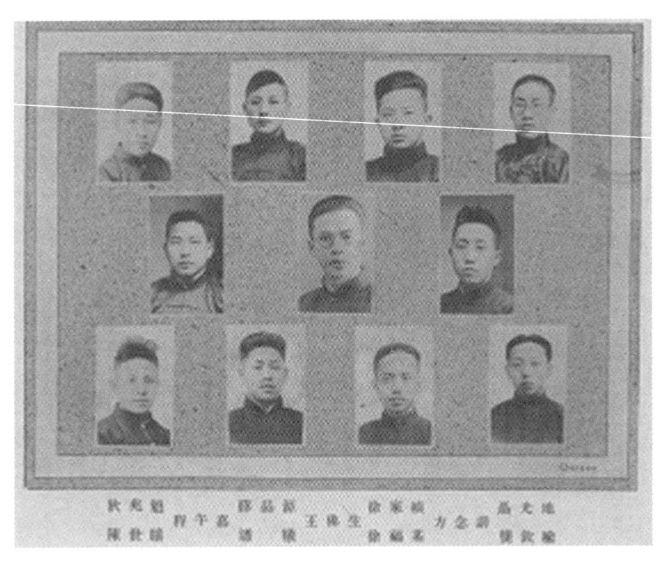

图2　陈世骧（下排左一）与复旦大学同学（1924年）

图3　20世纪50年代东亚飞蝗产卵习性插图

年，在注意到团队组建的重要性之后，陈世骧正式将中国科学院昆虫研究所分设12个研究方向。许多高度专业化的科学家活跃起来，包括朱弘复、刘玉素、马世骏、钦俊德、陆近仁、刘崇乐、蔡邦华、冯兰洲、熊尧、龚坤元、赵养昌，等等（中国科学院动物研究所所史编撰委员会，2008）。上述成员与其他国有研究机构成员一起从事蝗虫的基础研究。陈世骧与团队合作，坚持基础研究、跨学科合作和现有治蝗方法相结合的深刻理念，最后成功地控制了飞蝗肆虐，解决了千年蝗虫难题。此一系列创新突破得到周恩来总理的肯定，他批示为新成立的研究所建造一座办公楼（李晨阳，2020）。他们的深入研究“东亚飞蝗生态、生理学等的理论研究及其在根治蝗害中的意义”获得1982年国家自然科学奖二等奖。

在严谨治学的道路上，陈世骧勇于实事求是、坚持真理。1956年8月，为贯彻“百花齐放，百家争鸣”方针，由中国科学院与高等教育部在青岛联合召开遗传学座谈会（图4）。会上，陈世骧就李森科关于物种形成的新见解提出四点批评，纠正了新中国成立初期遗传学领域的偏差，给科研和教学发展树立了典范，也为遗传学领域的进一步研究奠定了坚实的基础。陈世骧对“争鸣”也有自己的理解：“必须以理论结合事实为基础，要‘言之有物，持之成理’，反对扣帽子，反对以概念批判概念，反对一切的清规戒律。”（陈世骧，1957）

图4　1956年8月遗传学座谈会，山东青岛

在科学实践中，昆虫学作为动物学分支学科，通常包括分类进化、行为学、形态学、生理学和遗传学等方面的研究。陈世骧始终热情洋溢、百折不挠地致力于鞘翅目、昆虫分类、生物进化等方面的前沿研究。在他看来，昆虫分类系统已经被概念化为进化史的缩影。他在《进化论与分类学》（图5）一书中系统讨论了三个基本问题：物种概念、系统原理和特征分析，由此创立进化分类学的理论体系，将生物分类学推到一个新的高度。他注重哲学和生物学的结合，提出物种“变与不变和适应与不适应”，总

结“生物进化史上的十件大事”，提出生物进化的三条路线，为丰富和发展达尔文学说作出贡献（陈世骧，1978）。他创造性地将生物进化分为非细胞总界、原核总界（细菌界、蓝藻界）和真核总界（植物界、真菌界、动物界）（陈世骧和陈受宜，1979）。其主编的《中国动物志·昆虫纲·鞘翅目·铁甲科》（图6），推动了我国生物学和动物学教育的发展，并于1988年获中国科学院科学技术进步奖一等奖，1989年获国家自然科学奖二等奖（陈世骧等，1986）。

图5 《进化论与分类学》（1978年）

图6 《中国动物志·昆虫纲·鞘翅目·铁甲科》（1986年）

陈世骧还是我国昆虫学体制化的先驱之一。1937年6月，以陈世骧为首的一批理想远大的昆虫学家预备发起成立中国昆虫学会，胡经甫、杨惟义等著名科学家均表支持，然而该计划因抗日战争全面爆发暂时中断（《科学》，1937）。1944年5月，陈世骧等30余位生物学家再次申请筹建，10月12日，中国昆虫学会最终成立。1960年，他担任中国农学会副理事长；1978年，担任中国昆虫学会第二届理事会理事长。1964年，亲手创办《动物分类学报》，曾担任《昆虫学报》《昆虫分类学报》和《科学通报》等重要刊物编委（卢嘉锡等，1994）。这些社团和期刊迄今为止仍然是生产性资源，具有深远的国际影响。陈世骧1952年加入九三学社，1979年加入中国共产党。1980年3月，陈世骧等38名代表在中国科学技术协会第二次全国代表大会上联名提出了一项关于恢复和加强生物教学的倡议。

陈世骧几十年来深知国际合作与学术交流不仅是中国科学的历史传统，也是全球科学不可阻挡的潮流。20世纪70年代前后，这一时期的国内外环境波动，促使研究的相关因素发生显著变化，更加广泛的科学交流也成为可能。因此，陈世骧迎来使其项目更加国际化的机遇。1973年，日本东京大学农学部Saburo Tamura教授来访中国科学院动物研究所。1975年5月，荷兰著名昆虫学家De Wilde教授应邀进行学术访问。同年7月，陈世骧和同事与来华访问的美国害虫综合防治代表团一行10人进行了有效沟通。1979年9月，陈世骧所长与马世骏研究员会见美籍华人、昆虫学家赵景熹博士（中国科学院动物研究所所史编撰委员会，2008）。1980年，美国昆虫学会科学家旅游团一行32人到访，陈世骧、朱弘复等人接待，双方就害虫综合治理、森林害虫防治及捕食性昆虫的饲养等议题举行多次学术讲座。1984年，年近80岁的陈世骧在德国汉堡举行的第17届国际昆虫学会议上分享“叶甲总科的演化与分类”报告，受到与会学者的一致赞赏，此次会议规模约为2500人（吕章申，2008）。

1988年1月25日，陈世骧院士与世长辞。直到生病住院的前一天，他还在和同事讨论《中国动物志》的工作。虽然他已离开，但他的爱国责任感、战略智慧、严谨态度、合作精神和开放意识并未消失，且值得所有科学家赞赏和学习。晚年，他曾简要写道（李方诗，1989）：

科研老卒，意在过河；
生命不已，壮志难磨；
涓流虽小，积少成多；
晚霞万里，照澈清波！

这也是他一生以诗抒发抱负的实践总结。

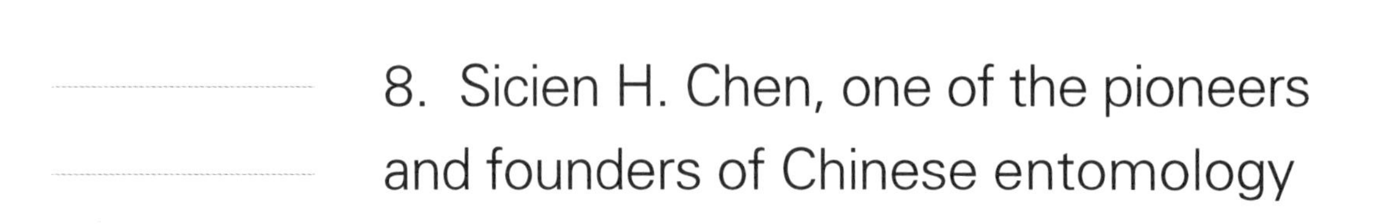

8. Sicien H. Chen, one of the pioneers and founders of Chinese entomology

Sicien H. Chen (陈世骧, 1905—1988) was an academician of the Chinese Academy of Sciences (CAS) and a distinguished biologist, entomologist, and evolutionary taxonomist. As a meticulous and diligent scientist, he discovered 76 new genera, 949 new species, and contributed more than 180 publications, including articles, monographs, and conference proceedings. Sicien H. Chen has been included in the *Who's Who* of various countries and continues to be highly regarded by the international scientific community.

Sicien H. Chen was born into a scholarly family in Jiaxing, Zhejiang Province, a fertile area rampant with rice borer infestation. To scientifically guard against the major threat of pests, his father, Chih-Kung Chen (陈志巩), initiated the first folk pest control organization, named the Committee of Borer Control. When the May 4th Movement broke out, Chen, a student at the Second Higher Primary School, performed a metaphorical play with his teachers in the street (Zhang, 1988). Thereafter, the seed of saving the nation with science was planted. Supported by this belief, he enrolled at the Biology Department of Fudan University in 1924, where he demonstrated great diligence and eagerness. Coincidentally, Auguste Savio, an entomologist and member of the French Entomological Society, taught here during his undergraduate period. After graduation in 1928, Chen continued with his academic career in entomology and became a doctoral candidate at the National Museum of Natural History in Paris as a result of his excellent performance. Notably, Sicien H. Chen and Yon-Chen Zia (谢蕴贞) were engaged in France on January 15, 1933. Subsequently, he finished his dissertation on Recherches sur les Chrysomelinae de la Chine et du Tonkin, which became a classic in the history of entomology and was awarded the Prix Passet in 1935, and he received his doctoral degree a year later (Chen, 1934). There were numerous opportunities to work in chief institutions such as the British Museum, National Museum of Natural History, and German Entomological Institute, but as an ardent patriot, he eventually returned to China and dedicated himself to entomology. He

Yejing Ge

Department for the History of Science and Scientific Archaeology, University of Science and Technology of China, Hefei 230026, China

Correspondence: geyejing@mail.ustc.edu.cn

successively served as a researcher at the Institute of Botany and Zoology in August 1934 and the Institute of Zoology in May 1944 at the Academia Sinica. After the founding of the People's Republic of China, he was designated as the director of the Laboratory of Entomology (1950—1953), the Institute of Entomology (1953—1962), and the Institute of Zoology (1962—1982). In 1955, he was elected as a member of the First Committee Member of Academic Divisions, now jointly known as the academician.

During the spring of 1952, many bacteria-carrying insects were discovered in Northeast China, which was inimical to the whole country both socially and economically. After learning of the outbreak, Sicien H. Chen undertook entomological research and identified a variety of specimens of these insects in a scientific investigation of anti-bacteriological warfare. Almost simultaneously, locust plagues broke out with great intensity. In early 1952, he had taken the initiative to establish a locust research group in Shanghai, where he bred East Asian migratory locusts to study their fertility. Meanwhile, in the following year, after noticing the significance of team construction, he formally divided the Institute of Entomology into 12 research directions. Many highly specialized scientists were active, including Hung-Fu Chu (朱弘复), Yu-Su Liu (刘玉素), Shih-Chun Ma (马世骏), Chun-Teh Chin (钦俊德), Chin-Jen Luh (陆近仁), Chung-Lo Liu (刘崇乐), Pang-Hua Tsai (蔡邦华), Lan-Chou Feng (冯兰洲), Yao Hsiung (熊尧), Kwen-Yuan Kung (龚坤元), Yung-Chang Chao (赵养昌), and so on (The History Compilation Committee of Institute of Zoology, Chinese Academy of Sciences, 2008). The members, as well as other state-owned organizations, were engaged in basic studies of locusts. As a corollary, he successfully controlled the ravages of the locusts and solved the millennium locust problem by collaborating with his team, based on the thoughtful and profound idea of combining basic research, interdisciplinary collaboration, and existing methods. A series of innovative breakthroughs by these elites were affirmed by Premier Enlai Zhou, who decided to construct a building for the newly established institute (Li, 2020), and their penetrating study on ecology and physiology of East Asian migratory locust and their significance in the eradication of locust pests won the second prize at the State Natural Science Award in 1982.

On the path of scrupulous scholarship, Sicien H. Chen had the courage to seek the truth from the facts and persevere. In August 1956, the Ministry of Higher Education and CAS jointly convened the Qingdao Meeting on Genetics to implement the policy of "letting a hundred flowers blossom and a hundred schools of thought contend" (双百方针). At the meeting, Chen offered four criticisms of Lysenko's new insights on species formation, which corrected the ideological bias, set a model for academic research, and laid a solid foundation for further research in the realm of genetics. Moreover, he insisted on a self-understanding of "zhengming" (争鸣); that is, it must be based on theory and facts and "said with solid judgment" ; opposed to labeling at random, criticizing "concepts" with "concepts" and all monastic rules (Chen, 1957).

As a zoological discipline, entomology typically consists of taxonomic evolutionary, behavioral, morphological, physiological, and genetic research in scientific practice. In a similar vein, Sicien H. Chen had always been an enthusiastic and indefatigable scientist, working on the frontiers of Chrysomelidae (Coleoptera), taxonomy, biological evolution, and so on. Particularly, the taxonomic system was conceptualized as a microcosm of evolutionary history. In the book *Evolution and Taxonomy*, Sicien H. Chen systematically described three fundamental issues, including the concept of species, phylogeny, and characterization, to create a theoretical system of evolutionary taxonomy that pushed biological taxonomy to a new level. Moreover, he concentrated on this meshing of natural philosophy with biology, proposed "change and unchanged, adaptive and non-adaptive" species, expounded "ten major events in the history of biological evolution" , and elaborated on the three paths of biological evolution, which prominently contributed to the enrichment and development of Darwinism (Chen, 1978). He creatively divided biological evolution into Superkingdom Acytonia, Superkingdom Procaryota (Kingdom Mycomonera and Kingdom Phycomonera), and Superkingdom Eucaryota (Kingdom Plantae, Kingdom Fungi, and Kingdom Animalia) (Chen and Chen, 1979). He was the editor-in-chief of the book *Fauna Sinica·Insecta·Coleoptera·Hispidae*, which won first prize at the Scientific and Technological Progress Award of CAS in 1988 and second prize at the State Natural Science Award in 1989, boosting biology and zoology education in China (Chen et al., 1986).

He was among the pioneer advocates of entomology institutionalization in China. In June 1937, a group of ambitious entomological scholars, led by Sicien H. Chen, were poised to establish the Chinese Entomological Society with the support of Chenfu Wu (胡经甫) and We-I Yang (杨惟义) (Science, 1937). Unfortunately, this project was temporarily interrupted due to full outbreak of the War of Resistance against Japanese Aggression. In May 1944, more than 30 biologists, including Sicien H. Chen, actively applied for the preparation of the society once again, and it was ultimately established on October 12. In the following year, Sicien H. Chen was the vice-chairman of the Chinese Agricultural Society in 1960 and chairman of the Chinese Entomological Society in 1978. Accordingly, he founded Zoological Systematics in 1964 and served as an editorial board member of journals such as *Acta Entomologica Sinica, Entomotaxonomia*, and *the Chinese Science Bulletin* (Lu et al., 1994). As such, these societies and journals still act as productive resources, and have had profound international influence so far. Sicien H. Chen joined the Jiu San Society (九三学社) in 1952 and the Communist Party of China in 1979. In March 1980, 38 members, represented by Sicien H. Chen, strongly urged for the restoration and reinforcement of biology teaching at the Second Congress of the National Association for Science and Technology.

Over the decades, Sicien H. Chen was deeply aware that international cooperation and exchange were not only the historical tradition of Chinese science but also the irresistible trend of global science. Around the 1970s, the research-related factors were significantly altered due to the national and international circumstances fluctuating in this period, making extensive scientific exchanges possible. Thus, Sicien H. Chen was able to make his projects more international. In 1973, Prof. Saburo Tamura of the Department of Agriculture, University of Tokyo visited the Institute of Zoology. In May 1975, Prof. De Wilde, a renowned entomologist in the Netherlands, was invited for academic visits. In July of the same year, he effectively communicated with a 10-member US delegation on integrated pest control, accompanied by his colleagues. In September 1979, Sicien H. Chen and Shih-Chun Ma met with Dr. Ching-Hsi Chao (赵景熹), a Chinese–American entomologist (The History Compilation Committee of Institute of Zoology, Chinese Academy of Sciences, 2008). In 1980, Sicien H. Chen received a 32-member delegation of the American Entomological Society. During this period, they gave several academic lectures on pest management, forest pest control, and feeding of predatory insects. In 1984, Sicien H. Chen, then nearly 80 years old, delivered the lecture, Evolution and classification of the Chrysomelid beetles, at the 17th International Congress of Entomology that was held in Hamburg, Germany with 2500 participants, which was appreciated by scholars at the conference (Lv, 2008).

On January 25, 1988, Sicien H. Chen passed away. He kept discussing *Fauna Sinica* with his colleagues until the day before his hospitalization. Although he has left us, his patriotic duty, strategic wisdom, rigorous approach, cooperative spirit, and opening consciousness have not disappeared and are worthy of appreciation and learning by all scientists. In his later years, he succinctly wrote (Li, 1989):

An old scientist advances cautiously (科研老卒，意在过河).
A lifelong career will always be (生命不已，壮志难磨).
Trifling as the trickle is, it makes a mickle (涓流虽小，积少成多).
The sunset shining in the sky illuminates the clear waves (晚霞万里，照澈清波).

It is likewise the summary of the practice of airing aspirations for all his life.

Figures

Fig.1 Sicien H. Chen (1905—1988)

Fig.2 Sicien H. Chen (first on the left bottom) and his classmates at Fudan University in 1924

Fig.3 An illustration of East Asian migratory locusts in the 1950s

Fig.4 The Qingdao Meeting on Genetics in August 1956

Fig.5 *Evolution and Taxonomy*(1978)

Fig.6 *Fauna Sinica·Insecta·Coleoptera·Hispidae*(1986)

References

Chen SH (1934) Recherches sur les Chrysomelinae de la Chine et du Tonkin. Paris: University of Paris.

Chen SH (1957) What are the reasons for the concerns? Guangming Daily 02: 2835. (陈世骧. 1957. 产生顾虑的原因是什么？光明日报，02: 2835.)

Chen SH (1978) Evolution and Taxonomy. Beijing: Science Press. (陈世骧. 1978. 进化论与分类学. 北京：科学出版社.)

Chen SH, et al (1986) Fauna Sinica·Insecta·Coleoptera·Hispidae. Beijing: Science Press. (陈世骧，等. 1986.

中国动物志·昆虫纲·鞘翅目·铁甲科.北京:科学出版社.)

Chen SH, Chen SY (1979) On the major classification of organisms. Acta Zootaxonomica Sinica 01: 1–12. (陈世骧,陈受宜.1979.生物的界级分类.动物分类学报,01: 1–12.)

Li CY (2020) Sicien H. Chen: Leaving a thick paper. Wisdom China 01: 59. (李晨阳.2020.陈世骧:留下一张厚重的考卷.智慧中国,01: 59.)

Li FS (1989) Yearbook of Who's Who of China. Beijing: China Society Press. (李方诗.1989.中国人物年鉴.北京:中国社会出版社.)

Lu JX, et al (1994) Biographies of Modern Chinese Scientists, Vol. 6. Beijing: Science Press. (卢嘉锡,等.1994.中国现代科学家传记,第6辑.北京:科学出版社.)

Lv ZS (2008) Biography of Modern Chinese Scholars Studying in France. Beijing: Palace Press. (吕章申.2008.中国近代留法学者传.北京:紫禁城出版社.)

Science (1937) Science news: Sicien H. Chen et al. initiated the Entomological Society. Science 21(6): 493. (《科学》.1937.科学新闻:陈世骧等发起昆虫学会.科学,21(6): 493.)

The History Compilation Committee of Institute of Zoology, Chinese Academy of Sciences (2008) A Brief History of the Institute of Zoology, Chinese Academy of Sciences. Beijing: Science Press. (中国科学院动物研究所所史编撰委员会.2008.中国科学院动物研究所简史.北京:科学出版社.)

Zhang YW (1988) A famous biologist: Sicien H. Chen. Bulletin of Biology 09: 45–46. (章有为.1988.著名生物学家——陈世骧.生物学通报,09: 45–46.)

9. 郑作新：中国鸟类学奠基人和中国动物地理学开拓者

郑作新1906年11月18日生于福建福州（图1）。1930年于美国密歇根大学获得博士学位，时年23岁，毕业后即回国报效祖国，此后毕生从事鸟类学和动物地理学研究，成为一代大师。

图1 郑作新（1906—1998）

新中国成立之前的近代鸟类学研究，多被外国人主导，以野外考察居多，且多为分类工作，涉及动物区系和动物地理等方面的内容很少。例如，R. Swinhoe在1863年记载了中国鸟类454种，这是对中国鸟类的最早报道。La Touche在1931—1934年撰写出版的《华东鸟类手册》记述了750种和亚种。我国学者对鸟类的研究工作始于20世纪早期，例如寿振黄、任国荣、郑作新等。其中，1944年郑作新发表的《三年来（1938—1941年）邵武野外鸟类观察报告》，对鸟类的分布、数量、居留类型等进行了较全面的报道，是我国关于野外观察鸟类数量统计的第一篇报道。1947年，郑作新在《中国科学社论文专刊》发表了英文版“Checklist of Chinese birds”，列有中国鸟类1087种、912亚种，首次建立了完整的中国鸟类区系和分类系统，是我国学者首次自行研究撰编的全国性鸟类名录（Cheng，1947）。在此基础上，同年在中国科学社主办的《科学》刊物上发表了《中国鸟类地理分布的初步研究》，开辟了我国鸟类动物地理学研究之首作。他将我国鸟类的分布格局划分为两个界（古北界和东洋界）、三个区（蒙藏区、华北区和华南区），建议自喜马拉雅山系以东，至陕西南部的秦岭山脉，经大别山，过长江至闽、浙山地为古北界与东洋界两大界的分界线。这是关于中国鸟类地理分布与地理区划最早的最完整的报道。Cheng和Chang（1956）发表的《中国动物地理区划》首次提出了“中国动物地理区划”的学科概念，确立秦岭为古北界和东洋界的分界线，并将我国划分为古北界和东洋界两界、7个动物

作者：雷富民，宋刚

中国科学院动物研究所，北京100101，中国

邮箱：leifm@ioz.ac.cn（雷富民）

地理区和16个二级区，这种区划在世界动物地理区划上具有重要意义，得到国内外专家的公认，并一直沿用至今，为中国动物地理区划的深入研究提供了总体框架。尤其Cheng（1987）出版的《中国鸟类区系纲要》首次对中国鸟纲所有物种的分类、分布、繁殖、种群、生态习性等做了全面而又系统的综合报道，提供了中国鸟类的完整资料，成为国际上鸟类学经典著作之一。

郑作新先生在鸟类系统分类、亚种分化与进化理论方面作出了重要贡献。他与合作者发现我国鸟类16个新亚种，同时对画眉类、雉类等的系统演化也有深入研究。例如，他通过对白鹇14个亚种的分布与特征比较研究发现，较原始的亚种类群被排挤到该种分布范围的边缘地区，认为较低等类型的亚种并不在种的起源地，而是被排挤后残存在这一个种分布范围的边缘，由此提出了物种分化的“排挤学说”，该观点体现了达尔文进化论之精髓，对我国鸟类的进化历史与分布格局研究产生了深远的影响。在系统成果的基础上，他负责了《中国动物志・鸟纲》系列巨著的编撰，并于1978 年出版了第一部著作《中国动物志・鸟纲・第四卷・鸡形目》，此后其他卷册相继出版。他还出版了其他鸟类地方志、图谱、图鉴以及大量的地方性区系调查报告，在鸟类资源调查、地理分布等鸟类学与动物地理学研究领域作出了重大贡献。

郑作新先生获得多项重要的奖励和荣誉，例如1989年获中国科学院科学技术进步奖一等奖；1990年获国家自然科学奖二等奖；1995年获“蔡冠深中国科学院院士荣誉基金会”奖；1996年获香港求是科技基金会杰出科技成就集体奖。1988年他还被美国野生动物联合会授予“自然保护特殊成就奖”，以表彰他对鸟类学研究与保护的卓越贡献，这是该联合会首次在美国以外举行授奖仪式，也是第一次颁发给一位中国的学者（图2）。

图2 郑作新（右二）获“自然保护特殊成就奖”

郑作新先生曾连任世界雉类协会副会长、会长、终身会长（图3）。他也是国际鹤类基金会首届顾问，第22届国际鸟类学大会名誉主席。郑作新先生是中国动物学会发起人之一，历任秘书长、副理事长、理事长、名誉理事长等职。他也是中国鸟类学会的发起人之一，被推选为第一任理事长、名誉理事长。1980年他当选中国科学院学部委员（院士）。为鼓励更多青年人献身于我国鸟类学事业，他拿出奖金，于1994年设立了“郑作新鸟类科学基金会”，以鼓励年轻人投入我国鸟类学研究与保护之中。截至目前已有28位获奖，这些获奖者目前都在我国鸟类学研究中发挥着引领作用，有些获奖者在国际鸟类学研究中也具有重要的影响力。

图3　郑作新在1989年世界雉类协会大会上致辞

郑作新先生于1998年6月27日离开了我们。为了纪念他对中国和世界鸟类学所作出的巨大贡献，Per Alström教授与雷富民教授等合作发现的中国鸟类一新种以郑先生的名字来命名——郑氏蝗莺（*Locustella chengi*）（Alström et al.，2015）。他的成果和精神一直存在，就像一座灯塔照耀着中国鸟类学事业不断发展。

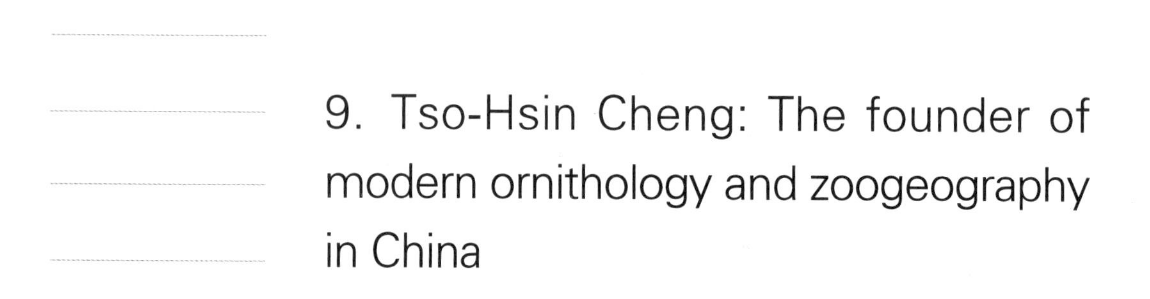

9. Tso-Hsin Cheng: The founder of modern ornithology and zoogeography in China

Tso-Hsin Cheng (Zuoxin Zheng, 郑作新) was born in Fuzhou, Fujian Province on November 18, 1906. He earned his Ph.D. from the University of Michigan in 1930, at the age of 23. After graduation, he returned to his homeland, where he devoted all his life to ornithology and zoogeography, and became a world-wide renowned scientist.

In the early time before the founding of the People's Republic of China, ornithological studies in China were mainly carried out by western naturalists, focusing on scientific expeditions, species description and taxonomy. For example, the earliest bird checklist of China was published by Swinhoe, in the *Proceedings of the Zoological Society of London* in 1863, documenting 454 bird species in China. La Touche (1931—1934) published "*A Handbook of the Birds of Eastern China*", in which a total of 750 species and subspecies of birds were descripted. However, they rarely focused on avian fauna and distributional concerns. Chinese scientists, such as Tsen-Hwang Shaw (Zhenhuang Shou, 寿振黄), Kwok-Yung Yen (Guorong Ren, 任国荣) and Tso-Hsin Cheng started ornithological research in the 1920's. Prof. Cheng is the most accomplished scientist in Chinese ornithology. He started the first fieldwork on wild birds and their population characteristics in China, "A report for three years' field survey (1938—1941) of birds in Shaowu", covering multiple aspects of birds in that area, including their distribution, abundance and residential status. In 1947, he published the "Checklist of Chinese birds" in the *Transactions of the Chinese Association for the Advancement of Science*, listing 1087 species and 912 subspecies of birds in China (Cheng, 1947). This is the first checklist of birds in China with completed classification and distribution to be compiled by a Chinese ornithologist. Based on this checklist, Prof. Cheng published "On the geographical distribution of birds in China" in the same year in the journal *China Science*. This is the earliest comprehensive research on the species distribution and biogeography of birds in China, dividing the avian fauna into two zoogeographical realms (the Oriental realm and the Palearctic realm)

Fumin Lei, Gang Song

Key Laboratory of Zoological Systematics and Evolution, Institute of Zoology, Chinese Academy of Sciences, Beijing 100101, China

Correspondence: leifm@ioz.ac.cn (F.-M. Lei)

and three zones (Mongolian Zone, North China Zone and South China Zone). This paper proposed a line beginning at the eastern edge of the Himalaya, along Qinling Mountains, Dabie Mountains, and across Yangtze River to the hilly areas in Fujian and Zhejiang as the boundary between the two realms. Prof. Cheng and Prof. Yung-Tsu Chang (Rongzu Zhang, 张荣祖) then firstly proposed the concept of "Zoogeographical Regions of China" in 1956 (Cheng and Chang, 1956), which suggested two zoogeographical realms in China with Qinling Mountains as the boundary, and additionally divided the two realms into seven regions and sixteen sub-regions in more detail. This zoogeographical regionalization has been recognized worldwide as one of the basic frameworks of zoogeography of China. In 1987, Cheng published "*A Synopsis of the Avifauna of China*" (Cheng, 1987). This is an encyclopedic monograph with a substantial compiling of multiple aspects of birds in China. For each of the 1186 species included in this monograph, nomenclature, breeding habitat, distribution and population status were also listed, which provided an inclusive ornithological reference for China, and is regarded as one of the classics of ornithological literature in the world.

Prof. Cheng is most famous for avian taxonomy and systematics. He and colleagues identified 16 new bird subspecies. They also studied the phylogeny and evolution of pheasants, laughing thrushes, and other bird families. After comparing the distribution and morphological variations of 14 subspecies of the silver pheasant (*Lophura nycthemera*), he found that more primitive subspecies were distributed at the periphery rather than the central area, and consequently proposed the "competitive exclusion" postulate to explain this pattern. This postulate is in accordance with the core idea of Darwin's species competition concept, which is crucial for explaining the evolutionary history and distribution pattern of birds in China. Based on the well-archived taxonomy and distribution of birds, he led the publication of the *Avian Fauna Sinica* series, and published the first volume (Volume 4, Galliformes) in 1978. He undertook great efforts in compiling the avian faunas, checklists, illustrations and a plethora of scientific reports about field expeditions across China. He made a tremendous contribution to ornithological and zoogeographical research with a number of fundamental reference works on species inventory and geographical distribution.

Prof. Cheng was awarded many honors for his great scientific achievements in ornithology, such as the First Prize of Science and Technology Progress Award of the Chinese Academy of Sciences in 1989, the Second Prize of the National Natural Science Award in 1990, the Forestry Ministry Lifetime Honor Award for China Wildlife Protection in 1993, the Award of Choi Koon-Shum foundation for Chinese Academician of Sciences in 1995, and the Award of Qiu Shi Science & Technologies Foundation in 1996. In 1988, he was awarded the "Special Conservation Achievement Award" in international animal resources protection by the National Wildlife Federation (NWF) for his great contributions to the research and conservation of birds. This is the first time that the NWF awarded a Chinese scientist outside of the United States of America.

Prof. Cheng was elected as the vice-president, president, and the lifetime honorary president of the World Pheasant Association. He was a consultant of the International Crane

Foundation, and the honorary Chairman of the 22nd International Ornithological Conference. He is a cofounder of the China Zoological Society, in which he actively served as the general secretary, vice chairman, chairman, and honorary chairman. He is also a founder of the China Ornithological Society, in which he was elected as the first chairman, and later on the honorary chairman. In 1980, Prof. Cheng was elected a member of the Academia Sinica. In order to encourage more young students and scholars to pursue a career in ornithology, Prof. Cheng donated his prize money to set up the "Cheng Foundation for Ornithological Science" in 1994. To date, 28 awardees were honored, all of whom have become leading ornithologists in China, and some of them are leading international scientists.

Prof. Cheng passed away on June 27, 1998. To memorialize his great contributions to the development of ornithology in China and the world, Per Alström and Fumin Lei, collaborating with other colleagues, nominated a new bird species—*Locustella chengi*, in name of Prof. Cheng (Alström et al., 2015). His scientific merits and spirit in pursuit of science are being passed on between generations, illuminating the ornithological development in China.

Figures

Fig.1 Tso-Hsin Cheng (1906—1998)

Fig.2 Tso-Hsin Cheng (2nd from right) was awarded the "Special Conservation Achievement Award"

Fig.3 Tso-Hsin Cheng gave a welcome speech during the International Conference of World Pheasant Association in 1989

References

Alström P, Xia CW, Rasmussen PC, et al (2015) Integrative taxonomy of the Russet Bush Warbler *Locustella mandelli* complex reveals a new species from central China. Avian Res 6:9.

Cheng TH (1947) Checklist of Chinese birds. Trans Chin Assoc Adv Sci 9:40–84.

Cheng TH (1987) A Synopsis of the Avifauna of China. Beijing: Science Press; Hamburg and Berlin: Paul Parey Scientific Publishers.

Cheng TH, Chang YT (1956) On tentative scheme for dividing zoogeographical regions of China. Acta Geographica Sinica 22(1):93–109.

10. 刘建康：脊椎动物性别决定研究的先驱

六七十年前，英国皇家学会会士W. S. Bullough博士在《自然》杂志上发表了专题评述《低等脊椎动物雌雄同体》，重点评论了一位年轻的中国学者刘建康（图1）所完成的关于低等脊椎动物性别机制的一个重要发现（Bullough，1947）。在这篇文章的开头，Bullough博士充满敬意地写道："在1944年发表的黄鳝性腺发育描述的论文中（Liu，1944a），刘提供了新的有趣的证据，打开了脊椎动物性别决定机制研究之门。"

图1 刘建康先生（1917—2017）

刘建康先生，1917年出生于江苏省吴江县。1938年毕业于东吴大学生物系，同年到中央研究院动植物研究所工作，在著名的鱼类学家伍献文教授（图2）的指导下开始了鱼类生物学的创新研究。抗日战争期间，中央研究院动植物研究所被迫从南京迁至阳朔、长沙、南岳，最后于1939年抵达重庆。在如此艰苦卓绝的条件下，刘建康先生作为一名年轻的学者，工作非常努力，首先发现了黄鳝不同寻常的性别现象，并在他27岁的时候就取得了他青年时期最突出的成就。刘建康先生仔细检查了体长分布在5.3厘米到57.6厘米的659尾黄鳝的性腺，揭示了低等脊椎动物的雌雄同体和性别反转的现象（Liu，1944a）。论文发表后，Bullough博士在荷兰乌得勒支大学的报告中介绍了这个有趣的发现，并在《自然》杂志上评论了它的科学意义（Bullough，1947）。同一时期，刘建康先生也关注鱼类分泌细胞的发育。他发现除了氯化钠外，硫酸钠也能诱导斗鱼氯化物分泌细胞的发育（Liu，1942）。1943年，剑桥大学的科技史专家Joseph Needham教授接受了英国文化委员会的委派，与其他国家进行文化交流，开始了与中央研究院动植物研究所的合作（图3）。在Needham教授的推荐下，刘建康先生的这项工作以致编辑信的形式发表在《自然》杂志上（Liu，1944b）。1939—1945年，刘建康先生发表了19篇研究论文（桂建芳，

作者：周莉，桂建芳

中国科学院水生生物研究所淡水生态与生物技术国家重点实验室，武汉430072，中国

邮箱：zhouli@ihb.ac.cn（周莉）；jfgui@ihb.ac.cn（桂建芳）

2007）。1946—1947年，他在N. J. Berrill教授的指导下在加拿大麦吉尔大学继续进行研究，获得了博士学位。在美国工作两年之后，他回到了上海，在中国科学院水生生物研究所获得了一个教职。从那时起，刘建康先生逐渐成长为著名的鱼类生物学家和淡水生态学家，并历任该研究所的所长和名誉所长。1981年当选中国科学院学部委员（院士）。

图2 刘建康先生（右）与导师伍献文先生（左）（1979年）

图3 1943年刘建康先生（一排右一）、Needham教授（二排左二）和中央研究院动植物研究所部分研究员合影

正如Bullough博士所说，黄鳝性别反转现象的发现的确为几十年后脊椎动物特别是鱼类的性别决定研究开辟了一个新的领域。到目前为止，我国学者已经在鱼类性别异形和性别决定的遗传基础研究和生物技术操作上取得了许多重大突破（Gui and Zhu，2012；Gui，2015a；Mei and Gui，2015）。由于一些水产养殖鱼类在生长速度和个体大

小上表现出明显的两性异形，因此生产和养殖全雌性或全雄性的群体会带来显著的经济效益。中国科学院水生生物研究所的几代鱼类遗传育种学家创建了一系列性别控制育种技术以获得单性群体。譬如，在多倍体银鲫中单性雌核生殖和有性生殖等多重生殖方式被发现和利用，并由此连续培育出异育银鲫（蒋一珪等，1983）、高体型异育银鲫（朱蓝菲和蒋一珪，1993）和异育银鲫“中科Ⅲ号”三代全雌性养殖新品种（Wang et al.，2011），在全国范围内大规模推广，促进了鲫鱼产业的快速发展（Gui and Zhou，2010；Gui and Zhu，2012）；通过人工雌核生殖和染色体组操作，培育了全雌鲤（Wu et al.，1986）；从黄颡鱼和乌苏里拟鲿中鉴定了一些性别特异或性染色体特异的遗传标记（Wang et al.，2009；Dan et al.，2013；Pan et al.，2015），研发出X和Y染色体连锁标记辅助的全雄黄颡鱼培育技术路线，大规模生产全雄黄颡鱼并用于商业养殖（Liu et al.，2013；Gui，2015b）。

作为脊椎动物性别决定研究的先驱和我国淡水生态学的奠基人，刘建康先生为鱼类生物学、湖泊生态学和可持续水产养殖业作出了重大贡献。作为他的后辈学者，我们对他带领我们领跑于鱼类性别决定研究表示最深切的感谢。

2017年11月6日，刘建康先生逝世，享年100岁。

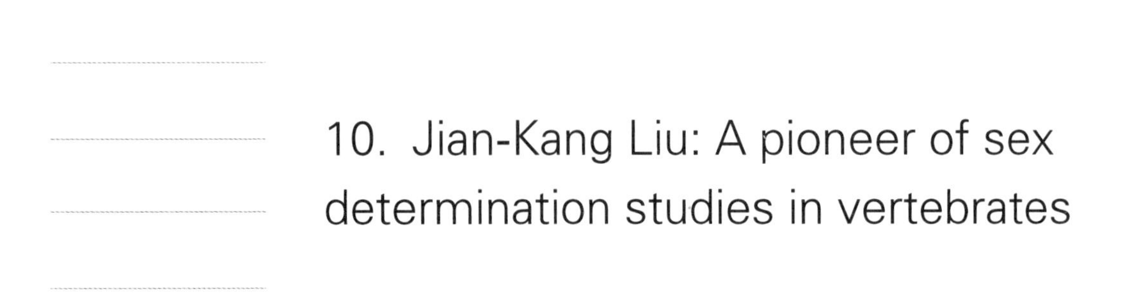

10. Jian-Kang Liu: A pioneer of sex determination studies in vertebrates

Sixty-eight years ago, Dr. W. S. Bullough, a Fellow of the Royal Society, published an article "Hermaphroditism in the lower vertebrates" in *Nature* (Bullough, 1947) and thereby commented a significant finding in lower vertebrate sex mechanism fulfilled by a young Chinese scholar Jian-Kang Liu (刘建康, C. K. Liu). At the beginning of the article, Dr. Bullough respectfully acknowledged that: "By the publication in 1944 of a description of the gonads of *Monopterus javanensis* Lac. (Symbranchii; Teleostei) (Liu, 1944a), Liu has furnished new and interesting evidence concerning the mechanism of sex determination in the lower vertebrates, and has opened a fresh field for research into this subject." (Bullough, 1947)

Jian-Kang Liu was born in Wujiang County, Jiangsu Province in 1917. In 1938, he graduated from the Biology Department of Soochow University, and accepted an offer from Institute of Zoology and Botany of Academia Sinica and began his innovative studies on fish biology directed by famous ichthyologist Dr. Xian-Wen Wu (伍献文). At that duration of the War of Resistance against Japanese Aggression, the Institute was forced to move from Nanjing to Yangshuo, Changsha, Nanyue, and finally to arrive Chongqing in 1939. Under the difficult conditions, as a young scholar, Jian-Kang Liu studied very hard and became the first to notice an unusual phenomenon about the sexuality of rice field eel (*Monopterus javanensis*) and obtained his early most remarkable achievement when he was only 27 years old. He carefully examined the gonads of 659 specimens ranging from 5.3 to 57.6 cm in total length and revealed hermaphroditism and sex reversal in the lower vertebrate (Liu, 1944a). After the paper was published, Dr. Bullough introduced the interesting findings in his lecture at Utrecht University in Netherlands and commented its scientific significance in *Nature* (Bullough, 1947). At the same time, Jian-Kang Liu also paid his attention to the development of fish secretory cells. He found that Sodium Sulphate can induce the development of "chloride-secreting cells" in *Macropodus* besides salinity (Liu, 1942). In 1943, Prof. Joseph Needham at Cambridge University accepted

Li Zhou, Jianfang Gui

State Key Laboratory of Freshwater Ecology and Biotechnology, Institute of Hydrobiology, Chinese Academy of Sciences, Wuhan 430072, China

Correspondence: jfgui@ihb.ac.cn (J.-F. Gui)

an assignment of the British Council for Cultural exchange with other Countries and cooperated with Institute of Zoology and Botany of Academia Sinica. Under Prof. Joseph Needham's recommendation, this work also published in *Nature* as a letter to editor (Liu, 1944b). From 1939 to 1945, young Jian-Kang Liu published a total of 19 research articles (Gui, 2007). From 1946 to 1947, he continued his research with Prof. N. J. Berrill at McGill University of Canada, and obtained his Ph.D. degree. After working for two years in USA, he returned to Shanghai, and was offered a faculty position at the current Institute of Hydrobiology, Chinese Academy of Sciences. Since then, Dr. Liu has developed into a well-known fish biologist and freshwater ecologist, and become a former and honorary director of the Institute. In 1981, he was elected as an Academician of the Chinese Academy of Sciences.

As Dr. Bullough said, the discovery of sex reversal in rice field eel indeed opened a fresh field for sex determination study after several decades in vertebrates, especially in fishes. Up to the present day, a large number of significant breakthroughs have been achieved from the genetic basis to the biotechnological manipulation of fish sexual dimorphism and sex determination (Gui and Zhu, 2012; Gui, 2015a; Mei and Gui, 2015). Because some aquaculture fishes exhibit significant sexual dimorphism in growth rate and body size, there are remarkable economic benefits by producing all-females or all-males for aquaculture, and several generations of fish genetic breeding scientists in the Institute have developed a series of sex control breeding biotechnologies to obtain mono-sex populations. Through heterologous sperm-induced gynogenesis and various reproduction mode utilization, three new unisexual all-female varieties, such as allogynogenetic gibel carp (Jiang et al., 1983), high dorsal allogynogenetic gibel carp (Zhu and Jiang, 1993), and allogynogenetic gibel carp "CAS Ⅲ" (Wang et al., 2011), were bred in gibel carp and applied to aquaculture practice throughout China (Gui and Zhou, 2010; Gui and Zhu, 2012). By artificial gynogenesis and chromosome set manipulation, all-female hybrid common carp was also produced (Wu et al., 1986). Moreover, some sex-specific or sex chromosome-specific genetic markers were identified from yellow catfish and other bagrid catfish (*Pseudobagrus ussuriensis*) (Wang et al., 2009; Dan et al., 2013; Pan et al., 2015), and all-males of yellow catfish had been massively produced and used for commercial aquaculture (Liu et al., 2013; Gui, 2015b).

As a pioneer of sex determination studies in vertebrates and a founder of freshwater ecology in China, Dr. Liu has made significant contributions to fish biology, lake ecology, and sustainable aquaculture industry. As his fellow scholars, we would like to express our deepest gratitude for his leading role as a front-runner of sex determination studies in fish and express our devout wishes for his health and for his long life.

Figures

Fig.1 Jian-Kang Liu (1917—2017)

Fig.2 Dr. Jian-Kang Liu (right) with his advisor Dr. Xian-Wen Wu (left) in 1979

Fig.3 Dr. Jian-Kang Liu (1st at the front raw from right), Prof. Joseph Needham (2nd at second raw from left), and the principal investigators in Institute of Zoology and Botany of Academia Sinica in 1943

References

Bullough WS (1947) Hermaphroditism in the lower vertebrates. Nature 160(4053):9–11.

Dan C, Mei J, Wang D, et al (2013) Genetic differentiation and efficient sex-specific marker development of a pair of Y- and X-linked markers in yellow catfish. International Journal of Biological Sciences 9(10):1043–1049.

Gui JF (2007) Genetic Basis and Artificial Control of Sexuality and Reproduction in Fish. Beijing: Science Press. (桂建芳. 2007. 鱼类性别和生殖的遗传基础及其人工控制. 北京: 科学出版社.)

Gui JF (2015a) Fish biology and biotechnology is the source for sustainable aquaculture. Science China–Life Sciences 58(2):121–123.

Gui JF (2015b) Scientific frontiers and hot issues in hydrobiology. Chinese Science Bulletin 22(60):2051–2057.

Gui JF, Zhou L (2010) Genetic basis and breeding application of clonal diversity and dual reproduction modes in polyploid *Carassius auratus gibelio*. Science China–Life Sciences 53(4):409–415.

Gui JF, Zhu Z (2012) Molecular basis and genetic improvement of economically important traits in aquaculture animals. Chinese Science Bulletin 57(15):1751–1760.

Jiang YG, Liang SC, Chen BD (1983) Biological effect of heterologous sperm on gynogenetic offspring in *Carassius auratus gibelio*. Acta Hydrobiologica Sinica 7(1):1–13. (蒋一珪, 梁绍昌, 陈本德. 1983. 异源精子在银鲫雌核发育子代中的生物学效应. 水生生物学, 7(1): 1–13.)

Liu CK (1942) Osmotic regulation and "chloride-secreting cells" in the paradise fish, *Macropodus opercularis*. Sinensia 13:15–20.

Liu CK (1944a) Rudimentary hermaphroditism in the symbranchoid eel, *Monopterus javanensis*. Sinensia 15:1–8.

Liu CK (1944b) Sodium sulphate also can induce the development of the "chloride-secreting cells" in *Macropodus*. Nature 153:252.

Liu HQ, Guan B, Xu J, et al (2013) Genetic manipulation of sex ratio for the large-scale breeding of YY supermale and XY all-male yellow catfish (*Pelteobagrus fulvidraco* (Richardson)). Mar Biotechnol 15:321–328.

Mei J, Gui JF (2015) Genetic basis and biotechnological manipulation of sexual dimorphism and sex determination in fish. Science China–Life Sciences 58(2):124–136.

Pan ZJ, Li XY, Zhou FJ, et al (2015) Identification of sex-specific markers reveals male heterogametic sex determination in *Pseudobagrus ussuriensis*. Mar Biotechnol 17(4):441–451.

Wang D, Mao HL, Chen HX, et al (2009) Isolation of Y and X-linked SCAR markers in yellow catfish and application in the production of all-male populations. Animal Genetics 40(6):978–981.

Wang ZW, Zhu HP, Wang D, et al (2011) A novel nucleo-cytoplasmic hybrid clone formed via androgenesis in polyploid gibel carp. BMC Research Notes 4(82):1–13.

Wu CJ, Ye YZ, Chen RD (1986) Genome manipulation in carp (*Cyprinus carpio* L). Aquaculture 54:57–61.

Zhu LF, Jiang YG (1993) A comparative study of the biological characters of gynogenetic clones of silver crucian carp (*Carassius auratus gibelio*). Acta Hydrobiologica Sinica 17(2):112–120. (朱蓝菲, 蒋一珪. 1993. 银鲫不同雌核发育系的生物学特性比较研究. 水生生物学报, 17(2): 112–120.)

11. 寿振黄：中国脊椎动物学的奠基人

“脊椎动物奠基者，鸟兽虫鱼无不通。分类生态相结合，生物统计开先声。”这是我国兽类学家夏武平先生在寿振黄先生逝世十五周年时写的悼念诗（寿先生于1964年7月5日逝世于北京）。

图1　寿振黄先生（1899—1964）

寿振黄先生（图1）1899年2月4日生于浙江省。他自幼在农村长大，所以对农村的贫困落后了解深刻，故立志学农以改变农村的落后面貌。1917年中学毕业后考入南京高等师范学校（即后来的东南大学）学农。1920年毕业后任中学生物教员，又认识到农业的基础是生物学，遂决定改学生物学。1921年在他22岁时考入东南大学生物学系。1925年毕业后赴美国留学，最初进入加利福尼亚大学伯克利分校学习，1926年转学到斯坦福大学研究院，在著名鱼类学家D. S. Jordan指导下从事鱼类分类学研究。同年8月，他又到霍普金斯海滨生物研究所，从事甲壳类生活史研究，并于当年完成论文，获得硕士学位，后又回到加利福尼亚大学，从事鸟类学和兽类学研究。1927年他先后到费城、芝加哥、纽约和华盛顿等地的大学和博物馆参观，学习博物馆陈列和标本制作技术。所以，寿振黄先生学习兴趣广泛，涉猎学科多，这些经历给他后来的发展奠定了坚实的基础。

他1928年回国，在清华学校生物学系分别任讲师和教授，一直到1936年。他自编讲义，讲授比较解剖学和鱼类学等课程。他一直兼任北平静生生物调查所动物部技师（图2）。

中华人民共和国成立后，他先后担任中国科学院标本管理委员会委员和研究员等职位。1953—1964年任中国科学院动物研究室（后为动物研究所）研究员。1957年任

作者：魏辅文[1,2]，王德华[1,2]

1　中国科学院动物研究所，北京100101，中国

2　中国科学院大学，北京100049，中国

邮箱：weifw@ioz.ac.cn（魏辅文）；wangdh@ioz.ac.cn（王德华）

图2 北平静生生物调查所成立时所内人员合影（1928年）。前排左起：何琦、秉志、胡先骕、寿振黄；后排左起：沈家瑞、冯澄如、唐进

兽类学研究室主任，1959年任动物生态学研究室主任。1962年任中国科学院动物研究所副所长。寿振黄先生发表的100多篇论文和相关著作中，涉及脊椎动物分类学、形态学、古生物学、生态学等多个领域。

他是我国最早研究鱼类的学者。寿先生在1927年与美国鱼类学家B. W. Evermann合作，发表了《华东鱼类及新种描述》（Evermann and Shaw，1927）。其后，他对我国华东和华北鱼类的分类学、形态学、生态学等都进行了开创性的研究，先后发表一系列相关研究论文（例如Shaw，1929a，1929b；Shaw and Tchang，1931；Shaw and Lee，1939）。

他是我国最早研究两栖类和爬行类动物的学者之一。寿先生在1929年曾调查北平两栖类动物（Shaw，1929b）；1940年对龟类的消化系统进行过研究（Shaw，1940）。

他是我国最早研究鸟类的学者。寿先生是我国现代鸟类学研究的奠基者之一。1927年在美国《科学》杂志上发表了《福建鸟类之记录》（Shaw，1927），这是中国学者在该杂志发表的第一篇鸟类学研究论文。1930—1934年，他分别在山东烟台、四川和浙江开展鸟类研究（Shaw，1930，1931，1932，1934）。1936年，他出版了《河北省鸟类志》（英文版）上下两卷（Shaw，1936），被誉为我国动物学家原创的具有国际水准的第一部鸟类志，也是以志书的形式出版的第一部地域性动物学专著，被视为我国地方动物志的重要典范以及我国脊椎动物区系分类研究的开端。

他是新中国兽类学发展的主要推动者。中华人民共和国成立前，我国的兽类学研究几乎是空白。1954年，年已65岁的寿先生根据国家学科发展的需要，将自己的研究方向转向兽类学。他组建了我国第一个兽类学研究室，建立了我国第一个动物生态学研究室，主持领导了一系列开拓性的研究，例如领导中青年科研人员，对东北三省及内蒙古东部几个盟进行了全面系统的兽类调查，指导年轻学者从事小兴安岭红松直播防鼠害的研究和大兴安岭流行性出血热宿主动物调查等。这些贡献体现在1958—1962

年完成的一系列著作中，如《东北兽类调查报告》是我国第一部地区性兽类调查报告（中国科学院动物研究所兽类研究组，1958）；《红松直播防鼠害研究工作报告》是理论结合生产实践方面的代表性成果（寿振黄等，1958）；《中国经济动物志·兽类》是新中国成立以后第一部有关全国性兽类分类、形态特征、习性、分布及其经济意义的专著并被广泛使用（寿振黄，1962）。通过这些科研活动，他为新中国培养了一支兽类学的专业队伍，影响深远。

他是我国野生动物保护和建立自然保护区的最早倡导者。寿先生于1956年赴德国参加“啮齿动物的生物学和防治会议”，并参观了自然保护区。1957年，寿先生与植物学家吴征镒先生向云南省政府提出在云南筹建自然保护区的建议，并提出了建立24个自然保护区的具体规划和方案，得到云南省政府的批复和采纳。寿先生最早记录了在我国广东发现儒艮，在云南发现大象和长臂猿等。

他是我国最早开展动物狩猎学研究的学者。寿先生转向兽类学研究后，在分类学和动物地理学的基础上，将研究领域拓展到动物生态学和应用生态学等领域，还开展了狩猎生产中合理利用资源动物的研究。他最早开展了我国毛皮兽的调查研究（寿振黄，1955），撰写了狩猎学的生物学基础（手稿），编写了《毛皮兽图说》（寿振黄，1958）等。

寿振黄先生在鱼类学、两栖动物和爬行动物学、鸟类学、兽类学等方面的这些工作都是奠基性的，为我国相关学科的建立和发展作出了重要贡献。他还特别重视人才培养，为我们国家培养了一批兽类学和动物生态学的研究骨干。他是当之无愧的中国脊椎动物学的奠基人。

致谢

感谢冯祚建先生和汪松先生在作者撰写本文时提出的意见和建议，感谢中国动物学会办公室提供照片。

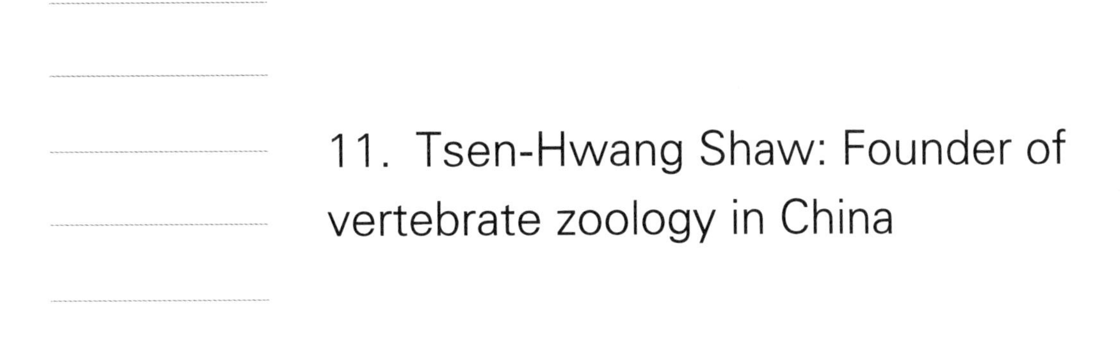

11. Tsen-Hwang Shaw: Founder of vertebrate zoology in China

"The founder of vertebrate zoology in China; an erudite scholar of insect, fish, bird and mammal research; combining taxonomy and animal ecology, and pioneering in biostatistics." These are memorial comments written by Chinese mammalogist Wuping Xia on the 15th anniversary of the death of Professor Tsen-Hwang Shaw (寿振黄，1899—1964), which offers an overview of the academic contributions of Professor Shaw.

Professor Shaw was born in Zhejiang Province on February 4, 1899. Growing up in the countryside, he was acutely aware of the constrained circumstances of life in rural areas. To change the backwardness of the countryside in China, he went to National Nanjing Higher Normal School (which later became Southeast University) to study agriculture in 1917. After graduation in 1920, he started to teach biology in a middle school, and he soon realized that biology is the foundation of agriculture and decided to switch from agriculture to biology. In 1921, he was admitted to the Department of Biology at Southeast University at age 22. After receiving a BS degree in 1925, he went to the United States for further study, and initially enrolled at the University of California, Berkeley and then Stanford University in 1926. He conducted his research on fish taxonomy under the supervision of the famous fish taxonomist D. S. Jordan. In August 1926, Shaw went to Hopkins Marine Station to conduct research on the life history of crustaceans. Later that year, he completed his thesis and obtained a master's degree. He then returned to the University of California and engaged in research related to ornithology and zoology. In 1927, he successively visited different universities and museums in Philadelphia, Chicago, New York, and Washington to learn the techniques of museum display and taxidermy. The wide range of learning interests and rich research experience in different fields laid a solid foundation for Shaw's future career.

Professor Shaw returned to China in 1928. He served at Tsinghua University for eight years as a lecturer and then a professor, teaching courses such as comparative anatomy and ichthyology

---

Fuwen Wei[1,2], Dehua Wang[1,2]

1 Institute of Zoology, Chinese Academy of Sciences, Beijing 100101, China

2 University of Chinese Academy of Sciences, Beijing 100049, China

Correspondence: weifw@ioz.ac.cn (F.-W. Wei); wangdh@ioz.ac.cn (D.-H. Wang)

using textbooks written by himself. At the same time, he also worked as a technician of the Animal Department of the Fan Memorial Institute of Biology in Beijing.

After the founding of the People's Republic of China, he served as a research fellow and professor in the Committee for Animal Specimens of the Chinese Academy of Sciences (CAS) and Institute of Zoology of CAS. In 1957, he was appointed as a principal investigator and the director of the Mammalogical Research Division of the Institute of Zoology of CAS. In 1959, he served as the director of the Department of Animal Ecology, and in 1962, deputy director of the Institute of Zoology, CAS. During Professor Shaw's academic career, he published more than 100 papers and related books on vertebrate taxonomy, morphology, paleontology, and ecology.

He was the first Chinese scientist to study ichthyology. Professor Shaw published the paper "Fishes from eastern China, with descriptions of new species" in collaboration with the American ichthyologist B. W. Evermann in 1927 (Evermann and Shaw, 1927). Later, he continued to carry out groundbreaking studies on the taxonomy, morphology, and ecology of fishes in eastern and northern China, and published a series of research papers (e.g., Shaw, 1929a, 1929b; Shaw and Tchang, 1931; Shaw and Lee, 1939).

He was one of the earliest researchers to study amphibians and reptiles in China. As far back as 1929, Prof. Shaw had investigated the amphibian species of Beiping (currently called Beijing) (Shaw, 1929b) and he also conducted research into the digestive system of turtles in the 1940 (Shaw, 1940).

He was one of the founders of modern ornithology in China. In 1927, Professor Shaw published the paper "A small collection of birds from Fukien" in *Science* (Shaw, 1927), making him the first Chinese scientist to publish an ornithological research paper in this journal. From 1930—1934, he conducted research into the birds of Chefoo (Yantai), Szechwan (Sichuan) and Chekiang (Zhejiang) (Shaw, 1930, 1931, 1932, 1934). In 1936, he published the English-language textbook "*The Birds of Hopei Province*" (Two volumes) (Shaw, 1936), which was the first ornithological book written by a Chinese zoologist with international influence and the first monograph on local fauna organized based on taxonomy. This monograph is regarded as a paradigm of local zoological study, as well as the beginning of vertebrate taxonomy in China.

He was also one of the main promoters of mammalogical research after the founding of the People's Republic of China. To fill in the gap in vertebrate zoology in China, Professor Shaw shifted his research focus to mammals at the age of 65 and established the first mammalogical research division and the first animal ecology laboratory in China. He led many pioneering studies, such as the survey of mammals, and rodent control in Korean pine forestry, and the survey of epidemic hemorrhagic fever in northeastern China. These profound contributions are reflected in a series of works completed in 1958—1962, such as "*A Report on the Mammals of Northeastern China*" (the first regional survey report on mammals), "*Rodent Control in the Direct Seeding of Korean Pine*" (a representative work combining theory and practice), "*Economic Fauna of China: Mammals*" (the first monograph on the taxonomy, morphological characteristics, habits, distribution and economic significance of animals nationwide), among other work (e.g., Mammalogical Research Division of the Institute of Zoology, Chinese Academy of Sciences, 1958; Shaw et al., 1958; Shaw, 1962). Through these scientific activities, he cultivated a professional team of mammalogists for the country and had a far-reaching influence on his successors.

He was among the first advocates of wildlife protection and the establishment of nature reserves in China. In 1956, Professor Shaw attended a conference on rodent biology and control in Germany and visited the nature reserves there, which inspired him to propose the establishment of nature reserves to the Yunnan provincial government, together with botanist Cheng-Yih Wu (Zhengyi Wu). Their specific proposals for establishing 24 nature reserves in Yunnan Province were approved and adopted by the government in 1957.

He was a pioneer of research on wildlife hunting in China. After he shifted to mammalogy, he expanded his research field to animal ecology, applied ecology on the basis of taxonomy and zoogeography, and carried out research on the rational utilization of animal resources in the game industry. He was the first to conduct surveys on fur animals in China (Shaw, 1955). He also wrote a manuscript on the biological basis of hunting, and compiled the book "*Illustrated Fur-bearing Mammals in China*" (Shaw, 1958).

Professor Shaw's research on fish, amphibians, reptiles, birds, and mammals was groundbreaking and made tremendous contributions to the establishment and development of the relevant research areas in China. He also paid much attention and displayed great enthusiasm for educating young mammal experts and animal ecologists, and some of them have become leading scientists in mammalogy and animal ecology in China. Accordingly, he is rightly deemed the founder and pioneer of vertebrate zoology in China.

Figures

Fig.1 Tsen-Hwang Shaw (1899—1964)

Fig.2 Group photo of Fan Memorial Institute of Biology in 1928 (Front row, from left: Qi He, Zhi Bing, Xiansu Hu, Tsen-Hwang Shaw; Back row, from left: Jiarui Shen, Chengru Feng, Jin Tang)

References

Evermann BW, Shaw TH (1927) Fishes from Eastern China, with description of new species. Proceedings of the California Academy of Sciences 16(4):97–122.

Mammalogical Research Division of the Institute of Zoology, Chinese Academy of Sciences (1958) A Report on the Mammals of Northeastern China. Beijing: Science Press. (中国科学院动物研究所兽类研究组. 1958. 东北兽类调查报告. 北京: 科学出版社.)

Shaw TH (1927) A small collection of birds from Fukien. Science 12:1289–1296.

Shaw TH (1929a) A new fresh-water goby from Tientsin. Bull Fan Mem Inst Biol 1(1):1–6.

Shaw TH (1929b) The amphibians of Beiping. Bull Fan Mem Inst Biol 1(5):77–78.

Shaw TH (1930) Note on some birds of Chefoo, China. Auk 47(4):542–545.

Shaw TH (1931) Note on some non-passerine birds from Szechwan. Bull Fan Mem Inst Biol 2(17):319–327.

Shaw TH (1932) Note on some passerine birds from Szechwan. Bull Fan Mem Inst Biol 3(15):217–233.

Shaw TH (1934) Note on the birds of Chekiang. Bull Fan Mem Inst Biol 5(5):285–338.

Shaw TH (1936) The Birds of Hopei Province (two vols). Beijing: Fan Memorial Institute of Biology.

Shaw TH (1940) The digestive system of the soft-shelled turtle. Bull Fan Mem Inst Biol 10:153–158.

Shaw TH (1955) The distribution of fur-bearing mammals in China. Acta Geographica Sinica 22(4): 405–421. (寿振黄. 1955. 中国毛皮兽的地理分布. 地理学报, 21: 405–421.)

Shaw TH (1958) Illustrated Fur-bearing Mammals in China. Beijing: Science Press. (寿振黄. 1958. 毛皮兽图说. 北京: 科学出版社.)

Shaw TH (1962) Economic Fauna of China: Mammals. Beijing: Science Press. (寿振黄. 1962. 中国经济动物志——兽类. 北京: 科学出版社.)

Shaw TH, Lee JS (1939) Age and growth in some food fishes. Bull Fan Mem Inst Biol 9(3): 251–262.

Shaw TH, Tchang TL (1931) Preliminary notes on the cyprinoid fishes of Hopei Province. Bull Fan Mem Inst Biol 2(15):289–296.

Shaw TH, Wang Z, Xia WP, et al (1958) Rodent Control in the Direct Seeding of Korean Pine. Beijing: Science Press. (寿振黄, 王战, 夏武平, 等. 1958. 红松直播防鼠害研究工作报告. 北京: 科学出版社.)

12. 朱弘复：世界著名昆虫学家

朱弘复教授是著名的生物学家、昆虫学家、分类学家。他研究领域广泛，涉及昆虫幼虫学、昆虫形态学、昆虫分类学和植物保护等多个学科。他一生做了大量开创性、奠基性的工作，是不少门类的创始人和学术带头人。他开创了中国蚜虫区系分类研究，系统地开展了中国蛾类区系分类研究。研究蛾类幼期形态学，创立了以数字命名毛序的方法。在我国率先引入和倡导应用数值分类学和支序分类学，对我国生物分类学的发展起到促进作用。他编写的《动物分类学理论基础》是我国学者第一本在此领域的理论著作。朱弘复先生一生致力于昆虫学研究，为我国昆虫学事业的发展壮大作出了巨大贡献，建昆虫分类之丰功，创中国昆虫学之伟业，在国内外昆虫学界享有盛誉。

朱弘复教授1910年生于江苏南通。1935年毕业于清华大学生物系并留校任教。1941年赴美国，在伊利诺伊大学理学院昆虫系知名昆虫学家W. P. Hayes教授及生态学家V. E. Shelford教授指导下攻读昆虫学，并以幼虫学为主修科目、生态学为副科。1942年获得理学硕士学位，1945年获得博士学位。1945—1946年在美国伊利诺伊州自然博物研究所与H. H. Ross教授专攻叶蜂分类学。1946—1947年在美国维思大学任动物学客座教授。由于朱弘复教授怀有返回祖国贡献所学、振兴中国昆虫学事业的夙愿，他于1947年偕眷毅然回国。

1950年，受郭沫若院长委托，朱弘复教授开始领导筹建中国科学院昆虫研究所。在原有分类学、昆虫形态学基础上，又添设资源昆虫学、昆虫组织学、昆虫生理学、昆虫生态学、昆虫毒理学与杀虫药剂学等研究室，并聘请了一批著名的专家，例如昆虫学家刘崇乐教授、蔡邦华教授，昆虫形态学家陆近仁教授，昆虫生理学家钦俊德教授，昆虫生态学家马世骏教授。经过三年左右的筹建，昆虫研究所诞生了，其中昆虫毒理学室与杀虫药剂室的建立是当时世界上首次以学科专门建室（图1）。

朱弘复教授历任中国科学院昆虫研究所、动物研究所副所长、所长，国家科学技术委员会农业组组长，中国农业科学院学术委员会委员，中国昆虫学会理事长，第十九届国际昆虫学大会主席，《中国动物志》《中国经济昆虫志》编辑委员会副主任、主任，《昆虫学报》《昆虫知识》《动物学集刊》《动物分类学报》、*Discovery and*

作者：薛大勇，韩红香

中国科学院动物研究所，北京100101，中国

邮箱：hanhx@ioz.ac.cn（韩红香）

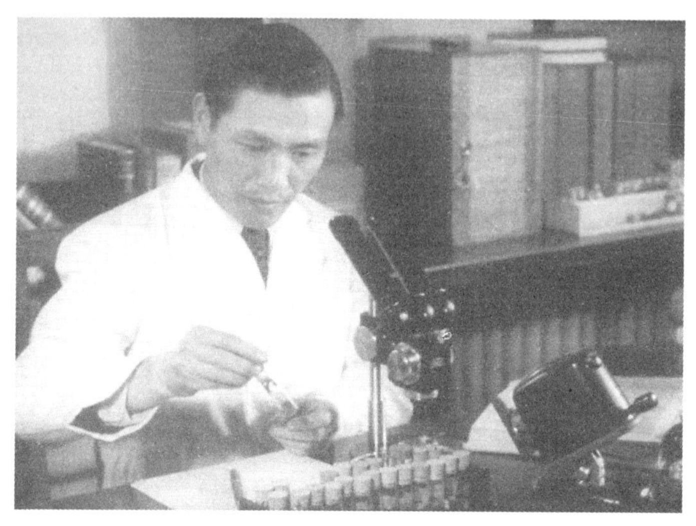

图1　朱弘复教授在实验室工作（20世纪50年代中期）

*Innovation*和*Annals of Entomology*主编或编委。

1970年以前，朱弘复教授的主要研究领域为幼虫生物学、昆虫形态学、害虫测报及防治，后来他主要从事昆虫分类学研究。他开创了昆虫幼虫形态学研究，1949年在美国出版了《如何认识幼虫》，迄今仍被列为美国大学昆虫学教学参考用书。这本著作描述了24个目的昆虫幼虫的形态学特征，并绘制幼虫图，编制了16个代表科的检索表。从那时起，国内对幼虫研究投入加大。1956年，朱弘复教授创立了以数字命名鳞翅目幼虫毛序的方法。一些重要的著作陆续出版，如《中国经济昆虫志·第七卷·鳞翅目》（夜蛾科Ⅲ幼虫）和《蛾类幼虫图册》。

朱弘复教授回国工作前期，主要从事重要农业害虫（以棉虫为主）的研究，包括鉴定、预测预报和治理。他在1957年发表了《蚜虫概论》，1959年发表了《中国棉花害虫》，准确鉴定了重要害虫如麦叶蜂和小麦吸浆虫，在植物保护方面作出巨大贡献。

朱弘复教授从20世纪50年代开始了鳞翅目分类学研究，成立了鳞翅目研究组，培养了十几个学生。他组织编写了《中国蛾类图鉴》（Ⅰ–Ⅳ），这在中国蛾类分类史上是一座里程碑。他共描述了7个新属，159个新种，21个新亚种。这些新的分类单元中，除重要农业害虫小麦叶蜂（*Dolerus tritici* Chu）外，其他都属于鳞翅目，包括虎蛾科、大蚕蛾科、蝙蝠蛾科、蚕蛾科、水蜡蛾科、钩蛾科、夜蛾科、凤蛾科、尺蛾科、天蛾科和网蛾科。朱弘复教授领导编写了系列巨著《中国动物志》和《中国经济昆虫志》，其中《中国经济昆虫志》曾获中国科学院自然科学奖（2000年）和国家自然科学奖二等奖（2001年）。

朱弘复教授为了我国昆虫学事业的发展，积极组织、主持和参加校译国外的重要昆虫学工具书、法规和理论著作，例如《国际动物命名法规》（第二版、第三版）和《英汉昆虫学词典》（第一版、第二版）。1975年他在国内率先倡导和研究数值分类学，20世纪80年代初开始教授支序分类学，并于1987年出版《动物分类学理论基础》专著，这是我国学者在这个研究领域的第一部理论专著（图2）。

图2　朱弘复教授在中国科学院讲述动物系统分类学（1984年）

朱弘复教授勤俭朴素，和蔼可亲，待人忠厚，治学严谨。他总是告诫学生、助手“学无止境”“珍惜时间”；而且他从不独揽成果，总是把具有初步成就的专题，交给年轻人独立去做。朱弘复教授以毕生精力致力于科研工作和人才培养，是大家永远怀念的老一辈科学家。

12. Hongfu Chu: World renowned entomologist

Professor Hongfu Chu (Hongfu Zhu) is a renowned Chinese biologist, entomologist and taxonomist who specializes in research fields such as insect immature, insect morphology, insect taxonomy and plant protection. He is a pioneer of taxonomist and the founder of Lepidoptera and aphid systematics. Hongfu Chu is the first person to erect the nomenclature of the chaetotaxy of lepidopteran larvae, and he introduced numerical taxonomy and cladistics to China. He also compiled the textbook *Theoretical Fundamentals of Animal Systematics*, which is the first theoretical monograph in animal systematics in China. Hongfu Chu devoted all of his time to the study of entomology, and he made magnificent contributions in establishing and developing entomology and insect taxonomy in China.

Born in 1910, Hongfu Chu received traditional education in his hometown, Nantong, Jiangsu province. He received a BS degree from the Department of Biology at Tsinghua University in 1935, and afterward, worked as an assistant professor in Tsinghua University. During his study at Tsinghua University, he showed great interests in zoology, and especially, entomology. In 1941, he was sent to Illinois University to study entomology and ecology under entomologist Prof. W. P. Hayes and ecologist V. E. Shelford. After receiving his Master's Degree in 1942, Hongfu Chu continued his graduate studies and received a Ph.D. degree in 1945. After graduation, Hongfu Chu accepted an offer from the Natural Museum of Illinois, where he researched the taxonomy of sawfly under Prof. H. H. Ross from 1945 to 1946. Shortly after his time at the Natural Museum of Illinois, Hongfu Chu worked as a guest professor at Wesleyan University from 1946 to 1947. In 1947, due to his commitment to the development of entomology in China, Hongfu Chu returned to China with his family.

In 1950, Hongfu Chu accepted commission from President Moruo Guo of the Chinese Academy of Sciences to initiate the establishment of the Institute of Entomology (which later became a major part of the Institute of Zoology, Chinese Academy of Sciences, Beijing). In addition to insect taxonomy and insect morphology, Hongfu Chu added resources on entomology,

Dayong Xue, Hongxiang Han

Institute of Zoology, Chinese Academy of Sciences, Beijing 100101, China

Correspondence: hanhx@ioz.ac.cn (H.-X. Han)

insect histology, insect physiology, insect ecology, insect toxicology and pesticide, and invited and hired many famous entomologists to join the institute, such as entomologist Chongle Liu, Banghua Cai, insect morphologist Jinren Lu, insect physiologist Junde Qin, and insect ecologist Shijun Ma. After three years of great efforts, the Institute of Entomology was established, including the first department of insect toxicology and pesticide in the world.

During his professional career, Hongfu Chu had held the positions of Vice Director and Acting Director of Institute of Entomology and Institute of Zoology, Chinese Academy of Sciences; Head of Agricultural Division of the National Science Commission; Member of Academic Committee of the Chinese Academy of Agricultural Sciences; Head of the Board of Directors of the Entomological Society of China; Chairman of the XIX International Congress of Entomology; Vice Chairman and Chairman of Board of Editors of *Fauna Sinica* and *Economic Insect Fauna of China*; Chief Editor or Editor of *Acta Entomologica Sinica, Chinese Journal of Entomology, Acta Zootaxonomica Sinica, Sinozoologica, Discovery and Innovation* and *Annals of Entomology*.

Before 1970, Hongfu Chu engaged in research regarding two main areas: (i) insect larvae, life history, and biology, and (ii) forecasting, prediction and control of pests. Later, he mainly engaged in researches of insect taxonomy.

Hongfu Chu initiated studies on the morphology of immature insects. In 1949, Chu and Cutkomp published *How to Know the Immature Insect*, which has been used as a textbook and reference book by students of entomology in America. The morphological characters of immature insects of 24 orders were described, and illustrations of larval morphological characters, and keys to 16 representative families, were provided. From then on, research on moth larvae in China greatly increased. Hongfu Chu erected the nomenclature of the chaetotaxy of lepidopteran larvae in 1956, thus prompting important publications such as *Economic Insect Fauna of China* (Fasc. 7, Lepidoptera Noctuidae Ⅲ larvae) and *Iconography of Larvae of Moths*.

During most of his life, Hongfu Chu worked mostly on important agricultural pests, especially cotton pests, and studied the identification, forecasting, prediction and control of said pests. He published several articles in these fields, such as An introduction to Aphidology and Cotton pests in China. Hongfu Chu's most significant contribution was that he identified many serious pests such as *Sitodiplosis mosellana* (Ghin) and wheat sawfly *Dolerus tritici*. He also greatly improved the development of plant protection in China.

Hongfu Chu mainly focused his research on the taxonomy of Lepidoptera in China from 1950s to his later days. He established the Lepidoptera workgroup and educated tens of graduated students. He also organized the monumental work of *Iconocraphia Heterocerorum Sinicorum* Ⅰ—Ⅳ, which was the milestone of moth taxonomy in China. Hongfu Chu described 7 new genera, 159 new species, and 21 new subspecies throughout his taxonomic career. All new taxa described by him, except the wheat sawfly *Dolerus tritici* Chu, were Lepidoptera, including Agaristidae, Saturniidae, Hepialidae, Bombycidae, Brahmaeidae, Drepanidae, Noctuidae, Epicopeiidae, Geometridae, Sphingidae and Thyrididae. Hongfu Chu had led the editing of the huge series *Fauna Sinica* and *Economic Insect Fauna of China*. The latter series of monograph was awarded the Nature Science Prize by the Chinese Academy of Sciences (2000) and National Science Prize (2001).

To facilitate the development of entomology in China, Hongfu Chu placed great emphasis on inducing important reference books, new theories, and methods. He presided over the translation of a series of the most important monographs in zoology and entomology, such as: *International Code of Zoological Nomenclature* (the Second and Third Editions), *English-Chinese Dictionary of Entomology* (the First and Second Editions). Hongfu Chu introduced numerical taxonomic methods to China in 1975, and he began to study and teach phylogenetic systematics in early 1980s, and published *Theoretical Fundamentals of Animal Systematics* in 1987, which was the first monograph on this subject in China.

Hongfu Chu often told students, researchers, and assistants, "Knowledge is infinite, and one need cherish the time." He was known to transfer the projects which had preliminary achievements to young students, and he never monopolized the outcome. He is the epitome of the old-generation of Chinese scientists, those who devote his or her whole life to science and country.

Figures

Fig.1 Prof. Hongfu Chu working in the laboratory (mid-1950s)

Fig.2 Hongfu Chu lecturing on cladistics in the Institute of Zoology, Chinese Academy of Sciences (1984)

References

Liang AP (1997) Entomologist: Hong-fu Zhu. In: Zhu GY. Biography Brief of Chinese Scientists and Technique Experts. Natural Science: Biology, vol. 1. Shijiazhuang: Hebei Education Press, 464–476. (梁爱萍. 1997. 昆虫学家: 朱弘复. 见: 朱光亚. 中国科学技术专家传略・理学编・生物学卷1. 石家庄: 河北教育出版社, 464–476.)

Xue D, Han H (2002) The entomological contributions of Prof. Hong-Fu Zhu (1910—2002). Fauna China 4:5–36.

Xue DY, Han HX (2013) Hongfu Chu. In: Qian WC. Overview of Academic Achievement of Renowned Chinese Scientists in 20th Century. Biology, Part 2. Beijing: Science Press, 190–197. (薛大勇, 韩红香. 2013. 朱弘复. 见: 钱伟长. 20世纪中国知名科学家学术成就概览・生物学卷・第二分册. 北京: 科学出版社, 190–197.)

二、植物学

13. 胡先骕：中国植物分类学奠基人

胡先骕（图1），字步曾，江西新建人，1916年获得美国加利福尼亚大学学士学位，1925年获得哈佛大学博士学位。胡先骕从1918年加入南京高等师范学校（东南大学）到1968年去世，在半个世纪的职业生涯中，为我国植物学的建立和发展作出了杰出贡献，是海内外公认的中国植物分类学奠基人。

图1 胡先骕（1894—1968）

杰出的科学家

作为学者，胡先骕在植物学研究上创造了大量的中国第一。在发现植物新类群方面，他是第一个命名中国植物新属（1928年发表的“捷克木”，即秤锤树）和植物新科（1934年发表的“鞘柄木科”）的中国学者。他发现的植物新种更是不计其数。他对古植物学也有很深的造诣，与R. W. Chaney合著的《中国山东省中新统植物群》（1938年，1940年）是中国新生代植物研究的第一部著作，也是亚洲新生代植物研究的划时代巨作。他最著名的植物学发现无疑是轰动世界的“活化石”水杉（1948年与郑万钧联名发表），被誉为20世纪最重要的植物学发现之一。

在植物系统研究领域，胡先骕在哈佛大学完成的博士论文《中国植物志属》（1925年）是世界上首部系统性整理全中国有花植物科属的专著，是中国植物分类研究的奠基性工作。这部论文在其后的30年间被我国植物分类学家作为标准参考文献广泛使用。胡先骕还更进一步提出了自己的植物分类系统。他于1950年提出了一个新的被子植物多元分类系统，在1965年又进一步提出了整个生命世界的分类系统。这些都是我国植物学家在该领域中的首创。值得注意的是，他在1944年第一次提议建设“中央植物园”。2022年，国家植物园正式在北京揭牌。

作者：胡晓江[1]，马金双[2]

1 北京师范大学社会学院，北京100875，中国

2 北京市植物园，北京100093，中国

邮箱：jinshuangma@gmail.com（马金双）

胡先骕一生发表学术论文150多篇和论著20多部，这些成就足以证明他是一位卓越的科学家。但仅有个人成就还不足以被称为“奠基人”。一项完整的科学事业，需要具备研究机构、教学机构、教科书、学术刊物、学术组织和文献积累等要素。只有这样，才可能大规模地培养人才和产生科学成果，最后才能惠及社会。胡先骕就承担起了在我国建立植物学事业的历史使命。

卓越的创业者

胡先骕对我国植物学事业的贡献是全方位的。中国是世界闻名的“园林之母”，从18世纪起就吸引了西方探索者的好奇心；但直到20世纪早期，真正意义上的植物学在我国仍然是一片空白。1918年，胡先骕任南京高等师范学校农科教授后，立刻开始筹划采集植物标本，并于1920年和1921年在浙江和江西进行了大规模采集。此后的20年间，胡先骕策划组织了多次更大规模的植物标本采集，采集范围覆盖了近半个中国。其中1930年代开始连续派出蔡希陶、王启无、俞德浚等人深入云南和四川采集标本，十多年采集标本十多万号，成为我国植物采集史的里程碑。与此同时，胡先骕在1930年代资助秦仁昌从欧洲各大标本馆拍摄了18337幅中国植物标本照片，使“中国植物学家不必再依靠西方人就可以鉴定自己的植物了”！ 这些植物标本和标本照片为我国植物分类学研究打下了不可或缺的实物基础。

植物学的建制基础也大都来源于胡先骕。胡先骕在1921年与动物学家秉志（1886—1965）创办了中国人所办大学中的第一个生物系——东南大学生物系；继而在1922年与秉志等创建了中国第一个生物学研究机构——中国科学社生物研究所，及其学术刊物*Contributions from the Biological Laboratory of the Science Society of China*（《中国科学社生物研究所论文丛刊》）；1923年，胡先骕与邹秉文、钱崇澍合作编写了首部中国人编著的大学教材《高等植物学》，被各大学植物学专业广泛采用；1928年胡先骕与秉志在北平创建北平静生生物调查所（图2），及其学术刊物*Bulletin of the Fan Memorial Institute of Biology*（《北平静生生物调查所汇报》）；1933年，胡先骕于重庆发起成立中国植物学会，次年发行会刊《中国植物学杂志》；1934年，胡先骕于江西创建庐山植物园；1938年，胡先骕于昆明创建云南农林植物研究所。

图2　北平静生生物调查所人员合影（1936年9月16日于北平文津街3号北平静生生物调查所大楼前），前排左八胡先骕（后排左六俞德浚，右二蔡希陶）

通过这些教学和研究机构、学术刊物和学术组织，我国植物分类学在1920年代早期至1930年代后期不到20年的时间内突飞猛进，快速地完成了自身建设，成为我国现代科学的众多学科中十分耀眼的一支。难能可贵的是，胡先骕是基于国际视野来建设我国植物学的。他创建的学术刊物都是英语或双语的，培养的人才也都中西兼顾。胡先骕与当时国际上主要的植物学机构建立了广泛的伙伴关系，使我国植物学成为国际植物学界里平等且极为重要的一员。

胡先骕以卓越的自身成就和领导才能成为我国当之无愧的植物学领袖，于1948年当选为中央研究院首届院士。更重要的是，胡先骕奠定了我国植物分类学的人才基础。我国第二代植物分类学家中大多数人都得益于胡先骕的直接培养和提携，第三代和第四代的植物分类学家至少有一半可以追溯到师承胡先骕。

胡先骕的一生与我国社会的起起伏伏紧密地联系在一起。他耿直的个性，使他能够在中青年时期取得巨大的成就，也给他的晚年带来了巨大的磨难。胡先骕从青年时期开始就积极参与社会事务，对我国的文学、科学与社会的走向发表了大量的议论。当1950年代李森科伪科学在我国遗传学界盛行一时之际，胡先骕又第一个站出来对其进行了公开批评（见《植物分类学简编》，1955年），这使得胡先骕在1955年和1957年两度落选中国科学院学部委员。1968年，胡先骕不幸逝世。之后几十年，胡先骕在我国植物学史上的地位渐渐不为人知。

然而，青山遮不住，毕竟东流去！随着1990年代越来越多的胡先骕研究成果问世，特别是1300万字的《胡先骕全集》（19卷）（2021年），胡先骕对于我国植物学事业的奠基性的功绩终究不可能被埋没。

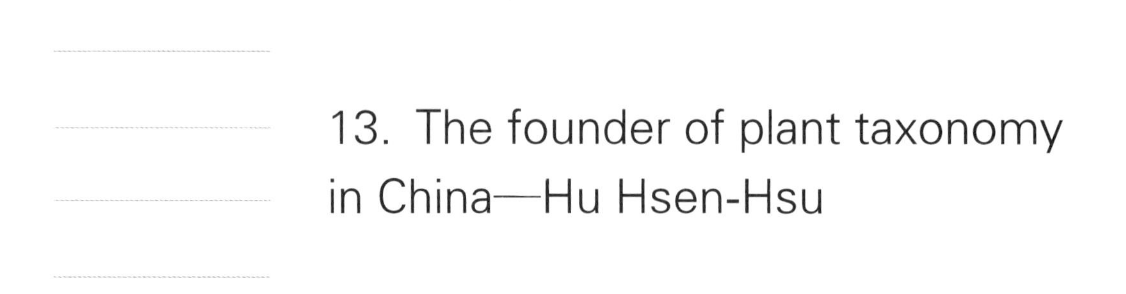

13. The founder of plant taxonomy in China—Hu Hsen-Hsu

Hu Hsen-Hsu (胡先骕, courtesy name 步曾, Buzeng) was born in Xinjian, Jiangxi Province of China in 1894. He received a bachelor degree of biology from the University of California in 1916, and a doctorate of applied biology from Harvard University in 1925. From 1918 when he started teaching in college until his death in 1968, in a half-century career Hu made foundational contributions to modern botany in China. He is widely regarded as the founder of Chinese plant taxonomy.

Accomplished Scientist

As a pioneer scientist, Hu created many "firsts" in Chinese botany. In terms of discovering new plant taxa, Hu was the first Chinese scientist to name a new genus (*Sinojackia*, in 1928) and then a new family (Torricelliaceae, in 1934). Several hundreds of new species of plants were named and described by him. In addition to modern plants, Hu was also a pioneer in paleobotany. *The Miocene Flora of Shandong Province, China* (1938, 1940), co-authored with Ralph W. Chaney, was the first work investigating China's Cenozoic fossil plants, the cornerstone of our knowledge of Asian Cenozoic plants. Hu's most famous discovery in botany is undoubtedly the "living fossil" *Metasequoia glyptostroboides* (published jointly with Cheng Wan-Chun(郑万钧) in 1948). The discovery of *Metasequoia* became a sensation throughout the world and was hailed as the most important botanic discovery in the 20th century.

In terms of systematic description, Hu's doctoral dissertation "Synopsis of Chinese genera of phanerogams" (Harvard University, 1925) was the first monograph that comprehensively surveyed and systematically described phanerogam plants in China. During the following 30 years since its completion, this work was widely used in China as the standard reference work for Chinese plants. In 1950, Hu proposed a new classification system for angiosperm plants. Furthermore in 1965, Hu proposed his own alternative classification system for the entire living

Xiaojiang Hu[1], Jinshuang Ma[2]

1 School of Sociology, Beijing Normal University, Beijing 100875, China

2 Beijing Botanical Garden, Beijing 100093, China

Correspondence: jinshuangma@gmail.com (J.-S. Ma)

world (published in *Taxon*). Both efforts were first of their kind from China. It is worth noting that he firstly proposed the construction of the "Central Botanical Garden" in 1944, and in 2022, the National Botanical Garden was officially unveiled in Beijing.

As a prolific scholar, Hu published more than 150 academic papers and more than 20 books throughout his life. But personal achievements alone are not enough to classify a person as a "founder". The Founder creates a whole disciplinary infrastructure of learning and research, including schools, textbooks, research institutions, academic organizations, academic journals, literature and resource accumulation, and other affine tasks, before the discipline can cultivate talent *en masse* and produce scientific results on a large scale, to the ultimate benefit of society. Hu's accomplishment in this area is what makes him the seminal figure in the establishment of botany in China.

Leading Entrepreneur in Science

Hu's contribution to the cause of botany in China is vast and comprehensive. With incomparable botanical variety, China is called "the mother of gardens" , and has attracted the curiosity of Western explorers since the 18th century; but until the early 20th century, the true science of botany was still nonexistent in China. After Hu became a professor of the Faculty of Agriculture at Nanjing Higher Normal College in 1918, he immediately planned to collect plant specimens. From 1920 to 1922, Hu personally carried out large-scale plant collections in Zhejiang and Jiangxi provinces. In the following 20 years, Hu continuously organized large-scale plant specimen collections, covering nearly half of China. Among these projects, the Yunnan and Sichuan collection carried out in the 1930s by Tsai Hse-Tao(蔡希陶), Wang Chi-Wu(王启无), Yu Te-Tsun(俞德浚) etc., which resulted in more than 100000 specimens, was regarded as the landmark of Chinese plant collection. At the same time, Hu sponsored Ching Ren-Chang(秦仁昌) to take high-quality photos of Chinese plant specimens from major herbaria in Europe. The project resulted in 18337 photos, which meant that "Chinese botanists no longer need to rely on Westerners to identify their own plants!" These new plant specimens and the photos of specimens laid an indispensable physical foundation for the study of Chinese plant taxonomy.

The human capital foundation of botany comes from the systematic organization of educational institutions. In 1921, Hu and zoologist Ping Chi (秉志, 1886—1965) founded the biology department in the Southeast University. This was the first biology department among Chinese-run universities, apart from missionary universities. Soon in 1922, Hu, Ping and others established China's first biology research institute—The Biology Laboratory of the Society of Science of China, and its periodical *Contributions from the Biological Laboratory of the Science Society of China*. In 1923, Hu, Chou Ping-Wen(邹秉文) and Chien Chong-Su(钱崇澍) compiled the first college textbook of botany in Chinese "*Advanced Botany*" . In 1928, Hu and Ping founded the Fan Memorial Institute of Biology in Peiping and its periodical the *Bulletin of the Fan Memorial Institute of Biology*. In 1933, Hu initiated the Botanical Society of China, and published its periodical *The Chinese Journal of Botany* in the following year. In 1934, Lushan Botanical Garden was founded in Jiangxi by Hu. And in 1938, he founded the Yunnan Botanical Institute in Kunming.

With this long list of teaching and research institutions, scientific periodicals, and academic organizations, China's plant taxonomy advanced rapidly from the early 1920s to the late 1930s. In less than 20 years, botany quickly completed its own disciplinary construction and became a modern science in China. It was a dazzling achievement, compared to the pace of many other disciplines. Moreover, Hu built Chinese botany with an international perspective. Most of the scientific periodicals he created were in English or bilingual, and the talent he cultivated were fluent in both in Chinese and English. Hu also established extensive partnerships with major botanical institutions around the world. With these efforts, Chinese botany became an equal and important member of the international botanical community.

With his outstanding achievements and leadership, Hu became a widely-respected leader of Chinese botany. He was elected as an academician of the Academia Sinica in 1948. The talent directly cultivated by him was very large. Almost all of the second-generation Chinese plant taxonomists, and about half of the third and fourth-generation Chinese plant taxonomists could be traced back to Hu.

Hu's life is closely intertwined with the ebbs and flows of Chinese history. His blunt and direct personality enabled him to make great achievements in youth and middle age, but also brought grave adversities to his old age. In addition to science, Hu was outspoken about all social affairs throughout his life, including literature, culture, education and politics. When the "Michurinism" of Lysenko of the USSR dominated China in the 1950s, Hu was the first to openly denounce it as a pseudoscience. This action resulted in his book *Plant Taxonomy Textbook* (1955) being banned, and he himself twice (1955 and 1957) failed to be appointed as an academician of the Chinese Academy of Sciences. Hu died in 1968. The situation eclipsed Hu's name after his death that time, and his public position in the history of Chinese botany was dimmed.

But as an old Chinese saying goes, "Thick mountains could not stop the river from flowing into the sea!" Since the 1990s, interest in his work has been reborn, and much research on Hu has been published. In 2021, 19 volumes of *H. H. Hu: Complete Works* will be published. Hu's foundational contributions to botany in China will not be buried after all.

Figures

Fig.1 Hu Hsen-Hsu (1894—1968)

Fig.2 Staff of the Fan Memorial Institute of Biology (in front of No.3, the Wenjin Street at Peiping, 1936.09.16). Front row: eighth from the left Hu Hsen-Hsu. Back row: sixth from the left Yu Te-Tsun, second from the right Tsai Hse-Tao

References

Hu XJ (2021) H. H. Hu: Complete Works, 19 volumes. Nanchang: Jiangxi People's Press.（胡晓江. 2021. 胡先

骕全集（19卷）. 南昌：江西人民出版社.）

Ma JS (2020) A Chronicle of Plant Taxonomy in China. Zhengzhou: Henan Science and Technology Press.（马金双. 2020. 中国植物分类学纪事. 郑州：河南科学技术出版社.）

14. 钱崇澍：中国近代植物学的奠基者之一

钱崇澍（图1）是我国著名的植物学家、教育家，我国植物分类学、植物生理学、地理植物学、植物区系学的创始人之一。他将毕生心血倾注于科研、教育工作中，是我国近代植物学的奠基者之一。钱崇澍开辟了我国近代植物学的多个研究领域：1916年发表了中国人用拉丁文为植物命名和分类的第一篇文献；次年发表中国人独立应用近代科学方法研究植物生理学的最早文献；1927年首次撰写我国植物生态学和地植物学方向的论文。他是系统研究分类工作难度较大的兰科、荨麻科等植物的先驱；1959—1965年担任《中国植物志》的主编并承担荨麻科部分的编写工作（俞德浚，1983）。钱崇澍开创了我国植物分类学、生理学和生态学研究的先河，为我国植被、植物区划以及植物专科专属研究打下了牢固的基础（钱崇澍等，1956），将科研与教学紧密结合，为我国培养了一大批植物学科技人才。

图1　钱崇澍（1883—1965）

求学与工作生涯

钱崇澍出生于浙江省，在书香家庭的熏陶下自幼勤奋好学，于1904年清朝举行的最后一次科举考试中考中秀才，后又顺应时代变革要求学习新科学，第二年考入上海南洋公学（西安交通大学、上海交通大学前身）学习。他于1910年考取清华学校（清华大学前身）留美公费生，与胡适、李四光、竺可桢等70 人一起赴美深造，先在美国伊利诺伊大学理学院学习农学，一年后转入伊利诺伊大学自然科学学院主攻植物学，

作者：刘欢[1,2]，黄凯靖[1]，袁雪凡[3]，程浩[4]

1　中国科学技术大学，合肥230026，中国

2　病毒学国家重点实验室，武汉430072，中国

3　南京艺术学院，南京210013，中国

4　中国科学院微生物研究所，北京100101，中国

邮箱：liuhuan520@ustc.edu.cn（刘欢）

于1914年7月获得理学学士学位。因对植物学有浓厚的兴趣，他选择到芝加哥大学和哈佛大学继续深造，进修植物生理学、生态学和分类学，获得芝加哥大学硕士学位（汪振儒，1984）。

1916年钱崇澍回国，先后在北京农业专门学校（中国农业大学前身）、清华学校、复旦大学等校担任教授。1923年，他与邹秉文、胡先骕合作编写中国第一部生物学教科书《高等植物学》。1926年，担任清华学校生物系首届系主任。1948年，钱崇澍当选中央研究院院士（图2）。1955年，钱崇澍当选中国科学院首批学部委员（刘昌芝，1981）。

图2　1948年9月钱崇澍（五排左六）参加中央研究院第一次院士会议

我国植物学研究奠基人之一

我国地大物博，植被类型丰富，珍稀品种繁多。但在19世纪，由于国内未建立系统的植物学研究方法，我国植物研究长期处于外国人的掌控之下，大批珍稀、模式标本散落国外，植物学论文稀缺，本土农业水平低下。这激起了青年钱崇澍极大的爱国情怀，立志建立属于中国的近代植物学，提高农业水平以建设祖国。

钱崇澍开创了国内在植物分类学、植物生理学、植物生态学等领域的研究，他早期的著述与翻译填补了我国植物学的空白，多篇论文受国内外学术界认可。1916年，钱崇澍在国外发表了第一篇由中国人用拉丁文为植物命名和分类的文献《宾夕法尼亚毛茛两个亚洲近缘种》，是近代植物分类学在我国诞生的标志（刘昌芝，1981）。在哈佛大学进修分类学的同时，他不忘苦学植物生理学，在该领域教授W. J. V. Osterhout的建议与指导下完成盐离子的拮抗现象和吸收机制的证明试验（Osterhout，1935），并于1917年在美国《植物学公报》发表首篇中国人独立应用近代科学方法研究植物生理学的文献《钡、锶、铈对水绵属的特殊作用》（Chien，1917），开启了我国植物生理学的研究（汪振儒，1984）；1927年，他首次完成了中国地植物学和区系植物学的论文《安徽黄山植被区系的初步研究》；1929年，他翻译了《细胞的渗透性质》《自养植物的光合作用》等植物生理学方面的论文（刘昌芝，1981）（图3）。

图3　钱崇澍在工作

珍稀物种植物分类屹立东方

钱崇澍对兰科、荨麻科、豆科、毛茛科植物的分类和系统研究作出了巨大贡献。兰科是单子叶植物中最大的科，目前全世界超过800属近25000种，我国就有约194属1388种。在自然界条件下，兰科植物能进行大范围的杂交，其中包括许多种间甚至属间杂交；同时兰科植物有极大的变异性，导致很多中间类型的产生，种的界限也不甚清晰（杨志娟等，2005）。荨麻科、豆科和毛茛科植物也具有种属多、种间交叉特征多等特性，为这些植物的分类和系统研究造成很大的困难。在国内相关标本、文献资料稀缺的20世纪初，钱崇澍迎难而上，毅然选择成为我国对这些科进行系统研究和分类的先驱。

独花兰是兰科、独花兰属的国家二级保护植物、我国的珍稀物种，由陈长年和邓世纬于1931年在广西宝华山首次发现并采集标本（图4）。1935年，时任中国科学社生物研究所教授兼植物部主任的钱崇澍在翻看植物标本时察觉并验证该物种为兰科新种乃至新属，同年在《中国科学社生物研究所论文丛刊（植物组）》上发表论文《来自中国东部的一种兰花新属》宣布新属新种的发现，并将其学名定为*Changnienia amoena*。论文记录了独花兰独特的生物特征，并指出该新属隶属于施勒赫特的顶孢科（Acrotonae-Calypsoeae of Schlechter），可根据叶片特征尤其是唇形和与其外观相似的布袋兰属区别开来：前者的唇形上有明显的刺状，后者则没有（Chien，1935）。

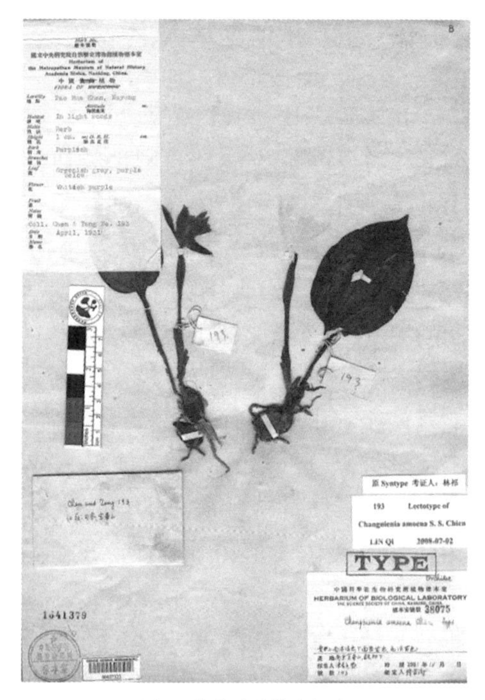

图4　独花兰的模式标本

主持编撰《中国植物志》

20世纪60年代，钱崇澍将自己投入主持编撰《中国植物志》这一庞大、艰巨却具有历史性意义与价值的工作中。植物志要包含植物的学名、形态特征、系统位置等重要的科学信息，需充分了解世界各国标本馆中的中国植物标本信息以及对全国各地植物标本进行详尽的考察和研究，这对植物物种丰富度名列世界前列但大量本土标本流失海外的中国来说难上加难。

钱崇澍在哈佛大学进修期间，时常到收藏有大量中国植物标本的格瑞标本室和阿诺德树木园标本室参观、学习祖国土地上生长的原生标本，累积了充足的植物分类学知识和本土植物信息（汪振儒，1984）。回国后，他在对浙江和江苏南部进行深入野外考察的过程中，采集一万多号植物标本用于分析研究，后又调查和收集了四川、安徽等地的许多植物标本，为编写植物志创造了条件。1959年10月，经过前后40多年的准备工作，以钱崇澍为主编的《中国植物志》编辑委员会正式成立，植物学科技史上的伟大工程拉开序幕。钱崇澍主持《中国植物志》的编撰工作至1965年，同时还承担了荨麻科部分的编写工作，他任主编期间共出版了3卷《中国植物志》。2004年，这一世界上最大的、凝聚四代中国植物学家的青春与心血、历经数十载编撰而成的《中国植物志》终于宣告完工，向世界彰显了我国近代植物学的巨大成就（图5）。《中国植物志》于2009年获得国家自然科学奖一等奖（骆洋和李德铢，2013）。

图5 《中国植物志》

第一部植物学教科书，第一个自然保护区

钱崇澍意识到要想在我国建立本土植物学，研究基地必不可少。1922年他与胡先骕合作在中国科学社生物研究所内组建植物部，建立符合科学标准条件的实验室、标本室和图书馆等，方便调查研究的开展，促进本土植物学的研究工作。1937年，钱崇澍与中国科学社生物研究所部分科技人员一同迁往重庆北碚，在最艰难的战争时刻也不曾放弃科研工作（俞德浚，1983）。

钱崇澍始终坚持科研与教育紧密结合，于1923年与邹秉文、胡先骕合编了《高等植物学》(图6)，是我国第一部植物学教科书，内容新颖体系完整，更正了过去相关教科书的错误，填补了该领域中文教科书的空缺，深刻地影响了我国植物学教育（俞德浚，1983）。

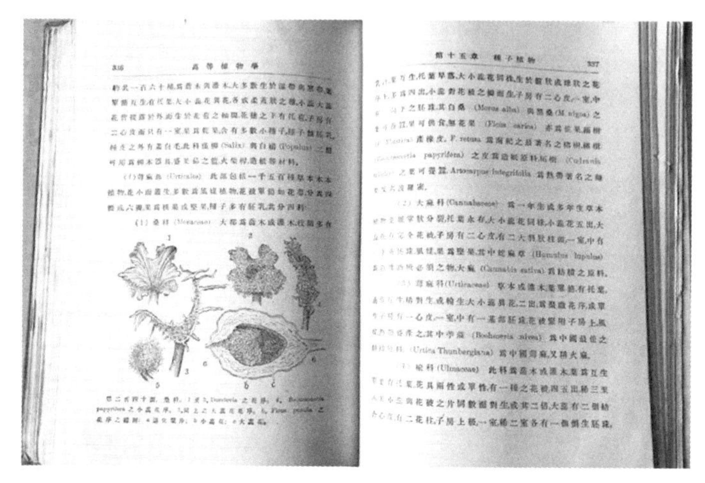
336 高等植物學

第十五章 種子植物 337

图6 《高等植物学》初版

通过深入研究植物生态学和实地考察，钱崇澍意识到我国自然资源的丰富性和被破坏情况，目光长远地提出要对森林等自然资源加强保护和合理利用，于1956年联名向全国人大提交在全国各地划分森林禁伐区的议案，并于当年促成在广东鼎湖山建立我国第一个国家级自然保护区，自此我国开启了自然保护区的建设事业（竺可桢等，1993）。

志在有恒

钱崇澍先生在文章《志在有恒》中自述：祖国内忧外患时，“我受到科学报国思想的影响，无意于追求高官厚禄，立志献身科学”；新中国成立后，“我深受鼓舞，把无比的喜悦全部浸透在忘我的工作中，为发展中国科学事业而努力”。他的真才实学和谦虚谨慎获得了业内同行的极高评价和敬佩。“他是一位具有高尚品德，热爱祖国，乐于助人，严谨治学，勇于献身的科学家和教育家”，植物分类学家、中国科学院植物研究所副所长俞德浚在钱老先生一百周年诞辰上这样评价道（俞德浚，1983）。钱崇澍作为我国近代植物学的开拓者、出色的教育家，他宝贵的学术成果成为我国近代植物学研究的里程碑，他开创的植物分类与研究方法将一直启迪后人将本土植物学不断推进，他为祖国科学事业奉献的一生将被国人永远铭记。

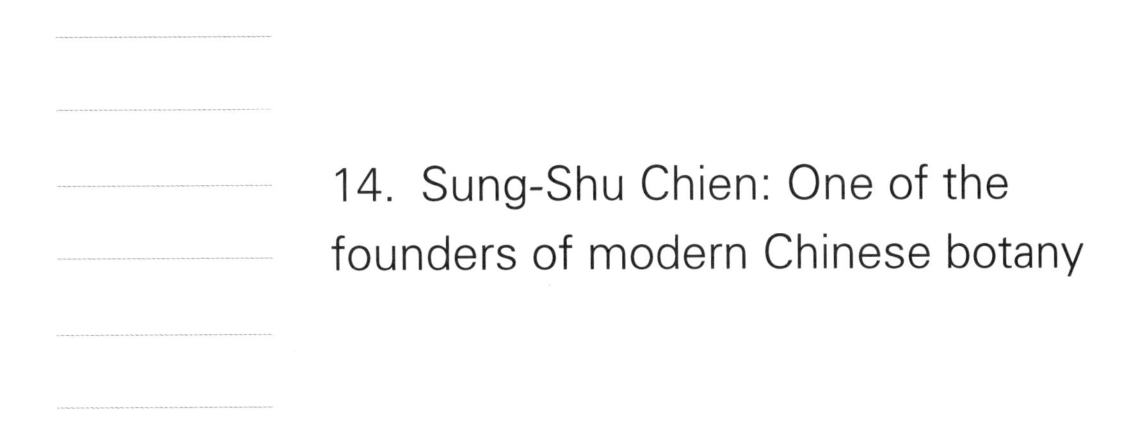

14. Sung-Shu Chien: One of the founders of modern Chinese botany

Sung-Shu Chien (钱崇澍, 1883—1965) was a famous botanist and educator in China, one of the founders of Chinese plant taxonomy, plant physiology, geobotany, and floristics. He devoted his whole life to scientific research and education, and was a founder of modern Chinese botany. Sung-Shu Chien pioneered several research fields in modern Chinese botany. In 1916, he published the first literature in Latin on the naming and classification of plants. In the following year, he published the earliest thesis on the application of modern scientific methods of plant physiology in China. He wrote the first paper in the area of plant ecology and geobotany in China. He was a pioneer in the systematic study of the families Orchidaceae, Urticaceae, Fabaceae, and Ranunculaceae in China. From 1959 to 1965, he was the chief editor of *Flora Republicae Popularis Sinicae* and was responsible for the section on Urticaceae (Yu, 1983). Sung-Shu Chien initiated the research of plant taxonomy, plant physiology and plant ecology in China, and laid a solid foundation for the extensive research of vegetation, plant zoning and plant specialties in China (Chien et al., 1956). By skillfully combining research and education, he cultivated Chinese scientific and technological talents in botany.

Studying and working life

Born in Zhejiang Province, Sung-Shu Chien was diligent since childhood under the cultivation of a well-educated family. He became a scholar in the last imperial examination held in the Qing Dynasty in 1904. Then, he studied new sciences in response to the demands of the times. In 1905, he was admitted to Shanghai Nanyang Public School (the predecessor of Xi'an Jiaotong University and Shanghai Jiao Tong University). In 1910, he was admitted to the Tsinghua School (the predecessor of Tsinghua University) as a state-funded student. He then

Huan Liu[1,2], Kaijing Huang[1], Xuefan Yuan[3], Hao Cheng[4]

1 University of Science and Technology of China, Hefei 230026, China

2 State Key Laboratory of Virology, Wuhan 430072, China

3 Nanjing University of the Arts, Nanjing 210013, China

4 Institute of Microbiology, Chinese Academy of Sciences, Beijing 100101, China.

Correspondence: liuhuan520@ustc.edu.cn (H. Liu)

went to the United States for further study together with almost 70 other students, including Shih Hu, Siguang Li, and Kezhen Zhu. He first studied agronomy at the College of Science of the University of Illinois. After one year, he transferred to the College of Natural Sciences of the University of Illinois, majoring in botany, and graduated with a Bachelor of Science degree in July 1914. Because of his strong interest in botany, he chose to continue his studies at The University of Chicago and Harvard University, where he studied plant physiology, plant ecology and plant taxonomy, and received his master's degree from The University of Chicago (Wang, 1984).

In 1916, Sung-Shu Chien returned to China and worked as a professor in multiple universities including Beijing Agricultural School (the predecessor of China Agricultural University), Tsinghua University, and Fudan University. In 1923, he collaborated with Bingwen Zou and Hsen-Hsu Hu to write China's first biology textbook, *Advanced Botany*. In 1926, he served as the first dean of the Department of Biology of Tsinghua University. In 1948, Sung-Shu Chien was elected as an Academician of Academia Sinica. In 1955, Chien was elected as one of the first Academicians of the Chinese Academy of Sciences (Liu, 1981).

One of the founders of botany in China

China is a vast country with rich vegetation and a wide variety of rare species. However, in the 19th century, Chinese botanical research was dominated by foreigners for a long time, a large number of rare and model specimens were shipped abroad, botanical literature was rare, and local agriculture was lagging behind the world. Reacting to this situation, Sung-Shu Chien was determined to establish modern botany in China and improve the agricultural development of the country.

Sung-Shu Chien initiated research in the fields of plant taxonomy, plant physiology, and plant ecology in China. His early works and translations filled the previously blank pages of botany in China. Many of his papers were praised by domestic and foreign academic communities. In 1916, Sung-Shu Chien published "Two Asiatic allies of *Ranunculus pensylvanicus*" , which is the first paper written by a Chinese author using Latin language for the plant name and classification. This paper marked the birth of modern plant taxonomy in China (Liu, 1981). During his taxonomic studies at Harvard University, he also showed a keen interest in plant physiology. He conducted experiments demonstrating the effects and mechanism of salt absorption by cells proposed at that time under the advice and guidance of Professor W. J. V. Osterhout (Osterhout, 1935). He published the first independent application of modern scientific methods to study plant physiology by a Chinese author in the *Botanical Gazette* in 1917 (Wang, 1984). His paper initiated plant physiology in China (Chien, 1917). In 1927, he completed the first paper on Chinese geobotany and zonal botany, "A preliminary study of the vegetation system of Anhui Huangshan" . In 1929, he translated literatures on plant physiology such as "Osmotic properties of cells" and "Photosynthesis in autotrophic plants" (Liu, 1981).

A new genus of orchids from China

Sung-Shu Chien contributed to the taxonomy of Orchidaceae, Urticaceae, Fabaceae, and Ranunculaceae. Orchidaceae is the largest family of Monocotyledons. At present, there are more than 800 genera and nearly 25000 species of orchids in the world, including approximately 194 genera and 1388 species in China. Under natural conditions, orchid plants are prone to hybridization, which has led to the occurrence of many interspecific and even intergeneric hybrids. At the same time, orchids have great variability, leading to the creation of many intermediate types and poorly defined species boundaries (Yang et al., 2005). Urticaceae, Fabaceae and Ranunculaceae also have multiple unclearly separated genera and interspecific hybridization, causing great difficulties for the classification and systematic study of these plants. In the early 20th century, when China lacked relevant specimens and literature, Sung-Shu Chien resolutely chose to devote himself to the systematic study and classification of these important plant families in China.

Changnienia amoena is a national second-class protected plant from the family Orchidaceae, and a rare species in China. It was first discovered and collected by Changnian Chen and Shiwei Deng in 1931 on Baohua Mountain, Guangxi, China. In 1935, Sung-Shu Chien, who was then a professor and director of the Botany Department at the Biological Laboratory of the Science Society of China, noticed and verified that this was a new species and even a new genus of orchids. The corresponding paper titled "A new genus of orchids from eastern China" was published in the *Contributions from the Biological Laboratory of Science Society of China, Botanical Series*, announcing the discovery of a new genus and species, which was named *Changnienia amoena*. The paper documented the unique biological characteristics of this species and pointed out that the new genus can be distinguished from the similar *Calypso* species based on the leaf characteristics, especially the labellum: The former has a distinctive spine-like shape on the labellum, while the latter does not (Chien, 1935).

Flora Republicae Popularis Sinicae

In the 1960s, Sung-Shu Chien devoted himself to the enormous and arduous but historically significant and valuable task of compiling the *Flora Republicae Popularis Sinicae*. This publication contains important scientific information such as nominal classification, morphological characteristics, and systematic locations in plant taxonomy. The compilation of this book required vast knowledge of Chinese plant specimens overseas, as well as detailed observation and study of domestic plant specimens, which were difficult at the time. The type specimens of many native species were overseas even though China has high plant species richness.

During his studies at Harvard University, Sung-Shu Chien often visited the Gray Herbarium and the Herbarium of Arnold Arboretum, which had a large collection of Chinese plant specimens. He closely observed the Chinese native specimens, acquiring knowledge of plant taxonomy and information on indigenous plants (Wang, 1984). After returning to China, he conducted a thorough field investigation of Zhejiang and southern Jiangsu, in which he collected more than 10000 plant specimens for analysis and research, and then investigated and collected many plant specimens from Sichuan, Anhui and other places, preparing for the book. In October 1959, after more than 40 years of accumulation, the editorial committee of *Flora Republicae Popularis Sinicae*, led by Sung-Shu Chien, was formally established, setting up a great project in the science and technology history of botany. Sung-Shu Chien led the compilation until 1965, and took charge of the edition on the Urticaceae. During his tenure as chief editor, three volumes were published. In 2004, the world's largest flora, which has been compiled by four generations of Chinese botanists over a period of decades, was finally completed, demonstrating to the world the great progress of modern botany in China. In 2009, the project received the first prize of The State Natural Science Award (Luo and Li, 2013).

The first botany textbook and the first nature reserve

Sung-Shu Chien realized that basic research is essential to establish botany in China. In 1922, he worked with Hsen-Hsu Hu to establish a botany department at the Biological Laboratory of the Science Society of China, setting up a laboratory, herbarium, and library that met scientific standards, which facilitated the development and promoted the research work of native botany. In 1937, Sung-Shu Chien and his colleagues in the Biological Laboratory of the Science Society of China moved to Beibei, Chongqing, and continued scientific work even in the hardest war time (Yu, 1983).

Sung-Shu Chien always paid attention to both scientific research and education. In 1923, he co-edited the book *Advanced Botany* with Bingwen Zou and Hsen-Hsu Hu. It was the first botany textbook in China, which corrected the mistakes of previous books and established a complete knowledge system, filling the gaps of Chinese botany textbooks and significantly promoting botanical education in China (Yu, 1983).

Through intensive study of plant ecology and fieldwork, Sung-Shu Chien realized the preciousness of China's natural resources and the threat they faced. Perspectively, he proposed to strengthen the protection and rational use of natural resources such as forests. In 1956, he jointly submitted a proposal to the National People's Congress of the People's Republic of China on the initiatives of non-logged areas of the country, which led to the establishment of China's first Nature Reserve on Dinghu Mountain, Guangdong Province. Since then, the establishment of Nature Reserves in China was initiated (Zhu et al., 1993).

A resolve makes a will

In a notable article, Sung-Shu Chien stated that "A resolve makes a will" : When China was suffering, "I was influenced by the idea of serving the country through science, and was not interested in pursuing authority or fortune, but was determined to devote myself to science" ; after the People's Republic of China was founded, "I was so inspired and worked hard for the development of China's scientific cause" . His true talent and modesty have won him high praise and admiration. "He was a scientist and educator with high moral character, love for the motherland, willingness to help others, qualities of rigorous study, and courageous dedication" , said Dejun Yu, a plant taxonomist and vice chairman of the Institute of Botany, the Chinese Academy of Sciences, in his speech on the 100th anniversary of Sung-Shu Chien's birth (Yu, 1983). As a pioneer and outstanding educator, Sung-Shu Chien's academic achievements have set milestones of modern Chinese botany. His pioneering approach to plant classification and research will continue to inspire future generations to advance botany, and his dedication to the development of modern science in China will be forever remembered.

Figures

Fig.1 Sung-Shu Chien(1883—1965)

Fig.2 Sung-Shu Chien(6th left from 5th row)attended The First Academician Conference of Academia Sinica in September, 1948

Fig.3 Sung-Shu Chien at work

Fig.4 model specimen of *Changnienia amoena*

Fig.5 *Flora Republicae Popularis Sinicae*

Fig.6 *Advanced Botany*(1st edition)

References

Chien SS (1917) Peculiar effects of Barium, Strontium, and Cerium on Spirogyra. Botanical Gazette 63:406–409.

Chien SS (1935) A new genus of orchids from Eastern China. Contributions from the Biological Laboratory of Science Society of China, Botanical Series 10(1):89–92.

Chien SS, Wu ZY, Chen CD (1956) Vegetation types in China. Acta Geographica Sinica 22: 37–92. (钱崇澍, 吴征镒, 陈昌笃. 1956. 中国植被的类型. 地理学报, 22: 37–92.)

Liu CZ (1981) The pioneer of modern botany—Sung-Shu Chien. The Chinese Journal for the History of Science and Technology: 35–39. (刘昌芝. 1981. 近代植物学的开拓者——钱崇澍. 中国科技史杂志:35–39.)

Luo Y, Li DZ (2013) *Flora of China* (English and revised Edition) was completed. Plant Diversity and Resources 35(6):742. (骆洋, 李德铢. 2013. Flora of China (《中国植物志》英文和修订版)全面完成. 植物分类与资源学报, 35(6):742.)

Osterhout WJV (1935) Mechanism of salt absorption by plant cells. Nature 136:1034–1035.

Wang ZR (1984) The initiator of plant physiology in China–Sung Shu Chien. Plant Physiology Journal 2:62–64. (汪振儒. 1984. 我国植物生理学的启业人钱崇澍先生. 植物生理学通讯, 2:62–64.)

Yang ZJ, Zhu GF, Zhang X (2005) Advances in the systematics and affinities of orchids. Acta Botanica Boreali-

Occidentalia Sinica 09:194–199. (杨志娟, 朱根发, 张显. 2005. 兰科植物系统学及亲缘关系研究进展. 西北植物学报, 09:194–199.)

Yu DJ (1983) Commemoration of the 100th anniversary of the birth of Professor Sung-Shu Chien, the founder of modern botany in China. Journal of Integrative Plant Biology 25(5): 495–496. (俞德浚. 1983. 纪念我国近代植物学的奠基人——钱崇澍教授诞辰一百周年. 植物学报, 25 (5):495–496.)

Zhu KZ, Chien SS, Bing Z, et al (1993) Views on the destruction of natural resources and suggestions on the future enhancement of rational utilization and protection. China Population, Resources and Environment 1:77–81. (竺可桢, 钱崇澍, 秉志, 等. 1993. 关于自然资源破坏情况及今后加强合理利用与保护的意见. 中国人口·资源与环境, 1:77–81.)

15. 陈焕镛：中国现代植物分类学开创者与奠基人之一，著名植物学家、分类学家

陈焕镛院士（1890—1971），字文农，号韶钟，是我国著名的植物学家和分类学家（图1）。他开创了我国现代植物科学研究的新纪元，对我国植物学，尤其是植物分类学作出了奠基式的贡献。他是中国科学院第一批院士（学部委员），第一、二、三届全国人大代表，《中国植物志》第一任主编。他也是最早进行植物标本采集的中国植物学家之一，在木兰科、樟科、壳斗科、绣球花科、苦苣苔科、桦木科、胡桃科等被子植物及裸子植物的分类研究方面有很深的造诣，一生共发表了11个植物新属、600多个新种及新组合，包括与匡可任教授联名发表的轰动世界的新种——“活化石”植物银杉（图2）。

图1　陈焕镛教授1956年在广州工作

他创建的华南植物园标本馆（原为中山大学农林植物研究所标本室），是国内最早的现代植物标本馆之一。他组织成立中国科学院华南植物园（图3）以及广西植物研究所，并担任首任所长，同时还建立了我国第一个自然保护区——鼎湖山国家级自然保护区。

作者：黄瑞兰

中国科学院华南植物园，广州510650，中国

邮箱：huangruilan@scbg.ac.cn

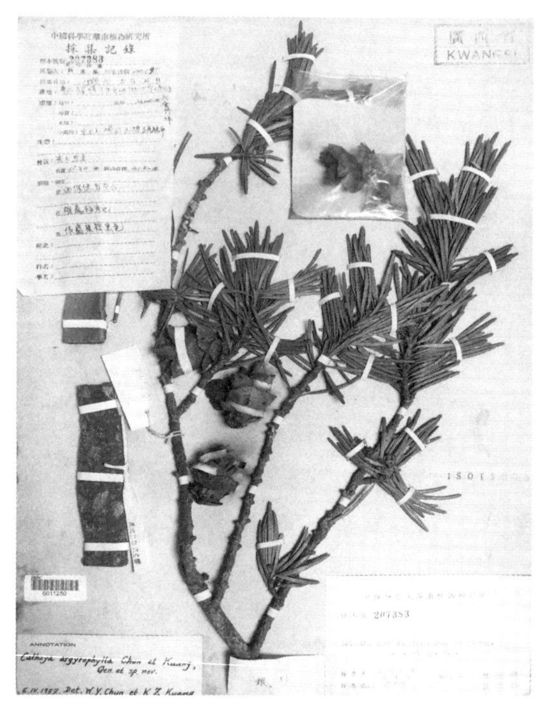

图2 陈焕镛教授和匡可任教授联名发表的银杉标本

图3 1956年，筹建华南植物园。前排左起：陈封怀、俞德浚、陈焕镛、赵同芳、程世抚、张肇骞

陈焕镛发行出版了我国第一份植物学英文杂志《中山专刊》。作为发起人之一，他创立了中国植物学会并担任会长，同时担任《中国植物学杂志》编辑。他是我国近现代最有建树的植物学家和分类学家之一。为了纪念陈焕镛对我国植物科学研究所作出的巨大贡献，有许多植物拉丁名以他的名字来命名，例如以他的姓氏陈（Chun）作为属名的山铜材属（*Chunia*）、琼棕属（*Chuniophoenix*）、乐东藤属（*Chunechites*），还有以他的名字“焕镛”（Woonyoung）来命名的焕镛木属（*Woonyoungia*），以及以他的姓

氏作为种名的鼎湖钓樟（*Lindera chunii*）和少花斑鸠菊（*Vernonia chunii*）等（陈德昭，1996）。

1890年7月22日，陈焕镛出生于香港，在广东、上海等地长大。1905年，年仅14岁的他远赴美国求学。1909年陈焕镛进入马萨诸塞州立阿默斯特学院学习树木学与昆虫学，1912年转赴纽约雪城大学林学院学习。1915年取得学士学位后，他继续前往哈佛大学伯西学院及阿诺德树木园学习森林学，并于1919年取得硕士学位（Hass，1988）。同年，他放弃了继续留美读博深造的机会，在获得谢尔顿旅游奖学金资助后，毅然选择回国到海南岛进行植物标本采集（图4）。他冒着生命危险，带领采集队深入五指山等腹地考察，待了9个月之久，收获甚丰，共采得数千份植物标本，而当时此地的植物还鲜为人知。

1920年，陈焕镛被聘为金陵大学森林学教授。1922年，他转到东南大学任教，并与胡先骕、秉志等知名动植物学家结识（图5）。同年，他与钱崇澍、秦仁昌首次组成一支中国植物学家考察队，奔赴湖北西部进行植物收集，共采得上千份标本。作为一名极富天赋又勤奋无比的年轻植物科学家，陈焕镛不仅有着远大的抱负和长远的眼光，同时还有扎实博学的知识积累。早在哈佛大学阿诺德树木园求学时，他就开始编写《中国经济树木》，1921年该书由上海商务印书馆出版，这是第一本由中国植物学家出版的英文植物学著作。

图4　1919年，陈焕镛教授在海南采集植物标本

图5　1925年，胡先骕、秉志和陈焕镛（第一排从左向右）与东南大学生物系的同事和学生

1927年，陈焕镛受聘为中山大学教授。随后第二年，中山大学农科设立植物学研究室，陈焕镛为研究室负责人，并兼任广州市动植物园植物学部主任。1928年秋，在他的努力下，创建了华南地区首个专业的植物标本馆。此外，早在20世纪20年代，陈焕镛就已经与60多个国家和地区建立起标本交换的良好关系，并获得3万多份珍贵的

国外标本，充实了标本馆的收藏。1929年，陈焕镛出任中山大学农林植物研究所（中国科学院华南植物园的前身）首任所长。自此以后，他把毕生的精力都奉献给了研究所，他担任所长直到去世，生前还把数百册珍贵书籍悉数无偿捐献给了植物所图书馆。1935年，他创建了广西植物研究所，并出任首任所长，参与了桂林植物园的园区规划建设（胡宗刚，2013）。

作为一名育人无数的教育家，陈焕镛可谓桃李满天下，他培养了许多知名的植物学家，包括我国蕨类植物研究开山鼻祖——秦仁昌，有“中国园林之父”美誉的陈封怀，著名植物分类学家张肇骞、蒋英等。为了更好地进行教学，陈焕镛精心编写了不少实用的教学材料，例如《植物分类属名指南》（拉丁文版）、《中国种子植物分类学》（英文版），以及专用于拉丁文教学的《植物学拉丁文解答》《中国植物学分类基础拉丁文》等。1930年，他创办了《中山专刊》并担任主编，这是我国历史上第一本专业的植物分类学英文杂志，历经战火颠簸，一直苦苦支撑到1948年，共出版了7卷26期（陈德昭，1996）。这本植物学杂志学术严谨、内容丰富、编排美观，获得了无数国内外同行的赞许与美誉。

20世纪30年代，作为知名的植物学家和分类学家，陈焕镛早已誉满国内外。1925—1937年，他与胡先骕合作，先后出版了共五卷的中英文版大型本《中国植物图谱》。自此，华南地区的植物科学研究事业，在陈焕镛的苦心经营下，逐渐发展起来，成绩斐然。1933年，他与钱崇澍、胡先骕等倡议创立了中国植物学会。同年，陈焕镛被选为《中国植物学杂志》学术评议员兼任编辑。1934年，他出任中国植物学会首任副会长，1936年又升任会长。

抗日战争期间，为了保存那些历经千辛万苦收集回来的数万份珍贵植物标本、文献和书籍等，陈焕镛冒着生命危险同时动用家族财力物力，将它们从广州用船运到香港。1938年，他和同事也一起搬迁到香港继续办公和做研究，一直工作到1941年香港沦陷。1942年，为了将标本、文献和书籍等顺利迁回广州，并得以继续工作，他被迫与日伪政府合作，也因此在1945年被人诬告，并身陷汉奸审查案（胡宗刚，2013）。虽然随后控诉撤销，但此段时间，农林植物研究所的研究工作和资助受到阻碍，进展缓慢。

1949年新中国成立之前，虽然处境艰难、前景难料，陈焕镛还是毅然选择留在祖国，放弃了去美国的机会。1954年，中山大学植物研究所改为隶属中国科学院，陈焕镛成为中国科学院华南植物研究所（园）首任所长。同年，他当选第一届全国人大代表，之后又连续两届当选全国人大代表。他同时被选为中国科学院生物学地学部学术委员。1955年，陈焕镛被选聘为中国科学院首批学部委员（院士），以表彰他在植物学和分类学上的杰出贡献，这是一位科学家在我国所能获得的最高荣誉与肯定。

陈焕镛崇尚自然之美，从早年起就不遗余力地向民众宣扬自然保护理念。1956年6月3日，在第一届全国人民代表大会上，陈焕镛与秉志、秦仁昌等专家学者一起，提交了建立国家自然保护区的议案，获得批准后，于当年建立起我国第一个自然保护区——鼎湖山树木园（国家级自然保护区）。

1951年，陈焕镛组织人员，对广州地区进行植物考察，并于1956年由他的得力助手侯宽昭主编出版了《广州植物志》。这是第一部由中国本土植物学家独立完成的地方植物志。随后他还主持了《海南植物志》四卷的编撰出版工作。1959年，陈焕镛当选《中国植物志》编辑委员会主编。1963年，陈焕镛被推选为在北京召开的中国植物学会成立三十周年纪念学术会议主席（陈德昭，1996）。

陈焕镛院士的父亲是清政府驻古巴总领事，母亲是西班牙籍的古巴人。在美国长达十数年的学习生活，使他拥有良好开阔的国际视野，同时将中西方文化融会贯通。他的拉丁文和英文造诣很深，因此他在我国植物学研究上拥有更为优越的语言和文化优势。事实上，在1949年之前，陈焕镛发表的所有学术论文和新种都是用英文或拉丁文。

凭借深厚的植物学涵养与优秀的语言造诣，陈焕镛院士多次代表中国参加重要的国际会议。1925年，他参加了在日本东京举办的第三届泛太平洋科学会议。1929年，在印度尼西亚雅加达（爪哇）举行的第四届泛太平洋科学会议上，陈焕镛提交了一篇关于广东植物的研究论文。第二年，第五届世界植物学大会在英国剑桥举办，这届大会上首次设立了中国植物讨论分会场，这也是第一次有中国植物学家参加的世界植物学大会。会上，陈焕镛做了“中国近十年来植物科学发展概况”的报告，回顾中国植物学发展的三个阶段（胡宗刚，2013）。也是在这次会议上，陈焕镛与胡先骕被公推为国际植物命名法规普通委员会委员；第二年，陈焕镛备选为专门委员，这标志着中国植物科学家在世界植物学研究上开始占有一席之地。1935年，陈焕镛出席在荷兰阿姆斯特丹举办的第六届世界植物学大会，他被选为大会植物分类学组副主席以及国际植物命名法规小组执行委员。1936年，陈焕镛被选为英国剑兰（唐菖蒲）协会名誉副主席以及美国马萨诸塞州园艺协会会员。1951年，陈焕镛院士作为中国代表团团长，出席在印度新德里举办的“南亚栽培植物之起源与分布学术讨论会”，会上他以中国古农学和遗传学观点讲到水稻的起源。1958年，陈焕镛院士又受邀出访苏联做学术交流，访问了当地多个植物园与研究所，并帮他们鉴定了不少中国的植物标本（图6）（陈焕镛1958—1963年日记）。

图6 陈焕镛教授（左二）与苏联院士共事（1958年）

终其一生，陈焕镛与世界顶级植物分类学家保持着良好的友谊关系。陈焕镛的导师——阿诺德树木园首任园长C. S. Sargent教授与J. G. Jack教授都对他早期事业发展给予支持。而著名的美国植物学家E. D. Merrill教授与陈焕镛志趣相投，自1920年认识后，两人便结为终身好友。Merrill教授为植物所标本馆捐赠了大量珍贵的植物标本，为图书馆捐赠文献资料，并为《中山专刊》首刊投稿。

除了植物学研究，生活中的陈焕镛还热爱文学诗歌。在哈佛大学求学期间，他从图书馆借阅了大量的欧美经典著作，并创作过不少诗歌。1948—1949年，陈焕镛旅居香港期间，曾写了两首英文诗表达爱国爱科学的心境，刊登于香港《南华早报》，其中一首《In high and mighty place》就传达了他对新中国成立的欢喜之情（陈德昭，1996）。

陈焕镛病逝于1971年，他虽然离开了，却为后人留下了许多珍贵的文化遗产。1978年，广东省科学院为他举办了骨灰安葬仪式和追悼会，以纪念这位伟大的植物学家，缅怀他为我国植物科学发展所作出的巨大贡献。1996年和2006年，陈焕镛的铜塑雕像分别树立在华南植物园的标本馆和展示区，凝视这片他倾注了无限心血的绿土。1996年，《陈焕镛文集》由华南植物园整理集结内部出版，记录了他为华南植物园（所）做出的硕果累累的科研成就。

作为20世纪最优秀的植物学家之一，我国现代植物分类学的先锋和奠基人之一，陈焕镛院士怀着科学兴国、创建中国现代植物科学研究的坚定信念，孜孜不倦，奋斗终生。1919年毕业前夕，恩师Sargent教授对临别的陈焕镛殷殷嘱咐："你是研究近代中国植物分类学的第一个中国学生，你的植物学生涯才刚开始，好好干，祝你成功。"天道酬勤，半个多世纪后，陈焕镛不负师恩，晚年的他，终于实现了年轻时的梦想与抱负，成功创建了中国自己的植物学研究、标本馆、图书馆、植物园（所）和国家级自然保护区，编写了《中国植物志》，他的丰功伟绩举世瞩目，他为国内外同仁所尊重敬爱，流芳千古。

致谢

感谢中国科学院华南植物园陈忠毅研究员、叶文博士、张林海研究员和廖景平研究员，上海辰山植物园马金双研究员，中国科学院微生物研究所Joel Haywood教授，美国拉霍亚过敏和免疫学研究所岳晓婧博士，美国Tom Wood老师以及编辑程浩博士对本文英文版的建议与校订。感谢中国科学院华南植物园陈忠毅研究员对本文中文版的修订。

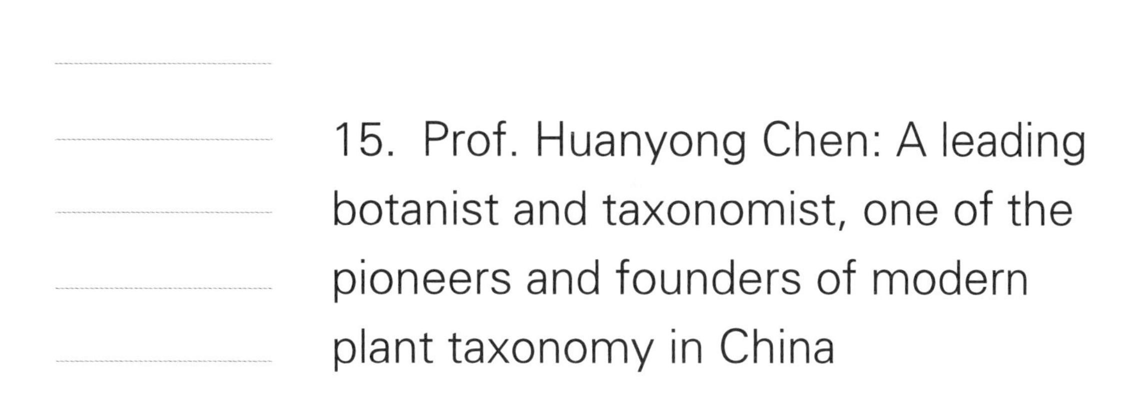

15. Prof. Huanyong Chen: A leading botanist and taxonomist, one of the pioneers and founders of modern plant taxonomy in China

Prof. Huanyong Chen (Woon-Young Chun, 陈焕镛, 1890—1971), a leading botanist and taxonomist, who started a new era for the modern plant science in China and made a great contribution to the plant taxonomy, botany and horticulture of China, especially in Southern China. He was one of the first academicians of the Chinese Academy of Sciences (CAS), a member of the first, second and third China's National People's Congress (NPC), one of the first editors of *Flora Republicae Popularis Sinicae* (*FRPS*, *Flora of China*). He was also a leading character in the history of botanical collection by Chinese botanists, and published 11 new genera and more than six hundreds new taxa, including the famous "living fossil" plant—*Cathaya argyrophylla* Chun et Kuang (Ko-Zen Kuang, 匡可任).

He built up one of the first modern herbariums in China, became the founder and first director of South China Botanical Garden (SCBG), CAS and Guangxi Institute of Plant Research. Moreover, he established the first national nature reserve Dinghushan Arboretum.

He launched the first English botanic journal *Sunyatsenia* in China, co-founded the Botanical Society of China, and served as the editor and president of *The Chinese Journal of Botany*. In honor of his great contributions, many plants were named after him, such as the genera *Chunia*, *Chuniophoenix*, *Chunechites*, and the species *Lindera chunii* and *Vernonia chunii*, etc(Chen, 1996). He is one of the most outstanding scientists in the field of botany and taxonomy in early modern China.

Born on July 22, 1890 in Hong Kong, Prof. Chen grew up in Guangdong and Shanghai, and then traveled to the US in 1905 when he was only 14 years old. He went to Massachusetts Agricultural College in 1909 to study forestry and entomology, and studied at Forestry College of Syracuse University in New York City in 1912. After he received his bachelor's degree in 1915, he continued his research career at Bussey Institution of Harvard University and obtained a Master's degree in forestry at the Arnold Arboretum of Harvard University in 1919(Haas,1988). He gave up the opportunity for further doctoral studies, instead, he accepted the Sheldon

Ruilan Huang
South China Botanical Garden, Chinese Academy of Sciences, Guangzhou 510650, China
Correspondence: huangruilan@scbg.ac.cn

Traveling Fellowship and completed a plant survey and collection in Hainan Island in China. He spent 9 months and risked his life to explore and collect large number of specimens at Wuzhi Mountain where the flora was barely known before.

The following year, he was appointed as a professor of forestry at the University of Nanking and started teaching at Southeast University in 1922, where he worked together with Xiansu Hu (Hsen-Hsu Hu, 胡先骕) , Zhi Bing(秉志) and other renowned biologists in China. In 1922, together with botanists Chongshu Qian (Sung-Shu Chien, 钱崇澍) and Renchang Qin (Ren-Chang Ching, 秦仁昌), Prof. Chen organized the first Chinese plant collection group to explore the west part of Hubei Province, and collected around one thousand specimens. As a talented and hard-working botanist with a broad view and advanced awareness for the development of Chinese botany, Prof. Chen complied the book *Chinese Economic Trees* when he was only 27 years old. It was the first botanical book in English authored by a Chinese botanist, which was translated into Chinese by Fenghuai Chen(Feng-Hwai Chen, 陈封怀)and published by Commercial Press in 1922.

In 1927, Prof. Chen was appointed as a professor in Sun Yat-sen University, and in the following year, he founded and directed a new institute of botany in the university. He helped to establish the professional and qualified herbarium in Southern China in 1928. Furthermore, he made specimens exchanges with more than 60 countries and obtained more than 30000 foreign specimens to enrich the specimens for herbarium in the 1920s. This herbarium is still considered as one of the top three herbariums in China till now. In 1929, Prof. Chen was appointed as the first director of Institute of Agriculture and Forestry, which was the predecessor of SCBG. Since then, he devoted his entire life as the director of SCBG and donated hundreds of publications to the library of SCBG. In 1935, he also founded the Economic Institute of Botany in Guangxi University and served as the first director, and he was also involved in the design and construction of the Guilin Botanical Garden(Hu, 2013).

As an educator, Prof. Chen trained numerous famous botanists including Renchang Qin, Fenghuai Chen, Ying Jiang (Ying Tsiang, 蒋英), et al. He also edited and compiled a number of textbooks and reference books for teaching, for example, the Latin version of *Deriration of Genera Names*, the English version of *Chinese Seed Plant Identification*, and books for teaching Latin: *Question from Botanical Latin* and *Eseential Latin for Chinese Taxonomy Botany*(Chen,1996). In 1930, he established and edited *Sunyatsenia*, the first Chinese professional plant taxonomy journal in English, and till 1948, 7 volumes 26 issues were published. This journal gained a great reputation and received awards from home and abroad by its high quality and professional articles.

Prof. Chen became well-known as the leading botanist in South China by both Chinese and foreign scientists in 1930s. He collaborated with Xiansu Hu and published five volumes of *Icones Plantarum Sincarum* from 1925 to 1937, in which they reviewed all the published illustrations and descriptions of Chinese plants. The scientific researches on the flora in South China gradually expanded under Prof. Chen's leadership. In 1933, Chongshu Qian, Xiansu Hu and Huanyong Chen were invited to found the Botanical Society of China, and in the same year, Prof. Chen was appointed as the editor and academic council member of *The Chinese Journal of Botany*. In 1936, he became the president of Batanical Society of China, after serving as vice president since 1934.

During the War of Resistance against Japanese Aggression, in order to preserve thousands of precious specimens and literatures, Prof. Chen risked his life to ship them from Guangzhou to Hong Kong. In 1938, with his staff, he managed to keep on doing researches in the Kowlong office until Hong Kong was also occupied in 1941. In 1942, he moved the stuff back to Guangzhou in order to continue his research. At that time, he was forced to collaborate with the Puppet Government (Japanese forces), which brought him into trouble in 1945 when the nationalist government accused him for collaborating with the enemy (Hu, 2013). Although the charges were squashed later, the development and the research work of the institute were badly affected by the lack of research funding and the war.

In spite of those adverse situations, Prof. Chen still chose to stay in China rather than US when The People's Republic of China was founded in 1949. He became the first director of SCBG in 1954. In the same year, he was elected as a member of the first National People's Congress (NPC), and was subsequently reappointed for second and third NPC. He was also, a faculty member of Biology and Earth Sciences department in Chinese Academy of Sciences (CAS). In 1955, Prof. Chen was elected as one of the first academicians in CAS for his outstanding contribution in the field of botany and taxonomy, which is the highest honor for a Chinese scientist.

Prof. Chen always admired the beauty of the nature and eagerly participated in conservation since his childhood. 3 June, 1956, He put forwards a joint proposal to set up natural reserves with Zhi Bing, Ren-Chang Ching et al. when he attended the first National People's Congress. The proposal was approved and he soon established the first national nature reserve Dinghushan Arboretum.

In 1951, Prof. Chen led a team to complete a survey of the Flora of Guangzhou, which was the first Chinese endemic flora survey completed by Chinese botanists. And later he also participate in the publication of the *Flora in Hainan*. In 1959, he was appointed as the Associate Editor-in-Chief of *Flora Republicae Popularis Sinicae* (FRPS, *Flora of China*).In 1963, Prof. Chen was elected as academic chairman of the 30th anniversary conference for Botanical Society of China in Beijing and a member of Academic Committee of Institute of Botany, CAS (Chen,1996).

Prof. Chen was born mixed-blood parentage, his father was a Chinese Consul in Cuba and his mother was a Spanish Cuban. In addition, he studied and lived in the US for decades, which broadened his international viewpoints and understanding of both eastern and western cultures. He was good at both English and professional Latin, which gave him more opportunities to present and improve the perception of Chinese botanical researches. In fact, before 1949, most of his thesis and academic articles were written in English or Latin.

From the 1920s to the 1950s, Prof. Chen attended several important international congresses, including the Third Pan-Pacific Science Conference in 1925, Tokyo, Japan. In 1929, at the Fourth Pan-Pacific Science Congress in Java, Indonesia, he presented a poster about the Flora of Guangdong. In 1930, at the Fifth International Botanical Congress in Cambridge, England, a symposium on the *Flora of China* was held for the first time in the history. The Chinese botanists attended this symposium and in the address to the symposium, Prof. Chen overviewed the development of botany in China(Hu, 2013). Xiansu Hu and he were elected as the board members of *Species Muscorum Frondosorum* (International Code of Botanical Nomenclature), which was a significant milestone of a worldwide appreciation of Chinese botanists. In 1935, Prof. Chen attended the Sixth International Botanical Congress as the Chinese team leader. In 1936, Prof. Chen was awarded as Honorary Vice President of British Gladiolus Society, and appointed as a member of Massachusetts Horticulture Society (MassHort, US). In 1951, at the symposium on Origin and distribution of cultivated plants of South Asia held in New Delhi, India, Prof. Chen presented his research work as the head of the delegation of CAS. He also visited several Plant Research Centers on behalf of CAS and helped identify plant specimens in Leningrad (Saint Petersburg), Soviet Union in 1958 (Diary of Huanyong Chen (1958—1963)) .

All through his life, Prof. Chen kept a good relationship with top botanists all over the world. The former director of Arnold Arboretum, Professor C. S. Sargent, supported him during his early career, and the famous botanist Dr. Elmer Drew Merrill (1876—1956) kept a life-long friendship with him ever since they meet in 1920. Dr. Merrill donated a lot of precious publications to the library and specimens to the herbarium, and also wrote an article for the first issue of the journal *Sunyatsenia*.

In addition to his botanic research, Prof. Chen was also interested in literature. he read lots of European and American novels from the library during his study at Harvard University and some of his poems were published in the Hong Kong newspaper *South China Morning Post*(《南华早报》)(Chen, 1996).

Prof. Chen passed away in 1971, and left countless treasures for the next generations. In 1978, a memorial meeting was held by Guangdong Academy of Sciences. Many government officials and scientists attended this meeting to greet Prof. Chen's outstanding contributions to the development of Chinese botanical sciences. In 1996 and 2006 respectively, his bronze statues were built in the Herbarium and the garden of SCBG. And the internal book of *Memorial Collections of Huan-Yong Chen* was published in 1996 to memorize all his endeavors to the development of SCBG.

As one of the most outstanding botanists in 20th century and one of the pioneers and founders of modern plant taxonomy in China, with a strong determination to establish the modern Chinese botanical research, Prof. Chen considered the development of Chinese botanical science as his life-long mission. He was blessed to do what he loved, and made his dream come true—to set up China's own botanical research, Herbarium, botanical library, Chinese flora, botanical gardens/institutions and natural reserves, which were deeply respected from all over the world.

Acknowledgements

I would like to thank Prof. Zhongyi Chen, Dr. Wen Ye, Prof. Linhai Zhang and Prof. Jingping Liao from SCBG, CAS, Prof. Jinshuang Ma from Shanghai Chenshan Botanical Garden, CAS, Dr. Joel Haywood from Institute of Microbiology, CAS, Dr. Xiaojing Yue from La Jolla Institute for Allergy & Immunology, USA and Tom Wood for the critical reading and helpful suggestions of the manuscript.

Figures

Fig.1 Prof. Huanyong Chen at work in 1956 in Guangzhou

Fig.2 Specimen of *Cathaya argyrophylla* published by Prof. Huanyong Chen and Prof. Keren Kuang

Fig.3 Prof. Huanyong Chen (in the middle) was discussing the design of the landscape for SCBG with colleagues in 1956. Front row, from left: Fenghuai Chen, Dejun Yu, Huanyong Chen, Tongfang Zhao, Shifu Cheng, Zhaoqian Zhang

Fig.4 Prof. Huanyong Chen was collecting plant specimens in Hainan Province in 1919

Fig.5 Xiansu Hu, Zhi Bing and Huanyong Chen (From left to right in the first row) together with colleagues and students in the Department of Biology, National Southeast University in 1925

Fig.6 Prof. Huanyong Chen (2nd left) worked with Academicians from the Soviet Union in 1958

Referencecs

Chen DZ (1996) Memorial collections of Huanyong Chen. Guangzhou: South China Botanical Garden (Internal publications). (陈德昭. 1996. 陈焕镛纪念文集. 广州：华南植物园(内部出版物).)

Chen HH. Diary of Huanyong Chen (1958—1963) from Herbarium of South China Botanical Garden and Documents from Archive of South China Botanical Garden. (陈焕镛1958—1963年日记和档案资料. 分别藏于华南植物园标本馆和华南植物园档案室.)

Haas WJ (1988) Transplanting botany to China: The cross-cultural experience of Chen Huanyong. Arnoldia 48:9–25.

Hu ZG (2013) The Early History of South China Institute of Botany. Shanghai: Shanghai Jiao Tong University Press. (胡宗刚. 2013. 华南植物研究所早期史. 上海: 上海交通大学出版社.)

16. 蕨类植物学家秦仁昌与“秦仁昌系统”

秦仁昌是我国著名的植物学家（图1），尤以蕨类植物研究见长，被誉为“中国蕨类学之父”，其建立的“秦仁昌系统”为世界蕨类植物分类学作出了巨大贡献。

图1　秦仁昌（1898—1986）

在江苏省第一甲种农业学校学习时，在陈嵘和钱崇澍的影响下，秦仁昌对植物学产生了浓厚的兴趣。从农校毕业后，他考入金陵大学，并获得林学学士学位。此后，他先是在东南大学担任助教，后受聘于中央研究院自然历史博物馆。1929年，秦仁昌到丹麦哥本哈根大学植物学博物馆，在世界蕨类学权威之一——C. Christensen教授的指导下开始从事蕨类植物分类学的研究。在欧洲学习期间，他详细考察了各国收藏的中国蕨类植物标本，为后来建立蕨类植物分类系统打下了坚实基础。回国后，他先在北平静生生物调查所任研究员，1934年被胡先骕派往江西建立庐山植物园并担任植物园主任。抗日战争时期，他在昆明成立了蕨类植物研究中心。1945年任云南大学林学系主任、教授，后任云南大学生物系主任。1949年兼任云南省林业局副局长。1955年

译者：付雷
浙江师范大学，金华321004，中国
邮箱：ful527@163.com

秦仁昌当选中国科学院学部委员（院士），并于次年调入中国科学院植物研究所，主持编写了《中国植物志》(图2)。

图2 《中国植物志》

秦仁昌对我国植物学尤其是蕨类植物学作出了巨大的贡献。早在东南大学工作时，在陈焕镛的支持下，他就开始对蕨类植物研究产生浓厚兴趣。秦仁昌在中央研究院任植物学技师时，认识到国内蕨类植物研究相当落后，要扭转这一局面，首先要对当时我国的蕨类研究进行梳理。而国外对于中国蕨类植物的研究已有180年的历史，内容多、分布散，原始材料和模式标本散落于国外，国内的蕨类植物的研究一时难以展开(吴兆洪，1986)。于是他自学英语、法语、拉丁语等多种语言，并到欧洲进修、考察。为了查清我国的蕨类植物模式，他奔赴英国皇家植物园标本馆及大英博物馆，拍摄蕨类植物模式照片18300张（王中仁，1998），收集了大量关于蕨类的珍贵资料，为中国蕨类植物分类学奠定了良好的基础。回国后，他修订了自己在1930年编写的《中国蕨类植物志》(初稿)，这是第一部比较完整的中国蕨类植物专著。秦仁昌从事植物学研究60多年，发表论文和著作160多篇（部），出版译著15部。他参与组织编写了《中国植物志》，这是世界上规模最大的植物志之一。1959年，他带领年轻人率先完成第二卷，又着手后几卷的编写。

秦仁昌提出了“秦仁昌系统”。在此之前，J. D. Hooker将真蕨分为里白科和水龙骨科两科，水龙骨科几乎包含85%的已知蕨类植物，但其内容庞杂，导致系统研究停滞不前。1940年，秦仁昌发表《水龙骨科的自然分类系统》，将水龙骨科分为33科、249属，提出了5条进化线索（邢公侠，1994）。他的分类体系结束了Hooker的保守分类体系，解决了当时蕨类植物学的最大难题，为世界蕨类植物分类学作出了卓越贡献。秦仁昌深知分类学需要依据形态学最新发现不断更新完善，便从解剖学、细胞学等方面深入研究，并于1954年发表了《中国蕨类植物科属名词及分类系统》，于1978年发表《中国蕨类植物科属的系统排列和历史来源》，使中国蕨类植物分类系统上升到新水平。“秦仁昌系统”将蕨类植物门划分为5个亚门，3纲，11目，63科；同时发表了5个新科，8个新属；阐明了科属的起源及其演化关系（吴兆洪，1984）。秦仁昌这一系统获1989年中国科学院自然科学奖一等奖以及1993年国家自然科学奖一等奖。“秦仁昌系统”得到了学术界的广泛采用。

除了开展自己的研究，秦仁昌还十分重视人才的培养。他大力倡导科技人员的培养，支持有前途的分类学工作者出国学习进修，与国际同行交流经验。全国各地的蕨类植物学工作者都是在他的指导下成长起来的。他十分关心后辈，对学生不论年龄、资历都一视同仁，竭尽全力给予帮助，例如冯国楣、王文采等都曾得到秦仁昌的指导。为了更好地培养国内的植物学学者，他翻译了80多万字的《植物学拉丁文》以及《近代科学技术大辞典》和《韦氏大辞典》的植物学部分（裘佩熹和邹安寿，1983），为我国植物学界的发展作出了巨大贡献。

秦仁昌一生兢兢业业，在科学事业上取得了卓越成就，在国内外享有盛誉。1988年，国际蕨类学会主席E. Hennipman在纪念秦仁昌诞辰九十周年大会上称："秦仁昌不仅是中国蕨类学之父，也是世界蕨类学之父。"

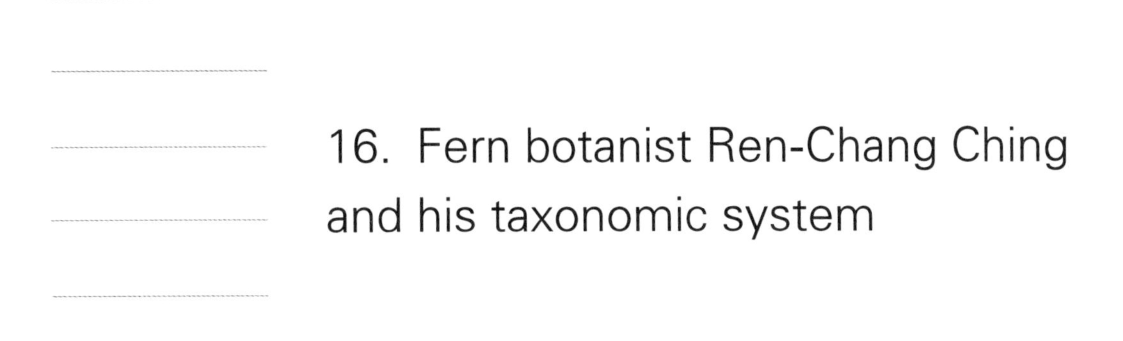

16. Fern botanist Ren-Chang Ching and his taxonomic system

Ren-Chang Ching (秦仁昌,1898—1986) is a famous botanist in China. He is known as the father of Chinese pteridology, having done extensive research on ferns. The Ren-Chang Ching System established by him has made great contributions to fern taxonomy all over the world. When he was studying in the First Agricultural School in Jiangsu Province, he became extremely interested in botany under the influence of Rong Chun (陈嵘) and Sung-Shu Chien (钱崇澍). After graduation, he was admitted to University of Nanking and obtained his bachelor's degree in forestry. He worked as an assistant at Southeast University and was employed later by the Natural History Museum of Academia Sinica. In 1929, he went to the Botany Museum of the University of Copenhagen in Denmark, and began to engage in research on fern taxonomy under the guidance of professor C. Christensen, who was one of the world's leading authorities on ferns. While studying in Europe, he inspected Chinese fern specimens kept in various countries in detail, which laid a solid foundation for the later establishment of his fern classification system. After returning to China, he worked as a researcher at the Institute of Hydrobiology in Peiping. In 1934, he was sent to Jiangxi by Hsen-Hsu Hu (胡先骕) to establish the Lushan Botanical Garden and served as its director. During the War of Resistance against Japanese Aggression, he established a fern research center in Kunming. From 1945, he was appointed successively as an associate professor at the Department of Forestry and Biology of Yunnan University. In 1949, he also served as the Deputy Director of the Forestry Bureau of Yunnan Province. He was elected as am academician of the Academic Department of the Chinese Academy of Sciences in 1955, and was transferred to the Institute of Botany of the Chinese Academy of Sciences in the following year, presiding over the compilation of *Flora of China*.

Ren-Chang Ching made a great contribution to Chinese botany, especially in the field of pteridology. While working at Southeast University, he became interested in fern research with the support of Woon-Young Chun (陈焕镛). When Ren-Chang Ching was a botanical technician at Academia Sinica, he realized that research on ferns in China was quite backward. To reverse

Xunfeng Xu, Lei Fu

Zhejiang Normal University, Jinhua 321004, China

Correspondence: ful527@163.com (L. Fu)

this situation, it was necessary to sort out the research on ferns in China available at the time. However, research on the ferns of China has a history of 180 years in foreign countries with abundant contents, with the protologue materials and type specimens scattered abroad. Therefore, it was difficult to carry out research on ferns in China (Wu, 1986). He learned English, French, Latin and other languages by himself, and went to Europe for further studies and investigations. In order to find out the patterns of ferns of China, he took 18300 photos of fern patterns (Wang, 1998), collected a lot of valuable information about ferns and laid a solid foundation for the fern taxonomy of China at the Royal Botanic Gardens Herbarium and British Museum. After returning to China, he revised the first draft of the *Monograph of Chinese Ferns* drafted in 1930, which was the first complete monograph of Chinese ferns. Ren-Chang Ching devoted himself to the study of botany for more than 60 years, and published over 160 articles and monographs, as well as 15 translations. He participated in organizing and compiling the *Flora of China*, one of the world's largest floras, and led young people in their challenging work to complete the second volume in 1959, immediately starting the compilation of the next several volumes afterwards.

Ren-Chang Ching proposed the Ren-Chang Ching System. Prior to this, Joseph Dalton Hooker divided the true ferns into two families, Gleicheniaceae and Polypodiaceae. The Polypodiaceae contained almost 85% of all known ferns, but their relationships were complex, which led to a stagnation of systematic research. In 1940, Ren-Chang Ching published the Natural classification system of Polypodiaceae, in which he divided plants of Polypodiaceae into 33 families and 249 genera, and proposed 5 evolutionary clues (Xing, 1994). His system ended the conservative classification system of Joseph Dalton Hooker and solved the biggest problem of fern botany at that time, which made great contributions to the worldwide fern taxonomy. He knew deeply that taxonomy needs to be constantly updated based on the latest morphological findings and he also conducted in-depth research on anatomy and cytology. He published the Systematic arrangements of families and genera of Chinese pteridophytes with corresponding names in Chinese in 1954 and The Chinese fern families and genera: Systematic arrangement and historical origin in 1978, which moved the Chinese fern classification system to a new level. This system divided the fern phyla into 5 subdivisions, 3 classes, 11 orders, and 63 families, while establishing 5 new families and 8 new genera at the same time, which clarified the origin and evolution of the family (Wu, 1984). The Ren-Chang Ching system won the First Prize for Natural Science of the Chinese Academy of Sciences in 1989 and the First Prize for National Natural Science Award in 1993. The Ren-Chang Ching system is widely applied by the academic community around the world.

In addition to his own research, Ren-Chang Ching attached great importance to the training of talents. He vigorously advocated for the training of scientific and technical personnel, supported promising taxonomists to study abroad, and shared ideas with international peers. Fern botanists all over the country grew up under his guidance. He was very concerned about the juniors, treated the students as equals, and did his best to help them regardless of age and qualifications. For example, Kuo-Mei Feng (冯国楣) and Wen-Tsai Wang (王文采) were trained by Ren-Chang Ching. In order to cultivate domestic botanical scholars, he translated more than 800000 words of *Latin of Botany*, the botany sections of the *Modern Dictionary of Science and Technology* and *Webster's Dictionary*(Qiu and Zou, 1983). He has made great contributions to the growth of the Chinese botanical community.

Ren-Chang Ching devoted his whole life to his studies and made great contributions to the development of science in the motherland. His distinguished achievements earned him a high reputation both at home and abroad. In 1988, at the 90th anniversary of the birth of Ren-Chang Ching, the president of the International Fern Society Hennipman said, Ren-Chang Ching is the father of pteridology not only in China, but also the world.

Figures

Fig.1 Ren-Chang Ching(1898—1986)

Fig.2 *Flora of China*

References

Qiu PX, Zou AS (1983) R. C. Ching, a pioneer of botany in China. Plants 2:41–43. (裘佩熹，邹安寿. 1983. 秦仁昌——中国植物学的一位拓荒者. 植物杂志，2: 41–43.)

Wang ZR (1998) To commemorate the 100th anniversary of the birth of Ching Ren-Chang(1898—1986), founder of Chinese pteridology. Acta Phytotaxonomica Sinica 36(3):286–288. (王中仁. 1998. 中国蕨类植物学的奠基人秦仁昌(1898—1986)——纪念秦仁昌先生诞辰一百周年. 植物分类学报，36(3): 286–288.)

Wu ZH (1984) The systematic scheme of Pteridophyta by R. C. Ching. Guihaia 4:289–307. (吴兆洪. 1984. 秦仁昌系统(蕨类植物门)总览. 广西植物，4: 289–307.)

Wu ZH (1986) The historical origin of the systematic scheme of pteridophyta by R. C. Ching. Guihaia 6(1–2):63–78. (吴兆洪. 1986. 秦仁昌分类系统(蕨类植物门)的历史渊源. 广西植物，6(1–2): 63–78.)

Xing GX (1994) Achievement of outstanding morality, dedicated to science career —Professor R. C. Ching, winner of the first prize of National Natural Science Award. China Award for Science and Technology 1:13–17. (邢公侠. 1994. 成就斐然，品德高尚，一生献给科学事业——国家自然科学奖一等奖获得者秦仁昌教授. 中国科技奖励，1: 13–17.)

17. 罗宗洛：中国20世纪30年代的海归

“海归”中文发音与“海龟”同音，已成为现代中国的热门词汇。受过海外教育的中国学者看到祖国经济的飞速发展，回国的想法会变得越来越强烈。截至2010年，近50万中国人从海外回国，这一趋势在生命科学学术领域也不例外。科学家回国能够获得有利的研究环境和上涨的薪资。许多海归人士表示，他们在中国获得的实验环境超过了他们在美国或欧洲的预期，为什么不回国呢？然而，早期的海归却没有这样幸运。早在20世纪20年代和30年代，科学家出于完全不同的原因回国。

随着社会框架和政治结构的调整，我国的学术体系在20世纪20年代发生了翻天覆地的变化，高等院校如雨后春笋出现在全国各地，使我国逐渐从传统的“崇尚孔子，轻视自然科学”的观念体系，进入相对现代的高等教育体系。所有新建立的高等院校都迫切需要熟悉现代学术体系和现代科学的教职员工，因此，许多受过国外培训的中国学者被聘请回国。他们从未被承诺会有更好的学术机会或者更高的薪资，但他们都毅然回到祖国，希望将现代知识带给自己的国家和人民。他们中的一些人比较幸运，尽管当时经济环境恶劣且政治情况不稳定，他们还是逐步开展了教学和科研工作，并在短时间内建立自己的实验室。然而还有一些人没有那么幸运，他们在艰苦的条件下奋斗了很多年才安顿下来，才能按照他们回国前的预期开展自己的工作。本文介绍的我国现代植物生理学主要创始人之一罗宗洛教授（图1）就属于后者。

图1　罗宗洛（1898—1978）

罗宗洛先生1898年出生在浙江一个商人家庭，在杭州安定中学和上海南洋中学接受早期教育后，他前往日本接受高等教育。他用了八年时间在北海道大学攻读本科和研究生课程，并于1930年获得了博士学位，成为第二位在日本获得博士学位的中国人。

译者：曲静

中国科学院北京生命科学研究院，北京100101，中国

邮箱：quj@biols.ac.cn

毕业后，他接受了中山大学的邀请，于1930年2月回国。

当罗宗洛先生第一次来到中山大学生物系时，惊讶地发现物资是那么匮乏，缺少各种教学的基本配置——教科书、实验室、仪器设备和动植物标本等。即使条件如此恶劣，他仍通过自己编纂、手写教科书等办法努力开展教学工作。他甚至建立了自己的植物生理学实验室，尽管它是那么简陋和破旧。但是，中山大学变得越来越不稳定，两年内更换了四位校长，每位校长对生物系都实行自己不同的政策。这样动荡的环境使得罗宗洛先生在中山大学成为教育家和植物生理学家的梦想破碎了，他毅然提交了辞呈，并于1932年离开了中山大学。

罗宗洛先生工作的第二所大学是位于上海的暨南大学。暨南大学校长承诺给予他教授的职位，并帮助他建立自己的实验室。然而，用于购买实验室设备、试剂和招聘助理的经费拨款承诺都未能实现。他在暨南大学一年期间所能做的一切就是教授普通生物学。罗宗洛先生感到非常失望，希望能够更好地运用自己的才能和知识，他于1933年离开暨南大学前往国立中央大学。

与前两所大学相比，国立中央大学在许多方面条件更加优越——更多的教职工、学生和经费预算。罗宗洛先生被分配教授三门课程，使他整年都忙碌起来。但他并不认为沉重的教学任务是一种负担，反而因如此忙碌而感到无比快乐。他还从大学争取到了相对充足的经费支持，这并不容易，因为教师之间的竞争非常激烈。凭借这笔资金，他于1934年建立了自己的植物生理学实验室，并在1935年发表了他回国后的第一篇科研论文。但是好景不长，抗日战争时期，国立中央大学于1937年迁至重庆，大学的基础设施被严重削弱，教职员工之间的竞争更加激烈。大多数教师在西方国家接受过培训，日本留学归来的罗宗洛先生则是大学里的少数派。他觉得这所大学已不适合他，便于1940年转到了浙江大学。

与战争期间的其他大学一样，浙江大学内迁至贵州省，生物系位于湄潭镇，这是一个安全而美丽的小镇，但资源十分匮乏。罗宗洛先生在湄潭镇教授两门课程的同时，还设法在一座废弃的小祠堂中建立了一个临时的植物生理学实验室，继续开展植物中微量元素的研究。他在湄潭镇的教学和研究一直持续到1944年，当时他被任命为中央研究院植物研究所所长，在那里他拥有了一个真正的实验室，至此他的职业生涯才真正起飞。

14年间辗转于4所大学，罗宗洛先生最终找到了理想的位置去做他真正感兴趣的事情，并逐步实现了他的梦想。有些人如果处于这种情况下，也许会放弃，也许会再次去海外寻求更好的机会；有些人则会面对现实，安于现状。但是罗宗洛先生一直没有停下来，直到实现了自己的理想。与此同时，无论征途有多么艰难，无论梦想离现实还有多远，他从不浪费任何时间。相反，在所待过的4所大学里，他一直全身心地投入自己的工作，以做出积极的改变。人们认为他是我国植物生理学的主要创始人，不仅基于他在中央研究院植物研究所里取得的学术成就，还因为他在大学里培养了大量年轻有为的学生，其中许多人后来成为我国植物生理学的骨干。

17. Zongluo Luo, a Chinese Haigui in 1930s

“Haigui” , which pronounces the same in Chinese as that of sea turtle, has become a trendy word in modern China. For those of you who are not familiar with the word yet, Haigui is a homonym for returnee from overseas. To overseas-educated Chinese scholars, the idea of returning home becomes more and more enticing because of China’s rapid economic growth. Until 2010, nearly half a million of Chinese have returned from overseas. And the trend has no exception in the life sciences academia. Scientists are attracted back home for promising research environment and rising salaries. Many returnees say that the lab environment they have got here in China definitely trumps what they could expect in the United States or Europe. Why wouldn’t they have returned? However, their earlier counterparts back in the old days were not as lucky as they are nowadays. In the 1920s and 1930s, scientists returned for completely different reasons.

Along with the transformation of the social framework and political structure, China’s academic system was going through dramatic changes in the 1920s—from the traditional system where Confucius meant everything and natural sciences was ignored, to the relatively modern educational system with colleges and universities being established everywhere. All the newly-established colleges and universities were in desperate need of faculty who were familiar with modern academic system and modern science; thus many foreign-trained Chinese scholars were recruited back home. They were never promised better academic opportunities or higher salaries, but they still returned in hopes of bringing modern knowledge to their home country and their people. Among these returnees there were the lucky ones, who managed to start teaching right away and/or set up their own labs in a short period of time, considering the poor economic environment and political instability at that time. Then there were the less fortunate ones, who struggled a long time before they could settle down and do what they had expected to be doing upon returning. Professor Zongluo Luo (Tsung-Lo Lo), a main founder of modern plant physiology in China, would qualify for the latter group.

Ming Li, Le Kang
Beijing Institutes of Life Science, Chinese Academy of Sciences, Beijing 100101, China
Contact: lkang@ioz.ac.cn (L. Kang)

Zongluo Luo was born in 1898 in a merchant's family in Zhejiang Province. After receiving his early education at Hangzhou Anding Middle School and Shanghai Nanyang High School, he traveled to Japan to receive higher education. He spent eight years at Hokkaido University for the undergraduate and graduate programs, and obtained his Ph.D. degree from Hokkaido University in 1930, which made him the second Chinese who got a Ph.D. degree in Japan. After graduation, Zongluo Luo accepted an offer from the Sun Yat-sen University, Guangzhou, and returned China in February 1930.

When he first came back to Department of Biology at the Sun Yat-sen University, Zongluo Luo was surprised to find how empty the department was. For a biology department, it was short of every essential element—textbooks, lab space, equipments, animal or plant specimens, etc. Even with the extreme lack of resources and funding, Zongluo Luo managed to start teaching, with hand-written textbooks made by himself. He even started his own plant physiology laboratory, no matter how simple and shabby it was. But the university itself became more and more unstable, and it went through four different presidents in two years, each having their own policy towards Department of Biology. With instability like this, it was meaningless for Zongluo Luo to continue pursuing his dream as an educator and a plant physiologist at the Sun Yat-sen University. Thus he submitted his resignation letter and left the university in 1932.

The second university Zongluo Luo worked at was Jinan University in Shanghai. The president of Jinan University offered him a professor position and promised to help him build his own laboratory. However, all the promises of funding for purchasing lab equipments and reagents and hiring assistants turned out to be empty ones after all. All he could do during his one year in Jinan University was to teach general biology. Feeling disappointed and hoping to put his talent and knowledge into better use, Zongluo Luo left Jinan University for the National Central University in 1933.

Compared to the other two universities he had stayed in, the National Central University was definitely more developed in many respects—more faculty, more students, and more budget. Zongluo Luo was assigned to teach three courses which kept him busy all year long. But he never thought of the heavy teaching task as a burden; he couldn't be happier that he could be so busy. He also managed to get relatively sufficient funding from the university, which was not easy as the competition among faculty was quite intense. With the money, he was able to set up his own plant physiology lab in 1934 and published his first article after his return in 1935. But the good time did not last long. The university moved to Chongqing, in 1937 during the War of Resistance against Japanese Aggression and the infrastructure of the university became fragile and the infighting among faculty became even more intense. Since most faculty members received their training in western countries, Zongluo Luo, a returnee from Japan, was always a minority in the university. In 1940, he felt that the university was no longer suitable for him and moved on to Zhejiang University.

Like other Chinese universities during war time, Zhejiang University had moved to Guizhou Province and the biology department was located in a small town, Meitan, a safe and beautiful town with nothing more to offer. While he was teaching two courses in Meitan, Zongluo Luo also managed to establish a make-shift plant physiology laboratory in a disused ancestral temple, continuing his research on trace elements in plants. His teaching and research went on in Meitan until 1944 when he was appointed the Director of the Botany Research Institute of Academia Sinica in Chongqing, where he could have a real laboratory and his career really took off.

Fourteen years and four universities, before Zongluo Luo finally found the right place to do what really interested him and began to fulfill his dream. Some people would give up and maybe go overseas again for better opportunities; others would face the reality and be happy with whatever they had got. But Zongluo Luo never stopped until he found what he was looking for. Meanwhile, no matter how difficult the journey was and how far the reality was from his dream, he never wasted any time. Instead, in all of the four universities he had stayed, he always devoted himself to his work and gathered all his energy to make a positive difference. People regard him as a main founder of Chinese plant physiology, not only based on the academic achievement he had made at the Botany Research Institute, but also because of the large number of young students he had trained in the universities, many of whom became the backbone of China plant physiology later.

Figures

Fig.1 Zongluo Luo (1898—1978)

18. 汤佩松：一位中国植物生理学家执着奉献的一生

汤佩松先生出生于1903年，他是我国现代植物生理学奠基人之一（图1）。也许你会对他的一些学术成就有所耳闻：他是第一个在植物中发现呼吸酶（细胞色素氧化酶）的人；他证明了水稻中呼吸代谢途径和电子传递系统的多样性；他首次通过实验证明了植物中碳酸酐酶的存在；他与王竹溪院士一起合作，利用热力学方法阐明了细胞内水分的运动机制，其后被称为“关于细胞水分化学势的汤－王理论”。你也可能听说过在20世纪40—50年代，他是第一位在《自然》和《科学》杂志上发表文章的中国生物学家。然而，仅仅从他所发表的200多篇学术论文中，并不能了解他的一切：作为一个中国人，他对国家和人民无私奉献；作为一名科学家，他在追求科学和真理过程中勤勉不懈。

图1　汤佩松（1903—2001）（选自《汤佩松论文选集》）

汤佩松先生于1917—1925年在清华学校接受了早期大学教育。当时的清华学校是用庚子赔款创立的，这让汤佩松感觉自己亏欠国家和人民，并立志在国外学成后要回国报效。在清华学校时，他就对生物学表现出了兴趣。当生物老师描述了种子胚乳中贮藏的淀粉如何转化为幼苗后，他问了一个老师也无法回答的问题：“无机物质到底是如何转化为有机的具有完整结构的物质，就像淀粉是如何成为幼苗的？”那时的他还不知道，这成为他科学生涯的起点。

从清华学校毕业后，汤佩松进入明尼苏达大学，获得了学士学位。他在大学的课程涵盖的学科范围很宽泛，这为他今后的研究提供了扎实的理论基础。在所有这些课程中，他尤其被生物物理学所吸引，这激发了他对生命体热力学的兴趣。在那之后，

译者：郝宁
中国科学院生物物理研究所，北京100101，中国
邮箱：haoning@ibp.ac.cn

他进入约翰斯·霍普金斯大学，师从Burton Livingston进行学习，但Livingston为他选择的毕业课题他并不感兴趣。与其他中国学生的谦逊、总是服从别人不同，汤佩松开始从事一些研究种子呼吸与光合作用的实验，同时也进行着毕业课题相关的工作。正是在Livingston的实验室，他关于呼吸代谢多条通路的理念开始形成。

在从约翰斯·霍普金斯大学获得博士学位后，1930年夏天，在前往哈佛大学W. J. Crozier的实验室工作之前，汤佩松到马萨诸塞州海滨小镇Woods Hole著名的海洋生物学研究所进行短期访问；在1931年夏天，他又到那里做了第二次短期访问。在那里，他与一些人建立了友谊，例如R. S. Lillie、Ralph Gerald以及W. J. V. Osterhout，受他们的影响，他决定选择细胞与植物的呼吸作用与光合作用的生物能作为他毕生的研究方向。

在哈佛大学的三年期间，他又遭遇了之前在约翰斯·霍普金斯大学遇到过的问题——Crozier希望他从事种子在萌发过程中呼吸系统的温度特性方面的研究，一个他并不感兴趣的课题。为了在温度特性课题上发表一系列文章，汤佩松比以往更加努力地工作，同时，他还进行着自己的实验，并于1932年第一次在植物中发现细胞色素氧化酶。在1933年，他实践了自己对国家的承诺，婉拒了一个在纽约大学任职的机会，接受武汉大学的邀请，回到了祖国。

在武汉大学最初的几年中，汤佩松就获得了许多学术成果。利用学校提供的2000元美金，他设法建立了一个小型的细胞与植物生理实验室，之后在细胞呼吸动力学研究方面发表了7篇系统性的研究论文。随后，1937年抗日战争全面爆发，关于细胞呼吸系统的研究不得不停滞，但汤佩松从未停止工作。他利用自己的知识和经验力所能及地支持国家抗战：首先，他将自己的实验室转而生产活性炭以用于制造防毒面具；武汉被占领后，他跋涉至贵阳，在那里建立了一个医学院；其后，与众多中国科学家一样，汤佩松转移到昆明。在相对安全稳定的昆明，他建立了一个小型植物生理学实验室（图2），那里也是青年生理学家聚会的场所，这些青年学者中的很多人在日后成为我国生理学界的中流砥柱。

图2　汤佩松在昆明的临时温室里
（选自《汤佩松论文选集》）

在新中国成立后，汤佩松终于得以继续从事植物呼吸代谢与光合作用方面的研究。我们不再细数他所取得的成就，但必须指出的是，在他的学术生涯中，他总是富于创造力并走在时代的前沿。一个很好的例子就是他与王竹溪在20世纪40年代发现的“关于细胞水分化学势的汤-王理论”，这个超前的理论直到60年代P. J. Kramer与他的同事发表了类似的理论之后才得到了广泛的重视。另一个例子是汤佩松与同事吴相钰发现水稻幼苗中硝酸还原酶的诱导形成，研究结果发表在1957年的《自然》杂志上。这不仅是学术界首次发现硝酸还原酶是诱导酶，而且也是学术界首次证实高等植物体内存在诱导酶。

甚至在离休后，汤佩松先生的远见卓识依然对我国植物生理学及生物化学的发展起到了重要作用。在20世纪80年代后期，他提出跨学科合作对研究光合系统膜蛋白是必须且非常重要的，并邀请了植物生理学家匡廷云、生物物理学家梁栋材和常文瑞到他家中讨论光合膜蛋白结构与功能研究合作的可能性，并决定了由中国科学院植物研究所和中国科学院生物物理研究所合作开展此项研究，这是一次世纪之约的会议。十余年后，汤佩松已经过世，常文瑞领导的研究小组与匡廷云领导的研究小组通过学科交叉、通力合作，在国际上首次解析了菠菜主要捕光复合物Ⅱ的三维结构，这是中国解析的第一个膜蛋白结构。这项研究结果作为当期封面文章发表在2004年的《自然》杂志上。

在1983年，为了庆贺汤佩松先生八十岁寿辰，国际著名刊物《植物生理学年评》邀请汤佩松撰写了一篇关于他自己学术生涯的文章，他将其描述为“漫步在植物生理学的迂回小径上”，并作为该文的题目。然而，正是他对祖国和人民的无私奉献、对科学真理的不懈追求，使他一直清楚前进的方向，并从未止步、始终前行。

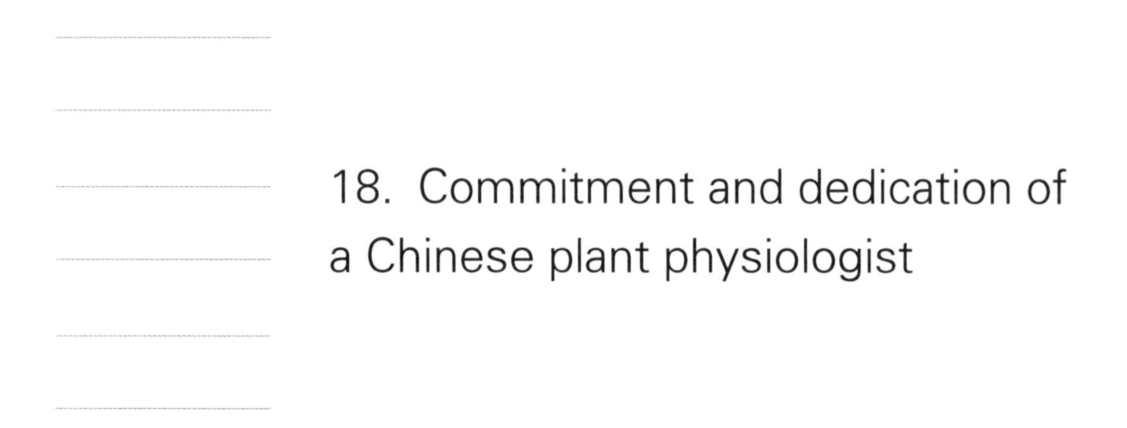

18. Commitment and dedication of a Chinese plant physiologist

Professor Peisong Tang (Pei-Sung Tang), born in 1903, is one of the co-founders of the modern Chinese plant physiology. You might be familiar with some of his academic achievements: He is the first one to discover respiratory enzymes (cytochorome oxidase) in plants; he proved that multiple respiratory metabolic pathways and electron transfers exist in the rice; he is the first one to experimentally prove the existence of carbonic anhydrase in plants; together with Prof. Zhuxi Wang, he used thermodynamical method to illustrate the mechanism of intracellular water movement, which later has been referred to as the "Tang-Wang theory of cellular water potential" . You probably also know that he is the first Chinese biologist who published in *Nature* and *Science* in 1940s and 1950s. But there is something about him that you will not learn from his over 200 publications: His commitment, as a Chinese, to his country and his people, and his dedication, as a scientist, in pursuing science and truths.

Peisong Tang received his early education at Tsinghua College from 1917 to 1925, financially supported by the "Boxer Indemnity Scholarship" . It was why Peisong Tang always felt indebted to his country and his people, and made a commitment of returning home after studying abroad for years. It was also at Tsinghua College where he first showed his interests in biology. After the biology teacher described how the starch stored in the endosperm of a seed converted into a seedling, he raised a question that the teacher could not answer: "How exactly do unorganized materials like starch become a seedling, something so well-organized and perfectly-structured?" Little did he then know that the question became the starting point of his scientific life.

After graduation from Tsinghua College, Peisong Tang went to University of Minnesota and graduated with a B.A. degree. His curriculums at the university covered a broad spectrum of courses, which provided a solid knowledge basis for his future research. Among the courses, he was particularly inspired by physical chemistry lectures, which strengthened his interests in the

Tingyun Kuang[1], Ming Li[2], Le Kang[2]

1 Institute of Botany, Chinese Academy of Sciences, Beijing 100093, China

2 Beijing Institutes of Life Science, Chinese Academy of Sciences, Beijing 100101, China

Contact: kuangty@ibcas.ac.cn (T. -Y. Kuang)

energetics of living matter. After that, he went to Johns Hopkins to study with Burton Livingston, but the thesis project that Livingston chose for him did not quite interest him. Unlike some other Chinese students who were humble and always did what they were told to do, Peisong Tang started to perform some experiments to study seed respiration and photosynthesis, meanwhile also working on his thesis project. It was in the Livingston lab where his concept of multiple pathways in respiratory metabolism started to form.

After getting his PhD degree at Johns Hopkins, and before he went to Harvard to work with W. J. Crozier, Peisong Tang spent a summer at the famous Marine Biological Laboratory at Woods Hole in 1930, and returned for another summer in 1931. It was at Woods Hole where he made quite a few friends such as Ralph S. Lillie, Ralph Gerald and W. J. V. Osterhout, whose influences made him decide to choose bioenergetics of cellular and plant respiration and of photosynthesis as his life work.

During his three years at Harvard, he encountered the same problem as he did at Johns Hopkins—Crozier wanted him to work on the temperature characteristics of seed respiration during germination, a project that did not interest him much. Working harder than ever, Peisong Tang managed to publish quite a few papers on the temperature characteristic project; meanwhile, he also set up experiments of his own, and for the first time discovered cytochrome oxidase in plants in 1932. In 1933, with his commitment to his country, he declined an offer for a position at a New York University, and accepted an invitation from Wuhan University to return to China.

Peisong Tang was quite productive during the first few years at Wuhan University. With the $2000 from the university, he managed to set up a small lab for cell and plant physiology, and later published a series of seven papers on the kinetics of cell respiration. Then the War of Resistance against Japanese Aggression outbroke fully in 1937. The research on cell respiration had to stop, but Peisong Tang never stopped. With his knowledge and experience, he did everything he could to serve the nation at war: First, he converted his lab into a factory making active carbon for gas masks; after Wuhan was occupied, he traveled to the city of Guiyang and helped establishing a medical school there. Later on, like most Chinese scientists, Peisong Tang moved to Kunming. With the relative safety and stability in Kunming, he built a small lab of plant physiology, which also served as an assembly place for young physiologists, many of whom later became the backbone of Chinese physiology.

After the founding of the People's Republic of China, Peisong Tang finally was able to continue his work on plant respiratory metabolism and photosynthesis. No need to repeatedly count his achievements, but it is important to point out that in his academic life, he was always creative and ahead of his time. One good example is the "Tang-Wang theory of cellular water potential" that he and Zhuxi Wang developed in 1940s. The forerunning concept did not get much attention until 1960s when P. J. Kramer and his colleges published similar theory. Another example is the study by Peisong Tang and Hsiang-Yu Wu on the adaptive formation of nitrate reductase in rice seedling, the result of which was published in *Nature* in 1957. It is not only the first report to show that the nitrate reductase is inducible, but also the first proved existence of inducible enzymes in plants.

Even after he retired, the far-seeing vision of Peisong Tang still played an important role in Chinese plant physiology. In the late 1980s, he suggested that inter-disciplinary collaboration was much needed to study the membrane proteins of photosynthesis, so he invited plant physiologist Tingyun Kuang, biophysicists Dongcai Liang and Wenrui Chang over to his house and the four of them had a small meeting to discuss the possibility of collaboration. Over a decade later, after Peisong Tang passed away, Tingyun Kuang's group and Wenrui Chang's group jointly solved the structure of the light harvesting complex Ⅱ from spinach, which was the first membrane protein structure from China. The result was published in *Nature* as a cover-story paper in 2004.

To celebrate his 80th birthday in 1983, *Annual Review of Plant Physiology* invited Peisong Tang to write an article about his academic life, which he described as "the devious trail of a roaming plant physiologist" in the title of the article. However, with his commitment to his country and his people, and his dedication in pursuing science and truths, he always knew where he was going and never stopped along the way.

Figures

Fig.1 Peisong Tang (1903—2001) (from *Peisong Tang Collections*)

Fig.2 Peisong Tang at his make-shift greenhouse in Kunming (from *Peisong Tang Collections*)

19. 俞德浚：爱国植物学家及其贡献

俞德浚是我国著名的植物分类学家、园艺学家以及植物园专家（图1）。他是一位爱国科学家，在1950年他拒绝了留在英国的机会，义无反顾地回到祖国，将他的余生都奉献给了植物学。而且，他的贡献远不止在植物学的发展上，还有植物园的建设和植物资源的开发，尤其是我国的果树资源开发。他将自己的科研和国家的经济发展紧密联系在了一起。

图1　俞德浚（1908—1986）

1931年，俞德浚从国立北平师范大学生物系毕业，随即加入了北平静生生物调查所，从事植物分类学研究。他第一次重要的采集任务是1932年到1934年在四川西部为期三年的植物采集，第二次采集活动是1937年到1939年在滇西北开展的。在20世纪30年代，他去的这些地方军阀混战、严重贫困、偏远闭塞、交通严重不便、自然条件复杂多变，但同时生物多样性很高。因此，这些采集活动一方面危险重重、充满挑战，另一方面对我国的植物分类和植物调查有着重要意义。俞德浚雇佣当地人当挑夫、向导和翻译，与他们相处得很好。独龙江的一位村民孔志清曾回忆说，俞德浚是第一个在独龙江地区采集植物的植物学家，为了采集标本他有时候甚至甘冒生命危险。在丛林里长时间艰苦穿梭后，依然不辞辛劳地整理标本，做好记录。他和当地人建立了深厚的感情，他教孔志清说普通话，帮他入学读书，让他成为独龙江地区第一个接受教育的人（孔志清和周元川，1986）。

作为一位植物学家，俞德浚既擅长野外调查，也擅长实验室工作，经常在两者间切换。他的分类学研究主要集中在蔷薇科、豆科、秋海棠科和山茶科，他和学生花了20多年在蔷薇科植物的分类研究上，整理和鉴定了在全国采集的30多万号标本。除了依靠标本，他们也经常到实地考察植物，鉴定和确立新种时尤其如此。1980年，俞德浚和学生李朝銮鉴定了一个新属——太行花属，他们远赴太行山找到了模式标本太行花进行确认后才完成了研究鉴定，发表了新属。他的团队最终完成了《中国植物志》

作者：姜虹
四川大学文化科技协同创新研发中心，成都610065，中国
邮箱：jh-iris@hotmail.com

中第36—38卷蔷薇科植物的编撰工作，对经济和学术都有重要意义（佚名，1986）。1947年到1950年，俞德浚得到邱园资助，在邱园和爱丁堡皇家植物园进修和从事研究，参与了高等植物分类学、园艺学和植物园建设，并因为《云南茶花及其园艺品种》一文被邀加入国际山茶花学会。从1978年到1986年，他先后担任《中国植物志》代主编和主编，期间共有35卷《中国植物志》完成。他在分类学上的另一个贡献是编著了《中国果树分类学》，收录了59科670种植物，包括形态学、染色体数、物候学、分布和栽培品种等信息。1982年，这部专著也为他赢得了全国优秀科技图书一等奖。

在访学英国期间，俞德浚意识到植物园在科研、经济发展和科普中扮演的重要角色。他一回到祖国就投身到北京植物园的建设中，为了给新植物园选址，俞德浚带领专家组在北京及周边十几个地方进行考察和比较，最终选择了香山脚下。作为北京植物园主任，他领导和参与了植物园的规划设计、植物种植、建设和日常管理工作。除了创办北京植物园，他也大力协助了我国其他几个植物园的创办，包括庐山植物园、武汉植物园、西双版纳热带植物园、杭州植物园等（图2）。结合植物园创办的实际经验，他还编著了《植物园工作手册》，成为我国建设现代植物园的重要参考手册；创办了《植物引种驯化集刊》，介绍这些新建成的植物园最新科研进展和建园经验；以及出版了画册《中国植物园》。就植物引种而言，俞德浚提出了记录植物的六条原则：① 引种年代、编号和原产地，② 正确的中文名和拉丁名，③ 种植图和名牌，④ 完整的物候记录，⑤ 详细的生物学特征记载和分析资料，⑥ 照片、种子标本和腊叶标本（佚名，1986）。这些要求和他对植物园建设的理念是一致的，尤其是隶属于中国科学院管辖的植物园，研究要放在第一位，然后是物种保护，以及开发经济植物以满足农业所需。

图2　中国科学院华南植物园内的俞德浚教授铜像

除了分类学和植物园建设，俞德浚的植物学兴趣还包括植物资源，尤其是果树和观赏植物的研究和栽培，两者都与他的分类学研究密切相关。在俞德浚的一生中，我国经历了战争和极端贫困，他对有经济价值或观赏价值的植物尤为敏感。在他所专长的蔷薇科和山茶科，有大量植物在农业和园艺中被开发和栽培。据他研究，中国有大量植物被引种到全世界，有200种以上的果树原产自中国，还有各种开花植物在欧洲那些植物园里绽放，所以中国被称为“园林之母”（俞德浚，1985）。在他的植物采集中，他尤为重视具有栽培潜力的野生植物资源的开发，考察全国各地的果树种类，在植物园培育新种，开展研究工作和发表果树分类学专著，还邀请苏联专家到我国开展研究，为果树开发献计献策。

俞德浚坚决保护祖国的植物资源，他曾撰文愤怒地控诉西方在中国的植物开发和掠夺。西方的植物猎人经常为自己的国家采集经济植物和观赏植物的种子、标本甚至活体植株，他们也窃取植物产品（如茶叶）的加工技术、自然资源和环境信息，直接买卖植株和种子。漂亮的中国植物至今还在欧洲的花园里生长开花，模式标本收藏在他们的标本馆，中国植物的拉丁名也经常以植物猎人的名字命名。中国的植物资源开发和植物学的发展极大地受到了西方植物学的影响（俞德浚，1952）。在这种情形下，俞德浚和其他中国植物学家在出访欧洲时都会竭尽全力研究收藏在那里的中国植物标本，尤其是那些模式标本。

俞德浚在1986年7月14日去世，他的遗愿是死后不要有任何告别仪式，遗体可以捐献给医院进行病理解剖。他的名字被刻在北京植物园（现国家植物园）的大理石墓碑上，作为植物园的创立者被铭记，他也被载入植物学史和植物园的历史中。

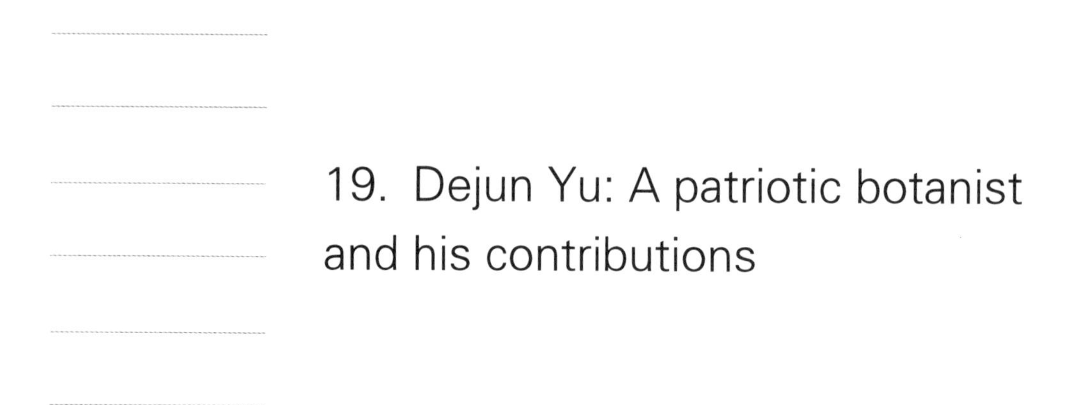

19. Dejun Yu: A patriotic botanist and his contributions

Dejun Yu (Te-Tsun Yu, Feb. 1, 1908—July 14, 1986) was a famous plant taxonomist and horticulturist in China as well as an expert on botanical gardens. As a patriotic scientist, Dejun Yu refused a position in Britain and returned to China without hesitation in 1950. He devoted his whole life to the botanical sciences. His contributions went far beyond advancing botany, however, extending to construction of botanical gardens and the exploration of plant resources, especially fruit trees in China. His research was closely connected to the scientific and economic development of his country.

Dejun Yu began his career in plant taxonomy at the Fan Memorial Institute of Biology upon his graduation from the Department of Biology at National Beiping Normal University in 1931. His first important collecting task was three-year plant hunting in western Sichuan Province from 1932 to 1934. He started his second hunting journey with his team to northwest Yunnan Province, which went from 1937 to 1939, three years later. The regions he went to in the 1930s suffered from warlord conflicts, serious poverty, isolation, poor transportation and complicated natural environment, but also had high biodiversity. Such explorations, therefore, meant a lot of dangers and challenges on one hand, and great contributions to the progress of plant taxonomy and investigation in China on the other hand. Dejun Yu often hired local people as carriers, guides and translators, and got along well with them. A villager from the Dulong River valley, Zhiqing Kong, once shared their stories. Dejun Yu was the first botanist collecting plants around Dulong River region. Sometimes he even risked his life to collect species. He also always worked hard on recording and organizing specimens after long trudges in the jungle. He established deep friendship with local people including Zhiqing Kong, teaching him mandarin and helping him enter school, which made Kong become the first educated Dulong person (Kong and Zhou, 1986).

Dejun Yu was both a field botanist and an armchair botanist, and often shifted between the two when he focused on his taxonomic research, including families of Rosaceae, Leguminosae, Begoniaceae and Theaceae. He and his students spent over 20 years on the taxonomy of

Hong Jiang
Center for Culture, Science and Technology, Sichuan University, Chengdu 610065, China
Correspondence: jh-iris@hotmail.com

Rosaceae plants, organizing and checking nearly 300000 specimens collected all over the country. However, this did not mean they completely relied on specimens. They often travelled to investigate plants in situ, especially to identify new species. In 1980, Dejun Yu and his student Chaoluan Li found a new genus named *Taihangbia* by identifying a specimen. They affirmed it by travelling to Taihang Mountain and finding its type species *Taihangbia rupestris*, and then published this genus. His team finally compiled three volumes of *Flora Republicae Popularis Sinicae* (FRPS) on this family, Vol. 36—38 (Anonymous, 1986). Supported by the Kew in London, UK, Dejun Yu did research and received advanced training in taxonomy of higher plants, horticulture, establishment and management of botanical gardens at the Royal Botanic Garden Edinburgh and the Kew from 1947 to 1950. In Britain, he was inducted into International Camellia Society because of his highly praised paper Camellia plants and their cultivars in Yunnan Province, China. From 1978 to 1986, he served as the acting editor-in-chief and then editor-in-chief of FRPS, as many as 35 volumes of which were published during that period. His another contribution to taxonomy was his research on fruit trees. In 1979 he compiled *Taxonomy of Chinese Fruit Trees*, collecting 59 families and 670 species, including detailed information on morphology, chromosome number, phenology, distribution, and cultivars. This monograph also won him the first prize of "National Excellent Monographies of Science & Technology" in 1982.

During his visit in Britain, Dejun Yu realized how important role botanical gardens played in research, economic development and science popularization. He devoted himself to the foundation of the Beijing Botanical Garden (BG) as soon as he returned to China. In order to find a proper place for the new garden, Dejun Yu led a team of experts to explore a dozen places all over Beijing city and compare them, and finally built it at the foot of Xiangshan Mountain. He was then appointed as the director of Beijing BG, engaged in the design, planting, building and daily management of it. Besides founding Beijing BG, he also greatly helped with the establishment of quite a few botanical gardens in China including: Lushan BG, Wuhan BG, Xishuangbanna Tropical BG, Hangzhou BG, etc. Along with the development of botanical gardens in China, he compiled *Handbook of Botanical Gardens*, an important guide to the building of modern botanical gardens in China, started the journal *Collections of Plant Introduction and Domestication*, to introduce research and progress achieved by these newly built gardens, and published the album of painting *Chinese Botanical Gardens*. As far as the introduction of plants is concerned, Dejun Yu proposed six principles of recording for each plant: 1. Introducing year, number and origin; 2. Accurate Chinese and Latin names; 3. Picture of planting and ID card; 4. Complete phenological records; 5. Detailed biological description; 6. Photo, seed samples and exsiccate (Anonymous, 1986). These requirements were identified with his ideas on the aims of botanical gardens, especially those under the administration of Chinese Academy of Sciences, namely research first, followed by protection of plants species and especially cultivation of economic plants for the need of agriculture.

Besides the taxonomy and botanical gardens, Dejun Yu's interests in botany also extended to plant resources, especially research and cultivation of fruit trees and ornamental plants, both closely related to his taxonomic research. During Dejun Yu's life, China experienced wars and deep poverty. He was very sensitive to species that may be cultivated as economic or ornamental plants. Many species of Rosaceae and Theaceae, the two families he specialized in, have been cultivated in agriculture and horticulture. According to his research, Chinese plants had been greatly introduced around the world. There were over two hundred fruit species around the world native to China. Various Chinese flower plants were blooming in European gardens, which won China a name of "Mother of Gardens" (Yu, 1985). Specifically, he contributed to the exploration of plant resource by means of collecting wild species with potential for cultivation during his plant hunting, exploring fruit species all over the country and trying to improve them, cultivating new species in the botanical garden, inviting experts from the Soviet Union to China to perform research and give advice on cultivating fruit trees, carrying research on fruit trees and publishing monography on the taxonomy, etc.

Dejun Yu also stood firm to protect the science and resources of his own country. He told the history of plant exploration and plunder by western countries in China, and expressed his anger in an article. He mentioned that, plant hunters often collected seeds, specimens and even living economic and ornamental plants for their own countries. They even stole processing technology of plant products (such as tea making) and information of natural resources and environment, and sold plants or their seeds directly. Beautiful Chinese plants still living in European gardens, type specimens kept in their herbariums and Latin names of Chinese plants named after plant hunters are outstanding examples. Because of the western botany, China was greatly influenced on the development of plant resources and the progress of botanical sciences (Yu, 1952). Given this situation, Yu and other Chinese botanists tried their best to study specimens of Chinese plants during their visits to Europe, especially type ones.

Dejun Yu passed away on the afternoon of July 14, 1986. His last wish was that there would be no farewell ceremony for him, and his body would be donated to a hospital for pathological anatomy. His name was carved on a marble tombstone as the founder of Beijing BG. He will also be remembered in the history of botany and botanical gardens.

Figures

Fig.1 Dejun Yu (1908—1986)

Fig.2 Bronze Statue of Prof. Dejun Yu located in South China Botanical Garden, Chinese Academy of Sciences (SCBG, CAS)

References

Anonymous (1986) In memory of Prof. Dejun Yu. Bot J 6:38–40 . (佚名 . 1986. 怀念俞德浚教授 . 植物杂志 , 6: 38–40.)

Kong ZQ, Zhou YC (1986) Deep friendship established on the bank of Dulong River—in memory of Mr. Dejun

Yu. Bot J 6:41–42. (孔志清, 周元川. 1986. 独龙江畔结深情——缅怀俞德浚先生. 植物杂志, 6: 41–42.)

Yu DJ (1952) How did imperialists plunder plant resource of China. Bull Biol 1:27–30. (俞德浚. 1952. 帝国主义者怎样掠夺我国的植物资源. 生物学通报, 1: 27–30.)

Yu DJ (1985) The contribution of Chinese plants to the world horticulture. Chin Bull Bot 3(2):1–5. (俞德浚. 1985. 中国植物对世界园艺的贡献. 植物学通报, 3(2): 1–5.)

20. 缅怀我的导师吴征镒院士

2013年6月20日凌晨，我完成在西藏的野外工作，刚刚乘机回到家中，就接到先生去世的噩耗，仿佛惊天霹雳。尽管先生在病榻已是一年有余，但当这一天来到的时候，我仍不愿意相信这是真的。就在几天前在藏东南崎岖的山路中，我试图辨认车窗外一晃而过的植物，发现十分困难。这让我不由想起先生西藏考察在吉普车上做笔记的事迹。看着车内昏昏欲睡的学生，我不由感叹："知道吗，你们的祖师吴先生60岁高龄参加青藏高原科考，随着吉普车的一路颠簸，他一路做笔记，一天下来竟是沿途植物分布的记录"。虽是寅夜，仍浮想联翩，和先生交往的一幕幕浮现心头（图1）。

图1　吴征镒教授（1916—2013）

我1985年从中国科学院南京地质古生物研究所硕士毕业，来到中国科学院昆明植物研究所工作。先生知道我的背景是古植物学专业，非常高兴，嘱咐我做一些古植物学和现代植物学交叉结合的工作。因为先生从事的植物区系地理和植物系统演化的研究，离不开古植物学证据。1986年先生开始招收博士研究生，德铢、建强和我有幸成为先生招收的第一批博士研究生（图2）。成为先生的博士生后，我又获得了公派出国攻读博士学位的机会。拿到这个机会以后，我既高兴，又有几分担忧，不知如何向先生开口。先生得知了我的顾虑，鼓励我出国深造，并亲自为我写了推荐信。最终我选择了中国科学院和英国皇家植物园邱园联合培养的方式完成了博士期间的学习，我始终为能师从先生而自豪。

1991年初，我获得博士学位，其时先生所主持的国家自然科学基金重大项目"中国种子植物区系研究"正如火如荼地展开。这个大项目是当时国家自然科学基金委员会投入经费最多的项目，全国50多个研究单位大约200名科技人员涉及其中。刚刚获

作者：周浙昆

中国科学院西双版纳热带植物园，西双版纳666303，中国

邮箱：zhouzk@xtbg.ac.cn

图2 1988年吴征镒教授和他的学生（一排左起：吴征镒、周浙昆、朱华；
二排左起：李建强、李德铢、杨亲二、唐亚）

得博士学位的我，理所当然地也投入其中。针对我的学科背景，先生要求我参加中科院植物研究所古植物学家陶君容老师主持的“中国植物区系形成与演变的地质背景研究”，这一要求使我走上了一条古今结合、研究我国植物区系形成演变和地球环境协同演化的道路。

在这个大项目中，有一项艰巨的任务——“中国植物区系中关键和薄弱地区植物区系的研究”。1990年孙航、德铢等完成了西藏阿里地区植物区系的考察。1991年，60多岁高龄的李恒老师主持完成了独龙江植物区系的越冬考察。更为艰苦的西藏墨脱的越冬考察，几经易人，最终交由孙航负责。西藏墨脱位于雅鲁藏布江大峡谷，是全国唯一不通公路的县，而且全年有大半年大雪封山，有“陆地孤岛”之称。得知孙航正在招兵买马，我跃跃欲试。先生知道我的想法后给予鼓励和支持。先生深知大自然是植物区系地理学的天然实验室，年轻人应该在大自然中吸取养分，锤炼自己。1992年9月，孙航、俞宏渊和我踏上了墨脱考察的征程，直到1993年6月才返回。西藏墨脱的9个月的野外考察中，我们徒步行程近万里，克服种种困难，采集各类标本近万份。先生非常满意，在欢迎我们考察归来的会上，先生高兴地说道：“年轻人担当重任，不负众望，又孚众望。”细心的先生甚至注意到，我们几个人的腰身都瘦了一圈。此后，以这些标本为材料，我们发表了20多篇论文和一部专著，为重大项目的完成作出了重要贡献。

1996年先生主持的“中国种子植物区系研究”项目在北京结题。已是耄耋之年的先生并未感觉轻松，反而觉得还有四座“大山”压在身上。先生指的是他想要完成的，代表他学术思想的四部专著：《中国被子植物科属综论》（Wu et al., 2003a, 2003b）、《中国植物志（总论）》（Wu and Chen, 2004）、《种子植物分布区类型及其起源和分化》（Wu et al., 2006）、《中国种子植物区系地理》（Wu et al., 2010）。先生相约德铢、孙航、彭华以及我共同搬这四座“大山”。

我协助先生搬的“大山”是《种子植物分布区类型及其起源和分化》。这部书的思想可以追溯到1965年先生在《科学通报》上发表的《中国植物区系的热带亲缘》(吴征镒，1965)，后又部分见于1977年先生和王荷生合作的《中国自然地理——植物地理》(上册)(Wu and Wang，1983)，其后1991年又见于《云南植物研究》发表的专辑《中国种子植物属的分布区类型》。在书稿中，先生从纷繁的植物分布现象中找寻植物分布的规律，娓娓道来，巧妙地将我国种子植物的属归为15种分布区类型。这种分析方法把地理成分和发生成分的研究联合起来，行之有效地揭示各分布区类型的特征及其相互关系，为进一步的区系分区奠定了坚实的基础，成为我国植物地理学研究的一个鲜明特色。其后，先生进行了大量的工作，将这一研究方法从中国逐步推向了世界。先生仔细钻研全球植物分类系统和植物区系学之间、植物区系和植物群落之间的区别与联系，从天、地、生这个更广袤的空间来研究植物的演替和分布的规律，研究植物分布与环境演变的相互关系。他依据自己多年的研究心得提出"生物演化和分布规律不但和地球演化规律同步进行，而且受到后者严格制约；生物演化总是一元多系，生物的演化韵律和地球的律动相合"的学术思想，建立东亚植物区和中亚植物区，提出世界科的分布区类型，并分析了各种分布区类型可能的发展演变过程。以书稿的核心思想总结而成的论文《世界种子植物科的分布区类型系统》，至今已经被引用几千次，是引用率最高的植物学文献之一。书中的许多学术思想，越来越多地成为后学选题的依据和研究的切入点。

修改和整理书稿的过程，是我向先生学习的良机，作为先生著作的第一位读者（承蒙先生提携，我也是专著的共同作者）常能和先生讨论学术，偶有拙见获先生赏识吸纳于书中，我会备受鼓舞，兴奋不已（图3）。历时8年，先生磨剑不止，书稿（图4）在不断修改和补充的过程中完善和升华，可见先生治学严谨，而我在此过程中饱纳养分，丰富了学养，仿佛又跟着先生读了一个博士学位。

书稿付梓前我请先生为书写自序，先生欣然允之。一个星期后，看着先生的自序，我不禁热泪盈眶，感慨万千。先生生于乱世，经历无数的战乱，新中国成立以后又是"树欲静而风不止"，60岁才逢盛世，方能专心学问，80岁以后息影林泉，著书立说。

图3 吴征镒教授和笔者正在讨论

图4　吴征镒教授的书稿

百万余字的书稿由先生一笔一画写就，其时先生目疾已久，医生嘱咐不可用眼过度，以留视力为生活所需。而先生不以为意，终日伏案奋蹄，终有兀兀穷年、呕心沥血的精心之作。

先生从事植物学研究70余年，认识中国植物的多样性、为中国植物编目是先生对中国乃至世界植物学的另一项重要贡献。根据《邱园索引》，先生定名和参与定名的植物分类类群有1766个（涵盖94科334属，其中新属22个），是中国植物学家发现和命名种子植物最多的一位。先生是《中国植物志》的第四任主编，他主持编撰完成了54卷《中国植物志》（全书80卷），他还主编了《云南植物志》《西藏植物志》等重要的植物学著作。为了让世界了解和认识中国植物，他还和美国植物学家Peter Raven共同主持了《中国植物志》英文版的编撰。25卷的《中国植物志》英文版，在先生去世一年后的2014年全部出版发行。

吴征镒院士先后获得国家自然科学奖一等奖2项，1999年获日本花卉绿地博览会纪念协会“COSMOS国际奖”，2003年获何梁何利基金“科学与技术成就奖”，2007年获国家最高科学技术奖。

斯人已逝，幽思长存。先生是一棵望天树，傲立林中；是一座灯塔，指引着植物区系地理学的发展方向。先生磨成的宝剑锋利无比，早已披荆斩棘。我虽不敢奢望“青出于蓝而胜于蓝”，但一定会殚精竭力，百折不挠；上下求索，不枉先生之谆谆教诲。

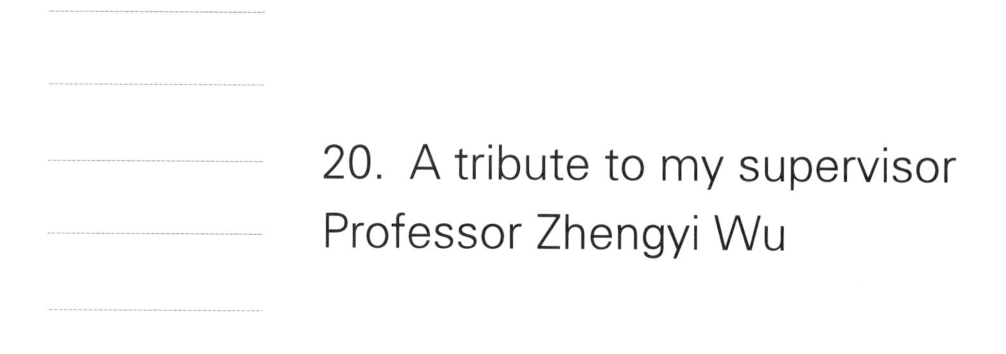

20. A tribute to my supervisor Professor Zhengyi Wu

I was saddened and quite shocked when I received the phone call informing me that Professor Zhengyi Wu had passed away. It was the evening of June 20, 2013. I had just got off a plane from Tibet and turned on my mobile phone. Prof. Wu had been unwell for nearly a year. I knew that this would happen one day, but refused to believe it would really come to pass. A few days ago, some students and I were on a field trip in Tibet. Sitting in our jeep, I watched the mountains and roadside as we drove, and tried to identify the plants along the route. The mountain roads in Tibet seem to always make my students sleepy. I woke them up. I told them that I had heard this first when Prof. Wu was on a field trip to Tibet. During the trip, he always kept an eye on the surroundings outside the vehicle, observing plants and taking notes. At the end of each day, he would prepare a check-list of plants for the region through which he had travelled. Back home it was the middle of the night, and I was awake. My thoughts had turned to my relationship with Professor Wu.

In 1986, one year after I had graduated with my Master's degree from Nanjing Institute of Geology and Palaeontology, Chinese Academy of Sciences, I, along with Dezhu Li and Jianqiang Li, became one of Professor Zhengyi Wu's first PhD students. Prof. Wu needed someone in his team to examine fossil plants for his research, and I was trained as a palaeobotanist. At the same time, I received a grant from the Chinese Academy of Sciences to pursue my doctorate overseas. I was very happy to have this opportunity but worried about how Prof. Wu would respond to the idea of me studying abroad. I was still a young graduate student, and did not know how to bring the subject up with my advisor. When Prof. Wu learned about my grant, however, he immediately encouraged me to study abroad. In the end, I was jointly supervised by Dr. Wilkinson from the Royal Botanical Gardens, Kew and Prof. Wu. I was at Kew for one year. I always felt honored to be one of Prof. Wu's students.

Zhekun Zhou[1,2]

1 Key Laboratory of Tropical Forest Ecology, Xishuangbanna Tropical Botanical Garden, Chinese Academy of Sciences, Xishuangbanna 666303, China

2 Key Laboratory for Plant Diversity and Biogeography of East Asia, Kunming Institute of Botany, Chinese Academy of Sciences, Kunming 650201, China

Correspondence: zhouzk@xtbg.ac.cn

I received my PhD in early 1991. At that time, Prof. Wu was leading a major project called "The Floristics of Chinese Seed Plants" . During the 1990s, this project was the largest of its kind at the National Natural Science Foundation of China. The project required a lot of fieldwork around China and many of my colleagues had already started to work in the field. According to Prof. Wu, the laboratories of floristic taxonomy, biogeography and ecology are in the field. Again and again, he told us, "As young students, you should work in the field, get your data first-hand in the field, and even contemplate the implications of what you find in the field." With his encouragement, I joined a field trip to Motuo, Tibet, which is at the big bend of the gorge of Yalu Tsangpo in Tibet, one of the most remote places in China. There, I worked in the field for nine months. We collected more than 7000 specimens from Motuo. Based on this work, we published over 20 papers, a book and received the National Prize for Natural Sciences from Chinese Academy of Sciences in 1998. With the great vision of Prof. Wu, my own scientific career benefitted immensely.

In 1996, Prof. Wu, who was then already eighty years old, completed his project on "The Floristics of Chinese Seed Plants" . But he was not content to rest. On the contrary, he said he still felt as if there were "four big mountains" on his shoulders. These "four big mountains" referred to books he was planning to write: *A Review of the Angiosperm Genera in China* (Wu et al., 2003a, 2003b); *Flora Republicae Popularis Sinicae* (FRPS), volume one, general overview (Wu and Chen, 2004); *The Areal-Types of Seed Plants and Their Origin and Differentiation* (Wu et al., 2006); *Floristics of Seed Plants from China* (Wu et al., 2010). In order to move these mountains off his shoulders, he asked four former students, Dr. Dezhu Li, Dr. Hang Sun, Dr. Hua Peng and me, for assistance. The plan was straightforward. Each student would help with one book. I was in charge of *The Areal-Types of Seed Plants and Their Origin and Differentiation*. The basic idea of this book originated from an article published in 1965 titled The tropical floristic affinity of the *Flora of China* (Wu, 1965). In this article, Prof. Wu analysed the distribution patterns of Chinese seed plants, classified them into 15 areal types and 31 sub-areal types, and then proposed the hypothesis of tropical affinity of Chinese floristics. He also proposed a hypothesis about the origin of East Asian floristics. Wu thought the regions of Indochina, South China and Southwest China, which have the richest primary families and genera, and have survived from the "Tertiary tropical" , are the core and cradle of East Asian floristics and have a tropical affinity (Wu, 1965). The classification of areal types of Chinese seed plants and the hypothesis that Chinese floristics have a tropical affinity were revised in the book *Chinese Physical Geography—Phytogeography* (I) (Wu and Wang, 1983). At the beginning of the project, I thought my job would be the easiest. The book had a well-developed framework and a lot of work had already been done. When we started preparing the manuscript, however, I found that Prof. Wu had compiled a substantial amount of new data. The content of the book had largely expanded. In the old book, he only provided an areal-type for Chinese genera of seed plants and a basic analysis of each areal-type. In the new book, Prof. Wu began with the basic concept of phytogeography. For instance, he addressed what floristic plant geography is, the difference between floristics and vegetation, and the research history of phytogeography. He continued by giving detailed explanations about the meaning of "areal" , and how different areal-types formed, particularly continental disjunction areal-types. He also addressed theories about dispersal and vicariance. Furthermore, he explained theories about the origin of disjunctions and proposed hypotheses on how continental drift impacted the formation of areal-types. In the third chapter, Prof. Wu employed his areal-type analysis method at the family level. In 2003, the part of this chapter that classifies 517 world families of seed plants into 18 areal types was published (Wu, 2003a, 2003b). This scheme has been widely used in analysing the national and regional floras of China at various levels and is helpful in understanding biogeographic issues, such as endemism, vicariance and disjunctive distributions. In fact, this article is one of the most cited papers in China.

In this new book, Prof. Wu reviewed the taxonomic status of all 3201 genera of Chinese seed plants and provided detailed information about the concept of every Chinese genus of seed plants in different systems and related references, synonyms, species number, detailed distribution information both in China and in the world and its areal-type. He summarised different systems of Gymnosperms, added distribution pattern data and proposed hypotheses for places of origin. This was a completely new book.

For me, helping to prepare the manuscript for this book was an incredible opportunity to learn from Prof. Wu. I had more time to discuss scientific issues with him. I was honored when Prof. Wu added any of my comments or suggestions to the manuscript. I felt like I was working on a second PhD. It took eight years for Prof. Wu to finish the book. The manuscript contains over 10 million words and Prof. Wu, who by that time was ninety years old already, wrote down every single one. He developed severe cataracts. His doctor advised him not to strain his eyes and to save his poor eyesight for his daily life. This did not make much of an impression on Prof. Wu. He continued to devote himself to his work.

The manuscript was finally completed, but I made one request. I asked Prof. Wu to write an autobiographical note for the book. He was happy to comply. One week later, Prof. Wu gave me his autobiographical note. When I read it, I was deeply moved. Prof. Wu was born in 1916. He developed an early interest in botany, inspired by his home garden and two books: *The Illustrated Book of Chinese Botany* (《植物名实图考》), written by Qing dynasty botanist Qiyong Wu, and *The Illustrated Book of Japanese Botany*. In middle school, he was fortunate to meet two Biology teachers: Mr. Shou Tang, his primary middle school biology teacher, and Mr. Yao Tang, his middle school biology teacher. They introduced him to fieldwork, and taught him to recognise plants and collect plant specimens. Profoundly inspired by his teachers, he decided to become a botanist. In 1933, he entered Tsinghua University in Beijing and received his BA in Biology four years later. By that time, the full outbreak of the War of Resistance against Japanese Aggression made it impossible to pursue research in botany. He moved, together with his University, to Kunming, Yunnan Province, which was home to the front during the war. He taught at National Southwestern Associated University and was a staff member at the Medicinal

Plant Institute of the National Ministry of Education. During the war, teaching and research conditions were very poor. Regardless, Wu still tried his best to use his botanical knowledge to serve his country. Prof. Wu and other colleagues published a book called *ICONES Plantarum Medicarum e libro Tien-Nan-Pen-Tsao Lanmaoano Tom 1* (《滇南本草图谱》) (King et al., 1945). They investigated medicinal plants from an ancient Chinese medicinal plant book called *Tien-Nan-Pen-Tsao* (《滇南本草》). At the time, medicine was badly needed but in extremely short supply. Their research, which identified medicinal plants around Kunming, was able to help. Twenty-six medicinal plants, including a new genus, *Psammosilene*, W. C. Wu et C. Y. Wu were described in the book, which included hand-drawn illustrations. Everything required for publication—writing, drawing and printing—all had to be done by themselves.

Since then, Prof. Wu started his scientific career which spanned over 70 years. Many articles have recorded his contributions to botanical studies already, here I only want to emphasise his contributions in plant taxonomy and phytogeography. Prof. Wu became the fourth editor in chief of the Editorial Committee of *Flora Republicae Popularis Sinicae* (FRPS) in 1987, and 82 books made up of 54 volumes (80 volumes in total) were published under his editorship. During the compilation of FRPS, he also chief edited another two great works: *Flora of Yunnan* and *Flora of Tibet*. In order that FRPS could be understood by non-Chinese speakers, Prof. Wu and Peter H. Raven became the co-chairs of the joint editorial committee. All 25 volumes of *Flora of China* were published in 2014, one year later Prof. Wu passed away unfortunately (Zhou and Sun, 2016).

By compiling the FRPS Prof. Wu and his colleagues answered the big question of how many plant species we have in China. His phytogeographcial research answered the next big question: "Where are these plants growing? " His contribution to phytogeography has been recorded in the above-mentioned book *The Areal-Types of Seed Plants and Their Origin and Differentiation*. He proposed about origin and differentiation of distribution pattern of many seed plants from China. This hypothesis became a big legacy to biogeographical research. Many research projects were proposed based on Prof. Wu's hypothesis (Zhou and Sun, 2016).

We hope he will be remembered in the history of botany forever.

Figures

Fig.1 Professor Zhengyi Wu (1916—2013)

Fig.2 Professor Zhengyi Wu and his students in 1988 (The first row from left: Zhengyi Wu, Zhekun Zhou, Hua Zhu; the second row from left: Jianqiang Li, Dezhu Li, Qiner Yang, Ya Tang)

Fig.3 Professor Zhengyi Wu and the author were discussing

Fig.4 The book and manuscript of Professor Zhengyi Wu

References

King LP, Wu CY, Kuang KZ, et al (1945) Icones Plantarum Medicarum e libro Tien-Nan-Pen-Tsao Lanmaoano. Tom.1., 25 spp., 26 plates.

Wu ZY (1965) The tropical floristic affinity of the Flora of China. Chinese Sci Bull 1:25–33. (吴征镒. 1965. 中国植物区系的热带亲缘. 科学通报, 1: 25–33.)

Wu ZY, Chen XQ (2004) Florae Reipublicae Popularis Sinicae. Beijing: Institutum Botanicum Kunmingense Academiae Sinicae.

Wu ZY, Lu AM, Tang YC, et al (2003b) The Families and Genera of Angiosperms in China—A Comprehensive Analysis. Beijing: Science Press.

Wu ZY, Sun H, Zhou ZK, et al (2010) Floristics of Seed Plants from China. Beijing: Science Press.

Wu ZY, Wang HS (1983) Chinese Physical Geography—Phytogeography (I). Beijing: Science Press.

Wu ZY, Zhou ZK, Li DZ, et al (2003a) The areal types of the world families of seed plants. Acta Bot Yunnanica 25(3):245–257.

Wu ZY, Zhou ZK, Sun H, et al (2006) The Areal-Types of Seed Plants and Their Origin and Differentiation. Kunming: Yunnan Science & Technology Press.

Zhou ZK, Sun H (2016) Wu Zhengyi and his contributions to plant taxonomy and phytogeography. Plant Diversity 38:259–261.

21. 蔡希陶：云南植物王国的拓荒者

1981年3月9日，我国著名植物学家蔡希陶先生（图1）因病在昆明去世，享年70岁。今天屈指算来，蔡先生离开我们已有37年了。1938年，蔡先生创立了云南农林植物研究所（简称“农林所”，是中国科学院昆明植物研究所前身）。在建所80周年之际，我们深切怀念这位研究所的创始人，对蔡先生一生扎根边疆、献身植物科学事业常怀敬意。

自学成才，与植物结缘

蔡希陶先生谱名中矩，名希陶，字侃如。1911年4月10日生于浙江东阳虎鹿镇蔡宅村乐顺堂。蔡先生于1929年9月考入上海光华大学物理系（图2）。在上海读书期间，他常到姐夫陈望道家，期间深受瞿秋白、李达、夏征农、鲁迅、冯雪峰、胡愈之等革

图1　蔡希陶先生（1911—1981）

图2　青年时期的蔡希陶先生

译者：王改变，杨永平
中国科学院昆明植物研究所，昆明650201，中国
邮箱：wanggaibian@mail.kib.ac.cn（王改变）；yangyp@mail.kib.ac.cn（杨永平）

命前辈和文学泰斗的影响。1930年9月，为避军警抓捕，经陈望道推荐，他进入北平静生生物调查所当练习生。工作期间，所长胡先骕先生常派他到北平附近采集植物标本。蔡先生刻苦钻研，自学成才，1931年即与胡先骕合作发表了《四川省唇形花科植物之研究》一文（旭文等，1993）。

云南植物王国的揭幕人

1932年2月，蔡希陶先生受北平静生生物调查所的派遣，到云南考察（图3）。他走遍云南的山山水水，历尽艰辛，冒着生命危险，深入边远山区开展植物调查和采集。1932年至1934年，共采集植物标本21000余号，其中有427个新种和不少云南新记录，揭开云南“植物王国”的面纱，为云南植物学研究作出奠基性贡献。

图3　蔡希陶先生（左）赴云南怒江地区考察（1933年）

后来，蔡希陶先生与俞德浚先生合译《农艺植物起源》（Jiang，2017），并发表豆科、蔷薇科和魔芋属等植物研究论文。考察期间，蔡希陶先生十分关注少数民族人情风俗，先后创作多篇被鲁迅先生誉为“很有气派”的文学作品。

植物科研机构的奠基者

1937年7月7日卢沟桥事变，平津危急，华北危急。北平静生生物调查所所长胡先骕与云南省教育厅厅长龚自知商定建立一个专事调查云南植物、促进云南农林经济事业发展的植物学机构。在各方支持下，蔡希陶先生受命负责具体筹办事宜，云南农林植物研究所于1938年7月在昆明北郊黑龙潭公园挂牌成立，蔡希陶先生兼任公园经理。

1940年春，云南农林植物研究所买下黑龙潭附近的一块土地建立办公室。龚自知命人书“原本山川　极命草木”之所训，刻于石碑，嵌入墙上。1941年办公楼的建成，结束了借住黑龙潭寺宇的局面，初奠云南农林植物研究所之永久基础。当时常带西南联合大学学生到农林所实习的吴征镒先生在《也是迟来的怀念》一文中把农林所誉为“那时的植物学最高学府”和“植物分类学活动中心”（吴征镒，1991）。

1945年抗日战争胜利后，迁滇的单位纷纷北归。后解放战争爆发，物价飞涨、民不聊生，云南农林植物研究所在风雨中飘摇。为维持所里十多名员工的生计，保护好十余万号植物标本，蔡希陶先生组织员工种菜、种花、种烟以自救，开设鹦鹉商店出售花木和观赏小动物等，苦撑至昆明解放。云南农林植物研究所，蔡希陶先生是见证者，也是实际上的负责人，唯一自始至终坚持到底的一个人。陈封怀先生在《忆述希陶同志一生》一文中认为，云南农林植物研究所的建立“以希陶同志为首创之功”。

1950年4月7日，云南农林植物研究所更名为中国科学院植物分类研究所昆明工作站，蔡希陶先生担任主任。他抓发展、搞建设，在周恩来总理的关怀下，工作站的园区面积得到扩大，并建设了研究所第一个集办公和科研于一体的“品字楼”。

在科学研究上，他创建了第一个以药用和香料植物为主要研究对象的植物资源化学研究室（植物化学与西部植物资源持续利用国家重点实验室前身）。后来工作站升格为中国科学院昆明植物研究所，吴征镒先生任所长，蔡希陶先生任副所长。

1959年，在年近半百的蔡希陶先生的带领下，一批被他的奉献科学的精神和高尚人格所感染的年轻科技人员和工人，自称“集体的鲁宾逊”，在一个由澜沧江一大支流——罗梭江环绕的“葫芦岛”上，创建了我国第一个热带植物园。在他的领导和身体力行的带领下，这个建立在热带林海和穷乡僻壤的植物园，克服了难以想象的诸多困难，很快在建园和科研上出成果、出人才，闻名中外。1978年，热带植物园升格为中国科学院云南热带植物研究所（中国科学院西双版纳热带植物园前身）（秦仁昌，1991）。

植物资源持续利用的先行者

蔡希陶先生在考察植物时深入民情，既向少数民族学习植物资源的利用知识，也痛惜他们过着“端着金饭碗要饭”的穷困生活。因此蔡希陶先生立志要“用植物学这门理论学科去为人民做一些有用的工作”，他成为我国植物资源持续利用的先行者，为国家和地方经济社会的发展作出了极其重要的贡献。

早在20世纪40年代，蔡希陶先生就一直关注云南烟草的生产。1945年，他通过陈焕镛先生从美国引进优良烤烟品种“大金元”，并驯化成功（Huang，2016）。后开设烟草推广培训班并提供烤烟良种，使“大金元”成为云南发展烟草生产的当家品种，为云南经济发展作出了重要贡献。

20世纪50年代，他急国家所急，组织了云南橡胶资源及其宜林地的考察研究，他提出“滇南西双版纳是发展三叶橡胶的最适宜地方”的建议，被国务院采纳，为我国建设橡胶生产基地作出了重大的贡献，他作为全国橡胶协作组成员于1982年获国家技术发明奖一等奖。

此外，他领导科技人员从野生植物资源中发掘和从国外引进了众多的重要药用植物、油料植物、香料植物和珍贵速生树种等经济植物，为我国热带区域经济社会发展提供了新资源及其发展技术。

青年植物学家的领路人

俗话说：十年树木，百年树人。蔡希陶先生在科技领导岗位上，不但运筹决策，为国为民开创了一个又一个的科研基地，推广了一项又一项的科研成果，形成巨大的社会物质生产力，而且呕心沥血培养了一批又一批的专业人才，造就了一支多学科的科研队伍。

他自己为年轻人教授专业课、英文和拉丁文，手把手指导植物考察、植物引种驯化、植物化学成分测试、植物栽培试验和论文写作等。他明确要求“科技人员到工作实践中去边干边学，不断积累和扩大知识，自己闯出一条路来”。

冯耀宗在《我的科学路上的引路人——蔡希陶老师》一文中写道：“蔡老不仅是我的启蒙老师，也是我不断前进的引路人。”蔡希陶先生一生亲贤爱才，善于引导、培养和提携后学。他是尊重知识、培养人才的模范，他的伯乐精神感人至深。

蔡希陶先生的一生是创造性的一生，艰苦奋斗的一生，也是富有成就的一生。他为“植物王国”——云南的植物学研究作出了卓越贡献。他的“献身科学”（方毅副总理题词）精神一直是西双版纳“葫芦岛”人和昆明“元宝山”人的精神支柱。

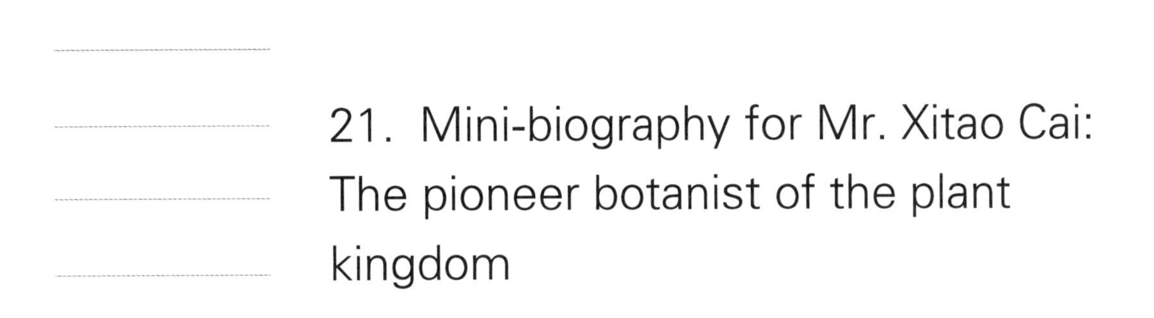

21. Mini-biography for Mr. Xitao Cai: The pioneer botanist of the plant kingdom

On March 9, 1981, Mr. Xitao Cai (Hse-Tao Tsai), one of the most famous pioneer botanists in China, passed away at his age of 70 in Kunming. It has been 37 years since he left us. Mr. Cai founded Yunnan Institute of Agricultural and Forestry Botany in 1938, the predecessor of Kunming Institute of Botany, Chinese Academy of Sciences. At the celebration of the 80th anniversary for the institute, we deeply cherish Mr. Cai, the founder of the institute, who has dedicated his entire life to the development of botanical science in China.

Mr. Xitao Cai was born in Zhejiang Province on April 10, 1911. He was admitted into the Department of Physics in Kwang Hua University in Shanghai in 1929. During his studies in Shanghai, he frequently visited his brother-in-law, Mr. Wangdao Chen and was influenced by many revolutionary predecessors and literary giants, such as Mr. Qiubai Qu, Mr. Da Li, Mr. Zhengnong Xia, Mr. Xun Lu, Mr. Xuefeng Feng and Mr. Yuzhi Hu. In September 1930, recommended by Mr. Wangdao Chen, Mr. Cai became a trainee in the Fan Memorial Institute of Biology. During this time, Professor Xiansu Hu (Hsen-Hsu Hu), the Director of the institute, often sent him to collect plant specimens near Peking. The next year, Mr. Cai published a research paper An study on Lamiaceae in Sichuan Province as a co-author with Prof. Hu (Xu et al., 1993). In February 1932, Mr. Cai was dispatched to Yunnan on a botanical expedition organized by the Fan Memorial Institute of Biology. He traveled through Yunnan, risked his life and conducted field survey and plant collection in the remote mountainous areas. During this expedition, he collected more than 21000 specimens including 427 new species, which unveiled the mask of Yunnan Province as a "Plant Kingdom" and contributed significantly to the development of the plant science in Yunnan. Later on, Mr. Cai translated the book *Origin of Cultivated Plants* written by de Candolle into Chinese, together with Mr. Dejun Yu (Te-Tsun Yu) (Jiang, 2017), and published a series of research papers on the family Leguminosaceae, Rosaceae and the genus *Amorphophallus* plants. His early works enabled the Chinese botanical community to have better understandings about the origins of crops and served as invaluable references for the following

Gaibian Wang, Quanxing Zhang, Yongping Yang

Kunming Institute of Botany, Chinese Academy of Sciences, Kunming 650201, China

Correspondence: yangyp@mail.kib.ac.cn (Y.-P. Yang)

studies till now. In addition, Mr. Cai was inspired by the customs of the ethnic minorities and wrote several essays which has been described as "very impressive" by famous Chinese writer, Mr. Xun Lu.

In 1937, the War of Resistance against Japanese Aggression broke out fully. Many institutions in Peking started to move to Southwest China. The Fan Memorial Institute of Biology planned to establish a botanical research base in Yunnan. Mr. Cai moved to Yunnan with his family. In July 1938, Yunnan Institute of Agricultural and Forestry Botany, the predecessor of Kunming Institute of Botany, was established in the Heilongtan Park, Kunming City. Mr. Cai was appointed as the park manager as well. In the spring of 1940, the institute bought a land site near the Heilongtan Park to construct the office building. And the construction of the office building was completed in 1941. The motto "Explore every mountain and river, name every grass and tree" was carved in stone and embedded into the wall. Mr. Zhengyi Wu (Cheng-Yhi Wu), who once took students from the National Southwest Associated University for internships in Yunnan Institute of Agricultural and Forestry Botany, described the institute as "the highest botanical institution in China at that time" and "a research center for plant taxonomy" in his commemorative essay to Mr. Cai (Wu, 1991).

In 1945, after the victory of the War of Resistance against Japanese Aggression, many institutions moved back from Yunnan Province. With the outbreak of the War of Liberation, prices soared and it became very difficult to live. In order to maintain the life of the employees and protect more than 100000 plant specimens, Mr. Cai organized the employees to grow vegetables, flowers and tobacco to save themselves. He opened a parrot shop to sell cutting flowers, ornamental plants and pets, and struggled till the liberation of Kunming. For the institute, Mr. Cai was not only the founder and the actual manager, but also the one who persisted from the beginning to the end. In the essay of Recalling Mr. Xitao's life by Prof. Fenghuai Chen (Feng-Hwai Chen), who worked together with Mr. Cai during 1930s—1940s. It is said that, "Mr. Cai made the most important contributions to the establishment of Yunnan Institute of Agricultural and Forestry Botany."

In April 1950, Yunnan Institute of Agricultural and Forestry Botany was merged and re-named as Kunming Station of the Institute of Plant Taxonomy, Chinese Academy of Sciences. Mr. Cai was appointed as the director of Kunming Station. With the special care from Premier Enlai Zhou, Mr. Cai focused on the development and construction. The campus of Kunming Station was expanded, three modern buildings used as administration office and research laboratories were constructed. In research fields, Mr. Cai established the first chemistry laboratory to study medicinal and aromatic plants (the predecessor of the State Key Laboratory of Phytochemistry and Sustainable Use of Plant Resources of Kunming Institute of Botany). Later on, Kunming Station was upgraded as an independent institute, Kunming Institute of Botany, Chinese Academy of Sciences. Mr. Zhengyi Wu was appointed as the Director and Mr. Xitao Cai as the deputy director.

In 1959, under the propose and leadership of Mr. Cai, Kunming Institute of Botany started to found the first tropical botanical garden of China in Xishuangbanna, South Yunnan. Mr. Cai made great contribution to the foundation of botanical garden, and all institute staff were deeply touched by Mr. Cai's scientific spirit and noble personalities. Mr. Cai and his team have overcome many unimaginable difficulties and made great achievements in the construction of the garden and scientific researches. They also cultivated many talents who later became famous scientists in China and abroad. In 1978, the tropical botanical garden was renamed as Yunnan Institute of Tropical Botany, the predecessor of Xishuangbanna Tropical Botanical Garden, Chinese Academy of Sciences (Qin, 1991).

As early as 1940s, Mr. Cai paid attention to tobacco production in Yunnan. In 1945, his team succeeded in the domestication and cultivation of Mammoth Gold in Kunming, a tobacco cultivar introduced by Mr. Huanyong Chen (Woon-Young Chun) (Huang, 2016) from Virginia, USA. Mr. Cai organized a series of training workshops for tobacco cultivar selection, demonstration and large-scale cultivation, and made earliest contributions for tobacco production in Yunnan. In the 1950s, he led an investigation team and explored the rubber resources in Yunnan. He proposed that Xishuangbanna was the most suitable place for rubber tree plantation, which was adopted by the State Council to build the rubber plantation farm in China. And as a member of national rubber research team, Mr. Cai was awarded the First Prize for State Technological Invention Award in 1982. Furthermore, he and his team explored and introduced many important medicinal plants, oil plants, aromatic and spice plants, valuable fast-growing species and so on, such as dragon blood trees, hodgsonia squash, camphor oil tree and mytenus plants. They have provided new resources and technologies for the economic and social development in the tropical areas in China.

As the saying goes, "it takes ten years to grow a tree but it takes a hundred years to rear people" . As a master in science, Mr. Cai organized many projects and researches, established lots of scientific research divisions and achieved a great deal of scientific research findings. As an educator, Mr. Cai has brought up numerous professionals and therefore built up a multi-disciplinary research team. He taught the younger generations English, Latin, botanical courses and guided them in botanical expedition, plant domestication, phytochemical analysis, plant cultivation and scientific paper writing. He clearly stated that the young scientists should learn in practice, accumulate and expand their knowledge and explore their own way to success. Mr. Yaozong Feng, one of Mr Cai's students, wrote in Teacher Cai Xitao—my guide in scientific exploration, "Mr. Cai is not only my enlightenment teacher but also the leader that guides me along all the time."

Mr. Cai has made outstanding contributions to botanical researches in Yunnan, the "Plant Kingdom" . His devotion to science will always be an inspiration for people in Xishuangbanna Tropical Botanical Garden and Kunming Institute of Botany, two botanical institutions founded by him.

Figures

Fig.1 Mr. Xitao Cai (1911—1981)

Fig.2 Mr. Xitao Cai at his young age

Fig.3 Mr. Xitao Cai (left) on the field trip to Nujiang Prefecture of Yunnan (1933)

References

Huang RL (2016) Prof Huanyong Chen: A leading botanist and taxonomist, one of the pioneers and founders of modern plant taxonomy in China. Protein & Cell 7(11):773–776.

Jiang H (2017) Dejun Yu: A patriotic botanist and his contributions. Protein & Cell 8(11):785–787.

Qin RC (1991) Trailblazer of Yunnan Botany Institute—Professor Cai Xitao. In: Wu ZY. Festschrift of Cai Xitao. Kunming: Yunnan Science and Technology Press, 13–14. (秦仁昌. 1991. 云南植物研究的拓荒者——蔡希陶教授. 见：吴征镒. 蔡希陶纪念文集. 昆明：云南科技出版社, 13–14.)

Wu ZY (1991) Belated missing. In: Wu ZY. Festschrift of Cai Xitao. Kunming: Yunnan Science and Technology Press, 10–12. (吴征镒. 1991. 也是迟来的怀念. 见：吴征镒. 蔡希陶纪念文集. 昆明：云南科技出版社, 10–12.)

Xu W, Wang ZH, Xiao G (1993) Brief Biography of Cai Xitao. Beijing: International Cultural Publishing. (旭文, 王振淮, 晓戈. 1993. 蔡希陶传略. 北京：国际文化出版公司.)

22. 邹秉文：中国农业高等教育的开拓者

图1　邹秉文（1893—1985）

邹秉文，字应崧，江苏吴县人，我国杰出的近代农业教育家、农业科教事业的奠基人、享誉国内外的著名农业问题专家（中华农学会和华恕，1993）（图1）。他是我国讲授植物病理学的第一位教授，一生致力于我国农业教育和科技的发展（Geng，2015），被周恩来总理誉为“东南三杰”之一。

1910年，邹秉文先生以驻美大使馆学习生名义赴美国纽约柯克中学读书，次年改入威里斯顿中学。1912年以优异的成绩毕业，补取为清华学校留美官费生，考入康奈尔大学学习机械工程。1913年暑假，他选读了农业课程，深受启发，便弃工从农，改读农科。1915年毕业，获农学学士学位，继续在研究生院专攻植物病理学。1916年归国，应金陵大学农林科主任芮思娄的聘请，担任金陵大学教授，主讲植物病理学。1917年应南京高等师范学校教务长郭秉文之邀，筹备南京高等师范学校农业专修科，并担任科主任。随着南京高等师范学校改组成立东南大学，邹秉文先生任首位农科主任，直至1927年离任。1929年任上海商品检验局局长，并发起成立中国植物病理学会。1931年任中央农业实验所筹备委员。在上海银行任职期间，协助民族企业家范旭东于1937年创办中国第一家硫酸铔厂。1943年起任联合国粮食与农业组织（FAO）筹备委员会副主席、FAO首任中方执行委员、南京国民政府农林部高等顾问兼驻美国代表、中美农业合作团中方团长。1945年与章之汶合编出版《我国战后农业建设计划纲要》，引起海内外广泛关注。1946年，美国密歇根大学授予其农学荣誉博士学位。1948年任美国纽约和昌公司（华侨经营）董事长，协助经营中美间农业物资贸易。1950年他从美国运回珍贵的岱字棉种，使得新中国的棉产量取得突飞猛进的发展。1956年8月，在周恩来总理直接关怀下，他返回祖国，并以一级教授身份受聘为农业部（后改为农林部、

作者：王源超

南京农业大学植物保护学院，南京 210095，中国

邮箱：wangyc@njau.edu.cn

农牧渔业部等）和高等教育部两部的顾问。

邹秉文先生是我国高等农业教育的主要奠基人。在主持南京高等师范学校至东南大学农科的10年中，确立了农科教结合的理念，编写《植物病理学概要》《高等植物学》（邹秉文等，1923）等我国近代植物病理学重要文献，发表《中国农业教育问题》《中国必须有农业工程》《中国农业建设方案》（图2）等多篇讲话、专著（邹秉文，1923，1945），对我国农业的发展提出许多重要的建议，先后为东南大学农科募得各项费用多达40余万元，培养出了金善宝、冯泽芳、邹钟琳等多位我国第一代现代农学家。

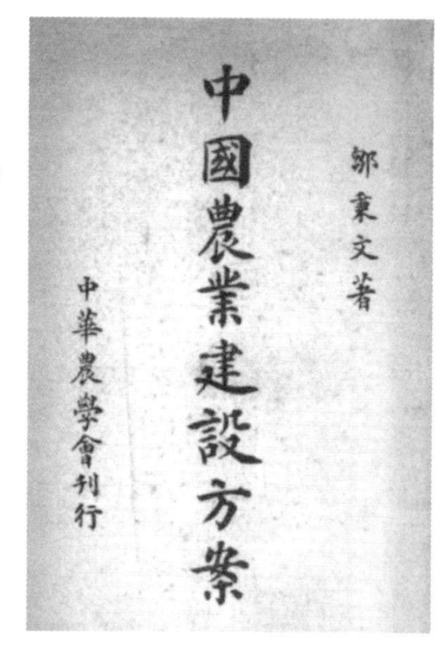

图2 《中国农业建设方案》

邹秉文先生是中美农业教育、科技交流的杰出组织者。20世纪40年代，他怀着战后复兴祖国农业的抱负，考察美国农业行政、教育及农业技术推广工作，他四处奔波，先后获得美国农业大学奖学金名额200余个，通过中华农学会选派中国各大学的农学院毕业生赴美进修农、林、牧、水利、气象、医卫等专业，其中有水利学家张含英、园艺学家章文才、农业工程学家陶鼎来等人，他们学成回国后都成为新中国建设的重要力量，影响深远。

邹秉文先生是我国近代植棉业的重要推动者。他率先在南京高等师范学校农科成立棉作改良推广委员会；开办暑期植棉讲习班，培养一批我国早期棉业改进骨干；为防治病虫害，他辗转磋商，设立我国第一个防治病虫害机构——江苏省昆虫局，开展昆虫研究和棉花、水稻害虫以及蝗虫等研究防治工作，在国内树立了科学治虫的新风，培养了邹钟琳等一批杰出的昆虫学家，拉开了我国近代农业昆虫研究的序幕。

邹秉文先生是中国科学社、中华农学会早期重要的参与者。早在1915年留美期间，与胡明复、任鸿隽起草《中国科学社总章》，成立中国科学社，编印《科学》月刊，这是我国最早的自然科学的学术团体和有影响力的学术杂志。他也是1917年成立的中华农学会（1951年更名为中国农学会）的创建人之一，1942—1948年出任该学会理事长，在战乱中让中华农学会面貌一新，在出版农学会会刊、培养人才、交流农业科学成果等方面发挥了重要的组织作用。

邹秉文先生1985年6月于北京病逝，享年92岁。他一生将振兴我国农业视为自己奋斗的目标，无论担任什么工作，始终尽心竭力，在筹办农业科研机构、培养农业人才、发展农业经济与贸易等方面，都作出了重大的贡献。1994年，遵从邹秉文先生的遗志，其子女及亲属将其收入捐赠，在南京农业大学设立“邹秉文奖学金”。2020年，南京农业大学教学楼树立了邹秉文先生铜像。

致谢

感谢南京农业大学邵刚老师、张岩老师、高俊老师和王燕老师在撰稿过程中给予的建议和对稿件的校正。

22. Bingwen Zou (Ping-Wen Tsou): Pioneer of agricultural higher education in China

Bingwen Zou, a renowned agronomist and educator, is one of the founders of agricultural higher education in China (Chinese Society of Agriculture and Hua, 1993). He is committed to set up modern agriculture education in China and supervised many students who later became well-known experts in Agriculture, including Shanbao Jin, Zefang Feng and Zhonglin Zou. Bingwen Zou made great contributions to the development of modern agriculture science in China, and was honored as one of the "Three Outstanding Scholars in Southeast" by Premier Enlai Zhou.

Bingwen Zou was born in 1893 in Jiangsu Province. He received high school education in the United States from 1910 to 1912 and continued higher education in Cornell University. During his study in Cornell University, Bingwen Zou was firstly enrolled into Mechanical Engineering. Soon afterwards, he realized that although China claims to established the country on agriculture, but the development of agriculture was declining gradually, and suffered from serious plant diseases. Therefore, he quitted from Mechanical Engineering to study Agriculture, majoring in Plant Pathology. In 1915, Bingwen Zou, together with his alumni in Cornell, initiated the first comprehensive Chinese scientific association—The Science Society of China, and compiled one of the most influential journal—*Kexue* (Science) monthly. Bingwen Zou is also one of the founders of the Chinese Society of Agriculture founded in 1917.

Bingwen Zou graduated from Cornell University in 1915 and returned to China in 1916. He was invited by Silou Rui, the director of the Department of Agriculture and Forestry at Jinling University, to teach plant pathology and botany courses at Jinling University. He emphasized the importance of quarantine and gave the earliest suggestion of plant quarantine in China in 1916. Bingwen Zou was the first person to teach plant pathology in China and made great efforts to become an excellent plant pathologist. With the deeper understanding of the problems in Chinese agricultural society, the deeper consideration to develop higher agricultural education in China. His efforts are not confined to the development of the plant pathology, but in the

Yuanchao Wang
College of Plant Protection, Nanjing Agriculture University, Nanjing 210095, China
Correspondence: wangyc@njau.edu.cn

whole construction of agricultural science by reforming and promoting Chinese agricultural education (Geng, 2015).

From 1917, Bingwen Zou became a director of Agricultural Science in Nanjing Higher Normal School. In November 1920, Bingwen Zou, together with Yuanpei Cai, Jian Zhang and Yanpei Huang, proposed the establishment of the Southeast University in Nanjing. Several departments of Nanjing Higher Normal School were transferred to Southeast University. In 1921, Southeast University was officially established with three branches: Engineering, Commerce and Agriculture. Bingwen Zou was the first director of the Agriculture department. He invited a number of famous professors such as Xiansu Hu, Chongshu Qian, Songzhou Yuan, Enlin Sun and Jubo Zhang to teach at the University. He questioned and refused to use Japanese, European and American textbooks as teaching materials for Agricultural science in Universities. He took the lead to compile the first college textbook of botany in Chinese—*Advanced Botany*—in 1923 (Zou et al., 1923). In addition, Bingwen Zou pointed out that lack of field practice was a major problem for Chinese agricultural education. He reformed the traditional teaching mode in the University from textbook based knowledge learning to a mode combining teaching with research and practice.

Bingwen Zou initiated the establishment of the Cotton Improvement and Promotion Committee under the Agricultural department of Southeast University, and took the lead in holding the Summer Cotton Planting Workshops in Southeast University. Bingwen Zou also initiated the establishment of Jiangsu Insect Bureau in 1922 to carry out research of insect and control of cotton, rice pests and locusts, which set up a new trend to scientifically control insect in China. He trained a group of renowed entomologists such as Zhonglin Zou, and opened the prelude to modern agricultural insect research in China.

By the time he left the director position at Southeast University in 1927, the agriculture science had been well developed, with seven departments containing Agronomy, Horticulture, Plant diseases and insect pests, Animal husbandry, Agricultural chemistry, Biology and Sericulture, and nine research stations for rice, sericulture, horticulture and cotton in Jiangsu, Henan, Hubei and Hebei Provinces, making the Agricultural department of Southeast University the center of agricultural science in China at that time.

Bingwen Zou raised the *Chinese Agricultural Construction Plan* in 1945 by virtue of his advanced scientific ideas and the practical problems in Chinese agriculture, which attracted extensive attention in China and abroad (Zou, 1945). Six months after the book was published, the University of Michigan awarded Bingwen Zou an honorary doctorate in 1946 in recognition of his contributions to Chinese agricultural development.

From 1943 to 1947, during Bingwen Zou served as the agricultural representative of the Government in the United States. He also served as the chief editor of the monthly magazine—*China Agriculture* (in English), where he openly discussed and collected suggestions on the development of Chinese agriculture. Moreover, he was engaged in persuading universities/institutions of the United States to provide scholarships and opportunities for agricultural graduate students in China. He obtained the opportunities for up to 200 graduate students for internships in universities/institutions in the United States. As the co-founder of early agricultural higher education in China, Bingwen Zou wrote the first monograph on agricultural education in China, *Agricultural Education in China* (Zou, 1923). This is based on his experience of teaching Agriculture in Jinling University and Southeast University, extensively investigating the advantages and disadvantages of agricultural education throughout the country, and referring to the precedents, such as Japan, the United States, Denmark and other countries. He pointed out a key to improve agricultural education is recruiting specialized teachers in agriculture. He mentioned that Agriculture field is so complicated that the experts can only be good at one or two aspects. The experts are often named as crop specialist, horticulturist, animal husbandry specialist, plant disease specialist, and so on, but not agricultural expert. Bingwen Zou emphasized that an expert must have expertise and be able to solve practical problems.

In 1994, following Bingwen Zou's legacy, his family donated all his savings, to set up "Zou Bingwen Scholarship" in Nanjing Agricultural University to support talent undergraduate students. Throughout his life, Bingwen Zou devoted himself to promote the development of Agriculture in China and serve the country.

Figures

Fig.1 Bingwen Zou (1893—1985)

Fig.2 *Chinese Agricultural Construction Plan*

References

Chinese Society of Agriculture, Hua S (1993) Zou Bingwen Memorial Collection. Beijing: Agricultural Press. (中华农学会,华恕. 1993. 邹秉文纪念集. 北京: 农业出版社.)

Geng X (2015) Serving China through agricultural science: American-trained Chinese scholars and "Scientific Nationalism" in decentralized China (1911—1945). PhD dissertation, University of Minnesota.

Zou BW (1923) Agricultural Education in China. Shanghai: The Commercial Press. (邹秉文. 1923. 中国农业教育问题. 上海: 商务印书馆.)

Zou BW (1945) Chinese Agricultural Construction Plan. Journal of Chinese Society of Agriculture. (邹秉文. 1945. 中国农业建设方案. 中华农学会刊.)

Zou BW, Hu XS, Qian CS (1923) Advanced Botany. Shanghai: The Commercial Press. (邹秉文, 胡先骕, 钱崇澍. 1923. 高等植物学. 上海: 商务印书馆.)

23. 辛树帜：中国生物学先驱、古农学家和教育家

图1　辛树帜（1894—1977）

辛树帜（图1）是我国生物学先驱，也是一位古农学家，他最早在广西和广东交界的大瑶山开展生物学考察，发现20多个新物种。他是中国植物学会和中国动物学会创始人和发起人之一。辛先生也是一位伟大的教育家。作为最早将科学方法引入农业史研究的学者之一，他参与筹建国立西北农林专科学校（现西北农林科技大学）和兰州大学。他将自己的一生都奉献给了西北地区的高等教育事业。

1894年，辛树帜出生在湖南省临澧县一个贫穷的农民家庭，9岁求学私塾。1910年，他考入湖南西路公立师范学堂（现常德一中）。1915年秋，考入国立武昌高等师范学校（现武汉大学），开始学习生物学，毕业后任教于长沙明德中学和湖南省第一师范学校。1924年至1927年，他自费赴欧洲留学。他先到伦敦大学学习，之后前往柏林大学，师从植物分类学家Diels博士。Diels博士对中国植物分类研究也很熟悉，他告诉辛树帜广东和广西这些区域的动植物分类工作还是空白，将来他可以做一些工作来填补这一空白。

1927年，辛树帜学成回国。在傅斯年推荐下，辛先生成为中山大学生物系教授和主任。受其老师Diels博士的影响，1928年5月辛树帜主持了广西大瑶山科考，期间他和同事共收集标本3万余份。同年11月，他和同事在大瑶山进行了更大规模野外考察，此次考察一直持续到1929年2月。此后不久，辛树帜组织了大瑶山第三次科学考察，开创了我国大规模野外科考和生物采集的先河。他们调查收集范围远超大瑶山地区，来到了贵州云武山、豆棚山、梵净山，湖南南部金童山，广东北江、永昌和瑶山，以及海南等地。

在野外考察中，辛先生和他同事收集超过6万份标本，包括3万份植物标本，40种

作者：谢群，程燕平，付雷
浙江师范大学教师教育学院，金华321000，中国
邮箱：xiequn@zjnu.cn（谢群）

哺乳动物100余份标本，大约40种爬行动物500份标本，超过2000份鸟类标本，以及大量的其他动物标本。除展示华南地区丰富的动植物外，他还发现近20个新物种，如鄂蜥（*Shinisaurus crocodilurus*）、辛氏木（*Sinia rhodoleuca*）、辛氏铠兰（*Corybas sinii*）等。基于他们的考察成果，中山大学建立了一个比较完整的动植物标本室，成为吸引和培养大批从事动植物研究的专业人才的重要基地，同时也提升了中国在国际生物学界的学术影响力。

在大瑶山的研究中，辛树帜和同事打破科学与人文之间的隔阂，突破性地将科学方法用于对瑶族生活习俗的调查。基于对村民实地走访和访谈，绘制村落地图。此外，辛树帜还撰写大量有关当地风俗习惯的笔记，并编撰《瑶山两月视察记》《正瑶舞歌》《甲子歌》《瑶山采集日程》等大量民族民俗学资料。通过他们的工作，瑶族人民及其生活被外面的世界所认识和接受。

1932年，辛树帜任国民政府教育部编审处处长。次年，编审处扩充为国立编译馆，辛先生任首任馆长。他认为科学术语规范化是我国自然科学发展的重要基础，因此编译馆开始组织翻译和编辑化学、天文学、地质学、医学、物理学和其他领域的术语。为促进这项工作，辛先生和同事积极招募了一批专家，并组织一系列学术会议。20世纪30年代初的科学术语规范化为我国科学发展作出至关重要的贡献。

辛先生认识到建立科学团体的重要性，他与很多学者都保持良好的关系。1933年仲夏，他参加了中国植物学会的筹备工作。作为植物学会的创始人和发起人之一，辛先生参加了在重庆举行的中国植物学会成立大会，并被推荐为学会理事（孙启高，2003）。中国植物学会的成立标志着我国现代植物学发展进入一个新阶段。1934年，辛先生、秉志先生、薛德育先生等人发起成立中国动物学会，辛先生被提名为副会长。中国植物学会和中国动物学会的成立为生物学家相互交流提供了重要的平台。

1932年，辛先生到洛阳和我国西北部其他城市调研，他发现那里没有高等农业教育相关的学校，这一局面对于西部发展很不利。通过调研，辛先生还发现陕西省武功县是我国古代农业的重要发祥地，他主张在武功县建立农业大学。1932年底，他在武功县参与创办国立西北农林专科学校。1936年7月，学校筹建工作结束。尽管西北地区条件恶劣，他仍毅然辞去国立编译馆馆长职务，来到武功县担任国立西北农林专科学校校长，后又担任西北农学院院长一职。任职期间，他每周定期为师生举办讲座，倾听他们的心声，早上与学生一起运动等。因此，他不仅对学校的教学和科研情况非常熟悉，而且与学校的师生关系融洽。他的许多学生传承着老院长的教诲，后来成为农业领域的中坚力量。

1940年，辛先生离开陕西前往重庆，担任国立中央大学教授兼主任导师。与此同时，他还被选为国民参政会参政员。后来由于母亲生病，他辞去各种职务回到家乡照顾母亲。在乡下的这段时间，他深刻感受到我国经济和文化的落后。他说服当地政府支持教育，并参与筹建湖南第十四中学、湖南诩武中学和九澧中学。1945年，辛先生当选湖南省教育会会长。

1946年3月26日，辛树帜被任命为兰州大学校长，负责兰州大学的筹建工作。他认为，在兰州办一所由五个学院组成的综合性大学对提高整个西北地区教育水平至关重要。为了让兰州大学成为西北教育之都，他首先关注学院的建立，强调开办兽医学院、藏语系和俄语系等院系，通过培养高素质的毕业生来满足西北地区经济和文化发展的迫切要求。其次，他成功吸引学者来兰州大学工作。由于当时西北地区交通和生活条件差，很少有学者愿意移居兰州。为此，他专程前往北平、上海和广东等地拜访各方学者，并邀请他们到兰州大学担任客座教授或短期讲师。由于辛先生的声誉以及与学者良好的关系，兰州大学吸引了来自全国各地的大量学者。辛先生第三个贡献是购买大量书籍和教学设备。在他的推荐下，兰州大学图书馆收集大量中外书籍和学术期刊（共10万多册），仅用三年时间，其藏书量就跃居西北高校前列。

中华人民共和国成立后，辛先生返回西北农学院担任院长一职。他深感调查我国农业遗产的重要性。1952年，他成立古代农学研究小组，组建了包括石声汉、夏纬瑛、周尧、崔允禔等学者在内的研究团队。虽然当时没有经费支持，但是团队所有成员都努力工作。1965年，他们的努力得到政府认可，西北农学院正式成立古农学研究室（图2）。他们研究整理古代农业文献，同时关注水土保持问题。他们研究了一批古代农业方面的文献并积累了丰富的经验，特别是对文献的真实性、来源考证、标点、分节、注释、翻译等方面进行了深入的探索（张曦堃和卜风贤，2012）。在辛先生主持下，研究室克服重重困难，鉴定了数十本古籍并出版数百万字专著，受到国内外学者的高度评价和赞扬。

图2　辛先生于1973年在古农学研究室工作

辛先生的研究硕果累累，其中包括《中国果树史研究》《禹贡新解》《易传的分析》和《〈农政全书〉159种栽培植物的初步探讨》等。根据西周至唐末有关果树记载，《中国果树史研究》对我国最早果树栽培成果进行了系统的分析和比较，考证了西周至唐末的果树种类和名称。在《禹贡新解》中，辛先生对《尚书·禹贡》中记载的“水和土壤”以及“贡品”如梳子、包等进行科学考证，树立古代农学研究典范。晚年，辛先生十分重视我国特别是西北地区的水土保持问题。发表《我国水土保持的历史研究》，主持编写《中国水土保持概论》。为研究植被破坏和土壤侵蚀积累第一手资料，他在陕西、云南等地进行实地考察。由于工作过度劳累，80岁的他从云南调研回来时生病了。1977年10月24日，辛先生与世长辞。

作为生物学家、古农学家和教育家，辛先生对工作和研究始终充满热情。他的工作总是富有远见和开创性。他深爱着祖国，并把自己的一切都献给了西北地区的科学和教育事业。

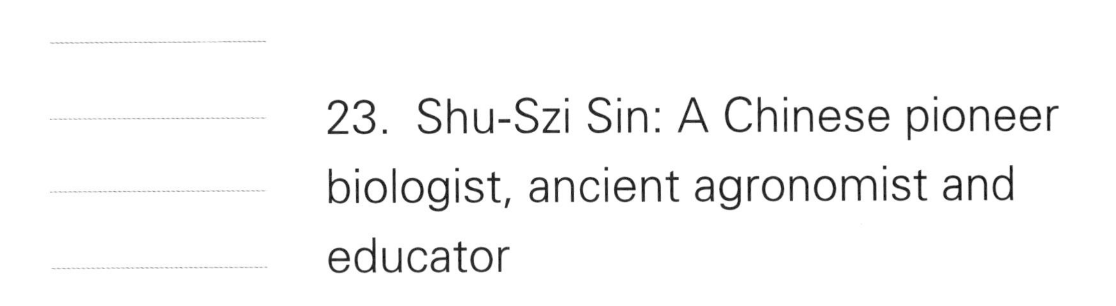

23. Shu-Szi Sin: A Chinese pioneer biologist, ancient agronomist and educator

Shu-Szi Sin (辛树帜, 1894—1977), a Chinese pioneer biologist and ancient agronomist, initiated the earliest scientific biological investigation of the Dayao mountain in both Guangxi and Guangdong, discovering more than 20 new species that have been acknowledged internationally. He was one of the founders and initiators of the China Botanical Society and the China Zoological Society. Mr. Sin was also a great educator. Among one of those who first introduced the scientific method into agricultural history research, Mr. Sin helped to set up the National Northwest Agriculture and Forestry College (now Northwest A&F University) and the Lanzhou University. He devoted himself to the development of higher education in Northwest China.

Shu-Szi Sin was born in a poor peasant family in Linli County, Hunan Province in 1894. At the age of 9, he went to study in a traditional Chinese private school.① In 1910, he entered the Hunan West Road Normal Academy (now Changde No. 1 Middle School). In the autumn of 1915, he was admitted to the National Wuchang Higher Normal School (now Wuhan University) and began to study biology. After graduation, he taught at the Changsha Mingde Middle School and the Hunan Provincial First Normal School. From 1924 to 1927, he went to Europe to study at his own expenses. At first, he studied at the London University. Then, he went to the University of Berlin and became a student of Dr. Diels, an expert in plant taxonomy who was also familiar with plant taxonomy in China. Dr. Diels told Sin that he could contribute to the classification of animals and plants in Guangdong and Guangxi, as no research studies had been conducted since then on this field.

In 1927, Shu-Szi Sin returned to China. In Ssu-Nien Fu's (傅斯年) referral, Mr. Sin became a professor and Director of the Department of Biology of the Sun Yat-sen University. Influenced by Dr. Diels, Shu-Szi Sin presided over the scientific investigation of the Dayao mountain in Guangxi together with his colleagues in May 1928, where they collected more than

Qun Xie, Yanping Cheng, Lei Fu

Teacher Education College, Zhejiang Normal University, Jinhua 321000, China

Correspondence: xiequn@zjnu.cn (Q. Xie)

① Traditional Chinese private school (私塾) is a kind of small scale school which was run and taught by educated persons. The schools are set by the teachers in their own homes. The students are mostly children aged six to eight nearby.

30000 specimens. In November of the same year, he and his colleagues conducted a larger field study at the Dayao mountain, which lasted until February 1929. Soon after, Mr. Sin led a third scientific investigation of the Dayao mountain, opening a precedent for large-scale scientific investigation and biological collection in China. The scope of its investigation and collection far exceeded the Yaoshan area, involving the Yunwu, Doupeng, and Fanjing mountains in Guizhou; the Jintong mountain in southern Hunan; Beijiang, Yongchang and Yaoshan in the Guangdong Province; and the Hainan Province.

Mr. Sin and his colleagues collected more than 60000 specimens, including 30000 specimens of plants, over 100 specimens of 40 species of mammals, about 4000 specimens of 210 species of birds, around 500 specimens of 40 species of reptiles, and more than 2000 specimens of birds. Apart from revealing the treasures of flora and fauna in southern China, he also identified nearly 20 new species, such as *Shinisaurus crocodilurus*, *Sinia rhodoleuca*, and *Corybas sinii* etc. Based on their investigations and collections, a relatively complete animal and plant specimen room was established in the Sun Yat-sen University, which became an important basis and resource to cultivate and attract a large number of specialized personnel engaged in animal and plant research, enhancing the academic influence of China in the international biology community.

In the Dayao mountain study, Mr. Sin and his colleagues broke the barriers between science and humanities, by using breakthrough scientific research methods in the humanities to investigate and report the life and customs of the Yao minority. Based on field visits and interviews with villagers, a map of the village was drawn. In addition, Mr. Sin also produced a considerable quantity of notes on local customs and habits, and compiled a large number of ethnic folklore materials such as the Yaoshan two-month inspection record, the Zheng yao dance song, the Ki zi ge, and the Yaoshan collection schedule. Thanks to their efforts, the people and the life of the Yao minority were introduced to, and accepted by, the outside world.

In 1932, Mr. Sin served as the head of the editorial department of the Ministry of Education of the National Government. The next year, the editorial office was expanded as the National Translator-Editor Center, of which Mr. Sin was the first director. He believed that standardizing the scientific terminology was an important foundation for the development of natural sciences in China. Accordingly, the National Translator–Editor Center began to translate and edit the terminology in chemistry, astronomy, geology, medicine, physics, and other fields. In order to facilitate this work, Mr. Sin and his colleagues also enrolled a number of experts and organized a series of academic conferences. The standardization of the scientific terminology in the first years of 1930s gave a crucial contribution to the development of science in China.

Realizing the importance of building a scientific community, Mr. Sin built good relationships with other scholars. He was one of the founders and sponsors of the Botanical Society of China, whose preparation and initiation began in mid-summer in 1933. As one of the initiators, Mr. Sin attended the inaugural meeting in Chongqing, when the Constitution of the Botanical Society of China was adopted, and Mr. Sin was recommended as a key member (Sun,

2003). The establishment of the Botanical Society of China marks the beginning of a new stage in the development of modern botany in China. In 1934, Mr. Sin, Mr. Zhi Bing (秉志), Mr. Deyu Xue (薛德育) and others initiated the establishment of the China Zoological Society, of which Mr. Sin was nominated as Vice Chairman. The establishment of both the Botanical Society and the Zoological Society provided important platforms for communication among biologists.

In 1932, Mr. Sin visited Luoyang and other cities of Northwest China, where he found the absence of higher agricultural schools. Therefore, he felt that he should do something to improve the situation in that region. Mr. Sin believed that Wugong, a city in the Shaanxi Province, was an important birthplace of ancient Chinese agriculture; for this reason, he advocated the establishment of agricultural colleges in that city. At the end of 1932, he participated in the founding of the Northwest Agriculture and Forestry College in Wugong. In July 1936, the preparation work for the Northwest Agriculture and Forestry College was over. In spite of the harsh conditions in the northwest, he resigned from the position of Director of the National Translator–Editor Center, serving as President of the Northwest Agriculture and Forestry College and, later, as Dean of the Northwest Agricultural College. During that time, he regularly held lectures with teachers and students every week, listening to opinions, and insisting on doing exercises with students in the morning. Therefore, he was not only familiar with the teaching situation, but also knew well teachers and even most students. Many of his students later became the backbone of the agricultural field. The teachings of the old Dean were appreciated.

In 1940, Mr. Sin left Shaanxi and became a professor and Director of the Central University in Chongqing. He was also elected as member of the National Political Participation. Due to the illness of his mother, he resigned his various duties and returned home village to take care of his mother. During his stay in the countryside, he was deeply impressed by its economic and cultural backwardness. He persuaded the government to support education, and participated in the preparation of the Hunan 14th Middle School, the Xuwu Middle School, and the Jiuli Middle School in the Hunan Province. In 1945, Mr. Sin was elected as President of the Hunan Provincial Education Association.

On March 26, 1946, Shu-Szi Sin was appointed by the government of the People's Republic of China as principal, responsible for the preparatory work of the Lanzhou University. He believed that running a comprehensive university, composed of five colleges, in Lanzhou was crucial to improve the education of the whole northwest. To make Lanzhou University the capital of northwest education, he first focused on the establishment of the faculty, emphasizing the need to run the veterinary college, the Tibetan language department, and the Russian department, among others, and to train qualified graduates that were urgently needed for the development of the northwest economy and culture. Secondly, he managed to attract scholars at the Lanzhou University. Due to the poor transportation and living conditions of Northwest China at that time, few scholars were willing to move to Lanzhou. Mr. Sin visited some scholars in Beijing, Shanghai, and Guangdong, and invited them to the Lanzhou University as visiting professors or short-term lecturers. Thanks to his good reputation and good relationships with other scholars, the Lanzhou

University attracted a number of scholars around China. His third contribution was the purchase of a large quantity of books and teaching equipment. With his recommendation, a large number of Chinese and foreign books and academic journals (totally more than 100000 volumes) were collected in the Lanzhou University Library, which in just three years jumped at the top of the northwest colleges and universities.

After the founding of the People's Republic of China, Mr. Sin returned as Dean of the Northwest Agricultural College. He deeply felt the importance of investigating the agronomic heritage of China. In 1952, he established the ancient agronomy research team, which included scholars such as Shenghan Shi (石声汉), Weiying Xia (夏纬瑛), Yao Zhou (周尧), and Yunti Cui (崔允禔). Although there was no funding support at that time, all team members were working hard. In 1965, their efforts were recognized by the government, and the Ancient Agricultural Research Laboratory was established at the Northwest Agricultural College. They studied and sorted the ancient agricultural literature, and addressed the issue of soil and water conservation. They gained experience in studying the ancient agricultural literature, especially in aspects such as the identification of authenticity, the examination of the origins of the literature, and the punctuation, subsection, annotation, and translation (Zhang and Pu, 2012). Under Mr. Sin's auspices, the laboratory overcame many difficulties, examining dozens of ancient books, publishing monographs of millions of words, and was highly valued and praised by scholars at home and abroad.

Mr. Sin's research outcomes include: The Study of Chinese fruit trees history; the New analysis of Yu Gong; the Study on the book of *Change*; and the Preliminary discussion on 159 cultivated plants in the book of *Agricultural Administration*. The Study of Chinese fruit trees history systematically analyzes and compares the achievements of the earliest fruit trees cultivation in China, according to the records of fruit trees from the Western Zhou Dynasty to the end of the Tang Dynasty, and carefully examines the types and names of fruit trees during this period. In the New analysis of Yu Gong, Mr. Sin conducts a scientific investigation of "water and soil" and "tribute" such as comb and bag which recorded in *Shang Shu* · Yu Gong, an ancient article describing the geography, establishing an example of ancient agronomy research. In his later years, Mr. Sin paid great attention to the issue of soil and water conservation in China, especially in the northwest region. He published the *Historical Research on Soil and Water Conservation in China*, and presided over the preparation of *The Introduction to Soil and Water Conservation*. In order to collect first-hand information on vegetation damage and soil erosion, he conducted field studies in Shaanxi, Yunnan, and other places. Having reached the age of eighty, he was overworked. When he returned from the Yunnan investigation, he fell ill. He passed away on October 24, 1977.

As a biologist, ancient agronomist, and educator, Mr. Shu-Szi Sin was always passionate in his work and research. His work was visionary and groundbreaking. He loved his motherland, and devoted all himself to the development of science and education, especially in Northwest China.

Figures

Fig.1 Shu-Szi Sin (1894—1977)

Fig.2 Mr. Shu-Szi Sin working in the Ancient Agricultural Research Laboratory in 1973

References

Sun QG (2003) Historical retrospect of the beginning years of the Botanical Society of China. In: Botanical Society of China. Compilation of Abstracts of the 70th Anniversary of the Botanical Society of China (1933—2003). Beijing: Higher Education Press. (孙启高. 2003. 中国植物学会成立初期的历史回顾. 见：中国植物学会. 中国植物学会七十周年年会论文摘要汇编(1933—2003). 北京: 高等教育出版社.)

Zhang XK, Pu FX (2012) Shu-Szi Sin and the Chinese agricultural history research. Agric Archaeol 6:273–277. (张曦堃, 卜风贤. 2012. 辛树帜与中国农史研究. 农业考古, 6: 273–277.)

24. 段永嘉：杰出的植物病理学家和农业教育家

段永嘉先生于1910年3月在吉林省四平市出生。1931年他收到日本北海道大学农学部农业生物学科植物病理专业的录取通知，随后获得庚子赔款留学。1937年，他怀着科技强国的心愿，冒着抗日战争的炮火，携日籍妻子——小学教师奥村里惠子，毅然离开日本回到热爱的祖国。他先后在江苏教育学院、广西大学、湖南省农业专科学校任教。

1944年底，段永嘉先生经云南大学农学院院长张海秋先生介绍，到云南大学农学院农艺学系任教。在此期间，段永嘉先生与随清华大学来昆明的植物病理学界老前辈戴芳澜、俞大绂教授，全国知名的小麦专家金善宝，生物统计专家汪厥明，林学家郑万钧，蚕桑专家陆星垣等在患难与共的教育生涯中结下了深厚的友谊。中华人民共和国成立前后，段永嘉先生担任云南大学农学院农艺学系系主任。1958年云南大学农学院独立建院成立昆明农林学院，1970年扩展为云南农业大学，段永嘉先生一直担任教授。1976年他出任云南农业大学植保系主任。1989年退休后，他仍没有离开教育和科研岗位，直至95岁逝世。他是云南高等农业教育的创立者，也是终身的建设者。

段永嘉教授是我国植物病理学著名教授，也是云南省植物病理学科的奠基人。1945年抗日战争胜利后，曾经云集于昆明的知名专家纷纷选择离开昆明，致使昆明高校教师奇缺，段永嘉先生临危受命，担任云南大学农艺学系系主任。为了广聘教师，他四处奔波，好事多磨，终于使云南高等农业教育得以恢复和发展。在这期间他在农艺学系开设了植物病理学课程，随后发展为现在的云南农业大学的植物病理学科。当时的教学条件极为艰苦，教材奇缺，1947年他编写出版了《植物病原菌学》，填补了当时国内植物病理教科书的空白（段永嘉，1947）。他还积极组织系里的师生采集病害标本（图1），创建了病害标本室，有的标本至今还保存完好。1957年，云南西双版纳的水稻暴发了一种水稻疑难病害，发病的水稻叶片发黄、生长明显迟滞，段先生通过深入调查和鉴定，不仅弄清了水稻黄矮病的病因，还为当地政府提供了防治措施，有效地控制了黄矮病的发生和蔓延。同年，他报道了30余种作物70种病原情况（段永嘉，

作者：杜飞[1,2]，邓维萍[1,2]，何霞红[1,2]，蔡红[1,2]，朱有勇[1,2]

1　云南农业大学云南生物资源保护与利用国家重点实验室，昆明650201，中国

2　云南农业大学教育部农业生物多样性与病害控制重点实验室，昆明650201，中国

邮箱：yyzhu@ynau.edu.cn（朱有勇）

图1 段先生在田间采集标本、调查稻瘟病

1957）。1958年，他第一次较全面系统地调查并报道了云南省主要作物病害（段永嘉，1958）。

段先生长期从事水稻白叶枯病和稻瘟病的研究，他带领的课题组报道的马唐等六种杂草根茎部可以带菌传播水稻白叶枯病（云南农业大学农学系病理教研组，1976a），这是国内首次发现白叶枯病传染源。这一成果开始不为某些“权威”承认，经过激烈的学术争论，随后江苏、浙江、安徽同行在研究中也得到相同结果，证实了其正确性。这一结果引起国内外植物病理学界的重视，编写进大学教材中。

当时国内外的研究学者对稻瘟病菌生理小种的研究还存在着不同看法，主要是根据菲律宾的一些研究，他们认为稻瘟病菌变异性太大，研究生理小种是毫无意义的，这种观点在植物病理学界引起很大的混乱。段永嘉先生有针对性地对稻瘟病菌的变异性进行了细致的研究，试验结果证明，稻瘟病菌和其他病菌一样虽有变异，但也有相对的稳定性，这种客观辩证的学术思想使稻瘟病菌生理小种研究这一重大课题得以肯定和继续，并取得了重大突破。他筛选出7个水稻鉴别品种和4个菌系，弄清了云南稻瘟病菌主要生理小种和分布，并对稻瘟病抗性基因分析及遗传规律的研究做了开拓性的工作（云南农业大学农学系病理教研组，1976b）。此外在植物病毒方面也做了大量研究（段永嘉等，1979）。

段先生很重视国际科技交流。1981年他受组织委托，邀请三位日本著名植物病理学家来昆明讲学，来自全国各地的244名农业科技人员参加学习。他还亲自翻译讲义，并担任口语翻译，受到一致好评。他还亲自联系组织各种大型国际学术交流活动7次，为中日植物病理学界搭起了友谊之桥。

他先后获得国家级和省部级科研成果奖7项，曾获全国农业劳动模范、国务院政府特殊津贴之殊荣。他曾担任中国植物病理学会顾问、西南区分会理事长，云南省植物保护学会理事长等职位。段先生为云南省高等农业教育和植物病理学科的发展倾注了全部心血，作出了杰出的贡献。

段永嘉教授非常重视学科建设。云南农业大学植物病理学科从创立至今，从实验室设备采购计划、发展规划、科研方向、学科梯队建设到教学和科研水平的提升无不包含着先生一生的心血。特殊时期，科研条件极其简陋，他还带着植物病理学科教师自己搭建简易温室开展科学研究，和大家一起到农村进行病害调查，借用云南省安宁市农科所农技站的剩余房间开展蚕豆茎枯病研究，连夏季超过40 ℃的元江和西双版纳都曾有过段先生的足迹。引进电子显微镜是植物病理学科师生多年的愿望，老先生在80岁高龄时还四处奔波寻求上级领导的支持，最终师生终于用上了日本生产的透视电镜（图2）。很多老师表示，他们在使用这台电子显微镜时，都会不由自主地想到段先生，都会倍加珍惜。

图2 段先生指导学生使用电子显微镜

段先生注重学术梯队建设，并为加强云南农业大学植物病理学学术团队付出了巨大的努力。他一贯治学严谨，严于律己，宽以待人，关心年轻人，爱惜人才，甘当人梯。在他的带动和影响下，植物病理学科成为团结、上进、充满生机的集体。这种传统已经传递了几代人（图3）。他的第一届硕士毕业生朱有勇就秉承了导师的优良思想作风和学术思想，开创生物多样性防治植物病害的理论，并在云南、贵州、四川等省一亿多亩农田上应用推广。段先生对朱有勇从本科到硕士，从助教到教授，从重点实验室主任到校长再到中国工程院院士的成长付出了太多。段先生在培养研究生工作中，以下几方面较为突出：一是重视思想教育，要求研究生德才兼备；二是言传身教，培养进取精神；三是科研教学密切结合，以广促深，以点带面；四是身先士卒，把好论文质量关（图4）（朱有勇和肖火根，1986）。朱有勇院士仅是一个代表，段先生还有许多在农业教学、科研战线上作出贡献的弟子。

段先生在植物病理学领域工作了54年，为我国植物病理学的发展作出了杰出贡献。虽然现在段先生已经离开了我们，可他那光辉的形象将永远留在我们云南农业大学植物保护专业的人的记忆中，无论是其高尚的道德情操、无私的奉献精神、严谨的治学作风，还是慈悲仁厚的包容胸襟，都永远值得我们学习！

图3　段先生（前排左三）与其指导的研究生在毕业论文答辩会后合影

图4　段先生（右一）指导学生研究

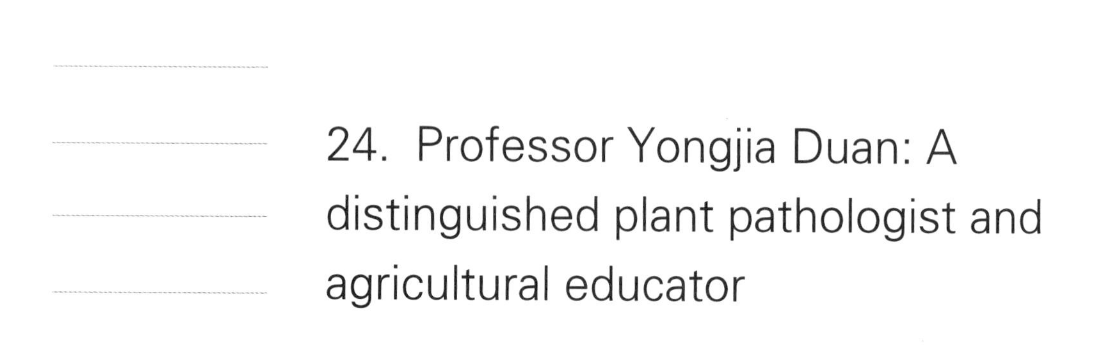

24. Professor Yongjia Duan: A distinguished plant pathologist and agricultural educator

Professor Yongjia Duan (段永嘉) is a distinguished plant pathologist and agricultural educator in China. He has devoted his entire life in developing and sharing his knowledge and experiences in Plant Pathology.

Prof. Duan was born in 1910 in Siping, Jilin Province. In 1931, he went to Hokkaido University in Japan to study plant pathology, where he obtained his bachelor's degree. In Japan, he met Ms. Hiroko Omura, a primary school teacher. They fell in love and got married. In 1937, Prof. Duan and his wife went back to China, hoping that he could contribute to the development of science and technology in China.

Prof. Duan had teaching experience in several universities, such as Jiangsu Educational College, Guangxi University and Hunan Agricultural College. At the end of 1944, he was introduced to the Agricultural College at Yunnan University by Prof. Haiqiu Zhang (张海秋), the Dean of the Agricultural College. During that period, Prof. Duan developed deep friendships and scientific relationships with Profs. Fanglan Dai (戴芳澜) and Dafu Yu (俞大绂), the pioneers in the field of plant pathology; Prof. Shanbao Jin (金善宝), a famous wheat expert; Prof. Jueming Wang (汪厥明), a biostatistician; Prof. Wanjun Zheng (郑万钧), a forestry scientist and Prof. Xingyuan Lu (陆星垣), a sericulture expert. After the founding of the People's Republic of China, Prof. Duan served as the dean of the Agricultural College at Yunnan University. In 1958, the Agricultural College became independent of Yunnan University as Kunming Agriculture and Forestry College, and later on developed as Yunnan Agricultural University in 1970. In 1976, Prof. Duan became the head of the Department of Plant Protection at Yunnan Agricultural University. As the founder and lifelong contributor to higher agricultural education in Yunnan, even retiring in 1989, he never left education and research positions until he passed away at the age of 95.

Fei Du[1,2], Weiping Deng[1,2], Xiahong He[1,2], Hong Cai[1,2], Youyong Zhu[1,2]

1 State Key Laboratory for Conservation and Utilization of Bio-Resources in Yunnan, Yunnan Agricultural University, Kunming 650201, China

2 Key Laboratory for Agro-biodiversity and Pest Control of Ministry of Education, Yunnan Agricultural University, Kunming 650201, China

Correspondence: yyzhu@ynau.edu.cn (Y.-Y. Zhu)

Prof. Duan is a well-known plant pathologist in China and the founder of the Department of Plant Pathology in Yunnan Province. In 1945, after the Second World War, lots of scientists and experts left Kunming, leading to a shortage of college teachers there. As the head of the Department of Agriculture, Prof. Duan actively looked for teachers to fulfill this shortage, and restored the development of higher education in Agriculture in Yunnan. He also started the course of plant pathology in the Department of Agriculture, which is now the plant pathology discipline of Yunnan Agricultural University. Due to the lack of teaching materials, he wrote and published *Phytopathogenic Bacteriology* in 1947, which was one of the earliest plant pathology textbook in China (Duan, 1947). He also encouraged the teachers and students to collect different disease specimens and set up places to preserve disease specimens. Till now, some specimens are still well preserved. In 1957, a puzzling disease caused yellow leaves and growth retardation in rice in Xishuangbanna, Yunnan Province. After a thorough investigation, Prof. Duan identified the cause of the disease, provided prevention and control methods to the government and effectively controlled the spread of the diseases. In the same year, more than 30 kinds of crops and 70 pathogens were reported (Duan, 1957). In 1958, the first comprehensive and systematic survey was reported on major crop diseases in Yunnan Province (Duan, 1958).

Prof. Duan was also engaged in the researches of rice bacterial blight and rice blast and found the transmission source of rice bacterial blight for the first time in China. He found that the roots and stems of six kinds of weeds, such as crabgrass, could carry and transmit rice bacterial blight (Pathology Teaching and Research Group of Agronomy Department in Agricultural University, 1976a). Later on, this research finding was confirmed by researchers from independent groups in Jiangsu, Zhejiang, and Anhui Provinces, attracted the attentions of plant pathologists in China and abroad, and has been written into the textbooks for Universities.

It was still controversial back then on the physiological races of rice blast fungus all over the world. Some researchers in the Philippines asserted that due to the variability of rice blast fungus, it was meaningless to study its physiological races. However, Prof. Duan conducted a detailed study on the variability of rice blast fungus and showed that despite of the variation, rice blast fungus displayed relative stability. He screened out seven rice identification varieties and four strains, and clarified the main physiological races and distribution of rice blast in Yunnan, and also performed pioneer work on the analysis of rice blast resistance genes and the study of genetic laws. In addition, a lot of research has been done on plant viruses (Pathology Teaching and Research Group of Agronomy Department in Agricultural University, 1976b; Duan et al., 1979).

Prof. Duan played important roles in organizing international scientific and technological exchanges. In 1981, he invited three famous Japanese pathologists to give lectures in Kunming. 244 students all over China attended these lectures. He personally translated the materials and served as a translator during those lectures. This event was highly praised by the teachers and the students. He also organized 7 large-scale international academic exchanges to strengthen the friendship and scientific connections for plant pathologists from China and Japan.

Prof. Duan has obtained 7 national, provincial and ministerial scientific research awards, and was also elected as the national agricultural labor model and received the special government allowance of the State Council. He also served as an advisor for the Chinese Society of Plant Pathology, chairman of the Southwest Branch of the Chinese Society of Plant Diseases, and chairman of the Yunnan Plant Protection Society. He made outstanding contributions to agricultural education and researches in the Yunnan Province.

Prof. Duan also paid great attention to discipline construction. Since the establishment of the discipline plant pathology at Yunnan Agricultural University, Prof. Duan was committed to laboratory equipment procurement, supervision of researches, and the improvement of teaching and scientific research levels. During the special time, the scientific research conditions were extremely rudimentary. Prof. Duan carried out scientific researches in a simple greenhouse with his colleagues. He also went to the countryside to conduct plant disease investigations, and performed researches on broad bean stem blight in the place borrowed from Agricultural Research Institute and Anning Agricultural Technology Station. Prof. Duan even went to Yuanjiang, where the temperature in summer could reach 40 degrees, and Xishuangbanna, to carry out his studies. The electron microscope has been the dreaming equipment for many teachers and students in the field of plant diseases. At the age of 80, Prof. Duan finally got the approval for the procurement of perspective electron microscope. The teachers said that they cherish the electron microscope since it reminded them of Prof. Duan.

Prof. Duan was always concerned about the construction of academic team and made tremendous efforts to strengthen the academic team in plant pathology at Yunnan Agricultural University. He was also a very kind person and was always willing to share his knowledge and experience with the young talents. Under his influence, the department of plant disease has become a united, progressive and ambitious group. This tradition lasted for generations. Mr. Youyong Zhu (朱有勇), the first student of Prof. Duan graduated with a master's degree, was deeply influenced by him and made great contributions to the pioneer work on the theory of biodiversity control of plant diseases in the academic community, which benefited more than 100 million acres of farmland in Yunnan, Guizhou, Sichuan and other provinces. Prof. Duan paid so much attention for Mr. Youyong Zhu from undergraduate to master's degree, from teaching assistant to professor and from key room laboratory director to principal and then to Academician of Chinese Academy of Engineering. Prof. Duan had his own criteria in training graduate students. Firstly, he emphasized the ideological education and required his graduate students to have both scientific ability and political integrity. Secondly, hc paid attention is "Words and deeds" , to cultivate his students with an enterprising spirit. Thirdly, he encouraged his students to not only focus on scientific researches but also participate in teaching (Zhu and Xiao, 1986). He also encouraged his students to lead and publish high quality research articles. Last but not least is to take the lead and set the quality of the good articles. He has cultivated lots of young talents, such as the Academician Youyong Zhu, who also made great contributions to the agricultural research, is only a representative of Prof. Duan's effective pedagogy. Many other of Prof. Duan's disciples have also gone on to contribute greatly to agricultural teaching and research.

Prof. Duan had worked hard in the field of plant pathology for 54 years and made outstanding contributions to the development of plant pathology in China. Prof. Duan has already left us, however, he has given us lots of precious spiritual wealth, such as his moral nobility, selfless dedication, rigorous style of scientific research, compassionate and generous tolerance.

Figures

Fig.1 Prof. Duan was working on collection and investigation of rice blast

Fig.2 Prof. Duan was instructing students to use the electron microscope

Fig.3 Prof. Duan (third from the left in the first row) in his student's master's thesis defence

Fig.4 Prof. Duan (right) was mentoring his students

References

Duan YJ (1947) Plant Etiology. Shanghai: The Commercial Press. (段永嘉. 1947. 植物病原菌学. 上海: 商务印书馆.)

Duan YJ (1957) The preliminary investigation on the main crop pathogens in Xishuangbanna Dai Autonomous Prefecture of Yunnan. (段永嘉. 1957. 云南西双版纳傣族自治州主要农作物病菌的调查初报.)

Duan YJ (1958) The preliminary report on major crop diseases in Yunnan. (段永嘉. 1958. 云南省主要农作物病害名录初报.)

Duan YJ, Wang YX, Chen HR, et al (1979) Studies on physiological race differentiation of *Pyricularia oryzae* (second report). Yunnan Agric Sci Tech 2:1–7. (段永嘉, 王英祥, 陈海如, 等. 1979. 稻瘟病生理分化研究 (第二报). 云南农大科技, 2: 1–7.)

Pathology Teaching and Research Group of Agronomy Department in Yunnan Agricultural University (1976a) The preliminary report on the weeds carrying pathogen and overwintering of *Xanthomonas oryzae* pv.oryzae. Yunnan Agric Sci Tech 2:25–29. (云南农业大学农学系病理教研组. 1976a. 水稻白叶枯病杂草带菌和越冬问题的研究初报. 云南农大科技, 2: 25–29.)

Pathology Teaching and Research Group of Agronomy Department in Yunnan Agricultural University (1976b) Primary report on physiological race differentiation of *Pyricularia oryzae*. Yunnan Agric Sci Tech 1:14–24. (云南农业大学农学系病理教研组. 1976b. 稻瘟病菌生理分化研究初报. 云南农大科技, 1: 14–24.)

Zhu YJ, Xiao HG (1986) How did Professor Yongjia Duan cultivate postgradates. Commun Postgrad. (朱有勇, 肖火根. 1986. 段永嘉教授是怎样培养研究生的. 研究生通讯.)

25. 涂治：新疆农业科学的开拓者

涂治，又名涂允治，我国杰出的植物病理学家、农学家和农业教育家（图1）。他的大半生深受革命情结的影响，扎根西部，与新疆的社会活动和科学事业，尤其是新疆的农业发展，结下不解之缘。他于1955年入选首批中国科学院学部委员（院士）。

图1　涂治（1903—1976）

1903年7月15日①，涂治出生在湖北黄陂的一个书香门第。祖父涂道泳是清朝秀才，以教私塾为业，却倡导新学；其父涂衡甫从商，叔伯兄弟中，多为学者，他还是后来中国科学院地学部院士涂光炽的叔父。在这样的成长环境中，涂治自幼接受了良好的家庭教育。1915年，涂治考上清华学校（清华大学的前身）的公费生，于是辞别父母，远离家乡，赴京求学。清华学校是当时的留美预备学校，学校倡导的学好科学、报效祖国的思想，对涂治影响颇深。五四运动爆发时，他像许多爱国青年一样，加入了呼吁民主与自由的热潮，并深受这一伟大思想解放运动的洗礼。

从清华学校毕业后，涂治与周培源等同班同学一道赴美留学。他抱着发展我国现代农业科学的理想，立志学农。在明尼苏达大学农学院，涂治师从Elvin C. Stakman教授，专攻植物病理学和作物育种学，他的毕业论文是《禾谷镰刀菌生理分化研究》。Stakman教授是美国著名的植物病理学家和农学教育家，他曾说，涂治是他最优秀的学生之一。涂治非常刻苦，经常晚上做试验到天亮，专心致志，不知东方之既白。涂治天赋过人，有着不同寻常的记忆力，遗传学家李先闻曾说他是学农的人中真正聪明的。

1929年，涂治怀着科学救国的理想回到祖国，先后在广州岭南大学、河南大学、

作者：张钫

清华大学科学技术与社会研究所，北京100084，中国

邮箱：fzhang1985@foxmail.com

① 涂治的出生日期存在争议。在他的后代所写的纪念文章中，他出生于1901年。但在其他一些文章中，他的出生日期是1903年8月20日。在明尼苏达大学的档案中，他的出生日期是1903年7月15日。本文采用这一记录。

武汉大学以及国立西北农林专科学校等学校任教。在岭南大学期间，他致力于农业科研与教学，希望能以此报国。在当时大家云集的河南大学农学院，他率先提出在河南大面积推广使用农业机械的主张。也是在这个时期，他便与中国共产党地方党组织成员乐天宇交往甚密，接受马列主义新思想，开始积极主动为党工作，参加抗日宣传活动。1934年，涂治应武汉大学之邀，回到家乡帮助筹建农学院，兼办湖北棉业试验场。1935年，他又来到刚成立不久的国立西北农林专科学校，他十分重视教学和科学研究，积极延聘名师来校任教。他在任期间，支持共产党和进步学生的活动，甚至拿出自己的薪水帮助进步学生奔赴革命圣地延安。

1938年11月，涂治欣然接受时任新疆学院院长杜重远的邀请，于1939年4月从兰州搭乘苏联运军火的卡车来到迪化（现乌鲁木齐）。此后，他便以孜孜不倦的忘我精神，夜以继日地工作着，数十年如一日，为新疆农林牧业生产和科学教育作出积极贡献。起初，涂治担任新疆高级农业学校的教务长，1941年该校并入新疆学院，改为农科，涂治任教务长兼农科主任。涂治在各族师生中均享有很高的威信，甚至在市民中还传颂着“涂博士”的动人故事。新疆学院前任教务长郭慎先后来回忆说：“我很是佩服涂治同志的，他博学多识，忠诚实在，很少说话，为人谦虚，大智若愚。”涂治支持马克思主义思想，参与中国共产党的诸多活动，后来因此被捕入狱。释放后，他又担任新疆学院副院长，并继续开展革命活动。他的革命热忱在党内外乃至全国都引起了广泛关注。新中国成立后，他应邀出席了中国人民政治协商会议第一届全体会议，并受到毛泽东等中央领导的亲切接见。

新中国成立后，涂治更全身心地致力于农业技术的发展与农学教育。1952年，八一农学院（新疆农业大学前身）成立，涂治担任院长。为了寻求优秀的专业技术人才，涂治四处奔波，走访了祖国的大江南北。有的专家为涂治的一片赤诚所感动，前一天谈好，第二天便赶赴新疆。涂治主张理论联系实际，并倡导教育结合生产。为了弥补农业技术干部的短缺，他采取措施促进学院与生产组之间的交流与沟通。此外，在他的建议和筹备下，1955年，新疆农业大学农林牧科学研究所成立，也就是后来的新疆维吾尔自治区农业科学院。

涂治对新生事物总是热情支持，推行先进技术，做到因地制宜。他积极支持成立原子能利用研究机构，应用核辐射选育新品种；他主张既搞杂交育种，也搞单倍体育种。鉴于新疆干旱缺水，他要求进行喷灌、滴灌试验；北疆无霜期短，他号召推广水稻塑料薄膜育秧；新疆地广人稀，他一直提倡实行农业机械化生产。1954年，涂治担任总指挥，在玛纳斯河流域获得2万亩棉花大丰收，创全国植棉高产纪录，推翻了西方学者“北纬45°地区不能植棉”的论断，成功地将中国棉区向北推移。此外，涂治还创办了期刊《新疆农业科学》。由于他在农业领域突出的贡献，1955年他当选中国科学院学部委员（院士）。

涂治很重视扎实的理论基础，倡导自由的学术讨论。在20世纪50年代，孟德尔遗传学与李森科主义之间的激烈争论在新疆也引发了一些影响。当时一些李森科主义的支持者破坏了孟德尔遗传学研究者的试验田和试验工具。这种粗暴的行为受到涂治的严厉批评，他还专门邀请一些专家到新疆来给大家讲授孟德尔遗传学。那个年代，小麦亩产万斤的浮夸风也吹到了新疆，涂治公开批评了这种不切实际的行为，并在“卫星田”不远的地方另外开辟一块丰产田作为对照，用事实来说服同志们，使大家心悦诚服。

涂治不仅在学术上造诣匪浅，还精通多国语言。他口译流畅，表达准确，语句典雅，发音标准；他笔译字句精当，意义准确，用词严谨。新疆学院曾经从苏联聘请了一批农牧业和水利专家讲课，但苦于没有合适的俄文翻译人员。在这种情况下，涂治把编译工作也承担下来了。1950年之后，涂治还将英国、法国、德国、苏联的一些农学著作翻译成了中文。

涂治对科学人才的关心更是令人难忘，他很重视年轻人的培养和发展。他努力帮助青年教师开展文献收集和研究工作，采集样本并建立育种基地。他对学生很和善，尽可能地帮助有需要的学生进一步深造。在20世纪30年代初，他为支持河南大学农学系年轻教员刘葆庆的小麦育种试验，将自己在美国读博期间的小麦优良品种资料提供给刘葆庆，使他取得了显著成绩。他还培养了许多优秀的科学技术人才，他们中的一些后来成为著名科学家，其中包括小麦育种专家赵洪璋和植物病理学家王鸣岐。

到了晚年，尽管健康状况一度恶化，涂治依旧坚持工作，直至他生命的终点。1976年3月30日，涂治病逝。他去世的消息使许多受到他热情和自由激励的人伤心不已。人们常常伫立在那里，向长眠在苍松翠柏中的这位“天山骄子”默哀致意。新疆农业大学专门在校园中铸造一尊涂治的雕像，以纪念这位新疆农业发展和农学教育的先驱（图2）。

图2 新疆农业大学中涂治的雕像

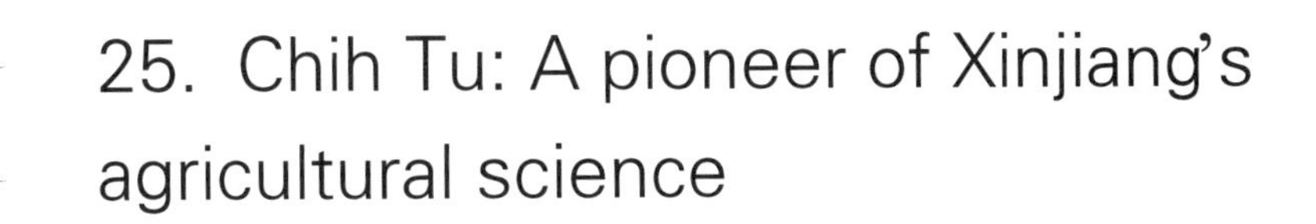

25. Chih Tu: A pioneer of Xinjiang's agricultural science

Chih Tu (涂治) was a pioneer of Xinjiang's agricultural science and technology, the founder of Xinjiang's agricultural education, and the first academician from Xinjiang to join the Chinese Academy of Sciences. In addition to being a distinguished plant pathologist and agriculturist, he was a communist fighter with a strong spirit and a revolutionary sentiment. Deeply influenced by revolutionary ideas and the Chinese Communist Party (CCP), he spent most of his life in the northwest part of China.

Chih Tu was born in a family of scholars in the city of Huangpi in Hubei Province on July 15, 1903.[①] His grandfather Tao-Yung Tu (涂道泳) was a scholar of late Qing Dynasty and a traditional private tutor who advocated new learning. His father Heng-Fu Tu (涂衡甫) was involved in business, but most of his uncles were scholars. Chih Tu himself was the uncle of Kuang-Chih Tu (涂光炽), a renowned geochemical scientist and expert on mineral deposits. Chih Tu received good education as a small boy. He attended Tsinghua School (the predecessor of Tsinghua University) in Beijing in 1915, which marked the start of his academic journey. During the period of the May 4th Movement, like many other young patriots, he joined the march calling for democracy and went through the great movement of ideological liberation. His motivation to save the nation through science originated from his experiences during that time.

After graduating from Tsinghua, Chih Tu went to the United States for graduate study with his classmate Pei-Yuan Chou (周培源). With the ambition of revitalizing China's agriculture by modern agricultural science, he planned to take agriculture as his major. He studied under the guidance of Professor Elvin Charles Stakman in the Department of Agriculture at the University of Minnesota. Stakman was a pioneering American plant pathologist and educator who established the methods for identifying and combating diseases of wheat and other important food crops. According to Dr. Stakman, Chih Tu was one of his most distinguished students. He was

Fang Zhang

The Institute for the History of Natural Sciences, Chinese Academy of Sciences, Beijing 100190, China

Correspondence: fzhang1985@foxmail.com

① His date of birth is controversial. In a commemorative article written by his son and grandson, he was born in 1901; while in some other papers, his birth date is August 20, 1903. However, in an archive from University of Minnesota, his birthday is July 15, 1903. Here follows this record.

extremely diligent and often spent entire nights conducting experiments. His memory, wisdom and talent were exceptional. Hsien-Wen Li (李先闻), a distinguished geneticist, described Chih Tu as an extraordinarily intelligent agriculturist. He earned a doctorate in 1929 for his dissertation titled "Physiologic specialization in *Fusarium* spp. causing headblight of small grains".

With the ambition of saving his nation through science, Chih Tu returned to China in 1929. From 1929 to 1938, he worked at several universities and institutes. At Lingnan University, he devoted himself to scientific research and education in service to his country. He was appointed as a professor of plant pathology and put in charge of all agricultural work at the Agricultural Department of Henan University where many notable agriculturists worked. Chih Tu was the first to advocate use of large-scale agricultural machinery in Henan Province. In 1934, with an invitation from Wuhan University, he returned to his hometown to help them establish a cotton experiment station at Wuhan University. In the following year, he took the position of the president of the Northwestern Agricultural School, which had recently been established at the time. It was during this period that he gradually realized that he should use practical techniques to save the nation. Accordingly, he established connections with Tien-Yu Le (乐天宇), an active member of the CCP from whom he learned much about Marxism–Leninism, and began to serve for the CCP and participated in anti-Japanese propaganda activities.

With the invitation of Chung-Yuan Tu (杜重远), the Director of Xinjiang College, Chih Tu arrived in Dihua (now Urumchi) from Lanzhou in April, 1939. From then on, he devoted himself to Xinjiang's agriculture and education for the rest of his life. At the beginning, he took the position of Provost at Xinjiang Senior Agricultural School, but later became the Director of the agricultural department at Xinjiang College and there after the Provost of Xinjiang College. He had high prestige among both teachers and students of all nationalities; the story of "Dr. Tu" was well known both at home and abroad. As a past Provost of Xinjiang College, Shen-Hsien Kuo (郭慎先) said, "I admire Chih Tu's talent and ability. He is knowledgeable and versatile, honest and trustful, modest and unobtrusive. He says less but does more; he seems to be ordinary but actually brilliant." He supported Marxism and the CCP and was imprisoned by the nationalist government for this reason. After his release, he was appointed as the Vice Director of Xinjiang College and continued to commit himself to revolutionary activities. His revolutionary zeal received much attention both within the CCP and across the country; as a result, he was invited to attend the First Chinese People's Political Consultative Conference.

After the founding of the People's Republic of China, he became firmly committed to agricultural development and education. In 1952, Bayi Agricultural College (the predecessor of Xinjiang Agricultural University) was established, and Chih Tu was appointed as the Director. In order to find good faculty and staffs, he traveled around most of China in search of talented individuals. Some experts were so deeply impressed by him that they arrived in Xinjiang the day after they received his invitation. He championed the idea of linking theory with practice and combining education with production. In order to address the shortage of agricultural technology leaders, he took measures to facilitate interactions and communications between colleges and production groups. Furthermore, under his proposal, Xinjiang Institute of Agriculture, Forestry and Livestock, which later became Xinjiang Academy of Agricultural Science, was established in 1955 and he acted as Director.

Chih Tu was always pleased to support new things and to advocate for new technology. He proposed to establish the Utilization of Atomic Energy Research Institute and to use radiation for generating mutant plant varieties. He also advocated both crossbreeding and haploid breeding. To address poor water resources, he proposed using sprinkling irrigation and drip irrigation. Due to the shortage of frost-free period, he promoted plastic sheeting for rice seedling. Since Xinjiang has a small population but covers a large geographical area, he determined to introduce mechanization into agricultural practices. In 1954, under his leadership, they had a bumper harvest of 20 thousand acres of cotton along the Manas River. This achievement not only set a new record but also dispelled the theory proposed by western scientists that cotton is not suitable for cultivation at northern latitudes. This achievement also expanded China's cotton growing areas to the north. Furthermore, Chih Tu launched the journal *Agricultural Science in Xinjiang*. For his outstanding contribution to agriculture, he was elected to an Academician of the Chinese Academy of Sciences in 1955.

He emphasized having a strong theoretical foundation and advocated free academic debate. During the 1950s, fierce controversy between Mendelian genetics and Lysenkoism caused some incidents in Xinjiang. At that time, some individuals who supported Lysenkoism destroyed the experimental plot and experimental instruments of researchers who advocated for Mendelian genetics. This rude, poor and terrible behavior was criticized by Chih Tu. He advocated for exchange of scientific results and civil academic debate. In particular, he invited several experts to Xinjiang to give lectures on Mendelian genetics. In the late 1950s, the current tendency of exaggeration also spread to Xinjiang, and some people claimed that per mu yield of wheat for an average year was ten thousands jin. Chih Tu denounced such unpractical behavior and conducted a comparative test to show that this was not the case.

Besides being an outstanding research scientist, he also knew several languages. Once, Xinjiang College had invited a group of agricultural and water conservancy experts from Russia to give lectures. However, no one was available for translation. Under this circumstance, Chih Tu took the responsibility of translation despite his busy schedule. After 1950, he took charge of translating foreign agricultural publications from English, French, German, and Russian into Chinese.

He also paid close attention to intellectual and career development of young individuals. He helped young teachers launch literature collection and research, collect specimen and build nursery gardens and herbariums. He was also very kind to his students and supported needy students for further studies. During the early 1930s, he supported the breeding experiment conducted by Pao-Tsing Liu (刘葆庆), who was a young teacher at Henan University, and supplied him with good varieties of wheat he had brought with him from America. This eventually helped Liu become successful in his breeding experiments. He educated many excellent science and technology students, some of whom later become leading scientists, including the wheat breeding expert Hung-Chang Chao (赵洪璋) and the plant pathologist Ming-Chi Wang (王鸣岐).

In his later years, Chih Tu continued to work even though his health condition deteriorated towards the end of his life. He passed away on March 30, 1976. The news of his death was received with great sorrow by many people who had been inspired by his enthusiasm and liberal ideas. He was buried in Urumchi Martyrs' Cemetery. People are often seen standing in silence, showing respect to this "Prideman of Tianshan" . In order to honor the founder of the university and the pioneer of agricultural education in Xinjiang, a statue of Chih Tu was erected in the campus of Xinjiang Agricultural University.

Figures

References

Chen ZW (1985) Chih Tu. In: Jin SB. The Biography of Chinese Modern Agriculturists 1. Changsha: Hunan Science & Technology Press. (陈之伟. 1985. 涂治. 见: 金善宝. 中国现代农学家传 (第1卷). 长沙: 湖南科学技术出版社.)

Deng LQ (2012) My initial experience in Xinjiang. Contemp China Hist Stud 19(2): 2–19. (邓力群. 2012. 初到新疆的历程. 当代中国史研究, 19(2): 2–19.)

Li HW (2007) Autobiography of Hsien-Wen Li. Changsha: Hunan Education Publishing House. (李先闻. 2007. 李先闻自述. 长沙: 湖南教育出版社.)

Shi L (1994) Professor Chih Tu and his Hui nationality students Zhanlin Yu and Wenxiang Ga. Changji Lit Hist Mater 17:118–122. (石磊. 1994. 涂治教授和他的回族学生禹占林、尕文祥. 昌吉文史资料, 17: 118–122.)

Tu GQ (2001) My two uncles. In: Jiang XY. Around China. Shijiazhuang: Hebei Education Press. (涂光群. 2001. 我的两个叔叔. 见: 蒋星煜. 走遍神州. 石家庄: 河北教育出版社.)

Tu GY, Tu ZD (2011) Zhi Tu. In: Shi YC. Overview of Academic Achievement of Renowned Chinese Scientists in 20th Century, Agriculture Part 1. Beijing: Science Press. (涂光一, 涂振东. 2011. 涂治. 见: 石元春. 20世纪中国知名科学家学术成就概览・农学卷・第一分册. 北京: 科学出版社.)

Wang DM (1997) The renowned agriculturist Chih Tu. Wuhan Lit Hist Mater 2: 122–135. (王东明. 1997. 著名农业科学家涂治. 武汉文史资料, 2: 122–135.)

26. 李博：中国生态学奠基人

“李博院士是我国以植物生理学为基础的植物生态学的第一代开创者，也是李继侗院士的弟子和继承人”（Wu，1980）。李博院士（图1）是一位优秀的植物学家和生态学家。他于20世纪70年代早期在我国创建了生态科学。他运用自己的生态学知识，同时利用传统和现代的研究方法，采用基础研究和实际应用相结合的模式，系统而综合地对干旱、半干旱地区的植物进行了研究。通过结合全球变化、可持续发展和生物多样性保护，李博先生将我国的生态学发展为大众热点。20世纪80年代他提出的关于遥感和地理信息系统在草场研究中的应用的创造性理论，将草地生态学提升到了一个新的水平。

图1　李博（1929—1998）

李博1929年4月15日出生于山东，1953年于北京农业大学毕业，后进入北京大学并成为李继侗先生的学生。李继侗先生是著名的植物生态学家，当时正致力于我国生态学的发展（李博，1979）。1955年，在李继侗先生的指导下，李博开始了自己的研究。当时，他对如何研究北京西山的植物几乎一无所知，他花了两年的时间学习基础知识和英语，这为他以后的研究奠定了坚实的基础。随后，他赴内蒙古的呼伦贝尔草原研究当地植被的生长，在那个时期，航拍、汽车等任何先进的研究工具和交通设备都没有，马车是唯一的交通工具。就是在这样艰苦的环境下，李博对草地生态学产生了浓厚的兴趣。李博自己都没意识到，从他踏上这片迷人土地的那一刻开始，他的一生就已经同草原紧紧地联系在了一起。

毕业两年后，有远大理想的李博开始了边境草地保护的建设工作。他的妻子蒋佩华一直给予他极大的支持，甚至放弃了自己的工作来陪伴他。1958年起，李博参加中国科学院治沙队的沙漠综合考察工作。1959年5月，李博及其团队抵达巴丹吉林沙漠，当时地表温度将近70 ℃，骆驼都难以生存。经过20多天的考察，科考队终于成功纵

作者：高志成

内蒙古大学，呼和浩特010000，中国

邮箱：gzcecology@163.com

穿巴丹吉林沙漠，获得了这一地区难得的动植物区系、植被、水文地貌等第一手资料，填补了这一地区研究的空白。60年代初，李博发表了几篇论文，阐明了我国内蒙古的植被和沙漠的类型和分布规律，提出了地带划分和分区方案。

70年代，《中国植被》一书开始组织编写，李博任该书“干旱、半干旱区植被”编写组副组长（Li，1990a）。李博在书中概述了我国草原植被的基本规律，提出了具有中国特色的分类系统，对该书的完成起到了重要的作用。

1981年，李博应美国爱达荷大学邀请，赴美进行了学术访问，并考察了美国21个州的草原，感受到了中美之间巨大的学术实力差距。回国后，他主持了科技攻关项目“遥感在内蒙古草场资源调查中的应用研究”，组织了全国9所高校和近百名专家，使我国草地资源的调查、评价与制图在方法上迈上了一个新台阶（Li，1990b）。

李博作为主要组织者之一，还成功地在呼和浩特举办了三次国际学术会议：国际草地植被会议、国际草地资源会议和蒙古高原草地管理国际学术会议。除会议外，内蒙古大学参与筹办的国际生态学协会（图2）在80年代影响巨大，当时内蒙古大学的生态学在全国乃至国际上都享有盛誉。李博在《退化草地的可持续发展》一文中详细阐述了退化草地的恢复措施，得到了国际同行的广泛认同和赞誉。

图2　李博院士（二排左四）在国际生态学协会

李博在内蒙古大学执教45年，建立了我国高校中的第一个生态学专业。为了追平我国生态学落后国际水平20年的差距，李博在教学和提高学生的科研能力方面付出了巨大的努力（图3）。他制定了新的专业教学计划，加强了外语、数学、生物统计和计算机应用等课程。此外，李博还强化了野外实习及大实验，组织翻译和编写了一系列教材。为提高教学效率，李博还邀请众多专家来做关于生态学教学方法的演讲。受益于李博高效的教学模式，他培养出的第一批毕业生就能很好地适应现代生态学工作的需要。接着，在他主持下，内蒙古大学先后建立了硕士点（1978年）、博士点（1990年），使内蒙古大学成为我国生态学人才培养的重要基地之一。

图3 李博院士在给生态学博士生做讲座

1993年他当选中国科学院院士之后说："国家培养了我，在有生之年，我一定努力拼搏，为祖国的科教事业、生态学和草原科学的发展，尽自己最大的力量。"李博，我国生态学的奠基人，把生态学教学和研究作为自己毕生的事业。他的著作《普通生态学》被当作高校教材沿用至今。

26. Li Bo: Pioneer of ecology in China

"Li Bo was the founder of Plant Ecology in China, the successor of Li Jitong and the pioneer of botany and phytophysiology of China." (Wu, 1980) Li Bo was a brilliant ecologist and botanist. He established the science of Ecology in China in the early 1970s. He conducted a comprehensive and systematic study on vegetation in arid and semi-arid regions using his knowledge of Ecology and combining basic research with applied study as well as traditional and modern research methods. By involving global change, sustainable development and biodiversity conservation, Li Bo linked Ecology in China to hot issues worldwide. His creative idea of applying remote sensing (RS) and Geographic Information System (GIS) to pasture resource investigation in 1980s promoted the Grassland Ecology to a new level.

Li Bo(李博) was born in Shandong Province on April 15, 1929. After graduating from Beijing Agricultural University in 1953, Li Bo worked in Peking University where he met his supervisor Li Jitong(李继侗), who was noted for his studies of plant biology and for devoting himself to the development of Ecology (Li, 1979). Li Bo started his research with Li Jitong in 1955. At that time, he knew almost nothing about how to investigate the vegetation on Xishan mountain of Beijing. He spent two years learning basic knowledge and English which laid a solid foundation for his following research career. Subsequently, he went to Hulun Buir grassland in Inner Mongolia to explore how the vegetation grew in this region. In that period, advanced research tools and transportation such as aerial photographs and vehicles were not available. The only means of transport was horse-drawn carriage. In such circumstances, with the guidance of Li Jitong, Li Bo began to take a strong interest in Grassland Ecology. From the first step on that fascinating area, although he wasn't aware of it, Li Bo had become closely tied to grassland.

As an ambitious youth, Li Bo tried his best to support the construction of border areas around the grasslands two years later after his graduation. His wife Mrs. Jiang Peihua, who sacrificed her career, strongly supported Li Bo's decision by following him wherever he went. The couple soon settled in Inner Mongolia. In 1958, Li Bo joined the team of Chinese Academy

Zhicheng Gao
Inner Mongolia University, Hohhot 010000, China
Correspondence: gzcecology@163.com

of Sciences combating desertification and started an integrated survey of desert ecology. In May 1959, Li Bo and his team arrived at Badain Jaran Desert, where even camels struggled to survive as the dune surface temperature rose close to 70 ℃. After more than 20 tough days, they obtained the primary data of flora, vegetation, hydrology and landforms that revealed the true features of the desert for the first time. Li Bo published several research papers illustrating the patterns of vegetation types and desert areas of Inner Mongolia. Furthermore, Li Bo put forward a zonal division and partition scheme in the early 1960s.

In the compiling of *Vegetation in China* in 1970s, Li Bo was the deputy leader of compilation group which described the vegetation conditions in arid and semi-arid regions of China (Li, 1990a). Li Bo summarized the basic laws of grassland vegetation which were essential to the book and created a brand new classification system of vegetation with Chinese characters.

Li Bo was invited to pay an academic visit to University of Idaho in October 1981. While there he observed and studied prairies in 21 States of the USA, and perceived a huge academic gap between China and America. When he came back to China, he presided at the program of Present Development in the Application of Remote Sensing to Grassland Resources Survey in China which updated the evaluation and map-making techniques in 1983 (Li, 1990b). This program involved almost one hundred specialists and nine universities in the ecological field.

As the major organizer, Li Bo presided over 3 international academic conferences in Hohhot: the International Grassland Vegetation Congress, the International Session on Grassland Resources and the International Conference on Grassland Management in Mongolia Highland. Besides these conferences, the International Association for Ecology led by Inner Mongolia University was highly influential in 1980s. At that time, the Ecology department of Inner Mongolia University had gained a significant reputation in China and even in the world. The paper "The sustainable development of deteriorated grassland" written by Li Bo expounded the measures of grassland restoration was widely recognized and appreciated internationally.

Li Bo was the founder of the first specialty of Ecology in China. He had been teaching in Inner Mongolia University for more than 45 years. To fill the 20 years gap between Ecology in China and international level, Li Bo made painstaking efforts to find how to teach students and help them develop better scientific studies. He made new plans and targets of inter-disciplinary teaching methods, including English, Mathematics, Biostatistics and even computer application. Field trips and labs were also enhanced under his supervision and many textbooks were written or translated by him. In order to be more efficient in teaching, Li Bo invited many experts to present lectures about teaching methods of Ecology. Due to his effective teaching model, the first graduates were well recognized in their work of modern Ecology. Li Bo set up the first Master Degree program in 1978 and the first Ecology PhD program of China in 1990. His contributions to Inner Mongolia University helped the university become one of the most important talent training bases of Ecology in China.

"I was cultivated by my country. So I will devote myself into Ecology, constructing my country and developing grassland science for the rest of my life." Li Bo said when he was selected as the academician of Chinese Academy of Sciences in 1993. Li Bo, the founder of Ecology in China, enjoyed teaching and researching Ecology all his life. The textbook of *General Ecology* written by him is still used by college students today.

Figures

Fig.1 Li Bo (1929—1998)

Fig.2 Li Bo (4th from left, line 2) in the International Association for Ecology

Fig.3 Li Bo was giving a brisk lecture to doctors of Ecology

References

Li B (1979) Basic characteristics of steppe vegetation in China. Grassland of China 1:2–12, 26. (李博. 1979. 中国草原植被的一般特征. 中国草地, 1:2–12, 26.)

Li B (1990a) The Vegetation of China. Beijing: Science Press, 156–168.

Li B (1990b) The Steppe of China. Beijing: Science Press.

Wu Z (1980) The Vegetation of China. Beijing: Science Press.

27. 钟扬：永远在路上的探索者

2017年9月25日上午，噩耗从遥远的内蒙古鄂尔多斯市传到复旦大学校园，正在当地出差的复旦大学研究生院院长、著名植物学家钟扬（图1）遭遇车祸，不幸逝世。

悲伤迅速地弥漫在整个复旦校园，每一个听到消息的人，无论与钟扬老师渊源深浅，都扼腕叹息，叹息我们永远失去了一位挚友、一位良师、一位永远充满激情与梦想的探索者。

图1　钟扬教授（1964—2017）

1964年，钟扬出生于湖北黄冈。如果要用一个字形容钟扬，最准确的莫过于一个“早”字。早慧的钟扬15岁就进入中国科学技术大学的少年班，学习无线电电子学。大学毕业后，钟扬进入中国科学院武汉植物研究所工作。在与所学专业几乎毫不相关的植物学领域，钟扬再次展现出他过人的天资。敏捷的思维和深厚扎实的数学物理功底，使他总能从全新的角度思考和发现问题（钟扬和何芳良，1986）。他受到“计算物理学”（computational physics）和“计算化学”（computational chemistry）等的启发，提出“计算生物学”（computational biology）的概念，并把这个尚属新生事物的边缘交叉学科作为自己努力的方向。1993年，他在中国科学院武汉植物研究所创建了第一个计算生物学青年实验室（图2），一群与钟扬志同道合的年轻人靠着他从美国带回的计算机系统，开始了一场全新的探索（钟扬和张晓艳，1990；钟扬，1995），由此开启了我国计算生物学的绚丽篇章。钟扬在计算生物学领域的探索即使在国际上也属先行，他领导下的青年实验室，在数量分类学、分支分类学、数量生态学、植物数据库等领域取得了大量成果。例如，他与陈家宽教授合作，在国内首先将数量分类和

译者：陈凡

中国科学院遗传与发育生物学研究所，北京100101，中国

邮箱：fchen@genetics.ac.cn

分支分类方法用于水生植物的研究（钟扬等，1990，1994）。有意思的是，当时“计算生物学”这个概念在国内显得太过超前，大家更习惯称之为“计算机在生物学中的应用”。然而，在他成立实验室后的一年，冠以《计算生物学》名称的国际学术刊物第一卷出版；三年后，国际计算生物学学会（International Society for Computational Biology，ISCB）成立。如今，计算生物学已经成为基础生命科学密不可分的一部分，深度融入了从分子尺度到生态系统尺度的不同领域。

图2　中国科学院武汉植物研究所计算生物学青年实验室部分成员（1993年）

钟扬出色地推进了计算生物学在植物学领域的国际合作。钟扬与美国密歇根州立大学和加利福尼亚大学伯克利分校等多家单位合作，针对国际上已有同类数据库所采用分类系统不一致的问题，提出了一种新的交互分类数据模型（UNIC结构）和一个检测分类树与系统树差异的新测度（Zhong et al.，1996，1997）；并设计和实现了基于分类本体论（taxonomic onotology）思想的交互分类数据库系统（HICLAS）（Zhong et al.，1999）。他建立的这些方法打破了不同机构的界限，为不同系统之间提供了一种“通行的语言”，为国际合作奠定了关键性基础。英国雷丁大学教授、生物分类及信息学权威Frank A. Bisby在为*Science*撰写的“生物多样性信息学”专题综述中引用了钟扬在*Taxon*上发表的数据模型（Bisby，2000），*Science*也于当年发表了钟扬的相关评述（Zhong et al.，2000）。

此外，针对植物分子进化分析中用不同基因重建的物种间系统发育关系有不同的拓扑结构，即产生不一致基因树的问题，钟扬与美国密歇根州立大学桑涛博士合作，提出了一种新的统计检测模型来区分产生不一致基因树的因素（杂交和谱系分选）。通过运用自展技术来检验杂交物种形成假说，并应用计算机模拟技术确定了该模型在基因树各分支进化速率相同和进化速率可变两种情况下的不同适用范围（Sang and Zhong，2000）。该工作被澳大利亚昆士兰大学M. A. Ragan教授列为该领域的前沿方向之一。

认识钟扬的人都知道，他学识渊博，绝不局限于植物学领域，他对生物学的各个分支都有涉猎，而且有自己独到的见解，这也使得他可以与多学科的专家合作，且硕果颇丰。近10年来，为了满足大规模基因组和蛋白质组分析的实际需求，钟扬在国内建立了一批分子数据库，包括多蛋白查询整合信息系统（Multi-Protein Survey System，MPSS），为同时提取多个蛋白质的相关信息提供了新的平台（Hao et al.，2005）；还有鉴定植物抗逆相关候选基因的数据库系统（PlantQTL-GE）（Zeng et al.，2007）以及PB转座子及小鼠插入突变的数据库系统（PBmice）（Sun et al.，2008）。这些数据库系统为生物信息学和分子进化分析的发展提供了新的工具。例如，通过建立C_3植物光合作用代谢的系统生物学分析模型和计算机模拟，发现环境扰动下代谢通路间协调性增强的规律，这无疑是植物系统生物学研究中的开创性工作（Luo et al.，2009）。他还参与了SARS冠状病毒分子进化分析和日本血吸虫全基因组的系统发育分析等工作（The Chinese SARS Molecular Epidemiology Consortium，2004；The *Schistosoma japonicum* Genome Sequencing and Functional Analysis Consortium，2009）。

钟扬的“早”还体现在他较早认识到藏族优秀人才培养的紧迫。科学家钟扬深谙教育的重要性，他连续三次申请援藏，在西藏教育科研第一线坚守了16年。在高原（拉萨，海拔>3600米）与海滨（上海，海拔<5米）之间的常年奔波使得钟扬的身体日益恶化，而西藏大学的教育和科研事业却迎来了翻天覆地的变化。他带领西藏大学理学院的老师成功申请并主持了第一个国家自然科学基金项目，培养了藏族第一个植物学博士，推动西藏大学获批了第一个理科硕士学位授予点（生物学）和第一个理科博士学位授予点（生态学）。他所领导的生态学学科最近也顺利入选了国家的“双一流学科”建设。他每年在西藏进行大量的野外科学考察和研究，在西藏野生资源的遗传多样性和化学多样性方面的研究已取得重要进展（图3）（Liu et al.，2006；Zhu et al.，2009）。功夫不负有心人，钟扬及其团队首次在西藏发现了分布于海拔4000米以上的拟南芥群体，并在其全基因组测序的基础上，检测了功能基因的适应性进化，表明西藏拟南芥是一个生长于最高海拔的生态型，为全世界的植物学基础研究提供了新的种质资源。在国家自然科学基金重大研究计划项目“青藏高原极端环境下植物基因组变异及适应性进化机制研究”支持下，钟扬及其团队获得了一系列的成果，例如，揭示了西藏特有植物西藏沙棘和山岭麻黄等的微进化特征，完成了青藏高原鱼腥藻的全基因组测序和转录组分析，发现了蓝藻高原适应的分子遗传学证据（Qiao et al.，2016）。在发生车祸时，钟扬教授已经预定了数日后再度进藏的机票。

图3　钟扬教授在从事植物资源的收集和调查工作

钟扬对近年来国内教育质量下滑忧心忡忡，在担任复旦大学研究生院院长后，他推动了多项教育相关的改革，启动了创新性的复旦大学夏季集中式授课（FIST）等项目，备受各方赞誉。他认为教育是场“接力赛”，学生的创新思维和批判性思维应该从基础教育抓起，在儿童阶段尽早开发。带着这份责任感，他在繁忙的工作之余，花费了大量精力在中小学生的科普教育和思维训练上。他热衷科普教育和科普宣传，翻译了包括《大流感》在内的近十本著作，通过分享这些他所读过的精品英文原著，将思考的种子传播开去，把教育的触角延伸在生活中的每一分钟，传达给每一个人。

斯人已去，音容犹存！钟扬教授的不幸离世，是植物学界的一大损失，也是教育界的一大损失！他永远离开了他热爱和为之奋斗的科研和教育事业，但是他的探索道路并没有终止，作为一个将分子进化分析模型和生物信息技术应用于进化基因组学与系统生物学研究的学者，作为一个永远不知疲倦、热情追求和探索着生命奥秘的奇人，钟扬教授和他的精神将永存于我们的心中。有更多缅怀他、崇敬他的后来者，将会追随着他探索的足迹不断前行，去圆他的梦想！他仍然行走在探索的路上！

安息吧，我们的挚友、良师和追梦者！

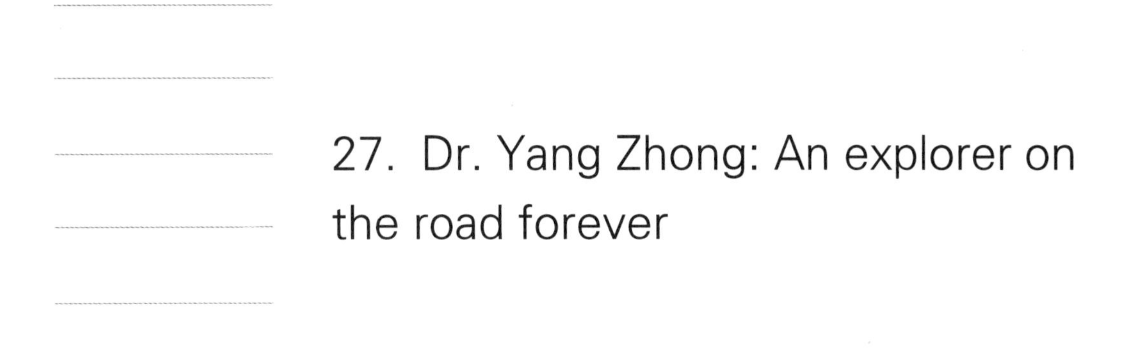

27. Dr. Yang Zhong: An explorer on the road forever

On the morning of September 25, 2017, grievous news spread from the remote Ordos region of Inner Mongolia to Fudan University campus in Shanghai. Professor Yang Zhong, a famous botanist and the Dean of Fudan University's graduate school, passed away in a tragic car accident while on a business trip.

Sorrow quickly spread throughout the entire campus of Fudan University, as well as beyond China. Whoever heard the news mourned with a heavy heart. We all have lost a close friend, a wonderful teacher, a mentor, a passionate dreamer and an excellent pioneer in many fields.

Born in 1964, in Huanggang, Hubei Province, Yang was marked with the word "early" throughout his life. He entered the University of Science and Technology of China when he was only 15 years old. After graduating from college with a major in radio-electronic engineering, Yang entered the Wuhan Institute of Botany of Chinese Academy of Sciences and started from zero knowledge in the field of botany. With his extraordinary talent and sharp thinking, he started to explore the succession and evolution of plants from a mathematical point of view (Zhong and He, 1986). Inspired by "computational physics" and "computational chemistry", he put forward the notion of "computational biology", which he dedicated his energy as a future direction. Yang founded the first computational biology lab at the Wuhan Institute of Botany of

Fan Chen[1], Baorong Lu[2], James C. Crabbe[3,4], Jiayuan Zhao[2], Bojian Zhong[5], Yupeng Geng[6], Yufang Zheng[7], Hongyan Wang[7]

1 Institute of Genetics and Developmental Biology, Chinese Academy of Sciences, Beijing 100101, China

2 Key Laboratory of Biodiversity and Ecological Engineering, Ministry of Education, Department of Ecology and Evolutionary Biology, Fudan University, Shanghai 200438, China

3 Wolfson College, University of Oxford, Linton Road, Oxford OX2 6UD, UK

4 Department of Life Sciences, Institute of Biomedical and Environmental Science & Technology, University of Bedfordshire, Park Square, Luton LU1 3JU, UK

5 College of Life Sciences, Nanjing Normal University, Nanjing 210046, China

6 School of Ecology and Environmental Science, Institute of Ecology and Geobotany, Yunnan University, Kunming 650091, China

7 Obstetrics & Gynecology Hospital, Institute of Reproduction & Development, Fudan University, Shanghai 200090, China

Correspondence: fchen@genetics.ac.cn (F. Chen)

Chinese Academy of Sciences in 1993.

Together with a group of young people who shared the same view, Yang started a new type of exploration in botany (Zhong and Zhang, 1990; Zhong, 1995) and opened up a brilliant chapter of computational biology in China. Under his leadership, the lab quickly made many great achievements in quantitative taxonomy, branch taxonomy, quantitative ecology, and plant databases which amount to the best in the world. For example, he and Professor Jiakuan Chen were the first ones in China to use quantitative classification and branch classification methods for aquatic plants (Zhong et al., 1990, 1994). However, the concept of "computational biology" was too far ahead of it's time to be accepted by everyone in China. Instead, people preferred to call it "the application of computers in biology" . Nevertheless, Yang's work was in line with trends internationally and the first volume of the international academic journal *Computational Biology* was published in 1994 and the International Society for Computational Biology (ISCB) was established in 1997. Computational biology has become an inseparable part of basic life sciences in both China and the world. It was entirely fitting that he was later appointed to the Editorial Board of the journal *Computational Biology and Chemistry*.

Yang also greatly advanced the international cooperation in computational biology in the field of botany. At that time, many different computational methods were developed by different scientists in the world, creating a problem of incongruence in taxonomic database systems. In collaboration with the partners from MSU and UC-Berkeley, Yang presented a new interactive classification data model (UNIC structure) to manage hierarchical classification data, and proposed a general comparison methodology for different classification trees and various types of dendrograms (Zhong et al., 1996, 1997). He also designed a prototype taxonomic database system called HICLAS (hierarchical classification system) based on taxonomic onotology (Zhong et al., 1999). Those methods proposed a solution to break the boundaries of different institutions, provided a "lingua franca" for different systems, and laid the foundations for international cooperation. Frank A. Bisby, professor at the University of Reading in the United Kingdom, a leading researcher in taxonomy and bioinformatics, quoted Yang's operative model in his review paper on biodiversity informatics in *Science* (Bisby, 2000), followed by Yang's commentary in *Science* in the same year (Zhong et al., 2000).

Another good example is Yang's collaboration with Prof. Tao Sang of Michigan State University in the field of plant evolution. They proposed a new statistical test model to distinguish hybridization and other biological processes as causes of topological incongruence. The computer simulation analyses they conducted not only supported the validity of the bootstrap test when each gene evolved at a constant rate, but also suggested that the model remained valid as long as the rate heterogeneity was occurring proportionally in the same taxa for both genes (Sang and Zhong, 2000). This work was referred to by Professor M. A. Ragan of the University of Queensland, Australia as one of the advanced aspects in phylogenetics. Collaborating with Professor Suhua Shi and her group in Sun Yat-sen University, Prof. Yang Zhong completed a series of studies on plant phylogenetics. They also detected evolutionary rate heterogeneity among the mangrove genera and their close terrestrial relatives of Rhizophoraceae, using phylogeny analysis and the relative-rate test. Basing on the results of the relative-rate test, the divergence times among genera as well as the average divergence time between the close related inland genus *Carallia* and the coastal Rhizophoraceae tribe were determined. These findings were published in *Ecology Letters*, which was the first work completed independently by Chinese scientists published in this internationally recognized peer-reviewed journal in ecology and biodiversity (Zhong et al., 2002).

Yang is well known for his profound knowledge and unique insights, which promoted his collaborations with multidisciplinary experts in China and abroad in the past 10 years. Yang's expertise and insightful thoughts allowed him to make great contributions to the new frontiers of large-scale genomic and proteomic analysis, molecular evolution analysis of the SARS coronavirus (The Chinese SARS Molecular Epidemiology Consortium, 2004) and genome-wide phylogenetic analysis of *Schistosoma japonicum* to name but two (The *Schistosoma japonicum* Genome Sequencing and Functional Analysis Consortium, 2009). He also established many early molecular databases in China, including a protein informatics system (the multi-protein survey system, MPSS) (Hao et al., 2005), a database system for plant resistance candidate genes (PlantQTL–GE) (Zeng et al., 2007), and a mouse mutagenesis database, generated with PB transposon methods (PBmice) (Sun et al., 2008). He pioneered computational simulation in plant systems biology to analyze the photosynthetic metabolism of C_3 plants, and his work revealed that metabolic pathways are synchronized and coordinated when under environmental perturbations (Luo et al., 2009). This is particularly relevant when organisms adapt to extreme environments, and potentially in understanding how organisms can mitigate against aspect of climate change.

Yang was also a great educator, and he recognized very early the urgency and importance of cultivating talented Tibetans. Yang was committed to this purpose, and applied for three consecutive periods of nine years as one of the cadres sent to support Tibet. He personally carried on this goal in Tibet for 16 years. Just before his tragic death, he had already booked tickets to go back to work in Tibet University. His health was deteriorating due to the harsh environment of Tibet and frequently traveling between high and low altitudes. There is no doubt that his efforts made tremendous changes in Tibet University. With his help, Tibet University successfully got the first NSFC funding in university history. He trained the first Tibetan PhD in Botany. Yang also made great efforts to establish the very first Masters in Science program (Biology) and the very first PhD program (Ecology) in Tibet University. The ecology department led by Yang also recently entered the pool of the nation's top disciplines. He conducted a large number of field scientific expeditions and studies in Tibet each year. Significant progress has been made in the study of the genetic diversity and chemical diversity of wild resources in Tibet (Liu et al., 2006; Zhu et al., 2009). Hard work pays off; Yang and his team found the first *Arabidopsis* population in Tibet at an altitude over 4000 m above sea level. As a unique ecotype, Tibetan *Arabidopsis* grows at highest altitude found for the plant, and provides new resources for botany research worldwide. Based on the whole genome sequencing data of this Tibetan *Arabidopsis*, Yang and his team analyzed the functional genes for high altitude adaptive evolution. Besides Tibetan *Arabidopsis*, Yang and his team also obtained a series of achievements in plant genome variation and adaptive evolution in the extreme environment of the Tibetan Plateau, including the micro-evolution of sea buckthorn and mountain ephedra, the WGS and transcriptome analysis of Qinghai–Tibet cordate houttuynia(Qiao et al., 2016). But it wasn't just plants that inspired him; his group also published a paper on why the Giant Panda eats bamboo (Jin et al., 2011).

Yang had great concerns for the declining quality of domestic education in recent years. While serving as the dean of the Fudan University graduate school, he promoted a number of education-related reforms and launched innovative projects such as the FIST program, which have won praise from all quarters. He believed that education is a "relay race" . Students' innovative and critical thinking should start from basic education and develop early in childhood. With this sense of responsibility, he devoted a great deal of energy to scientific training and thinking programs for primary and middle school students, adding a heavy burden to his already busy workload. Just before his death he arranged for Prof. James Crabbe, of the Universities of Oxford and Bedfordshire in the UK, and his research collaborator for over 10 years, to visit Qingdao and lecture to hundreds of young people—including one of his sons— about science and education. He was keen on public education and science popularization, and translated nearly a dozen books, including the popular seller *The Great Influenza: The Epic Story of the Deadliest Plague in History*. By translating these excellent books, he spread the seeds of thinking, extending the ideas of education to everyday life.

Professor Zhong Yang's tragic death is a great loss, not only for the field of plant biology, but also for education as a whole. He has left his beloved research and education career, however his spirits of exploration will never stop; they will carry on in his colleagues, in his students, and in people; his work will inspire in the future. As a scholar, an educator, a pioneer, Professor Yang Zhong will always be in our hearts. His passion will always encourage those who follow his footsteps! As he said, "It is not excellent people who have dreams, but the people who have dreams can be excellent."

Rest in peace, Dr. Yang Zhong, our best friend, teacher, mentor, and an excellent dreamer! "As long as our heart is flying, the road will continue to extend forwards."

Figures

Fig.1 Prof. Yang Zhong (1964—2017)

Fig.2 Some colleagues in Youth Laboratory of Computational Biology, Wuhan Insitute of Botany, CAS (1993)

Fig.3 Prof. Zhong is working on collection and investigation of plant resources

References

Bisby FA (2000) The quiet revolution: Biodiversity informatics and the Internet. Science 289:2309–2312.

Hao P, He W, Huang Y, et al (2005) MPSS: An integrated database system for surveying a set of proteins. Bioinformatics 21:2142–2143.

Jin K, Xue C, Qian J, et al (2011) Why does the giant panda eat bamboo? A comparative analysis of appetite-reward-related genes among mammals. PLoS One 6(7):e22602.

Liu J, Wang L, Geng Y, et al (2006) Genetic diversity and population structure of *Lamiophlomis rotata* (Lamiaceae), an endemic species of Qinghai-Tibet Plateau. Genetica 128:385–394.

Luo R, Wei H, Ye L, et al (2009) Photosynthetic metabolism of C_3 plants shows highly cooperative regulation under changing environments: A systems biological analysis. PNAS 106:847–852.

Qiao Q, Huang YY, Qi J, et al (2016) The genome and transcriptome of *Trichormus* sp. NMC–1: Insights into adaptation to extreme environments on the Qinghai–Tibet Plateau. Sci Rep 6: 29404.

Sang T, Zhong Y (2000) Testing hybridization hypotheses based on incongruent gene trees. Syst Biol 49:422–434.

Sun L, Jin K, Liu Y, et al (2008) PBmice: An integrated database system of piggyBac (PB) insertional mutations and their characterizations in mice. Nucleic Acids Res 36:D729–D734.

The Chinese SARS Molecular Epidemiology Consortium (2004) Molecular evolution of the SARS coronavirus during the course of the SARS epidemic in China. Science 303:1666–1669.

The *Schistosoma japonicum* Genome Sequencing and Functional Analysis Consortium (2009) The *Schistosoma japonicum* genome reveals features of host–parasite interplay. Nature 460:345–352.

Zeng H, Luo L, Zhang W, et al (2007) PlantQTL-GE: A database system for identifying candidate genes in rice and *Arabidopsis* by gene expression and QTL information. Nucleic Acids Res 35: D879–D882.

Zhong Y (1995) A survey of plant taxonomic information systems. Chinese Bulletin of Botany 12(Suppl.): 1–6. (钟扬. 1995. 植物分类信息系统概述. 植物学通报, 12(增刊): 1–6.)

Zhong Y, Chen JK, Huang DS (1990) The Methods and Programs for Numerical Taxonomy. Wuhan: Wuhan University Press. (钟扬, 陈家宽, 黄德世. 1990. 数量分类的方法与程序. 武汉: 武汉大学出版社.)

Zhong Y, He FL (1986) Prediction model of plant communities succession process. Prediction 6: 13–15, 18. (钟扬, 何芳良. 1986. 植物群落演替过程的预测模型. 预测, 6: 13–15, 18.)

Zhong Y, Jung S, Pramanik S, et al (1996) Data model and comparison and query methods for interacting classifications in a taxonomic database. Taxon 45:223–241.

Zhong Y, Li W, Huang DS (1994) Theories and Methods for Cladistic Taxonomy. Beijing: Science Press. (钟扬, 李伟, 黄德世. 1994. 分支分类的理论与方法. 北京: 科学出版社.)

Zhong Y, Luo Y, Pramanik S, et al (1999) HICLAS: A taxonomic database system for displaying and comparing biological classification and phylogenetic trees. Bioinformatics 15:149–156.

Zhong Y, Meacham CA, Pramanik S (1997) A general method for tree-comparison based on subtree similarity and its use in a taxonomic database. BioSystems 42:1–8.

Zhong Y, Zhang L, Su D (2000) Collaborations tailored for bioinformatics projects. Science 290(5499):2074.

Zhong Y, Zhang XY (1990) Studies on numerical cladistic taxonomy of Nymphaeales. Journal of

Biomathematics 2:156–161. (钟扬，张晓艳. 1990. 睡莲目的数量分支分类学研究. 生物数学学报，2: 156–161.)

Zhong Y, Zhao Q, Shi S, et al (2002) Detecting evolutionary rate heterogeneity among mangroves and their close terrestrial relatives. Ecol Lett 5:427–432.

Zhu Y, Geng Y, Tersing T, et al (2009) High genetic differentiation and low genetic diversity in *Incarvillea younghusbandii*, an endemic plant of Qinghai-Tibetan Plateau, revealed by AFLP markers. Biochem Syst Ecol 37:589–596.

三、微生物学

28. 戴芳澜：中国植物病理学的先驱和真菌学的开拓者

早年求学于中美

戴芳澜（图1）于1893年出生在湖北江陵一个没落的官僚地主家庭。祖父戴洪禧曾任浙江龙泉地方乡官。父亲戴经为晚清贡生。8岁时，戴芳澜母亲去世。12岁那年，就读于上海南洋公学附属高等小学。由于家庭经济窘迫，就读一年即辍学回家。后转入武昌文华学校，一年后再次肄业。1907年，戴芳澜和哥哥进入上海震旦中学读预科。这是一所以法语教学为主的教会学校，为其打下了坚实的学术基础，也激发了他继续求学的愿望。

图1　戴芳澜（1893—1973）

1911年，戴芳澜成功通过清华学校的招考，成为庚款留美预科班的第一批学生。1913年结业之后，国内秩序尚未恢复。与此同时，庚子赔款暂付阙如，他逗留一年后才得以出国（卢嘉锡等，1991）。期间回归故里，在一所商业学校担任临时英语教员。1914年赴美国威斯康星大学农学院学习。这段时间，他成为中国科学社第一批社员，还担任负责介绍新社员的"征求委员"。他于1916年转入康奈尔大学植物病理系，两年后获得农学学士学位。大学期间，戴芳澜在《科学》杂志发表了其最早的研究论文《说蝗》（戴芳澜，1916）。他在写给未婚妻的信中说："读了四年大学之后，我最大的感受就是认识到自己是无知的！学海无涯，我是永远不会毕业的。"（马春沅，1983）因此他选择进入哥伦比亚大学深造，在这里他结识了两位知音：美国真菌细胞学先驱R. A. Harper以及真菌学家和植物病理学家B. O. Dodge（刘作忠，2008）。然而，因父亲失业无法维持家庭生计，戴芳澜被迫提前回国。

作者：葛业静

中国科学技术大学科技史与科技考古系，合肥230026，中国

邮箱：geyejing@mail.ustc.edu.cn

我国植物病理学的先驱

1919年，26岁的戴芳澜获得了第一份正式工作——在南京第一农业专科学校任教授。为显严肃和年长，戴芳澜特意蓄起一小撮胡子（马春沅，1983）。因派系之争引发的混乱，他在第二年就离开了这里。暑假时只得在天津一家私人农场从事园艺工作过渡。不久后，他被聘为广东公立农业专门学校教授，承担植物学和植物病理学两门课程的教学，还在农业试验场调研病虫害。1921年，他与妻子邓淑媛完婚。他在广东结识了志趣相投的好友、后来的中国科学院院士丁颖。在广东任教期间，其最重要的贡献是在1922年夏天发现广东芋疫病。1923年发表的《芋疫病》一文是国人对疫霉属的第一篇研究报告（戴芳澜，1923）。这一时期军阀混战、物价飞涨，学校经常发不出薪水，生活非常艰难。

1923年，受康奈尔大学的同学邹秉文之邀，戴芳澜加入东南大学，任农科病虫害系教授。他在授课之余开展植物病害和病原真菌研究，此外还经常到中国科学社生物研究所查阅资料，完成了江苏麦类病害和全国植物病害问题等研究（吴汝焯等，2010）。1927年，由于“大学区制”的实行，原有农科体制被废除，戴芳澜再次遭到解聘。同年，金陵大学植物病理学教授R. H. Porter回美，戴芳澜被聘为金陵大学教授兼植物病理系主任。1920年代，各类科技社团和学会相继成立。戴芳澜敏锐地意识到，欲实现植物病理学在我国的体制化，必须要建立相应的学会和举办学术交流活动。1929年，他和邹秉文等共同发起成立中国植物病理学会，并于1936年继任会长。

我国真菌学的开拓者

在戴芳澜进入金陵大学不久，植物系主任A. N. Steward告知他，哈佛大学高等植物研究所委托采集所有的中国真菌标本。戴芳澜要求将所采标本一式两份，一份送往美国，一份留在中国。当A. N. Steward质疑中国没人鉴定真菌标本时，戴芳澜回答：“过去没有不等于以后永远没有！就从这次采集开始吧。”由此他开始了白粉菌的研究（马春沅，1983）。1930年，他发表了《三角枫上白粉菌之一新种》，这是国人用现代科学方法研究、发表的第一个中国真菌新种，被认为是中国真菌学创立的标志（戴芳澜，1930）。自18世纪中叶至20世纪上半叶，外国人持续在我国采集真菌标本。戴芳澜的爱国之心让他重视这个问题，并写了《外人在华采集真菌考》一文。1931年，他发表《竹鞘寄生菌之研究》，对多腔菌子囊双壁结构的形成方式和过程进行详细观察，这反映出他以后在真菌分类研究中逐渐形成的一个重要观点——不仅要注意形态还要注意细胞学、发育过程以及遗传学等所有生物学性状。该文得到了国际真菌学界的认可并被写进多种权威性真菌学教科书。直到1994年，国际公认的权威性教科书《真菌概论》还在引用这项工作。

1934年，清华大学成立农业研究所，拟聘请戴芳澜任该所植物病理研究室主任。适逢俞大绂从美国学成归来，接替了戴芳澜在金陵大学的工作。此时他已申请到中华教育文化基金会研究基金，预备赴纽约植物园和康奈尔大学进修真菌遗传学（吴汝焯等，2010）。访问期间，他与B. O. Dodge合作研究了脉胞霉的分类和细胞遗传学，其成果《脉胞霉的两个新种》和《脉胞霉的性反应连锁》先后发表在美国《真菌学报》(青宁生，2006)。这段时间他仍以通信方式关注着清华大学植物病害组的各项筹建工作。

抗战时期，清华大学被迫南迁，农业研究所迁到昆明西郊的大普吉村。在这段艰苦的岁月，戴芳澜每周进城一次到西南联合大学讲课，其余时间都在做研究。为了把项目继续做下去，戴芳澜及其学生动手办起了农场。利用昆明天然的生态条件，他在大普吉村临时设置的实验室里对云南当地丰富的地舌菌种类进行深入研究，还发现一种水稻新病害——稻一炷香病(真菌学报，1983；刘楠楠，2020)。抗战胜利之后，清华大学迁回北平，农业研究所扩大为农学院，戴芳澜改任植物病理系主任。他从昆明带回了全部科研资料，戴芳澜本人以及同仁在1945年至1948年所发表的近20篇论文都是从昆明到北平接续研究的成果。他把尽快恢复教学和研究工作放在优先位置，加紧编写教材、准备开课，还召开教授会议来商讨科研计划，夜深后仍在整理在昆明尚未完成的工作。1947年和1948年，他先后发表了《中国西部锈菌的研究》和《中国的尾孢霉Ⅱ》。1948年，戴芳澜当选中央研究院生物组院士。

从中国科学院植物研究所到微生物研究所

中华人民共和国成立之后，戴芳澜参与筹备全国第一次自然科学工作者代表会议，推动中国植物病理学会恢复和组建。1952年院系调整，北京农业大学成立，戴芳澜任教授。1953年，戴芳澜应中国科学院之邀承担植物研究所真菌植物病理研究室的筹建工作，同年任研究室主任。该研究室建立在北京农业大学校园内，以研究植物病害防治为主。他陆续聘请了北京农业大学俞大绂、林传光、沈其益、裘维蕃、王焕如五位教授为兼任研究员，后又聘请了几位副教授和讲师，进一步充实了研究室的科研队伍(中国科学院微生物研究所，2003)。同年，中国植物病理学会正式成立，戴芳澜当选理事长和《植物病理学报》主编。1955年，他当选中国科学院学部委员。此时，中国科学院拟聘沈阳农学院邓叔群教授到研究室工作。当时有人持不同意见，戴芳澜却没有门户之见。他从学术角度考虑，表示同意并保证搞好团结，还请邓担任真菌研究室副主任。

戴芳澜具有国际眼光，注重学术交流和合作。1954年，为了开展学科调查，加强国际间的业务交流，戴芳澜随着中苏友好协会苏联五一观礼团赴苏。期间参观了列宁格勒全苏科学院植物研究所和季米里亚捷夫农学院的植物保护系。他做了题为“中国植物病理学和真菌学发展情况”的报告，收到苏联植物病理学家赠送的许多珍贵书籍(马春沅，1983)。1955年，他与丁颖一起参加民主德国农业科学院成立四周年庆祝会和科学报告会，并被授予农业科学院通讯院士称号。

1955年12月，全国自然科学专门学会联合会、中国科学院和几个学会联合召开米丘林诞辰一百周年纪念会。会议筹备方拿着准备好的稿子请他在开幕式上作主报告。受李森科主义的影响，当时片面宣传米丘林学说，全部否定摩尔根学派的成果和理论。戴芳澜一口拒绝：“要我报告不合适，何况我又不是搞遗传的。”(边东子，2008)1956年，中国科学院一批科学工作者入党，戴芳澜是这批新党员中最年长的一位。

十二年远景规划制订期间，研究室正式扩建为应用真菌研究所。实际上在1954年，中国科学院高层领导和专家即已开始酝酿组建微生物研究所事宜。1955年10月6日，生物学地学部提出“建立微生物研究所筹备委员会”的意见。1958年，学部酝酿将应用真菌研究所与北京微生物研究室合并，扩建为微生物研究所。戴芳澜耐心劝说一些老同事，缓解其顾虑。事后，他从全局出发，加强来自两单位的人员的团结，采纳意见做规划，确定将扩大微生物的利用和有害微生物的防治为主要业务范围，带动微生物的形态、分类、生态、生理和遗传变异等研究（吴汝焯等，2010）。1958年12月3日，微生物研究所正式组建，戴芳澜任所长。戴芳澜脾气耿直，微生物研究所成立后由于实验室不够用，他几次给院里打报告。他的学术思想活跃，乐意接受新生事物。两单位合并不久后新成立的微生物物理室需要使用同位素，他对同事提议的成立公用同位素实验室率先表示同意。戴芳澜为人正直，凡未亲自参加工作的研究，在发表文章时从不署名，只在脚注中说明由他指导，以示负责（青宁生，2006）。

发光发热

20世纪60年代，戴芳澜着手整理几十年来搜集的资料，开始编写《中国真菌总汇》。他在夜深人静时还坚持，说“白天丢掉的时间，晚上补回来，工作要紧。”（中国科学院微生物研究所，2003）在他逝世前一个月，200余万字的初稿完成。《中国真菌总汇》于1979年出版，成为我国真菌学的巨著。

1973年1月3日，戴芳澜与世长辞。他一生为国家培养多位真菌学家和植物病理学家。他严谨的科学态度和高尚的道德风范值得后人永远学习与怀念。

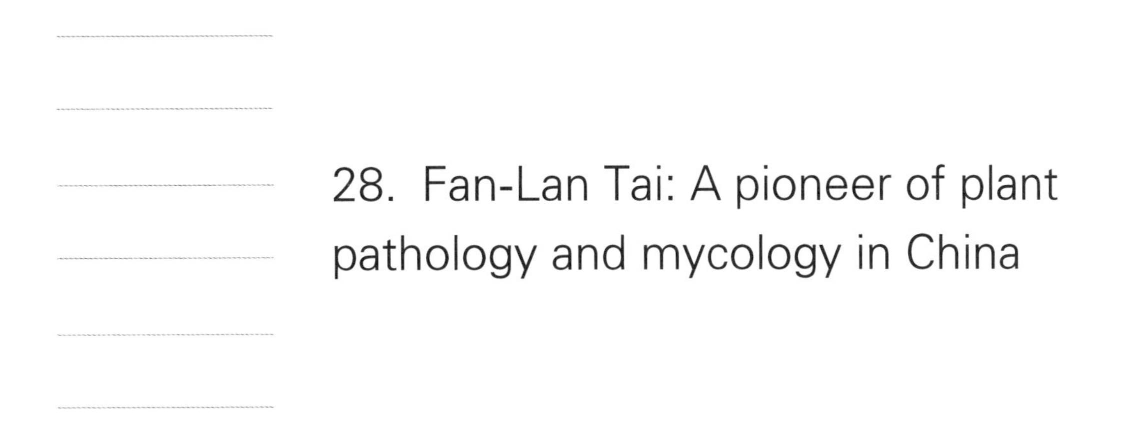

28. Fan-Lan Tai: A pioneer of plant pathology and mycology in China

Early studies in China and the United States

Fan-Lan Tai (戴芳澜, 1893—1973) was born in Jiangling, Hubei Province, to a dwindling family of bureaucratic landowners. His grandfather, Hongxi Tai (戴洪禧), was a local township official in Longquan, Zhejiang Province. His father, Jing Tai (戴经), was a senior licentiate in the late Qing dynasty. His mother died when he was eight years old. When he was twelve, he attended the Higher Primary School affiliated with Nanyang Mission College in Shanghai. He withdrew from school for a year because of his family's financial predicament. He was transferred to Wuchang Wenhua School, where he again withdrew the following year. In 1907, Fan-Lan Tai and his brother entered Shanghai Aurora School, a missionary school with a focus on French teaching, laying a firm academic foundation and reinforcing his desire to continue studying.

In 1911, he passed the entrance exam at Tsinghua School and became one of the first international students of the American Boxer Indemnity. Domestic order was not yet restored when he finished his courses in 1913. Meanwhile, because the Boxer Indemnity was unavailable, he lingered for a year before embarking on official duty abroad (Lu et al., 1991). During this time, he returned to his hometown and worked as a substitute English teacher at a business school. In 1914, he went to the University of Wisconsin to study at the School of Agriculture. During that time, he became one of the first members of the Science Society of China (SSC) and served in the "solicitation committee" of the university, responsible for introducing new members. He transferred to the Department of Plant Pathology at Cornell University in 1916 and received a bachelor's degree in agriculture two years later. During his college years, Fan-Lan Tai published his earliest paper, "Speaking of locusts" in *Kexue* magazine (Tai, 1916). In a letter to his fiancee, he said, "After four years of college, my greatest feeling is the realization that I am ignorant! Learning is endless, and I will never graduate." (Ma, 1983) He then went to Columbia

Yejing Ge

Department for the History of Science and Scientific Archaeology, University of Science and Technology of China, Hefei 230026, China

Correspondence: geyejing@mail.ustc.edu.cn

University, where he met two confidants, Robert Almer Harper (1862—1946), a pioneer in American fungal cytology, and Bernard Ogilvie Dodge (1872—1960), a mycologist and plant pathologist (Liu, 2008). However, Fan-Lan Tai had to return to China because his father was unemployed and unable to support the family.

A pioneer of plant pathology in China

In 1919, Fan-Lan Tai received his first job as a professor at Nanjing First Agricultural College. He deliberately grew a small beard to look serious and older (Ma, 1983). Due to the chaos caused by factional disputes, he left in the following year. During the summer vacation, he transitioned to gardening at a private farm in Tianjin. Soon after, he was employed as a professor at Guangdong Agricultural Special College, teaching botany and plant pathology, while also investigating diseases and insect pests in the agricultural experimental field. In 1921, he married Shuyuan Deng (邓淑媛). In Guangdong, he met Ying Ding (丁颖), a close friend with similar interests—later an academician at the Chinese Academy of Sciences (CAS). The most important contribution he made during his teaching in Guangdong was the discovery of Guangdong taro leaf blight in the summer of 1922. His paper "Taro leaf blight" published in 1923, was the first research report on *Phytophthora* in China (Tai, 1923). Due to conflicts between warlords, soaring prices, and the frequent failure of the college to pay salaries, life was difficult.

In 1923, at the invitation of Bingwen Zou (邹秉文), a classmate at Cornell University, Fan-Lan Tai became a professor of agricultural pests and diseases at National Southeast University. In addition to his lectures, he researched plant pathology and pathogenic fungi, usually going to the Institute of Biology of SSC to obtain data and complete research projects such as "Jiangsu wheat diseases and problems in Chinese plant pathology" (Wu et al., 2010). In 1927, "the university district" was reformed, and the original agricultural discipline system was abolished. Fan-Lan Tai was dismissed again. In the same year, R. H. Porter—a professor who had held a course in plant pathology at the Jinling University—returned to the United States and recruited Fan-Lan Tai as a professor and director of the Department of Plant Pathology at the Jinling University. In the 1920s, various scientific and technical societies and associations were established. He was keenly aware of the need to establish appropriate societies and organize academic exchanges to institutionalize phytopathology in China. In 1929, Bingwen Zou and Fan-Lan Tai co-founded the Chinese Society for Plant Pathology (CSPP), and Fan-Lan Tai succeeded as its president in 1936.

A pioneer of mycology in China

Shortly after Fan-Lan Tai entered the Jinling University, Albert N. Steward (1897—1959), the head of the Department of Botany, informed him that the Institute of Botany at Harvard University had commissioned the collection of all Chinese fungal specimens. Fan-Lan Tai requested that the collected specimens be sent in duplicate, with one specimen sent to the

United States and a duplicate left in China. When Steward quipped that no one could identify fungal specimens in China, Fan-Lan Tai replied, "No in the past does not mean no in the future! Let us start with this collection." and thus began research on Erysiphales (Ma, 1983). In 1930, he published "A new species of Erysiphales on *Acer buergerianum*" which was the first new species of fungi studied and published in China using modern scientific methods, denoting the founding of Chinese mycology (Tai, 1930). From the middle of the 18th century to the first half of the 20th century, Westerners continuously collected fungal specimens in China. Fan-Lan Tai was patriotic and attributed great importance to this, developing an article entitled "Collection of fungi in China by foreign explorers" . In 1931, he published "Observations on the development of *Myriangium bambusae* risk" which reported the detailed observation of the formation of the double-walled structure of Myriangiales, reflecting an approach to the study of fungal taxonomy that he developed—including the morphology and all biological traits such as cytology, developmental processes, and genetics. This article was recognized by the international mycological community and has been printed in a variety of authoritative mycology textbooks. *Introductory Mycology*, an internationally recognized authoritative textbook, still cited this work until 1994.

In 1934, Tsinghua University established the Institute of Agriculture and proposed appointing Fan-Lan Tai as the director of the Plant Pathology Laboratory. At the same time, Dafu Yu (俞大绂) returned from his studies in the United States and took over Tai's work at the Jinling University. At that time, Fan-Lan Tai had also applied for a research grant from the Education and Culture Foundation of China to study fungal genetics at the New York Botanical Garden and Cornell University (Wu et al., 2010). During his visit, he collaborated with Dodge on the taxonomy and cytogenetics of *Neurospora*, and his research papers "Two new species of *Neurospora*"and "Sex-reaction linkage in *Neurospora*"were published in *Mycologia* (Qing, 2006). At that time, he continued to follow the preparatory work of the Plant Pathology Group at Tsinghua University by correspondence.

During the War, Tsinghua University was forced to move south, and the Institute of Agriculture was relocated to the village of Dapuji in the western suburbs of Kunming. During these hard times, Fan-Lan Tai went to the city once a week to give lectures at the National Southwest Associated University and spent the rest of the time doing research. To continue their projects, Tai and his students started a farm. Taking advantage of the natural ecological conditions in Kunming, he conducted in-depth research on the abundant local species of *Geoglossum sinense* in Yunnan and discovered a new rice disease, *Ephelis oryzae*, in a makeshift laboratory (*Mycosystema*, 1983; Liu, 2020). After the War, Tsinghua University moved back to Beijing and the Institute of Agriculture was expanded to the School of Agriculture, with Fan-Lan Tai becoming the director of the Department of Plant Pathology. He brought back all the scientific materials from Kunming, and nearly twenty papers published by himself and his colleagues from 1945—1948 were the result of successive research materials brought from Kunming to Beijing. He prioritized resuming his teaching and research work as soon as possible, intensified the preparation of teaching materials, prepared to start classes, convened a meeting of professors to discuss research plans, and finished the Kunming work late at night. In 1947 and 1948, he published "Uredinales of western China" and "Cercosporae of China II" . In 1948, Fan-Lan Tai was elected as an academician in the Biology Section of Academia Sinica.

From the Institute of Botany to the Institute of Microbiology

After the founding of the People's Republic of China, Fan-Lan Tai participated in the preparation of the first national council of representatives of natural scientists, and promoted the revival and establishment of CSPP. In 1952, when the faculties were reorganized, Beijing Agricultural University was established, and in 1953, the CAS invited him to establish the Fungal Plant Pathology Laboratory of the Institute of Botany. In the same year, the CAS invited Fan-Lan Tai to be the director of the laboratory. The laboratory was established on the campus of Beijing Agricultural University and focused on plant disease control. He recruited five professors—Dafu Yu, Chuanguang Lin (林传光), Qiyi Shen (沈其益), Weifan Qiu (裘维蕃), and Huanru Wang (王焕如)—as part-time researchers, and later recruited several associate professors and lecturers to further enrich the research team of the laboratory (IM CAS, 2003). In the same year, CSPP was formally established, and Fan-Lan Tai was elected as the president and editor-in-chief of *Acta Phytopathologica Sinica*. In 1955, he was elected as an Academician of CAS. At that time, the CAS proposed hiring Professor Shuqun Deng (邓叔群) from Shenyang Agricultural College to work in the laboratory. When some members expressed disagreement, Fan-Lan Tai kept an open mind. From an academic point of view, he expressed his agreement with CAS, and promised to unify the department. The CAS invited him to be the deputy director of the laboratory.

Fan-Lan Tai cultivated an international perspective, paying attention to academic exchange and cooperation. In 1954, he went to the Soviet Union with the Soviet May Day Observation Group of the Sino-Soviet Friendship Association to carry out an investigation in the field and strengthen international exchange. He also visited the Institute of Botany of the Academy of Sciences in Leningrad and the Department of Plant Protection of the Gimiriazev Agricultural College. He gave a report entitled "The development of phytopathology and mycology in China" and received a gift of many valuable books from Soviet phytopathologists (Ma, 1983). In 1955, together with Ying Ding, he participated in the fourth-anniversary celebration and scientific presentation and was awarded the title of corresponding academician of the Academy of Agricultural Sciences of the DDR.

In December 1955, the China Association for Science and Technology, CAS, and several societies jointly held a conference to commemorate the centennial of the birth of Michurin. The organizers of the conference asked Fan-Lan Tai to give the main report at the opening ceremony with a prepared script. Influenced by Lysenkoism, the one-sided propaganda of the Michurin doctrine, and the total rejection of theories of the Morganian School at that time, Fan-Lan Tai refused and said, "It is not appropriate to ask me to report. Besides, I am not a geneticist." (Bian, 2008) In 1956, a group of scientists from CAS joined the Communist Party of China, and Fan-Lan Tai was the oldest member of this new group.

The laboratory was officially expanded to the Institute of Applied Mycology during the 12-Year Program for the Development of Science and Technology, and by 1954, the senior leaders and experts of CAS had already started to think about the establishment of the Institute of Microbiology. On October 6, 1955, the Department of Biology and Geology proposed "the idea of establishing a preparatory committee for the Institute of Microbiology" to the executive meeting of CAS. In 1958, the Department of Biology proposed merging the Institute of Applied Mycology with the Beijing Microbiology Research Laboratory and expanding it into the Institute of Microbiology. During this time, Fan-Lan Tai patiently persuaded some older colleagues to ease their concerns. Afterwards, he unified the staff from the two units by determining the opinions for planning, determining the expansion of the use of microorganisms, and research on the prevention and control of harmful microorganisms as the main scope of activities, driving the disciplines of morphology, classification, ecology, physiology, and genetic variation of microorganisms (Wu et al., 2010). On December 3, 1958, the Institute of Microbiology was formally established, and Fan-Lan Tai became its first director. After the establishment of the institute, he appealed to CAS several times because of the lack of laboratory space. He was undoubtedly an active academic thinker and willing to accept new things. When the new Microbial Physics Laboratory was established—soon after the merger of the two units—and needed to use isotopes, he was the first to agree to his colleagues' proposal to establish a common isotope laboratory. Fan-Lan Tai was also a person of integrity. He never signed his name when publishing research that he had not personally participated in, but stated in a footnote that he had supervised it, as a sign of responsibility (Qing, 2006).

Fan-Lan Tai gave light and heat

In the 1960s, Fan-Lan Tai started to compile the information he had collected over the decades and began to write the *Sylloge Fungorum Sinicorum*, which he insisted on writing in the dead of night, saying, "If I lose time during the day, I can make up for it at night. Work matters." (IM CAS, 2003) One month before his death, the first draft of more than two million words was completed. In 1979, the *Sylloge Fungorum Sinicorum* was published, and became a masterpiece of Chinese mycology.

On January 3, 1973, Fan-Lan Tai passed away. During his lifetime, he trained numerous mycologists and plant pathologists. His rigorous scientific attitude and noble character deserve to be studied and remembered by future generations.

Figures

Fig.1 Fan-Lan Tai (1893—1973)

References

Bian DZ (2008) Never Forget—Stories in the "Special Building" in Zhongguancun. Shanghai: Shanghai Education Press. (边东子. 2008. 风干的记忆——中关村"特楼"内的故事. 上海：上海教育出版社.)

IM CAS (2003) Collected Works of Xue Yugu: 80th Anniversary Celebration. (中国科学院微生物研究所. 2003. 薛禹谷文集：八十华诞纪念.)

Liu NN (2020) Fan-Lan Tai: Fanghua suiyue zhan jun zhang. China Archives 09:88-89. (刘楠楠. 2020. 戴芳澜：芳华岁月战"菌"章. 中国档案, 09: 88-89.)

Liu ZZ (2008) Ten Academicians from Jingzhou. Wuhan: Hubei Education Press. (刘作忠. 2008. 从荆州走出的十大院士. 武汉：湖北教育出版社.)

Lu JX, et al (1991) Biographies of Modern Chinese Scientists, Vol. 1. Beijing: Science Press. (卢嘉锡, 等. 1991. 中国现代科学家传记第1集. 北京：科学出版社.)

Ma CY (1983) Fan-Lan Tai, founder of Chinese Mycology. The Chinese Journal for the History of Science and Technology 01:35-44. (马春沅. 1983. 中国真菌学的奠基人戴芳澜. 中国科技史料, 01:35-44.)

Mycosystema (1983) Commemorating the 90th Anniversary of the Birth of Professor Fan-Lan Tai. Mycosystema 02:63-77. (真菌学报. 1983. 纪念戴芳澜教授诞辰九十周年. 真菌学报, 02:63-77.)

Qing NS (2006) The founding master of mycology in China—Fan-Lan Tai. Acta Microbiologica Sinica 02:169-170. (青宁生. 2006. 我国真菌学的开山大师——戴芳澜. 微生物学报, 02:169-170.)

Tai FL (1916) Speaking of locusts. Science 2(9):1030-1042. (戴芳澜. 1916. 说蝗. 科学, 2(9):1030-1042.)

Tai FL (1923) Taro Leaf Blight. Agriculture and Forestry Quarterly 1(3):29-37. (戴芳澜. 1923. 芋疫病. 农林季刊, 1(3):29-37.)

Tai FL (1930) A New Species of Erysiphales on *Acer buergerianum*. Proceedings of the Institute of Biology of the Science Society of China Plant Group 6(1). (戴芳澜. 1930. 三角枫上白粉菌之一新种. 中国科学社生物研究所论文集·植物组 6(1).)

Wu RC, Wang BZ, Xu ZH (2010) Remembering Our Mentors. Beijing: China Agricultural University Press. (吴汝焯, 王步峥, 许增华. 2010. 忆恩师. 北京：中国农业大学出版社.)

29. 邓叔群：勇往直前、信念坚定的菌物分类学先驱者

邓叔群（图1），字子牧，我国著名真菌学家，一生致力于真菌分类学、植物病理学和森林生态学研究，著有我国第一部真菌分类研究专著《中国高等真菌》，被誉为“才高行洁，德高望重”，是“中国分类学家之楷模”。他不仅是一位精于真菌研究的学者，更是一位正直、无私、忘我工作的爱国主义者，一位家人眼中的好父亲（程光胜；Deng et al.，2012；Korf，2002）。

图1　邓叔群教授（1902—1970）

邓叔群出生于福建省闽侯（今福州市）一个多子女的教师家庭，在当地接受了小学教育后，于13岁时通过了清华学校的招生考试。1915—1923年，他一直在清华学校庚子赔款留美预科班学习。

1923年，邓叔群进入康奈尔大学农学院林学系学习，并于三年内先后获得学士和硕士学位。他经常在课余时间到郊区农场从事木工和农业工作（图2）。在攻读植物病理学博士学位期间，他获得了全美荣誉学者学会的Phi Kappa Phi和Sigma Ⅺ奖章。1928年，为不耽误岭南大学开课时间，邓叔群在完成博士论文前回国，对于他来说，为国服务是最重要的事情，而获颁学位则轻如鸿毛。

1928年到1931年，邓叔群先后在岭南大学、金陵大学和国立中央大学任教，主要致力于棉花、小麦和水稻病害的研究工作，并教授真菌学和植物病理学课程。在1932年至1941年，邓叔群先后在中国科学社生物研究所、中央研究院自然历史博物馆、中央研究院植物研究所以及中央研究院林业实验研究所担任研究员和副所长等职务。我国虽然幅员辽阔，气候多样，林业和真菌资源丰富，但是当时这些方面的研究工作一直是个空白，在邓叔群回国的第一个十年（1928—1938年）中，他一共鉴定了由他本人采集的我国400多个属的1400多个分类单元的真菌标本，发表研究论文34篇，其中

作者：李国杰

中国科学院微生物研究所真菌学国家重点实验室，北京100101，中国

邮箱：liguojie.imcas@foxmail.com

包括新种121个，新属5个（图3）。1939年，他撰写的我国第一部真菌专著《中国高等真菌》出版，这本专著收录了我国子囊菌、担子菌和半知菌类23目、75科、387属共1391个种的形态、习性和分布信息。

图2　邓叔群教授在野外工作

图3　邓叔群教授的手稿

自20世纪30年代开始，邓叔群的研究方向转向森林生态学。应当时国民政府农林部的邀请和资助，他与12位年轻的同事一起，于1939年在我国西南部的西康、云南和四川的原始森林开展了数月的调查工作，初步查明了这一地区森林的组成、分布、蓄积量和病虫害。1940年，由于日军不断加紧对我国的侵略，邓叔群和同事将2000余份我国真菌标本的备份打包装箱，经中南半岛运往美国保管。这些标本一路先后换乘牛车、火车和轮船等多种交通工具。虽然这些标本在中国的部分几乎全部毁于战火，但在美国的部分一直被妥善保管。这些标本一共2278份，已于2009年被送还中国。1941年，邓叔群拒绝了担任国民政府农林部副部长的邀请，举家搬迁到甘肃卓尼的洮河林场开展林业工作。当时洮河上游流域尚未开发，人烟稀少，他不得不经常随身带着枪支工作，面部曾经严重受伤，最疼爱的小女儿也被洮河冲走，不知下落（Deng et al., 2012）。尽管遇到各种困难，他从未退缩过。在当时他写的一首诗中，他乐观地写道："让我们的热血燃烧起来。"在洮河林区工作的过程中，他提出了保证森林营造、木材生产、维护生态平衡、保证农牧业发展以及减少水土流失的一整套可持续发展策略，是我国林业史上的里程碑（程光胜）。

抗日战争胜利后的1946年，邓叔群回到南京，重建中央研究院植物研究所林业生态实验室，整理并发表了在甘肃期间的森林生态学、自然林管理和森林地理学研究成果。1948年，他被选为中央研究院院士。1950—1955年赴东北的沈阳农业大学和东北农业大学，领导并参与了学校的规划选址和建设工作。

1955年，邓叔群被选为中国科学院学部委员（院士）。1956年，邓叔群被任命为中国科学院应用真菌研究所副所长和真菌实验室副主任。尽管行政事务繁忙，但是他仍然坚持真菌研究工作，经常通宵达旦地工作。为了更好地让青年科研工作者学习真菌知识，他将1939年发表的真菌学专著翻译为中文发表，并增加了黏菌和霜霉等类群的400多幅插图，这本题为《中国的真菌》的专著收录的种类多达2400余种（程光胜）。目前亚洲保藏菌物标本最多的中国科学院微生物研究所菌物标本馆，也是当时由邓叔群主持建立的，他所设计的标本盒和标本柜尺寸，制定的标本编号、检索和借阅规章制度一直沿用至今（郑儒永，2002）。鉴于我国华南地区丰富的真菌资源，邓叔群还创建了中国科学院中南真菌研究室（今广东省微生物研究所）。

由于历史原因，邓叔群的部分论著和鉴定的标本已经散佚，但是他为我们留下的标本和著作，早已成为当代中国真菌分类研究的基石，他为中国真菌研究作出的贡献将会被永远铭记。

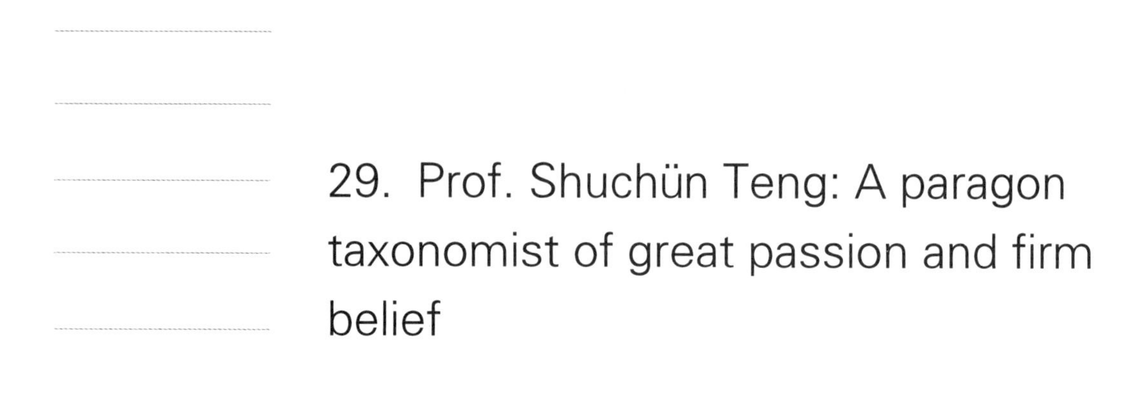

29. Prof. Shuchün Teng: A paragon taxonomist of great passion and firm belief

Shuchün Teng (or Shu-Chun Teng, 1902—1970), courtesy name Zimu (子牧), was one of the most famous Chinese mycologists. His main achievements were in fungal taxonomy, plant pathology and forest ecology. Being the author of *Higher Fungi of China*, the first monograph of fungi in China, he was regarded as "a paragon taxonomist" and "truly an immerse taxonomic legacy in which all China and all Chinese people can find pride" . In addition to being a studious mycologist in pursuit of truth and honesty all his lifetime, his diligence, thoroughness, optimism, and patriotism also made him a conscientious leader, a respected mentor, a brave mycology and forestry pioneer and a good father (Cheng; Deng et al., 2012; Korf, 2002).

Born and raised in a large and needy family in Min County (currently known as Minhou, Foochow or Fuzhou), China, Shuchün Teng's parents are both teachers. He received early childhood elementary education in primary school of Minhou. By taking part in the entrance examination of Tsinghua College, he got the admission qualification of Tsinghua in 13 years old. He was there as a student of preparatory course class for the Boxer Indemnity Scholarship Program (Cheng) in 1915—1923.

In 1923, Shuchün Teng entered the Department of Forestry, College of Agriculture in Cornell University, and got his bachelor's degree in Agriculture and master's degree in Forestry within 3 years. During that time, he also worked as a part-time farmer and carpenter in suburb farms. When taking his Ph.D. courses on phytopathology, he won two gold key badges for Phi Kappa Phi and Sigma XI, respectively, and was invited as the member of the two honor societies of science. With these honors, he went back to China in 1928, even before the completion of his Ph.D. The most important thing at that time was getting ready to work on the mycology in China, so the granting of the doctoral degree was as light as a feather for him (Cheng).

From 1928 to 1931, Shuchün Teng was professor in Lingnan University, Private University of Nanking, and National Central University. He mainly worked on the disease control and treatment of rice, wheat and cotton, and in the meantime, he also taught mycology and plant

Guojie Li

State Key Laboratory of Mycology, Institute of Microbiology, Chinese Academy of Sciences, Beijing 100101, China

Correspondence: liguojie.imcas@foxmail.com

pathology. From 1932 to 1941, he was appointed as the professor in Institute of Biology, Science Society of China, Central History and Nature Museum, Institute of Botany, Academia Sinica, and Institute of Forestry Experiment, Academia Sinica. China has a variety of forestry and fungal resources, because of the large territory which covers several climate zones and, but the investigations of such resources at that time were almost blank. During the first ten year after his returning to homeland (1928—1938), he identified 1400 taxa of 400 fungal genera based on specimens collected by himself, and published 34 research articles, in which 5 new genera and 121 new species were described. In 1939, his first monograph *A Contribution to Our Knowledge of the Higher Fungi of China* was published. This book covered the morphology, hosts, habits and habitats information about 1391 species of 23 orders, 75 families, and 387 genera in Basidiomycota, Ascomycota and Deuteromycota (viz. imperfect fungi).

From the early years of 1930s, Shuchün Teng switched his attention to forestry ecology. He organized a survey team with the grant support from the Ministry of Agriculture and Forestry. In 1939, he and his 12 young colleagues spent several months on investigating the virgin forests of Sichuan (四川), Yunnan (云南) and Xikang (西康) in southwestern China. They clarified the component, distribution, cumulation, diseases and insect pests of the forest in these areas. As the invasion of Japanese Aggression went on in 1940, Shuchün Teng and his colleagues packed up and sent over 2000 Chinese specimens to the US via Indochina Peninsula by ox-cart, train and ship. Although the specimens he collected remained in China were almost destroyed during that time, the duplicates kept in US are still in good conditions today, 2278 specimens of which were returned to China in 2009. In 1941, Prof. Teng declined the deputy minister of Ministry of Agriculture and Forestry, and moved to Jone (卓尼), Gansu Province, the central area of the Taohe River (洮河) forest with his family. At that time, full of danger, the upstream drainage of Taohe River was a remote, wild, inaccessible area. He travelled over the hills and through the rivers, suffered the rigors of living in the wilderness, and kept his gun in hand all the time. In the harsh natural environment, his face was severely injured one time, and the flooded Taohe River took away his nine-year-old third daughter forever (Deng et al., 2012). Despite all of the bitterness and hardships, he never shrank back. In one of his optimistic poems, it reads "Let your hot boiling blood keep you aglow" . He proposed a complete forestry management strategy to ensure the forest construction, and its regeneration greater than the amount of cutting. He brought forward the ecological balance theory of forest in order to keep the local ecological system beneficial to forestry agriculture, farming and animal husbandry, and to decrease the soil erosion caused by flood. His efforts on ecological forestry are good samples for his successors in Gansu Province, and he was regarded as "a commemorable milestone" (Cheng).

Shuchün Teng went back to Nanjing in 1946 after the winning of the War of Resistance against Japanese Aggression. He rebuilt the lab of fungi and established a lab for forest ecology in Institute of Botany, Academia Sinica. The data about forest and fungi that he obtained and collected in Gansu Province were reorganized, summarized and published. These publications were mainly about the forest ecology, forestation and natural forest management, forest management of eastern Qinghai-Tibetan Plateau, and forest geography in China. In 1948, he was elected as an academician of Academia Sinica. He was temporarily appointed as dean of the Shenyang Agriculture College and vice chancellor of Northeast Agriculture College from 1950—1955. He dedicated all of his time and energy to the construction of the two new colleges, such as making decisions on the campus location and planning, reviewing the capital blueprint, ordering experiment equipment, etc (Cheng).

In 1955, He was appointed as an academic committee member (known as academicians), Chinese Academy of Sciences. In 1956, Shuchün Teng was reassigned as deputy director of Institute of Applied Mycology, Chinese Academy of Sciences and the dean of the Mycology Lab. Despite his fully filled work schedule in administration, he held on to his own fungal research at countless nights until dawns and in holidays. In order to help the young people to improve their mycological knowledge, he translated and renewed his monography (published in 1939) in Chinese version, 400 plates including Phycomycetes and slime mold taxa were added, and the total number of the species in this book entitled "*Fungi of China*" reached up to 2400 (Cheng). He was the founder of Herbarium of Mycology, Institute of Microbiology, Chinese Academy of Sciences (HMAS), which is the largest collection of fungal specimens in Asia at present. The basal management of the herbarium, including exsiccatae cabinet design, accession code arrangement, index system and regulations for specimen loan, all drafted by him have been retained up to today (Zheng, 2002). According to the abundant sources of fungi in southern China, he also established the Central South Lab of Fungi, Chinese Academy of Sciences, which is currently known as Guangdong Institute of Microbiology, the mycological research center in South China.

Although many of his research findings are still not published, the specimen collections and publications of Shuchün Teng are regarded as the backbone for almost all the current fungal taxonomy work in China. His contribution to the development of the mycology in China is commemorated forever.

Figures

Fig.1 Prof. Shuchün Teng (1902—1970)

Fig.2 Prof. Shuchün Teng in field work

Fig.3 The manuscript of Prof. Shuchün Teng

References

Cheng GS. Biography of Shuchün Teng (submitted). (程光胜. 邓叔群传(未正式出版).)

Deng Y, Deng Z, Deng H, et al (2012) A capitalized man: Memoir of father Shuchün Teng. Information on Publication: 26–27.

Korf RP (2002) S.C. Teng, a Fitzpatrick student, a model taxonomist. Mycosystema 21(4):473–474. (Korf RP.

2002. 邓叔群:菲茨帕特里克的弟子,分类学家之楷模(英文). 菌物系统, 21(4): 473–474.)

Zheng RY (2002) Reminiscences of the days with Prof. S.C. Teng in the Mycology Department of the Institute of Microbiolgy. Mycosystema 21(4): 468–472. (郑儒永. 2002. 回忆邓叔群先生在微生物研究所真菌室的日子. 菌物系统, 21(4): 468–472.)

30. 俞大绂：人民的农业科学家和教育家

俞大绂（图1），著名的科学家、教育家，卓越的植物病理学家、微生物学家，我国植物病理学和农业微生物学的奠基人之一。他与大多数老一辈科学家一样，是早年抱着科学救国的理想而留学海外，又毅然回到疮痍满目的中华大地的祖国优秀儿女。自1924年从事植物病害的研究起，他一生坚持理想信念，潜心治学，竭力探索，无私奉献，不但身体力行，而且言传身教，为国家培养了几代栋梁之材，为发展我国的农业科学和教育事业作出了卓越的贡献。

图1 俞大绂先生（1901—1993）

俞大绂于1901年出生于南京的一个知识分子家庭，此时的中国内忧外患，国弱民穷，俞大绂的父母安排子女修习不同的专业，期冀成才后能以科学、教育、实业救国。壮志满怀的俞大绂眼见祖国贫瘠，农业落后，植物病害严重，便决定放弃最初的理化专业而选择了更为艰苦的农科。1924年毕业于南京金陵大学农学院，获学士学位后留校任助教。1928年，俞大绂怀着“科学救国”的理想，到美国艾奥瓦州立大学深造，1932年获得博士学位和金钥匙奖，成为美国植物病理学会会员，Sigma-Ⅺ荣誉会员，并获得美国“斐陶斐奖”。

毕业回国后，俞大绂历任南京金陵大学教授（图2），清华大学农业研究所教授，北京大学农学院院长、教授，中央研究院院士、评议员。新中国成立后，他一直在北京农业大学任教，并担任校长。1955年被选为中国科学院首批学部委员（院士），1956年当选苏联农业科学院通讯院士。俞大绂先生是第二、三届全国政协委员，第四、五、六届常务委员（青宁生，2007），先后任中国农学会副理事长、中国植物保护学会和中国植物病理学会理事长等多项职务。

译者：赵爽
北京市农林科学院农产品加工与食品营养研究所，北京 100097，中国
邮箱：shuangzhaow@163.com

俞大绂先生是我国现代植物病理学许多分支学科领域的开拓者和先驱。他开拓了我国植物细菌学、植物病毒学、植物抗病育种和病生理等多方面的研究领域，其研究的作物对象涉及禾谷类、豆类、果树和蔬菜等，致病菌涉及真菌、细菌和病毒，防治策略涉及种子处理、药剂防治和抗病育种等。他率先报道了小麦秆黑粉菌具有生理分化性，开创了我国生理育种研究的先河，是我国植物病理学和微生物学当之无愧的奠基人和开拓者之一。俞大绂先生在国内外学术刊物上发表研究论文110余篇，专著10部，译著8部，其中绝大部分是他本人研究结果的第一手资料。

早在20世纪20—30年代，俞大绂先生就致力于禾谷类作物抗病育种及种子消毒研究，发现大麦黑穗病和条纹病的侵染规律，育成抗黑粉病小麦、抗枯萎病的大豆、抗稻瘟病的水稻等品种，并在我国首先开创了对黑粉病菌生理小种的研究。40年代，俞教授应邀在西南联合大学清华农业研究所任教，他在抗日战争极其艰难的情况下坚持科学研究，率先报道了小麦秆黑粉菌具有生理分化性和我国若干果品贮藏病害，还开展了粟病害和蚕豆病害的全面研究，发表了多篇科技论文，其科研资料是我国此类病害的经典资料。

新中国成立后，俞大绂的满腹学识和一腔赤诚有了更多的施展空间，当时全国各地陆续出现大面积的苹果树腐烂病、马铃薯晚疫病、谷子红叶病、柑橘疮痂病等重大植物病害（俞大绂，1951），俞大绂受农业部的委托，带领防治专家组和学生，风尘仆仆奔走于大江南北的灾区，深入地调查研究（图3），提出了有效的防治措施，及时控制了病害的蔓延（俞大绂，1953），并在防治实践中培养了一批植物检疫人才，为我国的植物病害检疫工作奠定了基础。1958年俞教授和他的同事在北京农业大学首先开展了赤霉素的研究工作，培养出了我国特有的优良赤霉菌菌种，填补了我国赤霉素研究和生产的空白（Rong et al.，2017）。20世纪60年代，俞教授的研究逐步转向基础理论，主要集中在真菌异核遗传、镰刀菌的分类与鉴定、赤霉菌高产菌株的选育（俞大绂，1977），揭示了赤霉菌在自然界中的三种不同核型组成异核体，证实了各异核菌系在赤

图2　俞大绂在南京金陵大学实验室工作

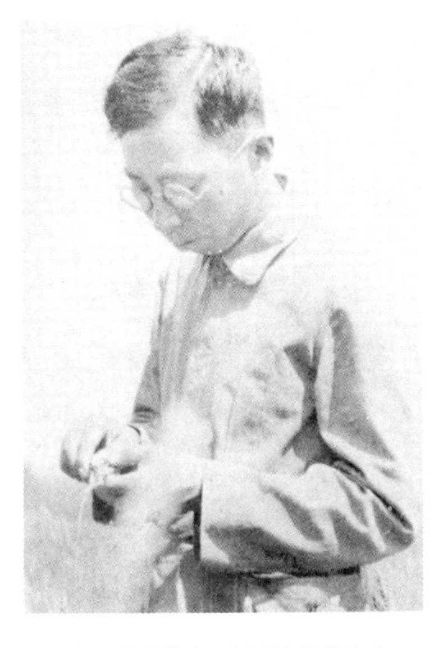

图3　俞大绂在田间观察植物病害

霉素产量和寄生力上的差异，有力地阐明了“异核现象在自然界中是普遍存在的”这一长期争论的问题，不仅证实和确认真菌异核体是病原真菌发生变异的重要途径，而且最终提高了抗病育种的针对性，把对植物病害的防治工作由被动治病转为主动防病，对农业生产具有重要的长远意义。此项研究当时处于世界先进水平，并在1980年获农牧渔业部科技成果一等奖。俞大绂先生的科研工作，一直是把解决农业生产问题和促进学科发展紧密结合起来，一生都在为解决我国农业生产中的病害问题而忙碌。他这一生在学术之路上取得了令人瞩目的成就，结合我国农业生产实践，极大推动了我国农业发展。

俞先生在致力于农业科研的同时，还投身于农业教育，是我国著名的农业教育家。抗日战争胜利后，俞大绂先生受命重建北京大学农学院，他殚精竭虑，不遗余力，从国外延聘了大批知名学者来校任教，建立起完整的学科框架和教学管理体制，主张民主、公开办学。俞先生的民主爱国思想，深受人们的敬佩。他旗帜鲜明地反对国民党反动派对学生爱国民主运动的镇压，保护师生的爱国活动，拒绝南京国民政府的邀请，毅然坚守岗位，保护校园财产。率留校师生携带贵重设备接受共产党的邀请前往解放区，继续科研，他的事迹收录在纪录片《共和国之恋》中。20世纪80年代，80岁高龄的俞先生再次出任北京农业大学校长，审时度势，从我国“农情”出发办高等农业教育，提出“农”与“工”的结合是农业院校的办学方向，农业大学要向着农业科学技术大学方向发展。俞大绂先生从教70年，以育人为己任，为国家培养了几代人才（图4）。众多弟子踏着恩师的足迹，在所从事的教育、科研及管理实践中，成为著名专家、教授和业务骨干，如方中达、林传光、裘维蕃院士、李季伦院士（Rong et al.，2017）、曾士迈院士、陈文新院士等。

图4　俞教授指导学生做研究

俞大绂一生为我国农业耕耘70年，为我国植物病理学和植物检疫、农业微生物学的发展奠定了基础，是受人敬爱的人民科学家。古话说“十年树木，百年树人”，俞先生集育树和育人于一身，科研硕果累累，桃李满天下，他的这种精神影响着一代代年轻的科研工作者，他对我国农业发展的伟大贡献将铭记于国人之心。

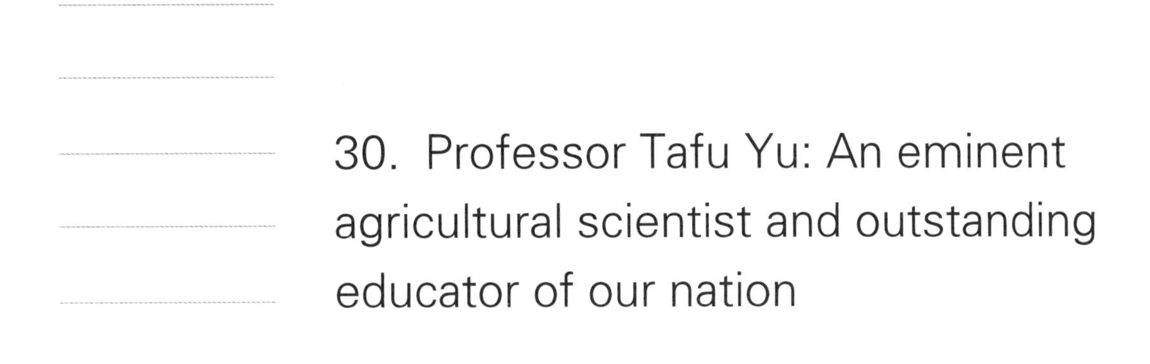

30. Professor Tafu Yu: An eminent agricultural scientist and outstanding educator of our nation

Prof. Tafu Yu (1901—1993) is a pioneering scientist of phytopathology and agricultural microbiology in China. Like most of the scientists of his generation, Prof. Yu gave up the attractive conditions abroad and went back to China with the determination to rescue the nation with his knowledge. He has devoted his entire life to the development of phytopathology in China since 1924. Prof. Yu is also a distinguished educator who has instructed and inspired several generations of talent students for the country and has made outstanding contributions to the development of the agricultural science and education in China.

Prof. Yu was born in 1901 in a well-educated family in Nanjing, Jiangsu Province. At that time, China was suffering from domestic strife and foreign invasion, with weakness and poverty being the major issues. Prof. Yu's parents hoped that their children could study different disciplines so that they can use their knowledge later on to rescue the country in different ways, such as science, education and industry. Prof. Yu gave up his original major in physics and chemistry and decided to study agriculture in order to solve the problems of poverty, backward agriculture and severe plant diseases for China. In 1924, he obtained his bachelor's degree from the College of Agriculture in Nanjing Jinling University. During that time, he also worked as a teaching assistant. In 1928, Prof. Yu continued his study in Iowa State University in the United States and received his doctoral degree and a Golden-Key award in 1932. Then he became a member of the American Phytopathological Society and Sigma-XI. He also received a "Phi Beta Kappa" award of the United States.

Prof. Yu returned back to China after graduation in 1932 and served as a professor at Nanjing Jinling University, and later on as a professor at the Agricultural Research Institute of Tsinghua University. He was then appointed as a director and professor of the College of Agriculture in Peking University, prior to taking up the positions of academician and senator

Shuang Zhao[1], Tiantian Xu[1], Hexiang Wang[2]

1 Institute of Plant and Environment Protection, Beijing Academy of Agriculture and Forestry Sciences, Beijing Engineering Research Center for Edible Mushroom, Beijing 100097, China

2 State Key Laboratory of Agrobiotechnology and College of Biological Sciences, China Agricultural University, Beijing 100193, China

Correspondence: hxwang@cau.edu.cn (H.-X. Wang)

of the Central Research Institute. Following the founding of the People's Republic of China, he started his career as an educator at Beijing Agricultural University as the President. In 1955, he was elected as one of the first generation of academicians of the Chinese Academy of Sciences and one year later, he was elected as the Communications Academician of the Soviet Academy of Agricultural Sciences. Prof. Yu was a member of the second and third CPPCC national committee and a member of the fourth, fifth and sixth standing committee (Qing, 2007). He also served successively as vice president of Chinese Association of Agricultural Science Societies, and chairman of China Society of Plant Protection and chairman of Chinese Society for Plant Pathology.

Prof. Yu was a pioneer of modern plant pathology and also the authority of plant bacteriology, plant virology, plant disease resistance breeding, disease physiology and other areas in China. He studied crops such as cereal crops, beans, fruit trees and vegetables, as well as crop pathogens including fungi, bacteria and viruses. He adopted the prevention and control strategies, including seed treatment, chemical prevention and control, cultivating disease-resistant varieties etc. Prof. Yu discovered that *Urocystis tritici* exhibited the characteristic of physiological differentiation and initiated the researches on physiological breedings in China. Prof. Yu published more than 110 research articles in international and domestic academic journals, 10 monographs and 8 translated articles.

In the 1920s and 1930s, Prof. Yu investigated the breeding of disease-resistant cereal crops and seed disinfection methods, clarified the infection patterns of barley smut and stripe disease, bred crop varieties such as smut-resistant wheat, blight-resistant soybean, and blast-resistant rice. He also launched a study on the physiological strains of the smut fungus.In 1940s, Prof. Yu was invited to teach in Tsinghua Agricultural Research Institute in Southwest Associated University in Kunming. At the time during the War of Resistance against Japanese Aggression, he conducted scientific researches under extremely difficult conditions. He discovered that *Urocystis tritici* displayed the characteristics of physiological differentiation and performed investigations on diseases that occurred during fruit storage. Concurrently, he also carried out a comprehensive study on the diseases that affected the millet and broad bean. Prof. Yu published numerous scientific papers which represent a classic literature of many critical plant diseases in China.

After the founding of the People's Republic of China, Prof. Yu had more freedom for his scientific researches. At that time, large area of farmland in the country was damaged by plant diseases, such as apple valsa canker disease, potato late blight disease, millet red leaf disease and citrus scab disease (Yu, 1951). Appointed by the Ministry of Agriculture, Prof. Yu led a research team to investigate the affected farmland, to prevent the spreading of the plant diseases, they try their best to conduct in-depth researches, execute effective prevention and control methods (Yu, 1953). He trained a number of plant quarantine personnel for the prevention and control of plant diseases, which laid the foundation of the plant disease quarantine work in China. In 1955, he was honored as a member of the first Chinese Academy of Sciences Committee and his work was highly praised by the Party Central Committee and the State Council. In 1958, Prof. Yu and his colleagues carried out researches on gibberellins in Beijing Agricultural University, where they developed the *Gibberella zeae* strains and filled in the gaps in gibberellin research and production in China (Rong et al., 2017). During the 1960s, Prof. Yu mainly focused on the basic researches and studied fungal heteronuclear genetics, classification and identification of *Fusarium* species as well as selected breeding of high-yield *Gibberella zeae* strains (Yu, 1977). His studies revealed that *Gibberella zeae* formed heterokaryons in nature from three different karyotypes, and proved that strains with different heterokaryons varied in gibberellin production and parasitism. His findings effectively demonstrated whether the heteronuclear phenomenon prevailed in nature, a long-term controversial question. It not only confirmed that fungal heterokaryon was implicated in mutations of pathogenic fungi, but also improved the targeting of breeding for disease-resistance. His achievements facilitated the upgrading of the methods for prevention and treatment of plant diseases from passive therapy to active prevention, which had a significant long-term impact on agricultural production. This achievement attained an advanced scientific level and was awarded the first prize of Science and Technology Achievements from Ministry of Agriculture, Animal Husbandry and Fisheries in 1980. Prof. Yu's researches implemented an intimate association between solving agricultural production issues and promoting the development of agricultural producing system. Prof. Yu spent his entire lifetime to tackle the diseases encountered in agricultural production in China. He has made spectacular accomplishments for China's agricultural production practice conditions, which profoundly enhanced the development of agriculture in China.

Besides devoting himself to agricultural researches, Prof. Yu was also dedicated to agricultural education. He was truly a prestigious and influential agricultural educator of China. After the victory of the War of Resistance against Japanese Aggression, Prof. Yu was appointed to rebuild the College of Agriculture in Peking University. He invited a large number of distinguished scholars from abroad to give lessons in the College of Agriculture, established a complete discipline framework and teaching management system and advocated a democratic and open minded education system. Prof. Yu's patriotism was highly respected. He opposed the Kuomintang (KMT) reactionaries who suppressed students' patriotic and democratic movement, and supported the patriotic activities of teachers and students. He rejected the invitation of Nanjing government, remained unwaveringly in his position and protected the campus and properties of the university. Meanwhile, he accepted the invitation of the communist party, stayed with the teachers and students in the university and then transported important equipment to the liberated areas and resumed scientific researches there.

His eminent deeds were recorded in the documentary "Love of The Republic" . In the 1980s, at the age of 80, Prof. Yu took the position of the President at Beijing Agricultural University again. He proposed the concept of combining "agriculture" and "engineering" for running the College of Agriculture and promoted the development of the Agricultural University toward a university of agriculture, science and technology. Being a teacher for over 70 years,

Prof. Yu educated several generations of talented students. Many of his students followed his footsteps and continued to make tremendous contributions in areas of education, science and management. Some of them are well-known, such as Zhongda Fang, Chuanguang Lin and the academicians Wenxin Chen, Jilun Li (Rong et al., 2017), Weifan Qiu and Shimai Zeng.

Prof. Yu, a highly respected scientist, has dedicated his entire life to the development of agriculture in China. He laid the foundation for plant pathology and agricultural microbiology in China. There is a famous Chinese proverb: It takes ten years to grow trees, but a hundred years to rear people. Prof. Yu not only imparted knowledge, but also educated his students how to behave. His spirit and exemplary roles influenced generations of young researchers and his prominent contributions to the development of China's agriculture will never be forgotten.

Figures

Fig.1 Prof. Tafu Yu (1901—1993)

Fig.2 Tafu Yu worked in laboratory at Nanjing Jinling University

Fig.3 Tafu Yu observed the plant diseases in the field

Fig.4 Prof. Yu supervised the students to do research

References

Qing NS (2007) Teaching in agricultural microbiology for 70 years—Tafu Yu. Acta Microbiol Sin 47(1):6–7. (青宁生. 2007. 执教农业微生物学70年——俞大绂. 微生物学报, 47(1): 6–7.)

Rong C, Liu Y, Wen Y (2017) Jilun Li and his profound impact on theoretical and applied agricultural microbiology. Protein & Cell 8(3):159–161.

Yu TF (1951) Grape brown blotch disease. Sci Agric Sin 12:7–8. (俞大绂. 1951. 葡萄褐斑病. 中国农业科学, 12: 7–8.)

Yu TF (1953) Implementing campestral sanitation work to eliminate the threat of diseases. Bul Agric Sci 10:411–414. (俞大绂. 1953. 展开田间卫生工作以减轻病害的威胁. 农业科学通讯, 10: 411–414.)

Yu TF (1977) Significance of *Fusarium* fungi systematics. Acta Microbiol Sin 17(2):163–171. (俞大绂. 1977. 镰刀菌分类学的意义. 微生物学报, 17(2): 163–171.)

31. 樊庆笙：中国农业微生物学、农业系统、农业教育和抗生素研究的先驱

地球生物圈中，微生物无处不在。它们对农业生态、营养循环、环境保护以及动物与人类健康都有着非常重要的作用。尽管我们已认识到微生物是物质代谢中的分解者，但对它们在其他方面的作用和开发利用价值的认识仍有待加强。本文纪念的科学家正是我国最早研发和利用微生物的先驱——樊庆笙先生（图1）。樊先生不仅为我国的青霉素生产事业作出了巨大贡献，还建立了利用微生物发展可持续农业的综合体系。无论是利用微生物从农业废弃物中分离转化新产品来提升农业综合生产力，还是使用微生物清理水体与农田环境污染物，提高土壤肥力，都证实了，他是一个极有远见的人。而在建立以微生物为中心发展可持续农业理论方面，以及坚持60年不懈努力研究各类微生物在众多生态位上的差异方面，更展现了他的前瞻智慧（图2）。

图1　樊庆笙先生（1911—1998）

樊先生于1911年8月4日出生在江苏省常熟县西周市镇（今属常熟市碧溪镇），地处江南水乡，濒临长江，并依托于上海北方。家中的九个孩子，他排行老大。自幼生活在农村，他一面跟随渔民在长江捕鱼，一面观察庄稼、蔬菜和林木等农作物的生长规律，采集果实为家中补贴生计，愉快地度过了他童年的大部分时光（樊真美，2011）。由于目睹了饥荒、贫困以及疾病对社会的影响，他深知农家生活之艰辛和改进

译者：张保元

中国科学院北京生命科学研究院，北京100101，中国

邮箱：zhangby@biols.ac.cn

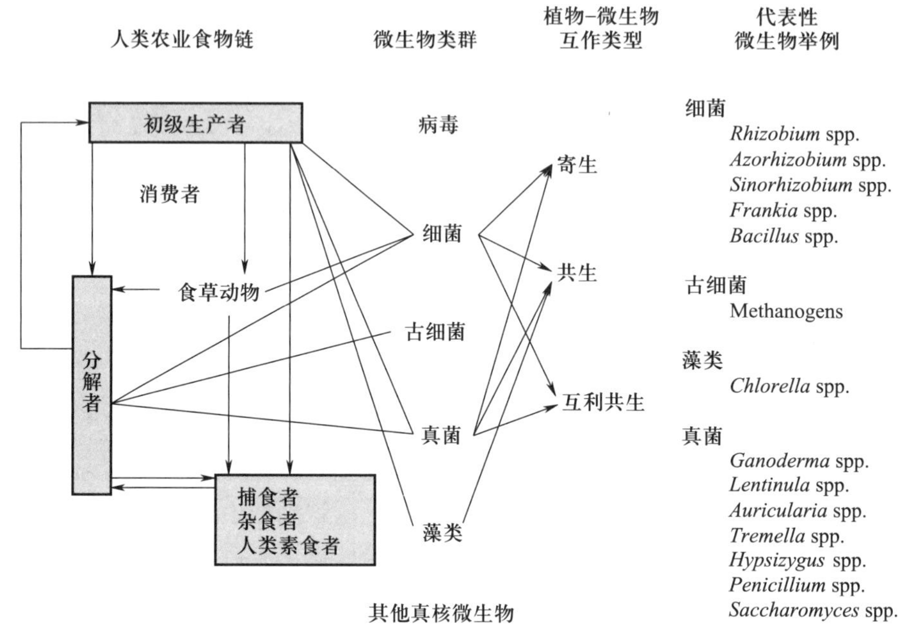

图2 以微生物为中心的农业微生物及其在广义系统农业框架中的作用。研究的生物实例如图所示。樊庆笙教授研究了这些微生物在提高作物生产力（如根瘤菌）、减少疾病（如青霉和链霉菌）、消除和/或将农业废物转化为有价值的物品（如产甲烷菌和食用菌）以及提高土壤肥力（如豆科植物和根瘤菌）方面的潜在作用

农业生产技术的必要，并对农业和农学产生了由衷的热爱。像大多数家庭一样，即使生活贫苦，为孩子提供最好的教育环境仍是家中的头等大事，樊先生的家庭也不例外。樊先生非常珍惜学习机会，从小学到高中，他的成绩一直名列前茅。1929年毕业于苏州萃英中学后，被保送进入金陵大学农学院森林系。由于缺乏资金，他的家人不得不借钱来支付他在金陵大学的首年学费。但由于成绩优异，他获得了全额奖学金，资助了他第二年到第四年的学习与科研。同时还被聘请为教学助理，帮助老师组织准备实验课程，这在当时的本科生中是非常罕见的。毕业的时候，他拿到了金陵大学应届生中的最高奖项“金钥匙奖”（樊真美，2011）（图3），并留校任助理讲师，负责植物学和植物分类学的教学工作。任教期间，他建立了当时我国最大的植物标本馆（李宗道，2003）。本科期间扎实的植物学基础，对樊先生建立植物-微生物相互促进农业可持续发展体系起到了重要的作用。

图3 樊庆笙1933年毕业于金陵大学，获林学学士学位

1937年，抗日战争全面爆发，政府和大批民族工业机构以及高校均被迫西迁，金陵大学迁至了四川省成都市华西坝。在那里，樊先生被提升为讲师并继续教授植物学（李宗道，2003）。1940年夏季，洛克菲勒基金会为金陵大学农学院的青年学者提供

一个留美学习的名额。尽管如此，农学院并没有只派遣一位学者去攻读三年的博士学位，而是决定充分利用这笔资金与机遇，申请并通过基金会的批准，派遣了三名青年学者到美国进行为期一年的学习，攻读理工硕士学位。樊先生就是三名青年学者中的一员，他在威斯康星大学麦迪逊分校农业细菌学系开始了新的学习生活。1941年，获得硕士学位后，樊先生的优异成绩和才华引起了微生物学家W. W. Umbreit教授的注意，他建议樊先生继续攻读博士学位，跟随自己研究海藻小球藻的光合作用生理学（图4）。1943年，樊先生顺利完成了博士学业，他的两篇研究成果分别发表在《普通生理学杂志》（Fan et al.，1943）和《细菌学杂志》（Fan and Umbreit，1943）上。

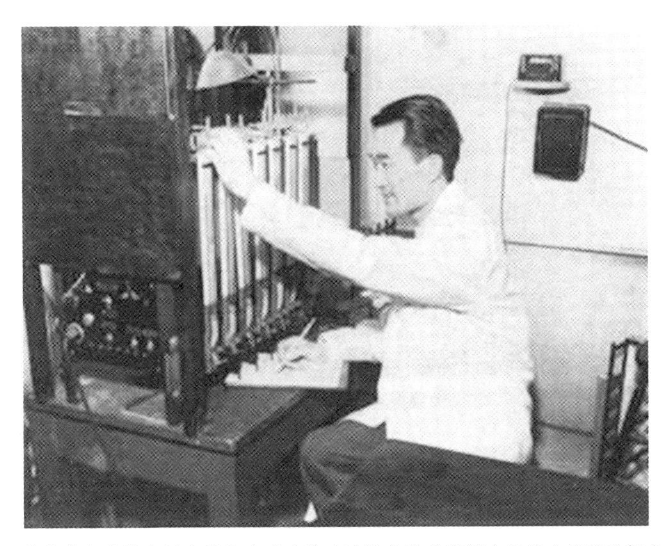

图4　樊庆笙在威斯康星大学麦迪逊分校攻读博士学位期间在气体交换试验机上工作

樊先生在威斯康星大学麦迪逊分校的三年时间里，认真学习了微生物学的各类理论知识，并进行了相关的探索研究。当时，该校农业细菌学系有两个国际领先的研究项目，分别是根瘤菌与豆科植物共生固氮项目和抗生素的生化应用研究项目（尤其是针对能够大量产生史上第一种抗生素——青霉素的工程菌株进行筛选研发）。樊先生对这两个研究领域投入了大量精力。由于当时正处于第二次世界大战时期，青霉素的生产引起他极大的关注。自1929年Alexander Fleming发现青霉素后，这种抗生素拯救了数百万人的生命，对于欧洲前线受伤的士兵尤其有效，抗生素能够帮助伤员免受传染病感染。而第二次世界大战期间，每年有数以千万计的中国人死于感染性疾病（Watt，2013），樊先生坚信青霉素能挽救更多中国人的生命。然而，由于在亚太地区的战火愈演愈烈，青霉素很难流通到中国。因此，樊先生决定利用自己的微生物学知识，在肯塔基州路易斯维尔的西格姆公司暂时从事发酵工作，借机寻找回国生产青霉素的机会。也正如下文将介绍的，这三个微生物学领域（青霉素生产研发、微生物发酵工程和根瘤菌生物固氮）都成为樊先生回国后研究的重要课题。

樊先生回国的机会出现在1943年年底，美国医药助华协会决定在昆明建立第一个血液库，将提供全部所需设备、人员及技术，以帮助盟国对抗日本在亚太地区的侵略（Watt，2013）。樊先生是细菌检验方面的人才，他可以确保血液和血液制品的安全。他向协会建议回国后进行青霉素的研制，该建议随即被接受。协会和他共同找到了极为珍贵的三株菌（图5），同时还准备了用于分离纯化抗生素的仪器和试剂。经过培训后，一个由八名专家组成的小组携带了全部的设备和医用品于1944年1月20日离开纽约，踏上了漫长而艰难的回国旅程。他们途经西大西洋、加勒比海、巴拿马运河、南太平洋后，绕道新西兰及澳大利亚南部海域，进入印度洋，到达印度孟买后换乘火车经加尔各答到达雷多，再搭乘美军运输飞机，沿着当时全世界最危险的、随时可能机毁人亡的“驼峰航线”，飞越喜马拉雅山，终于在1944年6月到达昆明（李宗道，2003；樊真美，2011）。

图5　樊教授1944年从美国带回中国的三株菌。这些菌株对我国开始青霉素生产至关重要。这些菌株目前在南京农业大学中国农业博物馆展出

在昆明，樊先生负责血库的检验工作，确保为中国在印度缅甸战区抗击日本侵略者的士兵提供安全的血液和血液制品。此时，他与汤飞凡（时任中央防疫处处长）和朱既明（我国著名的微生物学家）一起尝试开展了青霉素的生产和成品临床试验。他们成功研制了我国第一批青霉素试剂，由此我国成为世界上能够研制青霉素的七个国家（美国、英国、法国、荷兰、丹麦、瑞典和中国）之一（樊真美，2011）。这项工作挽救成千上万的生命，并为我国后续临床医学中的输血治疗和抗生素生产及使用积累了许多经验。

第二次世界大战结束后，樊先生回到金陵大学教授微生物生理学课程，并于1946年6月晋升为教授。除了在金陵大学开展日常教学工作外，他还定期往返于南京和上海之间，在上海的生化制品实验处进行青霉菌株的筛选改良工作。他与童村先生等人合作为青霉素的工业化生产进行了准备，最终，他们成功地实现了医用青霉素的大规模工业化生产。樊先生将英文的“penicillin”翻译成汉语的“青霉素”，也是在这个时期（樊真美，2011）。现如今，“青霉素”已家喻户晓。通过这个时期研究经验的积累，20世纪80年代至90年代，樊先生还领导了除青霉素之外的其他抗生素的研发，用以控制害虫和病原传播（例如，沈爱光等，1983；李顺鹏等，1993）。

在金陵大学农学院工作时期，樊先生就开始了对土壤微生物固氮的系统研究，开展的主要课题是共生固氮菌对豆科植物和土壤肥力的影响（曹正邦和樊庆笙，1957；樊庆笙，1963）。他和他的团队从大豆、花生、豌豆和紫云英等植物中分离出数百种根瘤菌，进行了大量的交叉接种试验，记录并测试了寄主的差异性状和共生固氮效率。

他们的这些工作不仅有助于我们学习土壤微生物学，而且还成功地将微生物应用到了农业生产中，研制出的产品使作物的生产能力和土壤肥力都有了显著的提高（樊庆笙，1985，1986）。

樊先生在生物固氮方面的研究中，最突出的贡献是将我国紫云英进行大规模种植与改良（樊庆笙等，1987）。紫云英是世界上最大的开花植物属，种类繁多，分布范围广。这些豆科植物不仅可以进行生物固氮，丰富土壤肥力，还可以用作动物饲料、蜜蜂酿制蜂蜜的花粉源、中草药等。紫云英生长在我国长江流域各省，主要分布在华南地区。几个世纪以来，我国农民一直在冬季将它种植于田间，作为稻田的“绿肥”。然而，尽管进行了许多尝试，我国中部和北部的农民却未能享受到“绿肥”带来的好处。樊先生和他的同事推测土壤中缺乏合适的固氮菌是主要原因。从1958年到70年代中期，樊先生被停职，派往农村进行劳动（李宗道，2003；樊真美，2011）。然而，他并没有因此放弃对紫云英种植的研究。利用这个机会，他和同事通过反复试验，检验了他们之前的推测，成功地筛选出了适合紫云英生长的根瘤菌组合，使得我国北部的黄河流域同样可以种植紫云英（樊庆笙等，1987；李宗道，2003）（图6）。这项工作为我国中部和北部地区的农业可持续发展作出了巨大贡献，并于1978年获得了全国科学大会一等奖。70年代末，樊先生在南京农业大学（原金陵大学农学院）的教研任职被恢复，他和他的研究团队继续开展了固氮菌的生态学、生理学、生物化学和遗传学研究，为我国进一步提升豆科作物产量并改善农田土壤肥力作出了巨大贡献（樊庆笙，1986；樊庆笙和娄无忌，1986；李宗道，2003）。

图6　樊庆笙（左）在紫云英田里

生态系统生产力通常由生物质的多少来衡量。而在农业、林业、畜牧业中，生物质只有很少一部分被人类直接使用或消耗。剩余的生物质在很大程度上不仅没有被充分利用，形成了浪费，还成为环境污染的来源。樊先生认为这种被浪费的生物质可以作为一种潜在的资源，通过微生物作用，转化为有价值的产品。为了实现这一目标，在70年代末，他积极推广了利用废弃生物质驯化培养食用和药用菌蘑项目（例如，李顺鹏等，1991）（图7）。他的这种推广是极富远见的，我国现已是世界上食用菌和药用菌的第一大生产国，产量约占世界总量的70%，可种植栽培的菌蘑多样性也远远超过其他国家。此外，他和他的团队还创建了我国第一座厌氧微生物实验室（王天光等，1984；李宗道，2003），该实验室为了实现生物燃料的创新，先后进行了厌氧微生物生理生态学研究，沼气发酵工艺调控和发酵残留物的综合利用研究，以及防治农作物病害和降解有机污染物的研究等工作（例如，李顺鹏等，1993；李宗道，2003）。厌氧发酵后的固体残留物被用作农田肥料，形成了有益于农业生产的循环利用。樊先生的这些工作表明，在我国建立健康高产的可持续农业系统是可行的，可食用菌能够为农民提供额外的收入和极好的营养，类似于甲烷的生物质燃料有助于降低能源成本并减少环境污染，厌氧发酵产生的固体残留废料可以改善土壤的肥力（李宗道，2003），这为形成以微生物为中心发展可持续农业的综合体系提供了可能。

图7 樊庆笙在蘑菇生产基地

樊先生一直倡导以微生物为中心发展可持续农业（李宗道，2003）。这一理论体系中，我们看不见的微生物扮演了多重关键角色。在农作物生产中，可持续的农业生态系统需要有经济划算的方法来保持土壤肥力。他证明这可以通过提供如下条件来实现：① 提供有利于固氮微生物生长的环境（实现方法可以是种植豆科植物），从而增加土壤中氮的含量；② 促进菌根真菌群落的形成，增加土壤中的磷含量；③ 将生产与消耗的生物质（如厌氧发酵和菌蘑栽培后的废料）返回农田，以增加农田中有机物、钾盐和其他元素的含量。他还指出，含有多种有益微生物群落的土壤可以帮助植物抵御传染病和农业害虫，与此类似的以微生物为中心的体系和方法可以广泛应用于林业、畜牧业和渔业。实践中，他具有将人类食物链中的所有成员结合在一起的战略远见，以整合的方式，充分发挥微生物的作用，提高生物质（由土壤环境中的植物和水生环境中的藻类产生）的利用效率。通过不断研究和积累，他证明了利用生态系统中关键因子改善农业生产的可行性。

樊先生不仅是农业微生物学领域富有远见的先驱，而且是农业教育领域极为优秀的前辈。他辛勤耕耘60个寒暑，培养了大批本科生、研究生及博士后，这些学生已成为我国农业领域科研前线的学术骨干、学科带头人或是技术人才。他曾亲自教授多门基础课程，包括植物学、植物分类学、植物病理学、土壤微生物学、细菌学、农业微生物学、微生物生理学、微生物分类学和固氮微生物学，同时还组织了各类专题讲座和研讨会。他的学术思想和工作作风不仅影响了与他一起工作的同事，还影响了他在全国各地的学生乃至国外的学者。他组织的重要研讨会主要有三次，成功帮助我国教育部门逐步开展了农业微生物学的教研工作。第一次是在1980年举办的微生物学讲习班，为期一个月，来自全国农业大学和学院的100多名教师参加了这次讲习班。这是我国高等教育第一个面向微生物教师的讲习班，来自全国各地的微生物学专家应邀就微生物学多样性主题汇报了最新研究进展，会后出版了名为《微生物学进展》的论文集，成为参会老师的参考书目（樊庆笙和陈华癸，1984）。第二次是1980年的全国厌氧微生物研讨会，会期同样是一个月；不同的是，此次会议的报告人是当时美国微生物学会的主席R. H. Hungate教授，通过交流研讨，此次会议帮助我国科研人员逐步开始了对厌氧微生物的研究。第三次是1985—1986年的全国微生物遗传学和生物技术培训会，培训为期一年，面向我国各个农业大学的研究生和科研人员，主要邀请了来自中国、日本和美国的微生物学家交流最新进展。这些会议的成功组织和举办，对我国高校培养微生物学和分子生物学方面的人才具有重要作用。即使在今天，我们仍能感受到这些培训会议带来的巨大影响。

樊先生在教授课程和报告演讲中，总是准备充分、逻辑清晰、生动简洁并且选材新颖。只要有可能，他就会把农业生产中的实际问题与课程中的重要概念进行关联讲解。教学中，他强调了实验循证对科学发现的重要性。虽然这在科研工作中是必不可少的，但并没有被广泛接受。例如，20世纪70年代末到80年代中期，一位大胆的“民间科学家”声称，他利用生物磁场能够改变植物的遗传性状，在许多非豆科作物（如小麦、水稻和棉花）中诱导产生固氮根瘤（王曼新，1981）。这一言论引起了全国各地媒体工作者的广泛关注。当时，许多人都在为这样一个“非凡的发明”喝彩，其中也不乏一些科学家（例如，苏云川，1981；吴思，1987）。然而，科研需要有事实依据，“非凡的发明”同样也需要有证据。作为我国权威的共生固氮研究专家，樊先生也被要求支持这种声明。他坚决要求对方提供固氮证据，但遭到拒绝和诽谤。所幸，追求“实验循证”的科学家在樊先生的支持下取得了胜利，最终证明了这些所谓的“根瘤”是由土壤线虫引起的，并且无法进行固氮（吴思，1987；李宗道，2003；樊真美，2011）。这一事件的发生，不仅肯定了以证据为依据的科学方法在农业生产中的重要性，也为其他类似的科学研究提供了经验和借鉴。

樊先生以微生物为中心的可持续发展农业的理论体系在他撰写的教科书中也有清晰的讲述（青宁生，2011）。他所撰写或参与撰写的专业书籍包括:《农业微生物学》《微生物学》《土壤微生物学》《固氮微生物学》和《微生物学和微生物生理学》等。他还是《中国农业百科全书·生物学卷》编委会主任，他与陈华癸合作主编的《微生物学》及其后续的修订本是50年代以来国内该学科最受欢迎的教科书之一，至今仍被广泛使用，该书的第四版获得1989年度国家级优秀教材奖（陈华癸和樊庆笙，1989）。

除了在教研方面的卓著贡献外，樊先生还是一位优秀的管理者和颇具影响力的领导者，他曾担任多项教研职务和学术任职（李宗道，2003）：1951年至1952年，任金陵大学教务长；1981年至1984年，任南京农业大学校长；他还曾任国际紫云英协会名誉会长、中国食用菌协会名誉会长等职。他曾多次组织和主持了与紫云英和土壤微生物相关的地区性、全国性和国际性的专业会议。为了推进南京食用菌生产的快速发展，他还组织了两次关于菌蘑栽培的大型会议，分别是：1989年的中国首届食用菌生物技术研讨会和1998年的国际食用菌科学与栽培研讨会。即便在生命的最后几年，他还担任了南京农业大学金陵研究院筹备委员会主任，主持了研究院的筹建工作。

樊先生去世于1998年，他充实而华丽的一生为我们留下了珍贵的遗产，并将持续鼓舞我们砥砺前行。有关他的事迹会经常出现在书刊、报纸和微信等媒体的报道中（例如，樊真美和黄为一，2013）。六十多年的科教生涯，樊先生完成了大量关于抗生素生产、农业微生物学和农业教育等方面的工作。他一贯的工作作风和工作方法对于现在比以前更具有指导意义。例如，由他提出的以微生物为中心开展农业可持续发展的战略思想，在如今的基因组学与宏基因组学时代受到了越来越多的关注和支持，微生物在地球生物圈的各类生态系统中的关键作用也越来越得到重视。再如，面对我国城市和农村环境的日益恶化，他强调利用生物“绿肥”构建农业体系，将生物质进行循环利用和再生，是行之有效的方法。与此类似，他坚持科研发现要有实验循证支持，更是在时刻提醒我们，无论什么学科都必须尊重事实依据。

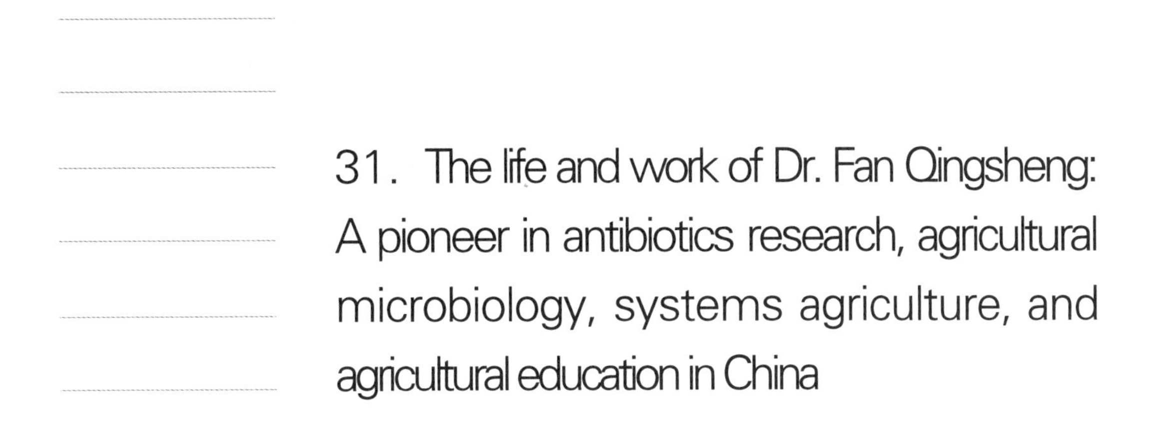

31. The life and work of Dr. Fan Qingsheng: A pioneer in antibiotics research, agricultural microbiology, systems agriculture, and agricultural education in China

Microbes are everywhere in the Earth's biosphere—they play crucial roles in agriculture, nutrient cycling, environmental protection, and animal and human health. However, despite the importance of microbes throughout human history, their roles remain to be fully recognized. In China, one of the key scientists who led to the recognition and utilization of microbes was Dr. Fan Qingsheng (樊庆笙博士). Dr. Fan not only contributed significantly to the production of penicillin in China but also established the broad framework for the effective use of microbes to develop sustainable agriculture by increasing agricultural production, creating valuable goods from agricultural wastes, cleaning up pollutants in aquatic and soil environments, and enhancing soil fertility through organic farming. Indeed, he was a visionary both in establishing the theoretical framework of a microbe-centric sustainable agriculture and in demonstrating the effectiveness of this approach through over 60 years of tireless work on a diversity of organisms across many ecological niches.

Dr. Fan was born on August 4, 1911, in the township of Xizhoushi, Changshu county, Jiangsu Province, along the south shore of the Yangtze River, northwest of Shanghai. As the oldest of nine children in the family, Dr. Fan spent much of his childhood in the countryside where he watched fishermen fishing along the Yangtze River, observed how crops, vegetables, and trees were grown in the traditional Chinese agricultural system, and helped his family obtain food (Fan, 2011). His passion for agriculture started during his youth when he witnessed the devastating effects of hunger, malnutrition, and diseases on the community. He believed that there must be ways to improve agriculture and alleviate the suffering. Like in most Chinese communities, even during politically unstable times, providing the best possible education for their children was a top priority for many families, including Dr. Fan's. Dr. Fan seized learning opportunities, earning some of the top grades in his classes throughout his elementary, middle, and high school years. Upon graduating from Cuiying High School in Suzhou, he was admitted to the Department of Forestry, Jinling University in Nanjing in 1929. Because of the lack of

Jianping Xu

Department of Biology, McMaster University, Hamilton ON L8S 4K1, Canada

Correspondence: jpxu@mcmaster.ca

financial resources, his family had to borrow money to cover his first-year's tuition at Jinling University. However, due to his excellent academic results, he was awarded full scholarships from the second to the fourth year of his study, including being hired as a teaching assistant to help organize laboratory classes—an extremely rare opportunity for undergraduate students. His top academic performance earned him the "Golden Key" award, the highest honor bestowed by Jinling University for its graduating students, in 1933 (Fan, 2011). Right after graduation, he was offered an assistant lecturer position for classes in Botany and Plant Taxonomy. In this role, he helped establish the largest herbarium in China at the time (Li, 2003). His solid training in botany during his undergraduate years played a vital role in developing his framework on the diversity of plant-microbe interactions and their importance in agriculture.

In 1937, the Japanese aggression caused many government agencies across eastern China, including universities, to move to western China. Jinling University was relocated to Huaxiba district in Chengdu, Sichuan Province. While there, Dr. Fan was promoted to Lecturer and continued to teach Botany (Li, 2003). In the summer of 1940, the Rockefeller Foundation offered one full graduate scholarship to a junior staff member in the Faculty of Agriculture at Jinling University to study at an American university. However, instead of sending one person for three years to study for a PhD degree, the Faculty of Agriculture decided, with permission from the Foundation, to make the best use of the opportunity and used the funding to send three young scholars to the US for one year study each to do their MS degrees. Dr. Fan was among those three young scholars and was admitted to the Department of Botany in the Faculty of Agriculture at the University of Wisconsin–Madison. However, after obtaining his MS in one year of study in 1941, his excellent academic performance and aptitude for learning caught the attention of the microbiologist Dr. W. W. Umbreit who suggested that he worked on his PhD degree on the physiology of photosynthesis in the alga *Chlorella pyrenoidosa*. Dr. Fan successfully completed his PhD studies in 1943 and published two papers from his thesis in the *Journal of General Physiology* (Fan et al., 1943) and *Journal of Bacteriology* (Fan and Umbreit, 1943).

During his three years of study at the UW-Madison, Dr. Fan devoted all his time to research and to learning the diverse aspects of microbiology. At that time, the Department of Agricultural Bacteriology had two world-leading research programs, one on symbiotic nitrogen fixation between rhizobia bacteria and legumes and the second on the biochemistry of antibiotics, especially on the development of Penicillium strains capable of producing high quantities of penicillin, the first antibiotic. Dr. Fan learned as much as he could about both frontiers. Due to the ongoing Second World War (WW Ⅱ), the production of penicillin especially attracted his attention. Since its discovery by Alexander Fleming in 1929, this antibiotic has saved millions of lives from infectious diseases, most prominently the injured soldiers at the front lines in Europe. During WW Ⅱ, tens of millions of Chinese were dying each year from infectious diseases (Watt, 2013). Dr. Fan believed penicillin could similarly save many lives in China. However, because of the difficulty of returning to China due to the ongoing WW Ⅱ across the Asian–Pacific region, Dr. Fan decided to use his microbiology knowledge and worked at Seagram & Sons, Inc. in

Louisville, Kentucky on fermentation technologies while searching for an opportunity to return to China. As we will see below, these three areas of microbiology (penicillin, microbial fermentation, and biological nitrogen fixation) all became important topics of Dr. Fan's research after his return to China.

The opportunity to return to China came in late 1943 when the American Bureau for Medical Advancement in China (ABMAC) decided to help China establish its first blood bank in Kunming, including providing all the required equipment, personnel, and technology, to help the Allied countries fight against Japanese aggression in the Asian–Pacific Region (Watt, 2013). Dr. Fan successfully obtained the position of diagnostic bacteriologist on the team, with the responsibility of ensuring the safety of blood and blood products. He also suggested to ABMAC that he would like to help produce penicillin in China. The suggestion was accepted and ABMAC helped procure the necessary strains (three strains total), materials and equipment for growing the fungi and for isolating, purifying, and testing the antibiotic. After finished their training, a team of eight experts and all their equipment and supplies left New York City on January 20, 1944 on a long and difficult journey. They sailed through the western Atlantic Ocean, the Caribbean Sea, the Panama Cannel, the southern Pacific Ocean around New Zealand and Australia, the Indian Ocean, and finally landed in Bombay, India. They then travelled by train from Bombay to Calcutta and finally to Ledo in Assam where they were air-lifted by a military transport plane over "The Hump" —the eastern Himalayan Mountains—to Kunming in June 1944 (Li, 2003; Fan, 2011).

While in Kunming, Dr. Fan was in charge of diagnosis for the blood bank, ensuring the safety of blood and blood products for soldiers fighting in the China–India–Burma Theater of the War of Resistance against the Japanese Aggression. In addition, he joined hands with Zhu Jiming and Dr. Tang Feifan, a prominent virologist and director of the Chinese Center for Disease Control and Prevention at the time, in penicillin research. They successfully produced the first batch of penicillin, making China the seventh country capable of producing this antibiotic (Fan, 2011). Their work helped save tens of thousands of lives and contributed to subsequent rapid medical developments in both blood transfusion and antibiotic research and production in China.

After the end of WW Ⅱ, Dr. Fan returned to Jinling University to teach "Microbial Physiology" and was promoted to Professor in June 1946. Aside from his regular academic responsibilities at Jinling University in Nanjing, he travelled regularly between Nanjing and Shanghai, supervising *Penicillium* strain improvement and working with Dr. Tong Cun and others on the production of penicillin at the Biochemical Products Research Laboratory of the Chinese National Institutes of Health in Shanghai. Their work successfully led to the large-scale industrial production of medical-grade penicillin. Dr. Fan translated "penicillin" to "青霉素" in Chinese, now a household word in China (Fan, 2011). Aside from penicillin, in the 1980s and 1990s, he also led the development of antibiotics to control agricultural pests and pathogens (e.g. Shen et al., 1983; Li et al., 1993).

After returning to the Faculty of Agriculture at Jinling University, Dr. Fan began his systematic investigations on soil microbes and biological nitrogen fixation, especially on the effects of symbiotic nitrogen-fixing bacteria on legumes and on soil fertility in general (Cao and Fan, 1957; Fan, 1963). He and his team isolated hundreds of rhizobia strains from soybeans, peanuts, peas, and Chinese milk vetch plants. They conducted numerous cross-inoculation experiments to test for host specificity and symbiotic nitrogen fixation efficiency. Their work contributed to our understanding of soil microbiology and the development of several highly successful products for agricultural application that have led to significant increases in both crop productivity and soil fertility (Fan, 1985, 1986).

Among the products and technologies associated with Dr. Fan's research on biological nitrogen fixation, the most notable was probably the improved productivity and range expansion of the Chinese milk vetch *Astragalus sinicus*, "紫云英" in Chinese (Fan et al., 1987). *Astragalus* is among the largest genera of flowering plants with different species having different ranges and distributed in different parts of the world. These leguminous plants can not only fix nitrogen and enrich soil fertility but also serve as animal feed, source of pollen for bees for honey production, and herbal medicine across the globe, among other uses. *A. sinicus* is naturally distributed in southern China, primarily south of the Yangtze River, and farmers have been growing it in winter months as a source of "green manure" for rice fields for centuries. However, farmers in central and northern China had not been able to enjoy this benefit, despite repeated introduction trials. Dr. Fan and his colleagues hypothesized that the lack of appropriate nitrogen fixing bacteria in the soil was the cause of the failed introduction. From 1958 to mid-1970s, Dr. Fan was not permitted to teach or conduct research at his university, and was sent to work as a farmer in the countryside (Fan, 2011; Li, 2003). However, he used the opportunity to test his hypothesis and through trial and error, he and his colleagues successfully selected appropriate rhizobia—*A. sinicus* combinations that allowed range expansion of *A. sinicus* all the way to the Yellow River basin in northern China (Fan et al., 1987; Li, 2003). This work has contributed enormously to agricultural sustainability in central and northern China and won him the First Chinese Science Congress's 1st Prize in 1978. With the restoration of his teaching and research responsibilities in the late 1970s at Nanjing Agricultural University, Dr. Fan and his group continued to investigate the ecology, physiology, biochemistry, and genetics of nitrogen-fixing bacteria and contributed to further increases in legume production and soil fertility in China (e.g. Fan, 1986; Fan and Lou, 1986; Li, 2003).

Productivity in ecosystems is often assessed by the amount of biomass production. However, only a small proportion of the biomass produced in agriculture, forestry, and animal husbandry is directly used or consumed by humans. The remaining biomass is largely under-utilized, completely wasted, or even becomes a source of environmental pollution. Dr. Fan saw such wasted biomass as a potential resource that could be converted to valuable goods through microbial action. To achieve this goal, he aggressively pushed for broad investigations on domesticating and cultivating edible and medicinal mushrooms using waste biomass since the late 1970s (e.g. Li et al., 1991). His vision for mushroom production was far-reaching. China now is the No. 1 producer of edible and medicinal mushrooms in the world, accounting for about 70% of the world's total mushroom production and with a cultivated species diversity far exceeding those in other countries. In addition, he and his team established the first anaerobic microbiology facility in China (Wang et al., 1984; Li, 2003) and initiated a series of studies that helped achieve several microbe-centric innovations in biofuel production, the elimination of plant pathogens and pests, and the degradation of organic pollutants from both agricultural and industrial runoff (e.g. Li et al., 1993; Li, 2003). The processed solid wastes were further used as fertilizers on agricultural fields, creating an extremely beneficial positive feedback loop. These areas of research by Dr. Fan showed that productive, healthy, and sustainable agricultural ecosystems were possible in China. The production of edible mushrooms serve as an additional source of income for farmers and an excellent source of nutrients; the production of biofuels such as methane gas also helps minimize energy costs and environmental pollution; and the spent mushroom substrates and solid wastes from anaerobic digesters further enrich soil fertility (Li, 2003).

Dr. Fan tirelessly advocated and tested the theoretic framework of a microbe-centric systems approach for sustainable agricultural and forestry management (Li, 2003). In this approach, invisible microbes play multiple critical roles. For example, in crop production, he believed that a productive and sustainable agriculture system required cost-effective ways to maintain soil fertility. He demonstrated that this could be achieved by creating conditions that (i) favor the growth of nitrogen-fixing microorganisms (e.g. by growing legumes), resulting in increased levels of nitrogen in the soil; (ii) facilitate colonization by mycorrhizal fungi to increase the level of soluable phosphorus in the soil; and (iii) return processed and spent biomass (e.g. from anaerobic digestion and mushroom growing) to agricultural fields to increase the levels of organic matter, potassium, and other elements. He also showed that a healthy soil microbial community containing various beneficial microbes could help plants defend against infectious diseases and agricultural pests. He suggested that a similar microbe-centric systems approach could be applied to forestry, animal husbandry, and fisheries. Indeed, his broad vision was to link all the components in the human food chain together, in a holistic way and through microbial actions, to make the best use of the biomass generated by the primary producers: plants on land and algae in aquatic environments. Through his own work, he has demonstrated the feasibility for many of the key components in this ecosystems approach to the improvement of agriculture.

Aside from being a visionary in agronomy and a pioneer in agricultural microbiology, Dr. Fan was also an excellent educator as evidenced by over 60 years of contribution to the training of tens of thousands of undergraduate students, graduate students, postdoctoral fellows, visiting scientists, and applied technical workers on the front lines of agriculture. He taught a diversity of courses including Botany, Plant Taxonomy, Plant Pathology, Soil Microbiology, Bacteriology, Agricultural Microbiology, Microbial Physiology, Microbial Taxonomy, and the Microbiology of Nitrogen Fixation. He also organized many workshops. His impacts were not limited to those who directly studied and/or worked with him, but extended to those from across

China and outside of China. The workshops that Dr. Fan organized included the following three very timely and important ones that helped re-launch microbiology education in the Chinese agricultural education sector. The first was the month-long Microbiology Training Workshop in 1980 involving over 100 teachers from agricultural universities and colleges across China. This was the first national workshop for training microbiology teachers after the Chinese higher education system was re-instated. Expert microbiologists from across China were invited to give presentations on a broad diversity of topics in microbiology. An edited book called *Advances in Microbiology* was subsequently published and used as a reference by teachers (Fan and Chen, 1984). The second was the month-long National Anaerobic Microbiology Workshop in 1980. Among the speakers at this workshop was the then-President of the American Society for Microbiology Dr. R. H. Hungate. This workshop helped start anaerobic microbiology research in China. The third was a one-year training program in Microbial Genetics and Biotechnology in 1985—1986 for graduate students and young researchers from all major agricultural universities, with lectures by invited experts from China, Japan, and the US. These workshops played a vital role in training microbiologists and molecular biologists in the agricultural university system across China. The impact of these workshops can still be felt even today.

In his own teaching and lecturing, Dr. Fan's presentations were always well prepared, logical, vivid, concise, and updated. Whenever possible, he linked real issues in agriculture to the key concepts in his lectures. In his teaching, he stressed the importance of the evidence-based approach to scientific discovery. While such an approach is essential in science, it has not always been accepted. For example, from the late 1970s to mid-1980s, an outspoken "citizen scientist" claimed that, using biological magnetism, he was capable of changing the genetic makeup of plants and inducing nitrogen-fixing root nodules in many non-leguminous crops such as wheat, rice, and cotton (Wang, 1981). His extraordinary claim attracted widespread attention from journalists working for a diversity of national and local newspapers and magazines. Many people, including some scientists, jumped to their feet to applaud such an "extraordinary invention" (e.g. Su, 1981; Wu, 1987). However, extraordinary claims require extraordinary evidence. As one of the most authoritative figures on symbiotic nitrogen fixation in China, Dr. Fan was asked to support such claims. He stood firm in his request for evidence of nitrogen fixation and was met with a smear campaign. Fortunately, science eventually triumphed in this case in Dr. Fan's favor, demonstrating that those "nodules" were caused by soil nematodes and unable to fix nitrogen (Fan, 2011; Wu, 1987; Li, 2003). This case played a big role in affirming the importance of an "evidence-based approach" in agricultural and other types of scientific research in China.

Dr. Fan's framework of microbe-based, systems view of agriculture is also evident in the textbooks and opinion pieces that he wrote (Qing, 2011). He authored or co-authored the general textbooks *Agricultural Microbiology*, *Microbiology*, and *Soil Microbiology*, as well as the specialized books *Microbiology of Nitrogen-Fixation*, *Advances in Microbiology*, and *Microbial Physiology*. He was the Editor-In-Chief for the Biology volume of the *Encyclopedia of Chinese Agriculture* book series. The book *Agricultural Microbiology* that Dr. Chen Huakui and he wrote has served as the standard textbook in the field since the 1950s and is still widely used today. Its fourth edition won the Chinese National Textbook Prize in 1989 (Chen and Fan, 1989).

Aside from his contributions to teaching and research, Dr. Fan was also an excellent administrator and an influential leader in a diversity of institutions and scholarly societies (Li, 2003). Among the many posts he held, he was a highly respected Provost of Jinling University (1951—1952) and President of Nanjing Agricultural University (1981—1984) during their critical transition periods. He served as the President of the International Astragalus Society and the Chinese Society of Mushroom Sciences. He chaired several national and international conferences on *Astragalus* research and Soil Microbiology. To help develop the Chinese mushroom industry, he organized two large international conferences on mushroom biology and cultivation: (i) the First International Symposium on Mushroom Biotechnology in 1989 in Nanjing; and (ii) the International Symposium on Science and Cultivation of Mushrooms in 1998. In the last few years of his life, he chaired the foundation in charge of establishing Jinling Research Institute within Nanjing Agricultural University.

Dr. Fan passed away in 1998, ending a long and decorative life and leaving a legacy that would have a lasting impact. Articles about him and his work continue to appear in books (e.g. Fan and Huang, 2013), newspapers, magazines, and the new media such as WeChat. Through over six decades of research and teaching, Dr. Fan left a large footprint on antibiotics production, agricultural education, microbiology, and agricultural science in China. Several of his perspectives are more relevant now than ever. For example, his emphasis on a microbe-centric systems approach to agriculture has been gaining increasing attention and support in this era of genomics and metagenomics where microbes are found to play crucial roles in all ecosystems in Earth's biosphere. His ideal of an agricultural system utilizing "green manure" and involving holistic biomass re-utilization and recycling should be emphasized in the face of rapid environmental degradation in our cities and rural communities. Similarly, his insistence on evidence-based science should serve as a constant reminder in our pursuit of knowledge, regardless of the discipline.

Figures

Fig.1 Dr. Fan Qingsheng (1911—1998)

Fig.2 Dr. Fan Qingsheng's work on agricultural microbes and their roles in the broad systems agriculture framework. Example organisms that Dr. Fan worked on are shown on the right. These microorganisms were investigated for their potential roles in enhancing crop productivity (e.g. *Rhizobia*), reducing diseases (e.g. *Penicillium* and *Streptomyces*), eliminating and/or bio-converting agricultural wastes into valuable goods (e.g. methanogens and edible fungi), and enhancing soil fertility (legumes and *Rhizobia*)

Fig.3 Graduating from Jinling University with an honor BS in Forestry in 1933

Fig.4 Working with a gas exchange testing machine at the University of Wisconsin–Madison for his PhD degree

Fig.5 Three strains of *Penicillium* brought by Dr. Fan from the US to China in 1944. These strains were critical for starting the production of penicillin in China. The strains are now on display at the Chinese Agricultural Museum in Nanjing Agricultural University

Fig.6 Dr. Fan (left) in a field of *Astragalus sinicus*

Fig.7 Dr. Fan at a mushroom fruiting facility

References

Cao ZB, Fan QS (1957) Effects of fertilizers on microbial community structure in red soil. Chin J Soil Sci 5(3): 206–208. (曹正邦，樊庆笙. 1957. 施用肥料对红壤中微生物区系影响的初步分析. 土壤学报，5(3): 206–208.)

Chen HK, Fan QS (1989) Microbiology. 4th edn. Beijing: Chinese Agriculture Press. (陈华癸，樊庆笙. 1989. 微生物学(第四版). 北京：中国农业出版社.)

Fan CS, Stauffer JF, Umbreit WW (1943) An experimental separation of oxygen liberation from carbon dioxide fixation in photosynthesis by *Chlorella*. J Gen Physiol 27(1):15–28.

Fan CS, Umbreit WW (1943) Oxygen production in green algae in the absence of carbon dioxide. J Bact 45(1):13.

Fan QS (1963) Review on Chinese research results and perspectives on nitrogen-fixing bacteria in soil. Chin J Soil Sci 11(2): 220–227. (樊庆笙. 1963. 我国土壤中固氮细菌的研究及其展望. 土壤学报，11(2): 220–227.)

Fan QS (1985) Ecology of rhizosphere microorganisms. In: Proceedings of the First Chinese National Symposium on Microbial Ecology, 7–13. (樊庆笙. 1985. 根际微生物的生态学研究. 见：第一届全国微生物生态学学术会议论文汇编，7–13.)

Fan QS (1986) Symbiotic nitrogen fixation and soil fertility. In: The Professional Committee on Soil Agricultural Chemistry, Soil Science Society of China. The Current Status and Future Perspectives on Soil Nitrogen in China. Beijing: Beijing Science and Technology Press, 212–216. (樊庆笙. 1986. 共生固氮和土壤肥力. 见：中国土壤学会土壤农业化学专业委员会. 我国土壤氮素研究工作的现状和展望. 北京：北京科技出版社，212–216.)

Fan QS, Chen HK (1984) Advances in Microbiology. Beijing: ChineseAgriculture Press. (樊庆笙，陈华癸. 1984. 微生物学进展. 北京：中国农业出版社.)

Fan QS, Lou WJ (1986) The ecology of *Rhizobia*. Chin J Microbiol 9(2): 50–54. (樊庆笙，娄无忌. 1986. 根瘤菌生态. 微生物学杂志，9(2): 50–54.)

Fan QS, Lou WJ, Cai DT (1987) *Rhizobia* inoculation and range expansion of *Astragalus sinicus* to the Huai River region. In: Fifty Years of Chinese Research Results on Symbiotic Nitrogen Fixation, 115–118. (樊庆笙，娄无忌，蔡大同. 1987. 接种根瘤菌紫云英北移淮北地区. 见：中国共生固氮研究五十年论文集，115–118.)

Fan ZM (2011) Our Father and Us. 2nd edn. Beijing: Contemporary Archives Press. (樊真美. 2011. 父亲和我们(第二版). 北京：时代文献出版社.)

Fan ZM, Huang WY (2013) Fan Qingsheng. In: Shi YC. Overview of Academic Achievement of Renowned Chinese Scientists in 20th Century, Agriculture Part 4. Beijing: Science Press. (樊真美, 黄为一. 2013. 樊庆笙. 见: 石元春. 20世纪中国知名科学家学术成就概览・农学卷・第四分册. 北京: 科学出版社.)

Li SP, Fan QS, Zhu JQ (1993) Effect of biogas fermentation fluids on rice diseases caused by microbial agents *Xanthomonas oryzae* and *Rhizoctonia solani* Kuhn. China Biogas 11(3):11–15. (李顺鹏, 樊庆笙, 朱家全. 1993. 沼气发酵液防治水稻白叶枯病和纹枯病的研究. 中国沼气, 11(3): 11–15.)

Li SP, Sheng B, Fan QS (1991) Nutritional physiology of a cultivated mushroom *Hypsizygus marmoreu*. J Nanjing Agri Univ 14(3):120–121. (李顺鹏, 沈标, 樊庆笙. 1991. 白玉菇的营养生理研究. 南京农业大学学报, 14(3): 120–121.)

Li ZD (2003) Commemorative Reflections on Professor Fan Qingsheng. Beijing: Agricultural Science and Technology Press. (李宗道. 2003. 樊庆笙教授纪念文集. 北京: 农业科技出版社.)

Qing NS (2011) Professor Fan Qingsheng—A pioneer and chief founder of Chinese agricultural microbiology. Acta Microbiol Sin 51(4): 566–567. (青宁生. 2011. 我国农业微生物之主要奠基人——樊庆笙. 微生物学报, 51(4): 566–567.)

Sheng AG, Li HJ, Zhou HM, et al (1983) *Streptomyces alloflavus* strain "New-1 ": Its characteristics and antimicrobial properties. J Nanjing Agri Univ 6(1):50–58. (沈爱光, 李惠君, 周惠民, 等. 1983. 链霉菌"新-1 "的特性及抗菌作用. 南京农业大学学报, 6(1): 50–58.)

Su YC (1981) New advancements in biological nitrogen fixation. J South China Agric Res 12(12):48. (苏云川. 1981. 生物固氮研究的新进展. 南方农业学报, 12(12): 48.)

Wang MX (1981) Artificial induction of nodules in non-legume crops through biological magnetism—mediated genetic modifications. Chin Nat Mag 4(3): 164. (王曼新. 1981. 作物根瘤的人工诱发——生物磁学工程的新成就. 自然杂志, 4(3): 164.)

Wang TG, Li SP, Liu MY, et al (1984) Microbial community change, biomass conversion and their effects on biogas production during anaerobic fermentation. J Nanjing Agri Univ 7(2):49–56. (王天光, 李顺鹏, 刘梦筠, 等. 1984. 沼气发酵过程中主要微生物的生理群的变化及物质转化对产气效率影响. 南京农业大学学报, 7(2): 49–56.)

Watt JR (2013) Saving Lives in Wartime China: How Medical Reformers Built Modern Healthcare Systems Amid War and Epidemics, 1928—1945. Boston: E. J. Brill Publisher.

Wu S (1987) Three years in Liu village: Field experiments on the validity and effects of "artificial induction of nitrogen-fixing root nodules in non-legume plants" . Lookout 49: 30–32. (吴思. 1987. 奇人王曼新在刘庄三年——"人工诱发非豆科植物结瘤固氮"试验记. 瞭望, 49: 30–32.)

32. 李季伦和他在农业微生物的理论及应用研究中的深远影响

微生物学家李季伦教授（图1）在微生物次级代谢产物和生物固氮领域作出了卓越贡献，创造了巨大的经济价值和社会效益。李季伦先生在微生物领域的杰出成就使其1995年当选中国科学院院士。

图1　李季伦教授

李季伦1925年出生于河北省乐亭县一个濒海的农村。1931年，他住读于邻村的一所小学，小学五年级时，在校园中见到了同乡孟昭英先生（清华大学教授，中国科学院院士）。孟先生当时刚从美国留学归来，应聘为燕京大学教授（后为清华大学教授），他非常崇敬孟先生，长大后想成为像孟先生那样的人，这个信念影响了他的一生。1937年李季伦小学毕业时，正逢日寇发动“七七事变”，北平被日军占领，家乡也成为日军沦陷区，他的求学之路也从此充满了艰辛。1942年太平洋战争爆发，他就读的中学被侵华日军查封，他不愿做亡国奴，想到抗日后方去读书，经过重重艰难险阻，冒着生命危险穿过日军封锁线，翻越秦岭，终于在1943年到达重庆（何志勇，2009），7月考入国立中央大学生物系，1948年7月毕业并留校任教。新中国成立后，在沈其益先生的帮助下，李季伦调入组建不久的北京农业大学（现中国农业大学）任教，从此开始了他卓有成就的科教生涯。

1958年李季伦和俞大绂先生等开始研制赤霉素（俞大绂等，1964）。赤霉素是水稻恶苗病病原真菌的次生代谢产物，当时是一种新型的植物生长素，国际上只有少数几个国家能生产。他们从全国各地稻田的恶苗病病株中分离出这种真菌，进行筛选诱变，获得一株不产孢子和色素的高产优良菌株，开始生产赤霉素GA3，产量达到当时最大值，填补了我国在这个项目上的空白。李季伦提取的赤霉素GA3结晶产品在1959年莱比锡国际博览会上展出，令国外厂商甚为震惊，为祖国争了光。此后李季伦组建了赤

译者：荣成博
首都医科大学附属北京世纪坛医院，北京100038，中国
邮箱：woshiboer@163.com

霉素生产工厂，经过多年的推广，赤霉素已经在我国大量生产，并在农业中得到广泛应用。他在研制微生物次生代谢产物为农牧业生产服务方面也是硕果累累，为国家创造了巨大的经济效益和社会效益。除早期研制的赤霉素外，他还开发出用于维持苹果高桩果型和克服锈果病的赤霉素GA4+7。

另外，李季伦先生和他的团队还研制出多种重要农用抗生素（图2）。1985年和1994年，李季伦等分别成功研制出了用于防治鸡球虫病的高效低毒多醚类抗生素——莫能菌素和马杜霉素，经国家有关部门批准，这两个产品均在国内生产和销售，解除了美国公司对产品的垄断，解决了我国养鸡业中的一大难题。

(a)

(b)

图2 （a）李季伦（左）正在为一名学生介绍工业发酵技术；（b）李季伦（左）与学生一起工作

阿维菌素是由阿维链霉菌产生的一组广谱、高效、低毒的大环内酯类农用杀虫抗生素。伊维菌素是阿维菌素B1的双氢还原产物，由于毒性更低，此产品主要作为针剂使用，用于杀灭畜禽的各种体内外寄生虫（吸虫和绦虫除外）和治疗人体盘尾丝虫病。阿维菌素和伊维菌素被称作近代超级抗生素，最早由美国Merk公司开发成功并垄断了国际市场。李季伦等从1986年起开始了阿维菌素的研制工作（Li et al.，2010），该研究一直被列为国家的科技攻关项目，经过“七五”“八五”“九五”和“十五”的连续攻关，完成了从菌种选育到工业化生产及作为农用杀虫剂的研究工作，使阿维菌素在国内实现了产业化，Merk公司产品也退出了中国市场。近几年他领导课题组又开展了应用基因工程技术改造阿维链霉菌的研究，构建了不产寡霉素仅产阿维菌素B组分的基因工程菌；通过新兴的组合生物合成手段成功构建了产伊维菌素的基因工程菌，为发酵法合成伊维菌素奠定了基础。有关阿维菌素的研究于2006年获得了国家科学技术进步奖二等奖。

李季伦在20世纪70年代后期开始了生物固氮的研究。1978年夏，我国组织代表团赴美参加第四届国际固氮大会，这是新中国成立后的第一个到美国参加学术会议的代表团，他有幸成为代表团的团员，第一次走出国门，结识了固氮生物化学研究的权威、美国科学院院士、威斯康星大学生化系的R. H. Burris教授。回国后，他于1979年启动了我国新疆地区豆科植物根瘤菌资源调查和分类的研究，为后来的研究奠定了基础。1980年，他受农业部派遣到Burris的实验室从事固氮酶的生化机制研究。早在20年前他就想从事固氮酶催化机制的研究，当时苦于国内没有条件无法进行，已过半百之年的他格外珍惜这次机会，夜以继日地工作，两年内完成了别人需要五年才能完成的工作，证实氘化氢（HD）形成是固氮酶催化的一个特性，而且是绝对依赖N_2的，否定了“不依赖N_2形成HD”的论点，并支持“N_2H_2是固氮过程中一个中间产物”的假说，受到国际同行的关注。Burris在给北京农业大学俞大绂校长的信中称：“他是你们国家的优秀代表。”1982年回国后，在农业部、国家自然科学基金委员会和学校的资助下，他得以继续进行生物固氮的相关研究。在研究固氮酶催化的机制方面，提出固氮酶可催化双位点H_2进化模式，目前已通过实验得到进一步验证（Guan et al.，2007）。在固氮螺菌的分子遗传学研究中，他建立了我国玉米固氮螺菌Yu62菌株的基因文库，构建了节约20%玉米氮肥的耐铵固氮基因工程菌株（Liu et al.，2003）。

李季伦一生热爱科研事业，在科学的道路上不懈追求。他还进行了聚β–羟基烷酸和1，3–丙二醇的发酵研究，以及趋磁螺菌的固氮作用和纳米磁小体生物合成机制等研究，也取得一系列突破性进展。

李季伦不仅在科学研究上取得了丰硕成果，在培养人才方面也是呕心沥血，兢兢业业，桃李满天下。他编译了四百多万字的高质量教材，与俞大绂教授合著的《微生物学》第二版，1988年被评为全国优秀科技图书一等奖；主编的《微生物生理学》被全国多数高校微生物专业采用为教材。他教育学生科研工作一定要实事求是，不能有半点虚假，不要怕出阴性结果。当学生的工作遇到困难挫折，彷徨不前时，他严而不厉，给予充分的信任和自由，辅以教导和鼓励，使学生有信心渡过难关。他为人正直、豁达，常和学生一起探讨人生哲学问题，启发学生去思索人生的价值，让学生懂得“人人为我，我为人人”的道理。他将一生都投入科研事业和培养人才之中，却从不计较个人得失。在他的言传身教之下，一批批学生知学奋进，他们从先生这里学到的不仅仅是专业知识，更重要的是一种爱国情操，一种敬业精神，一种为人和治学之道。我们有幸作为先生的学生，对他的卓越成就表示诚挚的敬佩，也献上最真挚的祝福，祝先生健康长寿（图3）。

图3　李季伦教授和他的夫人孟繁静教授

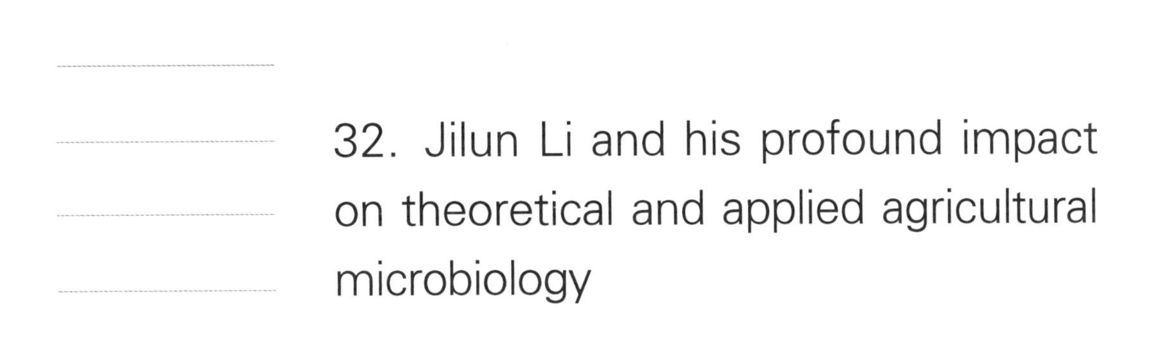

32. Jilun Li and his profound impact on theoretical and applied agricultural microbiology

The microbiologist Jilun Li's excellent work on microbial secondary metabolites and biological nitrogen fixation has made significant contributions to agriculture and has created enormous economic and social benefits in China. He has received many honors throughout his career and was elected as an academician of the Chinese Academy of Sciences in 1995.

Jilun Li was born on March 15, 1925 in Laoting country, Hebei Province, where he was deeply influenced by traditional culture. When he was in fifth grade in primary school, he met the famous scholar Zhaoying Meng, who was a professor at Yenching University. Li had tremendous respect for Prof. Meng and wanted to be a person like him, this belief remained with him all his life. The pursuit of knowledge in his youth was full of hardships, because China was suffering from the war. He risked his life to get through the Japanese blockade, and finally arrived in Chongqing (He, 2009). In 1943, he was admitted to the Department of Biology at the National Central University in Chongqing. He graduated and became a teacher at this university in 1948. Then he transferred to the Department of Plant Pathology at the Beijing Agricultural University (now China Agricultural University), and began his scientific research. In 1959, he assisted Prof. Yu to establish the Department of Microbiology, and gradually became a distinguished microbiologist, who developed new applications of microorganisms that greatly benefited agriculture and livestock husbandry.

In 1958, Prof. Li reached a marvelous milestone in his work on gibberellins (GAs), secondary metabolites produced by the pathogenic fungus *Gibberella fujikuroi*, the causative agent of rice bakanae disease, and used as phytohormones in seed germination and fruit development. At that time, GAs were a novel type of phytohormones and were only derived from a few plants abroad. Li and other researchers successfully isolated the GA-producing fungal strain, screened high-yield strains, and optimized fermentation conditions with the guidance of Prof.

Chengbo Rong[1], Yu Liu[1], Ying Wen[2]

1 Institute of Plant and Environment Protection, Beijing Academy of Agriculture and Forestry Sciences, Beijing Engineering Research Center for Edible Mushroom, Beijing 100097, China

2 State Key Laboratory of Agrobiotechnology and College of Biological Sciences, China Agricultural University, Beijing 100193, China

Correspondence: wen@cau.edu.cn (Y. Wen)

Dafu Yu (Yu et al., 1964). The fermentation yield of GAs produced by his *Gibberella* strains reached the world's advanced level at that period. A GA3 crystal weighing 100 g obtained with Li's strains was displayed at the Leipzig Trade Fair, shocked peer experts from other countries, and won great glory for our country. Thereafter GA3 was mass-produced and widely applied in the agricultural field in China. In addition to gibberellin GA3, Prof. Li had also great success in the research on other microbial secondary metabolites. In 1989, he successfully fermented and extracted gibberellins GA4+7, which had significant impact on increasing the index of fruit-length and overcoming rust disease.

In addition, Li and his group successfully developed several crucial antibiotics for veterinary and medical use. One example is Chicken coccidiosis, a terrible parasite that causes significant economic losses for the poultry industry. Prof. Li's group screened high-producing strains for monensin and maduramicin, polyether ionophore antibiotics that have a wide anticoccidial spectrum and good therapeutic effects on coccidiosis even in low dose, established industrial fermentation techniques, and made them suitable for mass production by 1985 and 1994, respectively. These accomplishments filled the gap in antibiotic discovery and solved a major problem of the poultry industry in China.

Another example is his outstanding work on avermectins (Li et al., 2010). Avermectins, a series of 16-membered macrocyclic lactones produced by *Streptomyces avermitilis*, are excellent pesticides with high efficiency and low side effects, and are widely used in medical, veterinary and agricultural fields. Avermectin B1a, one of the eight avermectin components, has the highest insecticidal activity. Avermectins were first produced by Merck & Co. which monopolized the international market. In the late 1980s and early 1990s, avermectin B1 (brand name is abamectin) was very expensive, almost 20000 yuan per kilogram in China. Prof. Li's group began to study avermectins in 1986. With their persistent efforts, the industrial production of avermectins in the late 1990s in China reached a competitive level and the moderate price forced Merck & Co. to retreat from the Chinese market. Due to this excellent work, Prof. Li and his group won the 2nd prize of the National Science and Technology Progress Award in 2006. In recent years, Li's group has focused on the elucidation of the complex regulatory network of avermectin biosynthesis and the rational design of new hyperproducer strains through genetic manipulation.

Prof. Li's research on nitrogen fixation started in the late 1970s. In the summer of 1978, he went to America for the first time to attend the international conference on nitrogen fixation. He was lucky to meet Prof. R. H. Burris from the University of Wisconsin, the leading scientist on the biochemistry research of nitrogen fixation. When he came back from America, Li started the investigation and classification of rhizobia resources in Xinjiang, which laid the foundation for later research on nitrogen fixation in China. In 1980, he returned to America to study in the laboratory of Prof. Burris at the University of Wisconsin. Prof. Li had been dreaming of studying the catalytic mechanism of nitrogenase for the past twenty years, and he cherished this opportunity very much and worked with great diligence, despite being already fifty-six years old. After two years of hard work, he demonstrated that HD formation is a characteristic of nitrogenase catalysis, and absolutely dependent on N_2. His findings refuted the hypothesis that HD formation is independent on N_2 and proved that N_2H_2 is the intermediate of the nitrogen fixation process; findings that were widely recognized by the international scientific community. He returned to China in 1982 and continued his research on nitrogen fixation. He proposed a two-site H_2 evolution model for nitrogenase and already verified it with scientific experiments (Guan et al., 2007). In addition, his group revealed the regulatory mechanism of nitrogen fixation of several nitrogen-fixing microorganisms (Liu et al., 2003), especially *Azospirillum brasilense*, a root-associated nitrogen fixing bacterium found on members of the Graminae family, including important agronomical crops like wheat, rice and maize. Li's group engineered an ammonium-resistant strain of *A. brasilense*, which could save 20% of nitrogen fertilizer in maize.

Prof. Li has very extensive research interests, including not only the fields mentioned above, but also the synthesis mechanism and application of other microbial metabolites such as fusarine, zearalenone, poly- β -hydroxyalkanoate (PHA), 1,3-propanediol, astaxanthin and megnetosome.

In addition to his great achievements in scientific research, he also devoted his life to teaching and educating students. He has compiled and translated about 4000000 words in books, and taught several courses on microbiology. He sets up examples for students with both precept and practice. Prof. Li's habit of working on experiments together with his students is his most prominent characteristic, and he has done this until the age of 80. He is an open-minded, righteous person, always enlightening students to explore the value of life, asking them to work for the need of our country, letting them know the truth that helping others is helping oneself. These teachings will never be forgotten by his students. As his students, we would like to express our deepest gratitude for his contributions and express our devout wishes for his good health and long life.

Figures

Fig.1 Prof. Jilun Li

Fig.2 (a) Jilun Li (left) was introducing industrial fermentation technology for a student; (b) Jilun Li (left) worked with students

Fig.3 Jilun Li and his wife, Prof. Fanjing Meng

References

Guan F, Zhao D, Pan M, et al (2007) Analysis of active sites for N_2 and H^+ reduction on FeMo-cofactor of nitrogenase. Chinese Science Bulletin 52(15):2088–2094.

He ZY (2009) The Life of Scientist. Beijing: China Agricultural University Press. (何志勇 . 2009. 科学人生 . 北京: 中国农业大学出版社 .)

Li M, Chen Z, Zhang X, et al (2010) Enhancement of avermectin and ivermectin production by overexpression of the maltose ATP-binding cassette transporter in *Streptomyces avermitilis*. Bioresource Technology 101:9228–9235.

Liu Y, Chen SF, Li JL (2003) Colonization pattern of *Azospirillum brasilense* Yu62 on maize roots. Acta Bot 45(6):748–752.

Yu DF, Li JL, Luo GG, et al (1964) Research on gibberellin in Beijing Agricultural University. Plant Physiol Commun 3:53–55. (俞大绂，李季伦，罗国光，等. 1964. 北京农业大学的赤霉素研究工作. 植物生理学通讯，3: 53–55.)

33. 追求真理，献身教育——陈华癸：中国土壤微生物学奠基人

“陈先生一生追求真理，追求进步，忠诚于党和国家的教育事业，献身科学，报效祖国，服务社会，桃李满天下。他的一生是革命的一生，奉献的一生，是为我国科教事业奋斗的一生。”华中农业大学党委在陈华癸先生的悼词中，对他给予了高度的评价。

陈华癸是我国著名的微生物学家、土壤学家，杰出的农业教育家（图1）。1944年，他首次揭示了紫云英根瘤菌是一个独立的互接种族，对根瘤菌剂的生产和大面积推广应用起了开创作用；1948年，他对水旱两作稻田的微生物区系和营养物质的生物循环进行了开拓性研究；1964年，他首次发现水稻土中兼气性硝化微生物进行亚硝化作用。陈华癸毕生致力于高等农业教育与管理以及科学研究（图2）。1980年因其学术上的卓越成就，当选中国科学院学部委员（院士），他为我国高等农业教育和科学研究事业，以及土壤微生物学科的发展作出了卓越的贡献。

图1　陈华癸教授（1914—2002）

作者：陈逗逗，张晗，刘欢
中国科学院武汉病毒研究所，武汉430071，中国
邮箱：liuhuan@wh.iov.cn（刘欢）

图2　陈华癸教授和他的美国学生

陈华癸祖籍江苏省昆山县，1914年1月11日生于北京市。他天资聪颖，1928年考入北京大学预科班，1935年毕业于北京大学生物系，留校任教一年。1936年7月由导师张景钺推荐，赴英国伦敦大学细菌及热带病学院学习1年。次年，年轻的陈华癸来到英国著名的洛桑试验站，在细菌学研究室Henry G. Thornton博士指导下攻读博士学位，研究方向为豆科植物与根瘤菌的共生固氮作用，也是从这里，开始了之后半个多世纪他与农田土壤和根瘤菌再不能分割的夙缘。在此期间，陈华癸单独和领衔发表研究报告4篇，第一篇研究报告《苜蓿根瘤菌生长物质的生产》于1938年发表在英国《自然》杂志上（Chen，1938），另外包括其博士论文在内的2篇文章在《英国皇家学会会刊》上刊登。他的博士论文在世界上首次阐明了根瘤组织大小、持续时间与共生固氮有效性的关系，在当时引起了英国学术界的轰动。1939年10月陈华癸取得博士学位，时年25岁。

1940年，国内正处于战火纷飞的抗日战争时期，几乎所有的通关口岸都被日军封锁，心系祖国的陈华癸毅然放弃国外优越的科研和生活条件与还未完成的论文，几经波折，辗转一年多的时间，终于途经美国回到祖国，来到西南联合大学汤佩松教授主持的清华大学农业研究所从事糖酵解方向的研究。时至今日，陈华癸的学生和同事程见尧回忆起先生的这段归国经历时仍感慨万分："1940年是抗战最艰难的时候，当时能够放弃国外舒适的生活冒着危险回国的人并不多。如果不是满怀报国的赤子之心，是很难在这样的历史节点做出归国选择的。"

1946年，他于北京大学筹建了我国第一个土壤学系；1947年，他来到武汉创建了武汉大学农业化学系；1952年全国院系调整，武汉大学农学院并入华中农学院，他也随之来到学院并建立了土壤微生物教研室。在这里，他招收了新中国第一批本科生和研究生。1956年参与筹建和成立中国科学院武汉微生物研究室（现为武汉病毒研究所）并任研究员（Chen et al.，2014）。

早在20世纪30年代初，当植物生长激素的研究尚处于开创阶段时，陈华癸就首先发现，作物根毛被根瘤菌感染之前发生伸长和弯曲的现象与根瘤菌分泌的激素有关。接着，他对有效根瘤和无效根瘤形态发育进行了比较研究，首次阐明了共生固氮有效性机制的一个重要方面，即结瘤植物的固氮量取决于根瘤数、每个根瘤中含菌组织的体积以及根瘤的寿命和含菌组织的持续时间。他发现，有效根瘤能够固定足够的化合态氮供宿主植物利用；无效根瘤不固氮或固氮量太少，植物得不到益处（Chen and

Thornton，1940）。有效菌株形成的根瘤寿命长，含菌组织体积大；无效菌株则相反，形成的根瘤寿命短，含菌组织体积小。最终这篇以《无效根瘤的结构及其对固氮作用的影响》为题的博士论文于1940年发表在《英国皇家学会会刊》上。

水稻是我国南方地区赖以生存的粮食作物，氮肥是水稻生长的第一重要肥料，但土壤中氮含量往往不足。紫云英是能形成共生固氮体系的豆科植物，主要用作水稻绿肥，对于提高中低产田土壤有机质含量、增加土壤肥力具有明显的效果。根瘤菌和豆科植物的共生关系非常复杂，在初次栽种紫云英的土壤中，接种有效根瘤菌是种植成功的关键因素。陈华癸实地调查了云南、四川、陕西、广西、湖南豆科绿肥的生产应用情况，并着重开展了紫云英共生固氮试验研究。他与同行一道首次分离并获得紫云英根瘤菌纯培养物，通过人工接种试验，发现紫云英根瘤菌不会入侵豆科植物的其他种属并产生根瘤，其他来源分离到的细菌也不能在紫云英上形成根瘤，除了异叶山蚂蝗，这表明紫云英根瘤菌是一个具有专一性的独立互接种族。研究结果于1944年5月在美国《土壤科学》杂志上发表（Chen and Shu，1944），为以后紫云英根瘤菌人工接种的大面积应用奠定了基础。为纪念陈华癸在紫云英根瘤菌研究方面的突出贡献，《国际系统细菌学杂志》于1991年将紫云英根瘤菌正式定名为“华癸根瘤菌”。

陈华癸还是我国稻田营养元素生物循环研究的开拓者，他早在1948年就发表了长江流域水旱两作稻田无论旱季还是水季都能进行氨化作用的论文。文章指出，在蓄水条件下，氨态氮是唯一重要的氮的来源。在冬季排水种旱作稻时，由于通气条件改善，硝化作用显著增强，硝态氮和氨态氮占有同样重要的地位。陈华癸研究了我国水稻田土壤物质的生物循环，揭示了稻田绿肥翻耕后氮和磷等元素的微生物转化规律。其突出贡献是首次发现了水稻土中兼厌气性硝化微生物进行亚硝化作用。陈华癸和他的同事从水稻土中发现了厌氧的亚硝酸细菌，并获得了亚硝酸细菌的纯培养物，这项发现记载在1964年举行的第八届国际土壤学大会上宣读的《兼厌气性的硝化作用和硝化微生物》论文中，并在美国《土壤科学》杂志上刊出（Zhou and Chen，1983），引起了各国学者的强烈反响。他于1957年主编出版的《土壤微生物学》是我国在这一领域的第一部专著；他主编的《微生物学》被各院校广泛采用，该教材的第四版荣获“国家级优秀教材奖”。

陈华癸知识渊博，治学严谨，他曾经这样说：“高等专业教育不可能提供给学生一生需要的猎物，而是提供给学生一支猎枪和使用猎枪的方法，而且即使是猎枪及其使用方法也是在不断更新换代的。”他的教学以此为基础，受到了学生的广泛好评。从1940年英国留学归国后，陈华癸在教育岗位上工作了60多个年头。在半个多世纪的教学、科研生涯中，他发表了大量论著和论文。

陈华癸一生淡泊名利，平易近人，即使身为华中农学院院长，也不配专车，甚至主动退掉了分配的住房，只与夫人住在武汉大学一栋老宿舍里；在学科建设和推广农业技术时，他始终站在国家的角度思考和谋划。时任农业部部长刘瑞龙曾这样评价：“陈华癸先生是在操国家的心，很珍贵。”

陈华癸一生追求科学真理，坚持求实创新，忠于教育事业。他献身科学，报效祖国，服务社会，在推动我国土壤微生物学的教学和科研工作中作出了卓越贡献。陈华癸斯人已去，但他的功德永存。

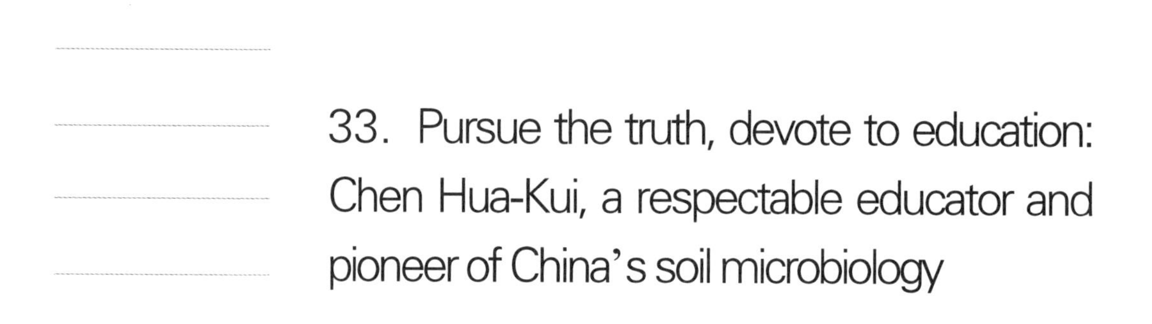

33. Pursue the truth, devote to education: Chen Hua-Kui, a respectable educator and pioneer of China's soil microbiology

"Professor Chen Hua-Kui dedicated his whole life to the pursuit of truth, the education of the nation, and services for society. The students of Prof. Chen blossom in the research field." The Party Committee of Huazhong Agricultural University appraised.

Prof. Chen was a renowned microbiologist and distinguished educator in China. In 1944, he first revealed that *Rhizobium astragali* belonged to a select cross-inoculation group, setting the basis for massive production and application of nitragin. Furthermore in 1948, he conducted pioneering research on microbiota and biological cycle of nutrients in wet and dry paddy fields, and in 1964, first discovered the nitrozation of facultatively anaerobic nitrification microorganism in paddy field. In 1980, Prof. Chen was elected as a member of the Academic Divisions of the Chinese Academy of Sciences (CAS) and continued to make great contributions for China's higher agricultural education and research, especially the research on soil microorganism.

Prof. Chen was born on January 11, 1914 in Beijing. He entered the preparatory class of Peking University in 1928 and graduated from the Biology Department in 1935. After graduation, he taught in the university for one year. In July 1936, with the recommendation of his tutor Zhang Jingyue, he went to England and studied in School of Bacteria & Tropical Medicine, University of London for one year. The following year, Chen began his postgraduate education in the bacteriology laboratory of Rothamsted Experimental Station, England, under the guidance of Dr. Henry Gerard Thornton. There, he undertook research into the symbiotic nitrogen fixation of leguminous plant and rhizobia, and began an over-a-half-century career of soil science and research. Whilst at the Rothamsted Experimental Station, Chen published "Production of growth-substance by clover nodule bacteria" in *Nature* (Chen, 1938) and his dissertation "The structure of 'ineffective' nodules and its influence on nitrogen fixation" was highly praised by the academic community of England. Chen was awarded a doctor's degree at the age of 25 in October 1939.

In 1940, during the War of Resistance against Japanese Aggression, Chen Hua-Kui

Doudou Chen, Han Zhang, Huan Liu

Wuhan Institute of Virology, Chinese Academy of Sciences, Wuhan 430071, China

Correspondence: liuhuan@wh.iov.cn (H. Liu)

embarked on a one-year-long arduous journey in the ruins and thorns, returned to China via the United States where he became a researcher in the field of glycolysis in Agricultural Institute led by Prof. Tang Pei-Sung in National Southwest Associated University. In 1946, he founded the first Soil Science Department of China in Peking University; in 1947 he founded the Agricultural Chemistry Department in Wuhan University, being a constitution of Huazhong Agricultural University later; in 1956, with Prof. H. Zanyin Gaw, Prof. Chen founded Wuhan Institute of Virology, CAS (Chen et al., 2014).

In the early 1930s, Prof. Chen first discovered that before the root hair of a crop was infected by rhizobia, a hormone secreted by rhizobia was responsible for extending and curling root hair. He conducted comparative research on the morphogenesis of effective and ineffective root nodules, indicating that the amount of nitrogen fixation carried out by a nodulated plant depended on several factors including: the number of nodules, the volume of bacteroid containing tissue in each, the lifecycle of nodule, and the time for which this tissue persisted. He also illustrated the mechanism of symbiotic nitrogen fixation, showing that an effective nodule could fix enough combined nitrogen to nourish the host plant; whereas the ineffective nodule could not (Chen and Thornton, 1940).

Prof. Chen investigated the applications of green manure in Yunnan, Sichuan, Shannxi, Guangxi and Hunan Provinces, and conducted researches on the symbiotic nitrogen fixation of *Astragalus* in 1940s. His group isolated for the first time in pure culture the microsymbiont of *Astragalus Sinicus*. In an artificial inoculation test, he discovered that the root-nodule bacteria of *Astragalus* did not produce nodules on other genera of leguminous plants, nor did bacteria isolated from other sources form nodules on *Astragalus* plants, except *Desmodium heterophyllum*, proving that *Astragalus* and its root-nodule bacteria must be considered as a select cross-inoculation group. The research result was published in *Soil Science* in May 1944 (Chen and Shu, 1944), which provided the theoretical evidence for large scale artificial inoculation of *Rhizobium astragali*. To commemorate his remarkable contributions, the *International Journal of Systematic Bacteriology* renamed *Rhizobium astragali* into *Mersohizobium huakuii* in 1991.

Prof. Chen was a pioneering researcher of nutrient bio-cycling in paddy fields of China. In 1948, he published a paper on ammoniation of paddy field in both wet and dry seasons along the Yangtze River valley. Ammoniacal nitrogen represents the only source of nitrogen when water is stored in contrast to the use of both nitrate nitrogen and ammoniacal nitrogen when paddy fields are drained in winter for improved aeration. Prof. Chen and his group were the first to discover a sample of anaerobic nitrate bacteria in paddy field, from which they were able to obtain a pure culture, and subsequently reveal the nitrozation effect of facultatively anaerobic nitrifying microorganisms in paddy fields. This discovery was included in the keynote speech Facultatively anaerobic nitrification and nitrate-forming organism delivered on the 8th World Congress of Soil Science (Bucharest) in 1964, and was published on the *Soil Science* (Zhou and Chen, 1983), receiving a widespread response in the world.

Prof. Chen was knowledgeable and rigorous in his research and strived to educate his students on the importance of higher education telling them that "Higher education provides you with the hunting gun and method of using it instead of the games" . He earned a great deal of respect from his peers with scientific achievement and moral integrity. Whilst serving as the director of Huazhong Agricultural College, he declined an exclusive car and house instead choosing to continue to live with his wife in an old dormitory of Wuhan University. The Minister of Agriculture Liu Ruilong said that "What precious concern Chen Hua-Kui has for the nation!" Prof. Chen spared no efforts to serve his homeland, and made great contributions for China's education and research into soil microbiology. His great contributions and virtues will be dearly remembered for generations!

Figures

Fig.1 Prof. Chen Hua-Kui (1914—2002)

Fig.2 Prof. Chen Hua-Kui and his American student

References

Chen HK (1938) Production of growth-substance by clover nodule bacteria. Nature 142:753–754.

Chen HK, Zhou Q (1964) Facultatively anaerobic nitrification and nitrate-forming organisms, The 8th World Congress of Soil Science (Bucharest), Ⅲ. Pedobiologia: 761–768.

Chen HK, Shu MK (1944) Note on the root-nodule bacteria of *Astragalus Sinicus* L. Soil Sci 58:291–293.

Chen HK, Thornton HG (1940) The structure of "Ineffective" nodules and its influence on nitrogen fixation. Proc R Soc B 129:208–229.

Chen HC, et al (2014) Festschrift for Mr. Chen Hua-Kui's 100th Anniversary. Wuhan: Science Press.

Zhou Q, Chen HK (1983) The activity of nitrifying and denitrifying bacteria in paddy soil. Soil Sci 135(1):31–34.

34. 超越源自积累——陈驹声：工业微生物学家

图1　陈驹声（1899—1992）

有句古话："不积跬步，无以至千里；不积小流，无以成江海"，告诉我们要想取得进步与成功，坚持与积累是绝不可少的。这个道理在一位老科学家——陈驹声先生身上得到了很好的印证。

陈驹声是我国近代工业微生物学的奠基人和开拓者之一，我国最早的发酵工业专家（图1）。他在改进我国传统酿造技术和建立近代工业微生物新体系方面作出了巨大贡献，尤其是在酒精生产技术上两次成功超越了聘来的外国专家，成为我国酿造历史上的佳话。

故事发生在20世纪初期，军阀混战，百业凋零，外商乘虚而入，在东北及华东地区以修路建厂为名，行掠夺之实。为了振兴民族工业，山东溥益酒厂于1922年成立，是第一个由中国人设立的酒精厂。该厂向日本定购了酒精蒸馏塔、蒸煮锅、糖化锅、冷却器等设备，并用水泥修筑了30个发酵槽（内涂沥青），聘请了日本酒精专家Watanabe主持酒精生产技术工作。而当时年仅24岁的陈驹声是该厂的技术员工，他到厂不久，凭着以往的知识储备，将储存在土坑中的糖蜜进行发酵试验，可惜发酵醪丝毫没有泡沫，闻起来也没有酒味。但他并没有因此而气馁，他将糖蜜中加入硫酸后加压蒸煮，闻到了恶臭气体。再将除去恶臭气体的糖蜜加水稀释，添加酵母进行发酵，最终得到了所需要的发酵产物——酒精。经过仔细分析，他发现厂里的糖蜜储藏过程中容易被丁酸细菌感染，该菌在代谢中可将葡萄糖分解成丁酸和氢气。糖蜜中的硝酸被氢气还原生成了一氧化氮存留在发酵醪中，阻碍酒精发酵，可使酒精发酵中途停止。这一重要发现阐明了酒精发酵停止、无法产出所需酒精产品的原因。这次意外的发现，不仅让陈驹声感到欣喜，也增强了他对自己发酵技术的信心。日本专家Watanabe对于谷类发酵颇有些经验，但是如何以甜菜糖蜜生产酒精，他也没有做过，但他认为甜菜发酵酒精是轻而易举的事。当时厂方迷信外国专家，看不起刚毕业的陈驹声，也没

译者：张保元，曲静
中国科学院北京生命科学研究院，北京100101，中国
邮箱：zhangby@biols.ac.cn（张保元）；quj@biols.ac.cn（曲静）

人相信他的实验结果。当Watanabe屡试屡败，厂方焦头烂额的时候，陈驹声挺身而出，用自创的方法对糖蜜进行预处理，结果每百斤糖蜜可出酒精24斤，发酵效率高达85%。生产调试取得成功，厂方特别举行了庆祝大会，当场宣布陈驹声跃升为工程师，代替Watanabe的技术主任职务，且加薪一倍。此时担任公司技术顾问的德国专家Lindermann和日本专家Hori Soichi也对他做出这项具有国际先进水平的成果表示祝贺，并在证明书上签字予以承认。

陈驹声先生另一次取得突破性进展、超越外国专家是在1933年，几位爱国华侨在上海浦东白莲泾建立了建源公司，用南洋输入的甘蔗糖蜜作为制造酒精的原料，从国外进口设备生产酒精。设定的产量为日产酒精2万加仑（约90922升），规模之大，堪称远东第一。在酒精生产初期，该厂高薪聘请英国工程师Brown负责生产，他未经小规模试验检测就盲目投产。几个月的大规模车间生产后，酒精发酵根本无法达到预期。陈驹声到工厂后，与一些技术人员对甘蔗糖蜜原料进行了成分分析，并开展小型发酵试验。他们通过试验发现甘蔗糖蜜中缺少氮源，所以必须补给氮源，否则酵母菌营养不足，发酵无法进行。同时他还进行多次反复的发酵比较试验，优选了对甘蔗糖蜜发酵效率高的酵母菌种。他们根据小规模试验的结果，进行了大规模糖蜜发酵，首次发酵便取得成功，酒精自蒸馏机的出口管滚滚流出，每日产量比预计还要高出一倍。建源公司总经理说："中国人真了不起。"

科研工作中经常会碰到与预期矛盾的试验结果，只有通过长期不懈的理论积累和实践探索才能接近最终的真理。有人问陈驹声先生："你研究工业微生物学数十年如一日，成果累累，你是怎样做到的呢？"他回答说："想要做好科学研究，有四个先决条件：第一，要有扎实的基本知识；第二，要善于总结前人的经验；第三，要有严谨的科研计划，并善于从研究过程中总结经验；第四，无论遇到怎样的困难，切不可中途而废，要总结失败的原因，坚持下去，必有成功的希望。"这番话对于现在的科研工作仍然有着重要的指导意义，如今新兴行业往往是高科技产业，而传统行业的基础研究大多偏于滞后，对于创新和超越的需求更高。陈先生还曾指出："如果有谁能把白酒的微生物研究透了，他能拿诺贝尔奖。"同样说明，在日常的科研工作中一定要重视基础研究，唯有在积累的过程中不断总结，才能够实现自我超越，取得领先时代的成果。

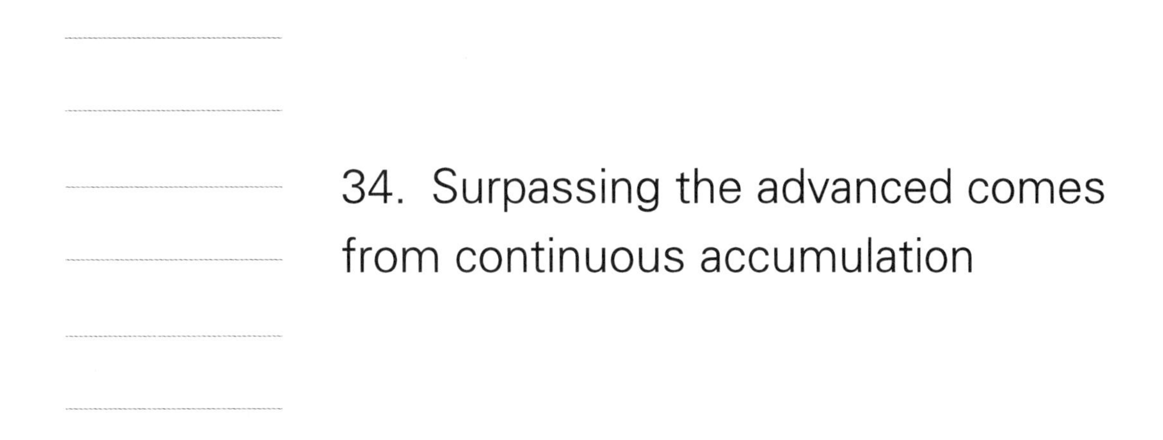

34. Surpassing the advanced comes from continuous accumulation

As a famous Chinese saying goes, "A journey of a thousand miles begins with a single step" . It teaches us that perseverance and steadfastness are absolutely essential for success and progress. This philosophy has been illustrated by an exemplary Chinese microbiologist, Taosheng Chen.

Taosheng Chen was one of the pioneers in modern industrial microbiology research in China. His pioneer work in improving the traditional Chinese brewing techniques and the establishment of modern industrial microbiology technology has been regarded as an outstanding contribution in the development of Chinese industry. Specifically in alcohol production technology, he had successfully surpassed international experts twice, marking them as commendable events in the history of Chinese alcohol fermentation.

The beginnings of Taosheng Chen's breakthrough began during the Chinese economic depression in the early 1920s. Shandong Pu Yi Winery was established in 1922, with the primary aim to revitalize the national industry. Shandong Pu Yi Winery was the first alcohol plant established by the Chinese. In its rudimentary stages, production equipment were ordered from Japan, which includes alcohol distillation towers, cooking pots, pots for glycosylation, coolers and other components. Therein, 30 cement fermentation pools (painted with asphalt) were built specifically for alcohol production. Notably, the plant owner hired a Japanese expert, Watanabe, for the operation of the alcohol production technology. At that time, Taosheng Chen, only 24 years old, was a technical staff of the plant. He examined the molasses fermentation stored in the pits and found that fermented mash produced no bubble, without the scent of alcohol. However, this result did not hinder him. By adding sulfuric acid into molasses, he smelled the stench of decomposing molasses after pressure cooking. After removing the stench, he diluted the molasses with water, added yeast for fermentation, and eventually achieved the desired

Baoyuan Zhang[1], Jing Qu[1], Nur Estya Binte Rahman[2]

1 Beijing Institutes of Life Science, Chinese Academy of Sciences, Beijing 100101, China

2 Temasek Life Sciences Laboratory, 1 Research Link, National University of Singapore, Singapore 117604, Singapore

Correspondence: zhangby@biols.ac.cn (B.-Y. Zhang)

fermented products—alcohol. Thorough analysis, he found that the molasses was easily infected by butyric acid bacteria during storage. The bacteria decomposed glucose into butyric acid and hydrogen in the metabolic processes. The hydrogen reduced nitrate in the molasses, resulting it to release nitric oxide which is an inhibitor of alcoholic fermentation. This important discovery elucidated the reasons the alcoholic fermentation ceased and was incomplete to produce the desired alcohol products. Taosheng Chen was quite delighted with this finding and gained more confidence in his fermentation technology methods. Although Japanese expert Watanabe was well-versed with cereal fermentation technology, he was not familiar with using beet molasses to ferment alcohol. The plant owners trusted Watanabe but did not recognize Taosheng Chen's potential in fermentation technology. As Watanabe experienced continuous failures in the beet molasses fermentation technology, Taosheng Chen decided to contribute his ideas confidently in refining the fermentation technology of beet molasses. He illustrated that the molasses could be pre-treated and fermented through his own methodology. Results indicated that one hundred pounds of molasses could produce 24 pounds of alcohol (with a fermentation efficiency as high as 85%). After the successful commissioning of the production, the plant held a special celebration meeting and announced that Taosheng Chen was promoted to be an engineer and technical director of the plant, rewarded with higher remuneration. Furthermore, the company's technical adviser, German expert Lindemann and Japanese expert Hori Soichi, also congratulated him for the internationally advanced achievement, and jointly signed the certification letter.

Another breakthrough made by Taosheng Chen that surpassed international experts was accomplished in 1933. Several Chinese patriots abroad decided to set up a company in Shanghai where they imported equipment abroad to produce alcohol using sugarcane molasses from Nanyang, as its raw material. The expected daily alcohol production was 20000 gallons (equivalent to an estimated 90,922 liters), and with that efficiency of alcohol production, the company aimed to be the first and leading alcohol production company in the Far East. In the initial stages of alcohol production, the plant hired a British engineer, Brown, to manage the technical work. However, Brown did the fermentation without a small scale trial. Several months passed, the production was still unable to achieve the desirable alcohol yield. After which, Taosheng Chen went to the factory, analyzed the component of sugarcane molasses with other technical staffs, and conducted small scale fermentation trials. They found that the sugarcane molasses lacked nitrogen sources. Sufficient nitrogen supplement was needed in the fermentation. Otherwise, yeast fermentation could not produce any alcohol. With this in mind, they repeated fermentation experiments many times in order to select a perfect yeast strain with high capacity in sugarcane molasses fermentation. With the results of these small scale trials, the first big scale fermentation thereafter proved to be successful and the daily production improved and doubled. When alcohol rolled out from the distillation machine pipe, the general manager of the company said, "The Chinese people are really great! "

Research work often encounters a paradoxical problem. Without long-term and persistent theoretical accumulation and practical exploration, people will not find the paths of truth.

Once, someone ever asked Taosheng Chen, "You have studied modern industrial microbiology for decades and accomplished numerous achievements. How did you do it?" He replied, "To someone who is engaged in scientific research, there are four prerequisites: first, he should have a solid basic knowledge; second, he must be good at summing up experiences of predecessors; third, he should have research plans and summarize the learning step by step; fourth, whatever the difficulties are, he must not give up. Persistent efforts lead to success." These remarks are still significant for today's research work. Nowadays, most of the emerging industries are high-tech ones, while traditional industries based on fundamental research often lag behind. They require more innovation to overcome difficulties. Taosheng Chen also pointed out, "If anyone can thoroughly study microbes in wine, he can be awarded the Nobel Prize." It also demonstrates that we must not forget and pay more attention on basic research. Only by continuous persistence and reflection can we realize self-transcendence and accomplish achievements in the future.

Figures

Fig.1 Taosheng Chen (1899—1992)

References

Chen TS (1992) My 70 years for engaging in the alcoholic fermentation. Chemistry 55:61–64. (陈驹声. 1992. 从事酒精发酵70年. 化学通报, 55: 61–64.)

Fu JQ (2006) Research data of wine-making microbes in China. Liquor-making Science & Technology 5:82–86. (傅金泉. 2006. 中国近代酿酒微生物研究史料. 酿酒科技, 5: 82–86.)

四、细胞生物学和遗传学

35. 贝时璋和他的细胞重建学说

发现细胞重建现象

在很长一段时间里，人们一直从传统视角研究细胞的繁殖增生，尤其关注细胞分裂，特别是有丝分裂。普遍认为分裂方式是细胞繁殖增生的唯一途径，很少注意其他可能性。

直到1932年春天，贝时璋（图1）在杭州郊区松木场稻田采集到的丰年虫（属于甲壳纲、叶足类）中，发现了中间性个体（intersex）；中间性丰年虫个体在发育过程中的某一时期，会进行性的转变，同时呈现雌雄性征，为雌雄同体。这种异常及其原因生物学界还没有做过研究与报道。根据呈现的性征偏于雌性或雄性的程度不同，中间体丰年虫可分为五种类型（图2）。通过研究，贝时璋还观察到在发生性转变时，生殖细胞也同时发生性的转变，转变的全部过程包括细胞的解体（即细胞解形，cell deformation）和细胞的重新形成（即细胞重建，cell reformation）。

图1　中国生物物理学会创始人贝时璋（1903—2009）

译者：李柳

中国科学院北京生命科学研究院，北京100101，中国

邮箱：liliu@ioz.ac.cn

图2　丰年虫中间性个体

1934年，在浙江大学生物系的一次讨论会上，贝时璋教授报告了这些观察到的现象和他的看法。但研究论文的发表却已经是1942年和1943年的事情了。迟迟未予发表，是因为抗日战争时期浙江大学屡次迁校，动荡不安。另外，没有一个合适的可以投稿的期刊也是一个原因。到1942年，《科学记录》杂志创刊，这项工作才得以投稿并正式发表。这些是客观的事实，而在主观上他多少也曾有所犹豫，因为要报道的是生物学上一个重大的细胞理论问题，是对细胞增殖传统概念的挑战。

在这两篇研究论文里，贝时璋分析了全部五种类型中间性丰年虫性转变过程中生殖细胞的解形和重建的情况，探讨了其中的机制。他之所以称它是细胞的“重建”，而非“新生”，因为重建是复兴，表现在卵黄颗粒具备组成细胞的一切原料。细胞解形产生了卵黄颗粒，卵黄颗粒反过来提供了重建细胞的材料。这说明，当组成细胞的物质基础存在以及环境合适的时候，可以不通过细胞分裂的方式形成细胞。而生物学界认为一切细胞来自细胞，认为细胞分裂是细胞繁殖增生的唯一途径。如果说可以通过细胞分裂以外的细胞重建的方式繁殖增生细胞，势将被看作对生物学的亵渎，是可笑的愚昧和狂妄。这样的疑虑当然不足取。好在他终究将论文发表了。论文发表后并未受到责难。可能是有人认为细胞重建即使存在，也不是那么容易见到和经常见到的现象，而“细胞以分裂产生细胞”“细胞分裂为产生细胞之唯一方法”的传统观点，被生物学界奉为金科玉律，它的影响毕竟是巨大而又深远的。因此，贝时璋的工作显得似乎只是一个“孤证”，不足以说明规律性的问题。既无响应，也无法展开讨论，当时又有许多事情等着去做，他只得把这项工作暂时放了下来。

惊风飘白日，光景驰西流。一直到1970年，贝时璋的细胞重建研究工作才又重新开展起来。与20世纪30年代相比，此时已经处在一个新的时代，研究工作也换了新的面貌。原先基本上是贝时璋一个人单独做，现在中国科学院生物物理研究所有了一个研究组；原先的研究材料只有南京丰年虫一种，现在除丰年虫外，还研究鸡胚、小鼠骨髓以及沙眼衣原体、大豆根瘤菌等；原来的研究方法只是光学显微镜观察，这时已经应用了电子显微镜、显微缩时电影、相差定位观察、放射自显影、荧光偏振、双荧光标记能量转移、荧光漂白恢复、拉曼光谱等以及生化方面的各种新技术和新方法。在原有工作的基础上，研究组获得了新的发现、新的知识。例如，发现卵黄颗粒内有DNA、组蛋白和染色质（图3）。卵黄颗粒内有染色质，这是生物学发展史上的第一次发现。卵黄颗粒的染色质和细胞核的染色质有同样的结构和行为，DNA分子的形状也与核DNA很相似。染色质一直被认为是细胞核所特有的，因此，卵黄颗粒也具有染色质的事实不能不引起生物学家的注意。

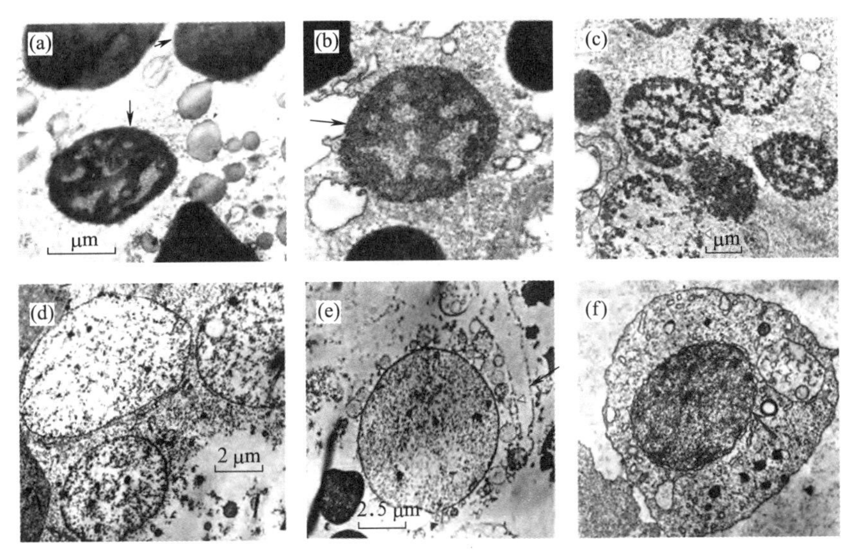

图3　丰年虫成熟卵母细胞内卵黄颗粒原位细胞重建的电子显微照片。(a) 卵黄颗粒的结构变化；(b) 卵黄颗粒的进一步开发；(c) 卵黄颗粒开始发育成核样结构；(d) 三个“裸核”，可见核膜和细胞核网状结构；(e) 重组的核，细胞质和一层质膜刚刚形成；(f) 一个相对完整的细胞，但整个细胞的形态结构相当原始

创立细胞重建学说

贝时璋和团队将实验研究结果和系统理论公之于世：1983年在《中国科学》(B辑) 正式发表5篇论文，1988年24篇研究论文收入《细胞重建》论文集 (第一辑)，2003年18篇论文收入《细胞重建》论文集 (第二辑)。贝时璋的这些认识已经形成了一个完整的细胞重建理论。这些认识可以概括为：

(1) 细胞重建是一个自组织 (self-organization) 的过程，只要具备组成细胞的物质基础和合适的环境，在生物体内，或在离体培养的不存在细胞的环境中，都有可能发生细胞重建或核重建。

(2) 细胞重建在自然界内广泛存在。不仅真核细胞能重建，原核细胞也能重建。不仅生殖细胞能重建，胚胎或成长个体的体细胞也能重建。

(3) 在鸡胚卵黄颗粒内有DNA、组蛋白和染色质，在合适的环境下能重建细胞。染色质不是细胞核独有的物质。卵黄颗粒也不是没有生命的细胞内含物 (图4)。

(4) 细胞和细胞核可以从细胞质重建，说明细胞质、细胞核之间本来就没有森严的壁垒。

(5) 细胞重建很可能是地球上细胞起源在今日生命世界的反映，是简单的生命形态发展为细胞的漫长过程的一个缩影。细胞重建的研究，有助于生命进化的阐释。

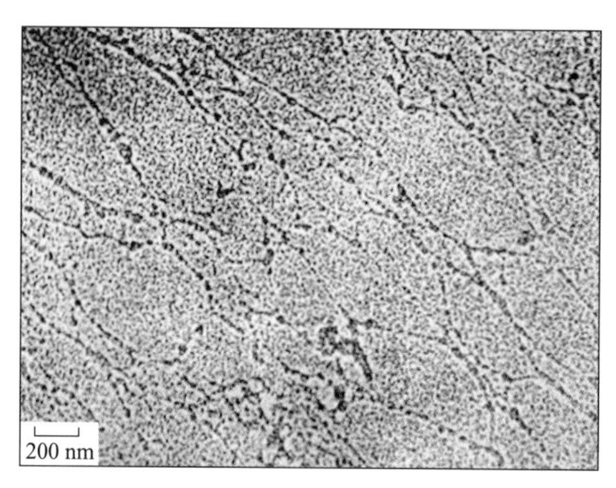

图4　鸡胚卵黄颗粒内的染色质

（6）细胞分裂是“闭锁性”的繁殖，细胞在分裂过程中和它的环境是以细胞膜隔离的。细胞重建是“开放性”的繁殖，在重建过程中细胞组分始终和周围环境打成一片。把细胞分裂和细胞重建结合起来研究，把模拟和诱导自组织结合起来研究，将为改变细胞的结构和性质、改造细胞的性状、选优汰劣、控制定向生产，即促进和发展细胞工艺和细胞工程，提供新的手段和途径。

也就是说，生物体内的细胞增殖不仅包括现有细胞的复制（细胞分裂），还可以通过自组织构成细胞成分的原料来繁殖新细胞（细胞重建）。在细胞重建中，新细胞可以从最初的非细胞形态开始，利用获得的母细胞细胞质原料逐渐重建繁殖。通常，繁殖过程一般先形成核，形成“裸核”状态，再包以细胞质和细胞膜，最后形成一个完整的重建细胞。

细胞重建是细胞起源的痕迹

贝时璋认为，细胞重建研究涉及细胞学的根本性问题，对细胞的起源、生命活动以及未来生物学的发展都会有重要的影响。

自组织是生命活动中十分重要的过程，形成了不同层次的生物结构。例如，在进化过程中，人高度发达的大脑与环境相互作用，改变自身以适应环境，是自组织过程；细胞重建是细胞的自组织过程，同样，细胞的起源也是一个自组织的过程。

作为生命起源与进化基本单位的细胞，是典型的自组织系统。在进化过程中，生命体通过完善自身结构来适应环境的变化。例如，在化学进化阶段，有机分子已经形成，它们会遵循热力学定律，通过自组织趋向于使分子结构稳定在某种状态，从而形成链、环、螺旋和折叠，呈现特殊生物活动。然后，这些分子形成一个相关的多分子系统。在原始生命出现后，它通过自组织进一步改善结构，并从非细胞形态演变为细胞形态。最后，原细胞发育为原核和真核细胞，具有复杂的结构和功能，如酶催化、DNA复制、转录和翻译。

贝时璋认为“细胞重建可能是地球上细胞起源的缩影”，也就是说，可能是地球上细胞起源过程在现存细胞生命活动中遗留的“痕迹”。可以想象，在地球原始生命出现之后，是以原始的非细胞生命形态“分子团聚物”中的生命物质为基础，例如美国科学家S. W. Fox在1965年主张的“微球体”或苏联科学家A. I. Oparin在1969年主张的“团聚体”，经历了类似细胞重建的自组织过程，从无到有，一步步产生了地球上最初的原始细胞。因此，细胞重建应该是细胞起源过程在现存生物体的繁殖增生中遗留的“痕迹”。深入研究现存生物体中的细胞重建现象，无疑是探讨细胞起源的一条重要途径。

现存生物体的细胞重建现象，就如同人的早期胚胎，其他哺乳动物、鸟类、爬行类、两栖类和鱼类同样保留着尾和鳃裂，人类大脑的新皮层之下保留着进化过程中形成的旧皮层和古皮层，人类基因组中保留着一系列低等动物祖先的基因一样，都是生命进化或起源过程遗留在现存生物体中的“痕迹”。深入研究现存生物体中这些起源与进化“痕迹”，在现存生物体内寻找非细胞形态转变为细胞的过程，无疑是探讨生命起源和进化的可行途径。

细胞重建是探索细胞起源的途径

细胞起源和细胞重建有着相似的过程，都是经由自组织从非细胞形态到细胞形态，因此研究细胞重建过程可为细胞起源提供线索：

（1）可以通过研究现存生物体在某些特殊情况下，特别是当细胞发生剧烈变化、损伤和大量新生时，如创伤愈合、骨髓造血、昆虫变态、癌细胞形成、辐射损伤与修复以及动物的性转变等，可能发生的核重建和细胞重建等由非细胞形态转变为细胞或者由细胞物质转变为细胞的现象，来寻找细胞起源的线索。

（2）可以模拟类似原始地球环境因素（重力场、电磁场等）对细胞重建进行人工诱导，以探讨细胞起源机制。

（3）可以依靠一价金属离子和二价金属离子浓度的改变等方法创造人工条件实现细胞重建的模拟，为细胞起源过程提供依据和资料。例如，用人工建立的微环境模拟核膜的消失和重建过程，以了解核膜的起源；进行细胞重建的模拟研究，可以为合成细胞探索条件和方法，而合成细胞的方法应该是进行细胞起源研究的重要方法。

（4）现在大都认为，地球上的细胞起源是先形成较为低等的原核细胞，而真核细胞由原核细胞进化而来。然而，现有的研究已经表明，在现代生物体中，原核细胞和真核细胞都可以经由自组织过程从无到有，通过细胞重建途径产生新的细胞。因此，可以推论，原核细胞和真核细胞可能具有并行的细胞起源途径；真核细胞的起源可能不经由原核细胞阶段。

（5）细胞和细胞的起源，是从头开始、从无到有，不一定如“内共生假说”（endosymbiotic hypothesis）以及“分隔假说”（compartmental hypothesis）等所主张的，要以某种结构为起源的基础。只要具备构成细胞的物质基础和合适的环境与条件，都有可能经由自组织过程，从无到有产生原始的细胞核和细胞。

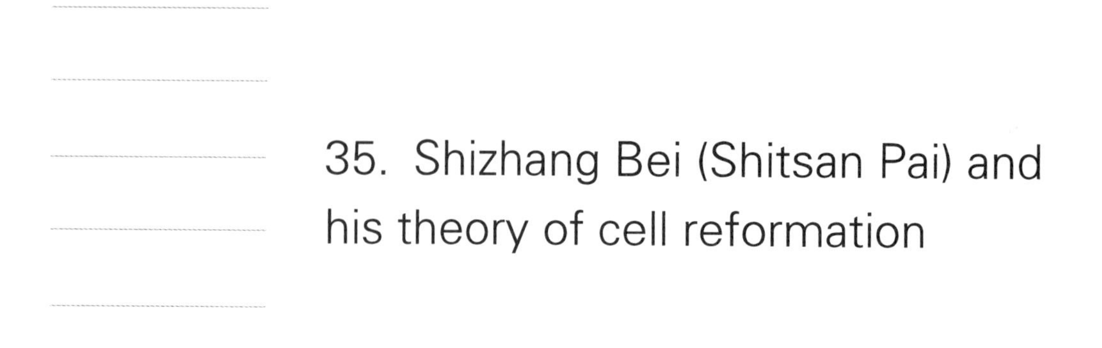

35. Shizhang Bei (Shitsan Pai) and his theory of cell reformation

The discovery of the phenomenon of cell reformation

For a long period of time, people always studied the cell reproduction and proliferation from the traditional viewpoints and in particular focused on cell division and mitosis. Cell division was thought to be the only way of cell reproduction and proliferation, little attention was drawn to other possibilities.

Until the spring of 1932, Professor Shizhang Bei (Shitsan Pai) found an intersex strain of *Chirocephalus nankinensis*, which belongs to Phyllopod of Crustacea, among the specimens collected in a paddy field of Songmuchang on the outskirts of Hangzhou. Individual of intersex has characteristics of both sexes at the same time during a certain period of development stage, and intersex in *Chirocephalus* has not been recorded before. The intersex of *Chirocephalus* could be divided into five types according to the extents of sexual inclination to male or female. The whole process of sex reversal of gonad was studied, and during its reversal, the transformations of germ cells including cell deformation and cell reformation were observed.

In 1934, Professor Bei reported the observation and his viewpoint of such phenomenon at a seminar in the Biology Department of Zhejiang University, where he held a position of professor. However, the report was not published until 1942 and 1943. The reason of the delayed publication of the report was multiple. Firstly, under the turbulent circumstances of the War of Resistance against Japanese Aggression, Zhejiang University was wandering from Hangzhou into inland and was moving from place to place many times. Secondly, there was no suitable journal to which the scientific report could be submitted for publication. Until 1942, the report was not formally published on a newly established periodical *Science Record*. These reasons were all the objective facts, and the subjective reason is that he also felt some hesitation in publishing the report to the public because his observation and viewpoint would be a challenge to the traditional theory of cell proliferation.

Guyan Wang
Institute of Biophysics, Chinese Academy of Sciences, Beijing 100101, China
Correspondence: wang_gy2000@yahoo.com.cn

In the two papers, Professor Bei analyzed the deformation and reformation of germ cells during the sex reversal process about all five types of intersexes, and discussed the possible mechanism. He called this phenomenon reformation instead of new formation because the reformation only means resurgence and the yolk granules possess all the raw materials of the cellular constituents. When the cells deformed, the yolk granules were yielded, which in turn provided the reformation of cells with the raw materials. It suggests that when the material basis for cell restitution and a suitable environment exist, the cell may be self-reconstructed not through the way of cell division. But in the circle of biology, it is a traditional view that a cell comes from its parent cell and the cell division is the only way of cell reproduction and proliferation. Therefore, the idea that the cell can be reproduced and proliferated by the way of cell reformation besides cell division would be considered as a profane conduct against the biological science. Some unfair words such as ridicule, ignorance or arrogance, etc. might be also imposed upon the idea. These doubts and misgivings should be beneath notice for the reports were published after all and did not incur censure. Perhaps somebody thinks that the cell reformation is not easy or frequent to observe if it exists. However, the traditional point of view is always accepted as an infallible law, the influence of which is wide and lasting. So, Professor Bei's research appeared to be an isolated example and was not enough to elucidate a general rule of cell proliferation. Since there was no response, his lonely feeling was hard to avoid and at that time he was busy at other work, so he could only put this research aside.

Time was slipping away fast and more than twenty years had elapsed when the study on cell reformation was started again in 1970. Since then, the research has entered a new period and had a new appearance as compared to that in the 1930s. In the past, the work was done mainly by one person, but now, they have established a research group in the Institute of Biophysics under the Chinese Academy of Sciences; in the past the experimental material was only one kind, i.e., *Chirocephalus nankinensis*, but now they use chicken embryo, mouse bone marrow, *Chlamydia trachomatis* and soy bean nodule bacteria besides *Chirocephalus nankinensis* as the studying materials. Light microscopic observation was the sole way used to study cell reformation in the 1930s, but now they have used many kinds of new techniques and methods such as electron microscopy, timelapse microcinematography, phase contrast microscopy, radioautography, fluorescence polarization, energy transfer with double fluorescence labels, fluorescence recovery after photobleaching, *Raman spectra* and many other biochemical methods. On the basis of the past research work, they have acquired new discoveries and understanding about cell reformation. For example, they have found DNA, histones and chromatin in the yolk granules of chicken eggs. A new discovery that the yolk granules of chicken eggs possess chromatin has never been reported before in the biological science. Furthermore, the chromatin of yolk granules and that of nucleus have the same structure and behavior. The shapes of DNA molecules from both origins are similar too. The chromatin is believed, according to the traditional biology, to exist only specifically in nucleus, so the fact that the yolk granules also possess chromatin cannot but attract the attention of the biological scientists.

The foundation of cell reformation theory

According to the results of experimental research, Professor Bei and his colleagues have published 5 papers in *Scientia Sinica* (Series B, 1983), 24 papers in compilation *Cell Reformation* (Series 1, 1988) and 18 papers in compilation *Cell Reformation* (Series 2, 2003). Professor Bei had obtained fairly systematic knowledge on cell reformation through experimental research and founded the theory of cell reformation. It can be summarized as follows:

(1) Cell reformation is a process of self-organization. It is able to take place in organism in situ or in cell-free preparations in vitro only if there exist a material basis for cell restitution and suitable conditions.

(2) Cell reformation exists extensively in the nature. It can be found not only in eukaryotic cells, but also in prokaryotic cells; not only in germ cells, but also in somatic cells of developing and adult animals.

(3) The superficial yolk granules beneath the chicken blastoderm contain DNA, histones and chromatin, and can reform cell under suitable environments. It should be pointed out that the chromatin is not a specific substance belonging to nucleus only and the yolk granules are not merely a non-living inclusion of cell.

(4) Cell and nucleus can be reformed from cytoplasm. It suggests that there is no strict barrier between nucleus and cytoplasm.

(5) Cell reformation is probably the reflection of cell origin on the earth in the present living world. It is an epitome of the long progress of a cell from the simplest living form. The research on cell reformation would help to elucidate the evolution of life.

(6) Cell division is a "closed" type of cell reproduction, when cells divide, they are isolated from their surrounding environment by a cellular membrane; while cell reformation is an "open" type of cell reproduction, during the whole process of cell reformation the constituents of cell and the surrounding environment are merged into one. So, if the study of cell division is combined together with that of cell reformation and the simulation of cell reformation is combined together with the induction of cell reformation, then they would provide a new approach of changing the cell structures and properties, selecting good qualities and eliminating harmful factors, controlling directive breeding, i.e., promoting and developing the cell technology and cell engineering.

That is to say, cell proliferation in the living organism includes not only the replication of present cells (cell devision), but also the reproduction of new cell by self-organization from the raw materials of cellular constituents (cell reformation). In cell reformation, new cells can start from the very beginning, non-cellular form, and be reproduced gradually from the materials in cytoplasm of mother cell. Generally, the process of reproduction includes the initial formation of nucleus as the naked nucleus state, then the cytoplasm and cellular membrane, and finally the intact reformed cell.

Cell reformation is a trace left behind by cell origin process

Professor Bei believed that cell reformation research involves the fundamental issues in cytology and will have an important impact on the knowledge of the origin and life activities of cell, as well as on the future development of biology.

Self-organization is an important process in living activities. The all biological structures at different levels can be formed by self-organization. For example, during evolution, the human cerebrum improves its structures by self-organization to adapt the environmental changes, cell reformation is a process of self-organization, similarly, the cell origin is also a self-organization process.

As the basic unit of life origin and evolution, cell is a typical self-organizing system. During evolution, life adapts the environment changes by improving its own structures. For example, in the process of chemistry evolution, once the organic molecule formed, it follows the rule of physics (thermodynamics), and tends to stabilize in a structure state by self-organization, then forms chain, ring, helix and fold, it presents special biological activities. Then, these molecules form a related multi-molecular system. After the original life appears, it further improves the structure through self-organization, and evolves to cellular form from the non-cellular form. Finally the original cell develops to be the prokaryotic and eukaryotic cell with sophisticated structure and functions, such as enzymic catalysis, DNA replication, transcription and translation.

Professor Bei's hypothesis on "cell reformation may be an epitome of cell origin on the earth" means that cell reformation is probably a trace of the progress of the cell origin in present living activities. We can imagine that the original lives exist in a non-cellular form of molecular aggregates, such as the microsphere proposed by S. W. Fox (1965), or the coacervate proposed by A. I. Oparin (1969). Then, gradually, the molecular aggregates reform to be the original cell in a process of self-organization, like cell reformation. Therefore, cell reformation should be a trace of the progress of the cell origin in reproduction and proliferation of the present life, while further knowledge of cell reformation in present organism is undoubtedly an important approach to explore the origin of cell.

Cell reformation is the trace left behind by the life origin in present organism, just like that tails and gill slits in early embryos of fish, amphibians, reptiles, birds, mammals and human, the human brain's the old cortex and the ancient cortex preserved in the evolution under neocortex of the human brain, the ancestral genes retained in the human genome. A thorough study on these traces and search for the process of changes of non-cellular form into cellular form are important approaches to explore the origin and evolution of life.

Cell reformation is an approach to explore the origin of cell

Cell origin and cell reformation are similar processes, both of which are self-organization processes from non-cellular form to cellular form. Therefore exploring the process of cell reformation could provide valuable hints for cell origin.

(1) The origin of cell by can be investigated through understanding nucleus reformation and cell reformation, the transformation from non-cell form to cell form, or the transformation from the cell products to cell. These phenomena may exist in present organism when cells are changing greatly, such as wound healing, marrow blood-forming, insect metamorphosis, carcinogenesis, radio injuring and repairing, or animal's sex translation.

(2) To determine the mechanism of cell origin, cell reformation can be artificially induced by electric, magnetic or gravitational field, which is similar to the ancient conditions on earth.

(3) Cell reformation can also be artificially induced by changing the concentration of one-valence and two-valence metal ions. For example, we can simulate the processes of disappearance and reformation of nuclear membrane under artificial micro-environment to understand the origin of the nuclear membrane, the simulation of cell reformation can be used to explore the techniques to synthesize cells, which is also important to study the origin of cell.

(4) Currently, it is widely accepted that eukaryotic cells are evolved from the prokaryotic cells. However, latest studies have shown that both the prokaryotic and the eukaryotic cells can be reconstructed from non-cellular living materials through self-organization or cell reformation. Therefore, we are not able to conclude whether the prokaryotic and eukaryotic cells are independently originated. That is to say, the eukaryotic cells are not necessarily evolved from the prokaryotic cells.

(5) The origin of nucleus and cell may be different from endosymbiotic hypothesis and compartmental hypothesis, which support that some structures are the basis of origin. In contrast, it is suggested that as far as there exist a material basis for cell restitution and suitable conditions, the original nucleus and cell may be self-reconstructed from non-cellular form through the processes of cell reformation.

Figures

Fig.1 Shizhang Bei (1903—2009), the founder of Biophysical Society of China

Fig.2 The intersex of *Chirocephalus*

Fig.3 Electron micrographs of the cell reformation of the yolk granules within the mature oocyte in situ of *Chirocephalus nankinensis*. (a) Structural change of the yolk granules; (b) Further development of the yolk granules; (c) The yolk granules beginning to develop into nucleus-like structures; (d) Three "naked nuclei" . Here the nuclear membranes and reticular structures of nuclei are obvious; (e) A reformed nucleus. Cytoplasm and a prat of plasma membrane just being formed. (f) A relatively full-developed cell, but the morphological structure of the whole cell is rather primitive

Fig.4 The chromatin contained in yolk granules beneath the chicken blastoderm

References

Bei SZ (1943) Yolk granule and cell reformation. Science 26: 38–49. (贝时璋. 1943. 卵黄粒与细胞之重建. 科学, 26: 38–49.)

Bei SZ (1988) Cell Reformation (Series 1). Beijing: Science Press. (贝时璋. 1988. 细胞重建(第一集). 北京: 科学出版社.)

Bei SZ (2003) Cell Reformation (Series 2). Beijing: Science Press. (贝时璋. 2003. 细胞重建(第二集). 北京: 科学出版社.)

Bei SZ, Cao MS, Chen CC, et al (1983) Time-lapse microcinematographic and phase contrast studies on the cell reformation of *Chirocephalus* yolk granules in vitro. Sci Sin B 26(5): 454–459.

Bei SZ, Chen CC, Cao MS, et al (1983) An Electron microscopy study on cell reformation of *Chirocephalus* yolk granules cultured in vitro. Sci Sin B 26: 592–597.

Bei SZ, Li YA, Li L, et al (1983) Chromatin and DNA of yolk granules beneath "blastoderm region" of unfertilized chicken egg. Sci Sin B 26: 823–827.

Bei SZ, Li YA, Li L, et al (1983) Electron microscopy of DNA molecules from blastoderm nuclei and yolk granules beneath blastoderm of fertilized and unincubated chicken eggs. Sci Sin B 26: 818–822.

Bei SZ, Li YA, Li L, et al (1983) Electron microscopy study on the chromatin of blastoderm nuclei and yolk granules beneath blastoderm of fertilized and unincubated chicken eggs. Sci Sin B 26: 708–715.

Pai SZ (1942) Diploide intersexen bei *Chirocephalus nankinensis* Shen. Sci Rec 1: 187–197.

Pai SZ (1943) Ueber die Transformation der Genitalzellen bei den *Chirocephalus*-Intersexen. Sci Rec 2: 573–583.

36. 陈桢及其金鱼遗传学研究

图1 陈桢（1894—1957）

陈桢是著名的遗传学家（图1），在金鱼的变异、演化和遗传上作出了杰出贡献，其相关成果发表在多种科学和教育期刊上。陈桢对于金鱼的研究在国际学术界获得了高度评价，为科学家和公众所熟知。

1894年，陈桢出生于江苏省。1914年从中国公学毕业后，陈桢被金陵大学农林科录取。1919年，陈桢考取清华学校留美官费生，先后在美国康奈尔大学和哥伦比亚大学动物学系攻读研究生，并于1921年获得硕士学位。毕业后他师从摩尔根开展了卓有成效的遗传学研究。1922年回国后，陈桢被东南大学聘为教授，并领导了中国科学社生物研究所的研究工作。1929年起，陈桢任清华大学生物系主任，1948年当选首批中央研究院院士，1955年当选中国科学院学部委员（院士）。1953年起，他担任中国科学院动物研究室主任。

陈桢从南京开始了他的金鱼遗传学研究。他认为金鱼是非常合适的研究材料，因为金鱼是我国原产的，且易于培养，其外部性状又有明显的变异性。同时，由于金鱼是体外受精的，这就为实验胚胎学研究提供了充足的卵和胚胎。为了对金鱼的遗传学有充分的了解，陈桢开展了杂交育种、胚胎学、细胞学、生物统计学和文献学研究。

陈桢认真研究了金鱼外形的变异，尤其是鱼鳍形状和肤色的变异（Chen，1925）。他将不同品种的金鱼杂交，让饲育品种金鱼和野鲫鱼杂交，研究了金鱼体色透明程度的变异，从而发现了“透明”这一性状，且发现这一性状也是符合孟德尔遗传规律的。他还进一步发现了遗传的多效性和不完全显性，并将成果发表在1928年的《遗传学》杂志上（Chen，1928）（图2）。此外，他在后来的实验过程中还研究了金鱼蓝色和棕色的遗传（Chen，1934a）。

陈桢还研究了金鱼在水外、蒸馏水、稀酒精等条件下的发育状况，并测量了野生

作者：付雷
浙江师范大学，金华321004，中国
邮箱：ful527@163.com

和驯化金鱼身体各个部分、器官、组织的比重。这些研究的成果发表在《中国科学社生物研究所汇刊》《科学》[①]《中国杂志》、*Biologia Generalis*、*Journal of Agriculture Science*等权威期刊上，对孟德尔遗传学的发展产生了重要影响（陈桢，1959）。他还将科研成果发表在《清华暑期周刊》上，由此他的清华大学学生还送他一个外号——金鱼店的老板（陈桢，1934b）。

图2 陈桢绘制的金鱼，从上往下：正常鳞鱼、杂合斑驳鱼、纯合透明鱼

通过大量整理研究我国古代文献中关于金鱼杂交的记载，陈桢指出金鱼是由鲫鱼培育而来，是从南宋时期开始家养化的。陈桢还发现，直到清朝人们才开始对金鱼进行定向杂交，而盆养发挥了重要作用（陈桢，1959）。这些发现对我们理解金鱼的遗传和变异有重要意义，也为我们了解我国古代的生物选择史提供了一个视角。

陈桢在中小学生物教育中也作出了重要贡献。1926年，他在《新教育评论》杂志发表了一篇介绍金鱼的文章，指出将金鱼运用到课程中可以提高学生的生物学学习兴趣，金鱼作为实验动物可以广泛用于动物的形体、生理、胚胎与发育、变异与遗传等研究中（陈桢，1926）。从美国回国以后，陈桢发现国内还没有关于生物学基本原理的教科书，因此，他在1924年给大学生写了一本《普通生物学》，被很多学校采用。1933年，他对这本书进行了修订，补充了最新的生物学研究成果，并将其中一些案例替换成了我国本土的材料，例如其中关于金鱼透明体色的遗传，更名为《复兴高级中学教科书：生物学》，由商务印书馆出版，作为高中生物学教科书（陈桢，1933）。这本教科书在国内颇受欢迎，一直使用到1951年。

陈桢因其在金鱼遗传学上的杰出贡献而获得国际上的高度评价，他是将科学研究与科学史研究融合在科学普及中的典范。

---

① 该期刊于1915年由中国科学社创办。

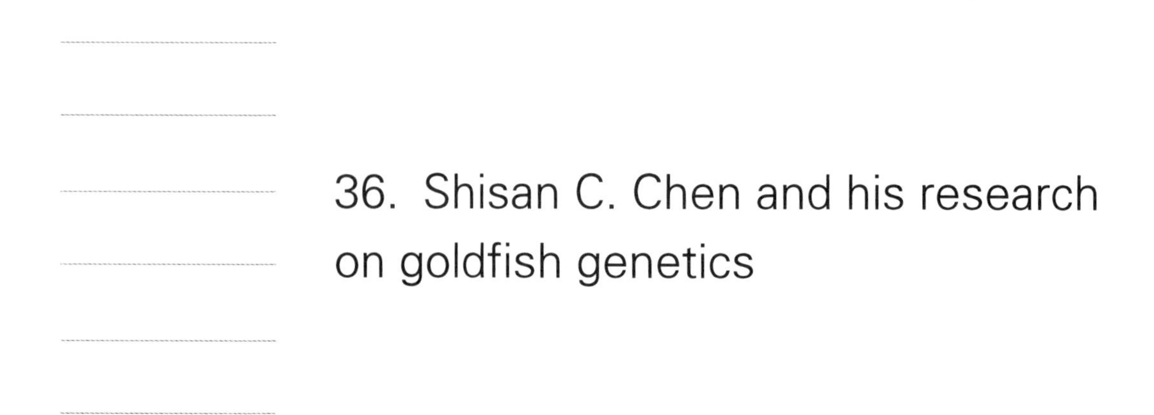

36. Shisan C. Chen and his research on goldfish genetics

Shisan C. Chen (1894—1957) was an excellent geneticist, who made outstanding contributions to research into the variation, evolution and heredity of goldfish and published many of his findings in scientific and educational journals. Shisan C. Chen's research into goldfish remains highly regarded by the international scientific community and resulted in the illumination of this research field to scientists and the general public alike.

Chen was born in Jiangsu Province in 1894. After graduating from China College in 1914, he was admitted to the University of Jinling, Jiangsu Province and majored in agriculture and forestry. In 1919 Chen passed an examination for overseas postgraduate studies held at Tsinghua University and subsequently carried out his postgraduate education in the United States at Cornell University and at the Department of Zoology in Columbia University, where he received his Master's degree in 1921. Following an informative and enlightening period of research into genetics at Thomas Hunt Morgan's laboratory, Chen returned to China in 1922. He was engaged as a professor by Southeastern University in Nanking and led research at the Biological Laboratory of the Science Society of China. In 1929 Chen was given the honor of the leadership of the Department of Biology at Tsinghua University and was among the first academicians of Academia Sinica in 1948 and of Chinese Academy of Sciences (CAS) in 1955. In 1953, he was appointed as a director of the zoology research unit of CAS.

As a geneticist, Chen carried out his research on goldfish and began his work in Nanking. He regarded goldfish as an appropriate study material as goldfish is native to China and can be easily cultivated from crucian based on their distinctive variations on the external characteristics. Moreover, the eggs and embryos of goldfishes provide convenient materials for experimental embryology as they can be fertilized in vitro. In his research, Chen undertook a comprehensive variety of genetic methodologies including cross breeding, embryology, cytology, statistical analysis and literature studies, etc., in order to gain a detailed understanding of goldfish genetics.

Chen studied variations on the external characteristics of goldfish, especially the shapes

Lei Fu
Zhejiang Normal University, Jinhua 321004, China
Correspondence: ful527@163.com

of fins and colors of skin (Chen, 1925). He investigated the inheritance of goldfish transparency scale by crossing various breeds of goldfish and mating domesticated breeds of goldfish and found a new characteristic called "transparent" , the inheritance of which was Mendelian. Furthermore, Chen proved the polypheny and incomplete dominant inheritance from this research and published his findings in *Genetics* in 1928 (Chen, 1928). Moreover, in later experiments, Chen studied the inheritance of blue and brown colours in goldfish (Chen, 1934a).

Chen also conducted research into the development of goldfish affected by being out of water, in distilled water and in solutions of alcohol and measured the specific gravity of various body-parts, organs and tissues of wild and domesticated goldfishes. These other research findings were published in *Contributions from the Biological Laboratory of the Science Society of China*, *Science*[①], *China Journal*, *Biologia Generalis*, *Journal of Agriculture Science* and other prestigious journals and had a major impact on Mendelian genetics (Chen, 1959). He also published his research findings in *Journal of Tsinghua University*. Thus, his students in Tsinghua University gave him the nickname: A Boss of Goldfish Store (Chen, 1934b).

By collating and researching a large number of historical documented observations of goldfish breeding in ancient China, Chen was able to propose that goldfish had originated from crucian and that their domestication began in the Southern Song Dynasty. Moreover, Chen found that directional breeding wasn't practiced until Qing Dynasty and pot raising played a key role in the course (Chen, 1959). These findings have contributed to our understanding of the variations and inheritance of goldfish and have provided an insight into the history of biological selection processes in ancient China.

Chen placed a great importance on biology education in secondary and elementary school. He delivered a speech on goldfish in *New Education Review* in 1926, in which he suggested that goldfish could be used in biology courses to increase students' interest in biology, and to demonstrate the morphology, physiology, embryo and development, variation and inheritance of animals as experimental materials (Chen, 1926). On his return from America, Chen found there were few textbooks concerning fundamental principles of biology and therefore resolved to write a book *General Biology* for college students in 1924, which became widely adopted in China. In 1933, he revised the textbook for high school students with a new name *Fuxing High School Textbook Biology* and published it by Commercial Press, in which he supplemented some advances of biology and replaced some materials with his own research, especially the inheritance of goldfish transparency scale (Chen, 1933). This textbook was very popular and adopted till 1951.

Shisan C. Chen won international acclaim for his work on goldfish genetics and created a model for integrating scientific research and scientific historiography research with the dissemination of science.

---

① This magazine was launched by the Science Society of China in 1915.

Figures

Fig.1 Shisan C. Chen (1894—1957)

Fig.2 Goldfish by Shisan C. Chen. From above: Normal scaled fish, heterozygous mottled fish, homozygous transparent fish

References

Chen SC (1925) Variation on external characters of goldfish, *Carassius auratus*. Cont Biol Lab Sci Soc 1(1):1–64.

Chen SC (1928) Transparency and molting, a case of Mendelian inheritance in the goldfish. Genetics 13:432–452.

Chen SC (1934a) The inheritance of blue and brown colours in the goldfish, *Carassius auratus*. J Genet 29:61–74.

Chen Z (1926) Speech on goldfish to secondary and elementary school biology teachers. New Educ Rev 1(11):9–12. (陈桢. 1926. 与中小学生物学教员谈金鱼. 新教育评论, 1(11): 9–12.)

Chen Z (1933) Fuxing High School Textbook Biology. Shanghai: Commercial Press. (陈桢. 1933. 复兴高级中学教科书: 生物学. 上海: 商务印书馆.)

Chen Z (1934b) Tsinghua Weekly in Summer Vocation, 8:465. (陈桢. 1934. 清华暑期周刊, 8: 465.)

Chen Z (1959) Domestication and Variation of Goldfish. Beijing: Science Press. (陈桢. 1959. 金鱼的家化与变异. 北京: 科学出版社.)

37. 谈家桢与中国现代遗传学

谈家桢是我国现代遗传学最重要的创立者之一，他为我国遗传学的国际化作出了巨大贡献（图1）。

1909年9月15日，谈家桢出生于浙江省。他先后在教会中学和公立中学读书，被免试保送至东吴大学学习。他选择了生物学专业，并对遗传学产生了兴趣。1930年毕业后，他进入燕京大学，师从李汝祺（图2）。李汝祺是我国第一个师从美国著名遗传学家摩尔根获得博士学位的人，也是燕京大学当时唯一的遗传学教授（Zhang，2017）。在胡经甫的建议和李汝祺的悉心指导下，谈家桢出色地完成了关于亚洲瓢虫遗传学的研究，发表了多篇优秀论文，获得了硕士学位。李汝祺将谈家桢的论文推荐给了摩尔根，当时摩尔根的实验室以果蝇遗传学而闻名。摩尔根注意到了他的才华并大加赞赏，由此开启了二人的合作。在东吴大学工作数月后，1934年谈家桢进入摩尔根的实验室进修。在摩尔根和杜布赞斯

图1　谈家桢（1909—2008）

图2　谈家桢（左）和李汝祺夫妇

作者：付雷

浙江师范大学，金华321004，中国

邮箱：ful527@163.com

基的指导下，1936年他获得加州理工学院博士学位。在接下来的几年里，他陆续在国际学术期刊上发表了十余篇论文，为国际学术界所认识。谈家桢婉言谢绝了摩尔根的挽留回到中国，受聘担任浙江大学教授。在抗日战争时期，他在贵州湄潭仍然继续关于瓢虫的研究，并培养出了第一批研究生。1952年，他开始担任复旦大学生物系主任，1961年担任遗传学研究所所长。1980年，谈家桢当选中国科学院学部委员（院士）。

谈家桢是杰出的多产的科学家，在遗传学的诸多领域都做出了举世瞩目的成就。即便是在战争年代，他也没有间断对瓢虫和果蝇的遗传学研究。在燕京大学时，谈家桢主要研究瓢虫色斑的遗传与变异。正是在这一时期，他在瓢虫色斑变异的镶嵌显性遗传上做出了举世瞩目的发现，在当今的教科书中仍被视为经典。20世纪40年代，谈家桢赴美继续研究瓢虫色斑变异的镶嵌显性遗传，其研究成果在国际遗传学界引起极大反响（Tan，1946），他认为瓢虫是研究微进化的模式生物。在摩尔根的实验室工作期间，谈家桢绘制了果蝇常染色体的遗传图谱。他研究了不同种果蝇染色体的本质，证明了基因是如何通过重复、分化而演化的（Tan，1942）。他从20世纪60年代开始关注辐射细胞遗传学。其研究团队将猕猴作为实验材料，检测X射线和γ射线的影响。此后他们又在国内率先将研究拓展到环境毒理学。晚年时期，谈家桢建议政府要大力支持人类基因组研究，以更好地保护中国人类遗传资源。在他的倡议下，我国建立了两大国家级基因组研究中心。在70余年的教学和科研生涯中，谈家桢发表了100余篇研究论文，这是宝贵的科学财富。谈家桢曾担任中国遗传学会理事长、中国环境诱变剂学会理事长、中国生物工程学会会长，对相关学科的发展也作出了突出贡献。

谈家桢积极参与国际合作。他是最早将“基因”这一术语介绍到国内的科学家。早在1945—1946年，他就作为访问教授在美国开展研究工作。1948年，他被选为国际遗传学会常务理事。几十年后的1978年，他在访问美国时，将中国的遗传学发展情况介绍到国外。此后，他又邀请了多位美国教授来华讲授分子遗传学。20世纪80、90年代，在他的邀请下，越来越多的欧美科学家来华开展学术交流。在他的力主下，1998年第18届国际遗传学大会在北京召开，他担任会议主席。此外，在他的帮助下，越来越多的中国学者也到海外进修。他被授予加州理工学院“杰出校友”荣誉称号、加拿大约克大学荣誉博士学位、日本遗传学会名誉会员。他是美国科学院外籍院士、意大利国家科学院院士、第三世界科学院院士（赵功民，1998）。在谈家桢和其他杰出科学家的努力下，我国遗传学和其他科学已经融入了世界科学界。

谈家桢热心于科学教育，培养了施履吉、盛祖嘉、刘祖洞、徐道觉等一大批杰出的遗传学家。他认为基础知识、基础理论和基本实验技术对学生来说是非常重要的。他对学生到海外进修是非常支持的。不仅如此，他还邀请他们回国参加工作。他热衷于科学普及，撰写了大量科普著作和文章，如《生命的密码》《基因和遗传》《孟德尔及其科学事业》《中国现代生物学家传》等，他在著作中还强调了科学技术史的重要作用。

谈家桢是世界著名科学家。为了纪念谈家桢的科学贡献，1999年，中国科学院紫金山天文台1964年发现的3542号小行星被命名为“谈家桢星”。

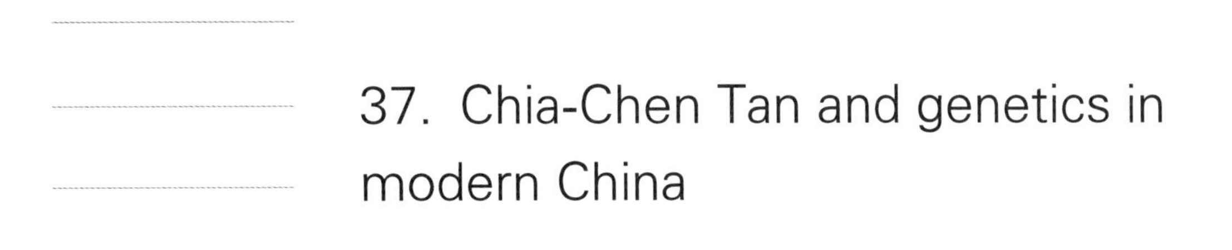

37. Chia-Chen Tan and genetics in modern China

Chia-Chen Tan (谈家桢, 1909—2008) was one of the most important founders of genetics in modern China and made great effort to the internationalization of Chinese genetics.

Chia-Chen Tan was born on September 15, 1909 in Zhejiang Province. He studied in missionary middle school and public high school successively before being admitted to Soochow University without examination. He majored in biology in the university and became interested in genetics. After graduating in 1930, he entered YenChing University and learnt from Ju-Chi Li (李汝祺) who was the first Chinese student to receive a doctor's degree from American famous geneticist Thomas Hunt Morgan and was the only genetics professor in YenChing University at that time (Zhang, 2017). Following Chenfu Wu Francis's (胡经甫) suggestion and Ju-Chi Li's instruction, Chia-Chen Tan completed his master research on lady-bird genetics with several excellent articles. Ju-Chi Li recommended his article to Thomas Hunt Morgan whose lab was famous for research on inheritance and variation in fruit flies. His outstanding work got Morgan's attention and appreciation, which promoted their cooperation later. After working at Soochow University for several months, Tan went abroad to Morgan's lab in 1934 for further study and he got PhD. from California Institute of Technology in 1936 under the supervision of Morgan and Dobzhansky. In the following years, he published dozens of articles in international journals, which made him known by the world. Then he declined Morgan's detainment and returned to China, being appointed professor in Zhejiang University. During the War of Resistance against Japanese Aggression, Tan continued his research on lady-bird and cultivated his first graduate students in Meitan, Guizhou Province. He then became the head of biology department in 1952 and director of Institute of Genetics in 1961 in Fudan University. He was elected as an academician of the Chinese Academy of Sciences in 1980.

As an outstanding and prolific scientist, Chia-Chen Tan worked on several fields of genetics and made a large number of world-famous findings. Even during war times, he continued his researches on genetics of lady-bird and fruit flies. When he was in YenChing University, Chia-

Lei Fu

Zhejiang Normal University, Jinhua 321004, China

Correspondence: ful527@163.com

Chen Tan worked on inheritance and variation in the color of patterns in the lady-bird beetles. It was in this period that he made world-famous discovery about mosaic dominant inheritance of color patterns in the lady-bird beetles, which is still a classical example in modern genetics textbooks. In 1940s, he went to USA again to continue his research on mosaic dominance in the inheritance of color patterns in the lady-bird beetles, which had a significant impact on international genetics (Tan, 1946). He considered lady-bird beetles as good model of micro-evolution. Tan figured out the genetic maps of autosomes in *Drosophila pseudoobscura* when he was in Morgan's lab. Then he proved how gene evolved through repetition and differentiation by investigating the nature of the race-differential chromosomes in *Drosophila montium* (Tan, 1942). From 1960s he paid attention to radiation cytogenetics. His team chose *Macaca mulatta* as experimental material and tested the effect of X-ray and γ-ray to *Macaca mulatta*. They expanded their research to environmental toxicology which was pioneering in China. Later in his life, Tan suggested the government supporting researches on human genome for the sake of protection of Chinese human gene resources. Thanks to his advice, China established two national genome research centers. In his 70 years of teaching and researching, Tan published more than 100 articles, which are valuable treasure for science. Tan served as the president of the Genetics Society of China, Chinese Environmental Mutagen Society and Chinese Society of Biotechnology and made substantial contributions to the development of the related disciplines.

Chia-Chen Tan participated actively in international cooperation. He was the first to introduce the term "gene" to China. As early as 1945—1946, he went to USA again as visiting professor. In 1948 he was elected as a member of council of the 8th International Congress of Genetics. After a few decades, he visited America in 1978 and introduced the progress of genetics in China. And then he invited several American professors to China to lecture on molecular genetics. From 1980s to 1990s more and more scientists from America and Europe came to China for academic communication at his invitation. He was the leading proponent and chairman of the 18th International Congress of Genetics held in Beijing in 1998. Besides, more and more Chinese scholars went abroad for further research under his aid. He was awarded outstanding alumni by California Institute of Technology, Honorary Doctoral Degree by York University, Canada and Honorary Fellow of Japanese Society of Genetics. He was the member of National Academy of Sciences of USA, Lincean Academy of Italy and The Third World Academy of Sciences (Zhao,1998). Under efforts of Chia-Chen Tan and other excellent scientists, Chinese genetics and other sciences became known to the world.

Chia-Chen Tan devoted himself to science education and cultivated several distinguished graduate students including Lvji Shi (施履吉), Zujia Sheng (盛祖嘉), Zudong Liu (刘祖洞), Daojue Xu (徐道觉) and so on. He considered the basics of knowledge, theories and experimental skills important for students. He supported his students to go abroad for further studies. More than that, He invited most of them to come back to work in China. He was continuously keen on science popularization and wrote several books and articles for public, such as *Secret of Life*, *Gene and Heredity*, *Mendel and His Science Career*, A *Biography for Biologists in Modern China* and so on, in which he laid stress on history of science and technology.

Chia-Chen Tan is a world-renowned scientist. In 1999 the No. 3542 asteroid discovered by Purple Mountain Observatory, Chinese Academy of Sciences in 1964 was named 3542 Tanjiazhen, which is in honor of Chia-Chen Tan's contribution to science.

Figures

Fig.1 Chia-Chen Tan (1909—2008)

Fig.2 Chia-Chen Tan (left) and Mr. and Mrs. Ju-Chi Li

References

Tan CC (1942) The nature of the race-differential chromosomes in *Drosophila montium* De Meijere. Sci Rec 1:178–187.

Tan CC (1946) Mosaic dominance in the inheritance of color patterns in the lady-bird beetles *Harmonia axyridis*. Genetics 31:195–210.

Zhang H (2017) Professor Ju-Chi Li, one of the pioneers and founders of modern genetics in China. Protein & Cell 10: 235–237.

Zhao GM (1998) Geneticist Chia-Chen Tan. J Dialect Nat 20(6): 60–72. (赵功民. 1998. 智者魅力学界楷模——遗传学家谈家桢. 自然辩证法通讯, 20(6): 60–72.)

38. 李汝祺：中国现代遗传学的奠基人之一

甲午中日战争中国战败的那一年，他降生在这个积贫积弱的国家；五四运动爆发的那一年，他怀着科学救国的志向远渡重洋赴美留学；北伐战争打响的那一年，他带着博士学位回国献身科学事业。他就是著名的生物学家李汝祺（图1），我国现代遗传学的奠基人之一。

李汝祺1895年3月2日出生于天津市，早年就读于清华学校。1919—1923年在美国普渡大学农学系学习畜牧学。毕业后，他进入哥伦比亚大学动物学系，在世界著名遗传学家摩尔根的指导下，从事果蝇发育遗传的研究。1926年，他以优异的成绩成为第一个在摩尔根实验室获得博士学位的中国留学生。同年，他回到中国，应蔡翘教授的邀请，去复旦大学任教，次年转任燕京大学生物学系教授。1952年全国高校院系调整后任北京大学生物学系教授直至退休。1956年被评为一级教授。曾任中国动物学会理事长、中国遗传学会首任理事长（图2）兼《遗传学报》主编。

图1　李汝祺教授（1895—1991）

图2　1978年李汝祺在中国遗传学会成立大会上讲话

作者：张翮
中国科学技术大学，合肥230026，中国
邮编：zhangheahu@163.com

李汝祺是果蝇发育遗传研究的先驱之一。早在1923年，他便开始在摩尔根教授的指导下进行染色体畸变对果蝇发育影响的研究。依据主要研究成果写成的博士论文，以《染色体畸变在黑腹果蝇发育中的效应》为题发表在《遗传学》杂志上，成为该杂志1927年的首篇文章（Li，1927）。这篇文章是果蝇发育遗传学领域最早的文章之一。十几年后，美国学者D. F. Poulson才发表果蝇发育遗传学的研究成果（Poulson，1940）。直至20世纪60年代，发育遗传学才正式创立。

李汝祺在动物染色体和胚胎发育领域也有过开拓性研究，他还是第一位把细胞遗传学介绍到中国的学者。1934年他首先报道了在中国马中发现的具有3对染色体的马蛔虫，不同于早先在国外发现的具有2对或1对染色体的马蛔虫（Li，1934）。在获取了新的更加充足的实验材料后，1937年，他将这一重要发现发表在国际权威学术期刊《科学》上（Li，1937），受到国际学术界的重视。他将生态学和胚胎学结合起来，研究了刺腹蛙、黑斑蛙和北方狭口蛙的胚胎发育及其对环境变化的适应性，取得了丰硕的研究成果。1935—1936年，他在加州理工学院从事细胞遗传学研究，回国后将果蝇唾腺染色体技术引入国内，开创了我国的细胞遗传学领域。他和李先闻、李景均、李竞雄并称民国时期遗传学“四大金刚”（Luo，2010）。

值得一提的是，李汝祺参加了1956年召开的青岛遗传学座谈会（Li and Kang，2011），在会上阐述了自己的观点，与许多遗传学家一道澄清了对摩尔根学派的错误批判。在此会议发言的基础上，他写出了题为《从遗传学谈百家争鸣》的文章，谈了自己参加会议的收获以及百家争鸣对于发展科学的重要性，在1957年4月29日的《光明日报》上发表。毛主席看后十分赞赏，把文章标题改为《发展科学的必由之路》，以原标题为副标题，还为之写了编者按，当年5月1日由《人民日报》转载，对于推动我国遗传学事业的健康发展起了积极的作用（吴鹤龄和戴灼华，2008）。

在长达60年的教学和科研生涯中，李汝祺培养了一大批遗传学界的骨干人才，为我国遗传学事业的发展奠定了坚实的基础。著名遗传学家谈家桢院士和动物分类学家刘承钊院士曾在他的指导下完成硕士学位论文。生物学家张作干、金荫昌、林子明和李肇特等都是他的得意门生。

李汝祺在长期的教学和科研工作中，积累了丰富的经验，著述甚多。1981年，他的《细胞遗传学的基本原理》（李汝祺，1981）出版，被国家教育委员会定为大学通用教材。1985年，90岁高龄的他出版了心血之作《发生遗传学》（李汝祺，1985a），该书将遗传学、胚胎学和细胞学的基本理论融为一体，被誉为我国的遗传学经典巨著。同年，收录了他40篇论文的《实验生物学论文选集》（李汝祺，1985b）出版。次年，他的《细胞遗传学若干问题的探讨》（李汝祺，1986）出版。

晚年，李汝祺捐出自己的积蓄，在中国遗传学会设立了“李汝祺优秀动物遗传学论文奖”，旨在鼓励年轻学者为我国遗传学的发展贡献力量。这是他一生中最后一次为中国遗传学事业作出重大贡献。

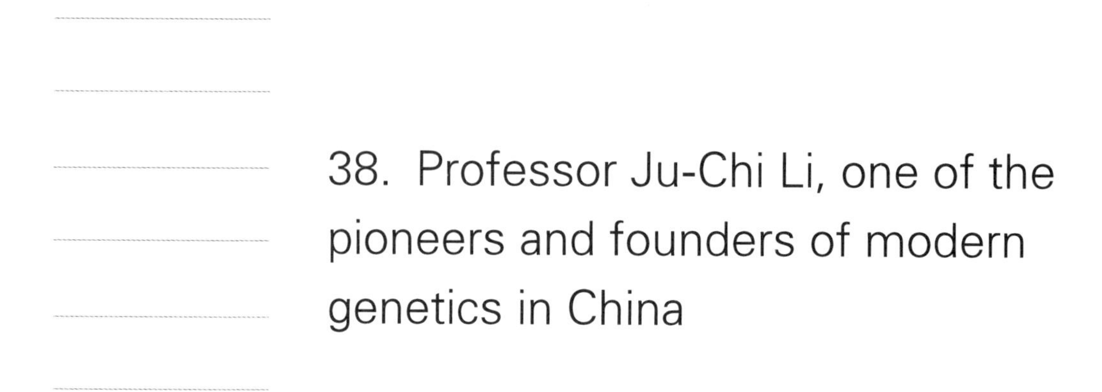

38. Professor Ju-Chi Li, one of the pioneers and founders of modern genetics in China

During the year China lost the Sino-Japanese War (1895), he was born in this poor and weak country. In the year the May 4th Movement broke out (1919), he traveled across the oceans to the United States with the ambition of saving the nation by science. It was in the year the Northern Expedition was declared (1926) that he returned to his motherland and dedicated himself to science. He was Prof. Ju-Chi Li, the famous biologist and one of the pioneers and founders of modern genetics in China.

Ju-Chi Li was born in Tientsin on March 2, 1895. He attended Tsinghua Imperial College from 1911 to 1919 before he went to the United States. From 1919 to 1923, he majored in animal husbandry at Purdue University. After graduation, he was admitted to the Department of Zoology at Columbia University, where he focused on the developmental genetics of *Drosophila melanogaster* under the supervision of world famous geneticist Prof. Thomas Hunt Morgan. In 1926, due to his remarkable achievement, he became the first Chinese student to receive a doctor's degree at Thomas Hunt Morgan's laboratory. In the same year, he went back to China to teach at Fudan University at Prof. Chiao Tsai's invitation. The following year, he moved to Yenching University to become a professor in the Department of Biology. In 1952, to follow the adjustment of colleges and departments in China, he was appointed as a professor in Department of Biology, Peking University until he retired in 1989. Li was promoted to the First Grade Professor in 1956. He was the former director-general of China Zoological Society, the first director-general of Genetics Society of China and the editor-in-chief of *Acta Genetica Sinica*.

Li was a pioneer in developmental genetics. As early as 1923, he initiated studies on the effect of chromosome aberrations on development in *Drosophila melanogaster*. Some of the findings in his dissertation for doctor's degree was published in *Genetics* (Li, 1927). This research article was one of the earliest articles in the field of developmental genetics. It was

He Zhang[1,2]

1 Department for the History of Science and Scientific Archaeology, University of Science and Technology of China, Hefei 230026, China

2 School of Marxism, Bengbu Medical College, Bengbu 233030, China

Correspondence: zhangheahu@163.com

a dozen years before the American geneticist D. F. Poulson published academic findings in the same field (Poulson, 1940). It was not until the 1960s that developmental genetics was established.

Li had also pioneered the study of chromosome and embryonic development of animals. He took the lead in introducing cytogenetics into China. In 1934, he first reported that *Ascaris megalocephala* having 3 pairs of chromosomes, unlike the classical *Ascaris megalocephala* found in other countries with 1 or 2 pairs of chromosomes, and named it "*Ascaris megalocephala* Trivalens" (Li, 1934). After obtaining new and more abundant experimental materials, in 1937, he announced this important discovery in the top-tier research journal *Science* (Li, 1937), which drew the attention of biologists all over the world. He combined ecology and embryology to study the embryonic developments of *Rana boulengeri*, *Rana nigromaculata*, and *Kaloula borealis* as well as their environmental adaptation, and made outstanding achievements. From 1935 to 1936, he carried out his research on cytogenetics in California Institute of Technology. After his return to China, he introduced the preparative technique of salivary gland chromosome of *Drosophila melanogaster* into China and initiated studies on cytogenetics in China. He and Hsien-Wen Li, Ching-Chun Li, Ching-Hsiung Li together are honored "Four Mars" in genetics area (1912—1949) (Luo, 2010).

It was worth mentioning that Li was invited to attend the Qingdao Meeting on Genetics in August, 1956 (Li and Kang, 2011). During the meeting, he elaborated his point of view and clarified the erroneous criticism of Morganian School along with many other geneticists. He also published an article titled "Talking about hundred schools of thoughts contend from genetics" in *Guangming Daily* on April 29, 1957 based on his speech at the meeting. In this article, he spoke of what he had gained in the meeting and the significance of Hundred Schools of Thoughts Contend for the development of science. Chairman Mao thought highly of this article, changed the title to "The only correct way for the development of science" with the original title as its subtitle and even wrote editor's note for it. It was reprinted in *People's Daily* on May 1 that year and played an active role in promoting the healthy development of science in China (Wu and Dai, 2008).

During his professional career spanned several decades, Li cultivated a large number of talents in the field of genetics to lay the solid foundation for development of modern genetics in China. The eminent geneticist Academician Chia-Chen Tan and animal taxonomist Academician Cheng-Chao Liu finished their master's dissertations under his supervision. Biologists Tso-Kan Chang, Yin-Chang Chin, Tsu-Ming Lin, and Chao-Te Li were all his brilliant students.

Li accumulated rich experience in his long-term professional career, and accomplished many seminal academic writings. In 1981, his *Principles of Cytogenetics* (Li, 1981) was published and widely used as a general textbook in Chinese colleges and universities before long. At the age of 90, he published his representative work with strenuous effort, which was titled *Developmental Genetics* (Li, 1985a). This book integrated the principal theories of genetics, embryology, and cytology and was regarded as the masterpiece of modern Chinese genetics. In the same year, the book *Selected Papers of Experimental Biology* (Li, 1985b), including forty previously published papers, was delivered. The next year, his *Discussion on Several Problems of Cytogenetics* (Li, 1986) appeared.

In his late years, Li donated his savings to the Genetics Society of China to set up a special fund named "Ju-Chi Li Animal Genetics Outstanding Paper Prize" , which aimed to encourage young scholars to make significant contributions to the development of genetics in China. This was the last great contribution he made for modern genetics in China.

Figures

Fig.1 Prof. Ju-Chi Li (1895—1991)

Fig.2 Ju-Chi Li delivering a speech at the founding conference of Genetics Society of China in 1978

References

Li JC (1927) The effect of chromosome aberrations on development in *Drosophila melanogaster*. Genetics 12:1–8.

Li JC (1934) A six-chromosome ascaris found in Chinese horses. Peking Nat Hist Bull 9:131–132.

Li JC (1937) A six-chromosome ascaris in Chinese horses. Science 86(2222):101–102.

Li JC (1981) Principles of Cytogenetics. Beijing: Science Press. (李汝祺. 1981. 细胞遗传学的基本原理. 北京: 科学出版社.)

Li JC (1985a) Developmental Genetics. Beijing: Science Press. (李汝祺. 1985a. 发生遗传学. 北京: 科学出版社.)

Li JC (1985b) Selected Papers of Experimental Biology. Beijing: Science Press. (李汝祺. 1985b. 实验生物学论文选集. 北京: 科学出版社.)

Li JC (1986) Discussion on Several Problems of Cytogenetics. Beijing: Peking University Press. (李汝祺. 1986. 细胞遗传学若干问题的探讨. 北京: 北京大学出版社.)

Li M, Kang L (2011) The 1956 Qingdao meeting on genetics: An important turning point of Chinese biology. Protein & Cell 2(1):5–6.

Luo GH (2010) The story of geneticist Hsien-Wen Li. Protein & Cell 1(8):709–710.

Poulson DF (1940) The effects of certain X-chromosome deficiencies on the embryonic development of *Drosophila melanogaster*. J Exp Zool 83:271–325.

Wu HL, Dai ZH (2008) Biography of professor Ju-Chi Li. Hereditas 30(7): 807–808. (吴鹤龄, 戴灼华. 2008. 李汝祺教授传. 遗传, 30(7): 807–808.)

39. 李景均先生逝世十五周年祭

李景均先生是世界遗传学的一位泰斗（图1和图2），是我国遗传学的先驱之一。在李先生逝世十五周年之际，让我们一起缅怀这位遗传学家。

图1　李景均（1912—2003）

图2　1982年李景均70寿辰照

如果试问一位学过遗传学或在相关领域工作的外国人：“你知道谁是世界上最著名的中国遗传学家吗？”答案会是：“李景均。”这令当今的很多人感到迷惑不解，其原因可能是：他（她）不了解中国的遗传学历史，或者我国的学者不了解全球遗传学家的历史。公平地说，这是不为我们几代遗传学家所知的一段特殊历史时期的遗传学奇事奇人。

李先生的一生充满坎坷。尽管在国际遗传学界尽人皆知，可是国内多数同事只有从那些曾与他交往并对他有较多了解的人那里才能听到关于他的些许故事（Petechuk，1989；杨焕明，2004）。本文拟按有关事件发生的先后顺序，主观地将先生的人生轨迹分为三个时间段，而这三个时间段皆出于他自己的三个人生选择，不管是己之所为或不得已而为之。

（1）1941年——第一个选择：回到祖国

1941年10月中旬，李先生携9月26日完婚的新婚妻子Clara Lem，乘荷兰远洋邮轮

译者：夏志，杨焕明

深圳华大生命科学研究院，深圳518083，中国

邮箱：xiazhi@genomics.cn（夏志）；yanghuanming@genomics.cn（杨焕明）

从美国加利福尼亚州的圣地亚哥启程经香港回到祖国（郭孙伟，2016）。此行的目的或许是蜜月旅行和传统的拜见长辈及亲朋好友，又或许是要回到祖国开启他的学术生涯。而当时，正是抗战最激烈的时候。

这是李先生做的改变他一生的第一个选择。

所有认识李先生，特别是曾与他相处较久的人都有完全一致的评价：他放眼世界，关注时局，涉猎广泛，阅典无数，无所不读，“无所不知”。当时，他的故乡因深陷战区而饱受创伤。据他的女儿Carol所告，李先生是了解中国时局的，对日军侵略一直义愤填膺，因而执意要去抗日后方。

此趟旅程，他和所有其他乘客一道亲历了战争带给人类的灾难：邮轮原定的航程为15天，结果在海上整整漂流了51天。到达香港那天已是1941年12月6日，正是“珍珠港偷袭”的前一天（郭孙伟，2016）。

在那段时间里，他的夫人已经怀孕。他夫人的家族在美国有着自己的产业，并希望他们回到美国（Spiess，1983，2005）。而更为可怕的是，无论是在船上还是在香港，他和夫人不只是饱受饥饿的威胁，而是实实在在地徘徊在饿死的边缘。

当时，折返美国已不现实，太平洋几乎停航，而奔赴抗战大后方的路途则困难重重。还没开始就知道那将是一段令人无法想象的漫长旅程。然而，李先生还是执意动身。他忍饥挨饿徒步走了38天，并花光了所有积蓄，终于从香港九龙到达了桂林。而他夫人由于怀上第一个儿子Jeff，不能走路而在轿子上颠簸了38天（Chen and Tai，1998；Spiess，1983，2005）。

这一段话足以证明当时是多么艰难，而他又是多么执着：

“在那段时间里，我知道了饥饿……当你饥饿时，你真的什么也做不了。除了食物你什么也不会想，你只能反应迟钝地躺在那里。当你看到一个东西，你的第一反应是它能吃吗？如果它不能吃，那就毫无用途……饥饿驱使我甚至期待所有的商店都应该卖食物……其余的事都是微不足道的和乏味的，连死亡也变得很平常。”当他到达惠阳时，他亲笔描述了这些感受。

在1942年到1948年期间，李先生随着战争辗转各地，但依然坚持他的工作——教学和研究。1942年他在广西大学农学院（位于郴州附近的沙塘）任教（Chen and Tai，1998），后到迁至成都的国立中央大学任职，1943年随校返回南京才稍稍安定下来。1946年他最终到了北平，并在北京大学任职（郭孙伟，2016）。

1948年，李先生的第一本书《群体遗传学导论》由北京大学出版社出版，这本书是在战火中锻造的他几年教学和研究的心血结晶。笔者之一也编过教科书，深知全靠记忆、缺少任何参考资料，坚持这一工作的艰难。李先生一直在考虑他的人生：做一个科学家。为了人类的科学，决不放弃自己钟爱的事业：研究与教学。

这一阶段（1942—1948年），他对我国遗传学作出直接贡献（高翼之，2005；Chakravarti，2004）。

（2）1949年——第二个选择：留在祖国

李先生做出的第二个选择是留在祖国以实现他的梦想。

说李先生是一位斗士，捍卫科学真理和自己人生理念的斗士，一点也不过分。

1949年1月31日北平才宣布解放，这几年时间他有足够的理由和机会回到美国去加入他与他夫人的家族。所有和他有着广泛联系的老同学、朋友和同事都曾劝他离开中国。实际上，早在1945年，他的夫人就去上海试图为他取得美国国籍（Spiess，1983）。他的胞弟Jerome那时已在美国俄勒冈大学任职。

下述三件事，虽尚无严格的文献支持，但或许可以解释他为何选择留在祖国：

第一，在北平宣布解放前后，他曾邀请共产党员到大学校园宣讲“解放区”的“一片蓝天”（叶笃庄，1997）。

第二，1949年1月，他曾向叶剑英先生表达过他希望继续为新中国的科学和教育事业作贡献的热切愿望（郭孙伟，2016）。这是一个非常重要的事件，也反映了李先生愿意为国效力的心迹。

第三，从1949年7月开始，李先生即与北京农业大学副教授陈延熙翻译了李森科的代表作《遗传及其变异》（是根据杜布赞斯基的英译本翻译的），说明他对我国遗传学的发展有着期许（郭孙伟，2016），因为如果去美国则绝对不需要这样的。据说，那些“李森科追随者”扬言李先生根本不知道什么是“米丘林学说”。李先生性格倔强，干脆来个“知已知彼”，所以才去“研究”李森科学说。经由胡乔木过问，该书才得以在1950年1月出版发行。也许正是通过翻译李森科的书，李先生更强烈地意识到“李森科学说”将会摧毁我国的遗传学。这令他失望至极。

可以肯定的是，他带着最美好的愿望投入新中国的怀抱，带着梦想与希望为我国的遗传学作出了贡献。

（3）1951年——第三个选择：别了，祖国！

他的第三个选择对其人生影响重大。当然，这也是他所做出的最为艰难的选择。

不难从他自己的叙述中了解到他当时的所思所想：“把一个纯学术的问题扣上政治的帽子是令人无法忍受的。”“即使有着再大的耐心，我和我的同事也无法把我国的遗传学从灭亡中拯救出来。一个人必须宣布他是忠于李森科学说的，否则能做的就只有逃离。”是的，他感觉到报国无门（郭孙伟，2008）。

不过，李先生拒绝放弃他的原则。他铁骨铮铮，堂堂正正，与当时整个遗传学界的李森科追随者抗衡。在1949年上半年，李先生在农学系师生大会上公开批评李森科学说是“完全错误的”和“缺乏科学性的”（叶笃庄，1997）。

他因此成为遗传学家中的第一个靶子。他的《群体遗传学导论》也旋即被批判。那些李森科追随者将李先生主张的遗传学、田间设计和生物统计都说成“唯心的”“伪科学”，李先生的课均被校方取缔，李先生无课可上，被晾在一边。

最终，李先生愤然“辞去”了农学系主任和试验田主任的职务（郭孙伟，2008）。至今我们仍然不知道这究竟是李先生的有意为之，还是实为无奈之举。他不能教授他所熟知的知识。他很清楚地明白李森科学说将误导年轻的一代。作为一名教师和科学家，他是不能接受这一切的。离开是最简单、最合理的方式。李先生的离开促使领导后来很快控制了局面，事实上也是保护了很多人。

到了美国后的几乎半个世纪里，他把自己看作一个“异乡人”。当李先生到达匹兹堡后，几乎是单枪匹马地开始了遗传学研究。他建立了美国最大、同时也是运行时间最长的遗传咨询师培训项目，为那些遗传病患者和家庭提供信息和咨询。从1953年到2000年的47年里，他发表了131篇科学论文和综述，其中25篇是在他退休后发表的。

他做了很多广受欢迎的演讲，例如“两个保温瓶的故事：人类智力遗传模型的特性”。他出版了10部著作（其中2部为译著），多部被用作教科书，《群体遗传学导论》更被奉为经典，在全球诸多一流大学被广泛使用，并被翻译成多种语言。这本书始终启发和激励着全世界的学生投身到遗传学研究中来。他的学生遍及全世界，其中包括12名博士生。他们与李先生的亲密关系“远超传统意义上的师生关系”（引自他的三个博士生的论文）。

李景均教授为我国的遗传学和生物统计学的发展作出了重要贡献，培养了许多在农业和生命科学领域中发挥了重要作用的杰出人才，被很多国内外同行称为“中国遗传学之父”。

他对整个遗传学的贡献可以这样总结：他是第一位具有全球影响力的美国人类遗传学会（ASHG）华裔主席；他发表了题为“现代人的下巴越来越小”的著名演讲；1998年他被授予ASHG最高荣誉“杰出教育奖”，时至今日仅有四位遗传学家获此殊荣。

李景均的主要研究领域是群体遗传学和人类遗传学以及它们的研究方法学之一——“统计学”。在群体遗传学方面，他的研究内容极为广泛，包括群体的遗传平衡及其稳定性，群体的适合度，基因型相关、配子相关和近缘系间相关，基因频率估计，以及突变和选择的效应等群体遗传和演化等问题。他对W. E. Castle群体平衡定律有新的发现，受到高度赞扬（郭孙伟，2008）。上述这些成就中，以遗传平衡和群体适合度的研究成果最为突出，集中表现在他所著的《群体遗传学》里。该书是首次向我国学术界介绍群体遗传学的论著，一经问世就被学术界公认为名著。

在人类遗传学方面，他的主要工作是对有关人类家系的孟德尔式分离的研究，这涉及人类许多遗传疾病的控制和防治，以及对诸如智力之类的复杂性状的基本科学认识。

除了对群体遗传学和人类遗传学方面的贡献，李景均对遗传流行病的早期发展也作出了突出的贡献。他创造了一种被称为“不计独子女方法”（method of discarding the singleton）或简称“单法”（the singles method）的分离分析方法，简单而实用（郭孙伟，2008）。在计算机普及之前，它是遗传流行病学研究中对疾病遗传传播规律进行分析的非常有用的方法。

在统计学方面，他的主要贡献是有关通径分析的应用和方差分析的数学模型。1964年，他的《试验统计学导论》一书出版。这本书深入浅出地介绍了实验的统计设计原理及其方法，受到读者欢迎。1975年，他的《通径分析入门》第一次系统论述了通径分析的原理、方法和应用。此书在遗传流行病学研究领域曾十分风行，只是随着个人计算机及统计软件的普及，通径分析才逐步让位于结构方程模型。1982年，李先生又出版了《不平衡数据的分析》。

或许李景均迄今为止影响最长久的一份学术精神遗产是他在20世纪50年代中期提出的临床试验的随机和双盲两个原则（郭孙伟，2008）。该思想虽在当时遭到一些美国医生的强烈反对，但李景均坚持己见，毫不让步。幸运的是，随着不懈努力，这两个原则逐渐被普遍接受。随机双盲，尤其是随机，再加上对照，是当今临床试验的基本原则，对客观评价治疗结果提供了方法学上的保证。

他的女儿Carol也曾与笔者之一谈到她的父亲："他是一位全身心投入的科学家。"

自从20世纪80年代，我国许多学者和留学生，包括笔者之一在内，尽管不一定是他的学生，但都有幸见到过李先生。作为ASHG年会的一部分，在美中国遗传学家举行的每一次学会工作午餐，李先生都会参加。席间，他从不提及自己的过往经历。他只是深切地关心祖国所发生的事情，并会坦诚地发表自己的观点，表达着他对祖国遗传学的关注和热忱。

从科学意义上讲，我们必须承认大多数时候他都是对的，例如，以他自己特有的坦诚和尖锐的表达方式，对"李森科学说"和"优生法"坚决批判。

1997年3月，他在给叶笃庄先生一封长达17页追忆往事的信中写道："这封信我写了两天，落了两次热泪。"叶笃庄先生说："有道是英雄有泪不轻弹，尤其是对一个85岁经历过人世沧桑的耄耋老人来说。写了两天，落了两次热泪——其委屈、其伤感、其悲哀、其遗恨，跃然纸上，真切、深沉、感人至深。"（叶笃庄，1997）

李先生对国内学术界一些现象真是"怒其不争"。李先生对于所谓"大跃进"和"优生法"的批评（Majumder，2004），从社会意义和科学意义上都是正确的。1997年提出的遗传资源管理，尽管在学术上，李先生对此颇有微词，但当时作为其中一位参与起草者，笔者之一特地向他征求意见，他的态度却非常积极，并特别高兴地答应修改英文稿。完全可以说，现在面世的英文稿里面就有他的贡献（Chen and Tai，1998）。这是我们不应忘记的。

我的诸多挚交师友，都是李先生的好友。每每提起先生，都"有口皆碑"（Chakravarti，1999；Majumder，2004）。

李先生安息吧！我们不会忘记您的。确实，已有好几次在国内举行的学术会议上，开设了"怀念李景均先生"的专场。

39. In memory of Prof. C. C. Li

Professor Ching Chun Li (C. C. Li), one of the greatest geneticists in the world and a pioneer of genetics in China, passed away fifteen years ago. Let us remember him together.

"Do you know who the greatest, world-renown Chinese geneticist is?"

The answer, "C. C. Li, of course!" would puzzle most, if not all, of us; many in our generation don't know Who is Who (Attention! Li, not Lee, the latter of which was generally used as the general spelling of that generation).

The reason for this answer is simple: Either the individual does not know the history of genetics in or outside China. To be fair, many of us really don't know so much about the history of genetics in those exceptional or specific historic periods in China.

C. C. Li's life was full of ups and downs. It has been reported by those who have known him and those who have known of him. This article, with the sole purpose of memorializing him, focuses on three choices which have made the most dramatic changes of his life in 1941, 1949 and 1951, respectively, based on a detailed Chronicle Vitae we try to make with reference to all the materials we are able to collect as Supplement to this article (Petechuk, 1989; Yang, 2004).

(1) 1941—The first choice: To return to his motherland.

In October 1941, Mr. C. C. Li, together with his bride, Clara Lem, took a Dutch ocean liner from San Diego to Hong Kong in order to reach China (Guo, 2016); the purpose of this trip was either for a honeymoon and family visit, or to start his academic career in his motherland.

It was the first choice C. C. Li made which changed his life.

All those who have known C. C. Li would have had the same impression: He was curious about everything. Ever since he learned how to read, he read about everything in the world, especially everything related to China. It would be wrong to conclude that he did not know the War of Resistance against Japanese Aggression was at a climax.

Immediately, he and his wife, together with all the passengers on the liner, experienced what that war brought to the people: The ocean liner, which was supposed to sail for 15 days, was

Zhi Xia, Juan Tian, Xiaoling Wang, Huanming Yang
BGI-Shenzhen, Shenzhen 518083, China
Correspondence: yanghuanming@genomics.cn (H.-M. Yang)

left drifting on ocean for 51 days—even by changing ships—before arriving in Hong Kong on December 6, 1941, just one day before the "Sudden Attack on Pearl Harbor" took place (Guo, 2016).

At that time, it was impossible for C. C. Li to get back to the US, mainly because his wife was pregnant. Her own family had business in the US and expected them to go back (Spiess, 1983, 2005). More horribly, he and his wife were still in the shadow of being "nearly starved to death" from their time on the ocean liner and in Hong Kong. No more starvation!

Unsurprisingly, the couple faced many difficulties on the way, or unoccupied part of China. It turned out, from the very beginning, to be a long journey which could still be impossible for us to imagine! However, C. C. Li was determined. He continued, walking and walking, with an empty stomach most of the time; his pregnant wife was carried on a sedan chair almost all the time, which emptied his small wallet. It took them 38 days to travel from Kowlong to Guilin on foot (Chen and Tai, 1998; Spiess, 1983, 2005).

The following quotations from him might help us understand how difficult it was and how determined he was:

"During this time, I learned about hunger... When you are starving, you really can't do anything. You can't think of anything but food. You just lie there like zombie. When you see a thing, the first question in your mind is: Is this edible? If it is not, it is useless... I thought all stores should sell food, that's what starvation does to your thinking... Everything else is trivial and boring. Even death is trivial." He described these thoughts after arriving in Huiyang."

Between 1942 and 1948, C. C. Li moved with the war but stuck to his profession as both a teacher and a researcher: From the Agricultural College of National Guangxi University (Shatang, near Chenzhou) in 1942 (Chen and Tai, 1998), to the National University of Nanking in Chengdu and then in Nanjing in 1943, finally to Peking University in 1946 (Guo, 2016).

C. C. Li's first book, *An Introduction to Population Genetics*, published in 1948 by the Peking University Press, is a full summary of his teaching and research from that Experimental Farm on the campus in Beijing.

The period between 1942 and the end of 1948 is the first and only period of his direct contributions to genetics in China (Gao, 2005; Chakravarti, 2004).

(2) 1949—The second choice: To stay in his motherland.

The second choice C. C. Li made was to stay in China again with his people in the hope for the future of genetics in his motherland.

Even before Beijing announced its liberation on January 31, 1949, there would be many reasons for him to join his wife's family in the US. His wide communication with his former classmates, friends and colleagues, might have advised him to leave. It was also true that his wife made an effort to visit Shanghai in order to resume her US citizenship as early as in 1945 (Spiess, 1983). In addition Jerome, his younger brother, was at the time already at Oregon State University, US.

Three events, not firmly documented yet, could help explain why he chose to stay in China:

First, it was he who invited a communist officer to the campus to give introduction to the "Liberation Region" around the time the liberation was announced (Ye, 1997).

Second, it was also true that "C. C. Li expressed a passionate wish to keep contributing to China's science and education to Mr. Jianying Ye, then Mayor of Beijing" in January 1949 (Guo, 2016).

Third, his motivation to translate Mr. Trofim Lysenko's representative book, *Heredity and Its Variability*, could be interpreted as a desire to know how genetics would develop in China (Guo, 2016).

It is absolutely sure: He embraced the best hope for China, as well as his dream to contribute to genetics in China.

(3) 1951—The third choice: To "leave" his motherland.

The third choice which had made a dramatic change in his life, as well as the most difficult choice he had ever made.

It is not difficult to find out what was in his mind from his angry remarks: "Taking a purely academic issue as a political 'hat' for an enemy cannot be tolerated." "Even with great patience, it is impossible for my colleagues and me to save genetics from extinction in China. In this case, one must declare that he is loyal to Lysenkoism, otherwise the only way is to escape." He felt that there was no way to serve his motherland (Guo, 2008). Perhaps just through the process of translation of Lysenco's book, C. C. Li realized that this would totally ruin genetics in China, which totally disappointed him.

However, he refused to abandon his principles. He stood up as a MAN against the "Lysenko's tide" which was flooding in the whole field of genetics. In the first half of 1949, C. C. Li publicly criticized Lysenkoism as "totally wrong" and "completely unscientific" several times at meetings with teachers and students at the Department of Agriculture (Ye, 1997).

He might have become the first target. Lysenko's followers, of course, further criticized "genetics" "field design" and "biostatistics" taught by C. C. Li as "ideal" "pseudo-science", leading to the official banning of all the courses he had been teaching. His book, "*Introduction to Population Genetics*" was immediately "criticized" by an official journal.

Finally, C. C. Li resigned from his positions in the Department of Agronomy and the Agro-experimental Farm (Guo, 2008). Really, we still don't know what C. C. Li did was an "action" or a "reaction". He was unable to "teach" something he did not know, or did not want to know; he understood very well that Lysenkoism was something to "mislead the younger generations". It was just an unacceptable fact by him as a teacher, let alone a researcher. C. C. Li's departure did disturb the leaders who took immediate measures to suspend this situation, actually protecting many geneticists.

In almost half of a century, he considered himself a "D.P. —a displaced person". It is well known that arriving in Pittsburgh, C. C. Li "almost singlehandedly started genetics research at Pitt". He established one of the nation's largest and longest-running programs for the training of genetic counselors—professionals who give information and advice to individuals with hereditary diseases—in the 1970s. He published 131 research and review papers in 47 years (from 1953 to 2000), with 25 of them published after his retirement. He delivered numerous speeches which received critical acclaim, such as "A tale of two thermos bottles: properties of a genetic model for human intelligence" . His 10 books (including 2 translated books), especially many textbooks, beginning with *Introduction to Population Genetics* published in Beijing in 1948, have become classics and been translated into several European and Asian languages and continued to inspire confidence and admiration of the students of generations all over the world to go into genetics.

More importantly, he has students all over the world, including 12 PhD students whose relationship with C. C. Li was even closer than that between the "traditional Teachers and Students" (paper by three of his PhD students).

He became the first Chinese American to be elected as a President of American Society of Human Genetics (ASHG), one of the most influential organizations in genetics in the world; he delivered the well-known inaugural speech, entitled "The diminishing jaw of civilized people" . He received the lifetime "Award for Excellence in Education" , the top honor by ASHG in 1998, which currently has been only awarded to 4 geneticists.

Many Chinese scholars and students from China, including the authors of this article, had been together with him many times since the 1980s. C. C. Li attended almost every one of the routine working dinner hosted by the Society of Chinese Geneticists in the US during the annual meeting of the ASHG. He generally did not talk about his own experiences, nor the past, but he always asked about and gave his own frank comments on what was happening in his motherland, demonstrating his passions and concerns for genetics in China.

In the sense of science, we also have to admit that, most of the time, he was correct, even when expressed in his unique frankness and sharpness; for example, his sharp criticism on Lysenkoism and "eugenics" .

Let's again take it as an example: No matter how he commented on it, he did help us with the translation of the Chinese version of the drafted *Provisional Regulations on the Genetic Materials* into English very carefully with many valuable and cordial suggestions. Now it could be said that he made great contribution to the English version of the Regulations (Chen and Tai, 1998).

In March 1997, he wrote another letter, as long as 17 pages, recollecting past events to Mr. Duzhuang Ye: "I wrote this letter for two days and fell into tears twice." "There is a saying that the hero is not tearful, especially for an old man who has been through the vicissitudes of life for 85 years. Written for two days, fell two tears—its grievances, its pain, its sadness, its resentment was really touching" (Ye, 1997).

May his soul rest in peace! He will live in our hearts forever.

Figures

Fig.1 Ching Chun Li (1912—2003)

Fig.2 C. C. Li on the occasion of his seventieth birthday in 1982

References

Chakravarti A (1999) Professor Ching Chun Li, courageous scholar and educator. Am J Hum Genet 64:14–15.

Chakravarti A (2004) Ching Chun Li (1912—2003): A personal remembrance of a hero of genetics. Am J Hum Genet 74:789–792.

Chen T, Tai J (1998) A conversation with C. C. Li. Stat Sci 13:378–387.

Crow JF (1950) An introduction to population genetics. Am J Hum Genet 2:279–280.

Gao YZ (2005) International genetics master C. C. Li. Life World: 76–79. (高翼之. 2005. 国际遗传学大师李景均. 生命世界: 76–79.)

Guo SW (2008) The geneticist and biostatistician Ching Chun Li—his life, his book, and spirit. Sci Cult Rev: 68–89. (郭孙伟. 2008. 遗传学家、生物统计学家李景均先生——其人其书及其精神. 科学文化评论: 68–89.)

Guo SW (2016) Deciphering: At the beginning of the founding of the PRC, a scientist's leaving shocked the central authorities. (郭孙伟. 2016. 解密: 建国初惊动了最高层的科学家出走事件.)

Majumder PP (2004) C. C. Li (1912—2003): His science and his spirit. J Genet 83(1):101–105.

Petechuk DA (1989) The life and times of C. C. Li. Health Sciences Review: 22–25.

Spiess EB (1983) Ching Chun Li, courageous scholar of population genetics, human genetics, and biostatistics: A living history essay. Am J Med Genet 16: 603–630.

Spiess EB (2005) Remembrance of Ching Chun Li, 1912—2003. Genetics 169:9–11.

Steele B (1998) In defense of scientific freedom: An oral interview with Li Ching Chun. University Times, University of Pittsburgh 31: 3–4.

Yang HM (2004) Commemorating the 1st anniversary of C. C. Li's passing away. J Genet 31(8):870. (杨焕明. 2004. 纪念李景均先生逝世1周年. 遗传学报, 31(8): 870.)

Ye DZ (1997) Why the prominent geneticist Li Ching-Chun left his country: The episode of forcing the acceptance of Michurinism in China. China Annu 7:35–41. (叶笃庄. 1997. 一代遗传学宗师李景均何故去国——记50年代初强制推行"米丘林学说"在中国引起的风波之一. 炎黄春秋, 7: 35–41.)

40. 李先闻：遗传育种学家

图1　李先闻（1902—1976）

著名的植物遗传学家和农作物育种专家李先闻（Hsien-Wen Li）（1902—1976），是我国现代生物学和农学发展史上一位非常重要的人物（图1）。他出生在四川江津（今属重庆）一个农民家庭，祖上是广东梅县客家人。李先闻既有农家子弟的勤奋节俭，又有客家人的坚毅和执着。自称是朴实认真的乡下人，却与李汝祺（Ju-Chi Li）、李景均（Ching-Chun Li）和李竞雄（Ching-Hsiung Li）三位同样姓李的遗传学家一道，被誉为民国时期遗传学“四大金刚”。

1915年，李先闻考入清华学校，从四川乡村来到大都市，开始了漫长的求学过程。前几年因年幼，不知用功，数学和物理等基础课都学得不好。到了毕业的前两年，班里已经分科，李先闻确定学农科时，才发奋努力，功课突飞猛进。在回首这段往事时，李先闻不无遗憾，认为如果数学和物理等课程都学好的话，应该能取得更大的成就。在清华学校学习时，他对一些老师的启发式教学法有着深刻的印象。像一位绰号“鸭蛋”的地理老师讲课时，“把某次某条约、割某地都连带讲出来”。李先闻认为“像这样有启发国家思想的教学法，对我们的学业实在有帮助”。而土壤学老师“育种学是现代最时髦的科学”的断言，深深地激励着他，使这位农家子弟后来义无反顾地投身于遗传育种学的研究，对我国育种学和农业生产的发展产生了很深的影响。

1923年从清华学校毕业后，李先闻到美国普渡大学学园艺。李先闻觉得那里的课程内容很简单，没有学到什么理论方面的知识。三年后转学到康奈尔大学攻读博士学位，主科是遗传学，指导老师是著名的植物遗传学家、当时植物育种系主任和研究院院长Rollins A. Emerson教授。在名师教导下，他的进步很快。尤其是他的细胞学老师Bababra McClintock（1983年诺贝尔生理学或医学奖获得者）生动直观的教学法，对他

作者：罗桂环

中国科学院自然科学史研究所，北京100190，中国

邮箱：guihuanl@ihns.ac.cn

帮助很大。他在给Randolph教授当助手的过程中，不仅学到很多实验技术，还深刻体会到学习生物学这种实验科学要“手脑并用”。与此同时，他与George W. Beadle、M. M. Rhoades和G. F. Sprague等美国同学很快成了好朋友。上述三个同学后来都成为美国科学院院士，Beadle还于1958年获得诺贝尔奖。在康奈尔大学上学，他既遇到一批良师，又结识一帮益友，学术环境非常好，心情十分畅快，学业迅速进步。Emerson教授言传身教，不辞辛劳和青年学生一起下玉米地里实验、研究，通过严谨的实验和缜密的思考解决科学问题，成为他的楷模。他决心一生都像老师那样努力，不敢懈怠。1929年，李先闻顺利通过论文答辩，获得博士学位。

获得博士学位后，李先闻为了更好地回国服务，曾去拜访系上的育种专家Love教授，想了解一下育种的专门技术，不料Love教授认为他是学遗传学理论的，对他想搞育种这种应用研究不以为然。刚刚离开美国时，李先闻充满豪情，心里默默地说：“再会吧！美国！我要回去救中国了。”不过，国内的现实却远不像李先闻想象的那样美好，找工作异常艰难。学术圈的人际关系非常复杂，其他人回国时有博士学位或者没有博士学位都给教授职称，而他有博士学位，回去只在国立中央大学谋到一个讲师的职位，工资比别人少不说，教的还是专业外的蚕桑课程，不久即失业。1930年好不容易才在东北大学找到一个教授职位，又因日本侵略东北很快丢失。有段时间甚至回到母校清华大学当体育辅导老师。际遇的坎坷和民族的深重灾难，让他倍感愤懑，但这个倔强的农家子弟从来不向困难低头。1931年8月，他去金陵大学看望康奈尔大学的老师Love教授和C. H. Myers教授。当时Love教授任政府实业部的顾问和中央农业实验所的总技师。谈话中Myers教授对学遗传学理论的李先闻一脸不屑，认为他只会在染色体上玩基因，别的一概不知。“而我们呢，实用的研究者，能为你们国家赚钱！”面对如此不公平的奚落，李先闻马上反驳他说：“假若给我机会，我会证明给你看。”

在同学好友的帮助下，李先闻从1932年开始，先后来到河南大学和武汉大学任教。在武汉大学期间还担任农学系主任。从那时起，逐渐开展小米和小麦等作物的育种研究，抽空做细胞遗传研究，逐渐使自己的生活和事业发展起来。因研究小米取得了成绩，在美国的农业杂志上发表了不少文章，他开始被称作“小米专家”或“狗尾巴草专家”。在武汉大学工作期间曾在珍珠小米中发现四倍体（当时叫四元体），为纪念这一发现，他还将自己的一个女儿起名“四元”。

抗日战争全面爆发后，李先闻一度想投笔从戎，参加抗日队伍，到前线杀敌立功。后来听从劝告，于1938年随学校撤退到成都，任四川农业改进所作物组组长。期间他对武汉大学进行的有关小米的遗传进化的研究成果进行了总结，搞清楚莠（狗尾巴草）是粟（小米）的祖先。在四川，他与李竞雄、鲍文奎两助手一起研究秋水仙碱诱导植物多倍体，在粟类远缘种间杂交及其进化、小麦矮生性状的遗传学分析等方面做出了卓越的成绩，在美国的遗传学刊物先后发表了十多篇文章。有一次他给美国一家刊物寄去了一篇长达25页的论文，文章发表后收到刊物负责人的来信，要求寄70多美元的出版费。李先闻回复说，他没法给这笔钱，一来自己实在太穷，二来也没渠道寄。刊物负责人只好权当把这笔钱送给他了。另外，他们培育出来的小麦良种，在川北绵阳等后方地区大面积推广后，普遍增产一至两成，对当时粮食产量的提高发挥了重要的作用。李先闻在一定程度上实现了儿时“建设国家，全国增产谷子”的理想。他还是最早进行玉米自交系育种研究的学者。

抗战胜利前夕，李先闻随一个学术代表团到美国考察农业，重新回到阔别15年的母校康奈尔大学。回到育种系讲述自己的工作成果时，Love教授已经对他刮目相看，在听讲过程中，很认真地做笔记。原因正如李先闻自己指出的那样："以他曾在中国的育种成绩看来，我们在短短的几年期间，成果远远超过他及他的伙伴在中国约十一年做出的成果。"可惜的是，当时Myers教授已经过世数年，李先闻无法将自己的育种成果展示给他看了。1948年，李先闻以自己杰出的研究业绩，当选首届中央研究院院士。

1948年，李先闻应友人的邀约，到台湾糖业公司的科学试验机构任负责人。后举家迁台，兼任台湾"中央研究院"植物研究所所长。在台湾期间，他在甘蔗育种方面下了很大的功夫。他的衣服和裤子很容易被试验田中如刀片一样的叶子割破，常常是补了又补，皮肉被割破也是常事。但他还是保持非常饱满的热情要把育种工作做好。研究成果发表后，也在国际上产生了广泛的影响。李先闻在良种推广方面做了大量的工作，在1953年推广良种的当年，就使台湾的糖产量由上年的55万吨增长到88万吨。他负责的甘蔗良种推广的成功，不但使公司业务大发展，也稳定了台湾的经济。当时台湾的外汇70%靠蔗糖的出口。在国外有人称他为"甘蔗李"（Sugar Li）。

李先闻不修边幅，衣着简朴，加上身体结实，皮肤黝黑，常常被人当成干苦力活的工人或农民。有一次有个农民看到他在试验田田埂上休息，就把他当成在糖厂打工的人，问他挣多少钱。在听说干活很苦而且挣钱很少的时候，这个农民立即对他表示了高度的同情。非常热心地告诉他养兔能挣钱，还花了半个小时向他传授养殖技术。

李先闻在有生之年一直期望有朝一日"四大金刚"能相聚，然而他没能等到这一天。1976年李先闻因心脏病发作而逝世，走完了自己充满传奇而丰富多彩的人生。

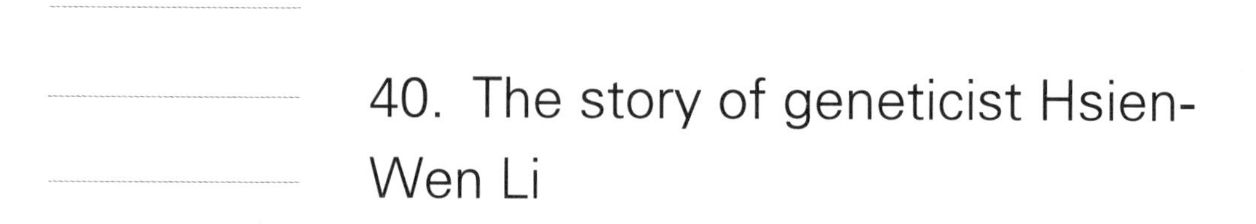

40. The story of geneticist Hsien-Wen Li

Hsien-Wen Li (1902—1976) is a distinguished geneticist and agricultural breeding specialist, and a significant figure in modern biology and agriculture development in China. He was born to a peasant's family in Jiangjin county, Sichuan Province, whose ancestors were Hakka (Kejia) people from Guangdong Province. Hence, He possesses the characteristics of farm boy, diligence and frugality, as well as of Hakka group, fortitude and persistence. Hsien-Wen Li called himself as "a simple and serious countryman" ; while he and Ju-Chi Li, Ching-Chun Li, Ching-Hsiung Li together are honored "Four Mars" in genetics area during Republic era.

At 1915, Hsien-Wen Li attended Tsinghua School (the predecessor of Tsinghua University) in Beijing and started his academic journey. Due to the young age, he was not very studious at the first six years in Tsinghua, and did not get good scores on basic courses, such as mathematics and physics. Until he chose agriculture as major at 1921, two years before graduation, he began to work hard and got significant improvement in academic studies. He felt very regretful about this experience—he believed that he would accomplish more if he had thorough knowledge of physics and mathematics. During the eight years of Tsinghua study, he was deeply impressed by the heuristic teaching mode. Words from the agrology teacher, "breeding is the most popular subject in modern science" , inspired him to continue the career on genetics and breeding.

After graduating from Tsinghua, Hsien-Wen Li went to the United States for graduate study. Initially he studied horticulture at Purdue University, but he felt the courses were too simple and contained limited theoretical knowledge. Therefore, three years later, he transferred to Cornell University to work on genetics. His advisor was a renowned plant geneticist, Prof. Rollins A. Emerson, who was also Chair of Department of Plant Breeding and Dean of Graduate School. He made rapid progress on academic studies under the instructions of many reputable teachers. For example, the cell biology teacher Bababra McClintock, who won the Nobel Prize in Physiology or Medicine at 1983, used direct and vivid teaching style to provide valuable assistance to young students; when he worked as an assistant for Prof. Randolph, he not only learned a lot

Guihuan Luo

The Institute for the History of Natural Sciences, Chinese Academy of Sciences, Beijing 100190, China

Correspondence: guihuanl@ihns.ac.cn

of experimental techniques, but also strongly felt the significance of experiments in biological science. During the graduate study, Hsien-Wen Li benefited a lot from the personal influence of Prof. Emerson, who tirelessly worked with young students in cornfields and investigated scientific puzzles through careful experiments and cautious thinking. He regarded Prof. Emerson as an example and followed his advisor's step during the rest of his life. At the same time, Hsien-Wen Li became a good friend with some American classmates, including G. W. Beadle, M. M. Rhoades and G. F. Sprague, who all became Academicians of Academys of Sciences in United States, among which George W. Beadle also won Nobel Prize in 1958. In the comfortable academic environment, Hsien-Wen Li's research went well and earned his PhD degree at 1929.

After graduate study, Hsien-Wen Li started to plan his return to China. Before the trip, he visited the breeding expert Prof. Love and hoped to learn techniques on plant breeding. Unexpectedly, Prof. Love showed little interest in Hsien-Wen Li's will to do applied research, because he thought Hsien-Wen Li knew only the theory of genetics. Despite the frustration, Hsien-Wen Li was still full of confidence and left US. However, the reality in China was more than tough—it was extremely hard to find a job and it was very complicated to deal with academic affairs. Some returnees, with or without PhD degree, were able to get full professorship; while he was offered only a position of assistant professor at Central University in Nanjing, teaching sericulture courses with low payment, and soon he lost the work. Later, he found a position of professor at Northeastern University in Shenyang, but it was quickly lost due to the invasion of Japan. He even went back to Tsinghua University to be a sport tutor for a while. In front of the difficulties from both country and himself, he never gave up or submitted to the situation. In August 1931, he went to Nanking University to visit his Cornell teacher, Prof. Love and Prof. C. H. Myers. At that time, Prof. Love was Consultant of the Department of Industry and Chief Technician of Central Agriculture Research Institute in China. During the talk, Prof. Myers said that Hsien-Wen Li was just a genetics theorist dealing with genes on chromosome, but not as an application researcher capable of make profits for the country. Facing this unfair statement, Hsien-Wen Li argued, "I will definitely show you when there is a chance."

With the assistance of colleagues and friends, Hsien-Wen Li joined Henan University at 1932 and then moved to Wuhan University, where he became Chair of Agriculture Department. Since then, he started working on millet and wheat breeding. Soon he made significant progress on millet breeding and published several papers on US agriculture journals; he was even called "millet expert" or "bristlegrass expert" . At Wuhan University, he discovered tetraploid in pearl millet, and his daughter was named "Siyuan" (the Chinese name of tetraploid) for memory.

During the War of Resistance against Japanese Aggression, Hsien-Wen Li summarized his research about genetics and evolution of millet, and indicated that bristlegrass is the predecessor of millet in Chengtu (Chengdu) city, Sichuan Province. Later, he worked with Ching-Hsiung Li and Wen-Kui Pao to investigate the polyploidy induction in *Euphorbis*, distant interspecific hybridization and evolution of millet, and genetics analysis of wheat dwarf characters. They got significant achievements, with over 10 publications on international genetics journals. Due to the war, the research condition was extremely tough. Once he mailed a 25-page manuscript to an American journal and was asked to pay $70 publication fee. He replied that he was too poor to afford that, and even if he had the money, he was not able to mail it out at all. Finally, the fee got waived. The improved species was applied to Mianyang at Northern Sichuan in large scale and produced 10%–20% higher yield. This had significant impacts on enhancing the food production, which partially realized his childhood dream of "work for the country, enhance grain production" . In addition, he was also the pioneer on corn inbred breeding studies.

In the mid-1940s, Hsien-Wen Li went to US with an academic group to study agriculture, and returned to the mother school Cornell after 15 years. When he presented his research, Prof. Love showed great surprise. Comparing to the work of Prof. Love and his colleagues during the 11 years in China, Hsien-Wen Li accomplished much more achievements within these short few years. Unfortunately, Prof. Myers had passed away and was not able to see such a miracle. At 1948, Hsien-Wen Li was elected to the Academician of Academia Sinica.

At 1948, Hsien-Wen Li was invited to Taiwan Sugar Company to be Manager of Scientific Research Department and Chair of Plant Institute of "Academia Sinica" . During that period, he devoted himself to the breeding of sugarcane, which was an extremely hard work. The clothes were easily cut by the sharp leaves of sugarcane, so did the skins. However, his motivation never faded in front of such difficulties. With the strong faith and passion, he got great success on sugarcane breeding and spread fine breed as well. At the first year of well-bred application, the sugar yield was enhanced to 880000 tons from the original 550000 tons, and this high yield remained stable in the following years. This achievement not only save the company, but also made significant contribution to Taiwan economy because 70% of foreign exchange at Taiwan relied on cane sugar export at that time. Therefore, he was also called "Sugar Li" by worldwide colleagues.

Although Hsien-Wen Li received Western education and peaked early, he still kept a simple lifestyle like a farm boy. He wore plain clothes and did not care for details; with strong body and dark skin, he was often mis-recognized as workers or peasants. Once, while he rested on the cane field, a person walking by took him for a worker in Sugar Company and asked how much he earned. After hearing about the high working load and low payment, the passerby showed great sympathy at once and suggested him to raise rabbits to make more money. This nice passerby even spent half an hour teaching him how to raise rabbits.

One of Hsien-Wen Li's wishes was reunion of "Four Mars" ; however, he was not able to see this moment. He passed away at 1976 due to heart attack and left his legend and colorful stories to the following people.

Figures

Fig.1 Hsien-Wen Li (1902—1976)

References

Li HW (1970) Autobiography of Hsien-Wen Li. Taipei: Taiwan Commercial Press. (李先闻. 1970. 李先闻自传. 台北: 台湾商务印书馆.)

Li HW, Pao WK (1943) Evolution of millets. Sichuan Agriculture Institute Bulletin 36: 1–17. (李先闻, 鲍文奎. 1943. 粟类之演化. 四川省农业改进所农业丛刊, 36: 1–17.)

Pao WK, Li CH, Cheng CF, et al (1943) Genetics of wheat dwarf characters. Scientific Agriculture 1: 1–12. (鲍文奎, 李竞雄, 陈之万, 等. 1943. 小麦矮生性之遗传. 科学农业, 1: 1–12.)

41. 黄翠芬：伟大的分子遗传学家和中国基因工程学奠基人

图1　黄翠芬教授出席全国英模大会（1987年）

黄翠芬教授是我国著名的分子遗传学家，我国基因工程学创始人之一（图1）。她一生致力于基因工程疫苗、基因工程多肽药物及分子肿瘤的研究，为我国生物技术和军事医学的发展作出了巨大贡献。她同时也是第一位获得尿激酶原（Pro-UK）基因的克隆及表达的中国科学家，该基因显示了对溶栓的高选择性而不会导致出血（中国工程院，2009）。

黄翠芬教授1921年出生于广东，在岭南大学（现中山大学）理学院获得化学学士学位后，到中央卫生实验院流行病微生物研究所工作，从事青霉素制备、鼠疫菌抗原提取和鼠疫菌的药物敏感试验。1948年，黄翠芬教授到美国康奈尔大学攻读细菌学硕士学位，师从时任《细菌学》杂志主编的J. Sherman教授。黄翠芬教授于1950年回到中国。起初，她在山东医学院（现山东大学医学院）担任细菌教研室副教授，负责微生物医学的教材编写、教学以及细菌学研究工作。1954年，黄翠芬教授调入军事医学科学院任研究员、副所长，主要从事微生物毒素的毒理机制和流行病综合防治的研究。经过多年的辛劳工作，她在国内首次成功研制四联创伤类毒素。1980年，黄翠芬教授筹建了中国人民解放军第一个分子遗传学研究室，专心"乙型肝炎核心抗原"遗传学研究。20世纪80年代中期，黄翠芬教授调至生物工程研究所，主要在基因工程领域进行研究。90年代末，黄翠芬教授从事分子肿瘤的相关研究。

黄翠芬教授先后在乙型肝炎核心抗原及e抗原试剂、腹泻疫苗、霍乱肠毒素、溶血栓制剂等基因工程研究领域取得显著成绩，分别于1987年和1995年获得国家科学技术进步奖一等奖，以及多项军队科技进步奖。1984年10月，黄翠芬教授被邓小平同志授予"模范科学工作者"荣誉称号，1996年当选中国工程院院士。同时，黄翠芬教授多次被党和国家领导人邀请，作为科技界的杰出代表登上北京天安门观礼。

译者：龚伟
浙江师范大学，金华321004，中国
邮箱：yzugw@163.com

黄翠芬教授参与多项国家项目，例如“七五”攻关项目和“863计划”项目，遗传工程是她工作的重中之重。黄翠芬教授是人类细菌腹泻基因工程疫苗的主要研究者，期间，她研究了引起人类腹泻的重要病原菌——产肠毒素大肠杆菌（ETEC）。ETEC包含两类有毒基因：热敏肠毒素（LT）基因和耐热肠毒素（ST）基因。黄教授和她的团队采用基因工程技术将热敏肠毒素亚单位（LTB）和ST基因进行融合，获得LTB/pro-ST融合蛋白。LTB/pro-ST融合蛋白不仅保留LTB的免疫性，还赋予ST免疫性，极大降低了ST的生物毒性。研究结果总结在《大肠杆菌热敏肠毒素和耐热肠毒素基因融合及其免疫原性研究》（张兆山等，2000）、《用基因工程技术研究预防细菌性腹泻疫苗的进展》（黄翠芬，1986），对日后的人用基因疫苗的制备起到奠基作用。另一件值得关注的事情是黄翠芬教授从1982年起开始指导研究Pro-UK的克隆和表达，通过基因克隆的方式在含有411个氨基酸的Pro-UK上加20个氨基酸的信号肽，首先在中国获得Pro-UK cDNA基因的克隆及表达（方继明等，1990）。此外，她还对人类组织型纤溶酶原激活剂结构进行改造，得到了溶血栓特效的基因多肽药物（周红等，1998），这些均达到了当时的领先技术水平。

黄翠芬教授一生从事科学研究，但由于国家科研课题的秘密性与严谨性，她放弃了很多成果的署名权。即便如此，黄教授还是取得了很大成就，其所带领的课题组相继在各学术杂志上发表数十篇高水平研究论文，并主编了《医学细菌分子生物学进展》《遗传工程理论与方法》等专著，编写教材《细菌学》，综合介绍了细菌的发展进化，成为新中国成立后我国医疗卫生教学的第一批教科书之一。此外，她还与丈夫周廷冲教授合著《药理学》（总政组织部，1990）。

黄翠芬教授还时刻关心着科研人才的培养。1950年回国后，黄教授始终坚持言传身教，她鼓励年轻人更多地在公共场合发表合理的研究见解，不要局限于所谓的学术权威，为国家培养出一大批生物技术领域的尖端人才，例如黄培堂、杨晓。黄培堂是中国“973项目”首席科学家、全国防治非典指挥部科技攻关组副组长，他在国内外学术刊物上发表论文50余篇，多次获得军队科技进步奖。黄翠芬教授的博士生杨晓是目前我国转基因重要课题组负责人。1996年，黄教授推荐她赴美学习基因敲除技术，随后杨晓发表多篇极具学术价值的论文，也被公认为国内掌握基因敲除技术第一人。

黄翠芬教授与丈夫周廷冲教授为军事医学科学院的第一对院士夫妻（郝成涛，2013），他们相濡以沫40多年，一起攻克了许多难关，为中国科研作出巨大贡献。1950年8月，他们克服了许多困难，冒险登上了一艘货船。经过56天的海上漂流，他们终于回到了中国（胡悌云等，1996）。1977年，周教授因病卧床不起，黄教授悉心照顾他，同时仍从事紧张忙碌的研究工作。不久，由于黄教授的精心护理，周教授重回工作岗位，夫妻俩一起为祖国的科研事业而奋斗着。当他们拥有私人时间时，黄教授会亲自为周教授下厨并一起在花园散步（图2）。

黄翠芬教授于2011年8月在北京逝世，她是我国伟大的基因遗传学家，其在学术研究和人才培养等方面为我国科学技术的发展作出了杰出贡献，尤其是在基因工程疫苗、基因多肽药物、分子肿瘤等领域。“直腰为人梯，弯腰为人桥”，这便是黄翠芬教授人生的真实写照。她脚踏实地的科研作风、成果共享的科研人格和忘我无私的科研精神，值得广大科研工作者学习与尊重。

图2　黄翠芬教授与周廷冲教授在中山公园（1995年）

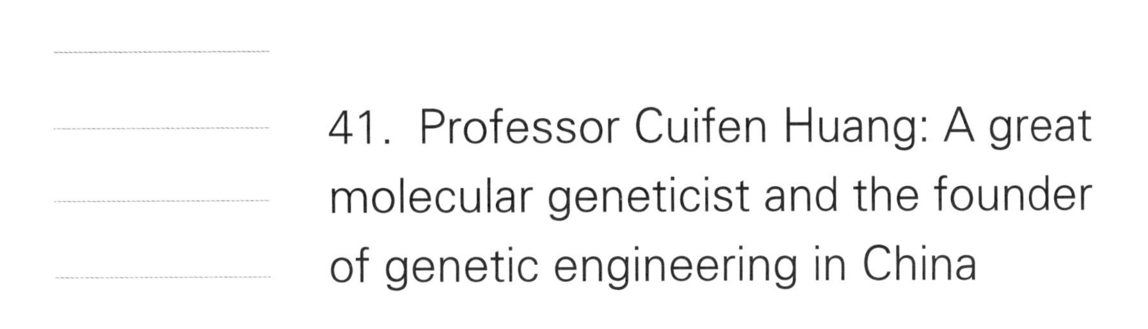

41. Professor Cuifen Huang: A great molecular geneticist and the founder of genetic engineering in China

Professor Cuifen Huang (黄翠芬, 1921—2011) is a famous molecular geneticist and one of the founders of Genetic Engineering in China. She devoted all her life to researches on genetic engineering of vaccines and polypeptide drugs and studies on molecular tumors. She made tremendous contributions to the development of biotechnology and military medical science in China. She was also the first scientist who cloned and expressed the pro-urokinase gene in China which showed high selective thrombolysis without systemic hemorrhage (Chinese Academy of Engineering, 2009).

Prof. Huang was born in Guangdong, China in 1921. After receiving her bachelor's degree in chemistry in Lingnan University (now known as Sun Yat-sen University), she went to the Microorganism Epidemiology Institute of Central Health Laboratory to work on the preparation of penicillin, the extraction of antigens and the drug sensitivity test of *Yersinia pestis*. In 1948, she studied in Cornell University in USA for her master's degree and was mentored by Professor J. Sherman who was also the editor-in-chief of the journal *Bacteriology*. Prof. Huang went back to China in 1950, shortly after the founding of the People's Republic of China. At the beginning, she was appointed as an associate professor in Shandong Medical College (now known as School of Medicine, Shandong University). There, she was responsible for the compilation and teaching of Microbiology, and also performed researches on Bacteriology. In 1954, Prof. Huang started her researches on the toxicological mechanism of microbial toxins and the integrated control methods of epidemics in the Academy of Military Medical Sciences. After years of hard work, she successfully synthesized Quadrivalent Traumatic toxoid for the first time. In 1980, Prof. Huang established the first molecular genetics laboratory of People's Liberation Army of China (PLA), and focused on the researches on hepatitis B core antigen. In the mid-1980s, Prof. Huang moved to the Institute of Bioengineering and studied in the field of Genetic Engineering. In the late 1990s, Prof. Huang started her researches on molecular tumors.

Prof. Huang accomplished remarkable achievements in genetic engineering, such as

Wei Gong, Leyi Cui, Yike Ying, Yijing Shen, Jiaqi Bao
Zhejiang Normal University, Jinhua 321004, China
Correspondence: yzugw@163.com (W. Gong)

her researches on hepatitis B core antigen and e antigen, diarrhea vaccines, cholera toxin, thrombolytic agents. Given Prof. Huang's outstanding achievements, she was honored the First Class Prize for Scientific and Technology Progress (STP) of China in 1987 and 1995, and obtained a number of Military Prizes for STP of PLA. In October 1984, Prof. Huang was awarded the national honorary title of "Advanced worker in Scientific research" by Deng Xiaoping. In 1996, Prof. Huang was elected as an academician of Chinese Academy of Engineering. In addition, Prof. Huang was invited for several times by the leaders of China to attend the National Ceremony at Tiananmen Square as an outstanding representative of the scientific and technological community.

Prof. Huang participated in multiple major national projects such as National Key Technology Research and Development Program of China during the "7th Five-Year Plan" and National High Technology Research and Development Program of China (National 863 Program), and genetic engineering was her top priority. Prof. Huang was the leading scientist of the National 863 Program—genetic engineering vaccine for human bacterial diarrhea. During this program, Prof. Huang studied the essential pathogen, enterotoxigenic *Escherichia coli* (ETEC), which caused human diarrhea. ETEC contain two types of noxious-gene: heat-labile toxin (LT) and heat-stable toxin (ST). Prof. Huang and her team used genetic engineering technology to fuse the B-subunits of LT (LTB) together with ST to obtain LTB/pro-ST fusion protein, which not only retained the immunity of LTB, but also endowed ST with immunity and reduced the biological toxicity. Prof. Huang also published several research articles to present these research data, such as "Fusion of genes encoding *Escherichia coli* heat-labile and heat-stable enterotoxins" (Zhang et al., 2000), "Research progress in the prevention of bacterial diarrhea vaccine by genetic engineering technology" (Huang, 1986), which laid a solid foundation for the production of humanized vaccines in the future. Another outstanding study is the cloning and expressing of pro-urokinase, which was supervised by Prof. Huang since 1982. She added a signal peptide containing 20 amino-acids to pro-urokinase that has 411 amino-acids using gene cloning, cloned and expressed the pro-urokinase for the first time in China (Fang et al., 1990). At the same time, she modified the structure of human tissue plasminogen activator and obtained one kind of anti-thrombolytic polypeptide drug (Zhou et al., 1998). All these research findings reached the leading levels in the world at that time.

Prof. Huang devoted her entire life to the scientific researches in China. Due to the confidential policy of national projects, she gave up the right to publish many research findings. Even though, Prof. Huang made outstanding achievements with her research team and published dozens of high quality research articles in multiple academic journals. Meanwhile, she edited several monographs including *Advances in Molecular Biology of Medical Bacteria* and *Theory and Methods of Genetic Engineering*, compiled the textbook *Bacteriology*, which comprehensively introduced the development and evolution of bacteria and was one of the first textbooks in the field. In addition, she coauthored *Pharmacology* with her husband, Professor Tingchong Zhou (周廷冲) (General Administration Department, 1990).

Prof. Huang was also a great mentor in the cultivation of young talents. After returning to China in 1950, Prof. Huang insisted on teaching by words and deeds, she encouraged young scholars to express their opinions in public places, not giving ways to so-called academic authority. Prof. Huang has mentored and trained a large number of outstanding talents in the field of biotechnology, such as Peitang Huang (黄培堂) and Xiao Yang (杨晓). Peitang Huang is the chief scientist of the Major State Basic Research Development Program of China (National 973 Program) and vice chairman of the scientific and technological research group of the National Severe Acute Respiratory Syndrome (SARS) Prevention and Control Center. He has published more than 50 articles in academic journals, and has won military prizes for STP of PLA for several times. Xiao Yang is the head of the transgenic group in China. In 1996, Prof. Huang recommended her graduate student Xiao Yang to go to USA to study gene knockout technology, who then published many high quality research articles and was recognized as the first person to master gene knockout technology in China.

Prof. Huang and her husband, Prof. Zhou were both academicians in the Academy of Military Medical Sciences (Hao, 2013). They have been together for more than 40 years, conquered many difficulties together and made tremendous contributions to the development of science and technology in China. In August 1950, they overcame many difficulties and ventured into a cargo ship. After 56 days of wandering at sea, they finally returned to China (Hu et al., 1996). In 1977, when Prof. Zhou was severely sick at home, Prof. Huang was busy taking care of him. Meanwhile, she was still engaged in her researches. Later on, Prof. Huang's meticulous care brought Prof. Zhou back to work, and they continued to work together for the development of science and technology in China. When they got some time for their personal life, Prof. Huang would cook for Prof. Zhou and they would take a walk in the park together.

Prof. Huang passed away in Beijing in August 2011. She was a great geneticist and made outstanding contributions to the development of science and technology in China in terms of academic researches and cultivation of young talents, especially in the fields of genetic engineering of vaccines and polypeptide drugs and studies on molecular tumors. "Straight waist to be a ladder, bending over to be a bridge for others" (直腰为人梯,弯腰为人桥)is the true reflection of Prof. Huang's life. Her down-to-earth research style, collaborating research personality and selfless research spirit are worthy of appreciation and learning by scientific researchers.

Figures

Fig.1 Prof. Huang attended the National Model Conference, 1987

Fig.2 Prof. Huang and Prof. Zhou at the Zhongshan Park, 1995

References

Chinese Academy of Engineering (2009) Academician of Chinese Academy of Engineering. Beijing: Higher Education Press. (中国工程院. 2009. 中国工程院院士. 北京: 高等教育出版社.)

Fang JM, Li XZ, Li FZ, et al (1990) Cloning of full-length cDNA gene of human pro-urokinase. Med J Chin People's Liberation Army 1:10. (方继明, 李秀珍, 李凤知, 等. 1990. 人尿激酶原全长cDNA基因的克隆. 解放军医学杂志, 1: 10.)

General Administration Department (1990) Contemporary Chinese Women Soldiers. Beijing: Chinese People's Liberation Army Publishing House. (总政组织部. 1990. 当代中国女兵. 北京: 解放军出版社.)

Hao CT (2013) RRC Health Guard. Beijing: Xinhua Publishing House. (郝成涛. 2013. 共和国健康卫士. 北京: 新华出版社.)

Hu TY, Zhang WB, Wang JX (1996) History of the People's Republic of China (1). Beijing: Contemporary Chinese Publishing House. (胡悌云, 张文彬, 王际欣. 1996. 新中国史话(一). 北京: 当代中国出版社.)

Huang CF (1986) Research progress in the prevention of bacterial diarrhea vaccine by genetic engineering technology. J Chin Biotechnol 6(2):9–18. (黄翠芬. 1986. 用基因工程技术研究预防细菌性腹泻疫苗的进展. 中国生物工程杂志, 6(2): 9–18.)

Zhang ZS, Xu B, Li SQ, et al (2000) The fusion of genes encoding *Escherichia coli* heat-labile and heat-stable enterotoxins and the study on the immunogenicity of the fusion peptides. High Technol Lett 10(1):15–18. (张兆山, 徐兵, 李淑琴, 等. 2000. 大肠杆菌热敏肠毒素和耐热肠毒素基因融合及其免疫原性研究. 高技术通讯, 10(1): 15–18.)

Zhou H, Wang GL, Huang PT, et al (1998) Cloning and expression of a novel tissue type plasminogen activator (t-PA) mutant. Sichuan J Physiol Sci 3:37. (周红, 王国力, 黄培堂, 等. 1998. 新型组织型纤溶酶原激活剂(t-PA)突变体的克隆及表达. 四川生理科学杂志, 3: 37.)

42. 杜若甫：中国人群体遗传学奠基人

杜若甫教授是享誉海内外的著名人类遗传学家（图1）。他为我国人类学和中国人群体遗传学的研究与发展奋斗了半个多世纪之久，使我国的人类遗传学研究跻身世界先进水平。杜教授作为中国人群体遗传学的奠基人，推动了我国人类基因组多样性计划，在北京建立了中华民族永生细胞库（我国保存人类遗传资源最丰富的细胞库），他编纂的多部人类遗传学基础理论著作，至今被广泛应用。鉴于他对我国人类学作出的卓越贡献，2014年杜教授获得人类学终身成就奖，该奖项是由上海人类学学会颁发，奖励在人类学研究领域取得巨大成就以及为推动学科发展作出杰出贡献的中国人类学学者。杜教授获得这一奖项可谓实至名归。

图1 杜若甫教授在办公室

1958年，杜教授从苏联列宁格勒大学获得博士学位后回到中国，时年28岁。当时他承担了一项科研任务，即评估电离辐射对哺乳动物和人类遗传的影响。他巧妙地在职业暴露于电离辐射的人群中开展调查，基于这些调查结果，他找到了电离辐射对人类遗传影响的直接证据（杜若甫，1964），据此，卫生部修改了我国接触辐射的专业工作人员的容许照射剂量。

译者：许崇凤

中国科学院遗传与发育生物学研究所，北京100101，中国

邮箱：cfxu@genetics.ac.cn

当他做此项调查的时候，他意识到我国人类遗传学研究数据非常匮乏。于是，他决定尽最大努力为这一领域作出贡献。难能可贵的是，即使在那动荡的10年，所有科研工作都被迫停止，他也没有停止对人类遗传学的研究，同时，他还翻译并出版了三本重要的科研工作指南：《作物辐射遗传与育种》《分子遗传学》（与杨纪柯先生合作）和《诱变育种手册》。

1978年改革开放，各项事业一片欣欣向荣。杜教授领导调查小组对澳大利亚进行学术访问，在访问后他意识到，中国的研究工作已经远落后于国际水平。当时有人问他："你回国后想要研究什么？"他回答说："在我国，我认为遗传学领域中最为落后的是人类遗传学。动植物领域已经有大量不错的研究，但是人类遗传学领域的研究却很少。我们有56个民族，还有许多特殊人群，有如此丰富的人类遗传学资源却没有人类遗传学研究，这是多么遗憾的事。"因此，杜若甫率先提出开展中国人群体遗传学的研究，正如他所说："开展人类群体遗传学研究是社会和科学的需要，也是我锲而不舍的奋斗目标。"

1979年，杜教授在中国科学院遗传研究所成立了人类群体遗传学研究小组，他和他的团队走访全国各地，其中有不少是偏远地区，对56个民族的体质人类学和文化人类学开展了调查及研究，并对人类群体遗传学的多项指标进行了广泛取样分析，获得了大量系统数据。经总结，杜教授编撰了两本专著，即《中国的民族》（杜若甫和叶传升，1994）和《中国人群体遗传学》（杜若甫，2004）。《中国的民族》广泛用于我国自治区和各级民族委员会的工作中，同时也是民族学、人类学和群体遗传学研究人员的常用参考书。《中国人群体遗传学》作为遗传学的基本理论书籍也被广泛应用。他强调，中国人群体遗传学旨在研究中国各民族之间遗传结构的相似性和差异，从而揭示群体之间的血缘关系，探索人类迁徙和进化规律，发现不同人群对各种疾病的易感性，以及对药物和不同的环境因素的反应。这是在我国首次开展的针对中国人各种遗传指标进行的广泛而详细的调查（杜若甫，1987，1997）。

杜教授首次提出在我国开展人类基因组多样性计划，提出建立中华民族永生细胞库。人类基因组计划（HGP）1990年正式启动。1991年，许多国家建立了人类基因组多样性计划（HGDP）执行委员会。我国人类基因组多样性研究计划由杜教授领导，几乎与国际同步开展工作。在此期间，杜教授担任国际人类基因组多样性委员会常务委员。在他的带领下，中华民族永生细胞库加入了国际人类基因组多样性计划。2002年，L. L. Cavalli-Sforza等在《科学》刊登的文章《建立人类基因组多样性细胞系计划》，杜教授是共同作者之一（Cann et al.，2002），由中华民族永生细胞库提供的细胞系保证了该计划数据的丰富性和完整性（图2）。

经过20年的数据采集，在几个少数民族学者的帮助下，杜教授编写了《中国少数民族姓氏》，该书约200万字，由民族出版社2011年出版，著名教授王泉根对这本书给予了高度评价（王泉根，2010）。

杜若甫热衷于人类遗传学的社会工作和学术活动，他曾担任过《遗传》《遗传学和基因组学》《人类学》《中国医学遗传学》《国外医学科学》（遗传学部分）等杂志的编委或副主编。

图2　杜若甫教授在北京

杜若甫在人类学的许多领域作出了杰出贡献。中华民族永生细胞库是他的成就之一。正如他所期待的，该细胞库正在逐步扩大，并在包括遗传学、进化、疾病的病理和药物敏感性等许多领域的基础研究中，发挥越来越大的支撑作用（杜若甫，1987，1997）。

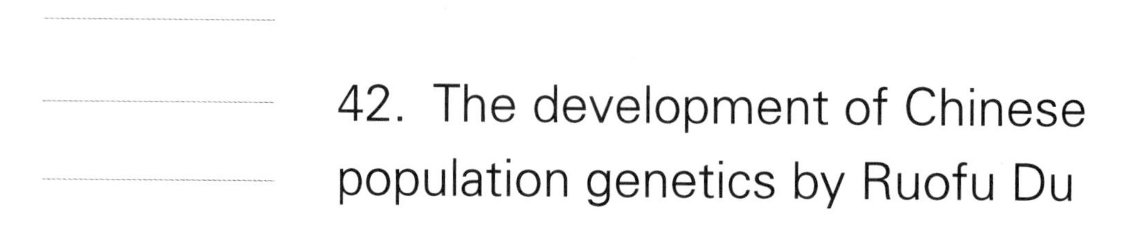

42. The development of Chinese population genetics by Ruofu Du

As a scientist, Professor Ruofu Du enjoyed an outstanding reputation both at home and abroad. For more than half a century, he devoted himself to the advancement of Chinese national anthropology and human population genetics, always keeping pace with global development trends of human genetics. As one of the founders of population genetics in China, Prof. Du promoted the Human Genome Diversity Project of China, and established the Chinese National Immortalized Cell Bank in Beijing, which is the most diverse population cell bank in China. Prof. Du compiled several books during his research career which have been widely used in the study of the basic theories of Chinese population genetics research. In 2014 Prof. Du won the "Anthropology of Lifetime Achievement Award" , which is sponsored by the Anthropological Association of Shanghai, to recognize those scientists in the field of anthropology who have made a significant and outstanding contribution the anthropology, for his 60 years of research in this field.

After receiving his PhD degree from the Soviet Union Leningrad College, Du returned to China in 1958, when he was 28 years old. There undertook a project to assess the effect of ionizing radiation on mammals and human genetics. Based on the research results of a survey into a population who were exposed to radiation in China, he provided direct evidence about the influence on human genetics of ionizing radiation (Du, 1964). Subsequently, the Health Department modified the permissible irradiation dose of exposure to professional radiation staff of our country according to Du's proposal.

Prof. Du realized the scarcity of research data on human genetics in our country during this survey, and resolved to try his best to contribute in this field. Remarkably, Prof. Du continued his research work during the 10 years even though most of scientific research was forced to stop at that time. Meanwhile, he translated and published three important guiding books, *The Radiation Genetics and Breeding of Crop*, *Handbook of Mutation Breeding* and *Molecular Genetics* during this period of unrest, with the cooperation of Mr. Jike Yang (杨纪柯).

Chongfeng Xu, Ziyuan Duan

Institute of Genetics and Developmental Biology, Chinese Academy of Sciences, Beijing 100101, China

Correspondence: zyduan@genetics.ac.cn (Z.-Y. Duan)

In 1978 China began to introduce social reform and opening-up policies, after leading an academic visit to Australia, Prof. Du realized that Chinese research work had lagged far behind their international counterparts. When someone asked him, "What are you going to do when you go back home?" He said: "I think human genetics is the most backward in the field of genetics in our country. There has been a great deal of research into animals and plants, however as for research into human genetics, there are 56 ethnic groups and many special populations in China, but there is few related research about that, it is pity that we leave such a wealth of population genetic resources alone" . Thus, Du first proposed to carry out the study of human population genetics in China. Just as he said, "research of human population genetics is the need of the society and sciences, and that is my persistent goal."

In 1979 Prof. Du and his human population genetics research group in Institute of Genetics, Chinese Academy of Sciences, which was established in Beijing, visited many remote regions all over the country and carried out a systematic investigation into the physical anthropology and cultural anthropology of the 56 ethnic groups in China, including extensive sampling and analysis on several indicators of human population genetics. Summing up the investigation, Prof. Du compiled two books, *Chinese Nations* (Du and Ye, 1994) and *Chinese National Population Genetics* (Du, 2004). *Chinese Nations* is not only widely used as a reference book in the National Committee at all levels of our country and autonomous region, but also for ethnology, anthropology and population genetics researchers. *Chinese National Population Genetics* has been widely used as an introduction into the theories of genetics. Prof. Du stressed that the aim of Chinese population genetics research is to study genetic structure, the similarities and differences, as well as changes among the population of all ethnic groups in China, so as to reveal the consanguineous relationship between populations to discover human migration and evolution routes, as well as susceptibility to various diseases and drugs and responses to different environmental factors. Prof. Du was the first to carry out such an extensive and detailed investigation into the various genetic indexes of the Chinese people (Du, 1987, 1997).

Prof. Du was the first to propose to carry out the Human Genome Diversity Project of China and to put forward the establishment of the Chinese National Immortalized Cell Bank. The Human Genome Project (HGP) was launched in 1990, and Human Genome Diversity Plan (HGDP) Executive Committee was established in many countries in 1991. The Human Genome Diversity Project of China which is led by Prof. Du is currently being conducted at a level equal to those of international peers. During this period in 1990s, Prof. Du served on the International Human Genome Diversity Standing Committee and the "Chinese National Immortalized Cell Bank" which is led by Prof. Du, joined the International Human Genome Diversity Project. Prof. Du is one of the authors in *Science* article "A human genome diversity cell line panel" which is corresponded by L. L. Cavalli-Sforza (Cann et al., 2002). The cell lines provided by Chinese National Immortalized Cell Bank ensure the richness and integrity of the panel.

With over 20 years of research experience and the help of several minority scholars, Prof. Du recently took the time to compile a new book *Family Names of Chinese Minorities*, published by the Ethnic Publishing House in 2011. Famous professor Quangen Wang of Beijing Normal University gave a high evaluation to this book (Wang, 2010), which is the first monograph in this field.

Prof. Du is keen on social work and academic activities in human genetics. He served as deputy editor in chief of *Hereditas* and *Journal of Genetics and Genomics*, also on the editorial board of *Journal of Anthropology*, *Chinese Journal of Medical Genetics* and *Foreign Medical Sciences* (Section of Genetics), *Human Biology* and so on.

Prof. Du has made outstanding contributions in many fields of anthropology. Just as he expected (Du, 1987, 1997), the immortalized cell bank of ethnic groups in China is gradually expanding, and plays an increasingly large role in supporting the basic research in many fields, including genetics, evolution, disease pathology, drug susceptibility, etc.

Figures

Fig.1 Professor Ruofu Du at office

Fig.2 Professor Ruofu Du in Beijing

References

Cann HM, De Toma C, Cazes L, et al (2002) A human genome diversity cell line panel. Science 296(5566):261.

Du RF (1964) The genetic effects of ionizing radiation on mammals and humans. Atomic Energy Sci Technol 3:286–305. (杜若甫. 1964. 电离辐射对哺乳动物和人类的遗传学效应. 原子能科学技术, 3: 286–305.)

Du RF (1987) Research on human population genetics in China. Bull Biol 32(7):9–11. (杜若甫. 1987. 我国的人类群体遗传学研究. 生物学通报, 32(7): 9–11.)

Du RF (1994) Chinese Nations (Chinese and English Edition). Beijing: Science Press.

Du RF (1997) Opinions about the research of human genome diversity in China. Bull Chin Acad Sci 6:398–402. (杜若甫. 1997. 对开展中国人类基因组多样性研究的思考. 中国科学院院刊, 6: 398–402.)

Du RF (2004) Chinese National Population Genetics. Beijing: Science Press.

Wang QG (2010) The development and deepening of the research on the surname of the minority nationalities in China. China Book Rev 1:116–119. (王泉根. 2010. 中国少数民族姓氏研究的开拓与深化. 中国图书评论, 1: 116–119.)

五、生物化学

43. 吴宪：中国生物化学及营养学的奠基者

图1　吴宪（1893—1959）

吴宪先生是世界知名的生物化学家及营养学家，在蛋白质化学、临床生物化学、免疫化学及营养学研究领域都有杰出的贡献（图1）。他是我国生物化学及营养学的奠基者，曾被美国学者J. Reardon-Anderson誉为“中国化学的巨人”，“毫无疑问，吴宪是20世纪前半叶中国最伟大的化学家，或者说是最伟大的科学家”。在寻求科学真理的事业中，吴宪先生是一位成绩卓然的学者，除此之外，他还具有广泛的兴趣爱好，富有同情心并严格自律，这使他同时又是一位爱国者、一位体贴的丈夫、一位开明的父亲，还是一位令人尊敬的同事。

吴宪先生出生于福州的一个书香门第家庭。他曾在私塾读过书，还参加过科举考试。1906年，他进入全闽高等学堂预科班，经过4年的学习，于1910年通过了清政府组织的庚款留美考试，获得了赴美留学的奖学金。

1911年，吴宪先生被派往美国，进入麻省理工学院学习。因为曾立志重建中国海军，他最初选择了造船工程专业。但不久，他阅读了包括赫胥黎的《生命的物质基础》在内的一些文章，受这些文章的影响，他转而攻读化学和生物学，于1916年获得学士学位。随后，他师从哈佛大学的Otto Folin教授，于1919年获得博士学位。

他的题为《一种血液分析系统》的博士论文已成为血液化学的经典文章。在文章中，他提出了一种只需10 mL血液就可定量测量血液成分的方法。尤其需要指出的是，用他提供的手段，人们只需取一滴血或尿，就可以测定其中的糖含量。学术界认为，如果没有吴宪先生的血糖测定方法，胰岛素的发现就会受到阻碍。这些测定方法后来被命名为“福林–吴”方法。

1920年，吴宪先生回到中国，在由美国洛克菲勒基金会创办的北京协和医学院任教。从此，他在这里奉献了大部分的时光，直到学院被日本军队占领而解散。在协

译者：蒋希萍

中国科学院生物物理研究所，北京100101，中国

邮箱：jiangxp@ibp.ac.cn

和医学院工作期间，他被任命为生物化学系主任，是该校的第一位中国籍主任。这段时间也是他科学生涯中的鼎盛时期，例如，他主持了一项大规模的研究工作——蛋白质变性的研究，他和同事共发表了关于蛋白质变性的系列论文16篇及相关论文14篇。1931年，他发表了题为《蛋白质变性的研究：XIII.蛋白质变性理论》的文章，认为天然球蛋白由肽链经规律折叠而成，若解折叠，蛋白质就将变性。这篇文章已成为国际蛋白质研究领域的一个里程碑。化学家Felix Haurowitz评论道："吴宪是首位用标记抗原的方法分析蛋白质的人，而且是第一个提出蛋白质变性学说的人。"

吴宪先生还发现，当时的中国有很多人营养不良、身体素质差，因此，他下决心要通过科学研究改善这一状况。在协和医学院期间，他开始系统研究人的健康与食物的关系。他建立了素食和杂食实验动物体系，经比较确证了杂食对健康的益处。通过观察大鼠摄入蛋白质的情况，他进一步指出蛋白质的摄入对健康是至关重要的。他编著了我国第一部食物成分分析表，并领导了第一次营养普查，还为不同的人群设计了特定的营养目录。20世纪40年代，他还担任了南京营养研究所的所长。

1948年，吴宪先生作为访问学者到哥伦比亚大学工作，由于战事，他终未能回到国内。1953年，因突发心肌梗死，他辞职退休，定居波士顿，直至1959年逝世。

诚如吴宪先生的儿子、杰出的生物学家吴瑞所说："吴宪先生是一位真正的科学家。"他的科学成就无法在此一一列举。他共发表科学论文163篇，撰写专著3部。他是美国化学会会员、德国自然科学院名誉院士、荷兰《生物化学与生物物理学报》顾问委员，并任联合国粮食及农业组织营养顾问委员会常务委员。D. D. Van Slykesaid博士曾这样评价他："只有亲眼看到他的工作时，你才能充分领略到他才智上的异彩和技术上的完美。他是当今生物化学领域伟大且具有真正独创性的领袖之一。"

吴瑞还提到，吴宪先生身上既有中国古典绅士的风范，又有哲学家的气质。他平和、从不急躁，注重自我约束。他曾经对子女说，良好的教育是对他们最有价值的投资，比物质上的给予更有益处。吴宪先生去世后，他的妻子决定不将钱财留给后代，而是捐赠于教育和科学研究事业。吴宪先生的子女也没有辜负他的期望，都在各自的研究领域中卓有建树。

尽管事业有成，然而吴宪先生绝不是一个冷漠无趣的人。他热爱家庭生活，与妻子感情和睦（图2）。他热切地期盼祖国强盛，为了国家科学事业的发展，有时甚至会不顾个人的安危。他关注世界和平及人类的幸福，曾写成《科学生活导论》一书，在书中，他论述了个人与国家的关系。他还设想通过科学方法解决社会问题。他充满生活激情，具有广泛的兴趣爱好：各种体育运动、建筑学、书法、园林设计，等等。吴宪先生不仅是一位伟大的科学家，而且是一个内心丰富的人，他演绎了一段多彩的人生。

图2　吴宪先生和他的家人

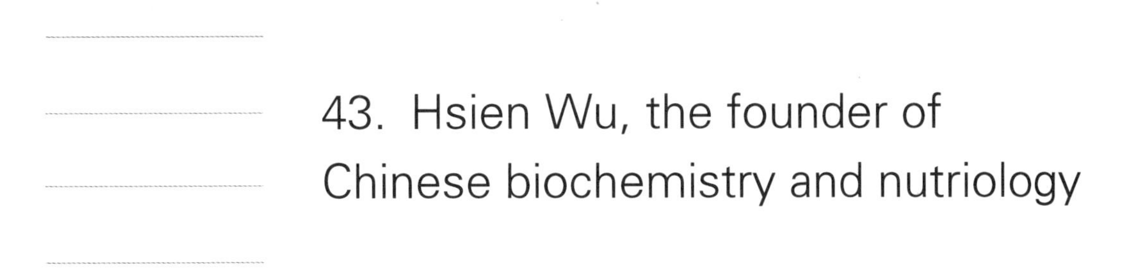

43. Hsien Wu, the founder of Chinese biochemistry and nutriology

Hsien Wu (1893—1959) was a world-renowned biochemist and nutrition scientist with excellent achievements in protein chemistry, clinical chemistry, immunochemistry, nutriology and so on. He was also the founder of Chinese biochemistry and nutriology, and was regarded as "a giant of Chinese chemistry" and "without doubt the greatest chemist and perhaps the greatest scientist in China during the first half of the 20th century" . Besides a productive scholar in pursuit of truth and perfection in career, his multiple interests, kindness and self-discipline also made him a patriot, a caring husband, a wise father, and a respected colleague.

Born in a scholarly family in Fuzhou, China, Hsien Wu once received the classical Chinese education and later took part in the civil service exam. In 1910, after 4 years of modern education in Fukien Provincial High School, he passed the government examination and won a Boxer indemnity scholarship to study abroad.

In 1911, Hsien Wu was sent to the United States, and entered the Massachusetts Institute of Technology. Dreaming of rebuilding Chinese navy, he first majored in naval architecture. But attracted by articles like Huxley's "On the physical basis of life" , he soon switched his major to chemistry and biology, and got his bachelor's degree in 1916. He then studied biochemistry with Professor Otto Folin at Harvard University, and obtained his doctor's degree in 1919.

His doctoral dissertation "A system of blood analysis" became a classic in the history of blood chemistry. It developed procedures to use only 10 mL blood samples to accomplish quantitative measurements of the major constituents of blood. Particularly, it's possible to measure the sugar content in blood or urine in a sample as small as one drop. In fact, scholars commented that without Wu's method, the discovery of insulin might have been hampered. These analytical procedures later were known as the "Folin-Wu" methods.

In 1920, Hsien Wu returned to China to work at the Peking Union Medical College (PUMC), a Rockefeller-endowed institution. From 1920 and on, he spent most of his time there, till the college was invaded by the Japanese army. During working at PUMC, he was appointed director

Shu Zheng
The Institute for the History of Natural Sciences, Chinese Academy of Sciences, Beijing 100190, China
Correspondence: zhengshu111@gmail.com

of the department of biochemistry, the first Chinese director of a department there. This period was also regarded as the prime time of his scientific life. For instance, he led a large-scale research at that college—the study of protein denaturation. He and his colleagues published 16 essays about "studies on denaturation of proteins" , along with other 14 related papers. In 1931, Hsien Wu published "Studies on denaturation of protein. XIII. A theory of denaturation" , a landmark contribution to the international study of protein. Chemist Felix Haurowitz commented that Hsien Wu was the first to use labeled antigens for determining proteins in the presence of antibodies and the first to propose that the peptide chains in globular proteins are folded and unfold on denaturation.

Hsien Wu noticed the malnutrition states of many Chinese people at that time and determined to improve this situation through scientific research. At PUMC, Hsien Wu began to conduct his systematic study on the relationship between food and human health. He raised experimental animals with purely vegetarian diet and omnivorous diet respectively, and confirmed the benefit of omnivorous meal. Through the observation of protein regimens in rats, he further pointed out that the nature of the protein was critical. He published the first analysis of food composition in China, and conducted the first nutrition survey, as well as set up specific nutrition indices for all kinds of people. In addition, Wu was the director of the Nutrition Institute in Nanking during the 1940s.

Hsien Wu went to Columbia University as a visiting scholar in 1948 and failed to return to China because of war. He retired in 1953 after the break of coronary thrombosis. He settled down in Boston and died there in 1959.

As his son, excellent biologist Ray Wu, pointed out, Hsien Wu "lived the life of a true scientist" . Scientific achievements of Hsien Wu are too many to be listed here. He wrote and published 163 scientific papers, as well as 3 books. He had membership in the American Society of Biological Chemists, and honorary membership in the Deutsche Akademie Naturforschor Leopoldina. He was on the advisory board of *Biochemica et biophysica acta*, and the Standing Advisory Committee on Nutrition of the Food and Agriculture Organization of the Untied Nations. Dr. D. D. Van Slykesaid once commented, "the brilliance of his mind and the beauty of his technique could be adequately appreciated only when one saw him in action. He was one of the great and truly original leaders in the creation of biochemistry today."

Professor Ray Wu also referred that elements of "classical Chinese gentleman (chun-tse) and philosopher" were combined in Hsien Wu's life. He was calm, never hurried, and self-disciplined. He told his children that good education was the best investment for them, and was much more important than material superiority. After Hsien Wu's death, his family decided not to leave money to their offspring, but to donate it to education and scientific research. All of Hsien Wu's children are experts in their own research areas.

Nevertheless, Hsien Wu was by no means a dull and cold person. He enjoyed warm family life, had a harmonious relationship with his wife all through his life. He was passionate for the development of his motherland, willing to help with China's scientific development despite life-threatening situation. He cared about peace and welfare of human being, and wrote "*Guide on Scientific Life*" in which he discussed the relationship between individuals and one's country. He imagined solving social problems with scientific method. His passion also resided in his multiple interests, such as all kinds of exercise, architecture, calligraphy, gardening, and so on. He was not only a great scientist, but a man with fully developed spirit and colorful life.

Figures

Fig.1 Hsien Wu (1893—1959)

Fig.2 Hsien Wu and his family

References

Bishop C (1982) Hsien Wu (1893—1959): A biographical sketch. Clinical Chemistry 28(2): 378.

Cao Y (1993) A biographical sketch of the father of biochemistry in China—in commemoration of the 100th birthday of Dr. Hsien Wu. Chin Historical Material Sci 14(4): 30–42. (曹育. 1993. 杰出的生物化学家——吴宪博士. 中国科技史料, 14(4): 30–42.)

Reardon-Anderson J (1991) The Study of Change: Chemistry in China, 1840—1949. New York: Cambridge University Press, 140–148.

Peking Union Medical College (1993) In Memory of the 100th Anniversary of Professor Hsien Wu. Beijing: Beijing Medical University/Peking Union Medical College Press. (中国协和医科大学. 1993. 纪念吴宪教授诞辰100周年. 北京: 北京医科大学/中国协和医科大学联合出版社.)

Wang Y (1991) Hsien Wu. In: Lu JX, et al. Biographies of Modern Chinese Scientists Vol 2. Beijing: Science Press. (汪猷. 1991. 吴宪. 见: 卢嘉锡, 等. 中国现代科学家传记第二辑. 北京: 科学出版社.)

Wu H (1931) Studies on denaturation of protein. XIII. A theory of denaturation. Chin J Physicol 5: 321–344.

44. 纪念曹天钦先生

2010年12月，“第四届纪念曹天钦蛋白质研究国际研讨会”在厦门举行。借此契机，能够写一篇回忆文章来纪念曹天钦先生（图1），我感到非常荣幸。

图1　曹天钦教授（1920—1995）

我进入曹先生实验室攻读研究生学位的时间比较晚。1948年，我从浙江大学化学工程系毕业后，发现很难在化工厂找到合适的工作。因此，我改学生物化学专业，在湘雅医学院生物化学系担任助教。随后，我便在北京和兰州从事了将近十年的生化教学。虽然我很希望在医学院做一些生化研究，但因为时间紧张和工作设施不完善而搁置。在得知中国科学院生理生化研究所是生物化学研究领域的科研殿堂后，我便希望有朝一日能在那里学习生物化学知识。1957年我终于梦想成真，国家放开了考生报名的限制，可以不需要原单位的批准证明，自由参加中国科学院研究生入学考试。非常幸运，我通过了生理生化研究所的入学考试，并且成为曹先生的研究生。

曹先生是一位平易近人的科学家，年龄只比我大5岁。1957年5月我从兰州来到上海时，曹先生告知我，抗日战争期间，他曾在兰州Rewi Alley创立的皮革加工厂工作，对皮革蛋白质很感兴趣。在上海，他主要从事肌肉蛋白的研究，此外还和他的学生彭加木研究兔皮中的胶原蛋白。1957年的夏天，新入学的硕士生和博士生必须参加3个月的高级生物化学强化课程，只有那些通过考试的人才可以在实验室里开展研究工作。3个月的学习非常艰苦，时间安排也非常紧张，上午学习理论课程，下午学习实验课程。我们非常喜欢听曹先生的课，他经常使用简单通俗的描述来解释复杂的理论。

译者：张保元

中国科学院北京生命科学研究院，北京100101，中国

邮箱：zhangby@biols.ac.cn

当他讲到烟草花叶病毒时，他将核酸比作朱丽叶，将蛋白质比作罗密欧，两者在病毒生命周期中都是不可或缺的。当他在讲授蛋白质的α螺旋和脱氧核糖核酸的双螺旋结构时，他引用了老子在《道德经》中所说的“玄之又玄，众妙之门”，意思是“交错螺旋结构蕴含了通向终极奥义的法门”。汉语中“螺旋”的“旋”谐音恰好是“玄妙”的“玄”。他幽默而融会贯通的讲授给我们留下了深刻的印象。

生物化学是一门实验科学，在研究生学习阶段做实验是很重要的。曹先生给我上的第一堂课便是如何杀死一只兔子。他紧紧抓住兔子的后腿，将它击昏，然后切开它的颈动脉进行放血，动作非常敏捷娴熟，给我留下了深刻的印象。他还教我们如何使用微电泳仪来检测蛋白质制备的纯度。水平黏度计是由他亲自设计后经机械厂加工而成的，他还帮我组装光散射仪测量蛋白质分子的大小和形状。更为复杂的精密仪器，如钛氏电泳仪、超速离心机和电子显微镜，则由一些有经验的工作人员进行操作和维护。曹先生精心地为实验室配备各种实验设备，这些被他称为“金刚钻”的设备是完成研究项目、获取关键数据所必需的。彭加木、潘家秀和任梅轩在这方面也都起到了很重要的作用。

在曹先生的指导下，我开始研究原肌球蛋白在不同溶剂中的构象。水溶剂是蛋白质的天然活性环境，有机溶剂可使蛋白质变性。尽管如此，一些蛋白质在有机溶剂中仍然是有活性的。我的结果表明，原肌球蛋白在67%的氯乙醇中会不可逆变性，它的螺旋含量略有增加。1958年，曹先生暂停了肌肉蛋白质的研究，全身心地投入国家的人工合成结晶牛胰岛素这项艰巨的科研任务，我的工作重心也从原肌球蛋白构象研究转移到了天然胰岛素的双链重组。1961年，我完成了研究生的学业，获得了由郭沫若院长签发的学位证书。

进入曹先生的实验室工作之前，我对他的家人了解很少（图2）。我甚至不知道他的夫人就是我久仰大名的物理学家谢希德先生。曹先生1944年毕业于燕京大学。1944年至1946年，他曾陪同李约瑟（Joseph Needham）在重庆中英合作办公室调研中国的科学和教育的发展状况。1944年12月，曹先生陪同李约瑟和他的夫人Dorothy Needham、讲师Lawrence Picken访问了在遵义和湄潭的浙江大学。在湄潭，李约瑟作了题为“科学与民主”的演讲。而此时，我正在遵义学习化学工程，所以没有见到他们。1946年，曹先生经李约瑟介绍获得英国文化委员会奖学金，赴英留学，在Kenneth Bailey实验室学习。1951年，他获取了博士学位，并被剑桥大学冈维尔与凯斯学院选为院士，这是该院历史上第一位获此殊荣的中国人。1952年，在李约瑟的帮助下，谢希德先生拿到了出国签证，途经美国抵达英国剑桥大学，与曹天钦举行了婚礼（图3和图4）。当时，我妹妹张友端和她的丈夫陈瑞铭也在剑桥学习，参加了他们的婚礼。1952年8月，这对

图2 20世纪50年代，曹天钦教授的全家福

图3 新婚照

图4 曹教授（左二）和李约瑟（右二）以及他的朋友们

新婚夫妇辗转香港，从南安普敦到上海，最终回到了祖国的怀抱。

在剑桥大学的学习过程中，曹先生对肌肉蛋白（包括肌球蛋白、原肌球蛋白和肌动蛋白）开展了系统的研究工作。他最重要的贡献是发现了肌球蛋白的小亚基，并将其称为肌球蛋白轻链。回到上海后，他继续与学生在肌肉蛋白上进行了很高效的研究，最重要的工作之一是他们开创性采用电子显微镜研究原肌球蛋白和副肌球蛋白晶体。1978年，曹先生被授予全国科学大会重大科技成果奖，表彰他对肌肉蛋白的研究。在这里，我们不应该忘记，虽然他的名字既不包括在工作者的名单中，也没有列在已发表的文章中，但曹先生是人工合成结晶牛胰岛素的积极倡导者和主要组织者。曹先生培养了许多硕士生和博士生，同时他还在复旦大学和上海科技大学教授蛋白质化学课程。曹先生的主要研究兴趣是肌肉蛋白，但他对其他蛋白质系统也很感兴趣。当学生希望研究其他蛋白质时，他总是毫不犹豫地支持。例如，在我研究生毕业后，我开始对植物病毒感兴趣。他不仅允许我自由开展工作，而且帮我联系了曾在英国学习植物病理学的周家炽先生，当时周先生在北京的中国科学院微生物研究所工作。我和微生物研究所的裴美云同时对周家炽先生实验室分离的烟草花叶病毒的不同菌株进行了比较研究，鉴定发现了一株特异性的新菌株。这项工作是早期的植物病毒分子生物学合作研究的典范。随后，彭加木和龚祖埙也对植物病毒的研究产生了兴趣，并将研究扩展到了感染重要作物的病毒和支原体上。类似的例子还有，戚正武先是跟随曹先生研究了一段肌动蛋白后，将研究兴趣转向蛋白酶和蛋白酶抑制剂，并一直在坚持他的探索。

曹先生一生致力于发展我国蛋白质科学的研究。如今，我非常高兴国家发展改革委员会已将蛋白质研究列为一个重点研究项目，并给予了前所未有的支持。我相信新一代的科研工作者必将继续完成曹先生的事业，推动我国蛋白质研究的快速发展。

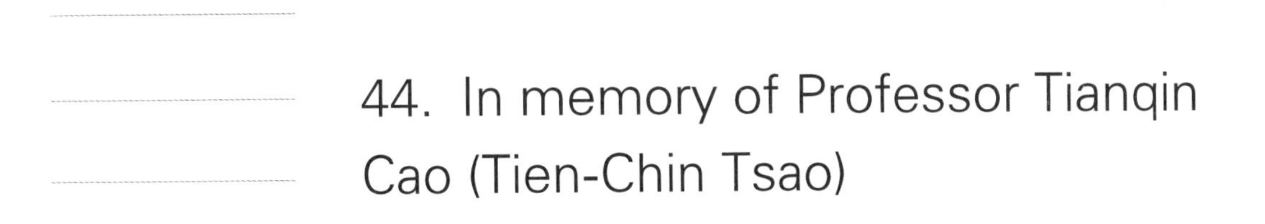

44. In memory of Professor Tianqin Cao (Tien-Chin Tsao)

In December this year, the 4th Tianqin Cao Memorial Symposium on Protein Research will be held in Xiamen (Amoy). On this occasion, it is my pleasure to write a recollection in memory of Professor Tianqin Cao.

It was rather late when I was able to study in Prof. Cao's laboratory as a graduate student in the Institute of Biochemistry. After graduating from the Department of Chemical Engineering of Zhejiang University in 1948, I could not find a job in a chemical engineering factory; therefore, I changed my profession to biochemistry and worked as an assistant in the Department of Biochemistry of Hunan Yale Medical College. Afterwards, I continued to teach biochemistry in Beijing and Lanzhou for nearly ten years. Though I wished to do some biochemical research in this medical college, yet I did not have enough time and working facilities. I knew that the Chinese Academy of Sciences (CAS) had a famous institute in Shanghai, the Institute of Physiology and Biochemistry. I wished someday I would be able to study biochemistry there. My dream did not come true until 1957, when people could freely participate the graduate student entrance examination of CAS without the approval of their affiliated institutions. Fortunately, I passed the examination of the Institute of Physiology and Biochemistry, and eventually became a graduate student of Prof. Cao.

Prof. Cao is an amiable scientist, only five years older than me. When I arrived in Shanghai in May 1957 from Lanzhou, Prof. Cao told me that he had been working in Lanzhou in a leather processing factory founded by Rewi Alley during the War of Resistance against Japanese Aggression and became interested in leather protein. In Shanghai, he and his student Jiamu Peng also worked on procollagen from rabbit skin along with his major work on muscle protein. In the summer of 1957, the newly arrived graduates and postgraduates must have a 3-month intensive course of advanced biochemistry; and only those who passed the examination were allowed to do research in the laboratory. Studies during these 3 months were very hard, with lectures in the morning and experiments in the afternoon. We enjoyed Prof. Cao's lectures a lot. He usually

Youshang Zhang

Institute of Biochemistry and Cell Biology, Chinese Academy of Sciences, Shanghai 200031, China

Correspondence: zys100@yahoo.com

explained complicated things with simple words. When he talked about tobacco mosaic virus (TMV), he compared nucleic acid to Juliet and protein to Romeo, both were indispensable. When he taught us alpha-helix in protein and double helix in DNA, he cited what Lao Tze said in *Dao De Jing*: "玄之又玄，众妙之门", which means "intriguing and intriguing, a gate to all mysteries". The "helix" [(螺)旋] in Chinese happens to be homophonic to "intriguing" (玄). His humorous speech deeply impressed us how subtle helix is involved in the living activities.

Biochemistry is an experimental science, and doing experiments is important in postgraduate studies. The first lesson I learned from Prof. Cao was how to kill a rabbit. What he did was to grasp the hind legs of the rabbit, hit its head, cut its carotid and evacuate the blood. I was deeply impressed with his swift and smart action. He also taught how to use the micro electrophoresis apparatus, which was needed to judge the purity of protein preparation. A horizontal viscometer was designed by him made by the machine shop. He also helped me to assemble a light scattering apparatus to measure the size and shape of protein molecules. Sophisticated instruments, like Tiselius electrophoresis apparatus, ultracentrifuge and electron microscope were operated and maintained by some experienced workers. Prof. Cao had paid special attention to equip his laboratory with experimental facilities, and called them "machine tools", which were essential to generating data for various research projects. Jiamu Peng, Jiaxiu Pan and Meixuan Ren had played important parts in this respect.

Under the guidance of Prof. Cao, I started my research on the conformation of tropomyosin in different solvents. Aqueous solution is the native environment of proteins while organic solvent may make protein denatured. However, it was reported that some proteins were still active in organic solvent. My results showed that tropomyosin was denatured irreversibly in 67% chloroethanol and its helical content was slightly increased. In 1958, when the formidable project of insulin synthesis started, Prof. Cao's project of muscle protein was suspended. My work was also shifted from tropomyosin to the recombination of native insulin from its two chains. In 1961, I finished my postgraduate studies and obtained a CAS diploma signed by President Moruo Guo.

Before I studied in Prof. Cao's laboratory, I knew very little about his family. I did not even know that his wife was Prof. Xide Xie, whom I knew long ago. Prof. Cao graduated from Yenching University in 1944. From 1944 to 1946, he worked with Joseph Needham in Chongqing at the Sino-British Cooperation Office to help investigate the development of science and education in China. In December 1944, Prof. Cao accompanied Joseph and Dorothy Needham, and Lawrence Picken to visit Zhejiang University in Zunyi and Meitan. In Meitan, Joseph Needham gave a lecture entitled "Science and democracy". At that time I was studying chemical engineering in Zunyi, so I did not meet Prof. Cao. In 1946, Prof. Cao went to Cambridge on a British Council Scholarship to study in Kenneth Bailey's laboratory with the recommendation of Joseph Needham. He obtained his PhD in 1951 and was elected a fellow of Gonville and Caius College, which was unusual for a non-British person. In May 1952, Prof. Xide Xie got her visa with the help of Joseph Needham and arrived in Cambridge from USA to join Prof. Cao, and they were soon married. At that time, my sister Youduan Zhang and her husband Ruiming Chen were also studying in Cambridge and witnessed their wedding. In August 1952, the newly married couple eventually came back to the motherland after travelling about one month from Southampton to Shanghai via Hong Kong.

In Cambridge, Prof. Cao performed systematic studies on muscle proteins, including myosin, tropomyosin and actin. His most important contribution was the discovery of a small subunit of myosin, known as myosin light chain. After coming back to Shanghai, he continued to work productively on muscle proteins in collaboration with his students, the most important being their pioneering electron microscopic studies of tropomyosin and paramyosin paracrystals. In 1978, Prof. Cao was awarded a prize of important scientific achievement by the National Science Congress in recognition of his research on muscle proteins. Here, we should not forget that Prof. Cao was an active advocator and a chief organizer for the total synthesis of insulin, while his name was not included in the author list of the published papers. Prof. Cao had educated many graduate students and postgraduates. He also taught protein chemistry in Fudan University and Shanghai University of Science and Technology. Prof. Cao's major interest was on muscle proteins, but he was also interested in other protein systems. When his students wished to study some other proteins, he always supported them without hesitation. For instance, after my graduate studies, I became interested in plant viruses. He not only let me start working freely, but also helped me to collaborate with the Institute of Microbiology in Beijing through his connection with Prof. Jiazhi Zhou, who used to study plant pathology in England. Meiyun Pei from the Institute of Microbiology and I did comparative studies on different strains of TMV isolated from Prof. Zhou's laboratory and identified a new strain with special characteristics. This collaborative work marked the earliest molecular biology studies of plant viruses in China. Afterwards, Jiamu Peng and Zuxun Gong worked on plant viruses as well and extended the research to viruses and mycoplasmas infecting important crops. Similarly, Zhengwu Qi first studied actin with Prof. Cao, but later switched his research to proteases and protease inhibitors, on which he has been studying continually.

Prof. Cao had devoted his whole life to the development of protein research in China. Now, I am glad to know that the National Development and Reform Commission has chosen protein research as a key project with unprecedented support. I firmly believe that the new generation of researchers will follow Prof. Cao's career and push forward rapidly the protein research in China.

Figures

Fig.1 Prof. Tianqin Cao (1920—1995)

Fig.2 Family photo taken in 1950s

Fig.3 Just married

Fig.4 Prof. Cao (2nd left) with Joseph Needham (2nd right) and old friends at home

45. 王应睐：心怀至善、情操高尚的生物化学大家

2001年7月5日，英国著名杂志《自然》以《王应睐（1907—2001）》为题刊载文章，追溯了王应睐的生平事迹和科学成就。用这样的方式缅怀一位中国学者，对于《自然》来说，并不多见。然而，作为一位心怀至善、情操高尚的生物化学大家，他的无私奉献精神无疑应当载入《自然》这样的国际学术期刊，供全世界广大科研工作者了解学习。

图1　王应睐（1907—2001）

王应睐，英国剑桥大学博士，中国科学院院士，国际著名生物化学家，我国近代生物化学科研事业的主要奠基人（图1）。他曾在世界上首次人工合成结晶牛胰岛素和酵母丙氨酸转移核糖核酸两项重大科研工作中担任首席领导，为我国生物化学科研事业的发展作出了杰出贡献。能够取得这样的成就，得益于他举贤育才的前瞻部署和淡泊名利的工作作风。

新中国成立初期，我国生物化学科研人才极为短缺。王应睐根据学科生长点的发展需要，有目的、有步骤地向国外留学生发出了一封又一封的邀请信。从1951年到1957年，邹承鲁、曹天钦、张友端、王德宝、钮经义、周光宇等一批优秀的生物化学科研骨干接受了王先生的邀请回到国内，成功组建了（上海）生物化学研究所（简称生化所）。形成了学科分类齐全并互为补充的研究梯队，具备了“啃坚克难”的科研攻关实力。他还分别把已跟随自己工作并具有培养潜力的伍钦荣、彭加木分别让给邹承鲁和曹天钦培养，使得全所上下成为一个民主、和谐的研究集体，为组织完成重大基础科研项目奠定了基础。

随后的工作中，为了解决氨基酸匮缺和价格昂贵的问题，改变生化试剂依赖进口的被动局面，王应睐不仅首创了“研产结合”之路，通过成立东风生化试剂厂，合成氨基酸和其他生化试剂，向全国科研院所和大专院校提供了多种生化试剂；而且先后组织了近10期高级生化培训班，通过系统讲授生化知识并辅以经典生化实验的方法，

译者：张保元

中国科学院北京生命科学研究院，北京100101，中国

邮箱：zhangby@biols.ac.cn

培养了大批科研攻坚力量。其中，1961年约400人参加，1979年和1983年的规模均在500人左右。这些学员后来成为国内各单位生化领域的骨干，李载平、许根俊、洪国藩、刘新垣等多位院士，都是培训班的学员。

作为人工合成结晶牛胰岛素和酵母丙氨酸转移核糖核酸工作的主要组织者和领导者，王应睐顾全大局，处处以国家利益和中国的生化事业为重，从不计较个人得失。可以说没有王应睐就没有这两项工作。但他在发表的有关文章上从不署上自己的名字，有人署上他的名字，他审核文章时还会删掉，他认为应该给一线工作人员署名。他淡泊名利，认为“吃亏是福”。这种全心奉献、甘为人梯的精神值得我们敬仰！

许根俊院士曾用《道德经》中的一句话——“上善若水，水善利万物而不争”来纪念王应睐。诚然，王应睐的一生无愧于这样的境界。他生前的很多故事，令人难忘！

1988年，为表彰中国科学家在人工合成生物高分子方面的功绩，在美国佛罗里达州迈阿密生物技术冬季讨论会上（图2），王应睐被国际学界授予“特殊成就奖”。随后在1996年，许多科学家推荐王应睐为香港“何梁何利基金科学技术成就奖”候选人，他推辞说：“这都是过去的事情，不必再提了。”当生化所通知他授奖的消息时，这位89岁的老人，很天真地问了一句：“拿奖有什么用啊？”等他领到100万元奖金后，立即在所里设立了奖学金。在他眼中，取得奖金是对他科研工作的激励，而不是获取个人名望和利益的渠道，治学严谨、奖掖后进是他的职责，自己的奖金无疑应当奖励成绩优秀的研究生！

图2 1988年王应睐在迈阿密生物技术冬季讨论会上

王应睐将“泽被万物而不争名利”的思想用到了他工作和生活的每一个角落。他集自己半个多世纪治学所得智慧题写的“献身、求实、团结、奋进”八字箴言不仅是上海生化所的所训，也是他一生上善若水的真实写照。他的作风和精神应当为我们后人所牢记！

致谢

中国科学院生物学部薛攀皋研究员通过口述为本文提供了许多回忆材料，为本文的撰写提供了极大帮助，在此表示感谢！

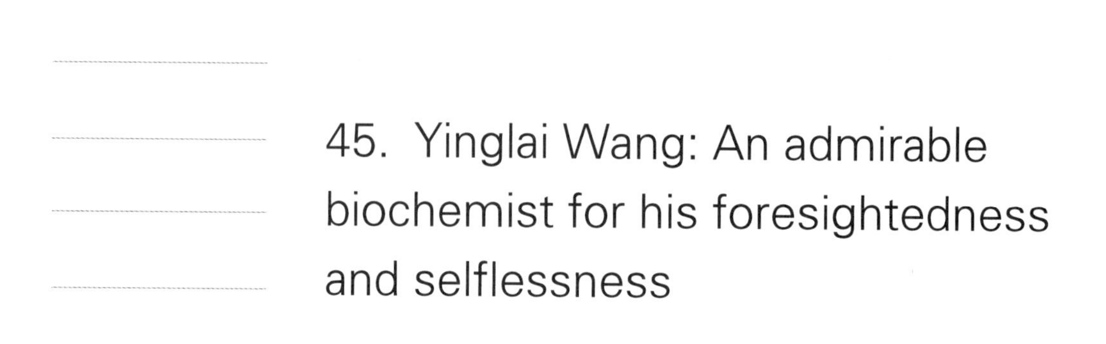

45. Yinglai Wang: An admirable biochemist for his foresightedness and selflessness

On July 5, 2001, *Nature* published an article "Wang Yinglai (1907—2001)" in honor of Dr. Yinglai Wang and his contributions to the advancement of total chemical synthesis of biological molecules. It is rather rare for *Nature* to describe a biochemist in this manner. However, Yinglai Wang is an admired and revered scientist, and as such, deserves such an honor! His scientific achievements and exemplary morality should be acknowledged by the scientific community and must be emulated by people from all walks of life.

Yinglai Wang, PhD of Cambridge University, UK, and Academician of the Chinese Academy of Sciences (CAS), was one of the founders of biochemical research in China. His most significant scientific contribution was the relevant leadership he provided in the first total chemical synthesis of crystalline bovine insulin and yeast alanine tRNA. His academic achievements were inseparable from his unique organizing ability and self-giving dedication.

In the early days of the People's Republic of China (1949—1957), biochemical research was unorganized. The country lacked qualified scientists and technicians. Realizing the need to change this situation, Yinglai Wang sent letters to some well-known Chinese scientists working overseas in a systematic way, and summoned them to return to China. His reasons were based on the disciplinary developmental trend of biochemistry. Consequently, from 1951 to 1957, he succeeded in recruiting prominent scientists from abroad including Chenglu Zou, Tianqin Cao, Yourui Zhang, Debao Wang, Jingyi Niu, and Guangyu Zhou, among others, and established the Shanghai Institute of Biochemistry at CAS. In order to support and highlight the scientific research careers of his best research assistants, namely, Qinrong Wu and Jiamu Peng, he introduced them to Chenglu Zou and Tianqin Cao, respectively. The esteemed biochemist's selfless spirit and untiring encouragement transformed the Shanghai Institute of Biochemistry into a strong and qualified institution capable of completing national key scientific research

Baoyuan Zhang[1], Pangao Xue[2], Enduo Wang[3]

1 Beijing Institutes of Life Science, Chinese Academy of Sciences, Beijing 100101, China

2 Bureau of Life Science and Biotechnology, Chinese Academy of Sciences, Beijing 100864, China

3 Shanghai Institutes for Biological Sciences, Chinese Academy of Sciences, Shanghai 200031, China

Contact: edwang@sibcb.ac.cn (E.-D. Wang)

projects. Under his guidance, the institute enjoyed a democratic and harmonious academic environment.

To solve the problem of amino acid scarcity for insulin synthesis, Yinglai Wang founded the Dongfeng Biochemical Reagent Factory in 1958. Under his leadership, the factory produced amino acids and various biochemical reagents to meet the whole country's need. Forty years before the advocacy for knowledge innovation and the acceleration of industrialization of scientific findings was initiated, he had already championed the model of and worked toward the integration of enterprises, universities, and institutes. Truly, Yinglai Wang was a visionary!

1958—1960, Yinglai Wang and other scientists had just succeeded in insulin synthesis. This inspired hundreds of scientific researchers to participate in protein synthesis research. They refocused their priorities, stayed in laboratories, and worked under stringent conditions as ordinary workers. However, these failed to take effect on the basic law of scientific research. Yinglai Wang reiterated that a capable contingent of scientists and meticulous and thorough work are vital to insulin synthesis research. Thanks to his proposal, the synthetic insulin research continued smoothly and at last, the mission was accomplished in 1965.

Yinglai Wang organized several national training programs to educate young scientists in addition to establishing a strong biochemistry research team for CAS. Hundreds of young scientists benefited significantly from the systematic knowledge and classical experiments rendered through the combination of classroom instruction and benchmarking. In 1961, approximately 400 people attended the training programs. In 1979 and 1983, the number of participants increased to more than 500. Many outstanding Chinese biochemists were trained under these programs, including Academicians Zaiping Li, Genjun Xu, Guofan Hong, and Xinyuan Liu, among others.

Aside from his endeavors as a leader and chief scientist, Yinglai Wang made significant contribution to the development of life sciences in China. He always gave top priority to the overall situation and national interests, and never considered personal gain or loss. Suffice to say, the achievements in biochemical research would not have been obtained if not for Yinglai Wang. However, he never affixed his name on the articles related to these achievements and even removed his name if it is cited in an article. In his opinion, "it is not that a penny saved is a penny earned" and only scientists whose name appears at the first line deserve to sign them. Such selflessness is worth the admiration!

Genjun Xu, a CAS academician, praised Yinglai Wang with a quotation from *Tao Te Ching*: "The top class of virtue is like water, which benefits ten thousand objects without any demands for return." This is an honest portrayal of the glorious life of Yinglai Wang.

In 1988, Yinglai Wang received the "Life Achievement Award" at the Miami Biotechnology Symposium held in America. In 1996, a significant number of scientists nominated him for the Prize for Science and Technology Achievement of the Ho Leung Ho Lee Foundation. Nevertheless, he considered that, his achievements need not be mentioned. When he was told that he had won a million RMB for his contribution, he naively asked where this money would be used! He immediately sponsored a scholarship for his students in CAS. For Yinglai Wang, winning an award is an encouragement rather than a means for gaining individual accolade or money, and that his prize should no doubt be a reward not for him but for deserving students!

Seemingly, "benefits ten thousand objects without any demands for return" was the basic philosophy that guided the life of Yinglai Wang. The maxim "dedication, realism, consolidation, courage" is not only the principle of instruction of the Shanghai Institute of Biochemistry at CAS, but also a summation of a lifetime of painstaking efforts of Yinglai Wang. Regardless of the education of scientific talents or the accomplishments in scientific research project, his contributions to the strategic and perspective development will never lose their significance in the context of biochemical research in China. He deserves the respect for his greatness and goodness!

Figures

Fig.1 Dr. Yinglai Wang (1907—2001)

Fig.2 Dr. Yinglai Wang at the Miami Biotechnology Symposium in 1988

46. 邹承鲁：积极倡导学术规范，为中国科研不懈努力

图1　邹承鲁(1923—2006)

与西方国家相比，我国现代科学发展历史较短。20世纪20年代，遗传学和进化论才首次被介绍到我国。早期的生物学家大多有海外留学经历，回国后，通过艰辛努力，为我国的动物学、植物学、生理学、神经生物学等学科奠定了基础。显然，我国现代生物学的发展与近代内忧外患的历史交织在了一起。

科学家不仅要“做”科学，还要“写”科学。原创性的科学成果应由同行评议审查后，发表到相关杂志，从而达到与国内外同行广泛交流的目的。然而，在相当长的一段时间内，我国生物学界曾经自我封闭，绝大多数研究成果都发表在国内期刊上，很少有与国外科学家进行国际交流的机会。因此，一方面，无论我们的科研成果多么新颖和重要，都很难介绍给国外同行并得到认可；另一方面，我国的科学家和青年学生没有接受过科技论文写作规范的训练，不了解国际科学界共同遵守的撰写、发表科技论文的规矩。这种情况持续了很多年，一直没有引起国内科学界的重视。直到20世纪80年代，邹承鲁先生（图1）开始提倡采用国际认可的科技论文写作规范和标准发表我国的科技成果，情况才得到较大的改观。近年来，我国科学家在国际重要学术期刊发表论文的数量和质量均有很大的提高，在国际学术界的地位也越来越重要。

邹承鲁院士出生于1923年，1951年于英国剑桥大学获生物化学博士学位。他长期从事蛋白质功能研究，是我国酶学研究领域的奠基人，也是我国人工合成结晶牛胰岛素的主要领导者和参与者，还是改革开放以后最早在英国《自然》杂志上发表科学研究论文的中国生物学家。

邹先生早在20世纪80年代初就开始呼吁开展“科研工作中的精神文明”的讨论，始终保持一颗讲真话的童心，对科技政策、成果评价、院士遴选、人才培养、学术普及、学术研究、学术弊端处理等问题提出自己中肯的看法，发表了一系列维护科学尊严、倡

译者：张保元
中国科学院北京生命科学研究院，北京100101，中国
邮箱：zhangby@biols.ac.cn

导科学道德的文章，对我国科学立足国内、走向世界发挥了重要作用。邹先生认为，中国科学要走向世界，必须要把自己的科研成果发表在国际重要刊物上，接受国际同行的检验和评价。当时，国内学术界对是否应该以及如何将研究成果在国际期刊上发表存在不同的看法。例如，很多国内期刊为了吸引稿源，常常会默许科研工作者把自己已经发表的科研结果稍加修改和补充后重复发表，甚至鼓励把在国内杂志上发表的论文再用英文重复发表到国际期刊上，这种一稿两投甚至一稿多投的陋习在国际上是绝对不被允许的。邹先生专门为《科学报》写了《一稿两投是违反科学道德的行为》一文。此外，很多作者在写论文时没有进行充分的文献检索，往往称自己的科研成果是“首次发现”，这不仅是对前人成果的无视和不尊重，也会被国际同行嘲笑为无知。再者，国内一些导师习惯在所有自己学生的论文上署名，却因为学生众多而无暇对论文质量一一把关，殊不知论文作者的署名代表着责任，一旦论文有任何疏漏、错失甚至造假，署名的导师必须为此负责。还有，引用文献时不加消化吸收而直接搬用，一些名词、度量单位、姓名的写法不符合国际惯例，等等。邹先生认为，中国科学家要成为国际学术界的一员，需要在国际刊物上发表论文，就必须改变国内学术界中存在的一些错误观念，必须遵循国际上通行的科学论文写作规范。这就像中国运动员需要参加奥运会，就必须遵循奥运会的比赛规则一样。邹先生认为，不仅项目/课题负责人需要认真纠正这些陋习，我们更要教育广大研究生学会按照国际标准和规范撰写学术论文，为他们今后的科研之路打下坚实的基础。邹先生身体力行，在中国科学院研究生院任教期间一直向学生宣讲这些理念。

邹先生不但提倡把科研成果发表在国际刊物上，与国际同行交流，扩大中国科学家在国际学术界的影响，还大力倡导中国期刊走向国际。他认为，我们应该多多创办自己的英文学术期刊，这是走向国际的第一步；既然要走向国际，中国期刊也应该像国际期刊一样，采用严格的学术标准和学术规范来管理和运行。这样做可以达到两方面的效果：中国的科研成果及期刊将为更多国外同行所了解，同时也能在国内学术界更广泛地推进学术规范。

近四十年过去了，我国在以上两方面都取得了巨大进步。学术论文的写作规范已经在国内科技界逐渐普及，越来越多的中国科学家在国际学术期刊上，尤其是高质量的国际学术期刊上发表论文。据统计，2010年，中国科技论文发表总数位居世界第二，被引用数排名已上升到世界第八位。同时，我们的英文学术期刊也越办越多，越办越好，在国际上初具影响力。追溯这些变化的源头，正是邹先生坚持不懈地提倡和呼吁，他的努力赢得了同行的尊重，对创建健康的学术生态具有重要意义。

尽管邹先生已经于2006年因病辞世，但他的教诲仍然影响着一代又一代年轻人。邹先生关于学术规范和学术道德的文章，被收录在中国科学院研究生院的《优秀研究生导师经验选编》中，一直作为向全体研究生普及学术规范的总则，提醒着每位科研工作者严格要求自己，遵守学术规范。

致谢

本文出版前夕，中国科学院生物物理研究所王志珍院士对本文进行了认真仔细的修改，在此衷心表示感谢！

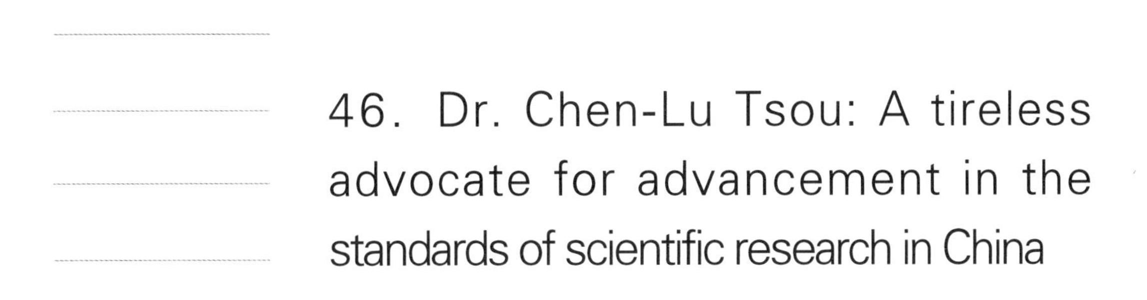

46. Dr. Chen-Lu Tsou: A tireless advocate for advancement in the standards of scientific research in China

The history of modern science in China is relatively short compared to that in western countries. For example, it was not until the 1920s that genetics and evolutionary theory were introduced to China. Most of the earliest Chinese biologists were trained overseas and returned to their beloved homeland to become pioneers of biomedical disciplines like zoology, plant biology, physiology, and neuroscience in China. Those who returned in the 1940s focused more on biochemistry, and after "Reform and Opening" in the late 1970s, molecular biology gradually thrived. With the continuous efforts of numerous Chinese biologists, modern biology started to flourish steadily despite a history of wars and political turbulence. However, even though Chinese biologists trained overseas brought back new knowledge and perspectives to China, the overall system of modern biology in China was still young, and many issues in scientific practice started to appear as time went by. One of them was lack of high academic standards.

For years, the lack of academic standards, especially the lack of scientific writing standards, had been ignored by many Chinese scientists. It is commonly accepted that scientists must not only "do" science, but also "write" science. "Writing" science allows original scientific research to be reviewed by peer scientists, and once approved, to be added to the existing database of scientific knowledge. Global visibility and accessibility of scientific work is particularly important for the basic sciences. For science conducted in China, what this means is that they should be published in English—the universal scientific language—in internationally distributed journals. However, the society of Chinese biologists was once a "closed box" and there was little intellectual exchange with western countries. Most research findings were published in Chinese in scientific journals in China. No matter how novel and important the work, the international scientific community did not get the chance to learn about, appreciate, and review them. Additionally, many scientists and young students never received formal

Chih-Chen Wang[1], Zhixin Wang[1,2], Baoyuan Zhang[3], Ming Li[3]

1 Institute of Biophysics, Chinese Academy of Sciences, Beijing 100101, China

2 School of Life Sciences, Tsinghua University, Beijing 100084, China

3 Beijing Institutes of Life Science, Chinese Academy of Sciences, Beijing 100101, China

Contact: liming@ioz.ac.cn (M. Li)

professional training on scientific writing standards, which were strictly followed in western scientific societies. This situation continued for years until a Chinese biochemist, Dr. Chen-Lu Tsou, spoke out and advocated applying internationally approved scientific writing standards in China.

Born in 1923, Dr. Chen-Lu Tsou, received a PhD degree at the University of Cambridge, UK, in 1951. He worked on protein functions during his entire academic career and pioneered the structure-function study of enzymes using partial proteolysis. He was also one of the few Chinese biologists who had published in *Nature* in the early years.

As early as the 1980s, Dr. Tsou started to publicly advocate scientific writing standards. He believed that Chinese scientists must publish their basic research results in internationally renowned peer-reviewed journals to make China's scientific research known to the world. Furthermore, Chinese scientists have to follow international standards of scientific publishing and change their unethical conducts to gain respect in the international scientific community. For example, in order to attract more papers, many domestic journals often encouraged authors to submit their previous publications with minor modifications which is known as "multiple submission" and strictly forbidden in the West. In addition, many authors often incorrectly claimed to be the first to find something without adequate literature research in the field. This not only showed disregard and disrespect towards the achievements of predecessors, but also led them to be ridiculed for ignorance by international colleagues. Furthermore, some supervisors signed their name on articles written by their students and sent them for publication without spending time and effort in ensuring proper quality. They neglected the fact that signing means responsibility, and advisors must be responsible for any omissions or errors in the articles they are co-authors in. Additionally, some researchers directly quoted others' results with the wrong interpretation and non-standard conversion of physical units. Due to such problems, Dr. Tsou suggested that we should change these attitudes in principal investigators (PIs) and promote academic writing standards to graduate students to lay a solid foundation for their research career. Dr. Tsou actively promoted these ideas not only to his peers but also to his students when he taught at the Graduate School of Chinese Academy of Sciences (CAS).

Dr. Tsou also insisted that Chinese journals should become more international, and this was as important as publishing manuscripts in internationally renowned journals. On one hand, China should publish more English journals independently, which is the first step to internationalization. On the other hand, it is necessary to apply internationally approved scientific writing standards and the peer-review process to journals published in China. In this way, China's scientific research and Chinese scientific journals can both get more attention from international scientists, which will in turn lead to the promotion of high academic standards in China.

In the past almost 40 years, China has made significant progress in these two fronts. The importance of high academic writing standards has been widely acknowledged in various scientific and technological fields in China. Increasingly more Chinese scientists have published highquality papers in international academic journals. According to some recent statistics, the total number of scientific publications from China sky-rocketed and ranked No. 2 in the world in 2010, and the total citation number of scientific publications moved up to No.8 in the world. In addition, the influence of international English journals published in China has been growing. Most if not all of these positive changes can be ascribed one way or another to Dr. Tsou's tireless advocacy of high scientific standards in China. His continued efforts won the respect of many researchers, and he has no doubt played a vital role in creating a favourable academic environment for scientific research in China.

Dr. Tsou passed away in 2006, but his contribution has left a profound impression on young people of every generation. His articles on academic ethics and norms have been included in *The Selection of Excellent Tutor Experience* in the Graduate School of the Chinese Academy of Sciences, and have always been viewed as the guideline for all the CAS graduates. Dr. Tsou's contribution to pushing China's academic standards several notches higher will be remembered forever. We are soberly aware of the gap between China and developed countries in science and technology, and there is a long way ahead to strive and persist in.

Figures

Fig.1 Dr. Chen-Lu Tsou (1923—2006)

47. 郭可信：杰出的科学家和人生导师

郭可信先生（1923—2006）是我国材料及电子显微学研究的先驱。他还为我国结构生物学研究引入冷冻电子显微学发挥了关键作用。郭先生是中国科学院院士、瑞典皇家工程科学院外籍院士、中国电子显微镜学会理事长，并获国家自然科学奖一等奖。半个世纪中，他培养了近130名博士及硕士研究生。郭先生对高分辨电子显微学的贡献，特别是对准晶材料及结构研究的贡献，以及他在培养和促进年轻科学家成长方面的成就，都对这些领域产生了深远的影响。作为曾经历过郭先生言传身教的学生，我们敬仰郭先生的杰出成就，并感念他的奉献精神。

求学年代

郭可信1923年生于北京。由于他父亲当时正在东北地区修筑铁路，郭可信的童年是在哈尔滨度过的。由于战乱频发，他在青少年时期学业常常被打断。最早遇到的是1931年的九一八事变。事变后日本很快就占领了整个东北。1936年春天，郭可信到天津进入南开中学继续学业。但好景不长，稍显宁静的南开求学之行很快又在1937年被全面抗日战争打断了。随后，少年郭可信同兄长历经艰辛来到重庆，在频繁的日军飞机轰炸中读了四年中学。在一次躲避空袭时，炸弹就在几米远处爆炸，在郭可信背上留下一块弹片烧伤的终生疤痕。

1941年郭可信考入抗战时从浙江杭州西迁到贵州遵义和湄潭继续办学的浙江大学攻读化工专业，并于1946年毕业。当1945年抗日战争胜利时，郭可信考取了公费奖学金准备赴瑞典学习冶金。

郭可信于1947年9月抵达瑞典斯德哥尔摩，并进入皇家理工学院师从Axel Hultgren教授学习金相学（图1）。在认真努力学习传统金相学的同时，郭可信也对新近发展起来的X射线衍射方法越来越感兴趣。1950年，郭可信决定放弃即将完成的博士论文，转赴乌普萨拉大学从事应用X射线衍射方法研究合金微结构的工作。在乌普萨拉大学

作者：王大能[1]，秦禄昌[2]

1 纽约大学医学院细胞生物系，纽约10016，美国

2 北卡罗来纳大学物理与天文系，教堂山27599，美国

邮箱：Da-Neng.Wang@med.nyu.edu（王大能）；lcqin@email.unc.edu（秦禄昌）

无机化学系Gunnar Hägg教授的实验室，郭可信的第一篇论文问世，于1952年发表在英国《自然》杂志上（Kuo and Hägg，1952）。在乌普萨拉大学，郭可信系统研究了碳化物的晶体结构，包括η相（Kuo，1953）。正是这个η相，其原子结构同郭可信回国多年后发现的准晶有密切关系。郭可信1954年又短暂回到斯德哥尔摩皇家理工学院从事研究工作，在这里郭可信发表了他应用刚刚起步的电子显微技术表征合金中的碳化物微结构的论文（Kuo，1956）。

(a)

(b)

(c)

图1　郭可信20世纪50年代初在瑞典（照片由郭桦女士提供）

郭可信1955年底转至荷兰代尔夫特理工学院，在W. G. Burgers教授的实验室从事白锡和灰锡的相变研究。在这期间，郭可信也发表了多篇金属间化合物的研究论文，包括Laves相和σ相。对这些金属间化合物的研究和理解，也为几十年后郭可信在准晶研究中得心应手的发展积蓄了雄厚的知识储备。

通过这些研究工作，郭可信已充分掌握了冶金学、X射线晶体学和电子显微学等近代冶金及材料研究的知识和技能，并展现了从事独立研究工作的卓越能力。1956年4月，周恩来总理号召海外留学生回国参加新中国建设。在离开祖国九年之后，郭可信决定响应召唤，启程回国。

开启电子衍射与电子显微术在材料研究中的应用

郭可信1956年回国时已经是一位卓有成就的物理冶金学家——他已经发表了二十多篇论文。他回到国内即担任在沈阳新成立的中国科学院金属研究所研究员。怀着为新中国建设和发展贡献力量的激情，郭可信教授到任当年即招收了他的第一个研究生，并组建了能力卓群的科研团队。虽然条件艰苦，郭可信教授却从未放弃对专业研究的追求。在金属研究所，他的实验室1965年获得了一台近代电子显微镜（JEM-150）。在这台电镜上，他和助手应用电子衍射和衍衬成像方法对镍基合金的显微结构与缺陷进行了系统研究。然而，1966年至1977年，他同外部世界的联系几乎完全隔断，基础研究和学术交流也陷于停滞。直到1980年，郭可信教授才重新开始在国际刊物上发表学术论文。然而，即便在那最无助的日子里，郭可信教授也从未放弃希望，坚信有一天能重新用他的知识和技能来促进国家科技的发展。所以他仍竭尽所能了解国际上相关研究领域的最新进展，并积极培养年轻一代研究工作者。例如，在20世纪70年代中期，郭可信教授便开始组织“个人学习班”，系统讲授电子衍射基础知识及相关研究领域的最新进展。这充分显示了郭可信教授的过人胆识和智慧。这些活动的一个珍贵见证是他的电子衍射学习笔记，当时曾在国内同行中广为誊抄流传。

1978年，郭可信教授又可以重新招收研究生，并开始将电子衍射与电子显微技术应用于材料科学研究。利用配有原位加热台的JEM-150，郭可信团队系统研究了镍基金属玻璃晶化过程的微结构演变。同时，基于多年应用电子衍射研究材料结构的成果和体会，他和叶恒强、吴玉琨1983年出版专著《电子衍射图在晶体学中的应用》（郭可信等，1983）。这部专著成为国内材料电子显微专业几代研究生和青年研究者的必读教科书。

在这一时期，郭可信教授开始关注世界上自20世纪70年代后期发展起来的高分辨电子显微术。利用他刚刚恢复的同国外科学界的联系，特别是通过同包括Hatsujiro Hashimoto、Sumio Iijima和Ryozi Uyeda等著名专家的日本电子显微镜学会代表团的交流，郭可信教授决定把固体材料高分辨电子显微学作为他的新的主要研究方向。

郭可信教授吸引了许多有才华的年轻学者加入他的科研团队。鉴于当时青年学者仍欠缺对最新科研技术与方法的专业训练，郭可信教授将他的年轻助理派往世界各地顶级实验室进修学习。这些开明做法为实验室的研究工作带来了显著的影响和水平的快速提高。从1982年末开始，通过购置一台在国内首次安装的新型JEM-200CX高分辨透射电子显微镜，郭可信团队开始将高分辨电子显微方法应用于各种材料结构与缺陷的研究，包括合金的缺陷分析、金属玻璃中的亚稳相转变、半导体和催化剂的微结构研究以及图像模拟和处理。这些研究内容也显示了他在课题选择上的高瞻远瞩。他还积极组织高级学术研讨会，邀请来自世界各地的知名科学家到国内讲学，并与他的团队成员和学生进行直接交流。为了推动电子显微学在材料研究中的应用，郭可信教授在金属研究所创立固体原子像实验室并担任首任主任。这也是我国第一个电子显微镜开放实验室，至今仍是材料电子显微学领域的一个重要实验室。1985年，他又在北京成立了中国科学院电子显微镜重点实验室（简称电镜室）。由于他对科学的杰出贡献，1980年郭可信教授当选中国科学院学部委员（院士）和瑞典皇家工程科学院外籍院士。

准晶研究

基于他对金属合金结构的深刻理解，以及在金属玻璃到亚稳相的晶化相变方面的大量研究工作的积累，郭先生团队迅速在材料微结构研究上得到了许多成果和新发现。这在准晶研究中特别突出。在金属间化合物中，金属原子通常形成与理想二十面体堆积相比稍微变形的对称团簇，例如Laves、σ、μ、H和C相，这些统称为Frank–Kasper相。在这些结构中，基本结构单元实际上是由沿着五次对称轴排列成列的变形二十面体（五角反对称棱面体）组成。具有这样的基本结构特征的结果是当这些晶体相有严重缺陷时，相应的电子衍射图将表现出单个拓扑单元的更多特征，也应该出现反映二十面体簇的基本结构单元的五次对称特征。而在实际研究中，确实很快就发现了在具有严重缺陷的Frank-Kasper合金相微区获得的电子衍射图具有明显的十次对称性（Ye et al.，1985）。在另一项对（$Ti_{0.9}V_{0.1}$）$_2$Ni合金金属玻璃晶化过程的研究中，当时正在郭先生和叶恒强研究员实验室攻读博士学位的研究生张泽很快获得了显示二十面体对称的电子衍射图和具有类似于准周期Penrose图的高分辨电子显微像。这项工作宣告在Ni-Ti合金中发现了准晶结构，并且这是一个全新的合金体系（Zhang et al.，1985）。与此同时，郭先生团队揭示了准晶与可以给出十次对称电子衍射图谱的五重晶体孪晶的根本结构差异（Jiang et al.，1985）。这项Ni-Ti新准晶合金的发现与研究独立于美国《物理评论快报》仅在几个月前报道的在Al-Mn合金中发现具二十面体对称的准晶结构的研究（Shechtman et al.，1984）。由于五次对称性在经典晶体学中是禁止的，所以这一发现彻底改变了传统晶体学。Shechtman教授也因此发现获得了2011年诺贝尔化学奖。

郭先生很快将全部注意力转到了准晶材料与结构的研究中（图2）。他在准晶研究上的主要贡献包括：① 发现了更多可以存在准晶结构的合金；② 发展并应用单元近似方法来逼近各种准晶体的原子结构（郭可信，2004）。郭先生团队不仅发现了具有二十面体对称性的更多三维准晶结构，而且还发现了几种具有其他类型对称性的一维和二维准晶（郭可信，2004）。作为该领域最高产和精深的研究团队之一，他的团队被国际准晶界称为"郭可信学派"。应用他具有的这些拓扑密堆相原子结构的渊博知识，推广到具有形变的二十面体结构的反棱面体单元，郭先生能够洞察在各种合金的稳定或亚稳结晶相之间许多结构的关系，从而判定形成准晶结构的可行性。这使他能够成功地指导他的学生在许多未知的金属系统中发现新的准晶结构。郭先生团队随后在近二十种其他的Al基和Mn基合金中发现了二十面体对称的准晶结构。郭先生团队也在Al-Co-Cu合金中首次发现十次对称的二维准晶，在Cr-Ni-Si和Mn-Si合金中发现八次对称二维准晶，以及在Cr-Ni-Si和V-Ni-Si合金中发现十二次对称准晶。此外，他的团队还

(a)

(b)

图2 （a）郭先生在东京都立大学的准对称壁画前；（b）郭先生（前排右二）参加1999年在德国斯图加特举行的第七届准晶国际会议

在Al-Co-Cu和Al-Ni-Si合金中发现了一维准晶。随着对这些准晶认识的积累，郭先生也开始寻找准晶原子结构的最终解决方案。为了测定准晶的原子结构，大多数科学家着迷于切割和投影高维立方晶格方法，但郭先生开创了一种全新的方法。从合金的玻璃态和稳定相之间的结构相似性得到启发，可以通过原子位置的微小变化产生具有准晶对称性的合理结构。这种方法在准晶原子结构的研究中取得了巨大的成功，现在通常被称为“郭可信学派的单元近似方法”。

郭先生一生总共发表了250多篇学术论文，其中约有170篇是关于准晶研究，而最后一篇论文是在他过世后才发表的（Cao and Kuo，2008）。郭先生在准晶结构方面的研究工作在国内和国际上都获得了广泛的盛誉。除经常受邀在国际学术会议上作专题报告外，郭先生也因对准晶研究的贡献获得了许多奖项，包括1987年国家自然科学奖一等奖（获奖人：郭可信，叶恒强，李斗星，张泽，王大能）。此外，他还被选为日本金属研究学会和印度材料研究学会的荣誉会员。

将生物冷冻电子显微学引入我国

1981年9月，在阔别欧洲25年之后，郭可信先生重访瑞典。在一个月的逗留期间，除了与老友新朋相聚，他还急于了解电子显微学和晶体学研究的最前沿。朋友送他一本1979年召开的第47届诺贝尔研讨会的会议文集《晶体和分子中原子的直接成像》。文集中，英国医学研究委员会分子生物学实验室（MRC-LMB）Aaron Klug博士的一篇文章引起了他的特别注意。文章描述了Klug本人在T4噬菌体图像重建方面的工作以及Richard Henderson和Nigel Unwin用电子衍射和成像测定细菌视紫红质结构的工作（Klug，1979）。郭先生将这本书长期放在沈阳的实验室里，并鼓励团队成员阅读。与此同时，他的团队建立了一个光学衍射系统来处理高分辨电子显微像并滤除噪声。当1982年诺贝尔化学奖授予Klug时，郭先生对大分子电子显微学和图像处理的兴趣变得更浓。

在中国《电子显微学报》1983年第2期，郭先生发表了《晶体电子显微学与诺贝尔奖》一文。文章中他系统地介绍了相位衬度、图像处理、三维重构等新概念，以及这些技术在各种生物大分子和病毒结构研究中的应用（郭可信，1983）。文章的最后一段清楚地展示了他在当时的思路：“Klug博士是一位物理学家，研究对象是生物大分子，得到的荣誉是诺贝尔化学奖……这正好说明物理、化学、生物这些基础科学之间的相互渗透，并在高分辨电子显微学这一新兴学科上联合起来。若不是Klug在这三个领域的渊博知识，很难设想他会在生物大分子的电子显微学研究方面获得这样显著的成就，这一点是很值得我们所有从事电子显微学的工作者深思的。”在他1985年出版的一本电子显微学著作中，他在第一章再次描述Klug、Henderson和Unwin的工作，并指出这个新领域的巨大潜力（郭可信和叶恒强，1985）。

郭先生不仅仅是感兴趣，这个新领域的前景也促使他采取行动。1985年初，他派一名学生到瑞典斯德哥尔摩大学学习图像处理。之后多年中，他鼓励若干位学生转向结构生物学，包括王大能、李慧林、施丹、隋海心、邢丽、张兴、唐敬华、谌东华、任罡、雷建林、吴渊、胡国彬、邓彬彬、阮娟芳等，其中大部分人至今仍然活跃在生物冷冻电子显微学领域（李慧林等，2003）。

郭先生想着手建立一个生物冷冻电子显微学研究组，而刚刚创办的电子显微镜重点实验室正好提供了一个好机会（图3）。他为此购买了Gatan 626冷台及专门用于生物研究的电子显微镜。1993年，他邀请刚刚从美国普渡大学学成回国的生物物理研究所徐伟研究员担任电镜室的生物冷冻电子显微学研究组组长，并安排自己的三名研究生与徐伟研究员一起工作。尽管当时结构生物学界的一些同行对冷冻电子显微学的潜力持怀疑态度，郭先生仍然获得了国家自然科学基金委员会的两项课题资助，以支持生物冷冻电子显微学的研究项目。徐伟研究员的研究重点是植物光捕获复合物二维晶体结构和二十面体病毒的结构。该小组在两个领域都取得了优秀的成果（张兴等，1997；Xu et al.,1998；郑东等,2000；Zheng et al.,2001）。为了进一步推动生物冷冻电子显微学，郭先生曾邀请清华大学隋森芳教授一道赴时任中国科学院院长周光召的晚餐，以说服中国科学院增加在该领域的投入。郭先生还邀请了海外各国研究人员，包括美国贝勒医学院的赵华教授和德国马普生物化学研究所的Wolfgang Baumeister教授访问电镜室。而旅美学者王大能、周正洪、李慧林等几位都先后在电镜室与郭先生的学生一起短期工作。正是这些早期的努力，为我国冷冻电子显微学研究的未来发展埋下了种子。

图3　郭先生（一排右六）1996年参加第一届结构生物学暨全国理论生物物理学会议（照片由美国印第安纳大学唐海旭博士提供）

培养青年科学家

郭先生是一位了不起的导师和长者。他用毕生精力培养学生，为他们事业的发展呕心沥血。即使学生离开他实验室十年、二十年之后，他仍然与他们保持联系、予以鼓励和支持。从1956年到1977年，他只有机会培养几名研究生（叶恒强等，2014）。后来，青年学子从四面八方蜂拥到他在沈阳及后来在北京的实验室。1982年至2000年，他每年接受5～10名硕士生和博士生。在整个学术生涯中，他培养了近130名研究生及大约10名博士后。

结合西方科学研究方法和东方的传统教育理念，郭先生用父亲般的关爱培养他的学生。而他对科学的热情也时刻感染着学生。独具一格的科学品位以及在材料科学、化学和晶体学方面的渊博知识，使他能够提出既前沿又有深度的论文项目。他常常给刚入门的学生讲德国著名物理学家Arnold Sommerfeld的故事：Sommerfeld有一个抽屉，里面放着许多小纸条；每当一个新学生加入实验室时，他都会打开抽屉拿出一张纸条给学生；纸条上写着论文题目，而这些题目常常带来诺贝尔奖！

一旦选择了论文项目，学生就在高年级学生或其他实验室成员的帮助下工作，从准备样品，到拍摄高分辨率电子显微照片。这有点像孔夫子开辟的传统私塾的教育方式。而彼此之间的密切来往使学生成为一群终生的朋友和合作者。郭先生本人则希望每个学生定期得到新的实验结果。每当见到学生时，他总会问："有什么好照片？"有时上午问过，下午遇到了仍然问。如果哪个学生在几星期都没有新结果，他（她）便知道自己需要更加努力。

郭先生不仅关注新的成果，他同样关心学生的生活。几乎每个学生都有一些终生难忘的经历，或者被先生邀请到家吃师母亲手做的佳肴，或者得到给自己新生孩子的礼物，或者在国外工作时收到他从中国寄来的包裹（图4）。郭先生去世前四天，就在他最后被送往医院抢救之前，他还给同事写了一封信，说自己病重无力，请这位同事帮助修改学生的文章。

图4　2003年8月23日，郭先生与在北京参加他80岁生日庆典的学生在一起

郭先生的许多学生在他的实验室中表现出色（叶恒强和王元明，2003）。学生对自己的科研项目充满热情，常常在实验室废寝忘食地工作。在20世纪80年代初期，实验室只有一台电子显微镜可用于高分辨率成像。这台电镜通常每天运行24小时，学生也排成4班，每次在电镜上工作6小时。学生经常得到一些新的结果，这总是让郭先生非常兴奋。几乎无一例外，每个学生毕业时都能够在国际顶级物理学或材料科学期刊上发表自己的论文。这在80年代和90年代的中国科学界中非常少见。有的学生甚至发表了多达15篇论文！

在开始自己独立的学术生涯之后，郭先生的许多学生在科学研究方面取得了很大的成就。其中三位成为中国科学院院士：叶恒强、张泽和万立俊。他的另几位学生也获得国际赞誉和奖项：2008年，邹晓冬被瑞典皇家科学院授予Göran Gustafsson化学奖，并在2017年被选为瑞典皇家工程科学院院士。贾春林则在2014年获得了国际显微学联合会颁发的Hashimoto奖。

领军人物和精神财富

除了在沈阳和北京建立和领导两个电子显微镜实验室，作为我国科学界的领军人物，郭先生还将大量精力用于建立和发展我国电子显微学组织（叶恒强等，2014）。中国电子显微镜学会成立两年后，他于1982年当选为理事长，任职到1996年。在此期间，郭先生在重建与国际社会的联系方面发挥了重要作用，使中国电子显微镜学会成为国际电子显微学联合会（IFSEM）的成员。他还曾出任亚太电子显微学联合会主席（1993—1996年），并担任多个国际科学委员会和专业期刊的编委会成员。他于1992年在北京组织了第五届亚太电子显微学会议，1993年在北京举办了国际晶体学联合会（IUCr）大会卫星会议。他发起的一年一度的中日双边电子显微学研讨会，极大地促进了中国电子显微学界与国际同行的联系与交流。他还积极寻求资助，使得更多的中国科学家有机会出席在京都（1986年）和西雅图（1990年）举办的国际电子显微学大会并且展示和交流研究成果。

为了纪念郭先生对科学的贡献以及他对培养年轻科学家的奉献精神，他的学生和朋友创立了郭可信教育基金会，以促进电子显微学研究和发展。作为非营利组织，基金会与中国电子显微镜学会合作，自2002年以来，两年一度颁发郭可信杰出科学家奖和郭可信青年科学家奖。

郭先生于2006年12月13日在北京去世。我们都为失去一位伟大的恩师、一位父亲般的长者和一位亲爱的朋友而感到难过。为了纪念他对电子显微学的贡献，郭先生的学生和朋友在2008年创办了郭可信电子显微学与晶体学暑期学校。正如郭先生35年前为我们所做的那样，暑期学校旨在将国际上一流科学家请到中国与年轻科学家交流。暑期学校每年举办一次，主题在生物冷冻电子显微学和材料科学电子显微学之间交替。十多年来，暑期学校已成为电子显微学领域最负盛名和最具影响力的活动之一（表 1）。2008年举办的第1届暑期学校不到100人参加；2018年的第11届已经吸引了400人与会，其中四分之一来自海外。未来三四年的暑期学校已经在筹备中。多年来，郭可信暑期学校吸引了世界上最顶尖的科学家参加和讲课。单粒子电子显微学先驱、2017年诺贝尔化学奖获得者Joachim Frank教授迄今为止已在4届生物冷冻电子显微学暑期学校（共6届）中讲学。而他的共同获奖者Richard Henderson博士也曾两次参加。这个暑期学校，以及受它培养或影响过的人，将使郭先生的治学精神和科学遗产代代相传。

郭先生开阔的科学视野、对研究主题的不拘一格的品位以及他热情、慷慨的人格魅力，都给我们留下了美好的回忆，让我们内心充满感激之情。他将被铭记为一位伟大的长者和一位纯粹的科学家。郭先生自己珍惜的座右铭最能准确地描述他的一生：清清白白做人，认认真真做学问。

表1 历届郭可信电子显微学与晶体学暑期学校

届/年	主题	组织者	主办单位	地点
第1届/2008	生物冷冻电子显微学	隋森芳 王大能	清华大学	北京
第2届/2009	材料科学电子显微学	张 泽 秦禄昌	郑州大学	郑州
第3届/2010	生物冷冻电子显微学	孙 飞 王大能	中国科学院生物物理研究所	北京
第4届/2011	材料科学电子显微学	马秀良 邹晓冬	中国科学院金属研究所	沈阳
第5届/2012	生物冷冻电子显微学	周正洪 蔡 刚 毕国强	中国科学技术大学	合肥/黄山
第6届/2013	材料科学电子显微学	张锦平	中国科学院苏州纳米技术与纳米仿生研究所	苏州
第7届/2014	生物冷冻电子显微学	李慧林 程易凡 丛 遥 何永宁	国家蛋白质科学中心	上海
第8届/2015	材料科学电子显微学	张 泽	浙江大学	杭州
第9届/2016	生物冷冻电子显微学	王宏伟 章佩君	清华大学	北京
第10届/2017	材料科学电子显微学	单智伟	西安交通大学	西安
第11届/2018	生物冷冻电子显微学	张 兴 刘 俊 隋海心	浙江大学	杭州
第12届/2019	材料科学	韩晓东 邹 进	北京工业大学	北京
第13届/2020 （因故取消）	生物冷冻电子显微学	沈庆涛 姜 文	上海科技大学	上海
第14届/2021	材料科学	黄晓旭 刘 庆	重庆大学	重庆
第15届/2022 （因故推迟）	生物冷冻电子显微学	王培毅 白晓晨	南方科技大学	深圳

致谢

衷心感谢以下各位分享他们与郭可信先生交往的经历、提供照片和提出修改意见：曹宝宝、郭桦、李斗星、李慧林、李明润、Jennifer Marden、潘明、彭练矛、隋海心、隋森芳、唐海旭、徐伟、叶恒强、张兴、章效峰、张泽。

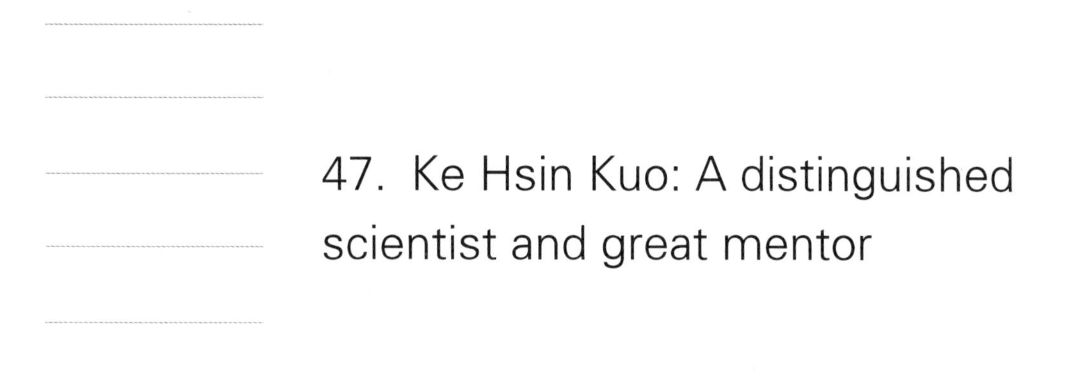

47. Ke Hsin Kuo: A distinguished scientist and great mentor

Early years

Professor Ke Hsin Kuo was born in Beijing in 1923, and his childhood was spent in Harbin in northeastern China where his father worked as an engineer to build the local railroad. His early education was at best characterized by frequent disruptions, the first of which being the "September 18th Incident" of 1931 when the Japanese Army launched an assault in Shenyang (Mukden) on the local Chinese troops and quickly took control of all the Northeastern Provinces. He moved to Tianjin in the spring of 1936, where he continued his education by enrolling in the famous Nankai High School. However, his school life at Nankai was soon disrupted again by full outbreak of the War of Resistance against Japanese Aggression in July of 1937 after Japan attacked and occupied Peking. Together with his brothers, he fled to join his parents who had previously relocated to Chongqing. He studied in Chongqing for four more years to complete his high school education, during which classes were constantly being disrupted due to the frequent bombings of the city by Japanese airplanes. These bombings not only left him with long lasting memories of the war, but also bore him a permanent burn scar on his back as a result of a bomb that exploded just meters away from him.

In 1941, He entered Zhejiang University, a national university that was evacuated to Guizhou Province from Hangzhou due to the advancement of the Japanese aggression. He graduated in 1946 having majored in Chemical Engineering. When an opportunity arose after the war ended in 1945, he took a competitive national examination and won a government scholarship to study metallurgy in Sweden.

Professor Kuo arrived in Stockholm in September 1947 to start his study of metallography with Professor Axel Hultgren at the Royal Institute of Technology (KTH). While actively engaged

Daneng Wang[1], Luchang Qin[2]

1 Skirball Institute of Biomolecular Medicine, and Department of Cell Biology, New York University School of Medicine, New York, NY 10016, USA

2 W.M. Keck Laboratory for Atomic Imaging and Manipulation, Department of Physics and Astronomy, University of North Carolina, Chapel Hill, NC 27599, USA

Correspondence: Da-Neng.Wang@med.nyu.edu (D.-N. Wang); lcqin@email.unc.edu (L.-C. Qin)

in research in traditional metallography, he also became increasingly interested in the newly developed X-ray diffraction methods. Later in 1950, after learning more about the usefulness and power of X-ray diffraction and crystallography in microstructural research of alloys, he decided to abandon his almost-completed doctoral thesis at KTH (although he eventually returned in 1980 to accept an honorary doctoral degree together with Foreign Membership in the Royal Swedish Academy of Engineering), and moved to Uppsala University. There he pursued X-ray diffraction studies of carbides in alloys in the Department of Inorganic Chemistry with Professor Gunnar Hägg, a world leading figure in X-ray crystallography at the time. With Hägg he soon published his first research paper in the journal *Nature* (Kuo and Hägg, 1952). He studied more structures of carbides, including η carbide (Kuo, 1953), which has a crystal structure closely related to that of a quasicrystal which would be discovered many years later in his laboratory in China. He returned to KTH briefly in 1954, where he carried out and published his first piece of work using another new technique—electron microscopy—for characterization of carbide precipitates in alloys (Kuo, 1956).

Professor Kuo briefly went to the Delft Institute of Technology in the Netherlands at the end of 1955 to work with Professor W. G. Burgers, to study the phase transformation of white tin to grey tin. In addition to his work on carbides, he also published several research papers on intermetallic phases including the structure of the Laves and σ phases, which would all be closely related to his work decades later on quasicrystals.

Professor Kuo was now well equipped with the knowledge and expertise in metallurgy, X-ray crystallography and electron microscopy needed to start his own lab. In April 1956, he left Europe after staying there for nine years in response to a call by the then Premier Zhou for overseas students to return to their homeland to build China.

Electron diffraction and microscopy for materials research in China

Upon his return to China, Professor Kuo was already an established physical metallurgist in his own right—he had more than twenty published papers to his credit. He joined the newly established Institute of Metal Research (IMR), Chinese Academy of Sciences, in Shenyang as a full professor. Bearing high hopes for using his expertise to contribute to the development of China, he accepted his very first graduate student in 1956 and began to assemble a capable research team. Despite of difficult conditions, he did his best to keep his research going. His laboratory acquired a modern transmission electron microscope (TEM) JEM-150 in 1965 that allowed them to perform electron diffraction as well as diffraction contrast imaging of crystal defects, a technique developed only in the early 1960s. With this TEM he and his associates studied crystal defects in Ni-based alloys. However, their work did not continue unabated from 1966 to 1977, his connections to the outside world were essentially severed, and all of his basic research and technical teaching came to a complete standstill. As a result, Professor Kuo did not begin publishing papers again in international journals until 1980. However, even during the darkest times he never gave up his dream of again applying his expertise and skills to advance science in China and, with whatever means available, he grasped every possible opportunity to refresh his knowledge and train young scientists. For instance, in the mid-1970s, he organized "personal" workshops or study groups to teach the latest developments in materials science to his junior colleagues, attesting to his personal resilience and intellectual stamina. A highly valued and demanded outcome of these endeavors was his personal notebook on electron diffraction, which was widely hand-copied as a high-level technical treatise within his small community.

Starting in 1978, Professor Kuo was once again able to accept graduate students into his lab, and he quickly instituted a research program in materials science using electron diffraction and microscopy techniques. Using the JEM-150 microscope equipped with an in situ heating stage, he directed his team to study the microstructural evolution during crystallization of various Ni-based metallic glasses. His group also systematically analyzed the use of electron diffraction geometry and initiated the use of computers to develop an automated program for indexing of electron diffraction patterns in China. As a result, he and his colleagues published a book on electron diffraction in 1983, which has trained and benefited generations of students and users throughout the country (Kuo et al., 1983).

The rapid development of high-resolution electron microscopy (HREM) in the late 1970s caught Professor Kuo's attention. With the re-establishment of connections with the outside world allowing him to learn the latest developments in electron microscopy, especially stimulated by intimate discussions with two high-level visiting delegations of electron microscopists from Japan including Drs. Hatsujiro Hashimoto, Sumio Iijima and Ryozi Uyeda, Professor Kuo soon chose the characterization of solid state materials using HREM as the major research area for his revitalized laboratory.

Professor Kuo attracted a number of talented investigators into his lab. Recognizing his programs' lack of expertise in recently developed techniques, he worked tirelessly himself and also sent his junior associates abroad to top labs to obtain advanced training and experience. These efforts reaped noticeable benefits and quick rewards to his research program. Beginning in late 1982, with the acquisition of a brand new JEM-200CX TEM, one of the first installed in China, his team started applying HREM to various studies including defect analysis of alloys, metastable phase transformation in metallic glasses, semiconductors, and catalysts, complemented with image simulations and processing, illustrating his eclectic taste in research topics. He also organized advanced workshops and invited prominent scientists from around the world to lecture in China and to interact with the members and graduate students in his group. To advance electron microscopy for materials research, Professor Kuo established and served as the founding director of the Laboratory for Atomic Imaging of Solids at IMR, a dedicated electron microscopy laboratory that to this day continues to make significant contributions to the science of electron microscopy. In 1985, he also founded the Laboratory of Electron Microscopy (BLEM), Chinese Academy of Sciences in Beijing. Due to his significant contributions to science, in 1980 Professor Kuo was elected as an Academician of the Chinese Academy of Sciences and the Royal

Swedish Academy of Engineering.

Quasicrystal research

Given his broad knowledge and deep understanding of metallic structures and the intensive effort he devoted to the study of the phase transitions of metallic glasses through a series of metastable phases, it came as no surprise, that Professor Kuo's team quickly made many discoveries, particularly in quasicrystal research. In intermetallic compounds, metal atoms often form clusters with slightly distorted symmetry from ideal icosahedral packing, such as the Laves, σ, μ, H and C phases, collectively called the Frank-Kasper phases. In these structures, the individual structural unit is actually composed of a distorted icosahedral cluster (pentagonal antiprism) arranged into a column along a five-fold axis. As a result, in a heavily faulted structure of these phases, the electron diffraction pattern will instead exhibit more characteristics of the individual topological unit, and distinct features reflecting the structure of icosahedral clusters, i.e., five-fold symmetry, should appear. Indeed, it was soon observed in heavily faulted alloys with nano-domains of the Frank-Kasper phases that an electron diffraction pattern with apparent ten-fold symmetry was present (Ye et al., 1985). In another investigation conducted in parallel examining the crystallization of metallic glasses in a $(Ti_{0.9}V_{0.1})_2Ni$ alloy, Ze Zhang, his doctoral student who at the time worked with him and Professor Hengqiang Ye, soon obtained electron diffraction patterns displaying icosahedral symmetry and high resolution TEM images of the samples that directly resembled quasi-periodic Penrose tiling. This work announced the discovery of quasicrystals in the Ni-Ti alloy, a totally new metallic system (Zhang et al., 1985). In the meantime, Professor Kuo's team also demonstrated side by side the fundamental differences in atomic structure from five-fold twinning that could result in an electron diffraction pattern with apparent ten-fold symmetry (Jiang et al., 1985). The entire project was carried out independently from the results published in *Physical Review Letters* only a few months earlier reporting the identification of icosahedral quasicrystals in an Al-Mn alloy with five-fold symmetry that is "forbidden" in classical crystallography (Shechtman et al., 1984), a discovery that revolutionized crystallography and later won the 2011 Nobel Prize in Chemistry.

Professor Kuo quickly turned his full attention to the research of quasicrystals. His contributions include (1) discovering more alloys where quasicrystals can exist and (2) developing an approximant method to approach the atomic structure of various quasicrystals (Kuo, 2004). Professor Kuo's team discovered not only more three-dimensional quasicrystals with icosahedral symmetry, but also one- and two-dimensional quasicrystals with several other types of symmetries (Kuo, 2004). As one of the most productive and proficient centers in this field, his team was referred to as the "Kuo School" by the international quasicrystal community. Using his knowledge of the atomic packing in these topologically close-packed structures down to the fundamental antiprisms with a distorted icosahedral symmetry, he was able to establish many structural connections between the stable or metastable crystalline phases of a variety of alloys having the potential to form quasicrystalline structures. This allowed him to guide his students successfully to discover new quasicrystals in many unexplored metallic systems. In particular, icosahedral quasicrystals were subsequently identified in nearly twenty other Al- and Mn-based alloys. His group was also the first to discover a stable phase of two-dimensional decagonal quasicrystals of ten-fold symmetry in an Al-Co-Cu alloy, two-dimensional octahedral quasicrystals in Cr-Ni-Si and Mn-Si alloys, and dodecagonal quasicrystals with twelve-fold symmetry in Cr-Ni-Si and V-Ni-Si alloys. Additionally, one-dimensional quasicrystals were also discovered by his team in Al-Co-Cu and Al-Ni-Si alloys. With the accumulation of knowledge on such quasicrystals, he began to search for an eventual solution to the atomic structure of quasicrystals. While most scientists were fascinated with cutting-and-projecting higher dimensional cubic lattices in order to obtain structural models for quasicrystals, Professor Kuo took a new approach. Recognizing the proximity between the glassy and stable structures of the forming alloys, he was able to produce rational structures with quasicrystalline symmetry through minor shifts in the atomic positions. This method led to tremendous success in the study of quasicrystal structures, and it is now often referred to as the Kuo School's approximant method.

In total Professor Kuo published more than 250 research papers; among them about 170 are on quasicrystals including one published posthumously (Cao and Kuo, 2008). His work on quasicrystals was well recognized, both domestically and internationally. He was frequently invited to speak at international conferences and was presented with numerous awards for his contributions to quasicrystal research, including the First Class National Science Medal in 1987 (with Hengqiang Ye, Douxing Li, Ze Zhang and Daneng Wang). In addition, he was also elected as a Honorary Member of both the Japanese Institute of Metals and the Materials Research Society of India.

Efforts in bringing biological cryo-electron microscopy to China

In September of 1981, twenty-five years after he left Europe and returned to China, Professor Kuo visited Sweden. During his one-month stay there, besides meeting with old friends and new, Professor Kuo was eager to learn what was at the forefront of electron microscopy and crystallography research. He was given a copy of the proceedings of the 47th Nobel Symposium *Direct Imaging of Atoms in Crystals and Molecules* that took place in 1979. One particular article in the volume caught his attention: the one by Aaron Klug from the Molecular Biology Laboratory of the Medical Research Council, in which he described both his own work on the image reconstruction of T4 bacteriophage and the structure determination of bacteriorhodopsin from electron diffraction and imaging by Richard Henderson and Nigel Unwin (Klug, 1979). Professor Kuo kept the book in his laboratory at IMR, and encouraged everybody in the group to read it. At the same time, his group set up an optical diffractometer to examine image quality and to filter out noises. When the Nobel Prize in Chemistry was awarded to Klug in 1982, Professor Kuo became even more excited about macromolecular electron microscopy and image processing.

In an article entitled "Crystallographic electron microscopy and the Nobel Prize" that Professor Kuo published in the second issue of the 1983 *Journal of Chinese Electron Microscopy Society*, he described the new concept and technical advances in phase contrast, image processing and reconstruction, as well as their applications with various macromolecular complexes and viruses (Kuo, 1983). His thinking at the time was most clearly revealed by the last paragraph of the article: "Trained as a physicist, Dr. Klug works on biological macromolecules, but has received a Nobel Prize in Chemistry... Such (success) shows that physics, chemistry and biology exchange ideas and have got merged at this novel field called high resolution electron microscopy." Without Klug's deep knowledge in all three areas, "it would be hard to imagine that he could have accomplished so much... — this is a point that we all electron microscopists should ponder about." In a book that he published in 1985, he again wrote extensively in the first chapter about the work of Klug, Henderson and Unwin and advocated for the great potential of the new field (Kuo and Ye, 1985).

Professor Kuo did not just reflect on it, this new field also spurred him into action. In early 1985 he sent a student to the University of Stockholm to learn image processing. This was followed over the years by his encouragement to more students to switch to structural biology including Daneng Wang, Huilin Li, Dan Shi, Haixin Sui, Li Xing, Xing Zhang, Jinghua Tang, Donghua Chen, Garry Ren, Jianlin Lei, Yuan Wu, Guobin Hu, Binbin Deng and Juanfang Ruan, many of whom are still active in the cryo-EM field to this day (Li et al., 2003).

Professor Kuo also wanted to start a biological cryo-electron microscopy group. His move to Beijing to establish the BLEM provided just such an opportunity. He purchased a Gatan 626 cryo-stage and dedicated one electron microscope to biological research. In 1993, he recruited Wei Xu, a professor at the Institute of Biophysics who recently returned from Purdue University, as a group leader at BLEM, and assigned three of his own graduate students to work with Xu. Despite skepticism from some members of the local structural biology community about the potential of cryo-EM, Professor Kuo was able to obtain two research grants from the Chinese Natural Science Foundation to support projects on biological cryo-EM. Xu's research there focused on structural studies of the plant light-harvesting complexes from two-dimensional crystals, and of icosahedral viruses, and the group was able to publish several papers (Zhang et al., 1997; Xu et al., 1998; Zheng et al., 2000, 2001). To promote the field, Professor Kuo once invited Senfang Sui, a professor at Tsinghua University, to join him at a meeting with the President of the Chinese Academy of Sciences in order to convince the Academy to invest more in the field. Professor Kuo also invited various investigators from overseas including Professor Wah Chiu of Baylor College, USA and Professor Wolfgang Baumeister of Max Planck Institute of Biochemistry, Germany to visit BLEM, and several of them including Daneng Wang, Zhenghong Zhou and Huilin Li spent a week or two there working with his students. Such early efforts planted the seeds for the future development of cryo-electron microscopy research in China.

Training young scientists

Professor Kuo was a great mentor. He was very dedicated to the training of his students and ensuring their career success. Even years after they left his lab he would remain in touch with them to provide support and encouragement. From 1956 to 1977, he trained several graduate students (Ye et al., 2014). Students flocked to his laboratory at the IMR and later at BLEM. Between 1982 and 2000, he often accepted five to ten students per year. Over the years, he had trained close to 130 graduate students and approximately ten postdocs.

Professor Kuo trained his students with father-like love, combining Western scientific research methodology and traditional Eastern teaching philosophy. His passion for science was infectious. His impeccable taste in science and broad knowledge in metallurgy, chemistry, and crystallography allowed him to suggest great thesis projects. He told every entering class the story of the renowned German physicist Arnold Sommerfeld: Legend said that Sommerfeld had a drawer in which a number of small pieces of paper were kept; each time a new student joined the lab he would open the drawer and give a slip to the student; a thesis project was written on the paper and the project often led to a Nobel Prize!

Once a thesis project was chosen, Professor Kuo's students would need to search for and get daily help on things ranging from preparing specimens to recording high resolution electron micrographs from senior students or other lab members, not unlike in a traditional Chinese private school in the Confucius style. Such close interactions with each other made Professor Kuo's students into a group of lifelong friends and collaborators. Professor Kuo himself would expect every student to deliver new experimental results regularly. He was well known for asking "What are your new results?" Every time he ran into a student, sometimes once in the morning and once again in the afternoon. If a student did not get new results in a few weeks, he or she knew they would have to work harder.

Not only did he care about obtaining new results, he cared equally as much for the well-being of his students. Almost every student has some personal stories about Professor Kuo inviting him to his home for dinner, or buying his newborn daughter a gift, or receiving presents sent by him from China while working abroad. Four days before he passed away—right before he was sent to the intensive care unit—he penned a letter to a colleague saying that he no longer had the strength and could this colleague please help to revise a manuscript for a student.

Professor Kuo's students did very well in his lab (Ye and Wang, 2003). The students were passionate about their projects, and they often worked long hours in the lab. In the early 1980s, when the lab had only one electron microscope for high resolution imaging, it was usually kept manually running nonstop for 24 h a day, with students working at the microscope in six-hour shifts. Professor Kuo directed the students to exciting areas, and they often discovered something new. Almost without exception, everybody was able to publish their thesis work in international top journals of physics or materials science—very uncommon for a laboratory in China in the 1980s and 1990s—some even published as many as 15 papers!

After training with Professor Kuo and starting their own independent careers, many of his former students accomplished a great deal in research, including three who became academicians of the Chinese Academy of Sciences: Hengqiang Ye, Ze Zhang, and Lijun Wan. Several of his former students also won international acclaim for their achievements in science and have received prestigious prizes: Xiaodong Zou was awarded the 2008 Göran Gustafsson Prize in Chemistry by the Royal Swedish Academy of Sciences and was elected to be a member of the Royal Swedish Academy of Engineering in 2017, and in 2014 Chunlin Jia received the H. Hashimoto Medal from the International Federation of Societies for Microscopy.

Leadership and legacies

In addition to founding and directing two major electron microscopy laboratories in Shenyang and later in Beijing, as a senior figure in the science and engineering community in China, Professor Kuo also devoted much of his energy to the establishment and development of the Chinese electron microscopy community (Ye et al., 2014). Two years after the founding of the Chinese Electron Microscopy Society (CEMS), he became its President in 1982 and served in this office until 1996. It was during this period that Professor Kuo was instrumental in re-establishing connections with the international community, including his efforts and success in making CEMS a member of the International Federation of the Societies of Electron Microscopy (IFSEM). He also served as President of the Federation of Asia-Pacific Societies for Electron Microscopy (1993—1996) and as a member of various international science committees and on the editorial boards of numerous professional journals. He organized the Fifth Asia-Pacific Electron Microscopy Conference in Beijing in 1992, the International Union of Crystallography (IUCr) Congress Satellite Meeting on quasicrystals in Beijing in 1993, and initiated the annual Sino-Japanese Bilateral Electron Microscopy Symposia, which helped tremendously to establish connections between the Chinese electron microscopists and the international community. As President of the CEMS, he solicited financial support to allow more scientists to participate and present results at international conferences during the time when international travel funds were scarce in China. As a result, noticeable Chinese attendance was made at the IFSEM Conferences in Kyoto (1986) and in Seattle (1990).

To celebrate Professor Kuo's contributions to science and his dedication to the training of young scientists, his students and friends established the K. H. Kuo Education Fund, a nonprofit organization to promote advances in electron microscopy. In collaboration with the Chinese Electron Microscopy Society, the K. H. Kuo Education Fund has awarded the K. H. Kuo Distinguished Scientist Award and the K. H. Kuo Young Scientist Award biennially since 2002.

Professor Kuo passed away on December 13, 2006. We were all saddened by the loss of a great mentor, a father-like advisor and a dear friend. As a memorial to honor his contributions to electron microscopy, a summer school series, the K. H. Kuo Summer School of Electron Microscopy and Crystallography, was launched in 2008 with the aim of bringing leading scientists to China to interact with young scientists, as Professor Kuo did for us over 35 years ago. The summer school is held annually and the topic alternates between structural biology and materials science. Over the years, the school series has become one of the most prestigious and influential meetings in electron microscopy. The first summer school, held in 2008, was attended by fewer than one hundred people; the 11th school in 2018 attracted about four hundred participants, a quarter of them coming from abroad; schools for the next three or four years have already been planned. Indeed, over the years the school series has been able to attract the very top scientists to attend and lecture. For example, Professor Joachim Frank, a winner of the 2017 Nobel Prize in Chemistry for developing single-particle cryo-EM, has lectured in four of the six biological Kuo summer schools so far, whereas Dr. Richard Henderson, his co-winner, has attended twice. This school series, along with the passion for science and the efforts of the people who were trained with or influenced by him, will help keep Professor Kuo's legacy alive.

Professor Ke Hsin Kuo's insightful scientific vision, eclectic taste of research themes, and his warm, generous and engaging personality, have left us with great memories and sincere gratitude. He will be remembered as a dear teacher, a great mentor and a devoted scientist, whose character is best described by the motto that he himself cherished dearly, Live in Immaculacy; Work with Dedication (清清白白做人，认认真真做学问).

Figures

Fig.1 Ke Hsin Kuo in Sweden in early 1950s (Photographs courtesy of Ms. Hua Guo)

Fig.2 (a) In front of a Penrose tiling at the Tokyo Metropolitan University; (b)Attending the 7th International Conference on Quasicrystals, Stuttgart in 1999 (Front row, second from the right)

Fig.3 Attending the First Chinese Conference of Structural Biology and Theoretical Biophysics, Qingdao in 1996 (First row, 6th from the right. Photograph courtesy of Dr. Haixu Tang, Indiana University)

Fig.4 With his students who attended his 80th birthday celebration in Beijing, August 23, 2003

References

Cao BB, Kuo KH (2008) Crystal structure of the monoclinic η-Al11Cr2. J Alloys & Compd 458: 238–247.

Jiang WJ, Hei ZK, Guo YX, et al (1985) Tenfold twins in a rapidly quenched NiZr alloy. Phil Mag A 52:L53–L57.

Klug A (1979) Image analysis and reconstruction in the electron microscopy of biological macromolecules. Chem Scripta 14:245–256.

Kuo KH, Hägg G (1952) A new molybdenum carbide. Nature 170:245–246.

Kuo KH (1953) The formation of η carbide. Acta Met 1:301–307.

Kuo KH (1956) Alloy carbide precipitated during the fourth stage of tempering. Electron microscopic examinations. J Iron Steel Inst 18:258–268.

Kuo KH (1983) Electron crystallography and the Nobel Prize. J Chin Electr Microsc Soc 2:1–5.

Kuo KH (2004) Quasiperiodic Crystals. Hangzhou: Zhejiang Science & Technology Press.

Kuo KH, Ye HQ, Wu YK (1983) Electron Diffraction Patterns for Applications in Crystallography. Beijing: Science Press.

Kuo KH, Ye HQ (1985) High Resolution Electron Microscopy. Beijing: Science Press.

Li H, Shi D, Ren G, et al (2003) Electron microscopy of biological macromolecules. In: Ye HQ, Wang YM. Recent Progress in Transmission Electron Microscopy. Beijing: Science Press, 114–148.

Shechtman D, Blech I, Gratia D, et al (1984) Metallic phase with long-range orientational order and no translational symmetry. Phys Rev Lett 53:1951–1953.

Xu W, Zhang X, Lou S, et al (1998) Two-dimensional crystallization and preliminary structure analysis of LHC-II from cucumber and spinach. Sci China Sci C41:265–271.

Ye HQ, Wang DN, Kuo KH (1985) Fivefold symmetry in the real and reciprocal spaces. Ultramicroscopy 16:273–278.

Ye HQ, Wang YM (2003) Progress in Transmission Electron Microscopy. Beijing: Science Press.

Ye HQ, Wang YM, Guo H (2014) Ke Hsin Kuo, A Biography. Beijing: Science Press.

Zhang X, Xu W, Lou S, et al (1997) Two-dimensional crystallization of the light-harvesting chlorophyll a/b protein complexes of spinach and cucumber. J Chin Electr Microsc Soc 16: 318–322.

Zhang Z, Ye HQ, Kuo KH (1985) A new icosahedral phase with the m35 symmetry. Phil Mag A52: L49–L52.

Zheng D, Chen D, Xue T, et al (2000) Visualization of RHDV particles packaging genomic and subgenomic RNAs by electron cryomicroscopy. J Chin Electr Microsc Soc 19:661–666.

Zheng D, Xue T, Chen D, et al (2001) Three-dimensional structure of the wild-type RHDV. Chinese Science Bulletin 46:1005–1008.

48. 沈家祥：中国制药工业的先驱者

在2007年中国药学会的百年庆典大会上，沈家祥院士等十位药学泰斗获得中国药学会突出贡献奖，这无疑是对他促进中国医药事业发展的巨大贡献的肯定和褒扬。沈家祥的主要学术贡献在于开发了氯霉素的全合成新工艺，实现了几种重要的甾体激素类药物的工业化生产，因此他也被认为是我国现代制药工业的奠基人之一。

图1　沈家祥24岁时在伦敦大学学习（1945年）

1921年11月11日，沈家祥出生于江苏省扬州市的一个普通家庭，后随父母迁居南京。中学时期，沈家祥就读于当时的南京市立一中，这所有着良好学风的学校使他受到很好的基础教育。1937年，日本发动全面侵华战争。国家处在风雨飘摇中，学校也难再维持正常的教学，此时沈家祥又因肺结核而吐血，不得已随父母迁往重庆。次年，由于身体原因而休学在家的沈家祥通过自修，以优异的成绩考入当时由南京迁至重庆的国立药学专科学校（今中国药科大学）。尽管个人的健康状况和时局动荡使这段求学经历显得极其艰难，但也坚定了沈家祥知识报国的信念，并奠定了他终生从事药学专业的基础。

1942年，从国立药学专科学校毕业的沈家祥在当时政府的陆军制药研究所短暂工作了一段时间，而后便通过考核获得奖学金资助前往英国伦敦大学药学院学习（图1）。沈家祥仅用四年时间就完成了从本科到博士的学习，于1949年获得了伦敦大学药物化学博士学位（Linnell and Shen，1949，1950）。在英国学习期间，沈家祥早已下定了学成回国的决心。为了尽早回国参加新中国建设，刚刚通过博士答辩的他等不及拿到学位证书，于1949年9月23日乘船经香港归国。沈家祥成为新中国第一批从海外回来的海归学者之一，自此，他一直工作在药物科研开发的第一线，成为新中国医药工业技

译者：郭翔海[1,2]

1　天津大学海洋科学与技术学院海洋科学系，天津300072，中国

2　系统生物工程教育部重点实验室，天津300072，中国

邮箱：guoxh@tju.edu.cn

术的开拓者之一。

在抗美援朝时期，沈家祥被指派负责氯霉素——一种志愿军战士急需的抗生素的合成方法研究。在短短几年的时间内，他完成了还原、水解等重大流程方面的技术革新，使氯霉素的生产工艺得到很大改善（图2）。1957年，新工艺成功用于大规模生产，这也被认为是新中国制药工业诞生的重要里程碑（Shen et al.，1950，1958a，1958b，1958c，1958d，1958e）。在20世纪50—60年代，他以国产资源为原料，成功地完成了结晶维生素A醋酸酯和维生素D2的合成方法研究。他还指导了多种甾体类药物（如氢化可的松和地塞米松）的合成和生产（Shen et al.，1964a，1964b，1964c），完成了雌性酮全合成方法的研究。雌性酮全合成方法的研究后来成为60年代后期投产的高诺酮（即18-甲基炔诺酮）及其他19-去甲基甾体激素类药物的研究基础。

70年代中期以后，沈家祥恢复了湖南医药工业研究所副总工程师职务，从此开始了以中药有效成分为基础的新药研究工作。1976年，他指导的驱绦虫药有效成分鹤草酚的全合成研究取得成功，并证明了它的独特化学结构（Shen et al.，1976a，1976b）。20世纪80年代中期，他被聘为北京医科大学（现北京大学医学部）药物化学专业的博士生导师（图3）。在那里他指导博士生进行丹参酮ⅡA和丹参新醌乙素等的全合成研究（1986—1988年）（Shen et al.，1988；Zhang and Shen，1988），并发现了某些衍生物的特殊药理活性。他还指导博士生进行丁公藤碱Ⅱ（包公藤甲素）及其类似物的全合成研究。

图2　沈家祥在图书馆寻找氯霉素的文献（1957年）

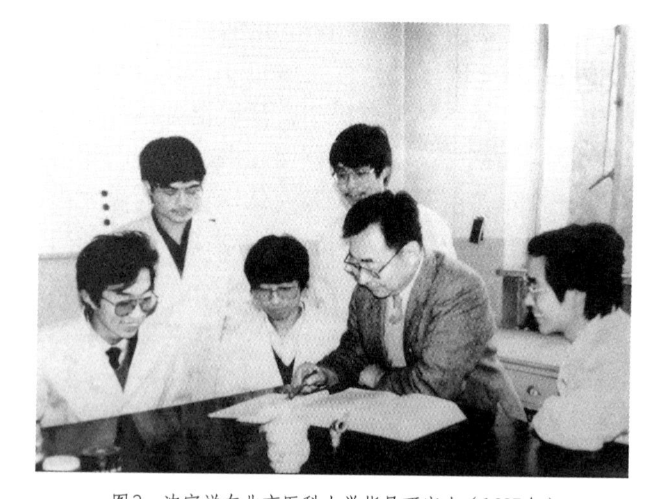

图3　沈家祥在北京医科大学指导研究生（1987年）

1992年，从国家中医药管理局副总工程师及中国医药研究开发中心主任的岗位上退休之后，沈家祥创办了北京市集才药物研究所，这也是我国第一家私营的药物研究机构。沈家祥把一生的心血都投入“为百姓做廉价好药”的事业中，他的多项科研成果不仅提高了药物疗效，还降低了生产成本，使其惠及百姓，实现了他年轻时“要强国，先要提高全民族身体素质”的梦想。他在集才药物研究所主持阿奇霉素的合成工艺研究时，成功获得含非结晶水的新晶型，并于1993年申请为专利。此项研究的成功，打破了辉瑞公司在我国获得的行政保护，实现了知识产权本土化，产品以远低于进口药价格供应市场并占压倒优势。此外，他还成功领导研究阿法骨化醇、替勃龙、布地奈德和坦索罗辛等合成方法有困难的新产品。

随着我国与国外科技交流的发展，沈家祥在国际药学界中的声望也日益提高（Shen and Zhuang，1984）。基于他在药物化学领域作出的突出贡献，并在国际药学界享有较高的声誉，沈家祥于1983年入选法国国家药学科学院通讯院士。20世纪80年代，沈家祥曾多次参加联合国工业发展组织和世界卫生组织的会议，并于1987年起，受聘为世界卫生组织疟疾化疗科学工作领导小组成员，为中国抗疟药——青蒿素的疗效得到国际公认作出了贡献（Luo and Shen，1987；Shen，1991）。1988年起，他担任中国-美国西尔研究中心基金会董事；1989年起，又受聘为*Medicinal Research Reviews*杂志编委和美国药学会主办的*Journal of Pharmaceutical Sciences*的编辑顾问委员会委员。

沈家祥以发展药学科学技术、促进医药事业发展、保障人民身体健康为己任，走过了一条不平凡的道路，见证了我国药学事业发展的风雨历程。他将智慧与忠诚奉献给祖国的医药科研事业，促进了我国药学基础研究、高新技术研究和重大关键技术研究的发展，推动了药学科研成果的转化和应用。他曾获得国家新产品奖2项（1964年），全国科学大会奖5项（1978年），国家发明奖三等奖1项（1982年），北京市科学技术进步奖二等奖1项（1999年）。1999年，他当选中国工程院院士。

沈家祥在青年时期目睹了祖国的落后，漂泊海外的求学生涯更使他亲身感受到被列强侵略的屈辱。这一切激发了他为民族振兴奋发图强和科学救国的思想，促使他用自己的努力证明中国人的志气。他在近半个世纪负责一线研究的工作中，提倡“任务带学科、浅入深出”的学风，主张脚踏实地、扎实深入地从事科研工作，做出不少独创性的成果。沈家祥还时时刻刻关心着国家的发展与进步，他曾多次对自己的学生说：“我们科研工作者，要坚持自力更生，靠自己的双手和大脑为国家创造财富，而不是依赖国家的扶持而生存！”

2001年，80岁高龄的沈家祥接受了天津大学的诚挚邀请，怀着对祖国医药事业的热爱，放弃了晚年闲适安逸的生活，来到天津与天津大学师生共同创建了药物科学与技术学院。在建院的前六年多时间里，沈家祥不顾年高体弱，始终坚持工作在第一线。他身上所凝聚的老一辈科技工作者的精神深深感染了天津大学的每一个人，这种精神已成为天津大学的宝贵财富！

致谢

本文资料搜集自中国科学技术协会资助的“老科学家学术成长资料采集工程”项目。对于沈家祥先生的儿子沈坚及女儿沈安在文章撰写过程中所提供的帮助和宝贵意见表示衷心的感谢。

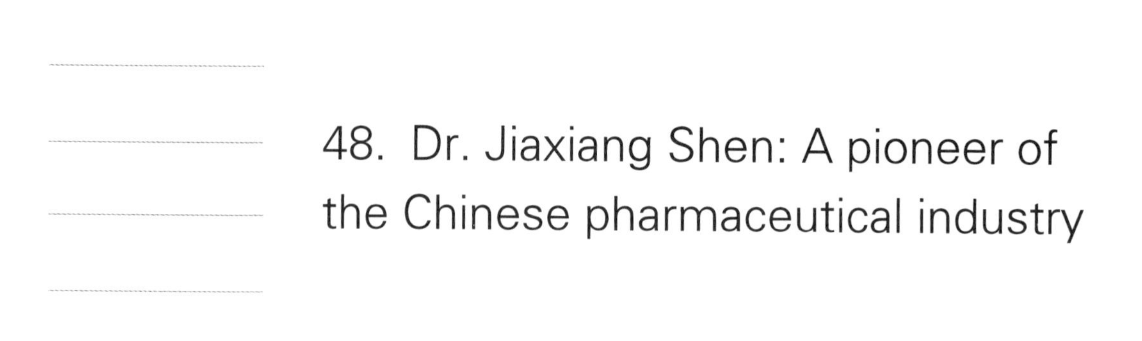

48. Dr. Jiaxiang Shen: A pioneer of the Chinese pharmaceutical industry

In 2007, Dr. Jiaxiang Shen received the "Outstanding Contribution Award" at the centenary celebrations of Chinese Pharmaceutical Association. Dr. Shen was acknowledged for his study on the new methodology for the total synthesis of chloramphenicol, the industrialization of several important steroid hormone drugs, and for being a co-founder of Chinese modern pharmaceutical industry.

Dr. Shen was born on November 11, 1921 in Yangzhou, Jiangsu Province, and later on moved to Nanjing with his parents. As a teenager, he studied at Nanjing Municipal High School where he was well trained. When the War of Resistance against Japanese Aggression fully broke out in 1937, the school was suspended. To make things worse, Dr. Shen got infected with tuberculosis, so he had no choice but moving to Chongqing with his family. One year later, with excellent exam scores, Dr. Shen was admitted to the National Advanced Pharmacy College (the predecessor of China Pharmaceutical University), which was just moved from Nanjing to Chongqing. During this time, Dr. Shen became extremely interested in Medicinal Chemistry with the influence of Prof. Xinghan Lei. This period of educational experiences, though turbulent but self-disciplined, laid foundation for his life-long path on pharmacy.

After graduation, Dr. Shen worked shortly in the Army Pharmaceutical Institute. Then, he went to England to study at the School of Pharmacy in University of London, U.K. It took him only four years to complete all the courses required for both bachelor and doctoral degrees. Dr. Shen received his PhD degree in Medicinal Chemistry from University of London in 1949 (Linnell and Shen, 1949, 1951). Immediately after graduation, he made up his mind to return back to China. He was so eager to devote himself to the development of the People's Republic of China that he couldn't even wait to receive the award of his PhD certificate. On September 23, 1949, Dr. Shen boarded on a ship and began the voyage back to his homeland via Hong Kong. He

Xianghai Guo[1,2], Baozhi Han[3]

1 Department of Pharmaceutical Engineering, School of Chemical Engineering and Technology, Tianjin University, Tianjin 300350, China

2 Key Laboratory of System Bioengineering, Ministry of Education, Tianjin 300072, China

3 Archives Department, Tianjin University, Tianjin 300350, China

Correspondence: guoxh@tju.edu.cn (X.-H. Guo)

became one of the first Western-trained Chinese scholars returning back to China. Since then, Dr. Shen has dedicated his entire life to the development of pharmaceutical industry in China and is renowned as one of the founding members of pharmaceutical industry.

During the Chinese People's Volunteers' departure for Korea to Resist US aggression and aid Korea, Dr. Shen took the task to lead the development of chloramphenicol, a badly-needed antibiotic for wounded Chinese soldiers. Under his guidance, a new synthetic method of chloramphenicol was developed. Within a few years, he achieved many significant technical innovations during the production processes, which significantly improved the manufacturing technique of chloramphenicol. In 1957, the new technique was successfully applied into large-scale production, which is hailed as an important milestone marking the beginning of modern pharmaceutical industry in China (Shen et al., 1950, 1958a, 1958b, 1958c, 1958d, 1958e, 1958f). In the 1950s and 1960s, he successfully synthesized crystalline Vitamin A acetate and Vitamin D2 using domestic resources, supervised the synthesis and production of multiple steroid medicines such as hydrocortisone and dexamethasone (Shen et al., 1964a, 1964b, 1964c), and accomplished the total synthesis of gestrinone, which laid a solid foundation for the industrial synthesis of 19-demethyl steroid drugs in China.

Late 1970s, Dr. Shen was reappointed as the Deputy Chief Engineer of Hunan Pharmaceutical Industry Research Institute. Shortly after that, he embarked on a new research topic based on effective constituents of Chinese Traditional Medicines (CTM). He demonstrated the unique chemical structure of agrimophol through the total synthesis method in 1976 (Shen et al., 1976a, 1976b), which was the active ingredient of an anti-tapeworm drug. In mid-1980s, he was appointed as the doctoral advisor in medicinal chemistry at Beijing Medical University (now part of Peking University). He supervised the PhD candidates on the study of the total synthesis of Tanshinone ⅡA and Danshenxinkun B (1986—1988) (Shen et al., 1988; Zhang and Shen, 1988), and discovered the special pharmacological activities in some derivatives. In addition, his group also studied the methodology of total synthesis of erycibe alkaloid Ⅱ (baogongteng A) and the analogues.

In late 1992, after his retirement, as the Deputy Chief Engineer of the State Pharmaceutical Administration of China and the Director General of the National Institutes of Pharmaceutical R&D, Dr. Shen established Beijing Jicai Pharmaceutical Research Institute. It was the first private pharmaceutical research institute in China, where he discovered a new crystal form of azithromycin. This discovery bypassed Pfizer's administrative restriction for azithromycin dehydrate in China. Later on, this product occupied the majority of domestic market share of Azithromycin with competitive price advantages, which substantially improved the access to the Chinese people. Additionally, his laboratory also overcame various technical challenges and developed for the first time in China various hard-to-synthesize specialty generics such as alfacalcidol, tibolone, budesonide, and tamsulosin.

With the development of China's science and technology, Dr. Shen became increasingly recognized in the international pharmaceutical field (Shen and Zhuang, 1984). Given his distinguished contributions and international reputation in the field of medicinal chemistry, Dr. Shen was elected as the Communication Academician of France Medication Academy in 1983. In the 1980s, Dr. Shen attended many conferences held by the United Nations Industrial Development Organization and the World Health Organization. In 1987, he was appointed as a member of the World Health Organization's leading group working on the chemotherapy for the treatment of Malaria and played a critical role in introducing the anti-malarial medicine artemisinine which was developed in China to the world (Luo and Shen, 1987; Shen, 1991). In 1988, he held the Directorship of Sino-American Searle Research Center Foundation. In 1989, he was invited to join the editorial board of *Medicinal Research Reviews* and the *Journal of Pharmaceutical Sciences*.

Dr. Shen took the development of pharmaceutical science, the promotion of China's pharmaceutical industry and the health of people as his own duty, and went through an extraordinary way along the stormy development path of Chinese pharmaceutical industry. He advanced the basic research and key technologies of Chinese pharmaceutical science, advocated the translation and application of research achievements. With his distinguished achievements, Dr. Shen won two prizes of National New Product Award (1964), five prizes of National Scientific Conference Award (1978), the Third Class of National Invention Award (1982), and the Second Class for Beijing Science and Technology Progress Award (1999). In 1999, he was elected as an Academician of Chinese Academy of Engineering.

Dr. Shen witnessed and experienced China's poverty during his youth, and he felt deeply humiliated by the aggressions of foreign powers. This inspired him to work hard for China's rejuvenation and empower the nation with modern science. He summarized his research style as "subject driven by mission" and "starting with easy things, but never leave without digging in great depth" . He always chose the research topics that are valuable to the development and progress of the country. He told his students many times, "As explorers of science, we must uphold the rule of self-reliance. Contributing to your country, rather than living on its support."

In 2001, at the age of 80, Dr. Shen gave up his quiet and comfortable life in Beijing and accepted the invitation to join Tianjin University with the hope of advancing pharmaceutical science and industry through education. He moved to Tianjin and co-founded the School of Pharmaceutical Science and Technology together with other colleagues at Tianjin University. During the school's early days, he always worked in the frontline regardless of his age and health. Everyone at Tianjin University was deeply inspired by his persevering spirit, which has become invaluable wealth to this institution.

Figures

Fig.1 Jiaxiang Shen at age 24, studying at University of London (1945)

Fig.2 Jiaxiang Shen was searching documents for chloramphenicol at a library (1957)

Fig.3 Dr. Shen was advising his graduate students at Beijing Medical University (1987)

References

Linnell WH, Shen CC (1949) Synthesis of the benzene analogues of vitamin A. J Pharm Pharmacol 1:971–986.

Linnell WH, Shen CC (1950) A note on Rupe's rearrangement. J Pharm Pharmacol 2(1):13–16.

Luo XD, Shen JX (1987) The chemistry, pharmacology and clinical applications of qinghaosu (artemisinin) and its derivatives. Med Res Rev 7(1):29–52.

Shen JX (1991) Antimalarial Drug Development in China. Beijing: China Medical Science Press.

Shen JX, Zhuang LG (1984) Current trends in new drug research in People's Republic of China. Med Res Rev 4(1):47–86.

Shen JX, Guo KY, Gao PM (1950) Synthesis of chloramphenicol (I). Chinese Science Bulletin 2:1167.

Shen JX, Zhang YQ, Zhou BW (1958a) Synthesis research of chloramphenicol II—VII. Acta Pharm Sin 6:207–209.

Shen JX, Guan JH, Yang QT (1958b) Synthesis research of chloramphenicol (III). New synthesis method of P-nitroacetophenone by the auto-oxidation of nitroethylbenzene. Acta Pharm Sin 6: 210–214.

Shen JX, Zhou BW, Pan FP (1958c) Synthesis research of chloramphenicol (IV). Research and improvement of epichlorohydrin aluminum reduction. Acta Pharm Sin 6: 218–219.

Shen JX, Xie K, Cai YZ (1958d) Synthesis research of chloramphenicol (V). Partition of DL-threo-1-parachloronitrobenzene-2-amino-1,3-propylene glycol. Acta Pharm Sin 6: 219–227.

Shen JX, Wang QF, Cai YZ (1958e) Synthesis research of chloramphenicol (VI). Racemization and reduction of L-α-dichloro acetyl-β-hydroxy-nitrobenzene acetone. Acta Pharm Sin 6: 308–311.

Shen JX, Cai YZ, Pan FP (1958f) Synthesis research of chloramphenicol (VII). Synthesis research of methyl dichloroacetate. Acta Pharm Sin 6: 312–315.

Shen JX, Li TS, Wang QF (1964a) Steroid hormone. Acta Pharm Sin 11(3):194–197.

Shen JX, Wang QF, Cai YK (1964b) Steroid hormone II. Acta Pharm Sin 11(3):156–161.

Shen JX, Chen YY, Zhang XD (1964c) Steroid hormone III. Acta Pharm Sin 11(4):242–245.

Shen JX, Ning DZ, Zhang LY (1976a) Complete synthesis of agrimophol. Chin Herb Med Commun 6:5.

Shen JX, Ning DZ, Zhang LY (1976b) Complete synthesis of agrimophol. Acta Chim Sin 34:313.

Shen JX, Zhang PZ, Qiao M (1988) New complete synthesis method of tanshinone IIA, the effective constituent in CTM salvia. Acta Pharm Sin 23(7):545–548.

Zhang PZ, Shen JX (1988) New complete synthesis of Danshenxinkun B, the effective constituent in CTM salvia. In: Proceedings of the Third National Conference of Natural Pharmaceutical Chemistry (Shanghai).

49. 陶慰孙：中国生物化学的先驱者

图1　陶慰孙（1895—1982）

陶慰孙，生物化学家和教育家，我国蛋白质化学研究奠基人之一（图1）。大同大学化学系、吉林大学化学系和生物系的主要创建者，吉林大学有机化学和生物化学学科的奠基人。曾任大同大学、东北工学院（现东北大学）和吉林大学教授，中国生物化学会第一、二届理事会名誉理事，第三届吉林省妇女联合会副主任，第三届全国人大代表，第五届全国政协委员，第四届吉林省政协副主席。

陶慰孙1895年2月20日出生于江苏省无锡市的一个大家庭中，1902年入上海务本女塾小学学习，1906年跟随在日本明治法律学校留学的父亲赴日本，在日本完成小学和中学学业后，于1914年进入日本东京女子高等师范学校学习，1918年毕业后回国，在北京女子高等师范学校任教，讲授物理和化学课程。1919年考取官费赴美国留学，1921年获哥伦比亚大学食品化学理学硕士学位，1923年获康奈尔大学教育学硕士学位（陶乃煌，2003）。后曾在英国、德国、法国、比利时、荷兰和瑞士等国著名化学研究机构参观学习。1923年冬季回国，在上海大同大学担任化学教授。1927年9月再次赴日本，入京都帝国大学理学部，在日本著名有机化学家Shigeru Komatsu指导下攻读博士学位，以东方各国主要粮食水稻为研究对象，系统研究了水稻发芽前后储藏淀粉的化学变化，不同温度下水稻发芽时糖类的变化，及不同温度下淀粉酶对淀粉的水解作用，相关结果以系列论文的形式于1930年在*Bulletin of the Chemical Society of Japan*上发表。之后又对大豆蛋白进行了研究。1931年秋，陶慰孙提前回国继续在上海大同大学任教授，1932年7月陶慰孙以《水稻淀粉的生物化学研究》为学位论文获理学博士学位（小松茂，1996），成为第一个在日本获得博士学位的中国女性。

1932年，陶慰孙在大同大学创建化学系，并任化学系主任，鉴于当时条件，工作

作者：贺天伟
吉林大学生命科学学院，长春130012，中国
邮箱：hetw@jlu.edu.cn

主要以有机化学、生物化学教学为主。她开课十余种，为人慈祥和蔼，有“好妈妈”之称。1934年，她对成熟的黄岩橘化学成分和福橘果实精油进行研究。1937年因为对大同大学的卓越贡献，她被选为大同大学核心管理层——立达学社社员。她还曾先后在上海中法大学药学专修科和上海交通大学兼职授课。1935—1944年兼任上海自然科学研究所研究员。1940年，研制出我国第一批注射用葡萄糖。20世纪40年代，她参与创建上海大同化学工业社和上海一心化学制造厂，先后生产纯盐酸、硫代硫酸钠、水银、液溴等化学试剂和天冬酰胺、胱氨酸、固体麦芽糖等生化试剂，为我国化学试剂工业作出了开拓性的贡献。

新中国成立以后，1950年暑期陶慰孙和丈夫关实之到东北旅行，夫妻二人耳闻目睹了东北地区在党的领导下大力建设工业基地以及取得的显著成就，深受感动，决定响应党的“支援东北地区教育与科学事业”的号召，放弃上海舒适的生活条件前往沈阳，9月任东北工学院化工系教授。1952年全国范围进行高校院系调整，国家决定在东北建立一所新型综合性大学，10月陶慰孙服从教育部安排来到长春，与蔡镏生教授、唐敖庆教授和关实之教授一起创建了东北人民大学（现吉林大学）化学系。化学系初创时设有无机化学、分析化学、有机化学和物理化学四个专业，陶慰孙任有机化学教研室主任（图2）。当时的东北人民大学是一所培养党政干部的学校，未曾开设过数理化课程，创系初始连实验室、实验仪器和药品都没有，房屋紧张就利用地下室作为实验室，没有实验台就用木板搭建实验台，用旧墨水瓶改装酒精灯，在较短的时间内就安排学生做实验。1958年8月东北人民大学更名为吉林大学。1960年6月，为了满足东北地区对生物人才的需求，陶慰孙在吉林大学创建了生物系，下设生物化学专业和生物物理专业。建系初始，面对师资缺乏的局面，陶慰孙一方面组织从北京大学、中山大学、厦门大学和山东大学等院校分配来的青年教师开设课程，另一方面还聘请吉林师范大学（现东北师范大学）的专业教师来校兼职，很快就使生物系的教学走上正轨。1962年9月，教育部决定撤销吉林大学生物系，应陶慰孙的要求生物化学专业被保留，成为化学系的一个专业，使吉林大学化学系成为国内最早建立生物化学专业的化学系。在50、60年代，陶慰孙针对东北地区植物资源，开展了向日葵种子、棉籽、南瓜种子、赤豆和大豆等植物蛋白质的研究工作，先后发表十余篇研究论文，1964年制成“赤豆

图2　陶慰孙教授在实验室指导学生

蛋白”和“大豆豆酪素”。陶慰孙1972年12月返回吉林大学，立即投入固氮酶的结构与功能研究的指导工作中，在她与同事的刻苦努力下，研究取得了新进展，发表了一批相关论文。

80多岁时，陶慰孙又满腔热忱地投入为国家培养人才的工作中。针对教师队伍水平不高的实际情况，她积极联系，送中青年教师去国内外高水平的科研机构攻读学位或进修，利用自己精通日语和英语的优势，辅导青年教师和研究生学习外语，帮助中青年教师确定科研方向，组织中青年教师为期刊撰写理论文章和编写教材，其中《蛋白质分子基础》于1981年6月由高等教育出版社出版，这是国内最早的一部蛋白质化学方面的专业教材，被国内同行广泛使用。陶慰孙教授积极参加学术活动，1979年和1981年还分别参加了在杭州和南宁召开的全国生物化学学术会议，并被中国生物化学会第一届和第二届理事会推选为荣誉理事。

1982年12月11日，陶慰孙因病在吉林省长春市去世，享年87岁。陶慰孙去世后，关实之教授按照她生前的遗愿，将二人多年的积蓄2.4万元捐给了吉林大学，吉林大学决定在化学系设立“陶慰孙奖学金”，奖励和资助品学兼优的本科生。

经过几代人的辛勤耕耘，吉林大学化学学科现已成为一级学科国家重点学科，生物化学与分子生物学现已成为二级学科国家重点学科。2012年8月，吉林大学举行了纪念化学学科建立六十周年的活动，四座新完成的蔡镏生、唐敖庆、关实之和陶慰孙雕像（图3）被安放在校园里，人们以这种形式来纪念这些为吉林大学作出突出贡献的人。

图3 吉林大学陶慰孙教授半身像

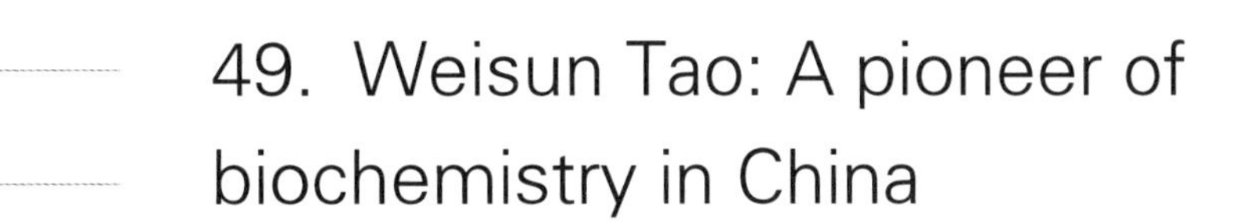

49. Weisun Tao: A pioneer of biochemistry in China

Professor Weisun Tao is a distinguished educator, a pioneering scientist of biochemical researches and one of the founders of the protein chemistry research in China. She is a main founder of the Department of Chemistry at Utopia University, the Department of Chemistry and the Department of Biology at Jilin University. She was appointed as a professor in Utopia University, Northeast Institute of Technology, Jilin University, and the honorary director of the Chinese Biochemical Society. She also established the disciplines of organic chemistry and biochemistry at Jilin University.

Prof. Tao was born in 1895 in Wuxi, Jiangsu Province. In 1902, she attended elementary school in Shanghai. She went to Japan with her father in 1906, who was then studying in Meiji Law School (now Meiji University). In 1914, after finishing her elementary and middle school studies, Prof. Tao was enrolled in Tokyo Women's Higher Normal School. After graduation, she came back to China in 1918, and started teaching chemistry in Beijing Women's Higher Normal School. In 1919, Prof. Tao went to America for further education. She obtained Master's Degree of Science from Columbia University in 1921 and Master's Degree of Education from Cornell University in 1923 (Tao, 2003). During that time, Prof. Tao has visited and studied at multiple famous chemical research institutions in the United Kingdom, Germany, France, Belgium, the Netherlands, and Switzerland. In the winter of 1923, she came back to China and was appointed as a professor in chemistry at Utopia University in Shanghai. In September 1927, Prof. Tao decided to go to Japan again for the doctoral study under the guidance of Shigeru Komatsu, a famous organic chemist in the College of Science, Kyoto Imperial University. There, Prof. Tao carried out a systematic study on the chemical change in the starch storage before and after rice germination, sugars change in rice germination at different temperatures, and hydrolysis of starch by diastase at different temperatures. In the autumn of 1931, Prof. Tao went back to China and continued to work as a professor of chemistry in Utopia University. In July 1932, Prof. Tao completed her doctoral dissertation entitled "Biochemical studies on rice starch" and received

Tianwei He

School of Life Sciences, Jilin University, Changchun 130012, China

Correspondence: hetw@jlu.edu.cn

her Doctoral Degree of Science (Komatsu, 1996). She was the first Chinese woman who obtained a doctoral degree in Japan.

In the same year, she founded and served as the head of the Department of Chemistry. During that time, Prof. Tao took education as her priority and taught more than ten courses. Besides teaching, she also conducted researches on chemical components of ripe Huangyan orange and the essential oil in the ripe Fu orange fruit. From 1935 to 1944, Prof. Tao also held the position as a researcher in the Shanghai Science Institute. In 1940, Prof. Tao developed the first batch of glucose that can be used for injection in China. In the 1940s, Prof. Tao participated in the establishment of Shanghai Datong Chemical Industry Factory and Shanghai Yixin Chemical Manufacturing Factory, which was a pioneering contribution to chemical reagent industry in China.

After the founding of the People's Republic of China, in September 1950, Prof. Tao gave up her comfortable life in Shanghai and went to Shenyang to join in the Department of Chemical Engineering in Northeast Institute of Technology as a professor in response to the encouragements to support the development of education and science in the northeastern area. In October 1952, in accordance with the arrangements of the Ministry of Education, Prof. Tao went to Changchun and founded the Department of Chemistry in the Northeast People's University together with Profs. Liusheng Tsai, Auchin Tang, and Shizhi Guan. Among the four majors available in the Department of Chemistry: inorganic chemistry, organic chemistry, analytical chemistry, and physical chemistry, Prof. Tao was mainly responsible for the establishment of organic chemistry major. At the beginning, it was very difficult. There were no laboratories, experimental instruments and reagents. Researchers have to conduct experiments in the basement: they used wooden boards to build the benches and used inkbottles as alcohol lamps. With all those efforts, students were finally able to do experiments. In August 1958, Northeast People's University was renamed as Jilin University. In June 1960, Prof. Tao established the Department of Biology with two majors: biochemistry and biophysics. To solve the problem of faculty shortage, Prof. Tao arranged young teachers from Peking University, Sun Yat-sen University, and Shandong University to give courses in Jilin University. She also invited teachers from Jilin Normal University (now Northeast Normal University) to work as adjunct teachers. In September 1962, the Ministry of Education decided to remove the Department of Biology in Jilin University. Upon Prof. Tao's request, the major of biochemistry was kept and transferred to the Department of Chemistry, making it the first chemistry department with biochemistry as a major in China. During 1950s and 1960s, Prof. Tao focused her researches on plants in the northeast regions and carried out researches on sunflower seeds, cottonseeds, pumpkin seeds, red beans, soybeans, and other plant proteins, and published more than ten research articles. She came back to Jilin University in December 1972 and immediately devoted herself into the study of the structure and function of nitrogenase. Together with her colleagues, Prof. Tao made great progresses on this study and published a series of research articles.

Prof. Tao at the age of 80 years old was still full of enthusiasm in education. She encouraged the young teachers to go for advanced studies and trainings in high-level research universities and institutions worldwide to increase the quality of the faculty. Prof. Tao also helped the young teachers with their foreign language studies, guided them to choose their research focuses and encouraged them to write articles and compile textbooks. *The Molecular Basis of Protein*, a textbook about protein chemistry, was published by Higher Education Press in June 1981 and was widely used by peers within the discipline. Prof. Tao played an active role in academic activity. In 1979 and 1981, she participated in National Biochemistry Conference held in Hangzhou and Nanning respectively, and was elected as honorary director in the first and second conferences.

On December 11, 1982, Prof. Tao passed away at the age of 87 in Changchun, Jilin Province. According to her will, her husband Prof. Shizhi Guan donated their savings of 24000 RMB to Jilin University for establishing the Weisun Tao Scholarship to sponsor the distinguished undergraduates in the Department of Chemistry.

After several generations of hard work, the chemistry discipline in Jilin University has now become Level I national key discipline, the biochemistry and molecular biology discipline in Jilin University has now become Level Ⅱ national key discipline. In August 2012, commemorative activities were held for the 60th anniversary for the Department of Chemistry in Jilin University. The busts for Profs. Liusheng Tsai, Auchin Tang, Shizhi Guan and Weisun Tao were built on campus in honor of their tremendous contributions to the university.

Figures

References

Komatsu S (1996) Doctor of science: Ms. Weisun Tao. In: Jilin University. In Memory of the 100th Anniversary of Shizhi Guan and Weisun Tao. Changchun: Jilin University Press. (小松茂. 1996. 理学博士陶慰孙女士. 见: 吉林大学. 关实之陶慰孙百年诞辰纪念文集. 长春: 吉林大学出版社.)

Tao NH (2003) Family biography of Tao Tingfang of Beitang, Wuxi. In: You XM, Zhao YL. Biographies of Renowned Families and Celebrities of Wuxi. Harbin: Heilongjiang People's Publishing House. (陶乃煌. 2003. 无锡北塘陶廷枋氏家传. 见: 尤学民,赵永良. 无锡望族与名人传记. 哈尔滨: 黑龙江人民出版社.)

50. 许豪文：中国运动生物化学的奠基人

许豪文教授是我国运动生物化学学科的奠基人（图1），一生致力于运动科学、实验生物学、运动医学等方面的相关研究。许教授曾任国务院学位委员会学科评议组成员，是华东师范大学首批终身教授之一（华东师范大学体育与健康学院，2015），为我国运动生物化学事业的发展作出了巨大贡献（俞立中，2006）。

图1　许豪文教授（1935—2004）

许豪文教授，浙江绍兴人，1935年出生于江苏南京。他的父亲是知名卫生统计学家、我国生命统计及卫生统计事业的奠基人许世瑾先生（上海医科大学公共卫生学院，1988）。由于生长在书香世家且受父亲的影响颇深，许教授自小对医学耳濡目染，产生了浓厚的兴趣。1954年，许教授选择进入上海第一医学院医疗系（现为复旦大学上海医学院）学习医学并于1956年加入中国共产党。1958年，为响应“苦战一月，基本消灭钩虫病和丝虫病”的号召，上海第一医学院的学生纷纷自觉组织起来进行支援。当时，许豪文担任孔桥乡突击队队长。由于在骑自行车时不慎跌伤了左手，大家劝说他休息，他却说：“完成了任务再说。”1959年，许教授在顺利完成了大学学业的同时，被送往北京外国语学院（现北京外国语大学）留苏预备班深造。1961年，因国情变化，许教授等多名毕业生被安排至北京体育科学研究所（现国家体育总局体育科学研究所）就职。那个年代，肾上腺皮质激素的测定技术十分复杂，在国内运动生理学研究中属首次应用，需要收集运动后48小时的尿液并测定总量。许教授就是在这种条件下不嫌脏臭，即使没有一副防护手套，也坚持与杨天乐、秦孝梅亲自完成研究，并将课题整理成文，于1964年全国体育科学报告会上进行报告，获得了一致好评。不仅如此，他与杨天乐、秦孝梅一起积极建设运动生化实验室，从事内分泌与生化研究。1978年，许教授被调至上海体育科学研究所工作，继续从事体育科研和竞技运动研究。期间，发表论文达数十篇，主持上海市体育局课题一项，并获国家

译者：龚伟
浙江师范大学，金华321004，中国
邮箱：yzugw@163.com

体育运动委员会体育科学技术进步奖四等奖和上海市科学技术进步奖三等奖。1984年7月，许教授被调至华东师范大学体育系工作，相继担任华东师范大学体育系副主任、学校体育研究所副所长，华东师范大学学位委员会委员、华东师范大学职称评审委员会委员等要职，此后许教授长居上海。1992年，他被批准享受国务院特殊津贴，直至2004年逝世。

20世纪80年代前后，许教授意识到我国在运动人体科学方面的研究相当薄弱。为此，他广泛涉猎国外优秀成果，积极开展有关的运动生物化学实验，并将研究成果汇集成一系列论文予以发表，使得运动生理学进入了一个崭新阶段，奠定了其在全国体育科研界的权威地位。血清磷酸肌酸激酶（SCPK）与尿素氮（SUN）分别是参与人体能量代谢的一种酶以及血浆中除蛋白质以外的一种含氮化合物，在经过不同强度与不同性质的运动后，其含量会有所变化。国外的研究只将其停留在实验室条件下观察这两个指标的变化，而许教授迈出了开拓性的一步，他监测真实的训练状态时SCPK与SUN的变化，形成了《不同项目运动应激后血清磷酸肌酸激酶和尿素氮的变化》一文并于1981年发表于《体育科研》，加深了国内外对SCPK与SUN变化规律的认识（许豪文等，1981）。90年代，对细胞、亚细胞结构的脂质膜在运动中受自由基的作用及功能的探索深化了学者对酶及非酶系统的直接防御作用的了解，但针对在运动中自由基与细胞酶蛋白及结构蛋白的关系阐述却相当缺乏，尤其是对谷胱甘肽转硫酶（GST）这样有各种功能的酶的观察较少。察觉到这一点，许教授以大鼠心脏、骨骼肌、肝脏为材料，观察了大鼠游泳320分钟后自身脂褐素和GST的活性，发现游泳使大鼠心肌、骨骼肌、肝脏的脂褐素呈下降趋势并以心肌下降最明显。另一方面，GST活性也呈下降趋势，以肝脏最为显著（丁树哲等，1992）。针对运动与蛋白质相结合，许教授还探究了运动与胰岛素受体及相关信号传导蛋白之间的关系，得出了“运动会引起骨骼肌胰岛素作用的提高与胰岛素信号传导能力的提高，并伴随胰岛素信号蛋白IRS-1/2表达的下降”这一结论，为运动训练作为治疗人胰岛素抵抗和糖尿病的方法找到了理论依据（孙清涛和许豪文，2002）。不仅如此，丝裂原活化蛋白激酶（MAPKs）信号系统作为存在于大多数原核生物和所有真核生物中的一组非保守的丝氨酸双重磷酸化蛋白激酶，在细胞生物信号传导中起着至关重要的作用。许教授针对运动对其的影响进行了研究，尤其对ERK1/2、JNK/SAPK、p38的研究最为深入，发现运动能激活骨骼肌中的MAPKs信号系统，并且不同运动方式与不同类型的肌肉会不同程度地影响MAPKs的激活以及激活后的时相性。此外，他还将这一成果撰文发表于《体育学刊》。当然，许教授将运动与蛋白相结合的研究不仅限于此。

许豪文教授十分关注民生问题，他认为健康的体魄必不可少。1982年，许教授同杨天乐、秦孝梅通过观察不同负荷的运动训练和不同情绪的比赛研究了运动对尿17-羟类固醇排泄量的影响并将研究成果发表于《生理科学》。荣获1985年国家体育运动委员会体育科学技术进步奖四等奖的“短跑运动员生化机能评定”也是最好的证明之一。为提高运动员身体健康水平与运动能力，许教授从短跑运动员大量运动训练及比赛后酶的变化入手，探讨了短跑运动量的合理性，并逐步向国民普及这些常识（许豪文等，1983）。另外，许教授还针对某些我们已知但无法为之正名的“事实”探讨研究，例如相较于女孩，男孩更喜欢或者擅长体育运动。但是，早期没有研究者运用科学的方法加以解释。直到1986年，许教授对其进行多方面实验分析并向世人揭示了其中真正的“奥秘”——雄性激素中的睾酮与运动能力的关系。研究阐明，青春发育期后，男孩血浆睾酮浓度迅速因年龄增长而增加，但女孩仍保持原来的水平，这对男女成人的生理和运动能力差别的形成起了相当重要的作用。同时，运动反过来又增加了雄性激素的分泌，增强了雄性激素的作用，使男孩的运动能力得到进一步的增强（许豪文，1986）。

此外，许教授还是一个敢于提出质疑的人。他提出了与当时的美国艾滋病防治中心相悖的观点，即适量参与体育运动对老年人某些疾病的预防与康复有一定的价值，但过度运动会适得其反，如增加感染风险、病情恶化等，并于1993年撰文发表在以实用为主、指导临床的综合性老年医学杂志《实用老年医学》中，旨在提倡国民根据具体情况制定合理的运动方案，开创了我国运动与自由基生物学、运动与衰老科学研究的先河（许豪文，1993）。冯炜权教授在《运动生物化学研究进展》一书中曾评价许教授关于衰老和体育锻炼方面的研究“对现实生活具有极大的意义”（冯炜权等，2006）。这又一次证实了体育锻炼有助于消除损害并延缓衰老。

许教授不仅自己毕生致力于运动生物化学的研究，他的夫人周红律同样也作出了重大的贡献。21世纪初，他们展开了针对“不同强度的运动与雌激素联合作用对去卵巢大鼠股骨上端的超微结构会产生何等影响”的探讨，并将其汇集成研究成果发表了论文。不仅如此，那个年代关于骨质疏松、维生素K与运动的关系一直都存在疑问，直到《骨质疏松、维生素K和运动》的发表揭示了真正的谜底，即运动可以通过多种途径来增加骨无机盐密度，但激烈的有氧运动会对骨密度产生负面影响，甚至影响下丘脑、垂体等；而维生素K的缺乏会导致骨钙蛋白中钙的结合能力有所下降，从而进一步影响骨代谢（许豪文和周红律，2003）。当然，他们两人的研究成果还有许多，填补了我国运动生物化学研究的空白，受到了人们的重视。

许豪文教授一生出版学术专著十余部，涉及运动生物化学研究的著作包括《运动生物化学概论》《运动生物化学》《人体生理学》《运动医学》，这些都成了运动生物化学专业研究者的重要读本。其中，2001年出版的《运动生物化学概论》是唯一获得教育部研究生工作办公室推荐的体育专业教学用书（中华人民共和国教育部研究生工作办公室，2000）（图2）。该书由许教授独自撰写完成，主要介绍了运动时体内的物质代谢与能量代谢的规律，以及运动性疲劳和竞赛中兴奋剂的禁用等，吸收了当时运动生物化学领域最新的学术研究成果，具有较强的理论深度和前瞻性，是研究生开展研究工作的重要指导用书。此外，由于我国运动人体科学的研究基础比较薄弱，许教授还积极向国内介绍国外运动生物化学的学术成果，为我国运动生物化学学科的发展起到了一定作用。例如，许教授在《山西体育科技》杂志连续发表了50多篇关于运动性疲劳的国外最新研究进展的文章；许教授参与翻译了《运动和训练的生理化学》（图3）。

许教授还是我国运动医学界第一本学术刊物《中国运动医学杂志》的编委。20世纪80年代，由于我国运动医学水平与体育事业的发展需求不协调，中国体育科学学会特于1982年4月创办了《中国运动医学杂志》，在国内公开发行，作为运动医学工作者发表各类科学研究成果和进行学术交流的园地，以深化研究如何有效防止运动损伤，大量运动训练后如何加快运动员身体恢复等问题。许豪文教授凭借自身扎实的知识功

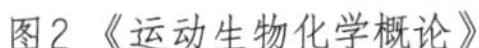
图2 《运动生物化学概论》

图3 《运动和训练的生理化学》

底担任编委。不仅如此，他还是中国体育科学学会理事，中华医学会运动医疗分会委员，中国康复医学会康复体育保健委员会荣誉主任委员。此外许教授还是武汉体育学院、北京体育师范学院（现为首都体育学院）、苏州大学等学校的兼职教授与科学训练顾问。1999年，河北师范大学为争取博士学位授权点做准备，为了扩大教师知识面，提高教学科研水平，于12月15日邀请了许教授等多位专家讲学。许教授针对“十五体育教育重点课题的展望”进行了概述，并与各教研室主任和研究生导师就“如何指导研究生”及“体育学科的学科建设与内容”开展了讨论。不仅如此，2000年5月15日，许教授到苏州大学体育学院开展题为“运动对恢复的作用”的讲座，获得了师生的好评。

许教授不仅是一位杰出的科研工作者，同时还是一名注重教育发展与人才培养的人民教师。在许教授的积极筹措与主持下，华东师范大学体育与健康学院的学科建设取得了全国领先的成就：1986年设立全国第一个运动生物化学硕士点，1993年设立全国第一个运动生物化学博士点（运动生物化学学科现更名为运动人体科学）（俞立中，2006）。许教授在华东师范大学工作期间，为我国培养了60余名硕士生、20余名博士生，于1999年荣获上海市育才奖。华东师范大学的丁树哲教授就是其中之一，他是许教授的第一个硕士生，现为中国生物化学与分子生物学会会员，中华医学会运动医疗分会委员，华人运动生理与体适能学者学会理事，坚持将线粒体调控作为自己的主要研究方向，积极推动我国运动生物化学事业继续发展（丁树哲等，1991）。

许豪文教授在学术钻研、人才培养和教育发展等方面的杰出贡献奠定了他在我国体育科研界的权威地位。与此同时，许教授求真务实的科研作风、一丝不苟的科研精神、成果共享的科研人格值得广大科研工作者学习。

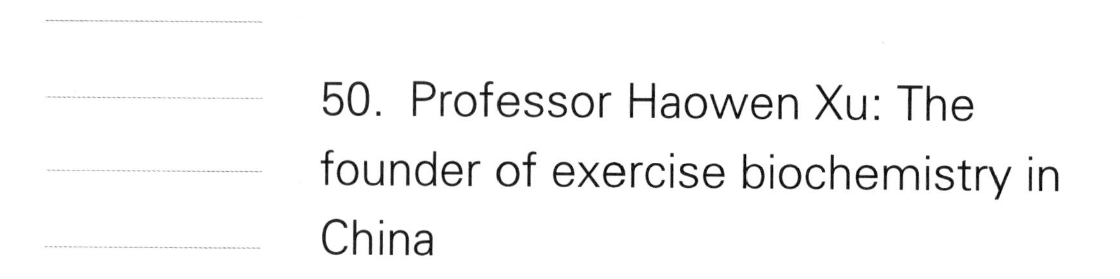

50. Professor Haowen Xu: The founder of exercise biochemistry in China

Professor Haowen Xu (许豪文, 1935—2004) was a famous sports scientist and the founder of Exercise Biochemistry in China. He dedicated his life to the study of sports science and made tremendous contributions to the formation and growth of the field of Exercise Biochemistry in China. He was the first to use theories of exercise biochemistry and sports medicine to guide athletes' training and competition in China (Yu, 2006).

Prof. Xu was born in 1935 to a family of scholars in Nanjing, China. While he was a child, Prof. Xu was an accomplished student and developed a deep interest in medicine. In 1959, Prof. Xu received his bachelor's degree as an outstanding graduate from the Department of Medicine and Therapeutics at Shanghai No. 1 Medical College (now known as Shanghai Medical College of Fudan University). Prof. Xu responded to a national call for the development of sports science and entered the Research Institute of Sports Science, part of the Physical Culture and Sports Commission of the People's Republic of China (now known as China Institute of Sports Science). There, he became involved in the brand-new field of real-time monitoring of athletes' training. During this time, Prof. Xu, in cooperation with Tianle Yang (杨天乐) and Xiaomei Qin (秦孝梅), studied the variation of urine 17-hydroxycorticosteroid (17-OH), which was affected by different training loads and emotional stress in the 48 h following exercise. In 1978, Prof. Xu was transferred to the Shanghai Research Institute of Sports Science. There he was responsible for the project Biochemical Assessment of Exercise Dose of Sprinters and reported several important findings. During this period, Prof. Xu was awarded the Fourth-Class Prize for Scientific and Technology Progress in Sports by the Physical Culture and Sports Commission of the PRC and the Third-Class Prize for Scientific and Technology Progress of Shanghai. In 1984, Prof. Xu continued his research at East China Normal University (ECNU). After that, he was successively appointed as the deputy director of the Department of Physical Education, the deputy director of

Wei Gong[1,2], Yijing Shen[1], Jiaqi Bao[1], Yike Ying[1], Han Zhou[1], Zhifeng Wu[2]

1 Zhejiang Normal University, Jinhua 321004, China

2 East China Normal University, Shanghai 200062, China

Correspondence: yzugw@163.com (W. Gong)

the Research Institute of Sports Science, and a member of the Academic Degree Committee and the Professional Title Evaluation Committee. As an academic leader at ECNU, Prof. Xu made significant positive contributions to the construction and development of the physical education curriculum. Given Prof. Xu's outstanding achievements, he was awarded the Special Government Allowances of the State Council in 1992.

During the 1980s, Prof. Xu published several high-quality research articles in academic journals such as *China Sports Science, Journal of Physical Education*, *Sports & Science*, among others. In 1981, Prof. Xu explored the variations in the laws of Serum Creatine Phosphate Kinase (SCPK) and Serum Urea Nitrogen (SUN) resulting from different exercise intensities, durations, and types in a real state rather than laboratory conditions, which represented a world-wide breakthrough in this area (Xu et al., 1981). In 1992, Prof. Xu compared the activity of lipofuscin and glutathione S-transferase (GST) after swimming for 320 min in a rat model. This research proved that the activities of lipofuscin as well as GST decreased in the heart, skeletal muscle, and liver after swimming, which provided more detail regarding the interactions between free radicals and cellular zymoprotein and structural proteins during exercise (Ding et al., 1992). During the early 21st century, Prof. Xu focused on the relationships between exercise, insulin receptors, signal transduction proteins. His research revealed that exercise resulted in the improvement of skeletal muscle insulin effects and insulin signal transduction ability, which were accompanied by decreases in the expression of IRS-1/2. This work helped show the great potential for exercise training as a therapeutic treatment for type Ⅱ diabetes (Sun and Xu, 2002).

Prof. Xu was also concerned with public health throughout his life. In the early 1980s, by monitoring enzyme changes in athletes during exercise, Prof. Xu discussed the rationality of exercise dose and shared his results with the public (Xu et al., 1983). In 1986, he found that the concentration of plasma testosterone increased as boys entered adolescence. In addition, androgenic hormones promoted the development of boys' exercise capacity, which promoted further secretion of androgenic hormones. In contrast, the level of androgenic hormone remained basically unchanged in girls as they entered adolescence, which helped to explain differences in exercise capacity between adolescent boys and girls. This important revelation laid the theoretical foundation for different gender groups participating in different forms of physical exercise (Xu, 1986). Since the 1990s, there has been a fervent belief that participating in sports could delay human aging and decrease causes of increased morbidity, such as tumors. In an attempt to rectify this onesided view, Prof. Xu released a paper entitled Rehabilitation and immunity of geriatric diseases, and made the point that excessive exercise could lead to unfavorable consequences, such as increasing the incidence of infection, enhancing the sensitivity of tumor onset, or aggravating a disease's course. Therefore, he proposed that people should choose reasonable exercises according to their personal situation (Xu, 1993).

Prof. Xu dedicated his entire life to the field of exercise biochemistry and contributed to more than ten monographs, such as *An Introduction to Exercise Biochemistry*, *Human Physiology*, and *Sports Medicine*, among others. *An Introduction to Exercise Biochemistry*, published in 2001, was the most representative monograph written by Prof. Xu independently. This book comprehensively introduced the laws of substance metabolism and energy metabolism in the human body during exercise, as well as sports fatigue, drug abuse in competition, and more. It was recommended as a Postgraduate Teaching Book by the Postgraduate Working Office of the Ministry of Education in China and was the only such book within the field of physical education (The Postgraduate Working Office of the Ministry of Education in China, 2000). Prof. Xu also played an energetic role in introducing advancements and achievements made abroad to China. For example, he consistently recommended dozens of foreign research papers on sports fatigue in the journal *Shanxi Sports Science and Technology* starting in 1989. Additionally, Prof. Xu co-translated the conference proceedings of the First International Physiological Chemistry Seminar on Sports Training in a collection entitled *Physiological Chemistry of Exercise and Training*, with Weiquan Feng (冯炜权), Ming Hua (华明), Kuisheng Yang (杨奎生) and Bubiao Wang (王步标).

Prof. Xu enjoyed a distinct status in the field of sports science and exercise biochemistry in China. He was elected as the director of the China Sports Science Society, an editor of the *China Journal of Sports Medicine*, an honorary director of the Chinese Association of Rehabilitation Medicine, as well as a member of the Discipline Evaluation Group of the Academic Degrees Committee of the State Council. In addition, Prof. Xu was also an adjunct professor and scientific training consultant at Beijing Sports Normal College (now known as Capital University of Physical Education and Sports), Wuhan Sports University, Guangzhou Municipal Sports Work Brigade (now known as Guangzhou Polytechnic of Sports), and Soochow University. Despite his outstanding achievements and standing within his field, Prof. Xu never regarded himself as a so-called "academic authority" and always insisted on academic freedom and sharing achievements with young scholars, promoting an atmosphere of creativity and academic prosperity.

Prof. Xu placed great importance on instruction within his discipline and the cultivation of talent, and was one of the first tenured professors at ECNU (The School of Sports and Health of East China Normal University, 2015). Due to Prof. Xu's efforts, ECNU established the first master's program in exercise biochemistry (now renamed as kinesiology) in 1986 and the first doctoral program in exercise biochemistry in 1993 (Yu, 2006). In total, he mentored and cultivated more than 60 master's students and 20 doctoral students at ECNU. Prof. Xu and his students worked together to overcome many difficulties in the field of exercise biochemistry, and became a significant academic force in the field of sports science in China. Professor Shuzhe Ding (丁树哲) of ECNU was Prof. Xu's first master's student, whose main research interests now are exercise adaptation and mitochondrial signal control. One of his outstanding contributions was being the first person to observe super oxygen free radicals (O_2^-) in rat myocardia after fatiguing exercise using Electron Spin Resonance (ESR) (Ding et al., 1991). He also reported other important findings about the effects of exercise on the structure and function of mitochondrial membranes, and the influences of aerobic training on mitochondrial DNA (mtDNA) and mitochondrial NO synthesis (Feng et al., 2006). Associate Professor Gang Zhou

(周刚) of the Physical Education Institute at Hunan University was Prof. Xu's last doctoral student. His research in exercise biochemistry focuses on exercise-induced oxidative stress, the mechanisms of sports fatigue, and physical fitness. His research results have been published in domestic and foreign journals (Zhou et al., 2009; Zhou et al., 2012).

Prof. Xu's remarkable achievements were made possible by the support and example set by his family. Prof. Xu's father, Professor Shijin Xu (许世瑾), was the founder of life statistics and health statistics in China. He devoted his attention to the health problems of other people throughout his life and adopted the methods of medical statistics to study the causes of death among Chinese residents (School of Public Health Shanghai Medical University, 1988). In 1935, he built the earliest reporting system for infectious diseases and parasitic diseases in China, with 204 hospitals nationwide. Consequently, the incidence and geographical distribution of 19 different kinds of diseases were investigated and counted, which provided important information for the prevention and control of infectious diseases and parasitic diseases in China (Xu and Ge, 1937). Prof. Xu's wife, Professor Honglyu Zhou (周红律), was also a well-known exercise biochemistry expert. Using advanced technology, the couple carried out multiple studies, such as osteoporosis, vitamin K and exercise. This study revealed that intense aerobic exercise had a negative impact on bone mineral density, and that lack of vitamin K could lead to a decline of calcium-binding capacity in osteocalcin, which helped to begin to interpret the relationships between osteoporosis, vitamin K, and exercise (Xu and Zhou, 2003).

Sadly, Prof. Xu passed away in Shanghai in December 2004. He was a ground-breaking exercise biochemist and made many influential contributions to the development of sports science in China, including academic research, disciplinary development and most importantly, the cultivation of young talents. Prof. Xu was a rigorous scholar with remarkable achievements and a modest, easy-going educator who was willing to evangelize his field. He is worthy of appreciation and will serve as a model for the next generation of scientific researchers.

Figures

Fig.1 Professor Haowen Xu (1935—2004)

Fig.2 *An Introduction to Exercise Biochemistry*

Fig.3 *Physiological Chemistry of Exercise and Training*

References

Ding SZ, Wang WX, Xu HW (1992) Observation of lipofuscin and glutathione S-transferase in different tissues of swimming rats. Sports Sci 1:67. (丁树哲，王文信，许豪文. 1992. 游泳大鼠不同组织脂褐素和谷胱甘肽转硫酶的观察. 体育科学，1: 67.)

Ding SZ, Xu HW, Cheng BJ (1991) Effects of exercise-induced endogenous free radicals on the rat myocardial mitochondrial membrane. Acta Biochim Biophys Sin 1:305–310. (丁树哲，许豪文，程伯基. 1991. 运动

性内源自由基对大鼠心肌线粒体膜的影响. 生物化学与生物物理学报(英文版), 1: 305–310.)

Feng WQ, Xie MH, Wang XS, et al (2006) Research Development on Exercise Biochemistry. Beijing: Beijing Sport University Press. (冯炜权, 谢敏豪, 王香生, 等. 2006. 运动生物化学研究进展. 北京: 北京体育大学出版社.)

School of Public Health Shanghai Medical University (1988) Deeply mourn Professor Shijin Xu. Chin J Health Stat 5:64. (上海医科大学公共卫生学院. 1988. 深切悼念许世瑾教授. 中国卫生统计, 5: 64.)

Sun QT, Xu HW (2002) Effects of exercise on insulin receptors and signal transduction proteins. Sports Sci 2:66–68. (孙清涛, 许豪文. 2002. 运动对胰岛素受体及相关信号传导蛋白的影响. 体育与科学, 2: 66–68.)

The Postgraduate Working Office of the Ministry of Education in China (2000) Notice on the examination and approval results of 1999—2000 postgraduate teaching books. (中华人民共和国教育部研究生工作办公室. 2000. 关于1999—2000年度“研究生教学用书”审定结果的通知.)

The School of Sports and Health of East China Normal University (2015) Xu HW. (华东师范大学体育与健康学院. 2015. 许豪文.)

Xu HW (1986) Exercise and androgenic hormones. Chin J Sports Med 3:158–161, 189. (许豪文. 1986. 运动与雄性激素. 中国运动医学杂志, 3: 158–161, 189.)

Xu HW (1993) Rehabilitation and immunity of geriatric diseases. Pract Geriatr 7(1):15–16. (许豪文. 1993. 老年病康复与免疫. 实用老年医学, 7(1): 15–16.)

Xu SJ, Ge JD (1937) Investigation of 19 infectious diseases and parasitic diseases. Natl Med J China 23(8):1067–1088. (许世瑾, 葛家栋. 1937. 十九种传染病及寄生虫病调查. 中华医学杂志, 23(8): 1067–1088.)

Xu HW, Zhou HL (2003) Osteoporosis, vitamin K and exercise. J Tianjin Univ Sport 2:51–53. (许豪文, 周红律. 2003. 骨质疏松、维生素K和运动. 天津体育学院学报, 2: 51–53.)

Xu HW, Zheng DQ, Liang PZ, et al (1981) Variation law of serum creatine phosphate kinase and serum urea nitrogen after exercise stress in different events. Sport Sci Res 12:2–6. (许豪文, 郑德倩, 梁佩珍, 等. 1981. 不同项目运动应激后血清磷酸肌酸激酶和尿素氮的变化. 体育科研, 12: 2–6.)

Xu HW, Zheng DQ, Liang PZ (1983) Biochemical assessment of exercise dose of sprinters. Sport Sci Res 7:7–11, 13. (许豪文, 郑德倩, 梁佩珍. 1983. 短跑运动员运动量的生化评定. 体育科研, 7: 7–11, 13.)

Yu LZ (2006) The Style of Teachers—Professors' Elegant Demeanour of East China Normal University. Shanghai: East China Normal University Press. (俞立中. 2006. 师・范——华东师范大学教授风采. 上海: 华东师范大学出版社.)

Zhou G, Ding SZ, Lu J, et al (2009) The effects of growth hormone administration on the circulation level of ghrelin and IGF-I of the trained rats. J Beijing Sport Univ 32(1):65–67. (周刚, 丁树哲, 卢健, 等. 2009. 外源生长激素对运动大鼠循环ghrelin和IGF-I水平的影响. 北京体育大学学报, 32(1): 65–67.)

Zhou G, Kamenos G, Pendem S, et al (2012) Ascorbate protects against vascular leakage in cecal ligation and puncture induced septic peritonitis. Am J Physiol 302(4):R409–R416.

51. 吴瑞：第五先生还是DNA测序之父？

第一次知道吴瑞先生（图1）的名字，是看了饶毅老师写的博文《君子爱“生” 得之有道》，这篇博文后来收录在《饶议科学 I 》里，过年期间又读过一遍。其中有一句写得很有意思：“1971年吴瑞的引物延伸，是测序的一个关键步骤，给奖是可以的。”但是，所有课本上讲的都是Sanger测序法，所以显然是Sanger的贡献最大，而且诺贝尔奖都发了。

图1　吴瑞先生于宾夕法尼亚大学（1954年）

一般来说，大家认为吴先生的贡献主要有三个：第一，发起中美生物化学联合招生项目（CUSBEA），这让当年很多杰出的中国学生有机会去国外读研究生，获得学位并成为当代的学术精英（Gu，2009）；第二，被誉为“植物遗传工程之父”（Jiang，2009），这个领域非常前沿；第三，培养了Jack Szostak，他在2009年因为端粒方面的研究拿了诺贝尔奖（Szostak，2009）。这些属于科研和社会服务方面的贡献。

本学期我要给本科生上“生物信息学”这门课，之前的内容有点陈旧，所以想讲讲第二代测序方面的数据分析。网上查了之后发现，中国科学院北京基因组研究所的于军老师等人翻译了《第二代测序信息处理》（Brown，2013）这本书，应该是这方面第一本也是目前唯一一本教科书，于是我买了回来看。书中提到，1971年吴瑞先生发表过类似Sanger法的测序方法（Wu and Taylor，1971）。众所周知，Sanger等于1975年在《分子生物学杂志》上发表第一篇有关DNA测序的有效方法的文章（Sanger and Coulson，1975）；1977年Gilbert等人在*PNAS*上发表文章，建立化学裂解法（Maxam and Gilbert，1977）；同年Sanger在*PNAS*上发表文章，改

译者：薛宇
华中科技大学，武汉430074，中国
邮箱：xueyu@hust.edu.cn

进了之前的方法，从而确立了DNA测序的主流方法——Sanger法（Sanger et al.，1977）。所以在DNA测序方面，吴先生有啥贡献？我又专门买了英文原版书来读。

根据书中内容以及查阅文献，可以肯定的是：第一，吴先生提出了第一个DNA测序方法（Wu and Kaiser，1968）。第二，从1968年至1972年，吴先生在DNA测序方面至少有9篇文章（Donelson and Wu，1972a，1972b；Padmanabhan and Wu，1972a，1972b；Padmanabhan et al.，1972；Wu，1970，1972；Wu and Kaiser，1968；Wu and Taylor，1971）。1973年吴先生发了3篇，1974年发了7篇，所以在Sanger1975年开始做测序的时候，吴先生已经发表了至少19篇文章！第三，吴先生1968年的第一篇文章只测定了DNA的碱基组成，没有测定顺序；但1970年的文章已经真正测定了DNA的序列。第一个既测定DNA碱基组成又测定出顺序的，是吴先生。

另外，吴瑞先生2008年2月10日去世后，2009年《中国科学（卷C）：生命科学》发表了一篇悼念文章，节选自康奈尔大学的官方讣告，其中写道："1970年，吴瑞先生发明了第一个DNA测序的方法……"肯定了吴先生在DNA测序方面的贡献，因此称吴先生为"DNA测序之父"，并不过分。

这样的话，你肯定有疑问：吴先生在DNA测序方面可能是最先做的，但并不是做得最好的，所以不肯定吴先生也许有道理？其实没有道理。这是因为，Sanger测序法最核心的是测序思想，而不是具体的技术。因为对技术革新有最突出贡献的是Leroy Hood，第一代测序仪也是根据他的方法发明的。并且现在第二代、第三代测序技术也陆续都发展起来或正在发展，要是比谁做得最好，那诺贝尔奖也不会颁给Sanger。况且诺贝尔奖一般不关心改进，而是关心原创。

吴先生是否意识到自己的贡献？答案是肯定的。2014年，新加坡科学史研究学者L. A. Onaga写了篇文章《吴瑞——第五先生：DNA测序历史上的重要回忆》，将吴先生描绘成"第五先生"（Onaga，2014），即很重要但是被忽略的关键人物。这篇文章对吴先生的生平和学术成就讲得非常详尽，并且讲到了吴先生的抗议。

2007年5月11日，《科学》杂志出刊的时候有个附带的夹页，描绘了从1865年孟德尔开始一直到宣布人类基因组的第一个序列的发展史。吴先生看到之后很生气，写信给《科学》。信中写道，虽然Sanger法是DNA测序中的重大突破，"然而，这个方法仍然是基于我的在序列分析之前标记DNA的位置特异性引物延伸原理。我认为应该写：吴发明了第一个DNA序列分析方法，即引物延伸策略"。

所以，吴先生很清楚自己的贡献，也非常清楚自己应该有的科学和科学史地位，并且也努力去争取获得认可，但是没有成功。

人类基因组计划、原子弹计划和登月计划并称为人类有史以来规模最大、最宏伟、最壮观的三大科研项目。基因组测序对生命科学和医学研究有直接的促进作用，影响深远，并直接促成了基因组、蛋白质组、生物信息学、系统生物学等多个领域的产生和发展，无论怎么突出其重要性都不为过，并且美国于2011年提出、2015年通过的"精准医学计划"，其研究基础正是1988年提出的"人类基因组计划"。吴瑞先生作为第一个提出DNA测序方法、确定了"引物延伸"的基本原理的人，可以称为"DNA测序之父"，也应该成为大家尊敬并崇拜的顶级学者。吴先生的科学史地位，至少应当与Sanger大致相当，后者因为蛋白质测序和DNA测序两次获颁诺贝尔奖，可称为"测序之王”。因此，未来我们介绍Sanger在DNA测序方面的贡献时，应当加上：基于吴瑞先生提出的“引物延伸”原理,Sanger做了重要改进。原创的贡献，既然是吴先生做出的，就不应该抹杀。

那么，吴先生这么大贡献，为什么没有获得诺贝尔奖？对于这个问题，哈佛大学公共卫生学院的刘小乐教授在信中是这么写的：

“作为一位20世纪50年代在美国的中国移民，吴瑞先生一定遇到非常多的来自社会和种族方面的挑战。没有社会支持和联系，他所取得的成就只能依靠自己的科学天分和意志。如果那时他是英国或美国的白人科学家，或是今日的华人科学家，他的科学成就也会被更为认可。很多时候，我们不能太看重奖励、表彰和头衔（例如吴瑞先生没有入选美国国家科学院院士），或者高被引次数（H指数）等具体指标，我们得客观评估学者对科学界的总体贡献。”

吴瑞先生，1928年8月14日出生于北平，后来在美国读书的时候，由于语言问题听不懂报告，所以付出了比一般人更多的努力。后来1964年，受到Robert Holley的RNA测序，以及噬菌体研究学者Max Delbrück、Alfred Hershey和Arthur Kornberg等影响，决心攻克DNA测序的难关，6年后取得成功，并直接开创了一个波澜壮阔的新时代：基因组时代。

DNA测序之父,实至名归。

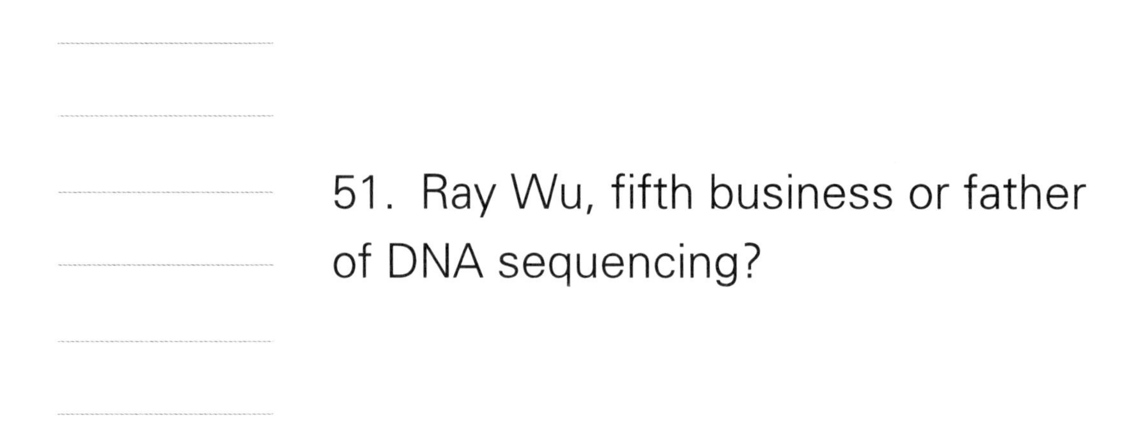

51. Ray Wu, fifth business or father of DNA sequencing?

The first author, Yu Xue, first read about Dr. Wu in the ScienceNet blog written in 2008 by Prof. Yi Rao, an eminent Chinese neurobiologist, who has made a profound and lasting influence on the new generation of Chinese scientists, including us. Later, this essay was included in one of his published books, and we carefully read it again during the Spring Festival of 2015. One sentence was quite confusing to us, "The primer-extension approach developed by Ray Wu in 1971, is a key step of DNA sequencing, and deserves a Nobel Prize." However, all textbooks that we had been aware of stated Sanger sequencing as the first and most important methodology. On top of everything the Nobel Prize in Chemistry was awarded to Dr. Frederick Sanger and Dr. Walter Gilbert "for their contributions concerning the determination of base sequences in nucleic acids" in 1980.

Generally, three major contributions of Dr. Wu have been widely recognized. The first and most well-known is the organization of the CUSBEA (China-United States Biochemistry Examination and Application, from 1981 to 1989) program. With this support, over four hundred excellent Chinese students got opportunities to pursue postgraduate studies in USA, achieved academic successes and have now largely become elite scientists in various fields (Gu, 2009). Second, Dr. Wu is deemed as "one of the founding fathers of plant genetic engineering" (Jiang, 2009), a field which looks wonderful but actually does not mean much to the public. Third, one of Dr. Wu's graduate students, Dr. Jack Szostak, won the Nobel Prize for Physiology or Medicine in 2009 for the discovery of telomeres (Szostak, 2009). These belong in the category of scientific and social service, and has nothing to do with Dr. Wu's scientific contributions per se.

Yu Xue needed to teach a Bioinformatics course to undergraduate students, and found that his previous materials were quite dated and decided to update the course with more cutting-

---

Yu Xue[1], Yongbo Wang[1], Hui Shen[2]

1 Department of Bioinformatics & Systems Biology, College of Life Science and Technology, Huazhong University of Science and Technology, Wuhan 430074, China

2 Center for Epigenetics, Van Andel Research Institute, Grand Rapids MI 49503, USA

Correspondence: xueyu@hust.edu.cn (Y. Xue)

edge content. During his research he came across a book entitled of *Next-Generation DNA Sequencing Informatics* (Brown, 2013). This book had been translated into Chinese by Dr. Jun Yu et al. at the Beijing Institute of Genomics (BIG) as a textbook for introducing computational processes in analyzing the NGS data. In the first paragraph of Chapter 1, he found the statement "An interesting approximation of the Sanger method was published in 1971 by Ray Wu of Cornell University (Wu and Taylor, 1971)" which reminded him of the comments made by Prof. Yi Rao. People knew that Dr. Sanger published the first paper introducing an efficient method for determining DNA sequences in 1975 (Sanger and Coulson, 1975), and two years later Dr. Gilbert developed a chemical procedure for DNA sequencing (Maxam and Gilbert, 1977). Then in 1977 Dr. Sanger greatly improved his strategy and developed the "Sanger Sequencing" approach, which earned him a Nobel Prize (Sanger et al., 1977). No other textbooks that he had read mentioned the contributions by Dr. Wu. So perhaps that Dr. Brown has made a mistake in his book?

So he decided to find the truth by doing a little bit research on PubMed. The literature showed several things. First, Dr. Wu was the first who attempted to develop the DNA sequencing approach, since his first paper on it was published in 1968 (Wu and Kaiser, 1968). Second, from 1968 to 1972, Dr. Wu's group at least published 9 papers in formal journals for DNA sequencing (Donelson and Wu, 1972a, 1972b; Padmanabhan and Wu, 1972a, 1972b; Padmanabhan et al., 1972; Wu, 1970, 1972; Wu and Kaiser, 1968; Wu and Taylor, 1971). Third, Dr. Wu continued to publish three papers in 1973 and seven papers in 1974. So before the Sanger's first paper on sequencing, Dr. Wu's group at least published up to 19 papers for DNA sequencing. Fourth, in Dr. Wu's first paper, only the nucleotide composition of the 5′-terminated strands of bacteriophage lambda DNA was determined with 13 nucleotides of dG, 13 of dC, 7 of dA, and 7 of dT. The orders of DNA sequences were not determined, and the methodology cannot be regarded as a successful sequencing approach. However, in 1970 but not 1971, Dr. Wu published a paper in *Journal of Molecular Biology* by himself, and reported a general method for determining the DNA sequence. In the abstract, he wrote: "When this terminal region is present as a single strand, as in bacteriophage lambda, *Escherichia coli* DNA polymerase can be used to repair the single-stranded region with the addition of radioactive nucleotides to the 3′-end copying the protruding 5′-terminated single strand. The partially labeled DNA can be degraded with nucleases, the radioactive oligonucleotides isolated, and their sequence determined." By this method, he successfully determined a short sequence of the first eight of twelve nucleotides as CGCCGCCC in the right-hand protruding strand of lambda DNA (Onaga, 2014; Wu, 1970). The method was continuously improved and collectively called as the location specific-primer-extension principle, or the primer extension method (Onaga, 2014; Padmanabhan and Wu, 1972b). Taken together, we can confirm that the first approach for determining both the composition and order of DNA sequences was reported by Dr. Wu (Wu, 1970).

After Dr. Wu passed away on Feb 10, 2008, the journal of *Science in China Series C: Life Sciences* published a special issue with 12 memorial essays in 2009. One essay derived from the official obituary of Cornell University Press stated, "In 1970, Wu developed the first method for sequencing DNA and some of the fundamental tools for DNA cloning (sequencing involves determining the base sequence in a DNA molecule)." So the contribution of Dr. Wu on DNA sequencing was fairly credited, and we can correct his title as "father of DNA sequencing" .

OK, we assume that you will have a question: Perhaps Dr. Wu was indeed the first that started to work on DNA sequencing, but his contributions were too trivial and others did much better and more important jobs. So is it reasonable to neglect this "humble" figure? Our answer is: No. While later adaptions and applications may have a more direct impact, the fundamental principles are at the root of everything. Obviously the Sanger sequencing was developed based on Wu's primer-extension method. If we only considered the technical innovations, we believe that Dr. Leroy Hood did the best job by developing the first automated DNA sequencer in 1986 together with Applied Biosystems. Although the Sanger sequencing was adopted in the first-generation sequencing, this approach was not used any longer in the second-generation and third-generation sequencing. However, the primer-extension principle, developed by Dr. Wu, has never been changed in all generations of sequencing techniques.

Whether Dr. Wu never realized his own contributions? Our answer is that he clearly knew. In an important historical research study, Prof. Lisa A. Onaga in Singapore analyzed and described Dr. Wu as a "Fifth Business" . The term "Fifth Business" was coined by the Canadian novelist Robertson Davies in 1970 (Onaga, 2014), to describe those who were "neither those of Hero nor Heroine, Confidante nor Villain, but which were nonetheless essential to bring about the Recognition or dénouement" . The lifetime and scientific careers of Dr. Wu were carefully described in details in Dr. Onaga's article, together with Dr. Wu's own protest on the dismissal of his seminal contributions.

So how did Dr. Wu defend himself? Prof. Lisa A. Onaga described an example. "On 11 May 2007, the journal *Science* published a colorful poster insert entitled 'The evolution of DNA sequencing technologies' . Starting with Mendel's analysis of inheritance in plants in 1865 and ending with the announcement of the first sequence of a named human being's genome, an unzipping double helix represented a timeline of the 'discovery process' leading to state-of-the-art DNA sequencing technologies." (Onaga, 2014) However, the name of Dr. Wu was not mentioned at all. So Dr. Wu felt being treated unfairly and wrote a letter to *Science*. Dr. Wu first acknowledged the Sanger sequencing as a great breakthrough, and continued to state the neglected fact: "However, the method was still based on my location specific-primer-extension principle in labeling the DNA before sequence analysis." Dr. Wu insisted, "If you agree to add my contribution to the chart 'The evolution of sequencing technology', you may add an entry for 1970 and write something like 'Wu introduced the first method for DNA sequence analysis by introducing the primer-extension approach' ." (Onaga, 2014) In this regard, Dr. Wu was clearly aware of his contributions, and knew that he should be remembered as a key figure in the history of life science. He tried his best to pursue a fair recognition, although not successful.

The Human Genome Project (HGP), the Manhattan Project and Apollo program, were considered as the three greatest scientific projects of all time. Genome sequencing, or DNA sequencing, has had a profound impact for researches in life science and medicine, and directly spurred the emergence of multiple new fields, such as genomics, proteomics, bioinformatics/ computational biology, and systems biology. So we can (perhaps arguably) regard the DNA sequencing as the most important technique in 20th century. Also, the fourth greatest project, the Precision Medicine Project, was started in 2011 with a solid foundation from HGP. Since Dr. Wu first developed the DNA sequencing methodology, and established the primer-extension principle, "father of DNA sequencing" should be a correct title for him. And his contributions can at least be comparative to Dr. Sanger, who made great contributions on determining both protein and DNA sequencing and is deserving of a title of "the emperor of sequencing" . Taken together, when we introduce the contributions of Dr. Sanger in DNA sequencing, we really have to add the information that based on the primer-extension approach developed by Dr. Wu, and Dr. Sanger further refined the methodology and played an important role in DNA sequencing. Dr. Wu's contribution should not be and hopefully will not be neglected.

So you will have a final question: Since Dr. Wu made a great contribution, why didn't he get the Nobel Prize? We can take a look at the comments from our friend, Prof. Xiaole Shirley Liu in Harvard School of Public Health:

"Being a Chinese immigrant in the US in the 50's, the social and racial challenges Ray Wu faced at that time must be tremendous. Without social backing and connections, he got where he did purely by his scientific genius and good heart. If he was a Caucasian scientist from UK or US, or even if he was in the current era, his scientific contributions would have been better recognized. Sometimes we don't have to pay too much attention to awards or *H*-index, but objectively evaluate someone by their overall impact to the scientific community."

Dr. Ray Wu was born on 14 August, 1928 in Beijing. Due to the language barrier, he had to make more efforts than others when he studied in USA. Later in 1964, he listened to the talk from Robert Holley on RNA sequencing, and was greatly influenced by Max Delbrück, Alfred Hershey, and Arthur Kornberg, who made great contributions on the genetics of bacteriophages. He decided to devote to resolving the problem of DNA sequencing, succeeded six years later, raising the curtain of a splendid new era: The Genomics Era.

So for Dr. Ray Wu, the title of "father of DNA sequencing" is more than befitting. He well and truly deserves this recognition.

This essay is an extending discussion on an outlook published in *Nature* by Dr. Chuanchao Wang, which immediately triggered a very broad discussion among the scientific societies in China, especially to the young generation (Wang, 2015). A lot of people participated in the online conversation launched by ScienceNet. All agree with Dr. Wang's opinion to give youth a chance. We shared our viewpoints to many friends, Prof. Xiaole Shirley Liu, Prof. Liangsheng Zhang, Prof. Pengyu Huang, and Prof. Kang Ning, and are thankful for their helpful comments, which considerably improved presentation. We are also grateful for Prof. Lisa Onaga, who believes, "Your comment reinforces my understanding that there is a different picture of this individual and I hope to continue to do more research to better understand the later part of his career." (Personal communication, 16 August, 2015) We also thank Prof. Le Kang for his encouragement, and Dr. Xiaoxue Zhang of *Protein & Cell* Editorial Office, for her kindness and patience. Yu Xue wrote the blog in Chinese by himself, but his student Yongbo Wang and his friend Prof. Hui Shen made great efforts on drafting and refining the English version.

Finally, it's the first time for Yu Xue to write an essay but not a formal scientific paper in English, or Chinglish. He used to write in Chinese, and is not sure whether Chinese and English-speaking people have a different sense of humor. When he attended conferences, a lot of foreign scientists frequently laughed at certain jokes the speakers were telling, when he was totally lost and confused. He pretended to laugh as well, of course, but in that light his efforts of being witty or humorous at certain part of the article might have been a total failure. We apologize for that, but hopefully it will not affect the message that we are trying to convey.

Figures

Fig.1 Mr. Ray Wu in the University of Pennsylvania in 1954

References

Brown SM (2013) Next-generation DNA Sequencing Informatics. Cold Spring Harbor: Cold Spring Harbor Laboratory Press.

Donelson JE, Wu R (1972a) Nucleotide sequence analysis of deoxyribonucleic acid. Ⅵ. Determination of 3′-terminal dinucleotide sequences of several species of duplex deoxyribonucleic acid using *Escherichia coli* deoxyribonucleic acid polymerase Ⅰ. J Biol Chem 247: 4654–4660.

Donelson JE, Wu R (1972b) Nucleotide sequence analysis of deoxyribonucleic acid. Ⅶ. Characterization of *Escherichia coli* exonuclease 3 activity for possible use in terminal nucleotide sequence analysis of duplex deoxyribonucleic acid. J Biol Chem 247: 4661–4668.

Gu X (2009) Ray Wu and the CUSBEA Program. Sci China Ser C 52:125–127.

Jiang C (2009) Commemorating the life of a legend. Sci China Ser C 52:97–98.

Maxam AM, Gilbert W (1977) A new method for sequencing DNA. PNAS 74:560–564.

Onaga LA (2014) Ray Wu as fifth business: Deconstructing collective memory in the history of DNA sequencing. Stud Hist Philos Biol Biomed Sci 46:1–14.

Padmanabhan R, Wu R (1972a) Nucleotide sequence analysis of DNA. Ⅳ. Complete nucleotide sequence of the left-hand cohesive end of coliphage 186 DNA. J Mol Biol 65:447–467.

Padmanabhan R, Wu R (1972b) Nucleotide sequence analysis of DNA. Ⅸ. Use of oligonucleotides of defined sequence as primers in DNA sequence analysis. Biochem Biophys Res Commun 48:1295–1302.

Padmanabhan R, Wu R, Bode VC (1972) Arrangement of DNA in lambda bacteriophage heads. 3. Location and number of nucleotides cleaved from lambda-DNA by micrococcal nuclease attack on heads. J Mol Biol 69:

201–207.

Sanger F, Coulson AR (1975) A rapid method for determining sequences in DNA by primed synthesis with DNA polymerase. J Mol Biol 94: 441–448.

Sanger F, Nicklen S, Coulson AR (1977) DNA sequencing with chain-terminating inhibitors. PNAS 74: 5463–5467.

Szostak J (2009) Ray Wu, as remembered by a former student. Sci China Ser C 52:108–110.

Wang CC (2015) Perspective: Give youth a chance. Nature 520: S36.

Wu R (1970) Nucleotide sequence analysis of DNA. Ⅰ. Partial sequence of the cohesive ends of bacteriophage lambda and 186 DNA. J Mol Biol 51: 501–521.

Wu R (1972) Nucleotide sequence analysis of DNA. Nat New Biol 236:198–200.

Wu RJ (2009) Ray Wu, Cornell's acclaimed pioneer of genetic engineering and developer of insect-resistant rice. Sci China Ser C 52:99–100.

Wu R, Kaiser AD (1968) Structure and base sequence in the cohesive ends of bacteriophage lambda DNA. J Mol Biol 35: 523–537.

Wu R, Taylor E (1971) Nucleotide sequence analysis of DNA. Ⅱ. Complete nucleotide sequence of the cohesive ends of bacteriophage lambda DNA. J Mol Biol 57: 491–511.

Yang NS, Lan CW (2009) For our founding father of ABRC, Prof. Ray Wu (1928—2008). Sci China Ser C, Life Sci / Chin Acad Sci 52: 128–129.

六、病毒学和免疫学

52. 伍连德：鼠疫斗士，中国公共卫生体系之父

著名学者、思想家、哲学家梁启超先生说："科学输入垂五十年，国中能以学者资格与世界相见者，伍星联（伍连德）博士一人而已。"1935年，伍连德（图1）成为第一个被提名诺贝尔生理学或医学奖的中国人。

1879年3月10日，伍连德出生于马来西亚的槟榔屿。他在17岁的时候远赴英国学习。在剑桥大学的伊曼纽尔学院，他以优异的成绩获医学博士学位。

1910年秋，一场致命的传染病在我国东北暴发。最初的疾病发生在满洲里，然后迅速传播到哈尔滨。在短短的四个月，6万人被传染。

伍连德博士到达哈尔滨不久，就做了首例尸检。他解剖了因感染致死的一位日籍妇女的尸体，发现其组织中有耶尔森菌。他认为这是一种肺鼠疫感染，可以通过空气或飞沫传播。这一观点与传统的"鼠疫通过鼠或跳蚤传播"的理论相矛盾，且传统观点认为鼠疫不会人传人。他的观点震惊了当时的科学同行，也遭到了普遍的质疑。

图1　伍连德博士（1879—1960）

法国著名医生Mesny是质疑伍连德观点的人士之一。他在视察疫区时拒绝戴口罩，几天后他死于肺鼠疫。他的死震惊了当时的国际同行。

与此同时，伍连德说服了俄国和日本的铁路当局停止运行所有列车。这些措施切断了交通，也因此阻止了疾病在东北进一步传播。但是，此时哈尔滨的死亡人数仍然在上升，因为很多病死者的尸体成了最大的疾病传染源。

伍连德博士上书申请焚烧尸体，大约3000具尸体及棺木被集中焚烧。至1911年3月31日，没有新发感染病例报告。这种致命的传染病在农历新年到来之后被消灭了。

在伍连德博士的建议下，1911年4月3日至4月28日，万国鼠疫研究会在奉天（今沈阳）举办（图2）。来自美国、英国、俄国、日本及法国等11个国家的知名流行病学

作者：马中良[1,2]，李艳利[1]

1　上海大学生命科学学院，上海200444，中国

2　伍连德研究所，黑龙江医学科学院，哈尔滨医科大学，哈尔滨150081，中国

邮箱：liyanli@shu.edu.cn（李艳利）

家、科学家参加了此次会议。伍连德博士当选为大会主席，他关于鼠疫防治的研究获得同行的高度赞扬。实际上，我国现代医学在该会议不久后建立。伍连德博士始终站在抗击鼠疫的前沿，在1921年他成功地阻止了鼠疫的再次发生。

图2 伍连德博士（一排左三）在万国鼠疫研究会

伍连德博士投入大量的精力在我国建立医院、医学院，并成立了中华医学会。1931年，日本侵占东北，伍连德博士离开东北到了上海。在上海，他建立了国家检疫体系。直到1937年，伍连德博士返回马来西亚。

在马来西亚，人生的最后岁月，伍连德博士用英语出版了他的个人自传《鼠疫斗士——一个现代中国医生的自传》。1960年1月20日，伍连德离开了这个世界，享年81岁。在世界科学和医学界，伍连德的去世引发巨大的悲痛。1960年1月27日，《泰晤士报》这样评论："伍连德的去世使得世界医学界失去了一位英雄，一位传奇人物，他做的远比我们知道的多，是值得我们纪念的人。"

那个年代，伍连德对现代医学在我国的发展作出的贡献至关重要，他被认为是建立我国现代医学服务和医学教育的第一人。为了纪念和铭记他的贡献，伍连德的铜像被安放在哈尔滨医科大学。他从没有被忘记，他的工作仍然对我们有指导作用。今天我们纪念伍连德，不仅因为伍连德博士拯救了很多人的生命，在我国建立了现代医学体系，而且希望这种精神激励我们今天做得更好。2015年12月24日，为了纪念他的伟大贡献，伍连德研究所成立了（图3）。伍连德研究所的目标就是聚焦传染性疾病的研究并将研究成果惠及每一个人。哈尔滨医科大学伍连德研究所所长、中国科学院院士高福博士也是抗击危险性病毒（包括埃博拉病毒）、中东呼吸综合征（MERS）及禽流感的斗士。

图3 2015年12月24日，伍连德研究所成立，高福院士（左二）任所长

致谢

感谢田波院士和高福院士，他们为我们打开了一扇通往现代生命科学的大门。本工作得到国家自然科学基金委员会（项目编号：31170750）和上海大学基础教育项目——科学史的支持。

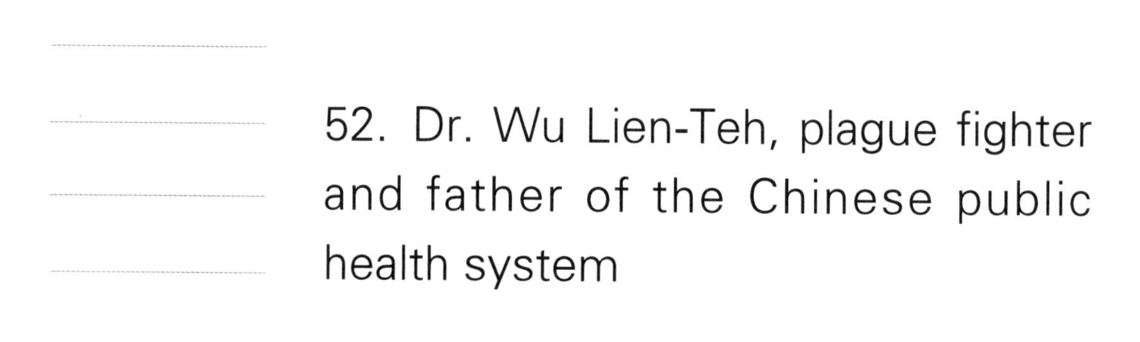

52. Dr. Wu Lien-Teh, plague fighter and father of the Chinese public health system

The Chinese scholar, journalist, and philosopher, Liang Qichao said, “About 50 years since science came into China, only Dr. Wu Lien-Teh can meet and talk with foreign scientists as a real scholar.” In 1935, he also became the first Chinese doctor nominated for the Nobel Prize in Physiology or Medicine.

Dr. Wu Lien-Teh (伍连德) was born in Penang on March 10, 1879 and, at 17, went to England to study at Emmanuel College, at the University of Cambridge. There, he was awarded the prestigious degree of M.D. from Cambridge University.

In the fall of 1910, the deadly epidemic broke out in the northeastern region of China. The first fatality was reported in the border town of Manchouli, and the epidemic quickly spread to Harbin. Within 4 months, it had claimed over 60000 lives.

After his arrival at Harbin, Dr. Wu performed the first-ever postmortem exam on a Japanese woman who had died from the epidemic. He discovered *Yersinia pestis* in the body tissues and further concluded that the epidemic was pneumonic plague, which could be transmitted by human breath or sputum. This was contrary to the general idea that plague could only be transmitted by rats or fleas and could not be transmitted from person to person. His idea surprised all of his scientific peers and was met with widespread disbelief.

Dr. Mesny, a prominent French doctor, was one of those who doubted Dr. Wu’s views. Dr. Mesny himself died of pneumonic plague several days later, after refusing to wear gauze and a mask and succumbing to the epidemic infection. His death shocked the international community.

At the same time, Dr. Wu had convinced the Russian and Japanese railway authorities to cease operation of all trains in 1911. These efforts cut off all transportation, and thus transmission of the disease in Northern-East China. Nonetheless, the death toll in Harbin continued to rise as the corpses of those who had died of the epidemic served as a perfect incubator for the plague bacillus.

Zhongliang Ma[1,2], Yanli Li[1]

1 School of Life Sciences, Shanghai University, Shanghai 200444, China

2 Wu Lien-Teh Institute, Heilongjiang Medical Academy, Harbin Medical University, Harbin 150081, China

Correspondence: liyanli@shu.edu.cn (Y.-L. Li)

Dr. Wu sent a petition to sanction the cremation of the deceased, and some 3000 corpses and coffins were gathered and cremated. No further infection was reported as of March 31, 1911. The deadly disease had vanished by the time the Chinese New Year had arrived.

On the advice of Dr. Wu, the International Plague Conference was held in Mukden from April 3 to April 28, 1911. Renowned epidemiologists and scientists from 11 countries, including the USA, UK, Japan, Russia and France were in attendance. Dr. Wu was elected as president of the conference and his work on plague prevention was highly praised by all. In fact, modern medical science was established in China soon after this conference was held. Dr. Wu always stood as fighter at the forefront in the battle to prevent plague and in 1921, he successfully stamped out the recurring epidemic.

Dr. Wu devoted many of his efforts to establishing hospitals and medical colleges and founded the Chinese Medical Society. In 1931, after the Japanese invaded the northeastern provinces of China, he left for Shanghai, where he set up the National Quarantine System. He returned to Malaysia with his family in 1937.

In his later years, after his homecoming to Malaysia, he wrote and published an autobiography in English, *Plague Fighter: The Autobiography of a Modern Chinese Physician*. Dr. Wu passed away at the age of 81 on January 20, 1960. He was deeply mourned by the scientific and medical communities. The *Times* London commented on January 27, 1960: "By his death, the world of medicine has lost a heroic and almost legendary figure and the world at large one of whom it is far more indebted to than it knows."

Dr. Wu's contributions to the development of modern medicine in China were of great importance during that era. He is regarded as the first person to modernize China's medical services and medical education. To honor him and to remember his contributions, bronze statues of Dr. Wu Lien-Teh, were erected at Harbin Medical University. He remains unforgotten and his work continues to serve us. We owe much to Dr. Wu, who saved many lives and whose contributions to modern medicine in China allow us to continue to do so today. On December 24, 2015, to thanks to his great contributions, Wu Lien-Teh Institute was opened, its aims are to research on infectious diseases and share the resulting knowledge to everyone. President of Wu Lien-Teh Institute, Harbin Medical University is Dr. George Fu Gao, an Academician of Chinese Academy of Sciences, is also a fighter to some dangerous viruses, such as Ebola virus, MERS (Middle East Respiratory Syndrome) and bird flu.

Acknowledgements

The authors would like to give great thanks to Prof. Po Tien, and Prof. George Fu Gao, who opened the door of Modern Life Sciences to us. This work was supported by the National Natural Science Foundation of China (Grant No. 31170750), the General Education Course of Shanghai University: The History of Science.

Figures

Fig.1 Dr. Wu Lien-Teh (1879—1960)

Fig.2 Dr. Wu Lien-Teh (3rd from left, line 1) in The International Plague Conference

Fig.3 Wu Lien-Teh Institute was opened on Dec 24, 2015 in Harbin. President of the Institute is Dr. George Gao Fu (2nd from left)

References

Wu LT (2014) Plague Fighter, the autobiography of a modern Chinese Physician. The Journal of Asian Studies 19(1): 75.

Wu LT (2014) The nomination database for the Nobel prize in physiology or medicine literatures of Dr. Wu Lien-Teh. Printed by Wu Lien-Teh Insitute.

53.“人类的朋友”汤飞凡：从“汤氏病毒”到沙眼衣原体再到“衣原体门”，一个关于沙眼病原体的故事

1979年，英国著名科学家Joseph Needham得知汤飞凡（图1）已经于1958年去世的消息后，给当时的卫生部生物制品研究所（现北京生物制品研究所）所长的信里写道：“汤飞凡正是英国谚语里说的‘人类的朋友’，他深爱着中国人民，是预防医学领域一位顽强的战士。我和所有的朋友一起向他致敬，我相信在中国，他将永远不会被忘记。”

图1　汤飞凡（1897—1958）

汤飞凡1897年出生于湖南醴陵汤家坪，1921年他以第一名的成绩成为长沙湘雅医学院的首届毕业生。他到北京协和医学院进修三年，随后到美国哈佛大学医学院细菌学系深造，在Hans Zinsser教授指导下进行病毒学研究。汤飞凡1929年回到上海，在国立中央大学医学院细菌学系任职，同时兼任上海雷士德医学研究院细菌学系主任。1937年，汤飞凡被任命为中央防疫处处长，并于1947年当选国际微生物学会联合会理事。1949年新中国成立后，他被任命为卫生部生物制品研究所所长，并主持组建中央生物制品检定所（现中国食品药品检定研究院），他一度兼任该所所长。1952年，汤飞凡任中国微生物学会第一届理事会理事长，5年后被增选为中国科学院学部委员（生物学部）。

汤飞凡是我国第一代医学病毒学家，1929年从美国回国后开始从事沙眼病的研究，这是一种当时严重威胁我国民众健康的疾病。他首先针对日本科学家野口（Noguchi）提出的“颗粒杆菌是沙眼的病原体”假设进行了一系列实验，将“颗粒杆菌”接种在包括他自己在内的12名志愿者的眼睛里，实验证明野口的结论是错误的。在随后几年的沙眼病原学研究期间，他在第一代超滤膜的发明者英国科学家W. J. Elford邀请下，于1936年到英国进行合作研究，他在那里获得了科研灵感，他确信自然界存在的微生物是从小到大一个长长的系列，在已知的病毒和细菌之间应该存在着“过渡的微生物”，

译者：程浩

中国科学院北京生命科学研究院，北京100101，中国

邮箱：chengh@biols.ac.cn

图2 汤飞凡（左起第五位）和他的同事

如立克次体和支原体等。他认为，引起沙眼的病原体应该是比牛痘病毒更大的、接近立克次体的“大病毒”。由于战争，他不得不暂停这些研究，直到1949年后才得以继续研究（图2），而此时科学界还没有确定引起沙眼的真正病原体。为了保证病理材料的可靠，汤飞凡特别邀请北京同仁医院眼科专家张晓楼合作收集和鉴定了200余份沙眼患者病例，并阐明了沙眼病原体的形成和演变过程。1955年，他详细描述了沙眼病原体侵入宿主细胞后的发育周期，澄清了此前50年对沙眼病原体的混乱认知。汤飞凡还进行了沙眼包涵体研究、猴体感染试验和病毒分离试验。同样在1955年，猴体感染沙眼的试验也获得了重大进展，他采用研究立克次体常用的卵黄囊接种技术，并用链霉素和青霉素等抗生素作为抑制剂，分离出了一株引起猴体沙眼的病原体——“病毒TE”，后来许多外国科学家将它称为“汤氏病毒”。他又将这株“病毒”接种到自己眼内，导致自己出现典型的沙眼症状，并详细记录了40多天病程，进一步确认了这个“病毒”对人类的致病性。

这些突破性的研究进展发表在1956年的《微生物学报》上，汤飞凡的研究成果很快被全世界众多相关实验室证实。1958年，英国李斯特研究所的L. H. Collier教授用汤氏方法在冈比亚分离出沙眼病原体。不久，其他国家的科学家也相继分离出“沙眼病毒”。随后，国外学者又确定鹦鹉热和淋巴肉芽肿等的病原体与沙眼病原体属于同一类。1973年，世界卫生组织（WHO）将它们命名为“衣原体”。现在，这一类微生物被《伯杰氏系统细菌学手册》归为衣原体门，衣原体学也成为医学微生物学的一个重要分支。

尽管汤飞凡从事沙眼研究的近30年正值我国处于动荡不安的时期，他仍然在临床研究和基础研究领域都作出了重要的贡献。如果他没有被任命去从事麻疹和脊髓灰质炎疫苗的研究，而是继续在他擅长的科研领域探索，他将会取得更多成就。1958年，性情耿直的汤飞凡采取自尽的方式来保护自己的尊严。直到23年以后，因为汤飞凡在沙眼研究中作出的杰出贡献，他被国际沙眼防治组织授予金质奖章。1992年，我国发行“中国现代科学家纪念邮票”，四位科学家之中就有汤飞凡（图3），他的铜像屹立在北京生物制品研究所前。非常遗憾的是，沙眼的研究在我国暂停了一段时间，如今我国进入了创新研究的新纪元，我们永远不能忘记先辈的经验和教训。

图3 1992年发行的纪念汤飞凡的邮票

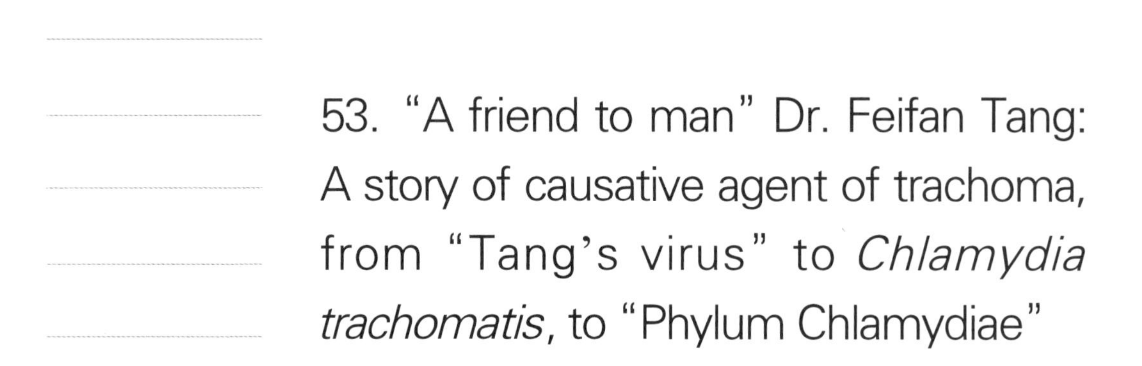

53. "A friend to man" Dr. Feifan Tang: A story of causative agent of trachoma, from "Tang's virus" to *Chlamydia trachomatis*, to "Phylum Chlamydiae"

"Dr. Feifan Tang (F. F. Tang) was indeed 'a friend to man' as our old English phrase has it; he loved the Chinese people and was a doughty fighter in the fundamental field of preventive medicine. With all friends here, I salute his memory and I am sure he will never be forgotten in China." The famous British scientist Joseph Needham wrote this in 1979 in his letter to the then director of the National Vaccine & Serum Institute of China, when he received the obituary notice of Dr. F. F. Tang who had died in 1958.

Dr. F. F. Tang was born in Tangjiaping, Liling County, Hunan Province in 1897. In 1921, he became the first session of graduates from the Hsiang-Ya School of Medicine (Changsha, the capital city of Hunan Province) ranking the first in academic grades among his peers. After working at the Peking Union Medical College for three years, he continued his research on virology with Prof. Hans Zinsser at the Department of Bacteriology of the Harvard School of Medicine, Harvard University, USA. In 1929, he returned to Shanghai, and was offered a faculty position at the Department of Bacteriology, Medical College, National Central University, as well as director of the Bacteriology Department of the Henry Lester Institute of Medical Research (Shanghai). In 1937, Dr. F. F. Tang was appointed director of the National Epidemic Prevention Bureau, and in 1947, he was elected Executive Committee Member of the International Union of Microbiology Societies. After the founding of the People's Republic of China in 1949, he was appointed director of the National Vaccine & Serum Institute of China and later founded the National Institute for the Control of Pharmaceutical and Biological Products. In 1952, he became chairman of the Chinese Society for Microbiology. Five years later, he was elected as a member of Academic Division of Biology, Chinese Academy of Sciences, Beijing.

Dr. F. F. Tang was a pioneer virologist in China and had been conducting virological research on trachoma since 1929, when he returned from the USA and when trachoma was severely threatening the health condition of Chinese people. He first performed a series of

---

Guangsheng Cheng[1], Ming Li[2], George F. Gao[1,2]

1 Institute of Microbiology, Chinese Academy of Sciences, Beijing 100101, China

2 Beijing Institutes of Life Science, Chinese Academy of Sciences, Beijing 100101, China

Contact: guangshch@vip.163.com (G.-S. Cheng); gaof@im.ac.cn (G.-F. Gao)

inoculation experiments in human eyes including his own, and disproved the Noguchi hypothesis that bacteria were the etiologic agent of trachoma. During the following years of etiological research on trachoma, he was once invited to work with Dr. W. J. Elford in England, who then had developed the ultra-filtration technology. Dr. F. F. Tang was greatly inspired by this experience. He was convinced that natural microorganisms consist of a long series of microbes, from small to large in size, and there should be "transition microbes" between the known virus and bacteria, such as rickettsia and mycoplasma. The etiologic agent of trachoma should be a "big virus" whose size is larger than vaccinia virus and similar to rickettsia. However, his etiological study of trachoma was interrupted due to the war, and did not resume until 1949, when the real agent of trachoma remained yet unknown. To ensure the reliability of pathological material, he carefully collected samples from over 200 trachoma patients with the help of ophthalmologist Dr. Xiaolou Zhang, and investigated the formation and evolution of trachoma pathogen. In 1955, he elucidated the development cycle of trachoma pathogen in host cells, which clarified the confusion in the past 50 years. In the same year, his experiment of trachoma infection in monkeys also made great progress. By the yolk sac inoculation approach commonly used in the rickettsia research and using antibiotics streptomycin and penicillin as inhibitors, he successfully isolated the virus strain TE, which was confirmed to be the etiologic agent of trachoma in monkey eye. The virus has been referred to as "Tang's virus" by many researchers later on. He even inoculated the isolated virus into his own eye and got typical symptoms of trachoma. He recorded the cause of trachoma for over 40 days and, further proved the pathogenicity of this virus.

This breakthrough was published in *Acta Microbiologica Sinica* in 1956, and was quickly confirmed by many laboratories all over the world. In 1958, L. H. Collier in the Lister Institute isolated the trachoma pathogen in Gambia using Tang's method, and later isolated the "trachoma virus" in other countries. It was also found that the trachoma virus belonged to the same category as the pathogens of psittacosis and of lymphogranuloma. In 1973, the category was named Chlamydia by the WHO. Today, these microorganisms are classified into Phylum Chlamydiae in *Bergey's Manual for Systematic Bacteriology*, and chlamydiology has become an important branch of medical microbiology.

It took Dr. F. F. Tang nearly 30 years in his turbulent homeland to identify the trachoma agent, which made significant contribution to both clinical practices and basic science. His research might have been continued for greater achievements if he had not been appointed to study vaccines of measles and poliomyelitis. In 1958, he gave up his life to protect his integrity and dignity. Until 23 years later when Dr. F. F. Tang was awarded the gold medal by the International Organization against Trachoma, his great contribution became to be treasured. In 1992, the Chinese government issued a postage stamp in his honor. A bronze statue of Dr. F. F. Tang stands in front of Beijing Institute of Biological Products. It is a great pity that the trachoma study was discontinued in China for a while. Today, when we are in the new era of innovation, the lessons from our predecessors should never be forgotten.

Figures

Fig.1 Mr. F. F. Tang (1897—1958)

Fig.2 Mr. F. F. Tang (5th from the left) and his colleagues

Fig.3 The postage stamp in memory of Mr. F. F. Tang in 1992

54.“努力在我，评价在人”——高尚荫：敢为天下先，中国病毒学开拓者和奠基人

“高尚荫教授是一位学术造诣很深、在国内外享有很高声誉的科学家和教育家。在半个多世纪的科学、教育生涯中，他呕心沥血，锲而不舍，奋斗不息，为我国科学、教育事业的发展作出了重要贡献……高尚荫教授学识渊博，却仍虚心好学，手不释卷，密切注视世界科学的新进展，始终使自己站在当今世界生物科学的前沿……他的逝世，是我国科学界和教育界的不幸……”我国著名数学家、时任武汉大学校长齐民友在高尚荫教授的追悼大会上对这位杰出的科学家做出了以上崇高评价（Hu et al., 2002; 高尚荫, 2007）。高尚荫教授（图1）是我国病毒学的奠基人之一，创办了我国最早的病毒学研究机构、第一个微生物专业和第一个病毒学专业。他于1958年完成的“家蚕脓病病毒的组织培养方法研究”是无脊椎动物组织培养和昆虫病毒研究中的开创性工作，他的著作《昆虫病毒理论及应用研究》（*Theory and Applied Research on Insect Viruses*）在国内外产生了重要影响。1980年，高尚荫被选为中国科学院学部委员。在他56年的科研与教育生涯中，他锲而不舍地奋斗，对我国微生物学和病毒学事业的发展产生了重要影响（图2）。

图1　高尚荫教授（1909—1989）

1909年3月3日，高尚荫出生于浙江省嘉善县陶庄镇的一个书香世家。7岁那年，他进入父亲办的一所乡间小学接受启蒙教育。1926年中学毕业后考上了东吴大学，主修专业是生物学，选修化学。他学习努力刻苦，对任何问题都喜欢追根溯源。他博览群书，常进入图书馆如饥似渴地读书，对于生物学科的书籍更是如获至宝。这为他尔后献身于生命科学，成为国内外著名的病毒学家奠定了牢固的基石。1930年他完成了大学学业，获得东吴大学理学学士学位。同年，21岁的高尚荫由一位旅美亲戚介绍，获得了美国佛罗里达州劳伦斯大学的奖学金赴美国学习。在劳伦斯大学，他各科学习

作者：刘欢[1,2]，张晗[1]，陈逗逗[1]

1　中国科学院武汉病毒研究所，武汉430071，中国

2　病毒学国家重点实验室，武汉430072，中国

邮箱：liuhuan@wh.iov.cn（刘欢）

图2　高尚荫教授在美国耶鲁-中国交流会

成绩优秀，免修了很多课程，一年后就获得了文学学士学位。1931年秋，高尚荫进入美国耶鲁大学读研究生。前两年通过在实验室协助教授工作以获得维持生活的费用。1933年，在美国著名原生动物学家L. L. Woodruff教授的指导下攻读博士学位。1935年初，他的毕业论文《草履虫伸缩泡的生理研究》提前完成，在答辩过程中受到导师和专家的好评，获得了耶鲁大学理学博士学位。他的几位美国朋友希望他在美国工作，可他想得更多的是自己贫穷落后的祖国需要掌握科学知识的儿女。1935年2月，高尚荫等不及5月底举行的毕业典礼，就提前离开耶鲁大学来到了欧洲，为的是利用回国前的宝贵时间接触和学习更多的先进技术，更全面地了解西方发达国家的科技发展现状，以便回国后更好地开展工作。他在英国伦敦大学研究院开展了短期科学研究。

1935年8月，年仅26岁的高尚荫回到了祖国，受聘任教于武汉大学，成为该校当时最年轻的教授。1935—1945年，他先后讲授过普通生物学、原生动物学、无脊椎动物学、微生物学和土壤微生物学等课程。除担任教学工作外，他还积极从事科研工作。在环境极度困难的条件下，他不知疲倦地工作，先后在《中国生理学杂志》《科学》（*Science*）等国内外刊物上发表有关原生动物生理学和微生物固氮菌方面的论文20余篇。1945年，高尚荫再次赴美，在洛克菲勒医学研究所任客座研究员，在美国著名生物化学家、病毒学家、诺贝尔奖获得者W. M. Stanley的实验室从事病毒学研究工作，从此开始了他在病毒学领域近半个世纪的奋斗。1947年，风华正茂的高尚荫，怀着报效祖国的一片赤子之心，推辞了研究所的高薪聘用，谢绝了英国皇家科学院的盛情聘请，毅然返回祖国，在武汉大学创办了我国第一个病毒学研究室。1955年，在高尚荫的主持下，武汉大学创立了国内大学中第一个微生物学专业，为国家培养了大量的微生物学专业人才（Hu，2004）。

1956年1月，作为当时我国病毒学研究的领军人物，高尚荫响应“向科学进军”的伟大号召，看到了在我国开展病毒学研究的重要性和紧迫性。于是，他主动承担重任，与陈华癸等科学家一起，建立了国内第一个病毒学基础研究机构——武汉微生物研究室筹备委员会（1958年更名为武汉微生物研究室，1961年更名为中南微生物研究所，1970年更名为湖北省微生物研究所，1978年更名为中国科学院武汉病毒研究所，后简称“病毒所”），重点开展病毒学基础研究，服务于工业、农业和国防建设，提高我国整体生命科学研究水平。建所初期，为了能在较短时间内组织与学科发展相匹配的科技人才，高尚荫等科学家一方面选拔一批最优秀的大学毕业生作为研究工作的新生力量，另一方面依靠高等院校人才优势撑起门户，使病毒所拥有了各级研究人员共30余人组成的队伍，并成立了病毒研究组、土壤研究组、遗传研究组以及植保研究组四个研究组，面向国家需求，开展病毒学基础研究。在担任病毒所所长的28年间，高尚荫夜以继日，不知疲倦，他带领一批科研人员，共同承担科学研究课题370余项。在研究方向上，高尚荫等科学家紧跟国家需求，不断调整学科布局，将最初的四个研究组发展成三个大的研究室，即病毒学研究室、工业微生物室和农业微生物室，涵盖了病毒学研究的多个领域，并在昆虫病毒的基础研究、乙型肝炎病毒及其诊断方法、根瘤菌的固氮探索和工业污水的微生物处理等领域取得了重大进展，其中有些工作当时在国内是开创性的或占领先地位的。在做科研的同时，高尚荫一直重视在研究工作中培养科研人员。在他的努力下，病毒所的各级人员发展到100余人，这不仅使病毒所有能力培养硕士和博士研究生并有条件承担国家和地方的相关科研任务，也为病毒所今后的发展奠定了人才基础。

高尚荫一直立足于病毒学研究前沿。在Stanley的实验室工作期间，他发表了《从土耳其烟草和福禄草分离出来的两株烟草花叶病毒的比较研究》一文，阐述了病毒理化性质的稳定性，证明病毒性质，特别是理化性质不以宿主的不同而存在差异。这一研究成果在苏联病毒学家Surkov的《病毒本质》一书中和其他专业文献中被广泛引用。高尚荫从美国回国后继续在武汉大学从事这一研究，他克服了研究条件、实验经费等多重困难，取得了重要研究进展（Gaw，1953）。当时国内外有关流行性感冒病毒的研究一直是在孵育的鸡胚蛋中进行的，而他首次成功地在孵育的鸭胚蛋中培养出流感病毒并比较了鸡胚和鸭胚中流感病毒的理化性质，再次通过动物病毒证实病毒理化性质不以宿主的不同而异。这一研究成果得到了病毒学研究工作者的高度评价。

此外，他还在无脊椎动物组织培养方面取得重大突破。1935年Trager曾用悬滴法培养家蚕的各种组织，结果除卵巢外，其余的组织（如神经、脂肪等）都不能生长。此后的20余年，国际上在家蚕组织培养这方面的工作都没有取得大的进展。1957年，高尚荫带领研究小组开展“家蚕脓病病毒的组织培养方法”课题研究，并于1958年在捷克斯洛伐克第六届国际病毒学研讨会上宣读了《用单层组织培养法培养家蚕的各种组织》的论文，介绍了他们研究小组首创的单层组织培养方法并将该方法用于家蚕脓病病毒的研究，引起了与会科学家的强烈反响，这是无脊椎动物组织培养和昆虫病毒研究的重大突破。

家蚕传染病暴发会给丝绸产业带来严重损失。与饲养蚕的方式相比，用组织培养的方式繁殖病毒可不依赖季节。此前的研究认为脓病病毒仅在卵巢中增殖。高尚荫和他的同事培养了来自雌性和雄性家蚕生殖腺、气管、肌肉、肠和丝腺组织中的单层细胞。为了制备单层培养物，他们采用两种方法：细胞悬浮培养（cell suspension culture）和通过胰蛋白酶消化获得细胞（cells obtained by trypsinization）。最终，他们不仅将家蚕的各种组织培养成功，并对家蚕卵巢细胞体外培养22代，获得了昆虫细胞传代培养的成功经验；更为重要的是，他们在单层细胞培养上观察到多角体病毒在细胞内的形成和致细胞病变作用，产生自带病毒和形成病毒包涵体。这项研究成果被国外病毒工作者誉为“具有国际领先水平的开创性工作”“最重要的突破”和“经典实验”。

高尚荫常引用马克思说的话“科学绝对不是一种自私自利的享乐。能有幸致力于科学研究的人，首先应拿出自己的知识为人类服务”来教导学生。在科学研究的过程中，他遇到了很多困难和挫折。“行路难，多歧路”，但他以百折不挠的献身精神持续探索，坚韧不拔，始终站在生物科学研究的前沿。他虽然为祖国为人民作出了重大贡献，却常常说：“努力在我，评价在人。”高尚荫虽已离我们而去，但他留下了极为宝贵而丰富的精神财富。他以“敢为天下先”的创新精神带领青年向科学前沿进军，成为后辈心中的楷模。

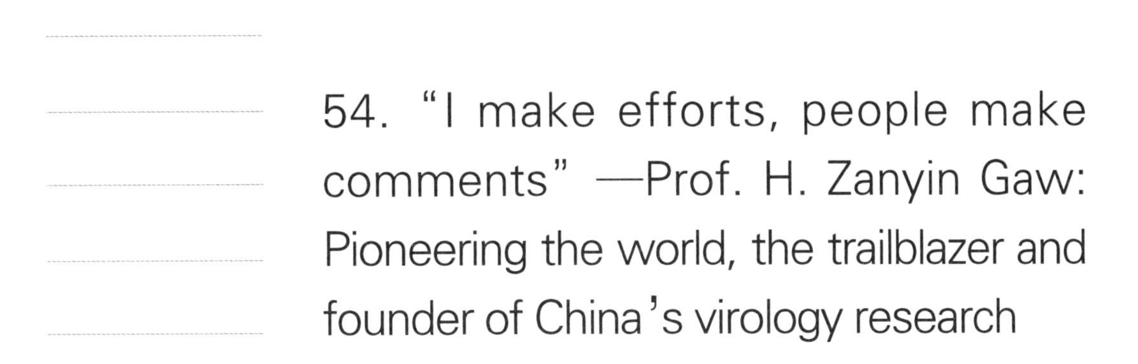

54. "I make efforts, people make comments" —Prof. H. Zanyin Gaw: Pioneering the world, the trailblazer and founder of China's virology research

"As a scientist and educator, Professor Harry Zanyin Gaw (H. Zanyin Gaw) enjoyed an outstanding reputation both at home and abroad. For more than half a century, he was always erudite and studious, devoting himself to the advancement of science and education of China, always keeping pace with global development trend of biology." Prof. Minyou Qi, the former president of Wuhan University appraised (Hu et al., 2002; Gaw, 2007) . Prof. Gaw was one of the founders of virology in China. He established the first virology research institute, the first microbiology major and the first virology major of China. In 1958, his research on the silkworm pus virus tissue culture represented a breakthrough of invertebrate tissue culture and insect virus research. His book *Theory and Applied Research on Insect Viruses* was received to worldwide widespread acclaim. In 1980, he was elected as an Academician of the Chinese Academy of Sciences (CAS). Throughout his 56 years of research and education, he made tremendous contributions to the progress of Chinese microbiology and virology.

Prof. Gaw was born on March 3, 1909 in a family of scholars of Taozhuang Township, Jiashan County, Zhejiang Province. In 1926, majoring in biology, he became a college student of Dongwu University. In 1930, he was awarded a Bachelor of Science Degree. When he was 21 years old, he was offered a scholarship by Lawrence University and obtained a Bachelor of Letters in only one year. In 1931, he began his postgraduate education in Yale University. His dissertation "Physiology of the contractile vacuole in ciliates" won praise from his tutor and experts, and in early 1935, Prof. Gaw was awarded a doctor's degree. Though being invited to stay in the United States, he was concerned more about his impoverished motherland and its lack of virology knowledge and committed himself to returning home.

In August 1935, Prof. Gaw came back to China and at the age of 26 he became the youngest professor of Wuhan University. Between 1935 and 1945, he taught general biology, microbiology, and other courses at Wuhan University. In 1945, as a visiting researcher, Prof. Gaw went to the

---

Huan Liu[1,2], Han Zhang[1], Doudou Chen[1]

1 Wuhan Institute of Virology, Chinese Academy of Sciences, Wuhan 430071, China

2 State Key Laboratory of Virology, Wuhan 430072, China

Correspondence: liuhuan@wh.iov.cn (H. Liu)

Rockefeller Institute for Medical Research and engaged in virology research in the laboratory of the Nobel Prize Laureate W. M. Stanley. In 1947, after declining an offer of Rockefeller Institute and the invitation of the Royal Society of Britain, he returned China and founded the first virology research laboratory of China. In 1955, led by Prof. Gaw, Wuhan University established the first microbiology major domestically (Hu, 2004).

In January 1956, Prof. Gaw answered the Call to "March towards Science" . He took the initiative to establish the Wuhan Institute of Virology, CAS. He worked with Prof. Huagui Chen and other scientists to establish the Wuhan Microbiological Research Preparatory Committee, later renamed to the Wuhan Institute of Virology in 1978, which was the first research institute of virology in China. The institute was dedicated to basic virology research and was aimed at serving the demands of industry, agriculture and national defense. Prof. Gaw provided outstanding leadership for the institute for 28 years, and enabled the institute to make significant progresses and several groundbreaking discoveries in China.

While working in the lab of W. M. Stanley, Prof. Gaw published several influential academic papers including a paper entitled "A comparative study of the properties of two strains of tobacco mosaic virus prepared from the sap and from the leaf residues of diseases Turkish tobacco plants" , where he described the stability of physicochemical properties of virus and proved that the virus' properties, especially its physicochemical properties remained the same despite separate hosts. This research result, which was highly appraised by peer scientists, has been widely referenced in professional literatures throughout the world. After returning from the U.S., Prof. Gaw continued to develop this research in Wuhan University, proving the physicochemical properties of influenza virus successfully (Gaw, 1953), despite the difficulties in experimental budgets and research conditions.

In 1957, he led a research team to commit to Research on Cultivation of Virus of Grasserie in Silkworm Tissue Cultures, and presented the thesis of "Culturing all types of silkworm tissues using monolayer culture" at the 6th International Invertebrate Virology Conference in Czechoslovakia in 1958, which aroused a strong response among the attending experts. The use of the monolayer tissue culture method for research into silkworm pus disease virus by Prof. Gaw and his research team represented a major breakthrough in invertebrate tissue culture and insect viruses research.

Outbreaks of silkworm infectious diseases could cause severe loses in silk production. Compared with feeding of silkworms, tissue culture is season-independent for virus propagation; the pus virus was considered to be proliferated only in ovaries. Prof. Gaw and his colleagues were able to culture the monolayers cells obtained from male and female gonads, trachea, muscle, intestine and silkgland tissue. To prepare monolayer cultures, they used two methods: cell suspension culture and cells obtained by trypsinization. Ultimately, they succeeded in maintaining subcultures of male and female gonad cells for twenty-two generations, which can maintain a normal phenotype and represent the typical cytopathic effect after virus infection. This elegant technique, permitting continuous studies of the virus disease without raising silkworms, is now regarded by international virologists as a pioneering work in this field.

Prof. Gaw always quoted Karl Marx, who once said that "Science must not be a selfish pleasure. Those who have the good fortune to be able to devote themselves to scientific pursuits must be first to place their knowledge at the service of humanity" , to inspire students. During his scientific research, he came across difficulties and frustrations, but Prof. Gaw always had unswerving dedication and stood on the leading edge of advanced biological science. Making significant achievements for China, Prof. Gaw always said "I make efforts, people make comments" . Prof. Gaw has left a valuable and generous spiritual fortune for us. His entrepreneurship of "Pioneering the World" will continue to lead youth to march towards the advancement of science, the lessons from Prof. Gaw should be everlastingly cherished and never forgotten.

Figures

Fig.1 Prof. H. Zanyin Gaw (1909—1989)

Fig.2 Prof. H. Zanyin Gaw at the U. S. Yale-China Association

References

Gaw ZY (1947) A comparative study of the properties of two strains of tobacco mosaic virus prepared from the sap and from the leaf residues of diseases Turkish tobacco plants. Arch Virol 167:347-355.

Gaw ZY (1953) Properties of purified PR influenza virus from inflected duck embryo. Acta Microbiol Sin 1:36-41.

Gaw ZY (2007) The Anthology of Shangyin Gao. Wuhan: Wuhan University Press.(高尚荫. 2007. 高尚荫文选. 武汉: 武汉大学出版社.)

Hu Y (2004) Insect virus research and prospects in China. Virol Sin 3: 303-308.

Hu Y, Qi Y, et al (2002) A Festschrift for Shangyin Gao. Wuhan: Wuhan University Press.

Liu N, Xie T, Gaw ZY (1958) Culturing all types of silkworm tissues using monolayer culture. Chin Sci Bull 7: 219-220.

55. 黄祯祥：病毒体外培养技术的创新者

1954年，美国科学家J. F. Enders、T. H. Weller和F. C. Robbins由于首次在人体组织试管培养中成功培养了小儿麻痹病毒，为防治严重威胁儿童健康的神经性疾病奠定了理论基础，共同获得诺贝尔生理学或医学奖。当年12月11日，3位获奖者在受奖演说中谈及他们采用的方法是由前人创建的，特别是黄祯祥（图1）。

图1　黄祯祥博士（1910—1987）

1941年，黄祯祥到美国洛克菲勒医学研究所从事病毒学研究，1943年，他在美国《实验医学杂志》上发表论文《马脑炎病毒西方毒株在组织培养中滴定和中和作用的进一步研究》。在这篇论文中，他成功建立了一种新的病毒培养技术，从而在一个制约着病毒学发展的关键问题上取得了突破。在此之前，检测病毒必须通过注射感染，观察动物发病或死亡情况来判断。黄祯祥创造的方法则是用人工方法将动物组织消化成单层细胞，并添加一定的营养成分使其在试管内存活，然后将病毒接种在细胞内，经过一段时间后，如果发生了病毒侵染和增殖，用光学显微镜即可观察到细胞病理改变。这一新技术使病毒培养从实验动物和鸡胚的活体水平到了细胞水平。Enders等在他们的受奖演说中提到，以往直接观察脊髓灰质炎病毒引起的细胞病变非常麻烦，往往需

作者：程光胜

中国科学院微生物研究所，北京100101，中国

邮箱：guangshch@vip.163.com

要2周以上的时间。他们试验过许多改进方法，而最后采用黄祯祥的方法获得了成功。后来，全世界许多实验室应用这一技术，阐明了许多病毒性疾病的病原学，分离出许多新病毒，解决了当时还鲜为人知的一些疾病的病毒病因问题。正因如此，在病毒学发展历史中，黄祯祥被认为为现代病毒学作出了重要贡献。

黄祯祥出生在我国厦门鼓浪屿的一个信仰基督教的医生家庭，从小就立志学医。1934年毕业于北京协和医学院，1941年9月至1943年10月在美国从事病毒学研究，1943年12月回到祖国参加抗日战争。归国前，他在发表的论文中毫无保留地将自己的成果公开，使后人能够在探索科学时少走弯路。新中国成立后，他在北京一直从事病毒学研究，1980年当选中国科学院学部委员（生物学部）。他是美国实验生物医学会会员，美国传染病学会名誉会员，国际比较病毒学组织咨询委员会委员。在20世纪50年代曾任苏联与其他东欧国家合办的《病毒学问题》编委，还担任过美国《国际病毒学杂志》和《传染病学论丛》编委。

黄祯祥在50多年的学术生涯中，主要从事的是医学病毒学理论和病毒病的研究。他主导的对流行性乙型脑炎的系统研究，于1978年获得了全国科学大会奖，又在他逝世后的1988年获得了卫生部科学技术进步奖一等奖。他还对麻疹病毒的致病性和免疫性进行过深入研究，对麻疹病毒的血凝素、麻疹疫苗的佐剂和疫苗生产工艺等进行了研究。

他在晚年开始从事病毒免疫的研究，先后发表了《被动免疫对活病毒主动免疫的影响》等论文。他在国内首先将病毒干扰现象应用到医学病毒学研究中，他提出了通过病毒免疫治疗肿瘤的设想，认为利用病毒感染肿瘤细胞，不但有可能直接杀伤肿瘤细胞，而且在病毒感染后会改变肿瘤细胞膜的抗原性，有利于调动机体免疫系统识别肿瘤细胞、控制肿瘤的发展。经过几年的探索，这个课题在实验研究方面已经取得了可喜的进展。他还在研究别人不注意的非致病性病毒，在虫媒病毒病的防治方面开拓了新的前景。在他的指导下，研究人员对病毒免疫治疗肿瘤进行了大量探索性工作。他不幸罹患白血病后，曾在自己身体上进行试验，用他的病体为学科发展作出了最后的贡献。

黄祯祥在国际上享有较高的学术地位。1983年，他率中国微生物学会代表团赴美国参加第13届国际微生物学大会，美国得克萨斯州丹顿市曾为他颁发金钥匙，授予他“荣誉市民”称号（图2）。1986年，他曾在北京接待过美国著名病毒学家、国际病毒命名委员会创建人J. L. Melnick教授（图3）。

黄祯祥生前说过：“走老路永远不能超过别人，一定要注意创新。我一生所有的研究工作都遵循这个原则。”

图2　1983年得克萨斯州丹顿市授予黄祯祥的金钥匙

图3　J. L. Melnick教授和黄祯祥教授1986年在北京

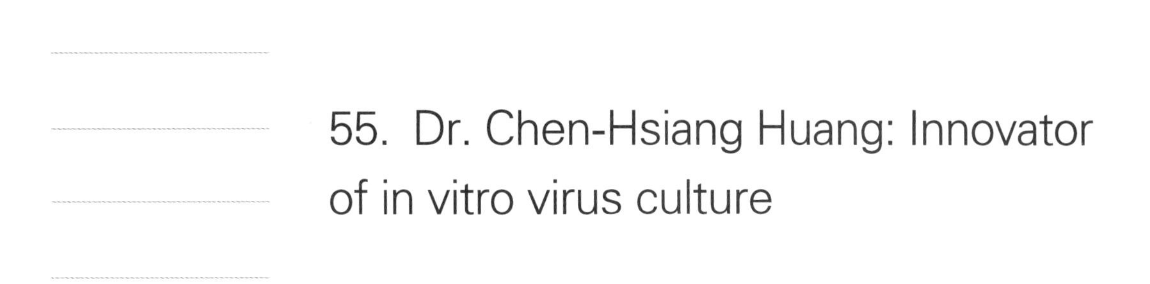

55. Dr. Chen-Hsiang Huang: Innovator of in vitro virus culture

In 1954, American scientists John Franklin Enders, Thomas Huckle Weller and Frederick Chapman Robbins were awarded the Nobel Prize in Physiology or Medicine due to their contribution to the development of the technique to grow polio virus in cultures of human tissues. Their technique later became a powerful tool for the prevention of poliomyelitis in children. At the Nobel lecture on December 11, 1954, when they talked about how they developed this technique, the three scientists mentioned that the methods had been used in the past by other workers, in particular Dr. Chen-Hsiang Huang.

Dr. Chen-Hsiang Huang had been studying virology at the Rockefeller Institute for Medical Research since 1941. In 1943, he published an original research article "Further studies on the titration and neutralization of the western strain of equine encephalomyelitis virus in tissue culture" in *Journal of Experimental Medicine*. In the paper, he reported on his successful establishment of a new technique to culture viruses, thus making a breakthrough in virology. Prior to his studies, viruses had to be examined by injection into animals and observation of the incidence or mortality of infected animals. Dr. Chen-Hsiang Huang's method is to digest animal tissues into a single layer of cells, and then inoculate the virus in the cultured cells. If there is a virus infection and proliferation, pathological changes of the inoculated cells can be observed under a light microscope. This technique brought virus culture from animal level to cellular level. In their Nobel lecture, Enders, Weller and Robbins said that in the past, observation of cellular pathological changes caused by the poliovirus was very troublesome and often took more than two weeks. They tested a number of methods to improve it, and finally developed the technique with Dr. Huang's method. Later on, many laboratories around the world applied this technique in pathogenic virus studies and virus isolation, especially in the virus etiology of some little-known diseases at that time. For his innovative achievement, Chen-Hsiang Huang was later considered to have made important contributions to modern virology.

Dr. Chen-Hsiang Huang was born in a Christian family in Kulangsu, Xiamen, China and

---

Guangsheng Cheng
Institute of Microbiology, Chinese Academy of Sciences, Beijing 100101, China
Contact: guangshch@vip.163.com

grew up determined to study medicine. He graduated from Peking Union Medical College in 1934, and was engaged in virology research in the United States from 1941 to 1943. Before he returned to his motherland in December 1943, he unreservedly published his findings in his paper, so that other virologists could avoid some detours in seeking the truth. After the founding of the People's Republic of China, he continued devoting himself to virology studies in Beijing. In 1980, he was elected a member of the Academic Divisions of Chinese Academy of Sciences. He was also a member of American Society for Experimental Biology and Medicine, an honorary member of Infectious Diseases Society of America, and a member on the Advisory Committee of the International Comparative Virology Organization. In the 1950s, he served as a member of the editorial board of the journal *Issues of Virology* published by the former Soviet Union and other Eastern European countries, as well as the editorial boards of the American journals *International Journal of Virology* and *Reviews of Infectious Diseases*.

In over 50 years of his academic life, Dr. Chen-Hsiang Huang conducted extensive research on the theory of medical virology and viral diseases. On account of his contribution to the systematic study of encephalitis, he won the National Science Conference Prize in 1978 and, after he passed away, the Science and Technology Award of Ministry of Health in 1988. He also studied the pathogenicity and immunity of measles virus, as well as the measles virus hemagglutinin, measles vaccine adjuvants, vaccine production techniques, etc.

In his later years, he began to focus on viral immunity studies, and published research articles such as "The effects of passive immunity on active immunity by live virus" . He was the first one in China to apply viral interference into medical virology research. He presented a vision of cancer immunotherapy through virus, in which virus was used to infect tumor cells, not only to directly kill the tumor cells, but also to change the antigenicity of the cell membranes of the tumor cells, so that they can be recognized and targeted by the patient's immune system. After years of exploration, this project had made encouraging progress. He also studied nonpathogenic viruses that others paid little attention to at that time, as well as the insect-vector viral diseases. Under his guidance, a lot of exploratory work were carried out on virus-mediated immunotherapy of tumor. After he was diagnosed with leukemia, he tested the immunotherapy on himself, making a final contribution to the research.

Dr. Chen-Hsiang Huang had a high academic status internationally. In 1983, when he led a delegation of the Chinese Society for Microbiology to the United States to attend the 13th International Congress of Microbiology, he was awarded the honorary citizenship and the "Key to the City" by Denton, Texas. In 1986, he met Prof. Joseph Louis Melnick, American virologist and the founder of the International Committee on Taxonomy of Viruses, during Melnick's visit in Beijing, and introduced him to the Chinese community of virologists.

"One can never surpass others if always taking the old paths. Innovation is the key in exploring truth. This is the principle that I've been following all my life." said Dr. Chen-Hsiang Huang to summarize his academic life.

Figures

Fig.1 Dr. Chen-Hsiang Huang (1910—1987)

Fig.2 The Golden Key to the City of Denton, Texas that was awarded to Chen-Hsiang Huang in 1983

Fig.3 J. L. Melnick and Chen-Hsiang Huang in Beijing (1986)

56.“糖丸爷爷”顾方舟：只为挺起中国的脊梁

图1　顾方舟教授（1926—2019）

顾方舟教授（图1），1926年6月出生于上海。医学家、病毒学家，第三世界科学院院士，中国医学科学院北京协和医学院院长、一级教授。2019年9月17日，顾方舟教授被授予“人民科学家”国家荣誉称号。

童年家庭的悲惨境遇和屈辱的亡国奴生活，使顾方舟在思想上和民族感情上受到极大的冲击，他刻苦学习，自立自强。1944年，顾方舟考入北京大学医学院，立志当一名医生，救民众于危难。1950年毕业分配，他毅然选择了从事公共卫生事业，让更多的人受益，1950年10月，他怀着这样的信念，欣然接受学校的分配，到大连卫生研究所工作，从此，顾方舟和病毒结下了不解之缘。

1951年8月，顾方舟作为新中国成立后派往苏联的第一批留学人员，被分配到苏联医学科学院病毒学研究所（图2）。

图2　顾方舟教授在苏联学习

四年的学习为后来开展的科学研究奠定了坚实的基础。回国后，29岁的顾方舟被任命为卫生部微生物流行病研究所脑炎室副主任，1958年被调入中国医学科学院病毒研究所任脊髓灰质炎研究室主任，从此，他就把毕生精力投入消灭这一可怕的儿童急性病毒传染病的战斗中。为利用高等动物猿猴进行医学研究和研制脊髓灰质炎疫苗，1958年，中国医学科学院沈其震副院长和顾方舟到云南选址筹建医学生物学研究所（简称“生物所”）。在昆明西郊海拔2100米的玉案山一片荒芜的山坡地上，无水无电无道路，从零开始艰苦创业，最终建成（图3）。

译者：吴俊
中国医学科学院医学生物学研究所，昆明650118，中国
邮箱：wujun@imbcams.com.cn

图3 中国医学科学院医学生物学研究所成立初期

图4 顾方舟教授（二排左一）和同事在苏联考察疫苗生产情况

在1953年南通发生脊髓灰质炎流行之后，卫生部将此病列为法定报告传染病，自此疫情报告日趋增多，发病地区不断扩大，严重影响儿童健康。1959年3月，国家派出由顾方舟任组长的四人小组（顾方舟、董德祥、闻仲权、蒋竞武）去苏联考察疫苗生产情况（图4）。顾方舟从学术会议上了解到，美苏两国正在合作研制“脊灰活疫苗”。

他查阅了所有能获得的公开资料，比较了两种疫苗的优劣，根据我国人口众多、经济不发达等具体情况，大胆地提出了走活疫苗技术路线的建议，这项建议得到了卫生部的赞同和支持。这一决策对我国最终消灭脊髓灰质炎具有决定性意义，也是顾方舟一生中所作的最大贡献。

于是，顾方舟立即提前回国，开始了脊灰活疫苗的研制。1959年12月，卫生部召集中国医学科学院及北京和成都生物制品研究所组成协作组，由顾方舟任组长，在北京生物制品研究所进行试生产。1960年初，协作组用Sabin Ⅰ、Ⅱ、Ⅲ型原始毒种试制出我国首批500万人份的Ⅰ、Ⅱ、Ⅲ型减毒活疫苗（顾方舟等，1961）。这时需要进行三期临床试验，第Ⅰ期临床试验需有10名易感小儿服苗进行安全性观察，顾方舟带头报名给他刚出生几个月的儿子服苗，带动了实验室同仁纷纷参与，从而顺利验证了疫苗的安全性。第Ⅲ期临床试验主要观察流行病学效果，与北京、上海、青岛等11个城市的卫生防疫站合作，组织近400万名7岁以下儿童试服疫苗，结果发病率明显降低，削平了季节高峰。三期临床试验结果证明活疫苗安全有效，具有良好的免疫学和流行病学效果。

1961年继续由此三个单位组建专业团队前往昆明医学生物学研究所扩大试生产规模，建立生产基地。团队由顾方舟领导的四人小组为骨干，培训年轻技术人员熟练掌握技术操作，又以苏联技术为基础，结合我国实际，编写制定疫苗生产、检定规程及操作细则，保证了疫苗安全生产及质量控制。

1959年决定活疫苗技术路线后，由于活疫苗对低温的严格要求，必须改进剂型保证疫苗质量。1960年，顾方舟提出研制糖丸疫苗，并由董德祥具体负责。中国医学科学院沈其震副院长亲自挑选与上海信谊药厂协作，采用中药制丸技术将病毒液包裹在糖丸中制成糖丸疫苗。历经3年上百次反复试验，不断改进糖丸配方和滚丸工艺，终于在1963年成功研制出可在室温及4～8 ℃条件下延长保质期的糖丸疫苗，1964年推广到全国使用，受到广大防疫人员及家长和儿童的欢迎与喜爱，又为预防、控制和消灭脊髓灰质炎增添了一个有力工具。

1962年，疫苗生产专业团队完成扩大试生产任务后返回原单位。1963年，董德祥被调回生物所具体负责疫苗生产；1964年，顾方舟调任生物所副所长，并举家定居昆明。随着疫苗需求增长，生物所疫苗生产任务不断加大，从最初的每年500万增至6000万～7000万，最高达到1亿多人份。随着经验的积累，每一到两年修订一次疫苗生产及检定规程，不断提高疫苗质量。

在顾方舟教授主持下的生物所始终坚持研究、生产和应用相结合的原则，使活疫苗真正为控制、消灭脊髓灰质炎服务。生物所每年组织科技人员到全国各地开展病毒学、血清学、流行病学调查，及时掌握病毒传播动态、免疫效果及存在问题，为不断改进免疫方案、修订免疫规划、开展计划免疫提供科学依据。生物所为提高各型疫苗的免疫效能，减少接种次数，经多年反复试验，调整型间剂量配比，成功研制出三价糖丸疫苗。为了发动各省（自治区、直辖市）防疫站积极参与免疫效果及流行病学调查，生物所在卫生部支持下，分别于1971年、1975年、1978年组织了三次“全国预防、控制脊髓灰质炎经验交流会”，交流预防经验，推动计划免疫；还举办了多期脊髓灰质炎及其他肠道病毒实验技术培训班，推动病毒学、血清学监察。

1971年，顾方舟教授因工作需要调往中国医学科学院，但他对脊灰活疫苗生产、服用情况及存在问题始终十分关注，经常给予指导和帮助。董德祥教授回忆，每次到北京开会或汇报工作，顾老都会详细询问和讨论有关消灭脊髓灰质炎的方方面面的问题，研究解决方案。

2000年，世界卫生组织确认中国已成功阻断了本土脊髓灰质炎野毒株的传播，实现了消灭脊髓灰质炎的目标。顾方舟作为代表之一在《中国消灭脊髓灰质炎证实报告》上签上了自己的名字，完成了他一生的夙愿。这是继全球消灭天花之后，世界公共卫生史上的又一创举，顾方舟作为这一创举在中国的领头人，功不可没。他尽职尽责，用自己的一生完成了国家托付的使命。他将永远被铭记在人民心中。

在顾方舟教授的回忆录中，他写道：“我活这一辈子，不是说从别人那里得到什么，而是我自己给了别人什么。”

顾方舟教授的一生，是奉献于国家免疫事业的一生，默默无言，大爱坚守，他的故事留在共和国的历史中，闪耀着不褪的荣光。

致谢

感谢李琦涵教授和董德祥教授对本文翻译工作的指导和对稿件的修改！

56. Prof. Fangzhou Gu, father of "sugar pills" : Just for Chinese people's stood up

Professor Fangzhou Gu (顾方舟) was born in Shanghai in June 1926. He is a medical scientist, expert virologist, member of the Third World Academy of Sciences, the former president and first-level professor of Peking Union Medical College of Chinese Academy of Medical Sciences. On September 17, 2019, Professor Fangzhou Gu was conferred the "People's Scientist" national honorary title.

The family tragedies of his childhood and the humiliating life as conquered people had a tremendous impact on Professor Fangzhou Gu, including his thoughts and national sentiments. He was independent, self-support and hard working. In 1944, Professor Fangzhou Gu was admitted to the Medical School of Peking University and inspired to be a doctor to rescue the people in suffering. After graduating in 1950, he resolutely chose to devote to public health to benefit more people. In October 1950, with such a belief, he readily accepted the school's assignment and worked at the Dalian Health Research Institute. Since then, Professor Fangzhou Gu developed a deep infatuation for virology.

In August 1951, Professor Fangzhou Gu, as one of the first batch of overseas students who were sent to the Soviet Union after the founding of the People's Republic of China, was assigned to the Institute of Virology of the Academy of Medical Sciences of the Soviet Union.

Four years of study laid a solid foundation for his later scientific research. After returning home, Professor Fangzhou Gu, at the age of 29, was appointed as the Deputy Head of the Encephalitis Laboratory of the Institute of Microbiology and Epidemiology of the Ministry of Health. In 1958, he was dispatched to join the Institute of Virology of the Chinese Academy of Medical Sciences as the Head of the Polio Laboratory. Since then, he has devoted his whole life into the battles to eradicate this terrible acute viral infection in children. In order to use higher primates in medical research and develop polio vaccine, Qizhen Shen (沈其震), Vice President of the Chinese Academy of Medical Sciences, and Professor Fangzhou Gu went to Yunnan to find a place for Institute of Medical Biology in 1958. On the barren hillside of the Jade Mountain

Jun Wu, Dexiang Dong, Ying Li
Institute of Medical Biology, Chinese Academy of Medical Sciences, Kunming 650118, China
Correspondence: wujun@imbcams.com.cn (J. Wu)

in the western suburbs of Kunming, where no water, no electricity, no roads was connected, the Institute of Medical Biology was built from scratch.

After the epidemic of polio in Nantong in 1953, the Ministry of Health listed the disease as a legally reported infectious disease. Since then, the epidemic reports have been increasing and the infected area of the disease has been expanding, severely affecting children's health. In March 1959, the four-member group (Fangzhou Gu, Dexiang Dong (董德祥), Zhongquan Wen (闻仲权), and Jingwu Jiang (蒋竞武)) led by Professor Fangzhou Gu was sent to the Soviet Union to investigate the vaccine production. Professor Fangzhou Gu learned from the academic conference that the United States and the Soviet Union are cooperating to develop a live vaccine for polio.

He reviewed all the public available materials and compared the advantages and disadvantages between the live and inactivated Polio vaccines. According to the specific situation on China's huge population and underdeveloped economics at that time, he boldly proposed the strategy of employing the route of the live vaccine. This proposal was approved and supported by the Ministry of Health. This proposal was decisive meaningful for the elimination of polio in China eventually, which also becomes the greatest contribution made by Professor Fangzhou Gu in his life.

Therefore, Professor Fangzhou Gu immediately returned to China and started the development of a live vaccine for polio. In December 1959, the Ministry of Health called for a collaborative team of the Chinese Academy of Medical Sciences, the Beijing Institute and Chengdu Institute of Biological Products to conduct trial production. The collaborative team was again chaired by Professor Fangzhou Gu and conducted at the Beijing Institute of Biological Products. In the early 1960s, the collaborative team used the original attenuated strains, Sabin Ⅰ、Ⅱ and Ⅲ, to produce the very first Ⅰ、Ⅱ and Ⅲ attenuated vaccines for 5 million people in China (Gu et al., 1961). At this time, three phases of clinical trials were required. In the phase Ⅰ trial, 10 susceptible children were required for safety observation. Professor Fangzhou Gu took the lead role to use the test vaccine on his newborn son, which inspired the other colleagues in the laboratory to participate in the clinical trial. Thus the safety of the vaccine has been successfully verified. The phase Ⅲ of the clinical trial was mainly for testing the epidemiological effect. In cooperation with the health and epidemic prevention stations in 11 cities including Beijing, Shanghai and Qingdao, nearly 4 million children under the age of 7 were tested for vaccines. The results significantly reduced the incidence and flattened the season peaks. The results of phase Ⅲ clinical trials demonstrated that live vaccines were safe and effective, with good immunological and epidemiological effects.

In 1961, the collaborative team continued to build a professional team to go to the Institute of Medical Biology in Kunming to expand the scale of trial production and establish the manufacturing base. A four-person team led by Professor Fangzhou Gu was regarded as the backbone scientists, through the training of young technicians to master the technical operations, the production and quality-control instructions and detailed operations procedures were compiled upon the technics of Soviet Union further with the consideration of Chinese situation, which guaranteed the safe production and quality control of the vaccine.

After the decision on choosing the route of live vaccine immunization in 1959, due to the strict requirements for low temperature of the live vaccine, it is necessary to improve the dosage form to ensure the quality of the vaccine. In 1960, Professor Fangzhou Gu proposed to develop a "sugar pill" vaccine, and Dexiang Dong was responsible for the implementation of this project specifically. Qizhen Shen, vice President of the Academy of Medical Sciences, selected Shanghai Xinyi Pharmaceutical Factory to collaborate with Institute of Medical Biology in person, and they used the preparation technology of traditional Chinese medicine pill to wrap the viral liquid in the sugar pill to make a "sugar pill" vaccine. After hundreds of repeated experiments in three years, the "sugar pill" formula and the preparation technology were continuously improved. Finally, the "sugar pill" vaccine which can prolong the shelf life at room temperature and 4~8 ℃ was successfully produced in 1963. In 1964, the vaccine was spread in the whole country and was welcomed by the majority of personnel in epidemic prevention, parents and children, becoming a new powerful weapon for the prevention, control and elimination of polio.

In 1962, the professional team of vaccine production returned to their original organizations after completing the expansion of trial production. In 1963, Dexiang Dong was dispatched back to the Institute of Medical Biology to be responsible for vaccine production. In 1964, Professor Fangzhou Gu was appointed to the deputy director of the Institute of Medical Biology, and his family moved to Kunming. With the growing demand for vaccines, the task of vaccine production continued to increase year by year, from an initial 5 million to (60~70) million per year, up to more than 100 million. The more experience was obtained, the greater improvement on vaccine quality was reached through the revision on production and quality-control instruction every one or two year.

Under supervision of Professor Fangzhou Gu, the Institute of Medical Biology has always adhered to the principle of integration of research, development and utilization, and prompt live vaccines to control and eliminate polio indeed. Every year, scientists and technicians were organized to conduct investigations of virology, serology and epidemics all over the country. They were timely informed of the dynamics of viral transmission, immunological efficacy and existing problems in order to provide scientific basis for continuous improvement of immunization methods, revision of immunization strategies, and implementation of planned immunization. To improve the immune efficacy of each type of vaccines and reduce the number of inoculations, after years of repeated trials and adjustment of the dosage ratio of the types, the trivalent vaccine in sugar pill was successfully developed. In order to encourage the epidemic prevention stations in various provinces, cities and municipalities to actively participate in the investigations on immune effect and epidemiology, with the support of the Ministry of Health, "Experience Exchange Conference of National Prevention and Control of Polio" were organized in 1971, 1975, and 1978. The experience of polio prevention was spread and planned immunization was promoted; also, a few training courses on polio and other enterovirus experimental technologies were provided to promote virology and serology surveillance.

In 1971, Professor Fangzhou Gu relocated to join the Chinese Academy of Medical Sciences for his work. However, he always paid close attention to the production, utilization and existing problems of live polio vaccines, and often gave his guidance and help. Professor Dexiang Dong recalled that every time he went to Beijing, Professor Fangzhou Gu would inquire and discuss all details of research of polio.

In 2000, the World Health Organization confirmed that China had successfully blocked the spread of native wild poliovirus strains and achieved the goal of eliminating polio. As one of the representatives, Professor Fangzhou Gu signed his own name on the "Certification Report for Eliminating Polio in China" and realized his lifelong dream. This is another pioneering work in the history of public health in the world after the global eradication of smallpox. Fangzhou Gu, as the pioneer initiative in China, has made an invaluable contribution to this work. He dedicated his life to fulfilling his duty and the national mission. Professor Fangzhou Gu will be remembered in people's hearts forever.

In the memoir of Professor Fangzhou Gu, he wrote as:

"I regard as import what have I done for the people rather than what can I obtain from them all through my life."

This is also the promise claimed by all the health guardians, never fading with the passage of time. We will succeed and continue the ideal of Professor Fangzhou Gu, safeguard public health and public health safety, and stay true to the medical mission.

Figures

Fig.1 Professor Fangzhou Gu (1926—2019)

Fig.2 Professor Fangzhou Gu studied in the Soviet Union

Fig.3 The Institute of Medical Biology of Chinese Academy of Medical Sciences was established in the early stage

Fig.4 Professor Fangzhou Gu (first from left, the second row) and his colleague investigated the vaccine production in Soviet Union

References

Gu FZ, Dong DX, Jiang JW, et al (1961) Experience in manufacturing and verifying polio oral live vaccine. In: Department of Virology, Chinese Academy of Medical Sciences. Compilation of Polio Live Vaccine Research Data. Beijing: Science Information Laboratory, Chinese Academy of Medical Sciences, 1–23. (顾方舟，董德祥，蒋竞武，等. 1961. 制造和检定脊髓灰质炎口服活疫苗的几点经验. 见：中国医学科学院病毒学系. 脊髓灰质炎活疫苗研究资料汇编. 北京：中国医学科学院科学情报研究室，1–23.)

57. 朱既明：中国分子病毒学的奠基人和生物制品研发的先驱

“我非常敬重朱既明教授，他在中国的困难时期仍然以高标准进行科学研究，在流感病毒基因表征研究方面取得了卓越成就。”这是美国国家科学院院士、RNA病毒专家Peter Palese教授在1998年收到朱既明教授讣告后所写的悼词。

“他不仅是一位优秀的科研工作者和老师，同时也是一个了不起的国际知名学者，为促进科学发展和国际合作作出了重要贡献。”这是美国圣犹达儿童研究医院病毒学与分子生物学系的著名病毒学专家Graeme Laver教授和 Robert Webster教授在1999年为朱既明教授撰写悼词时对他的评价（Laver and Webster，1999）。

图1　朱既明教授（1917—1998）

朱既明教授是国际著名病毒学家，我国分子病毒学重要奠基人之一，我国生物制品研发和产业化的重要先驱（图1）。从经典病毒学到分子病毒学时代，他都是我国病毒学研究的领军人物。他在流感病毒结构方面的研究，对国际上流感病毒鉴定和阐明各类病毒结构影响深远，也为亚单位疫苗的开发提供了理论依据和技术方法。在青霉素、病毒减毒活疫苗、重组乙型肝炎病毒（HBV）疫苗和痘苗病毒载体的研究方面，他同样取得了卓越的成就，开拓了我国生物制品研究领域。

朱既明教授1917年9月12日出生于江苏省宜兴县，1939年毕业于上海医学院，1948年获剑桥大学博士学位。1948年至1950年，他在英国国家医学研究所（位于伦敦米尔山）工作。1950年回国后，他先后在卫生部生物制品研究所、长春生物制品研究所、中国医学科学院病毒学研究所从事分子病毒学研究和生物制品研发工作（阮力和陆德敏，1998；张嘉铭，1998）。1998年1月6日，朱既明教授因病在北京逝世。他是中国科学院院士，英国皇家内科医学院院士，美国微生物学会名誉会员，中国预防医学科学院病毒学研究所名誉所长。1964年至1974年，他当选全国人大代表；1978年至

作者：严伟峥，黄保英，阮力，谭文杰

中国疾病预防控制中心病毒病预防控制所，北京102206，中国

邮箱：tanwj28@163.com（谭文杰）

1992年，他当选第五届、第六届和第七届全国政协委员（阮力，2013）。

朱既明教授对我国生物制品的研发贡献卓著

青霉素在第二次世界大战期间挽救了无数生命，因而得到了全世界的广泛关注。朱既明教授是我国青霉素研制的先驱（阮力，2015）。1940年至1945年，朱教授在昆明中央防疫处任技师。在汤飞凡教授的指导下，他与著名的农业微生物学家樊庆笙教授合作，从昆明当地分离收集了30余株青霉素菌种，通过比较成功筛选出了产量最高而且能够稳定传代的22号菌种（Chu et al.，1945；Xu，2013）。通过不断改进培养和提纯条件，成功研制出了一批注射用青霉素冻干粉末，并开展了少量临床注射研究。这是我国抗生素事业的开端。

早在20世纪50年代初，朱既明教授就开始研究病毒减毒活疫苗，他先后研究了森林脑炎、麻疹和流感减毒活疫苗。在研究流感疫苗时，他发现不同时期流感病毒减毒速度不同，并设计了“鸡胚传代法”和“遗传重组法”两种获得疫苗株的方法。1959年，全国麻疹大流行，朱教授与同事在长春开始了麻疹疫苗研究。经过6年的努力，最终选育出副作用低、抗体阳性率和抗体滴度都较高的安全高效的麻疹减毒活疫苗毒株“长–47”（Chu et al.，1964），该毒株于1966年投产使用，一直沿用至今，为我国麻疹的防治作出了重要贡献。

1951年至1963年，朱既明教授历任卫生部生物制品研究所主任技师，第二研究室主任兼检定科长，生物制品人员训练班班主任，长春生物制品研究所副所长（张嘉铭，1998）、病毒研究室主任。在这段时间，他起草了我国第一部生物制品制造与检定规程，举办了生物制品人员训练班，培养了一批优秀的青年科技骨干（阮力，2013）。

朱既明教授推动了我国痘苗病毒载体的研究。1984年初，他和同事进行了以痘苗病毒（天坛株）为载体的基因工程研究，目的是开发一种多价疫苗表达系统。他建立了能够高效表达乙型肝炎病毒、EB病毒、单纯疱疹病毒抗原的痘苗病毒载体多价疫苗表达系统（Tsao et al.，1988）。该系统广泛应用于甲型肝炎病毒、麻疹病毒和呼吸道合胞病毒的抗原表达研究。该项工作1986年得到国家高技术研究发展计划（863计划）的资助。在他的带领下，我国在世界上首次使用痘苗病毒疫苗株研制出外源基因表达系统，并用该系统完成了针对甲肝、乙肝、麻疹、EB病毒感染性疾病的5株重组痘苗病毒疫苗的临床Ⅰ期研究。此外，他还发起并指导了具有我国自主知识产权的痘苗病毒（天坛株）载体的研发计划，该载体在我国疫苗研究中得到广泛应用。

20世纪80年代，乙型肝炎已成为严重危害我国人民健康的病毒病，研制新一代疫苗已成为乙型肝炎防控的当务之急。年过花甲的朱既明教授提出并领导了中国仓鼠卵巢（CHO）细胞表达的乙肝基因工程疫苗研究课题。经过10多年的努力，我国自行研制的基因工程乙肝疫苗成功上市，这也是世界上第一个使用CHO细胞研制成功的基因工程乙肝疫苗（阮力等，1983）。该疫苗自1991年上市以来，已在我国广泛应用至今。

朱既明教授在流感病毒研究方面硕果累累

从1945年到1984年的近40年，朱既明在从事其他研究工作的同时，从未中断对流感病毒的研究。1949年，他在国际上首次发现流感病毒杆丝状形态可向球状形态转变，这项工作率先在*Lancet*发表（Chu et al.，1949），并被后来的科学研究所证实。在此期间，他还发现了β抑制素和γ抑制素可以抑制流感病毒对红细胞的凝集（Chu and Coombs，1947；Chu，1948），β抑制素作为新的病毒抑制素也被称为“朱氏抑制素”。他还发现，乙醚和其他去污剂的处理可以大大降低流感病毒的感染性，同时病毒被裂解成不同大小的亚单位：血凝素和可溶性抗原。他在此基础上提出“病毒由类脂膜包裹，外层是血凝素、内部是可溶性抗原”的设想，成为20世纪50年代后阐明各种病毒结构的先导，为后来研究亚单位疫苗奠定了理论基础。

朱既明教授发现流感病毒有多个亚型，是我国1957年和1977年流感大流行病毒的确认者（Laver and Webster，1999）。1949年，朱教授证实了流感病毒抗原变异（漂移）后会传播到其他地方（Chu，1951）。1957年2月，朱教授从流行病学上追踪，明确H2N2病毒首次出现于贵州省贵阳市，不同于之前的H1N1毒株。该病毒于1957年3月底向全国散播。这种流感病毒是一种新的亚型，它的血凝素抗原发生了质的变异（即抗原转移），这是我国首次有病毒学证据的甲型流感病毒新亚型引起世界大流行。此后，他于1977年2月在鞍山确认了1950 年H1N1流感的再次暴发，比该毒株到达世界其他地区早9个月（Laver et al.，1984）。流感病毒多个亚型的发现为流感监测和疫苗株筛选提供了指导原则。

朱既明教授在国际上首次发现自然界存在流感病毒温度敏感株。他首次证明在自然界流感病毒不仅会发生抗原变异，而且会发生毒力变异。1977年，他发现新分离的流感病毒株可分为正常株和温度敏感株（简称温敏株）（Chu et al.，1982）。温敏株指只在33 ℃以下才能繁殖的毒株，对人类无致病性。遗传研究表明，温度敏感性可以通过基因重组传递到正常株中（Chu et al.，1992），为分离减毒株用于活疫苗的研究提供了理论可行性。

朱既明教授博学多才，极具大师风范

朱既明教授的格言是“生活朴素，少物质欲望，方能立理想志气；头脑冷静，无急躁情绪，方能有远大目光”（图2）。他强调学生和导师都应诚信治学。他为人正直，生活简朴，不受名利驱使。工作中他坚持原则，痛恨不正之风；研究中他科学严谨，反对弄虚作假。他不仅是一位伟大的科学家，也是一位了不起的人生导师。朱教授的重要贡献和高尚品格应该被后人铭记，他的事迹激励着我们继续为科研事业而奋斗。

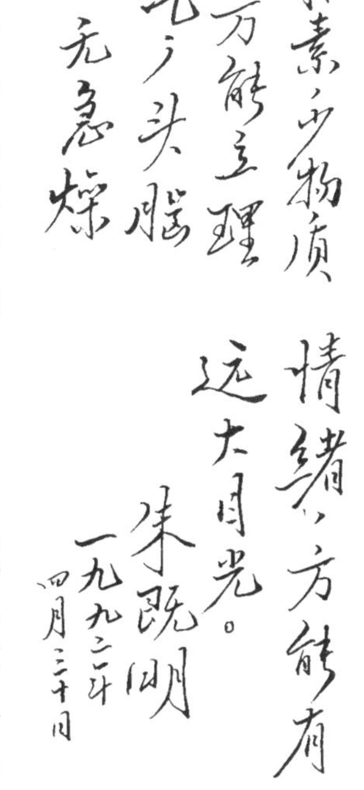

图2　朱既明教授的格言手稿（1992年）

朱既明教授非常重视国际合作与关注学科前沿发展动态。他鼓励学生和同事从战略角度进行创新思考；他勇于将科研灵感付诸实践，并鼓励大家克服困难，开展长期稳定的国际合作。他多次为我国病毒性疾病防控中的关键问题提出开拓性的解决方案。20世纪70年代初，他回顾了全球病毒学研究的发展历史，意识到分子病毒学时代正在到来。1978年，他赴美开会之际有意考察了美国一些分子病毒学机构后，回国在中国预防医学科学院病毒学研究所建立了病毒遗传研究室（中国疾病预防控制中心病毒病预防控制所应急技术中心前身）。1981年，他发表了《病毒与重组DNA》一文，提出了两个研究方向：CHO细胞中表达外源基因和利用病毒作为基因工程载体。这标志着他的学术思想从经典病毒学向分子病毒学转变。

作为《病毒学报》的创办人，对该杂志接收发表的每一篇论文，朱既明教授都做最后的审读，并就学术、文字、英文图注和摘要等细节提出具体意见（金冬雁，2009）。朱教授始终以学术规范、学术质量以及同行评议结果作为接收或拒稿的依据。他亲自指导学生和同事开展科研实践，经常参加实验室和研究所的学术活动，以严谨的态度对待科学问题，鼓励创新（图3）。这种无私的治学态度和原则对年轻科研工作者是很好的鼓励。

图3 朱既明教授（右二）1994年在中国预防医学科学院病毒学研究所（现更名为中国疾病预防控制中心病毒病预防控制所）举行的实验室会议上与同事讨论科学问题

两位著名的流感病毒学家Graeme Laver和Robert Webster描述了他们与朱既明教授的首次会面："我们第一次见到朱教授是在1972年。凌晨4点，他亲自到北京火车站迎接我们澳大利亚医学代表访问团，给我们留下非常深刻的印象。这样一位有声望的科学家会抽出时间欢迎我们访问中国，我们为他的友好谦和所打动。"（Laver and Webster，1999）朱教授像一位英国绅士：他总是穿着整洁，精心打理头发，以谦逊和善良对待每一个人。

回顾朱既明教授的一生，他在病毒学和生物制品领域作出杰出贡献，荣获全国科学大会奖、全国卫生科技大会奖、国家科学技术进步奖一等奖和二等奖、国家自然科学奖三等奖、卫生部科技进步奖一等奖以及何梁何利基金医学药学界科学与技术进步奖等奖项。他是《病毒学报》总编辑，《中华微生物学和免疫学杂志》《中华预防医学杂志》《热带医学》副总编辑，《中华医学杂志》英文版、《微生物学报》《中华流行病学杂志》等期刊的编委，国际杂志*Archive of Virology*顾问（1982—1988年）。

朱既明教授是国际知名的病毒学家，他为我国病毒性疾病防控作出了重大贡献。逝者长已矣，但他的精神和风骨永存。我们作为他的同事和学生，在朱教授诞辰一百周年之际写下了这篇纪念文章，我们将践行他的遗志，学习他淡泊名利、为科学献身的优秀品德和崇高精神，为疾控事业继续奋斗。

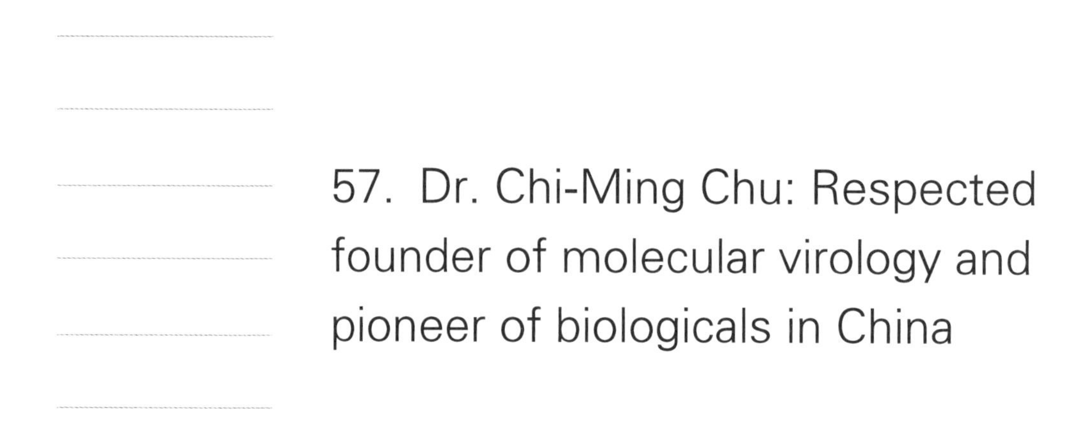

57. Dr. Chi-Ming Chu: Respected founder of molecular virology and pioneer of biologicals in China

"I have the greatest respect for Dr. Chu, who upheld great standards of science during very difficult times in China. His contributions to the genetic characterisation of influenza viruses were extraordinary."

—Professor Peter Palese, a member of the National Academy of Sciences and expert in the field of RNA viruses, wrote this when he received the obituary notice of Dr. Chu in 1998.

"He was not only a great scientist and teacher, but also a remarkable human being who contributed significantly to openness in science and international relations."

—Professors Graeme Laver and Robert Webster, famous virologists at the Department of Virology and Molecular Biology (St. Jude's Children's Research Hospital, USA), wrote an eulogy for Dr. Chu in 1999 (Laver and Webster, 1999) .

Dr. Chi-Ming Chu (Ji-Ming Zhu, 朱既明) was an internationally renowned virologist, one of the most important founders of molecular virology and a pioneer of biologicals in China. He was a leader in virology research in China, from classical virology to the era of molecular virology. His seminal observations and research on influenza viruses laid the foundation for worldwide research on influenza virus structure, surveillance and subunit vaccine development. He was a pioneer in the field of biologicals in China due to his extraordinary contributions to research on penicillin, attenuated viral vaccines, recombinant hepatitis B virus (HBV) vaccines and vaccinia viral vectors.

Dr. Chu was born on September 12, 1917, in Yixing County, Jiangsu Province. He graduated from Shanghai Medical College in 1939 and was awarded a PhD by Cambridge University (UK) in 1948. He worked for the World Influenza Centre at the National Institute for Medical Research at Mill Hill (London) from 1948 to 1950. After returning to China in 1950, he worked successively at the National Vaccine and Serum Institute of China, National Institute of Biologicals in Changchun and the Chinese Academy of Preventive Medicine (CAPM) (Ruan and Lu, 1998; Zhang, 1998). At the time of his death on January 6, 1998, Dr. Chu was a member of

Weizheng Yan, Baoying Huang, Li Ruan, Wenjie Tan

National Institute for Viral Disease Control and Prevention, China CDC, Beijing 102206, China

Correspondence: tanwj28@163.com (W.-J. Tan)

Academic Division of the Chinese Academy of Sciences, an Academician of the British Royal Institute for Internal Medicine, an Honorary Member of the American Society for Microbiology and Honorary Director of the Institute of Virology (CAPM).

Dr. Chu made an extraordinary contribution to the development of Chinese biologicals

Penicillin was a magical drug saved thousands and millions people's life during the Second World War. Dr. Chu contributed significantly to the development of China's first antibiotic of penicillin (Ruan, 2015). From 1940 to 1945, Dr. Chu worked at the National Epidemic Prevention Bureau (NEPB) in Kunming under Professor Tang Feifan's direction, cooperated together with Dr. Fan Qingsheng, a prominent agricultural microbiologist at the time, Dr. Chu and his colleagues compared 30 local strains and some foreign strains, and finally selected the local No. 22 strain for the extraction and purification of penicillin (Chu et al., 1945; Xu, 2013). A seed strain of lyophilized penicillin for injection was successfully developed and the preliminary clinical observation was performed by Dr. Chu's group. This work contributed significantly as the beginning of antibiotics research and application in China.

Beginning in the 1950s, Dr. Chu conducted research on live attenuated vaccines. Specifically, he studied forest encephalitis, measles, and influenza live vaccines. During his research on the influenza vaccine, he found that the rate of attenuation of influenza viruses differs yearly, and he developed two methods for producing live vaccines: Chick embryo passage and genetic recombination. In 1959, a measles pandemic occurred in China, and Dr. Chu and his colleagues in Changchun began to study live attenuated measles vaccines. After 6 years of hard work, he, together with his colleagues, selected the Chang 47 measles strain; the vaccine produced using this strain had mild adverse reactions, a high seroconversion rate and induced high-titre antibody responses (Chu et al., 1964). The vaccine produced from this strain in 1966 made an important contribution to the control of measles in China.

From 1951 to 1963, Dr. Chu was appointed Chief of the Second Research Laboratory, Chief of the Quality Control Department of the National Vaccine and Serum Institute, and Chief of the Biological Product Training Class conducted by the Ministry of Health of China. Subsequently, he became Deputy Director of the National Institute of Biological Products in Changchun (Zhang, 1998), and Deputy Director and General Director of the Institute of Virology. He wrote the first biological product regulation draft in Chinese and trained groups of young scientists (Ruan, 2013).

Dr. Chu promoted research on vaccinia viral vectors in China. At the start of 1984, he directed the genetic engineering of a polyvalent vaccine vectored by the Tiantan vaccinia virus strain, with the aim of developing a vaccine effective against several diseases. He designed a polyvalent vaccinia vaccine expressing antigens from HBV, Epstein-Barr virus, and herpes simplex virus (Tsao et al., 1988). This repertoire was later expanded to include antigens of hepatitis A virus, measles virus, and respiratory syncytial virus. This work was supported by the National High-Technology Research and Development Program (863 Program) in 1986. Under his direction, the Tiantan vaccinia virus strain was used for the genetic engineering of antigen expression, resulting in clinical trials of several recombinant vaccine candidates. In addition, he initiated and guided the R&D program of a non-replicating vaccinia virus vector (Tiantan strain) with the intellectual property rights for Chinese, which was subsequently used extensively in vaccine research in China.

Hepatitis B has had a major impact on the health of the Chinese. In the 1980s, it became clear that a new HBV vaccine was needed. Dr. Chu contributed significantly to the first genetically engineered recombinant HBV vaccine. He undertook research into a recombinant hepatitis B vaccine expressed in Chinese hamster ovary (CHO) cells, assuming directorship of the project. After more than 10 years of hard work, he and his colleagues finally succeeded in developing a genetically engineered hepatitis B vaccine expressed in CHO cells, which was the first genetically engineered HBV vaccine produced in CHO cells (Ruan et al., 1983). The vaccine has been produced since 1991 and has played an important role in the control of HBV transmission in China.

Dr. Chu was one of the greatest influenza research scientists in the world

Dr. Chu was engaged in influenza virus research from 1945 to 1984. Among his contributions, Dr. Chu discovered the filamentous form of influenza virus and was the first to report its variability in *Lancet* in 1949 (Chu et al., 1949); his findings were later confirmed by other scientists. During the same period, he found that beta- and gamma-inhibitors could inhibit the agglutination of red blood cells by influenza virus (Chu and Coombs, 1947; Chu, 1948); the beta-inhibitor is also termed Chu's inhibitor. Subsequently, he discovered that influenza virus could be divided into two subunits by treatment with ether or detergents: Haemagglutinin and soluble antigens. This led to the determination of the structure of the virus in the 1950s and subsequently to the development of a subunit vaccine.

Antigenic shifts in influenza viruses were discovered by Dr. Chu. He also discovered the origin of the 1957 and 1977 influenza pandemics (Laver and Webster, 1999). In 1949, Dr. Chu demonstrated that a qualitative change in influenza virus antigens (drift) subsequently spreads to other locations (Chu, 1951). In February of 1957, the H2N2 Asian pandemic strain appeared in Kweiyang, Kweichow Province, Southern China, and spread to all of China by the end of March, replacing the previously prevalent H1N1 strains. Dr. Chu showed that this pandemic influenza virus was a new subtype resulting from a qualitative change in the haemagglutinin antigen (i.e., antigenic shift). This was the first pandemic caused by a new subtype that was proven virologically. Following this, he showed the re-emergence of H1N1 flu in 1950 in Anshan in February of 1977, 9 months before it reached the rest of the world (Laver et al., 1984). The concept of quantitative changes in influenza virus antigens subsequently became a guiding principle for flu monitoring and vaccine seed selection.

Influenza temperature-sensitive strains in nature were firstly discovered by Dr. Chu, who was the first scientist to show that influenza virus undergoes not only antigenic variation but also virulence variation. In 1977, he found that the newly isolated influenza virus strains could be divided into normal and temperature-sensitive types (Chu et al., 1982). Temperature-sensitive strains can reproduce only below 33℃ and are not pathogenic to humans. Genetic research showed that temperature sensitivity could be transferred to a normal strain by genetic recombination (Chu et al., 1992). This suggested the possibility of isolating attenuated strains in the field for use in live vaccines.

Dr. Chu was always courteous as well as a master of knowledge and a worthy mentor

This is the personal motto of Dr. Chu: "Only by simplifying your everyday life and suppressing your desire for material things can you reach your goals. Only a calm mind and absence of anxiety will grant you broad vision." He stressed the importance of integrity to his students and to himself. He was upright and lived simply, and was not motivated by fame or wealth. He was rigorous in both scientific research and in his life. He was not only a great scientist but also a remarkable human being. Dr. Chu will be remembered for his great contributions and nobility of character, and we who knew him will miss him forever.

Dr. Chu was always keen to keep up with international developments in science and technology. He encouraged his colleagues and students to adopt a strategic perspective and think innovatively; he emphasised putting ideas into practice, and he encouraged collaboration and long-term efforts to overcome major difficulties. He repeatedly proposed creative solutions to key issues in the prevention and control of viral diseases in China. In the early 1970s, he reviewed global developments in virology research and realised that the era of molecular virology was approaching. After visiting several American laboratories engaged in viral molecular genetics and genetic engineering research in 1978, he proposed and actively supported genetic engineering research in China and established the laboratory of viral genetics (The predecessor of Biotech Center for Viral Diseases Emergency, China CDC now) at the Institute of Virology. In 1981, he published a paper titled "Virus and recombinant DNA" and proposed two research directions for the laboratory: Gene expression in CHO cells and the virus as a vector for genetic engineering. This marked a change in his research focus from classic to molecular virology.

As the founder of the *Chinese Journal of Virology* (Bing du xue bao), Dr. Chu made specific comments on each article accepted by this journal, such as on academic issues, English usage and illustrations. He made the final decision to accept or reject a manuscript after peer review based on academic quality (Jin, 2009). He personally instructed colleagues and students, often attending academic activities in laboratories and institutes, discussing issues in a rigorous manner and encouraging innovation. He took the initiative to encourage the younger generation, for entirely selfless reasons.

The internationally famous influenza virologists Graeme Laver and Robert Webster described their first meeting with Dr. Chu (Laver and Webster, 1999): "We first met Professor

Chu in 1972. He had come to the Beijing railway station at 4:00 AM to meet our visiting Australian medical delegation. We were most impressed that such a prestigious scientist would find the time to welcome us to China. This courtesy was typical of Dr. Chu." Dr. Chu also lived like a British gentleman; he was unfailingly neatly dressed with meticulously looked-after hair. He treated everyone with modesty and kindness.

Based his distinguished contributions to virology and biologicals, Dr. Chu was honoured with the National Science Conference Award, the First-Class Award for Scientific Advancement, the Third-Class Prize in Natural Science, and the First-Class Award for Scientific Advancement by the Ministry of Health, and the Ho Leung Ho Lee Scientific Advancement in Medical Science Award. He was invited to be an adviser to the international journal *Archive of Virology* from 1982 to 1988, served as chief editor of *Chinese Journal of Virology*, as deputy editor of *Journal of Microbiology and Immunology*, *Chinese Journal of Preventive Medicine* and *Journal of Tropical Medicine*, and he was on the editorial boards of *the National Medical Journal of China* (English Edition), *Acta Microbiologica Sinica* and *Chinese Journal of Epidemiology*.

Dr. Chu was a truly international Chinese virologist who made significant contributions to solving key problems in viral disease control in China. Dr. Chu has died, but his spirit and style are eternal. We, his former colleagues and students, penned this memorial on the occasion of the centennial of Dr. Chu's birthday. We try always to emulate Dr. Chu's spirit, courage, enthusiasm, meticulousness and dedication, and to be worthy recipients of his legacy.

Figures

Fig.1 Prof. Chi-Ming Chu (1917—1998)

Fig.2 Motto written by Prof. Chi-Ming Chu in 1992

Fig.3 Prof. Chi-Ming Chu (2nd from right) discussing scientific topics with colleagues at a laboratory meeting in the Institute of Virology, CAPM (now renamed as National Institute for Viral Disease Control and Prevention, China CDC) in 1994

References

Chu CM (1948) Enzymic action of viruses and bacterial products on human red cells. Nature 161:606–608.

Chu CM (1951) The action of normal mouse serum on influenza virus. J Gen Microbiol 5(4):739–757.

Chu CM, Coombs RRA (1947) Modification of human red cells by virus action. Lancet 1:484–487.

Chu CM, Dawson IM, Elford WJ (1949) Filamentous forms associated with newly isolated influenza virus. Lancet 253(6554):602–603.

Chu CM, Ku YF, Chang YH (1964) Observations on the further attenuation of measles virus and on live vaccines prepared from highly attenuated lines of virus. Chinese Med J 83(8):483–496.

Chu CM, Tian SF, Ren GF, et al (1982) Occurrence of temperature-sensitive influenza A viruses in nature. J Virol 41(2):353–359.

Chu CM, Wei BY, Tian SF, et al (1992) Host dependence of naturally occurring temperature-sensitive influenza A viruses and location of their genetic lesions. Virus Res 22(3):207–214.

Chu CM, Wong YW, Fan CC, et al (1945) Experimental production of penicillin in China. Chin Med J 64(5 and 6): 89–101.

Jin DY (2009) Impressed memory on Dr. Chu CM. Personal blog. (金冬雁. 2009. 追忆朱既明先生两三事. 个人博客.)

Laver WG, Webster RG (1999) In Memoriam: Chu Chi Ming (1917—1998). Virology 225(1):1.

Laver WG, Webster RG, Chu CM (1984) Summary of a meeting on the origin of pandemic influenza viruses. J Infect Dis 149(1):108–115.

Ruan L (2013) Zhu Jiming: Chinese famous medical virology and biological expert. Chin J Biotechnol 4: 540–543. (阮力. 2013. 我国著名医学病毒学和生物制品学专家——朱既明. 生物工程学报, 4: 540–543.)

Ruan L (2015) Simple for explicating one's ambition, fair and softly go far in a day: Chu Chiming, a famous virologist, founder of Chinese virus biotechnologist. J Microbs Infections 3:130. (阮力. 2015. 淡泊明志, 宁静致远——著名病毒学家,我国病毒生物技术的奠基人朱既明. 微生物与感染, 3:130.)

Ruan L, Lu DM (1998) The Life History of Dr. Zhu Jiming: 80th Anniversary Celebration of Dr. Zhu Jiming, famous virologist(1917—). J Microbiol Immunol 18(1):4–7. (阮力, 陆德敏. 1998. 朱既明教授传略——庆贺著名病毒学家朱既明教授八十寿辰(1917—). 中华微生物学和免疫学杂志, 18(1):4–7.)

Ruan L, Ren GF, Ruan WQ, et al (1983). Expression of hepatitis B virus surface antigen in mammalian cells using co-transformation technique. Chinese Med J 5:269–274. (阮力,任贵方,阮薇琴,等. 1983. 应用共转化技术在哺乳动物细胞中表达乙型肝炎病毒表面抗原. 中国医学科学院学报, 5: 269–274.)

Tsao H, Liu GQ, Ruan L, et al (1988) Construction and application of plasmids containing bilateral promoters of vaccinia virus. J Virol 62(12):4832–4834.

Xu D (2013) Experimental production of penicillin in Kunming in the early 1940s. Protein & Cell 4(5):323–324.

Zhang JM (1998) A high prestige scientist in career and life: In memory of Professor Chu Chiming. Chin J Biol 1:1–2. (张嘉铭. 1998. 事业宗师,人生楷模——缅怀朱既明教授. 中国生物制品学杂志, 1:1–2.)

58. 周健：伟大的宫颈癌疫苗发明者

2005年11月，科学家聚集在纽约时代广场，宣布了一个令人振奋的消息：世界上第一个宫颈癌疫苗研制成功。宫颈癌疫苗可以保护未患病的女性抵抗四种人乳头瘤病毒（HPV）毒株的感染，能够预防近70%的宫颈癌疾病。美国食品药品监督管理局（FDA）等均批准了这种疫苗的使用。随后，《纽约时报》将宫颈癌疫苗的研制评为2006年世界十大科技成果之一。2006年8月，在昆士兰州州长Anna Bligh的见证下，Ian Frazer教授实施了首剂疫苗的注射。Bligh说："今天，我们正在创造新的医学史，对于这个国家的科学来说是一个伟大的时刻。"而Frazer教授告诉所有人，没有周健博士（图1）的贡献，这一历史性突破是不可能实现的。令人遗憾的是，这位伟大的疫苗发明者，早在1999年就离开了我们，无法亲眼见到疫苗惠泽众生的这一天。

周健博士1957年出生于杭州市，彼时父亲已有50岁高龄。他的出生，为家庭带来了希望和欢乐（图2）。儿童时期，周健博士很活泼，经常由于自顾自地玩耍而被要求离开幼儿园教室。然而，这样的惩罚似乎并没有影响到年幼的周健博士，在教室外他依旧自己玩得不亦乐乎。从幼童伊始，他似乎就有着无穷无尽的精力，而这种能量伴其一生，并成为他成功的重要因素。

图1　温州医科大学校园内的周健博士雕像

图2　周健博士童年时期照片

译者：程浩
中国科学院北京生命科学研究院，北京100101，中国
邮箱：chengh@biols.ac.cn

周健博士在杭州度过了愉快的童年。高中毕业后，他被派到农村，那里无学可上，还要辛苦劳作，这让他对农民的艰辛深感同情。在农场工作一年后，他返回杭州成为六一针织厂的一名临时工。在那里，他每天要用手推车运送超过50千克重的包裹，却只有少得可怜的8毛钱。他吃了很多苦，但从未因此而意志消沉。不久，周健博士去一家无线电工厂成为一名焊接学徒。因为他年纪尚小，身材瘦削，工厂的车间组长不想要他，但其中一个组长注意到了他的聪明伶俐和心灵手巧，他能够快速掌握焊接技术并生产出高质量产品。周健博士在这位组长的带领下成为团队的核心成员，并一直工作到他离开工厂去上大学。他在农场和工厂的经历培养了他后来在科学工作中吃苦耐劳和默默奉献的精神。

1977年，周健博士进入温州医学院（现温州医科大学）学习，并获医学学士学位。本科毕业后，他在浙江医科大学进一步深造并获医学硕士学位。也正是在硕士就读期间，他开始接触人乳头瘤病毒并对人体病理学产生了浓厚的兴趣。1985年，他在河南医科大学开始了对人体病理学的博士研究，继续开展人乳头瘤病毒的研究工作，并前往中国预防医学科学院病毒学研究所（现中国疾病预防控制中心病毒病预防控制所）学习了分子病毒学和分子克隆技术。在硕士和博士学习阶段，周健博士发现食管癌与人乳头瘤病毒感染有关，并获国家级奖励。1988年，他在北京医科大学（现北京大学医学部）开始了他的第一次博士后研究工作。在此期间，他以痘苗病毒为载体在体外成功表达出特定的蛋白，也为他后续的疫苗开发奠定了技术基础。同年，他受邀前往英国剑桥大学病理系的肿瘤病毒实验室，在Lionel Crawford教授的指导下继续从事人乳头瘤病毒的分子生物学研究（图3）。他刻苦勤奋，勇于创新，对人乳头瘤病毒研究的现有状况和发展方向有着清晰的理解，并很快在《普通病毒学杂志》《病毒学》和《病毒学杂志》等国际知名期刊上相继发表了多篇有影响力的学术论文。

图3 周健博士（左一）在剑桥大学肿瘤病毒实验室

在剑桥大学工作期间，周健博士结识了来自澳大利亚昆士兰大学的免疫学家Frazer教授。出于对周健博士创新理念和科研兴趣的认可和尊重，Frazer教授劝说他前往澳大利亚与其一起深入开展人乳头瘤病毒的研究。1990年，周健博士抵达昆士兰大学，揭开了他研究生涯最重要的篇章。

经过无数次实验之后，周健博士最终以痘苗病毒为载体，利用DNA重组技术，成功在体外哺乳动物细胞中表达出乳头瘤病毒L1和L2衣壳蛋白。这些表达的衣壳蛋白还可以自我组装成病毒样颗粒。同年，上述结果在《病毒学》顺利发表（Zhou et al., 1991a）。1991年6月，昆士兰大学为这项发明申请了临时专利。7月，周博士和Frazer教授在美国西雅图举办的国际人乳头瘤病毒会议上公开了他们的研究成果，随后被公认为人类医学史上的重大突破。澳大利亚CSL医药公司和美国默克公司先后进行了大规模的动物和人类临床试验，最终验证了人乳头瘤病毒样颗粒在预防宫颈癌中的作用。

周健博士是一位才华横溢的科学家，他为乳头瘤病毒和癌症生物学研究作出了史无前例的贡献。在昆士兰大学癌症和免疫学研究中心（现为癌症、免疫学和代谢医学研究所）的前四年，他以第一作者的身份在病毒学前沿性期刊上先后发表了11篇研究论文，对于从事生物医学研究的每一位科学家来讲，这都是难以置信的成就。在研究生涯中，他创造了多项闻名世界的发明，并曾在8年内获得12项发明专利，其中就包括对乳头瘤病毒衣壳蛋白L2在DNA结合和衣壳化中关键作用的研究，揭示其在人乳头瘤病毒生命周期里的重要功能（Zhou et al., 1991b）。周健博士还发现，L1蛋白的C末端不影响衣壳的形成，这让他可以人工整合C末端截短的L1蛋白和其他早期病毒蛋白的T细胞表位组成的嵌合体衣壳，用于开发预防性和治疗性的疫苗。另一项创新性的研究成果是遗传密码（基因密码子）的优化。他发现，因为哺乳动物细胞的转移核糖核酸（tRNA）的限制作用，人乳头瘤病毒的L1和L2衣壳基因在大多数哺乳动物细胞中不能表达，却能在酵母中成功表达（Zhou et al., 1999）。利用哺乳动物稀有密码子，在感染的末端分化上皮中有效地表达衣壳蛋白。密码子优化和变更可以有效地提高蛋白表达效率和疫苗的免疫原性，开辟新的基因治疗技术（Zhou et al., 1999）。

周健博士于1994年在昆士兰大学获医学博士学位，随后在芝加哥洛约拉大学医学院任副教授一职。他组建了自己的研究团队，并开始指导博士生和博士后从事不同的医学项目研究。1996年，他回到昆士兰大学，任首席研究员，并成为癌症和免疫学研究中心乳头瘤病毒结构蛋白实验室主任（图4）。1998年，周健博士获得三项澳大利亚国家医疗卫生研究委员会基金资助，一项美国国家健康研究院基金资助，一项美国癌症基金会资助，两项昆士兰癌症委员会资助，以及其他各种来源的项目基金和多项专利使用费。作为一名研究人员，周健博士成为昆士兰大学历史上获得基金和科研经费最多的科学家。

图4 周健博士（前排左一）在昆士兰大学癌症和免疫学研究中心

周健博士以身为温州医科大学的毕业生为荣，并时刻关注母校的教学和科研发展。他经常返回母校开课讲学，指导科研工作（图5），并无私地贡献出个人的时间和资金用来支持母校的学者出国留学。1996年周健博士还在芝加哥的时候，他得知一位来自母校的学者正在波士顿学习，他立刻为其购买波士顿和芝加哥之间的往返机票，以便聚在一起讨论如何支持温州医科大学开展科研工作。在周健博士加入昆士兰大学之后，他专门创建温州医科大学与昆士兰大学的合作项目，通过进一步地开展科研合作和学术交流，大力支持母校的发展。他的家人在温州医科大学设立了“周健博士纪念奖学金”，支持母校的学生前往国外从事博士研究。周健博士为温州医科大学如今的闻名中外作出了重大贡献。

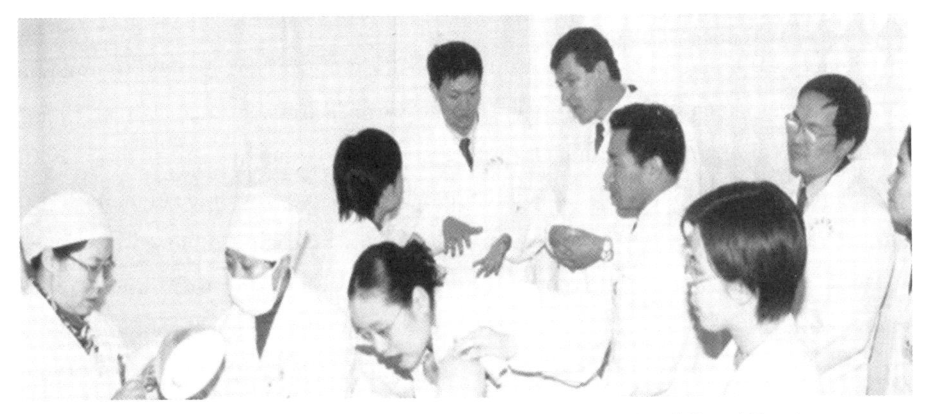

图5 周健博士（左五）和Frazer教授（左六）在温州医科大学第一附属医院

周健博士在科学和人文方面的杰出贡献受到了广泛的认可并值得纪念。昆士兰大学的癌症、免疫学和代谢医学研究所举办了周健纪念演讲年会，世界著名的科学家纷纷前来演讲缅怀英才。研究所还将一间会议室命名为周健会议室。自2000年以来，每年的国际人乳头瘤病毒会议都会为他举办系列的纪念活动。2006年，亚洲–大洋洲生殖感染和肿瘤研究组织（AOGIN）设立了AOGIN周健博士奖，作为三项最佳报告奖之一。同年，昆士兰政府创立“周健博士智慧国家奖学金”。昆士兰政府、澳大利亚华人协会和澳大利亚中国友好协会共同组织了“周健博士纪念仪式”，并出版了《英才济苍生》周健博士纪念册（图6）。周健博士和Frazer教授凭借共同研发出全球首个人乳头瘤病毒疫苗，最终在2015年荣获欧洲发明奖。

图6 周健博士纪念册封面照片

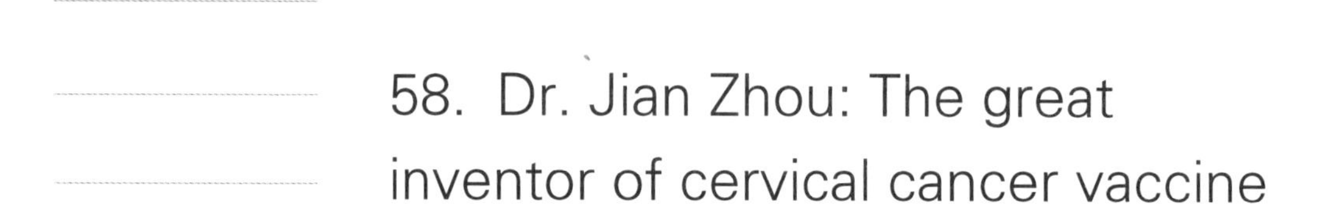

58. Dr. Jian Zhou: The great inventor of cervical cancer vaccine

In November 2005, scientists gathered at Times Square in New York City announced the exciting news of the successful development of the world's first cervical cancer vaccine. The cervical cancer vaccine, which is able to protect unexposed women against the infection with four HPV strains responsible for 70% of cervical cancers, was approved for human use by the US Federal Drug Administration (FDA) and so on. The *Times* magazine reported the development of the cervical cancer vaccine as one of the world's top ten scientific and technological achievements of 2006. In August 2006, the first dose of the vaccine was administered by Professor Ian Frazer in the presence of Anna Bligh, the Premier of Queensland. Anna Bligh said "Today we are making medical history. It is a great moment for science in this country." Professor Frazer said that this historic breakthrough could not have been achieved without the contribution of Dr. Jian Zhou. Sadly, Dr. Zhou, the great inventor of the vaccine, passed away in 1999 and was unable to see the benefits brought by the vaccine.

Dr. Zhou was born in 1957 in Hangzhou, Zhejiang Province, China. At the time of his birth, his father was already 50 years old and the birth of a boy brought the family hope and joy. As a child, Dr. Zhou was active and playful and was often asked to leave his kindergarten classroom because he was busy in doing and playing by himself. However, the punishment did not seem to affect the young Dr. Zhou and he was often seen having a good time playing by himself outside of the classroom. Even as a child, he seemed to have inexhaustible energy and this energy, an important factor that contributed to his later success, stayed with him all of his life.

Dr. Zhou spent his childhood in Hangzhou. After graduating from high school, he was sent

---

Kong-Nan Zhao[1,2], Lifang Zhang[1], Jia Qu[3,4]

1 Institute of Molecular Virology and Immunology, Wenzhou Medical University, Wenzhou 325035, China

2 Centre for Clinical Research, Translational Research Institute, The University of Queensland School of Medicine, 37 Kent Street, Woolloongabba, Brisbane QLD 4102, Australia

3 Laboratory of Neurovascular Biology, School of Ophthalmology and Optometry and the Eye Hospital of Wenzhou Medical University, Wenzhou 325003, China

4 The State Key Laboratory Cultivation Base and Key Laboratory of Vision Science, Ministry of Health, Wenzhou 325027, China

Correspondence: k.zhao@uq.edu.au, knzhao@wzc.edu.cn (K.-N. Zhao); jqu@wz.zj.cn, jia.qu@163.com (J. Qu)

to work in a rural village. There was no school to attend. During his time working in the village, Dr. Zhou gained profound empathy for the hardships experienced by farmers. After working on farms for a year, he returned to Hangzhou as a temporary worker at the "June-One Knitting Factory" . There, he pulled carts and carried packages weighing more than 50 kilograms for a measly eighty cents a day. Although the work was extremely hard, Dr. Zhou never lost his spirit. Later, Dr. Zhou worked as a welding apprentice at a radio factory. Because of his youth and lanky build, a number of team leaders at the factory were reluctant to accept him as their apprentice. However, one team leader noticed that he was very smart, more thoughtful than others and had skilful hands—he was able to grasp the welding technology and produce high quality products quickly. This team leader took the young Dr. Zhou on and relied on him as a core team member until he left the factory to attend university. Dr. Zhou's experiences of working on farms and in factories cultivated his ability to endure hardships and his dedication to the development of science.

In 1977, Dr. Zhou was admitted to study at the Department of Medicine at Wenzhou Medical University where he obtained his Bachelor of Medicine degree. After graduation, he went on to study at the Zhejiang Medical University and obtained a Master's degree in Medical Sciences. It was at the Zhejiang Medical University that Dr. Zhou began to study the human papillomavirus and developed his interest in the field of human pathology. In 1985, Dr. Zhou began his PhD studies in human pathology at the Henan Medical University where he continued his research in human papillomavirus. During that time, he also attended the Institute of Virology, Chinese Academy of Preventive Medicine and studied molecular virology and molecular cloning technology. During his Master's degree and PhD studies, Dr. Zhou discovered that the development of esophageal cancer is associated with human papillomavirus infections. He received a national award in China for this discovery. In 1988, after he had obtained his PhD degree, Dr. Zhou undertook his first postdoctoral study at Beijing Medical University. During this period, he used vaccinia virus as a vector to express a specific protein in vitro, thereby laying the technological foundation for his later invention. In the same year, he was invited to work at the Tumor Virus Laboratory, in the Department of Pathology of Cambridge University to continue molecular biology research in the human papillomavirus under the guidance of Professor Lionel Crawford. Dr. Zhou's hard work, diligence and innovation allowed him to develop a clear understanding of the state of the research and the developmental direction for human papillomavirus research. He was soon publishing influential papers in journals such as the *Journal of General Virology*, *Virology*, *Journal of Virology* and other well-known international journals.

Dr. Zhou first met Professor Frazer, an immunologist from the University of Queensland, at Cambridge. Impressed by Dr. Zhou's creative research ideas and bonded by mutual respect, Professor Frazer convinced Dr. Zhou to join him in Australia to pursue their common interest in human papillomavirus research. In 1990, Dr. Zhou joined Professor Frazer at the University of Queensland, Australia, starting the most important chapter of Dr. Zhou's research career.

After multiple experiments, Dr. Zhou successfully used vaccinia virus as a carrier to express papillomavirus L1 and L2 capsid proteins in mammalian cells in vitro using recombinant DNA technology. The expressed capsid protein can be self-assembled into virus-like particles. In the same year, Dr. Zhou and Professor Frazer published these results in the *Virology* journal (Zhou et al., 1991a). In June 1991, the University of Queensland applied for a provisional patent for the invention. In July of that year, Dr. Zhou and Professor Frazer reported their results at the International Conference of Human Papillomavirus in Seattle in the United States. This achievement was considered a major breakthrough in human medical history. Australian medical company, CSL and later Merck Company, USA conducted large-scale animal and human clinical trials which ultimately verified the effect of human papillomavirus-like particles on the prevention of cervical cancer.

Dr. Zhou was an incredibly talented scientist. His contributions to papillomavirus and cancer biology research were exceptional. During his first four years at the Centre for Cancer and Immunology Research (later renamed The Diamantina Institute for Cancer, Immunology and Metabolic Medicine) at the University of Queensland, he published 11 first-author scientific papers in leading virology journals. This is an incredible achievement for any scientist involved in biomedical research. During his research career, Dr. Zhou made a number of world-renowned inventions and obtained twelve invention patents within eight years. One of these involved the discovery that the minor papillomavirus capsid protein L2 plays a crucial role in DNA binding and capsidification, revealing the importance of this capsid protein in the HPV life cycle (Zhou et al., 1991b). Dr. Zhou also found that L1 protein C-terminus does not affect the formation of capsids. This discovery led to the use of a chimeric capsid consisting of the C-terminal truncated L1 protein, and other T-cell epitopes of the early proteins of the virus, to produce both prophylactic and therapeutic vaccines. Another very novel study was genetic code (gene codon) optimization. In this study, he found that the HPV L1 and L2 capsid genes could not be expressed in most mammalian cells, but it can be expressed in yeast because the transfer ribonucleic acid (tRNA) in mammalian cells limits the expression of the capsid genes (Zhou et al., 1999). Dr. Zhou discovered that a capsid gene using the mammalian rare codon effectively expresses capsid protein in the infected terminal differentiation epithelium. Codon optimization and change could be used to improve significantly the efficiency of protein expression and improve the immunogenicity of the vaccine and opened up a new gene therapy technology (Zhou et al., 1999).

Dr. Zhou received his MD from the University of Queensland in 1994. After this he took on an associate professorship at the Loyola University School of Medicine in Chicago. He set up his own research group there and supervised PhD students and postdoctoral fellows working on different research projects in medical sciences. In 1996, Dr. Zhou returned to the University of Queensland and was appointed the principle Lions Research fellow and the Head of the Papillomavirus Structural Protein Laboratory at the Center for Cancer and Immunology Research. In 1998, Dr. Zhou was awarded with three research grants by the Australian National Medical and Health Research Council, a grant by the American National Institute of Health, a grant by the American Cancer Foundation, two grants by the Cancer Council Queensland, and received funding for several other research projects from other resources as well as royalties from his patents. In that year, Dr. Zhou was the well-funded researcher receiving the largest amount of research funding in the history of the University of Queensland.

Dr. Zhou was always proud of being a graduate of the Wenzhou Medical University and was always concerned with the development of teaching and scientific research at his alma mater. He travelled to Wenzhou Medical University to give lectures and guide scientific research and was selfless in spending his personal time and money inviting scholars from his alma mater to further their studies overseas. One such example occurred in 1996 when Dr. Zhou was based in Chicago. When he found out that a scholar from his alma mater was studying in Boston, he immediately sent that person return flight tickets between Boston and Chicago so that they could get together to discuss how to support and carry out scientific research at the Wenzhou Medical University. Later, after Dr. Zhou joined the University of Queensland, he established a joint program between the Wenzhou Medical University and the University of Queensland to further scientific research collaboration and facilitate the exchange of medical students to accelerate the development of his alma mater. His family established the "Dr. Jian Zhou Foundation" to provide scholarships to medical students from Wenzhou Medical University, Dr. Zhou's alma mater, to undertake PhD studies outside of China to train as medical research scientists. Dr. Zhou made a significant contribution to the development of Wenzhou Medical University as a well-known university in China and the world now.

Dr. Zhou's outstanding contributions to science and humanity have been recognised and commemorated. The Diamantina Institute for Cancer, Immunology and Metabolic Medicine of the University of Queensland established the annual Jian Zhou Memorial Oration at which world renowned scientists give lectures to commemorate Dr. Zhou. The Diamantina Institute has also named one of its conference rooms the Jian Zhou room. Since 2000, the annual international HPV Conference has held a number of memorials for Dr. Zhou. The Asian-Oceania Research Organization on Genital Infection and Neoplasia (AOGIN) in 2006 established the AOGIN Dr. Jian Zhou award as one of the three awards for best oral presentation. In 2006, the Queensland Government established the "Dr. Jian Zhou Smart State Fellowships Program" . The Queensland Government, the Australian Chinese Society and Australia China Friendship Society organized the "Dr. Jian Zhou Memorial" and published a book entitled *Dr. Jian Zhou's Brilliant Mind — The Inventor of Cervical Cancer Vaccine* (《英才济苍生》). Lastly, the vaccine against human papillomavirus (HPV) jointly invented by Dr. Zhou and Professor Frazer won the European Inventor Patent award in 2015.

Figures

Fig.1 Statue of Dr. Jian Zhou on the campus of Wenzhou Medical University

Fig.2 Dr. Jian Zhou's childhood photo

Fig.3 Dr. Jian Zhou (1st from left) at the Tumor Virus Laboratory of Cambridge University

Fig.4 Dr. Jian Zhou (1st from left in the front row) at the Center for Cancer and Immunology Research of The University of Queensland

Fig.5 Dr. Jian Zhou (5th from left) and Professor Ian Frazer (6th from left) at the first affiliation hospital of Wenzhou Medical University

Fig.6 Cover photo of Dr. Jian Zhou memorial book

References

Zhou J, Sun XY, Stenzel DJ, et al (1991a) Expression of vaccinia recombinant HPV 16 L1 and L2 ORF proteins in epithelial cells is sufficient for assembly of HPV virion-like particles. Virology 185(1):251–257.

Zhou J, Sun XY, Louis K, et al (1991b) Interaction of human papilloma virus (HPV) type 16 capsid proteins with HPV DNA requires an intact L2 N-terminal sequence. J Virol 68(2):619–625.

Zhou J, Liu WJ, Peng SW, et al (1999) Papilloma virus capsid protein expression level depends on the match between codon usage and tRNA availability. J Virol 73(6):4972–4982.

59. 超越课堂——王良：中国防痨先驱

结核病是由结核分枝杆菌引起的最为致命、流行最广泛的传染病之一，目前全球约有三分之一的人口感染结核分枝杆菌（Zumla et al.，2015）。因此，结核病也成为医学生课堂教学的一个重点内容。结核分枝杆菌是一种需氧的细小杆菌，细胞壁中含有大量的脂质，在细菌致病过程中发挥关键作用。尽管卡介苗（BCG）的保护效果存在争议（Villarreal-Ramos，2009），但目前它是预防结核病的唯一疫苗。自1921年问世以来，卡介苗在世界范围内广泛使用，特别是发展中国家。自20世纪70年代中期以来，我国已将新生儿和15岁以下儿童接种卡介苗纳入国家免疫规划项目。这一伟大成就的实现，要归功于一位内科医生——王良，是他最早于20世纪30年代初将卡介苗引入我国并率先带领我国人民抗击结核病（Yan，2003；Dai and Xiao，2013）。

虽然在医学生的课堂上，王良的伟大成就是一个永恒的话题，但是很少有人知道他传奇的一生和在防痨过程中遭遇的种种艰辛。1891年5月5日，王良出生于四川成都。他幼年丧父，1913年从越南的河内医学院毕业回国后，在云南和四川等地从医。20世纪初，结核病仍位于各种传染病致死原因之首。由于当时缺乏有效的治疗方法，结核病对人类的威胁犹如今日之癌症，王良的一兄一妹皆死于结核病。作为一个青年医生，他立志献身于祖国的防痨事业。

1924年，法国巴斯德研究所的Albert Calmette和Camille Guérin发表了关于卡介苗的研究成果，证明该疫苗能有效地预防结核病。1925年王良得知此消息后，便开始筹资到法国学习。在法国外交部的帮助下，王良于1931年来到了巴斯德研究所，并得到了Guérin的亲自指导，他的虚心好学很快受到了老师的赞赏。在研究所学习的两年间，王良共完成了4篇研究论文，其中有3篇都是跟卡介苗有关的。1933年夏天，王良带着卡介苗菌苗和仪器设备回到了祖国。在重庆一家私立医院担任内科医生期间，王良创建了自己的微生物实验室，并将业余时间全部投入到卡介苗的培养和制备国内首批疫苗上。1933年10月至1935年8月，王良用自己制造的卡介苗接种了248名婴幼儿，没有出现一例不良反应。该疫苗对预防结核病起到了很好的作用，甚至对一般流行病都具有一定的抵抗力。

译者：李明
陆军军医大学基础医学院微生物学教研室，重庆400038，中国
邮箱：sotx7080@163.com

正当王良准备进一步推广卡介苗接种时，1937年抗日战争全面爆发了，王良的实验室被迫关闭。此后，我国的卡介苗发展事业一度被中断10年以上。

1949年11月，西南军政委员会卫生部在重庆成立，王良得以重续他的卡介苗事业。1950年，王良受邀参加第一届全国卫生工作会议，并受卫生部指示在重庆筹建西南卡介苗制造研究所。自1951年8月起，卫生部举办了3期卡介苗应用培训班，卡介苗的应用被迅速推广。1956年，西南卡介苗制造研究所并入成都生物制品研究所（图1），在那里王良将其大半生的职业生涯都倾注在卡介苗的研制上。

图1　成都生物制品研究所成立之初

在之后的30多年里，王良始终坚持以临床应用为主、理论联系实际的原则，以其广博的知识和精湛的技术，集中开展了以下3个方面的研究工作，为我国卡介苗事业作出了不可磨灭的贡献。

第一，注重提高卡介苗的质量。王良改良了原来卡介苗的多种成分，将培养基中的麸酸钠含量由8%改为6%，甘油含量由6%改为5%，使之更适合活菌苗的培养；他还用优选法确定冻干卡介苗的保护液中蔗糖的最佳含量为8%。此外，为了测定卡介苗的贮存寿命，他还分析了10批次生产的卡介苗在4℃冰箱保存中活力丧失的情况，确定6周后活菌数量仍能维持在26%左右，足以引起机体的免疫反应。

第二，选育了优良的卡介苗生产菌种。由于各国卡介苗菌种长期传代的方法和培养条件有所不同，因此各国使用的菌种也有所差异。王良曾会同各个国家和地区卫生部门的技术人员，进行了一次历时两年的卡介苗选种工作。他们比较了大量的国内外菌株，建立了许多有效的检测方法。选种工作的成果为后来我国选用来源于丹麦的上海D2株作为全国卡介苗生产菌种提供了重要的实验依据。

第三，积极探讨卡介苗的免疫机制。王良对卡介苗免疫动物的血液成分进行了多次试验分析发现，动物一般于免疫后第3天白细胞总数明显上升，第5天达到高峰，然后于第12天恢复正常，白细胞分类中以淋巴细胞增加最为显著。他还多次注意到，经免疫后动物的网状内皮细胞和吞噬细胞的吞噬能力显著提高，血浆中的各种免疫球蛋白也明显增加。当免疫动物的变态反应消失后，卡介苗的免疫保护力在一定时间内依然存在。这些试验证明了卡介苗不仅能刺激细胞免疫，也能刺激体液免疫，从而验证了他在1935年首次接种婴幼儿时观察到的非特异性免疫现象。王良还发现在接种卡介苗后，动物对金黄色葡萄球菌以及伤寒菌和链球菌的抵抗力明显提高。这些重要的实验结果为后来卡介苗及其有效组分用于肿瘤及其他疾病的免疫治疗提供了依据，也为后来积极开发治疗型卡介苗系列产品指明了方向。

王良治学严谨，一丝不苟，一生都奉献给了祖国的防痨事业。他为人谦虚谨慎，平易近人，注重培养年轻人。年近90岁高龄，还亲自编写讲稿为青年讲课，为防痨事业鞠躬尽瘁。1983年，为纪念王良从事卡介苗工作50周年并庆祝他的92岁寿辰，成都生物制品研究所及中国防痨协会成都分会联合邀请中华医学会、中国防痨协会及全国各生物制品研究所的专家、学者及王良的好友、学生汇聚一堂，举行了一场学术论文报告会，鼓励大家学习王良同志为科学事业献身的精神。

1985年8月31日，王良病逝于家乡成都，享年94岁。

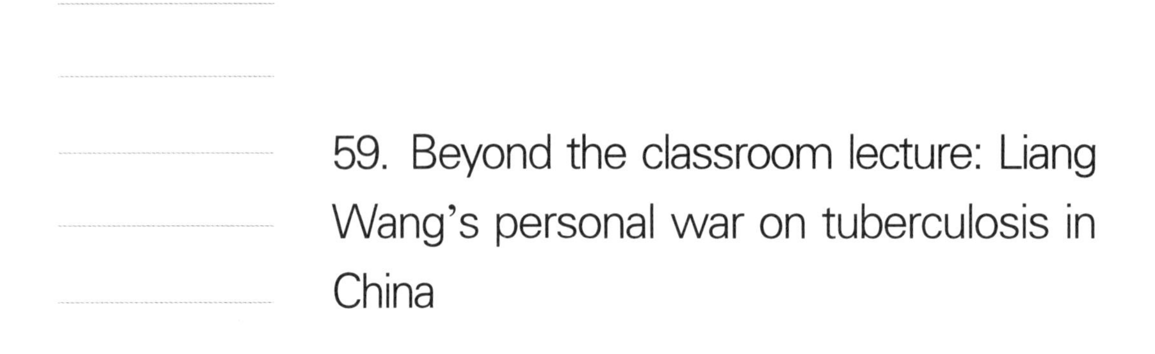

59. Beyond the classroom lecture: Liang Wang's personal war on tuberculosis in China

Tuberculosis, historically one of the deadliest and most prevalent infectious diseases, is caused by the bacterium *Mycobacterium tuberculosis*. It is estimated that one-third of the world's population is infected by this agent (Zumla et al., 2015), and tuberculosis is therefore a major topic for medical students. *M. tuberculosis* is a small aerobic bacillus, with a remarkably high lipid-content cell wall that plays a critical role in its pathogenicity. Although the efficacy of the Bacillus Calmette-Guérin (BCG) vaccine for preventing tuberculosis continues to be debated (Villarreal-Ramos, 2009), live attenuated BCG has been the only approved vaccine against *M. tuberculosis* infection since its introduction in 1921, and is still administered routinely around the world, particularly in developing countries. In China, BCG has been included in the national vaccination program for newborns and children under 15 since the mid-1970s. After decades of intensive effort, remarkable progress has been made in the fight against tuberculosis in China. This great achievement was made possible by Liang Wang, a physician who pioneered the introduction of the BCG vaccine to China in the early 1930s and spearheaded China's fight against tuberculosis (Yan, 2003; Dai and Xiao, 2013).

Although his accomplishments are frequently discussed in the medical school classroom, most students are unfamiliar with Liang Wang's personal story and the significant challenges he faced in his battle against tuberculosis. Liang Wang was born on May 5, 1891 in Chengdu, Sichuan Province. He lost his father at very young age. After graduating from the Hanoi Medical School in Vietnam in 1913, he entered private practice in the Yunnan and Sichuan regions of China. In the early 20th century, tuberculosis was the leading cause of death. Because there was no known cure, the severe health threat posed by tuberculosis was considered to be as serious as that posed by cancer today. Wang's older brother and younger sister both died of tuberculosis. As a young doctor, Liang Wang determined to devote himself completely to the fight against tuberculosis.

Ming Li, Xiaomei Hu, Fuquan Hu, Xiancai Rao

Department of Microbiology, College of Basic Medical Sciences, Army Medical University, Chongqing 400038, China

Correspondence: raoxiancai@126.com (X.-C. Rao)

In 1924, Albert Calmette and Camille Guérin from the Pasteur Institute in France published their research results on the BCG vaccine, demonstrating its efficacy in the prevention of tuberculosis. One year later, Liang Wang learned about BCG and began fundraising to support a trip to France. With the help of France's foreign ministry, Wang was able to visit the Pasteur Institute in 1931, where he began his research career under the personal guidance of Guérin. Wang's modesty and eagerness to learn were quickly appreciated. During his two years at the Pasteur, Wang completed four research papers, three of which focused on BCG and the culture of *M. tuberculosis*. In the summer 1933, Liang Wang returned to China with the BCG seed strain and lab equipment. He established a microbiology laboratory while working as a physician in a private hospital in Chongqing. He committed much of his spare time to BCG culture preparation and developed the first batch of BCG vaccine in China. Beginning October 1933 through August 1935, 248 infants were inoculated with the BCG vaccine manufactured by Liang Wang. No adverse reactions were observed. Crucially, BCG vaccination conferred significant protection against the development of tuberculosis, and also offered some degree of immunity against other common infectious diseases.

However, just as Liang Wang was planning to expand and promote BCG vaccination, in 1937, Wang's lab was forced to close, and the development of the vaccination project halted for more than a decade.

The new Department of Health, opened by the Southwest Military Administrative Committee in Chongqing, provided Liang Wang with the opportunity continuing his work. He was invited to participate in the first national health working conference in 1950, and was authorized to found the Southwest Chinese BCG Production and Research Institute in Chongqing under his leadership. Beginning August 1951, the Health Ministry held three training classes to promote the use of Wang's BCG vaccine, and BCG vaccination quickly became popularized. In 1956, the Southwest BCG Production and Research Institute was incorporated into the Chengdu Institute of Biological Products, where Liang Wang devoted the later stages of his professional career to BCG research and manufacturing.

For the next 30 years, Wang insisted on the priority of clinical application and the principle of combining theory with practice. By focusing on three specific areas of research, he contributed significantly to the refinement of BCG-based immunization strategies to control tuberculosis in China.

First, Wang emphasized the improvement of BCG vaccine quality. Wang optimized the chemical composition of BCG by decreasing the content of monosodium glutamate from 8% to 6% and glycerol from 6% to 5%, making it more suitable for live attenuated BCG. He also determined the most effective concentration of sucrose (8%) for the protection of lyophilized BCG preparations by using the optimum seeking method. To determine the shelf life for the vaccine, Wang measured the loss in potency for 10 different batches of BCG vaccine after storage at 4 ℃ for 6 weeks, and confirmed that 26% of bacteria remained viable, a level sufficient to induce the body's immune response.

Second, Wang began a process to isolate a superior BCG strain for vaccine production. Due to different passage methods and culture conditions, the BCG vaccine used in different countries varied. In collaboration with others, both locally and nationally, Liang Wang performed a two-year-long BCG breeding program in which he compared a large number of domestic and international BCG strains and established efficient detection protocols. His achievements provided the experimental basis for the selection of the Shanghai D2 strain from Denmark as the foundation for BCG vaccine production.

Third, Wang actively investigated the immunologic mechanisms underlying BCG efficacy. In an analysis of various blood constituents in BCG-immunized animals, he observed that the number of leukocytes increased significantly 3 days after vaccination, peaked on day 5, and returned to normal on day 12. Lymphocytes were the most markedly increased cell type. Through many experiments, he determined that the cytophagic capacity of reticuloendothelial cells and phagocytes were remarkably enhanced in BCG-vaccinated animals, and serum immunoglobulins were also significantly elevated. Wang also noted that the protection conferred by BCG vaccination persists even after allergies have disappeared in immunized animals. His work demonstrated that BCG induces not only cell-mediated immunity but also humoral immunity, consistent with the non-specific immunity observed in infants enrolled in Wang's first vaccination project in 1935. Furthermore, Wang also found that BCG-immunized animals exhibited an enhanced ability to resist infection caused by *Staphylococcus aureus* and other species from the genus *Salmonella* and *Streptococcus*. This remarkable discovery provided the foundation for the clinical application of BCG and its active ingredients as immunological treatments for cancer and other diseases, and defined the future direction for development of a series of therapeutic BCG-associated products.

With his strong work ethic and rigorous attitude toward research, Liang Wang devoted himself wholeheartedly to his anti-tuberculosis career. He was invariably courteous, amiable, and easy to approach, and paid close attention to the cultivation of young scientists and physicians. He remained active as a writer and lecturer even when nearly 90 years old. In 1983, an academic symposium was organized by the Chengdu Institute of Biological Products to celebrate the 50th anniversary of Wang's professional career and his 92th birthday.

On August 31, 1985, Liang Wang died at the age of 94 in his hometown of Chengdu.

Figures

Fig.1 The Chengdu Institute of Biological Products in its early days

References

Dai Z, Xiao D (2013) A History of Tuberculosis. Shelton: People's Medical Publishing.

Villarreal-Ramos B (2009) Towards improved understanding of protective mechanisms induced by the BCG vaccine. Expert Rev Vaccines 8:1531–1534.

Yan Z (2003) The founder of China's BCG career—Liang Wang (1891—1985). Chin J Microbiol Immunol 1:1–2.

Zumla A, George A, Sharma V, et al (2015) The WHO 2014 global tuberculosis report—further to go. Lancet Glob Health 3:e10–e12.

60.“朱老夫子”朱关福：病毒学家

朱关福教授1926年6月出生于浙江省定海县，1946年10月起就读于同济大学医学院（图1），1954年9月起就读于北京协和医学院细菌免疫学系，师从微生物学家和免疫学家谢少文教授，获得硕士学位。1951年7月，朱教授担任第二军医大学微生物教研室助教，1958年1月起就职于军事医学科学院微生物流行病研究所，历任研究实习员、助理研究员、副研究员、研究员。朱教授于2015年11月7日因心脑血管疾病逝世，享年89岁。

图1　青年时代的朱关福教授

朱教授是国内最早从事病毒学研究的开拓者之一，几十年辛勤耕耘、矢志不渝，为我国病毒学发展作出了重要贡献。早在20世纪70年代初期，朱教授自我国感冒患者鼻分泌物和咽液标本中，成功分离出多株冠状病毒、鼻病毒和腺病毒（黄如统等，1979；中国人民解放军某部队感冒防治小组，1975；朱关福，1976），并调查明确了冠状病毒的人群感染率，为我国呼吸道传染病病因研究提供了重要依据。在80年代初期，朱教授系统建立了多种虫媒甲病毒与黄病毒的组织培养技术（朱关福等，1982；刘紫慧等，1982），并于90年代初采用淋巴细胞共培养的方法第一次从我国艾滋病患者血液中分离获得了人免疫缺陷病毒（HIV）（孙中和等，1991）。此外，朱教授还较早地将单克隆抗体技术应用于多种常见病毒性疾病的实验室检测和诊断。他先后主编出版了《病毒名称》（朱关福，1987）、《现代临床病毒学》（杜平等，1991）、《常见新发传染病防治手册》（邵一鸣等，2005）等多部专著，发表了多篇研究论文和综述。

朱关福教授为人谦和，“学究味”十足，被同行称为“朱老夫子”。他为中国微生物学会发展和国际交流作出重要贡献，受到业内人士的爱戴和尊敬。朱教授曾经连任三届中国微生物学会副理事长，并获荣誉理事长称号；1984—2001年，朱教授任第二

译者：秦成峰，叶青

军事科学院军事医学研究院微生物流行病研究所，北京100071，中国

邮箱：qincf@bmi.ac.cn（秦成峰）；yy.0526@163.com（叶青）

届至第五届中国微生物学会病毒学专业委员会主任委员。他发起的由中国微生物学会病毒学专业委员会主办的全国病毒学学术研讨会，自1986年7月15—19日在天津举办首届会议以来，至今已经连续举办11届，成为国内病毒学交流的盛会，2015年在武汉举办的第十一届全国病毒学学术研讨会与会人数达到1200人。退休后，朱关福教授仍然不遗余力积极组织、筹办和参加学会的学术活动，并担任第六届病毒学专业委员会名誉主任委员和第五届中华医学会微生物学与免疫学分会名誉主任委员。2009年，83岁高龄的朱关福教授仍然出席并参加了在北京举办的第八届全国病毒学学术研讨会，与众多病毒学专家开展学术交流和讨论。

朱关福教授积极推动我国与国际病毒学界的交流（图2）。1990年8月27—31日，国际微生物学会联合会（IUMS）病毒学部在柏林召开第八届国际病毒学会议，朱关福教授以中国微生物学会病毒学专业委员会主任委员的身份，代表中国病毒学界参加会议，建立了我国与国际病毒学界的正式联系。1992年，朱关福教授在北京友谊宾馆与日本同行共同倡导发起中日病毒学国际研讨会（图3），每两年举办一次并延续至今，促进了国际合作交流。

图2 朱关福教授（左一）1996年在第十届国际病毒学会议上

图3 朱关福教授（中）在第七届中日病毒学国际研讨会上

朱关福教授因其卓越贡献和高尚品格被尊称为“朱老夫子”，犹如“孔老夫子”（孔子）作为我国几千年来的文化象征受到世人尊敬一样。我们深切缅怀朱关福教授。

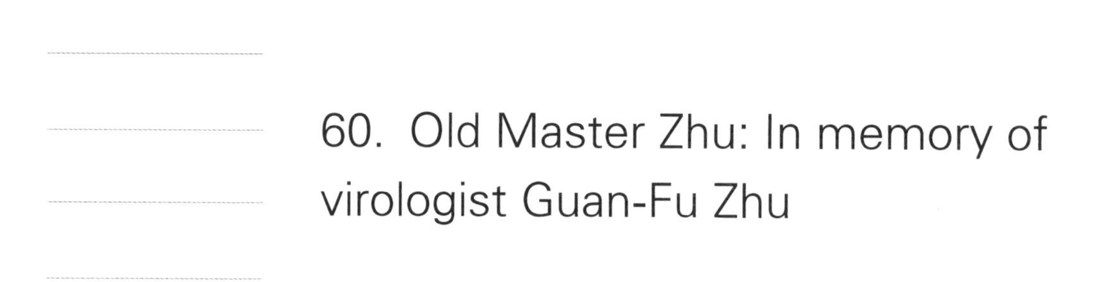

60. Old Master Zhu: In memory of virologist Guan-Fu Zhu

Born in Dinghai County of Zhejiang Province in June, 1926, Guan-Fu Zhu began to study at School of Medicine, Tongji University, in October, 1946. Then he moved to the Department of Bacteriology and Immunology of Peking Union Medical College in September, 1954, where he got his master degree under the supervision of Prof. Shao-Wen Xie, a renowned microbiologist and immunologist. In July, 1951, Mr. Zhu served as an assistant in teaching and research section of the Second Military Medical University of PLA. In January, 1958, he began to work at the Institute of Microbiology and Epidemiology, Academy of Military Medical Sciences, successively serving as a research assistant, assistant professor, associate professor and professor. Due to cardiovascular and cerebrovascular diseases, Mr. Guan-Fu Zhu unfortunately passed away on November 7, 2015, at the age of 89.

Prof. Zhu made significant contribution to the development of virology in China and was deemed as one of the pioneer virologists in China. In the early 1970s, Prof. Zhu successfully isolated dozens of viral strains of coronaviruses, rhinoviruses, and adenoviruses from nasal discharge and pharyngeal fluid samples from patients with cold (Huang et al., 1979; A group of cold prevention and control of PLA, 1975; Zhu, 1976), and investigated the infection rate of coronavirus in population, providing basic foundation for the study of the etiology of respiratory infectious diseases in China. In the early 1980s, Prof. Zhu systematically established virus propagation techniques in tissue culture including a variety of viruses such as vector borne alphavirus and flavivirus (Zhu et al., 1982; Liu et al., 1982). Then at the beginning of 1990s, he and his colleagues isolated the first Chinese HIV strain from the blood samples of a Chinese patient using co-culture of lymphocyte (Sun et al., 1991). In addition, Prof. Zhu had tried to utilize the monoclonal antibody technology to the laboratory detection and diagnosis of common viral diseases at the very early stage. These brilliant works lead to numerous publications in scientific journals and textbooks. He also co-edited several keystone monographs, including

---

Qing Ye, Tao Jiang, Chengfeng Qin

Department of Virology, State Key Laboratory of Pathogen and Biosecurity, Beijing Institute of Microbiology and Epidemiology, Beijing 100071, China

Correspondence: qincf@bmi.ac.cn (C.-F. Qin)

Virus Names (Zhu, 1987), *Modern Clinical Virology* (Du et al., 1991), and *Prevention and Control of Common Emerging Infectious Diseases* (Shao et al., 2005).

Prof. Guan-Fu Zhu spent all his energy and life time on virology research. His personality affects everyone he meets, and his old friends called him "Old Master Zhu" . Prof. Zhu had made great contributions to the development of Chinese Society for Microbiology and the international communication and collaboration in the field of virology. Prof. Zhu served as three-term vice president for the council of the Chinese Society for Microbiology, and was then awarded the title of honorary chairman. Since 1984, Guan-Fu Zhu was elected as the chairman (1984—2001) of the Virology Committee of the Chinese Society for Microbiology. Prof. Zhu had launched the National Conference on Virology, held by the Virology Committee of the Chinese Society for Microbiology. It has been held biannually since the first meeting in Tianjin, China in July 15–19, 1986, and has become a grand gathering for all the Chinese virologists. The 11th National Conference on Virology held in Wuhan in 2015 attracted more than 1200 participants. Prof. Zhu was active in organizing and participating the academic activities even after retirement, and served as the honorary chairman for the sixth session of Virology Committee and the fifth session of the Microbiology and Immunology Branch of Chinese Medical Association. In 2009, Prof. Zhu, at the age of 83, still attended the 8th National Conference on Virology in Beijing, and communicated with next generation of Chinese virologists.

Prof. Zhu had been actively promoting the international communications in the field of virology. On August 27–31, 1990, as the chairman of Virology Committee of the Chinese Society for Microbiology, Prof. Zhu attended the 8th International Conference on Virology in Berlin, held by the Virology Division of International Union of Microbiological Societies (IUMS). In 1992, Prof. Zhu initiated and launched the China-Japan International Conference of Virology with Japanese scholars in Beijing Friendship Hotel, which has been held biannually until now.

"Old Master Kong" (Confucius) has been alternately idealized as a culturally symbolic figure of China over the millennia. Prof. Guan-Fu Zhu will be remembered for his great contribution and nobility in character, and we will miss Old Master Zhu forever.

Figures

Fig.1 Prof. Guan-Fu Zhu in his youth

Fig.2 Prof. Guan-Fu Zhu (1st from the left) in the 10th International Conference on Virology, 1996

Fig.3 Prof. Guan-Fu Zhu (middle) in the 7th China–Japan International Conference of Virology

References

A group of cold prevention and control of PLA (1975) Pathogenic examination of patients with cold in Beijing from 1971 to 1973. Natl Med J China 10:728–731. (中国人民解放军某部队感冒防治小组. 1975. 北京地区 1971—1973 年感冒患者的病毒病原检查. 中华医学杂志, 10: 728–731.)

Du P, Zhu GF, Liu XY (1991) Modern Clinical Virology. Beijing: People's Military Medical Press. (杜平，朱关福，刘湘云. 1991. 现代临床病毒学. 北京：人民军医出版社.)

Huang RT, Qu CM, Chen XR, et al (1979) Microdiagnosis of the pathogens of adenovirus pneumonia. Bull Acad Mil Med Sci 1(5):121-126. (黄如统，渠川玫，陈香蕊，等. 1979. 腺病毒肺炎的病原微量诊断. 中国人民解放军军事医学科学院院刊, 1(5):121-126.)

Liu ZH, Huang RT, Zhu GF (1982) Microculture of arbovirus in cells. Bull Acad Mil Med Sci 6(22):697-700. (刘紫慧，黄如统，朱关福. 1982. 甲组虫媒病毒的微量细胞培养. 军事医学科学院院刊, 6(22):697-700.)

Shao YM, Zhang XY, Zhu GF (2005) Prevention and Control of Common Emerging Infectious Diseases. Hangzhou: Zhejiang University Press. (邵一鸣，张晓燕，朱关福. 2005. 常见新发传染病防治手册. 杭州：浙江大学出版社.)

Sun ZH, Xu WP, Li H, et al (1991) The first isolation of HIV strain in China from the blood samples of AIDS patients. Bull Acad Mil Med Sci 15(1):79. (孙中和，许万平，李红，等. 1991. 从中国艾滋病病人血液分离出首株人免疫缺陷病毒. 军事医学科学院院刊, 15(1):79.)

Zhu GF (1976) Human coronavirus: An emerging respiratory virus. Microbiology China 1:22-24. (朱关福. 1976. 新发现的呼吸道感染病毒——人冠状病毒. 微生物学通报, 1: 22-24.)

Zhu GF (1987) Virus Names. Beijing: Science Press. (朱关福. 1987. 病毒名称. 北京：科学出版社.)

Zhu GF, Huang RT, Wang J (1982) Microculture of dengue virus in cells. Med J Chin People's Lib Army 7(3): 139-141+5. (朱关福，黄如统，王津. 1982. 登革病毒微量细胞培养. 解放军医学杂志, 7(3): 139-141+5.)

七、生理学和医学

61. 林可胜：中国生理学的奠基者

林可胜先生是我国现代生理学与医学领域的一位传奇式人物（图1）。他不仅是杰出的生理科学研究者和学科领导者，是中国现代生理学的奠基人；同时，他还是一位赤诚的爱国者，他投身抗战前线，创建和领导战时军队医疗救护系统，为抗战胜利作出无私奉献。

图1　林可胜先生（1897—1969）

林可胜先生1897年10月15日出生于新加坡，祖籍福建。他的父亲林文庆先生毕业于爱丁堡大学，不仅是一位医术精湛的医生，还是一位社会活动家，为推进我国社会和教育改革作出重要贡献。母亲黄端琼女士是华侨领袖黄乃裳的女儿，是最早赴美留学的女性之一，也是一位杰出的新时代女性。著名的鼠疫斗士伍连德是林先生的姨父。

林可胜8岁时就被父亲送到苏格兰。很快，他就进入了爱丁堡最好的一所学校——乔治·沃森学校学习。从沃森学校毕业后，他考入爱丁堡大学学习医学。不久，第一次世界大战爆发，林可胜应征入伍到一所印度军医院担任外科助理，主要进行新兵的战地救护培训。1916年，林可胜重新返回爱丁堡大学继续学业，于1919年获得了医学学士和化学学士学位（Davenport，1980），并留校担任生理学系讲师。由于才能出众，林可胜深受生理学教授Edward Sharpey-Schafer的赏识。在授课之余，他继续跟随Sharpey-Schafer从事生理学研究，于1921年获得生理学博士学位。1923年，年仅26岁的林可胜当选爱丁堡皇家学会会员。

在此期间，林可胜曾于1922年向美国中华医学基金会申请过专项奖学金并因此获得赴欧洲或美国学习的机会，并表达回国效力的愿望。他的学术背景引起了基金会负责人R. S. Greene的关注。后来在洛克菲勒基金会的支持下，他在芝加哥大学工作了一年。

1924年，林可胜原本计划回国去厦门大学建立医学院，也曾计划在厦门实验室建立的过渡期先去北京协和医学院（简称“协和”）待半年，积累一些经验。但是后来他

作者：张钫
清华大学科学技术与社会研究所，北京100084，中国
邮箱：fzhang1985@foxmail.com

接受了Greene的建议，先以客座教授的身份到协和工作一年。

那时，协和的生理学系人手不足。林可胜一到这里，就和他的同事及学生着手开展各种研究项目。很快，在林可胜28岁时，他被任命为协和生理学系主任兼教授。从1924年到1937年，林可胜在协和生理学系工作了13年之久，并且在生理学科学研究和学科建设上作出了一系列突出的贡献。

为了带动协和生理学系的发展，林可胜在团队建设、科学研究、教学活动、基础设施建设以及人才培养等各个方面都倾注了大量心血。协和的生理学系工作局面打开之后，林可胜又开始将精力向全国拓展。开展学术交流是生理学发展的一个重要途径（曹育，1998），因此，林可胜借鉴欧美学会的经验，筹划创立了中国生理学会，并担任学会第一任会长。第二年，他又创办了《中国生理学杂志》，并担任主编（罗桂环等，2018）。学会和杂志的创立在我国生理学的发展中发挥了重要作用。

学科建设之余，林可胜也一直活跃在生理学研究前沿领域，他发表了90余篇研究论文。在1937年抗日战争全面爆发之前，林可胜主要集中于消化和循环生理的研究，并且在这两个领域都取得了突出的成就，处于国际领先地位。

在消化生理领域，林可胜凭借其精湛的动物实验技术，以犬作为研究对象，制成了全胃瘘、活体灌流胃、移植胃等模型，和他的同事及学生成功地进行了有关胃的多方面研究，他们明确揭示了激素作用于胃分泌和脂肪抑制胃运动的机制，引起国际生理学界的高度关注。20世纪30年代，他发现并命名的“肠抑胃素”，是我国科学家发现的第一种激素（曹育，1998）。

从1936年到1939年，林可胜从事循环生理学的研究。林可胜对血管的中枢定位进行系统研究，他证明了交感反射可以通过刺激大脑的髓质而诱发。这个领域的研究工作由于战争的爆发而暂时中断。后来，他的同事和学生沿着他的思路和方法继续进行研究，之后协和在该领域的工作一直受到国际生理学界的重视。

20世纪30年代末，林可胜辞去了协和的职务，开始更大规模地为国家服务。战争开始时，林可胜组织并领导了中国红十字会医疗救援队，为士兵提供现代化医疗服务。他建立了战时卫生人员训练所，训练医生、护士和医疗技术人员的医疗急救服务技术，为军队和民间救援机构培训了大量的医务人员。到1939年，他们已经带领众多救援队，奔赴全国各大战区（图2）。战争期间，林可胜在贵阳建立了当时我国最大的医学中心。1941年，他被任命为医疗服务总监（Davenport，1980）。

图2　林可胜先生（左三）等人在贵阳

战后，林可胜担任军医署署长，重建了我国的医学教育和军医体系，为我国军队医疗系统的发展铺平道路。1943年，他获得罗斯福总统授予的荣誉勋章，1946年获得自由勋章，以表彰他在战时的工作。他被认为是“中国最伟大的人之一”。

1948年，林可胜当选中央研究院院士。1949年，他前往美国，受邀在伊利诺伊大学任临床医学的客座教授。1950年到1951年，他在克莱顿大学担任生理与药理学系主任和教授。1952年，他接受了印第安纳州迈尔斯实验室的邀请，负责生理学和药理学系，并开展了一些研究。由于在战争期间的经历，他对止痛药产生兴趣，对痛觉和止痛药的神经生理机制进行了研究。

林可胜突出的成就为他赢得国际学界的尊重。他曾当选美国、英国和德国许多研究机构和学术团体的会员或荣誉会员（Ivy，1970）。1942年，他当选为美国国家科学院外籍院士。

1967年，林可胜被诊断出患有食道癌。1969年，他和妻子搬到牙买加的儿子家中，女儿也从英国赶来。在那里，他与家人一起度过了最后时光。1969年7月8日，林可胜平静地去世了，享年72岁。

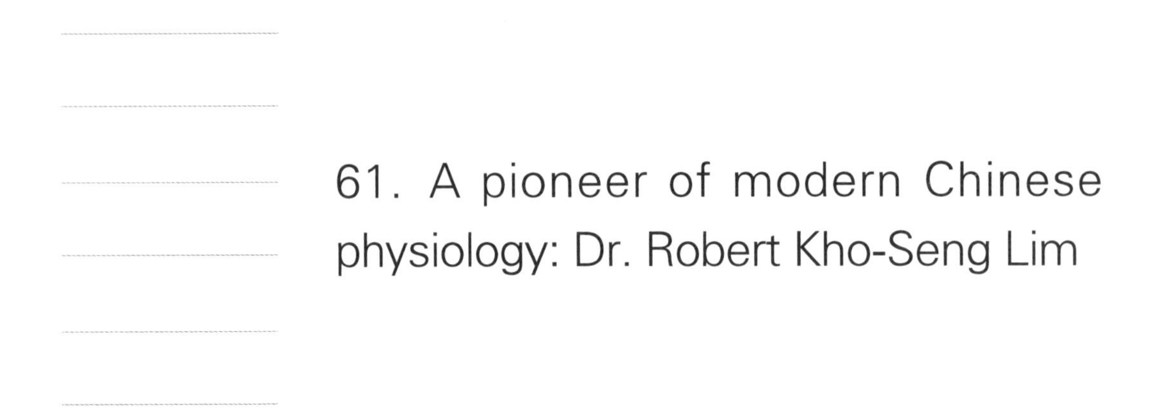

61. A pioneer of modern Chinese physiology: Dr. Robert Kho-Seng Lim

Dr. Robert Kho-Seng Lim (林可胜) was a legendary figure in modern Chinese physiology and medical fields. He was a famous physiologist and was considered as a pioneer of modern Chinese Physiology. He was also an ardent patriot and organized medical relief corps and trained various medical workers to meet the needs of China during the war. He was a man of modesty, wisdom, fortitude and tenacity.

Dr. Lim was born of Chinese parentage in Singapore on October 15, 1897. His nickname was Booby. His father, Lim Boon-Keng (林文庆), was a distinguished physician graduated from the University of Edinburgh, who promoted social and educational reforms in China. His mother, Wong Tuan-Keng (黄端琼), the daughter of Wong Nai-Siong (黄乃裳), was one of the first generation of Chinese women who received educations in the United States. His uncle, Wu Lien-Teh (伍连德), was a plague fighter and the father of Chinese public health system.

Dr. Lim went to Scotland when he was eight years old. Later, he attended George Watson's School in Edinburgh. In 1911, he was admitted to the University of Edinburgh for medical studies after the graduation from Watson's school. At the outbreak of the First World War, he volunteered and served for the Indian army medical service to train recruits for the field rescue. In 1916, Dr. Lim returned to Edinburgh, where he received his M.B. and Ch.B. degrees in 1919 (Davenport, 1980). In the Medical School, due to his excellent performance, Dr. Lim got appreciated and supported by Sir Edward Sharpey-Schafer, the professor of physiology. After his graduation, he was appointed as lecturer in the department of physiology to teach histology. At the same time, he continued his researches and obtained his Ph.D. degree in 1921. Aged 26, he was elected as fellowship of Royal Society of Edinburgh.

In 1922, Dr. Lim applied for a fellowship from China Medical Board of America for the opportunities to study in Europe or USA. His application caught attention from Roger S. Greene, the Board's Secretary. And, he was granted a Rockefeller Foundation Fellowship to work at the University of Chicago for one year.

Fang Zhang
China University of Political Science and Law, Beijing 102249, China
Correspondence: fzhang1985@foxmail.com

In 1924, Dr. Lim planned to return to China with his original thought to set up a medical school for Xiamen University. He also planned to work in the Peking Union Medical College (PUMC) for half a year during this transition. At the end, he accepted Greene's advice and worked as a visiting professor in the PUMC for a year.

At that time, it was relatively lackluster of the department of physiology at PUMC. Upon his arrival, he started to establish a vigorous research program in collaboration with his colleagues and students. Soon, Dr. Lim was appointed as professor of physiology and head of the department at PUMC when he was only 28. Dr. Lim worked in the department of physiology for 13 years from 1924 to 1937 and made outstanding contributions to the development of physiological researches and the construction of the research programs.

Dr. Lim spent all his efforts on team construction, scientific researches, teaching activities, infrastructure construction and talent training for the development of the department of physiology in PUMC. After opening new avenues for the department of physiology in PUMC, Dr. Lim then began to extend the subject of physiology to the whole country. Bring all of the professionals together and carrying out academic exchanges were important ways for the development of physiology in China (Cao, 1998). Dr. Lim and his colleagues founded the Chinese Physiological Society, and Dr. Lim served as the first president of the society. In the following year, Dr. Lim founded the *Chinese Journal of Physiology* and served as editor-in-chief (Luo et al., 2018). Both the Society and the journal had played an essential role in promoting the development of physiology in China.

In addition to the subject construction, Dr. Lim had always been an active scientist in the frontier of physiology, and had around 90 publications. Before 1937, Dr. Lim mainly focused on digestive physiology and circulatory physiology and gained remarkable achievements in both fields.

In the field of digestive physiology, Dr. Lim had successfully conducted various researches on gastric secretion with his colleagues and students with his skilled animal experimental techniques. They had explicitly demonstrated the hormonal mechanism for gastric secretion and the inhibition of gastric motility by fat, which attracted great attentions from scientists all over the world in the field of physiology. In 1930s, he coined the word "enterogastrone" to designate the "gastric inhibitory agent" , which is the first hormone discovered in China (Cao, 1998).

Dr. Lim worked on circulatory physiology from 1936 to 1939 and showed that sympathetic reflexes could be evoked by stimulation of the medulla of the brain. The studies in this research field were temporarily interrupted due to the outbreak of the war. And later on, his colleagues and students continued the researches in this field, which attracted great attentions from the international physiological community.

In the late 1930s, Dr. Lim resigned from his position at PUMC, and started serving his country on a larger scale. As the war began, Dr. Lim founded and directed the Chinese Red Cross Medical Relief Corps, and provided modern medical treatments to the soldiers. He organized the Emergency Medical Service Training School for physicians, nurses, and medical technicians and trained hundreds of medical workers. The graduates served in the army and civilian relief agencies. By 1939, they had led many rescue teams to cover the major battlefields all over the country. Dr. Lim built the largest medical center at Kweiyang (贵阳) in China during the war, and he was appointed as Inspector General of the Medical Services in 1941 (Davenport, 1980).

After the war, as Surgeon General, Dr. Lim rebuilt China's medical education and military medical system, paving the way for the development of military medicine in China. He received the Legion of Merit by President Roosevelt in 1943 and the Medal of Freedom in 1946 given his medical support during the war. He was considered as "one of the greatest men of China" .

In 1948, Dr. Lim was elected to be a member of the Academia Sinica. In the same year, he was offered the position of the Minister of Health, but he rejected this job offer. He went to the United States in 1949 and was invited as a visiting research professor of clinical science at the University of Illinois. From 1950 to 1951, he severed as professor and the head of department of physiology and pharmacology at Creighton University. In 1952, he accepted a position at Miles Laboratories in Indiana, where he was responsible for physiological and pharmacological researches and the guidance of medical scientific research. From his military experience during the war, he became interested in pain relief medications. He found that pain receptors were chemo-sensitive, in which pain-producing agents such as bradykinin peptides were produced when tissues of the body were injured.

Because of his outstanding achievements, Dr. Lim was a member or an honorary member of numerous research institutions and academic societies in the United states, Great Britain and Germany (Ivy, 1970). He was elected a Foreign Associate of the National Academy of Sciences of the United States in 1942. When he became a United States citizen in 1955, he became a regular member of the Academy.

Dr. Lim was diagnosed with esophageal cancer in 1967. In 1969, he moved to his son's home in Jamaica together with his wife, and his daughter also came back from England. There, he enjoyed his last moments with his family. Dr. Lim passed away on July 8, 1969 at the age of 72.

Figures

Fig.1 Mr. Robert Kho-Seng Lim (1897—1969)

Fig.2 Mr. Robert Kho-Seng Lim (3rd from the left) and others in Kweiyang

References

Cao Y (1998) Dr. Robert Kho Seng Lim: A pioneer of modern Chinese physiology. Chin Hist Mater Sci Technol 19(1):26–41. (曹育. 1998. 中国现代生理学奠基人林可胜博士. 中国科技史料, 19(1): 26–41.)

Davenport HW (1980) Robert Kho Seng Lim: A biographical memoirs. Nat Acad Sci 51:280–306.

Ivy AC (1970) Robert Kho-Seng Lim, M.B., Ch. B, Ph. D, D. Sc, 1897—1969. Gastroenterology 58(4):580–581.

Luo GH, et al (2018) History of Chinese Biology (modern volume). Nanning: Guangxi Education Publishing House, 69–73. (罗桂环, 等. 2018. 中国生物学史(近现代卷). 南宁: 广西教育出版社, 69–73.)

62. 陈克恢：中国现代药理学研究先驱

陈克恢（图1），药理学家，一生为医药学研究奋斗不息，是现代中药药理学研究的创始人（钱伟长，2014）。

他系统研究了麻黄素的药理作用，发表了世界上关于麻黄素药理作用的第一篇论文；发现亚硝酸钠和硫代硫酸钠静脉注射可有效地解除急性氰化物中毒（Chen and Rose，1952）；从蟾酥中分离到华蟾蜍精和华蟾蜍毒素，为蟾蜍毒素等研究丰富了药物化学宝库。陈克恢开辟了中药研究的药理学道路，是我国中药药理研究划时代的奠基人，对麻黄素的研究成果是国际药理学的里程碑。

图1　陈克恢（1898—1988）

求学生涯

陈克恢幼年时父亲病逝，由舅父周寿南代为照顾。因舅父是中医，陈克恢受他影响自幼即对中草药感兴趣。1916年，考入清华学校（清华大学前身）。1918年，陈克恢从清华学校毕业。1920年，陈克恢在美国威斯康星大学药学系进行桂皮油的研究，并以这项研究完成了他的学士论文，获理学学士学位。1923年，获威斯康星大学生理学博士学位（图2）。

图2　陈克恢1921年摄于威斯康星州

1923年至1925年，陈克恢任北京协和医学院药

作者：刘欢[1,2]，刘兆祺[1]，龚雪[1]，程浩[3]

1　中国科学技术大学，合肥230026，中国

2　病毒学国家重点实验室，武汉430072，中国

3　中国科学院微生物研究所，北京100101，中国

邮箱：liuhuan520@ustc.edu.cn（刘欢）

理学助教。1927年，他获得约翰斯·霍普金斯大学的医学博士学位，并晋升为药理学副教授。1929年，任美国礼来药厂研究部主任。1937年至1968年，陈克恢任印第安纳大学医学院药理学教授。1968年，陈克恢从印第安纳大学退休（钱伟长，2014）。

麻黄素药理作用的重大发现

陈克恢最著名的工作是对麻黄素的研究。麻黄素药理作用的重大发现在医药史上具有划时代意义，真正认识、了解并应用于临床治疗则是从陈克恢开始的。

1923年，陈克恢在北京协和医学院任助教，除了教学之外，他希望继续研究中草药，并得到了系主任C. F. Schmidt教授的支持。陈克恢从舅父处得知麻黄有治疗哮喘的作用，在舅父的建议下，他从数百种常用中药中选了麻黄作为第一个研究对象。在短短几周内，便用氨氯仿方法从麻黄中成功地分离出左旋麻黄碱（Chen and Schmidt，1926）。在此之前，日本学者长井长义于1887年从麻黄中分离出左旋麻黄碱，命名为麻黄素（ephedrine），并发现麻黄素具有放大瞳孔的作用。陈克恢和Schmidt详细研究了麻黄素的药理作用，发现麻黄素可使颈动脉压长时间升高，心肌收缩力增强，对内脏血管、黏膜和皮肤有收缩作用，使支气管舒张（钱伟长，2014）；并能使离体子宫很快收缩，对中枢神经有兴奋作用，滴入眼内引起瞳孔散大（Chen and Schmidt，1926）。这些作用都和肾上腺素相同，不同的是麻黄素口服有效、作用时间长、毒性低。1924年，陈克恢与Schmidt在美国实验生物与医学学会北京分会上作了初步报告。同年，经过系统的实验研究后，二人将研究结果写入论文《中药麻黄有效成分麻黄素的作用》，发表在美国权威杂志《药理学与实验治疗学杂志》上，这是世界上关于麻黄素药理作用的第一篇论文。文章指出，麻黄的有效成分麻黄素的生理作用与肾上腺素类似且较持久，主要通过刺激交感神经系统发挥功效。至此麻黄素一跃成为国际瞩目的一个拟交感神经新药（图3）。这项研究是从天然产物中寻找先导化合物进行优化、开发新药的一个典范。

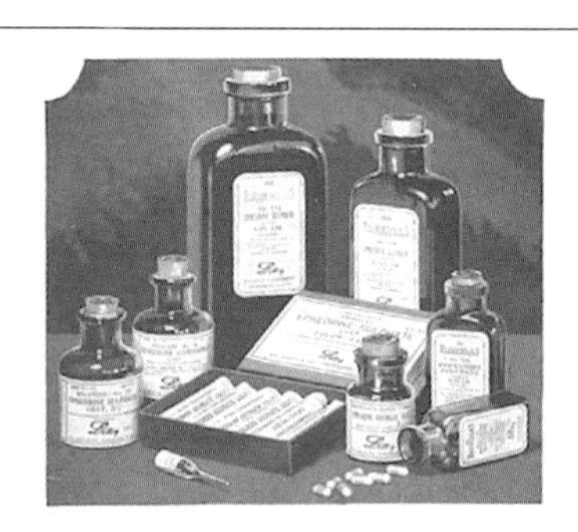

图3 礼来药厂介绍硫酸麻黄素的印刷品

急性氰化物中毒解救方法的发现

20世纪30年代早期，陈克恢和他的同事的另一项重要研究成果是关于急性氰化物中毒的解救方法。陈克恢与他的科研团队在前人的研究基础上发现，先后静脉注射亚硝酸钠和硫代硫酸钠可有效地解除急性氰化物中毒。亚硝酸盐和硫代硫酸盐的结合不仅表现出协同效应，而且超过了它们各自效应的总和（Chen and Rose，1952）。这一发现临床意义重大。目前，在抢救急性氰化物中毒的病人时，陈克恢等的研究成果仍是主要解毒方法之一，即先静脉注射亚硝酸钠溶液，随后注射硫代硫酸钠溶液。

对蟾蜍毒素的研究

陈克恢的一生有大部分时间在研究蟾蜍毒素。陈克恢在阿贝尔实验室时（1927年）即对蟾蜍毒素的研究很感兴趣。蟾蜍毒素即著名中药蟾酥。陈克恢是第一个成功地从蟾酥晶体中获得肾上腺素和含氮化合物的人（Chen et al.，1931）。他很快从蟾酥中分离到两种成分，即华蟾蜍精和华蟾蜍毒素，并发现这两种成分都有类似洋地黄的强心作用，与洋地黄毒苷比较，华蟾蜍精的作用时间持续较短，而且口服无效。1929年，陈克恢来到礼来药厂后，继续蟾蜍毒素的研究达40多年，对400多种强心苷和甾类化合物进行了构效关系的研究，发现除了蟾蜍外，其他动物也能够从植物性食物中储存卡烯内酯类型的糖苷（Chen，1970）。陈克恢发表了大量文章，丰富了药物化学宝库，并为其他药物的研究提供了宝贵经验。

其他药理研究

20世纪40年代，陈克恢发现中药常山中提取到的常山碱丙有抗疟疾的作用，其抗疟作用为奎宁的148倍（Henderson et al.，1949），但会导致呕吐，而且容易引起肝脏水肿变性，最终常山碱丙未能推向市场。第二次世界大战后不久，陈克恢得到缴获的美沙酮（Methadone）样品，并肯定了它的镇痛作用。在此基础上，他和同事在礼来药厂合成并开发了丙氧芬（钱伟长，2014），此药临床效果虽仅与可待因相当或稍差，但其成瘾性小，而且与阿司匹林有协同作用。

国际药理学界的一座丰碑

陈克恢从事药理学事业50余年，发表论文和综述350多篇，研究领域广泛、深入，对新药开发贡献巨大。陈克恢从中药中分离、提取有效成分，发展为化学药物。遵循这一思路，我国的医药学家和化学家对此进行了大量的研究，并取得了很多标志性成果和重大突破，如治疗肝炎的新药联苯双酯、双环醇和抗疟药青蒿素、双氢青蒿素等。陈克恢的学术思想对中草药等天然药物与合成药物的研究都具有深远的指导意义，他的科学思路和研究方法开创了近代中医药物药理研究的先河。

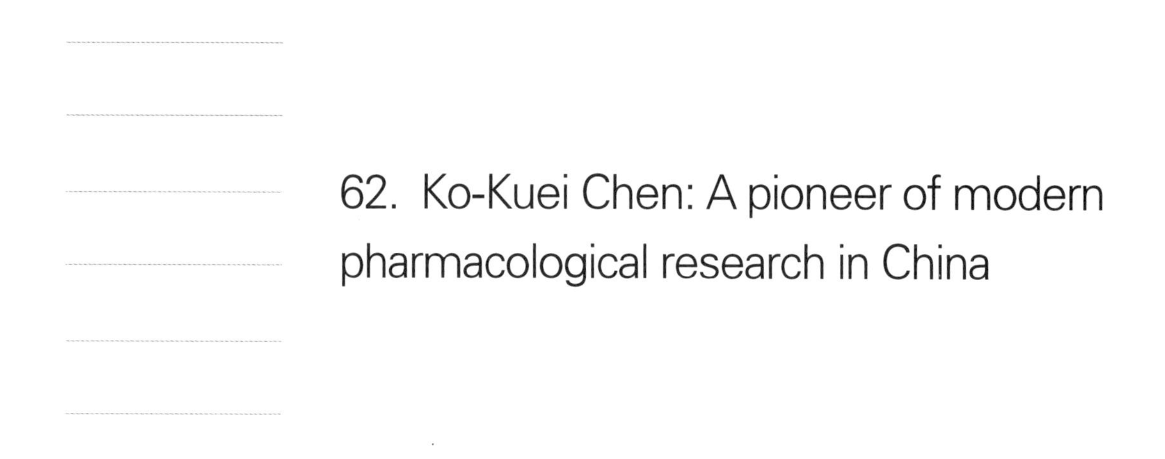

62. Ko-Kuei Chen: A pioneer of modern pharmacological research in China

Ko-Kuei Chen (陈克恢,1898—1988) was a pharmacologist who ceaselessly strove for medical research all his life. He was a pioneer of modern pharmacological research in Chinese medicine (Qian, 2014).

He systematically studied the pharmacological action of ephedrine and published the first paper on this topic in the world. He discovered that injecting sodium nitrite and sodium thiosulfate intravenously could effectively treat acute cyanide poisoning (Chen and Rose, 1952). He isolated cinobufagin and cinobufotoxin from Chan Su, the dried venom of the Chinese toad. His research in pharmacology enriched the treasure trove of medicine. Ko-Kuei Chen was a pioneer of pharmacology in Chinese medicine, and his research had epoch-making significance in the field. His pharmacological research on ephedrine set a milestone in the world medicine history.

Ko-Kuei Chen's Education

When Ko-Kuei Chen was young, his father died and he was raised by his uncle Shounan Zhou (周寿南), a Chinese physician. Due to his influence, Ko-Kuei Chen became interested in Chinese herbal medicine at an early age. In 1916, he was admitted to "Tsinghua College" (the predecessor of Tsinghua University). In 1918, Ko-Kuei Chen graduated from Tsinghua College. In 1920, Chen conducted research on cassia oil in the School of Pharmacy at University of Wisconsin-Madison, where he completed his thesis and received a bachelor of science degree.

In 1923, he received a Ph.D. in Physiology and Pathology Sciences at University of Wisconsin-Madison. From 1923 to 1925, Chen served as an assistant professor in the Department of Pharmacy at Peking Union Medical College. In 1927, he received his M.D. from Johns Hopkins University and was promoted to an associate professor in pharmacology. In 1929, he became the director of the research

Huan Liu[1,2], Zhaoqi Liu[1], Xue Gong[1], Hao Cheng[3]

1 University of Science and Technology of China, Hefei 230026, China

2 State Key Laboratory of Virology, Wuhan 430072, China

3 Institute of Microbiology, Chinese Academy of Sciences, Beijing 100101, China.

Correspondence: liuhuan520@ustc.edu.cn (H. Liu)

department of Eli Lilly and Company. From 1937 to 1968, Chen was a Professor of Pharmacology at Indiana University School of Medicine. In 1968, Chen retired from Indiana University (Qian, 2014).

Pharmacological Action of Ephedrine

Ko-Kuei Chen is best known for his research on ephedrine. The discovery of ephedrine's pharmacological action was of epoch-making significance in the history of medicine. Ko-Kuei Chen was the first to reveal the pharmacological action of ephedrine and apply it in clinical treatment.

In 1923, when Ko-Kuei Chen was at Peking Union Medical College, in addition to teaching, he wished to continue his research on Chinese herbal medicine, and received support from Prof. Carl Frederic Schmidt, the dean of the department. Ko-Kuei Chen learned from his uncle Shounan Zhou that Ma Huang could relieve asthma. At his uncle's suggestion, he chose Ma Huang among hundreds of commonly used Chinese medicinal herbs as the first research object. Ephedrine was successfully isolated from Ma Huang using the ammonia-chloroform method in just a few weeks (Chen and Schmidt, 1926). Previously, Nagai Nagayoshi had separated ephedrine from Ma Huang in 1887 and named it ephedrine, noting that it can cause pupil dilatation. Chen and Schmidt studied the pharmacological action of ephedrine in detail, and found that ephedrine could increase the carotid pressure for a long time, enhance cardiac contractions, constrict splenic and renal vessels, as well as mucous membrane and skin, and relax bronchial muscles (Qian, 2014). Ephedrine could stimulate the isolated uterus and central nervous system, and cause pupil dilatation after dripping it into eyes (Chen and Schmidt, 1926). These effects of ephedrine are qualitatively identical with those of epinephrine, while ephedrine can be taken orally, with longer efficacy duration and lower toxicity. In 1924, Chen and Schmidt made a preliminary report at the Beijing branch venue of the Society for Experimental Biology and Medicine. In the same year, after systematic experiments, they wrote their research paper "The action of ephedrine, the active principle of Chinese drug Ma Huang" , and published it in the magazine *Journal of Pharmacology and Experimental Therapeutics*, which was the first paper on the pharmacological action of ephedrine in the world. The physiological effects of ephedrine, the active principle of Ma Huang, are similar to epinephrine and last longer, primarily acting via the stimulation of the sympathetic nervous system. Since then, ephedrine has become an important sympathomimetic drug, which attracted international attention. The study was a model for identifying lead compounds from natural products and optimizing them to develop new drugs. Ko-Kuei Chen is reputed as a pioneer of modern pharmacological research in Chinese medicine.

Technique for Treating Acute Cyanide Poisoning

In the early 1930s, Ko-Kuei Chen and his colleagues discovered a technique for treating acute cyanide poisoning. On the basis of previous research, Chen and his research team found that the combination of sodium nitrite and sodium thiosulfate, consecutively injected intravenously, detoxifies cyanide to a marked degree (Chen and Rose, 1952). Furthermore, they also found that the combination of nitrite and thiosulfate has a synergistic effect that surpasses the sum of their individual values (Chen and Rose, 1952). At present, nitrite-thiosulfate therapy is still an effective detoxification method in the treatment of acute cyanide poisoning, based on the intravenous administration of sodium nitrite, followed by the administration of sodium thiosulfate.

Research on Chan Su

Ko-Kuei Chen spent much of his life studying Chan Su, the dried venom of Chinese toad. Since 1927 when he was in Abel's Laboratory, he had shown great interest in Chan Su. Ko-Kuei Chen was the first to have succeeded in obtaining epinephrine and the N-containing compound from Chan Su in crystalline form (Chen et al., 1931). Soon after, Ko-Kuei Chen isolated the two pure compounds cinobufagin and cinobufotoxin from Chan Su, and found that these two components had a digitalis-like cardiotonic effect. Compared with digoxigenin, cinobufagin had a shorter duration of action and was ineffective when taken orally. In 1929, after Ko-Kuei Chen came to Eli Lilly and Company, he continued his research on Chan Su for more than 40 years. He studied the structure-activity relationship of more than 400 cardiac glycosides and steroids. Furthermore, he also found that animals other than toads were capable of storing glycosides of the cardenolide type from their plant foods (Chen, 1970). Ko-Kuei Chen published a number of articles that have enriched the treasure trove of medicinal chemistry and provided valuable reference for the study of other drugs.

Other Pharmacological Studies

In the 1940s, Ko-Kuei Chen found that γ-dichroine extracted from the Chinese medicinal herb Chang Shan (*Dichroa febrifuga*) is anti-malarial, and its effect is 148 times stronger than that of quinine (Henderson et al., 1949). Since it causes vomiting, hepatic edema and degeneration, γ-dichroine wasn't introduced into clinical practice. Shortly after World War Ⅱ, Ko-Kuei Chen obtained a sample of methadone and confirmed its analgesic effect. On this basis, he and his colleagues synthesized and developed propoxyphene in Eli Lilly and Company (Qian, 2014). Although its clinical effect is equal to or slightly worse than that of codeine, it is less addictive, and it exhibits synergism with aspirin.

Master of International Medicine

Ko-Kuei Chen engaged in pharmacology for more than 50 years and published more than 350 papers and reviews. Due to his extensive and in-depth research interests, he made great contributions to the development of new drugs.

Ko-Kuei Chen separated and extracted active ingredients from traditional Chinese medicinal materials and developed them into chemical drugs. Following this approach, Chinese medical scientists and chemists have carried out numerous studies and achieved major breakthroughs, such as the discovery of the new antihepatitis drugs bifendate and bicyclol, as well as the antimalarial drugs artemisinin and dihydroartemisinin, etc. Ko-Kuei Chen's academic thoughts have far-reaching significance for the research of Chinese herbal medicine as well as other natural and synthetic drugs. His scientific thoughts and research methods have walked broad road and greatly promoted pharmacological research in modern Chinese medicine.

Figures

Fig.1 Ko-Kuei Chen(1898—1988)

Fig.2 Ko-Kuei Chen in Madison, Wisconsin, 1921

Fig.3 Introduction of ephedrine sulphate.

References

Chen KK (1923) Phytochemical notes No. 91 cassia oils from leaves and twigs. Journal of the American Pharmaceutical Association 12:294–296.

Chen KK (1948) Pharmacology of methadone and related compounds. Annals of the New York Academy of Sciences 51:83–97.

Chen KK (1970) Newer cardiac glycosides and aglycones. Journal of Medicinal Chemistry 13:1029–1037.

Chen KK, Jensen H, Chen AL (1931) The pharmacological action of the principles isolated from Chan Su, the dried venom of the Chinese toad. Journal of Pharmacology and Experimental Therapeutics 1:244–251.

Chen KK, Rose CL (1952) Nitrite and thiosulfate therapy in cyanide poisoning. The Journal of the American Medical Association 2:113–119.

Chen KK, Schmidt CF (1924) The action of ephedrine, an alkaloid from Ma Huang. Proceedings of the Society for Experimental Biology and Medicine 6:351–354.

Chen KK, Schmidt CF (1926) The action and clinical use of ephedrine: An alkaloid isolated from the Chinese drug Ma Huang. The Journal of the American Medical Association 11:836–842.

Ding GS (2009) Ko-Kuei Chen, an international pharmacologist. Progress in Physiological Sciences 4:289–291. (丁光生. 2009. 陈克恢——国际著名药理学家. 生理科学进展, 4:289–291.)

Henderson FG, Rose CL, Harris PN, et al (1949) γ-dichroine, the antimalarial alkaloid of Chang Shan. Journal of Pharmacology and Experimental Therapeutics 95(2):191–200.

Qian WC (2014) Overview of Chinese renowned scientists' academic achievements in the 20th Century. Beijing: Science Press. (钱伟长. 2014. 20世纪中国知名科学家学术成就概览. 北京: 科学出版社.)

Song ZY (1996) Biography Brief of Chinese Scientists and Technique Experts·Medicine, Vol. Pharmacology. Beijing: China Science and Technology Press. (宋振玉. 1996. 中国科学技术专家传略·医学编·药学卷. 北京: 中国科学技术出版社.)

63.“在病人面前，我永远都是一名小学生。”——张孝骞：现代中国胃肠病学创始人

张孝骞教授是一位卓越的临床医学家、医学教育家，他是中国科学院学部委员，是中国胃肠病学创始人（北京协和医院，1988）（图1）。作为医生，他在血容量、胃泌素功能、消化性溃疡、胃溃疡和胃癌等方面进行了深入的研究。作为一名教育家，他重视对学生临床技能的培养，鼓励学生将自己的知识应用在学习和工作中。他为我国医学界培养出了一大批优秀的青年才俊。

图1　张孝骞博士（1897—1987）

1897年12月28日，张孝骞出生于湖南长沙的一个教师家庭。1914年，他选择进入湘雅医学院（简称“湘雅”），因为他认识到我国不仅需要发展工业来改变贫穷的状况，我国的“病”同样危害至深（张孝骞，1983）。在现代内科学奠基人William Osler的故事的影响下，张孝骞沉浸在广阔而深奥的内科学研究领域，并为此奋斗了一生。1921年，他从湘雅医学院毕业并获得美国康涅狄格州政府颁发的硕士学位证书。毕业后，他顺利成为湘雅的一名住院医师（北京协和医院，1988）。

1923年，他获得了到北京协和医院（简称“协和”）进修的机会。1927年，进修期间的出色表现为他赢得了到约翰斯·霍普金斯医院学习一年的机会。在Harrop教授的指导下，他致力于人体血容量的研究。1928年，他在美国《临床研究杂志》上发表了两篇研究论文，受到同行科学家的高度评价，在世界各地的专业文献中得到了广泛的引用（Chang and Harrop，1928；Chang et al.，1928）。尽管Harrop教授一再挽留，希望他留在美国继续从事血容量的相关研究，但他毅然返回了自己魂牵梦绕的祖国。回到中国后，他主持开展了一系列的研究工作。1931年，他发表了《甲亢血容量》，受到世界各国专家的高度赞扬（Chang，1931）。在20世纪30年代，他在北京协和医院创建了第一个胃肠病学组，并发表了关于胃分泌功能的论文（Chang，1933）。从此张孝骞开始

译者：刘旭东
北京协和医院，北京100730，中国
邮箱：liuxudong@pumch.cn

在医学界崭露头角。

九一八事变后，日本侵略者步步逼近华北。当时已经在北京协和医院工作13年的张孝骞不愿在侵略者的控制下工作。1937年，他接受了母校邀请，回到湘雅医学院继续他的研究。不幸的是，随着日本侵略势力向南推进，湘雅已无法在长沙立足。师生纷纷离开学校，学校陷入危机。在最危险的时刻，张孝骞接任校长，经过一番思考，张孝骞决定：西迁。“湘雅培养的人才属于中国，我不能眼看着湘雅毁在日本人手里。”1938年10月11日，全校40多吨教学仪器和图书资料，260多名学生、教职员工及家属，辗转一个星期，胜利到达目的地贵阳。两周后，湘雅师生在这里上起了第一堂课。1944年冬，日本侵略者入侵贵阳，师生又陷入恐慌。他再次带领师生长途跋涉到重庆并继续他们的课程学习直到战争结束（荣印林和刘笑春，1987）。凭着毅力和勇气，他成功为我国医学界保留了一个重要的源泉，这生动地诠释了他的爱国情怀：“生命的泉，即使拌和着血和泪，也应该在自己的祖国流淌。”（北京协和医院和湘雅医学院，2008）

新中国成立后，在担任湘雅医学院院长11年后，张孝骞再次被邀请回到北京协和医院承担内科的复建工作。在他的感召下，来自南方甚至海外的许多精英汇聚起来（图2）。他负责建立消化、传染病、血液、呼吸、免疫、遗传病等专业组，并写信给政府恢复协和八年医学教育。他还帮助恢复了协和的优良传统，如住院医师制度和查房制度，并有条理地制定了内科学教育计划，改进了协和医学教育体系。

图2 原内科普通住院医师（1949年10月9日）。从左至右依次为：张孝骞、刘士豪、谢少文、吴朝仁、朱宪彝、邓家栋、马万森、朱贵卿、张安、方圻

重视临床是协和历来所坚持的传统，这也是张孝骞所坚持的。他一生致力于内科学的临床实践，在内科学领域多有建树。在对病人进行诊断时，张孝骞总是能在意见复杂多变时找到切入口，甚至在相当长的一段时间之后，准确回忆起病人的每一个症状，这很大程度上依赖于他特殊的“小本子”。在门诊或查房期间，他总是随身携带一

个小本子，记录重点和疑点（图3）。他认为，只有在没有主观性或任意性的情况下，才能准确分析具体的病例，才能更好地解决临床问题。他的小本子里的每一个字都是他严谨求精的展现，每一句话都有可能从生死边缘抢救回一条生命（张孝骞，1984）。

图3 张孝骞的笔记

20世纪60年代中期，张孝骞教授的门诊来了一位女病人，她一患感冒就休克。张教授对该患者有种似曾相识的感觉，询问病史得知，30年前她曾因难产大出血来医院住院。但因战争原因，很多资料已被毁坏。张教授仅凭着一大堆已经变了色的小本子和自己的片段记忆寻找到了蛛丝马迹，原来30年前那场大出血引起了患者脑垂体坏死，导致脑垂体机能减退，造成甲状腺和肾上腺等内分泌物不足和应激反应的缺陷，在受到紧急感染时，就会发生休克。据此，患者被诊断是席汉综合征。经过治疗后，病情很快好转（北京协和医院，1988）。

他的许多座右铭直到如今也一直被人们高度传颂，例如“如临深渊，如履薄冰”（讴歌，2016）。在1981年冬季，他收到了一份病历，邀请他进行书面诊断。在仔细研读病历后，他要求患者进行骨髓抽吸和淋巴结检查。两天后，因为没有收到检查的结果，已经80多岁的张孝骞不顾距离和严寒，亲自去病人家中为他看病。即使是在晚年，他对病人也有着特殊的感情。1987年3月的一个晚上，生病住院的张孝骞教授在病房里醒来，要求内科的副主任到他床边。他一看到副主任就问：“医院太大了，病人太多了，在紧急情况下，他们能在夜间找到医生吗？”得到满意的回答后，他才没有任何负担地入睡了。像这样的故事不胜枚举。医德医风在他这里始终是统一的。“一息尚存，仍当继续努力”一直是他的人生哲学（北京协和医院和湘雅医学院，2008）。

张教授医学知识渊博，临床经验丰富，理论功底深厚，在追求新理论、新知识方面也从不懈怠，孜孜不倦地读书学习。他几乎每个星期天上午都是在图书馆度过的。做完白内障手术以后，刚能看东西时，他就想看书了。眼科大夫一再劝阻，可他早就心急如焚。怕别人阻拦，张孝骞教授自己叫了车到了图书馆。被人撞见后，他不由自主地笑了，诙谐地说：“看来一个人不能做一点亏心事呀！不过，你知道我好久不看书，感到落后了，医学科学发展得太快了，紧着赶还赶不上呢！”就在那天，他一直看到闭馆的铃声响了好几次（北京协和医院，1988）。

1987年8月8日，张孝骞在北京逝世，享年90岁。临终前，他有两件憾事，一个是《临床拾遗》刚刚开始，另一个是他一直出诊到90岁的愿望永远都无法实现了。正如《希波克拉底：医学誓言》中所写的那样：“在我来的每一个房子里，我都会为了病人的利益而进入，远离所有的潜意识里想做的坏事和所有的诱惑，尤其是与女人或男人的爱的乐趣，无论他们是自由的还是被奴役的。”张孝骞用“戒、慎、恐、惧”作为他的人生座右铭，并为此坚持了一生（张孝骞，1982）。

张孝骞教授毕生致力于医学救国，密切关注民生，把病人视为“教师”（讴歌，2016）；他是现代医学先驱，却从未忘记过“我是医生”。他不仅是一位医生，而且是一座跨越了两个世纪的丰碑，深深地影响着当代社会（图4）。

图4 中国现代科学家纪念邮票（第三辑）。单枚邮票分别是：数学家熊庆来（左上）、微生物学家汤飞凡（右上）、医学家张孝骞（左下）、建筑学家梁思成（右下）

30年后，张孝骞的音容仍然浮现在我们的脑海中。他所热爱的祖国经历了巨大变化，已经拥有更为健全的医疗保健体系和更加康健的人民，而这也正是张孝骞一生奋斗的目标。

63. "In front of patients, I will always be a pupil." Dr. Xiaoqian Zhang: The founder of the modern Chinese gastroenterology

Dr. Xiaoqian Zhang (张孝骞, Hsiao-Chien Chang) was an outstanding clinician, therapist and medical educator, a member of the Academic Divisions of the Chinese Academy of Sciences, he is considered to be the founder of the modern Chinese gastroenterology (PUMCH, 1988). As a doctor, he made a systemic and deep research on human blood volume, gastric secretion function, peptic ulcer disease, gastric ulcer and gastric cancer, celiac tuberculosis, amoebic dysentery, ulcerative colitis and so on. He diagnosed and treated multiple intractable diseases. As an educator, he emphasized the training of clinical basic skills, urged students to grasp the utilization of science in study and work. He trained a large number of talents for Chinese medicine.

Dr. Zhang was born on December 28, 1897, in a family of teachers, Changsha, Hunan Province. In 1914, he chose to enter Hsiang-Ya Medical College studying medicine, for he realized that "poverty" which leaned to industrial salvation was not the only contradiction at that time, the destructiveness of "disease" was as deep as it was (Zhang, 1983). Under the influence of William Osler's story, the father of Modern Medicine, Dr. Zhang had been indulging in the extensive and profound internal medicine research field for a few decades and had struggled for a lifetime. In 1921, he graduated from Hsiang-Ya Medical College and received his M.D. from the Connecticut government. After that, he became a resident doctor in Hsiang-Ya Medical College (PUMCH, 1988).

In 1923, he was offered a chance to further study at Peking Union Medical College Hospital (PUMCH). During the residency period in PUMCH, because of his excellent performance, he got an opportunity to study in the Johns Hopkins Hospital for one year in 1927. Under Prof. Harrop's supervision, he immersed in the study of human blood volume. In 1928, he published two research articles in the *Journal of Clinical Investigation* on progress in blood volume, which was highly appraised by peer scientists and had been widely referenced in professional literature

Lu Wang, Xudong Liu, Wenli Duan, Shuyang Zhang

Peking Union Medical College Hospital, Chinese Academy of Medical Sciences and Peking Union Medical College, Beijing 100730, China

Correspondence: shuyangzhang103@163.com (S.-Y. Zhang)

throughout the world (Chang and Harrop, 1928; Chang et al., 1928). Though being invited to stay in the United States, he was concerned more about his impoverished motherland. After returning to China, he hosted a series of research work. In 1931, he published another research article entitled "The blood volume in hyperthyroidism" on *Journal of Clinical Investigation*, which was also received extensive attention from peer scientists worldwide (Chang, 1931). In the 1930s, he created the first gastroenterology group in PUMCH and published several research articles on gastric secretion function (Chang, 1933). Since then, Dr. Zhang developed into a well-known clinician gradually.

After the "September 18th Incident" , the Japanese aggressors pressed in step by step. Dr. Zhang, who had been working in PUMCH for 13 years at that time, decided not to work under the control of the invaders. In 1937, he accepted the invitation from his alma mater and moved back to Hsiang-Ya Medical College to continue his study. Unfortunately, as the Japanese aggressors advanced southward, Hsiang-Ya could no longer survive in Changsha. Teachers and students left school one after another, and the college was deep in crisis. At the most dangerous moment, Dr. Zhang took over the role of dean and decided to move the whole college westward to Guiyang. They overcame various difficulties, taking 40 tons of teaching equipment and books with them. Students, faculty staff and their families, approximately 300 people, finally reached Guiyang in one week. In the winter of 1944, the Japanese aggressors invaded Guiyang and the college was in panic again. At that time, Dr. Zhang once again led the teachers and students to set foot on a long journey to Chongqing and continued their classes until the end of the War (Rong and Liu, 1987). With perseverance and courage, he avoided the Hsiang-Ya from being destroyed by war and reserved an important source for the Chinese medical science, which was a vivid interpretation of his patriotic feelings that "spring of life, even if mixed with blood and tears, should flow in one's own country" (PUMCH and Xiangya Hospital of Central South University, 2008).

After the founding of the People's Republic of China, Dr. Zhang, the dean of Hsiang-Ya Medical School for 11 years, was invited to return to Peking Union Medical College Hospital again to assume the rehabilitation of internal medicine. Under his inspiration, many elites from the south area and even overseas gathered at PUMCH with the same mission. He took charge of the establishment of the professional groups of digestion, infectious diseases, blood, respiration, immunity, genetic disease and so on; He wrote to the Central Government to resume the eight-year medical education of Peking Union Medical College. He restored the PUMCH's old traditions like the resident system and ward rounds; He developed the disciplined planning of the internal medicine and improved the medical education system for PUMCH.

PUMCH has always attached great importance on clinical practice which was also what Dr. Zhang insisted on. He devoted all his life to the clinical practice of internal medicine and made many achievements in the field of internal medicine. When it comes to the diagnoses for patients, Dr. Zhang could always hit the mark when opinions varied and could recall every symptom of a patient accurately even after quite a long period which relied on his special "notes" to a great extent. He always carried a small notebook during outpatient service or ward rounds and would record everything he deemed important or doubtful. He believed that clinical issue could be well completed only when one could analyze specific issue case by case without subjectivity or arbitrariness. Every word in his small notebooks is the culmination of his painstaking effort; each sentence may retrieve life from the verge of death (Zhang, 1984).

In the middle of 1960s, a female patient who would shock once catching a cold came to PUMCH. Dr. Zhang had a familiar feeling with the patient and asked about her medical history and found out that she had been hospitalized in PUMCH due to dystocia hemorrhage 30 years ago. However, the original medical records and data were completely gone because of war. After thoroughly digging into a pile of his old special "notes" and fragments of memories, he was suddenly enlightened. The original hemorrhage caused necrosis of the pituitary gland, leading to hypopituitarism, resulting in deficiencies of endocrine thyroid glands, adrenal glands and lacks in emergency response. When the patient was subjected to an emergency infection, the shock would occur. Based on this, she was diagnosed as Sheehan's syndrome. She was given thyroid and adrenal cortex hormones as an alternative treatment, and then her condition improved quickly (PUMCH, 1988).

Many of his lifelong mottoes has been praised until now like "Tread, as if on the thin ice; Walk, as if on the brink of a deep gulf " (Ou, 2016). In the winter of 1981, he received a medical record requiring him to give a written consultation. After reading the medical records, he demanded the bone marrow aspiration and lymph node examination for the patient. Two days later, having not received the results of the tests, Dr. Zhang, though in his 80s, went to visit the patient in person regardless of the distance and severe coldness. He had a particular affection on the patients, even in his late years. One night in March 1987, Dr. Zhang who just weakly woke up in the ward, asked for the vice director of the internal medicine department to come to his bed. He cast his doubt directly once he saw the vice director: "The hospital is so large that there are so many patients. Can they find a doctor at night in an emergency?" After getting a satisfactory answer, he fell asleep without any burden. Stories like these are too numerous to enumerate. Medical ethics and medical skill were always completely unified for him. "I would die before I give up" was Dr. Zhang's life philosophy all along (PUMCH and Xiangya Hospital of Central South University, 2008).

Dr. Zhang was erudite in medical knowledge and clinical experience and was well known for his profound theoretical foundation, but he never slacked in the pursuit of advanced knowledge and studied sedulously during his lifetime. He spent almost every Sunday morning in the library. After cataract surgery, he wanted to read once seeing things. In case other people might stop him, he ordered a car to the library by himself regardless of his ophthalmologist's dissuade. When he was caught by others, he laughed and said humorously: "It seems that one cannot do anything wrong but you know I haven't read books for such a long period." On that day, he kept reading until the closed bell rang for several times. When Dr. Zhang was already an octogenarian, his students advised him not to go to the library on Sunday. He replied: "Medical science nowadays is developing so fast and I could hardly catch up with it without continued learning." (PUMCH,

1988)

On August 8, 1987, Dr. Zhang passed away in Beijing, at the age of 90. There were two regrettable things for him: One was *The Clinical Supplements* had just started; the other was his desire of continuing outpatient service until the age of 90 never came true. As what was written in *Hippocrates: The Oath of Medicine*: "In every house where I come I will enter only for the good of my patients, keeping myself far from all intentional ill-doing and all seduction and especially from the pleasures of love with women or with men, be they free or slaves." Dr. Zhang used "caution, prudence, dread and fear" as his life motto and carried out the oath all along (Zhang, 1982).

He devoted all himself to save the nation via medical science; he paid close attention to people's livelihood and regarded the patients as his "teacher" (Ou, 2016); he was the modern medical pioneer, but never failed to remember that "I'm a doctor" . He is not only a mere doctor but also a monument that has stretched over two centuries that deeply influence the contemporary society.

Thirty years later, Dr. Zhang's edification, forthright admonition and his voice and expression still emerge in our mind. China has been undergoing a radical change after Dr. Zhang's period, and PUMCH, the hospital he dedicated his whole life to, is now also heading into a new era. China has already grown up with a more advancing healthcare system and healthier people which were just the destination Dr. Zhang had been struggled for throughout his life (Zhang, 1983).

Figures

Fig.1 Dr. Xiaoqian Zhang (1897—1987)

Fig.2 Previous internal medicine general residents (Oct. 9, 1949). From left to right: Xiaoqian Zhang, Shihao Liu, Shaowen Xie, Chaoren Wu, Xianyi Zhu, Jiadong Deng, Wansen Ma, Guiqing Zhu, An Zhang, Qi Fang

Fig.3 Dr. Xiaoqian Zhang's Notes

Fig.4 "Modern Chinese Scientists" stamps (3rd series). The stamps are respectively: Qinglai Xiong (up left), mathematician; Feifan Tang (up right), microbiologist; Xiaoqian Zhang(down left), physician; Shih Cheng Liang (down right), architect

References

Chang HC (1931) Blood volume in hyperthyroidism. J Clin Investig 10:475–487.

Chang HC (1933) Gastric secretion in fever and infectious diseases. J Clin Investig 12:155–169.

Chang HC, Harrop GA Jr (1928) Determination of circulating blood volume with carbon monoxide. J Clin Investig 5:393–405.

Chang HC, Harrop GA Jr, Schaub BM (1928) Circulating blood volume in diabetic acidosis. J Clin Investig

5:407–474.

Ou G (2016) The Medical Story of PUMC. Beijing: SDX Joint Publishing Company. (讴歌. 2016. 协和医事. 北京: 三联书店.)

PUMCH (1988) Zhang Xiaoqian. Beijing: Central Party Literature Press. (北京协和医院. 1988. 张孝骞. 北京: 中央文献出版社.)

PUMCH, Xiangya Hospital of Central South University (2008) Biography-in-photo of Zhang Xiaoqian. Beijing: Peking Union Medical College Press. (北京协和医院, 湘雅医学院. 2008. 张孝骞画传. 北京: 中国协和医科大学出版社.)

Rong YL, Liu XC (1987) He is passionately devoted to his Alma Mater. Hunan Medical University Newspaper 63. (荣印林, 刘笑春. 1987. 他, 对母校一往情深. 湖南医学院院报, 63.)

Zhang XQ (1982) Professional ethics of medical workers. Health News 1982–11–24. (张孝骞. 1982. 医务工作者的职业道德. 健康报, 1982–11–24.)

Zhang XQ (1983) How did my dream come true. In: Bo J, Qing S. Scientists Talk about Ideals. Hefei: Anhui Peoples Publishing House. (张孝骞. 1983. 我的理想是怎样实现的. 见: 卜汲, 青韶. 科学家谈理想. 合肥: 安徽人民出版社.)

Zhang XQ (1984) Discussion on clinical thinking. Medicine & Philosophy 2. (张孝骞. 1984. 漫谈临床思维. 医学与哲学, 2.)

64. 蔡翘：中国生理学和中国航空、航天与航海医学的奠基人

2019年2月16日，蔡翘院士博物馆在广东省揭阳市举行了小行星（207681）落户启动仪式。这颗小行星于2011年10月14日被命名为蔡翘星（图1）。蔡翘星于2007年8月16日在中国科学院紫金山天文台盱眙观测站被发现。那么，这颗小行星被冠以的姓名是谁呢？

蔡翘（Tsai Chiao）（图2），字卓夫，乳名（族名）义忠。作为我国生理学的奠基人之一，他引导、促进了我国生理科学事业的发展，也促进了国际神经解剖学的繁荣。同时他也是我国航空、航天与航海医学的奠基人和创始人，初步建立了我国的航空生理体系。

图1　以我国生理学家蔡翘命名的小行星题字

图2　蔡翘（1897—1990）

1897年10月11日，蔡翘出生于广东揭阳县揭东区新亨镇仙美村。7岁时，他进入本村私塾学习，后又至蓝田小学。1913年，蔡翘到广东潮安县金山书院上学，1917年毕业。1918年，他到上海复旦大学附中补习英语，同年到北京大学中文系当旁听生。

译者：王亚兰[1]，钱燕燕[2]

1　南京师范大学心理学院，南京210097，中国

2　荷兰莱顿大学社会与行为科学学院，莱顿2333AK，荷兰

邮箱：psyylw@126.com（王亚兰）；psyqyy@163.com（钱燕燕）

1919年秋，受五四运动“科学救国”的思潮影响，蔡翘自费留学美国。

在美国，蔡翘结识了行为心理学家郭任远，在郭任远的影响下（Qian et al., 2020），蔡翘先是进入郭任远所在学校所学专业——美国加利福尼亚大学心理学系学习，后来又转到印第安纳大学的心理学系，总共耗时两年修完本科课程。1921年冬，蔡翘进入哥伦比亚大学读研究生。1922年至1925年，他又转学至芝加哥大学文理学院生理系读研究生，在比较心理学教授H. A. Carr担当导师的情况下，以心理学为主修专业，兼修生理学和神经解剖学两门副科。1924年，蔡翘凭借论文《运动习惯记忆保持曲线的比较研究》获博士学位，论文对白鼠在迷宫问题上的记忆保持和人类记忆保持之间的差异进行了比较（Tsai, 1924）。1925年，蔡翘在芝加哥大学赫尔解剖实验室探究了负鼠的视域和视觉运动的中枢部位（Tsai, 1925a），并描述了负鼠的下行束及其丘脑与中脑的相关结构（Tsai, 1925b）。在探究负鼠的视束及视觉中枢时，蔡翘详细地描述了顶盖前核所在的腹侧中脑被盖层（图3），这一区域因此被称为“蔡氏区”（Tsai's area）或“蔡氏被盖区”（ventral area of Tsai）。另外，针对弗洛伊德提出的“性驱力在人类所有动机中是力量最强的”这一观点，蔡翘用白化大鼠的性和饥饿动机的比较研究证明了饥饿对白鼠的驱力远强于性对它的驱力（Tsai, 1925c）。同年，蔡翘获芝加哥大学金钥匙奖（Gold Key Award），并被推荐为美国解剖学会会员。

图3　蔡翘（左一）和技术人员发现腹侧中脑被盖层

1925年秋，蔡翘回国。郭任远时任复旦大学副校长，蔡翘在其支持下始创复旦大学心理学院生物学科，教授生物学和生理学。他培养了一批生理学研究者，包括徐丰彦、冯德培、朱鹤年。1927年，蔡翘到上海的国立中央大学医学院任生理学教授。期间，蔡翘担任主编编写了我国第一本大学生理学教科书《生理学》，该书于1929年作为大学系列丛书之一出版。在这之后，他又编订了《生理学实验》《运动生理学》《生理学常识》等10本生理学教材。编写《生理学》期间，蔡翘主要关注甲状旁腺，发表了8篇相关文章。首先，他和助教徐丰彦关注犬类甲状旁腺痉挛的发病机制，研究了大肠切除（Tsai and Hsü, 1929a）、结扎胆管（Tsai and Hsü, 1929b）、肠梗阻（Tsai and Hsü, 1929c）和肠道腐烂（Tsai and Hsü, 1929d）四种可能的影响因素与甲状旁腺痉挛之间的相关性。其次，他们研究了静脉注射不同物质对甲状旁腺的血浆钙和无机磷的影响（Tsai and Hsü, 1930a, 1930b）。再次，他们研究了甲状旁腺切除术和甲状旁腺素注射后骨骼肌的钙含量（Hsü and Tsai, 1930）。最后，他研究了甲状旁腺切除术和甲状旁腺素给药对犬的胃运动的影响（Tsai, 1930）。

1930年秋，蔡翘由美国洛克菲勒基金会资助再度出国进修。蔡翘先到伦敦大学生理学系实验室跟C. L. Evans教授研究了不同情况下的猫肝脏中的糖原（Evans et al., 1931a, 1931b）以及肾上腺素对猫体内糖原分布的影响（Evans et al., 1931c）。之后，蔡翘到剑桥大学E. D. Adrian教授的生理实验室研究麻醉剂对神经纤维动作电位的影响（Tsai, 1931），这一研究中的神经电位图后被《人体生理学原理》及《实验生理学》所引用。

1931年冬，蔡翘在德国法兰克福大学及其他大学的著名生理实验室短期访问。1932年春至1936年末，蔡翘受聘于由英国人主办的上海雷士德医学研究院，任生理学研究员。在这期间，他共发表11篇论文，刊于1933—1937年的《中国生理学杂志》。他主要对动物肝脏进行了研究，探索了禁食兔的肝组织中总碳水化合物的含量（Tsai，1933a），并在助手易见龙的协助下，以断头猫（Tsai，1933b；Tsai and Yi，1934a，1934b，1934c）和正常猫（Tsai and Yi，1936a，1936b）的肝脏为研究对象，探索了在血糖浓度正常的前提下肝在碳水化合物代谢中的作用。接着，他探索水提取物用于估计肝脏的糖原和总碳水化合物的有效性问题（Tsai，1934），并基于肝脏糖原的研究尝试对肝脏糖原和肌肉糖原进行联合研究（Tsai，1937a，1937b）。最后，蔡翘尝试改进血管造口术（Tsai，1935）。

1936年秋，蔡翘提前一年中断与上海雷士德医学研究院的合同，到国立中央大学新成立的医学院任生理学教授。半年后，“七七事变”爆发。蔡翘随国立中央大学医学院迁到成都。1937年至1945年，蔡翘做的工作可以简要总结如下。首先，制造实验仪器。蔡翘于医学院内建立机械室，自造土车床、钻床等生理学教学及科研所需仪器，并提纯药品试剂用于教学和研究。其次，建立学会。1938年秋，蔡翘与华西大学生理学教授O. L. Kilborn领导并成立中国生理学会成都分会。同年，蔡翘组织成立生理学研究所，培训进修生，其中有延安军医学校秘密派来的学员，方怀时（我国航空生理学家）就是其一。再次，设立学刊。为使西南地区生理科学的研究成果与国内外的研究进行交流，1941年6月，蔡翘主持创刊《中国生理学会成都分会简报》并担任主编。这一刊物自1941年6月延续至1945年6月，共出版13期，是当时抗战后方唯一的生理学学术刊物。最后，开展血液研究。蔡翘以第一作者参与撰写11篇论文，指导进修生发表20余篇论文，主要包含三个主题：①红细胞脆性增高的原因（Tsai et al.，1940，1941，1942；Puh et al.，1945）。②溶血物质与抗溶血物质的研究（Lee and Tsai，1942a，1942b，1943）。③中国人的血液学标准的探索（Wu and Tsai，1940）和中国人的感觉测量方法的标准化（Wu and Tsai，1943）。在此基础上，1948年，美国医药援华会会长Gregerson带专业人士来华，以蔡翘领导的实验室为基地，举行了研究班教授血液研究相关的技术，并组织研究班成员测定了数十名我国青壮年的血液系数。

1943年，受美国政府邀请，蔡翘作为交换教授与费孝通等六人到美国讲学一年。蔡翘根据当时我国的经济状况、农业发展以及人民的健康习惯和教育问题向美国民众介绍了我国的营养问题（Liu et al.，1946）。讲学期间，他与哥伦比亚大学医学院生理学系的两位学者共同研究血清中的血管收缩物质（Tsai et al.，1944），该研究促成了5-羟色胺（即血清素）的发现（张衡，2013）。

1924年至1948年是蔡翘的研究生涯前期，研究主题为基础生理学。1948年后，因为我国国防建设的需要，蔡翘的研究转向航空生理和航海生理。

1948年，中央研究院评选出第一届院士，共81位，蔡翘是生物组25位院士之一（图4）。1949年，蔡翘被正式任命为南京大学医学院院长。1952年，蔡翘任第五军医大学校长。作为带头人，蔡翘领导建成了我国第一台混凝土人用低压舱，为高空低氧实验的开展争取了时间（张立藩，1997）。1952年至1954年，蔡翘指导中青年科技人员及大学生开展了有关睡眠、劳动代谢、飞行员营养等一系列理论意义与军事作业相结合的生理研究课题，形成了我国第一份《航空生理研究总结初步报告》。

1954年至1968年，蔡翘接受中央军委命令担任军事医学科学院副院长，兼任军事劳动生理研究所（1964年后更名为航空宇宙医学研究所）所长（图5和图6）。1957年至1966年，蔡翘领导建成多项大型设备，主持制定适合我国军队的防护制度和装置要求。同期，蔡翘指导研究人员开展航空生理的相关科学研究。他们就此写了12篇相关论文，涉及飞行疲劳的实验性模拟、鉴定人体对高温和低温的耐受性等问题，这些论文被收入《中国人民解放军军事医学科学院论文汇编》。蔡翘也积极参加国际学术会议：1956年7月，他出席国际生理科学会布鲁塞尔会议；8月至次年2月，他作为中国军事医学代表团副团长赴苏联考察半年（图7和图8）；后又赴捷克斯洛伐克和波兰参加学术会议。蔡翘还赴华沙参加国际空间科学会议，赴爱尔兰参加国际航空空间医学会议，并顺访了英国伦敦等地的研究实验室。1966年，我国根据蔡翘等研究者起草的规划意见制定了《宇宙医学规划》。航天医学工程研究所根据规划成立，他为我国航天医学事业的发展作出了贡献。

图4 1986年，蔡翘（右）和1980年当选为中国科学院院士的生理学家王志军（图片来源：中国科学院院士文库）

图5 1985年，蔡翘（前排左二）及军事医学科学院的同事（图片来源：中国科学院院士文库）

图6 军事医学科学院成立初期师生合影（蔡翘在二排右六）（图片来源：中国科学院院士文库）

图7 1956年，蔡翘（前排左一）与苏联列宁格勒基洛夫军事医学科学院生理学系的苏联生理学家合影（图片来源：中国科学院院士文库）

图8 1956年，蔡翘（右一）访问苏联，在列宁格勒实验医学研究所留影（图片来源：中国科学院院士文库）

1966年至1976年，蔡翘失去医学工作，然而这期间他坚持写作《航空与空间医学基础》，该书于1979年出版。1978年，蔡翘重新担任军事医学科学院副院长，指导神经生物学实验室的研究工作并培训研究生（图9）。1981年，蔡翘辞去中国生理科学会（即中国生理学会）理事长职务，告别科学研究。1990年，蔡翘于北京病逝。

图9　1986年9月，蔡翘（前排左四）作为博士论文答辩的评委，在答辩会后与评审团和博士研究生合影留念（图片来源：中国科学院院士文库）

致谢

钱燕燕的研究获得国家留学基金管理委员会的研究生（博士）奖学金支持。

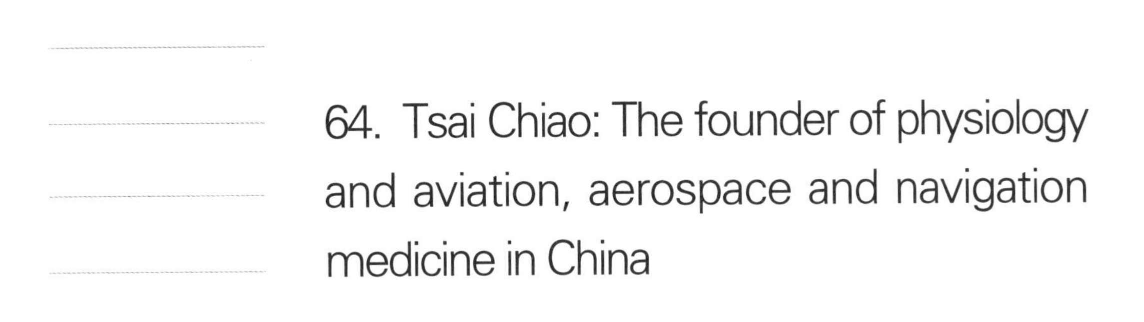

64. Tsai Chiao: The founder of physiology and aviation, aerospace and navigation medicine in China

On February 16, 2019, Tsai Academician Museum held a launch ceremony of a minor planet (207681)'s settling in Jieyang County, Guangdong Province. This minor planet was given a new name Cai Qiao on October 14, 2011. Cai Qiao was discovered at Xu Yi Station in the Purple Hills Observatory of the Chinese Academy of Sciences on August 16, 2007. So who is the planet named after?

Tsai Chiao (蔡翘) styled name was Zhuofu (卓夫), infant name was Yizhong (义忠). As one of the founders of Chinese physiology, Tsai promoted the development of Chinese physiology and also furthered the prosperity of international neuroanatomy. He was also the originator of China's aviation, aerospace and navigation medicine, who initially established China's aviation physiological system.

October 11, 1897, Tsai was born in Xianmei Country, Xinheng Town, Jiedong District, Jieyang City, Guangdong Province. Tsai went to an old-style private school when he was 7, and subsequently transferred to Lam Tin Government Elementary School. In 1913, he received education from Jinshan Academy located at Chaoan County, Guangdong Province, and graduated in 1917. In 1918, he received further education from Shanghai Fudan University Attached Middle School to improve his English. The same year, he became an external student of the Chinese Department of Peking University. In the autumn of 1919, deeply influenced by the cultural trend "Saving the Country through Science" of May 4th Movement, he went to America at his own expense.

Affected by behavioral psychologist Zing-Yang Kuo (郭任远) (Qian et al., 2020), Tsai studied a regular college course for two years, firstly at Psychology Department, California University, and then transferred to the same department at Indiana University. In the winter of 1921, he entered the University of Columbia as a postgraduate. From 1922 to 1925, he transferred to the Department of Physiology, College of Arts and Sciences, University of Chicago

Yalan Wang[1], Zijian Li[2], Yanyan Qian[2], Benyu Guo[1]

1 School of Psychology, Nanjing Normal University, Nanjing 210097, China

2 Social and Behavioral Sciences Faculty, Leiden University, Leiden 2333 AK, The Netherlands

Correspondence: psyqyy@163.com (Y.-Y. Qian)

to take more postgraduate courses. During this period, Professor Harvey A. Carr of comparative psychology supervised Tsai's study and Tsai majored mainly in psychology, minored in physiology and neuroanatomy. In 1924, Tsai gained a philosophic doctor degree with the thesis "A comparative study of retention curves for motor habits" in which he further investigated the differences of memory retention between white rats and humans on a stylus maze problem (Tsai, 1924). In 1925, at the Hull Laboratory of Anatomy of Chicago University, Tsai explored "the optic tracts and centers of the opossum" (Tsai, 1925a) and described its "descending tracts and related structures of the thalamus and midbrain" (Tsai, 1925b). In his exploration, Tsai described in great detail the ventral mesencephalic tegmentum where the nucleus pretectalis is located. This area is called Tsai's area or ventral area of Tsai. In addition, Tsai proved the fact that "the normal food-seeking impulse in the albino rat as measured by the choice method is stronger than that of sex" (Tsai, 1925c) against the Freudian conception that the sex urge is the strongest of all human motives. The same year Tsai was awarded Chicago University's Gold Key Award and recommended as a member of the American Association of Anatomists.

In the autumn of 1925, Tsai returned back to China. With the support of Zing-Yang Kuo, the vice-president of Fudan University, Tsai constructed biology discipline at Psychology College, Fudan University, taught biology and physiology, and cultivated a bunch of researchers like Fengyan Hsü (徐丰彦), Depei Feng (冯德培), Henian Zhu (朱鹤年). In 1927, Tsai served as a Physiology Professor in the Medical Colleague of National Central University in Shanghai. Tsai acted as editor of the textbook called *Physiology*, the first textbook of physiology in China, which was published in 1929 as part of a college book series. After that, Tsai also compiled 10 more Chinese physiology books including *Physiology Experiment*, *Exercise Physiology* and *The General Knowledge of Physiology*. During this period, Tsai also published 8 papers concerning the issue of parathyroid. Firstly, co-operating with his teaching assistant Hsü, Tsai explored the relationship between parathyroid tetany and the removal of the large intestine (Tsai and Hsü, 1929a), ligation of the bile duct (Tsai and Hsü, 1929b), intestinal obstruction (Tsai and Hsü, 1929c) and intestine putrefaction (Tsai and Hsü, 1929d). Secondly, they studied plasma calcium and inorganic phosphorus following intravenous injections (Tsai and Hsü, 1930a, 1930b), and the calcium content of skeletal muscles after thyroparathyroidectomy and parathormone injections (Hsü and Tsai, 1930). Finally, he studied the effect of thyroparathyroidectomy and parathormone administration on the gastric motility in dogs (Tsai, 1930).

In the autumn of 1930, with funding from the Rockefeller Foundation, Tsai went to the Psychophysiology Laboratory in London. Under the supervision of Prof. C. L. Evans, he studied the changing situation of glycogen in cat livers under different conditions (Evans et al., 1931a, 1931b), as well as the action of adrenaline on glycogen distribution in the cat (Evans et al., 1931c). Afterwards, Tsai studied the action of narcotics on the conduction of nerve impulses from a single end-organ (Tsai, 1931) at E. D. Adrian's physiological laboratory, Cambridge University, the diagram of neuropotential of which cited by Starling's *Principles of Human Physiology* and Harris' *Experimental Physiology*.

In the winter of 1931, Tsai briefly visited Goethe University Frankfurt and the other famous physiological laboratories in Germany. From the spring of 1932 to the winter of 1936, Tsai was employed as a researcher in the field of physiology at the Henry Lester Institute of Medical Research, a British-owned institution in Shanghai. During this period, he published 11 research articles in the *Chinese Journal of Physiology* from 1933 to 1937. Tsai mainly studied the liver of animals. He explored the question of total carbohydrate content of the liver tissue in the fasting rabbit (Tsai, 1933a), and, with the cooperation of assistant Chien Lung Yi (易见龙), taking the liver of the decapitated cat (Tsai, 1933b; Tsai and Yi, 1934a, 1934b, 1934c) and normal intact cat (Tsai and Yi, 1936a, 1936b) as the research object, explored the role of the liver in carbohydrate metabolism on the premise of maintaining the normal blood glucose concentration, which were published in Volumes 8–10 of the *Chinese Journal of Physiology*. Then he explored the question of the validity of using aqueous extracts for estimating glycogen and total carbohydrate of the liver (Tsai, 1934). Based on the study of liver glycogen above, Tasi attempted to study the combined glycogen of liver and muscle (Tsai, 1937a, 1937b). Finally, Tsai tried to improve angiostomy (Tsai, 1935). Finally, Tsai tried to improve angiostomy (Tsai, 1935).

In the autumn of 1936, Tsai waived his contract with the Henry Lester Institute of Medical Research one year ahead of schedule, and became a professor of physiology in the newly established Medical College of National Central University in Nanjing. Six months later, the July 7th Incident happened, and Tsai moved to Chengdu together with the Medical Department of Central University. His contributions in Chengdu from 1937 to 1945 can be summarized as below: Firstly, Tsai manufactured experimental apparatus. He set up mechanical rooms in Medical College, built equipment, such as soil lathes and drilling machines, which are a pivotal issue to physiological teaching and scientific research and purified pharmaceutical reagents for teaching and research. Secondly, he established academy. In the autumn of 1938, the Chengdu Branch of Chinese Association for Physiological Sciences was founded by Tsai and O. L. Kilborn, a physiology professor at West China Medical University. In the same year, Tsai organized the establishment of the Institute of Physiology to train trainees, among whom were secretly sent from the Yan'an Military Medical School, such as aeronautical physiologist Huishih Fang (方怀时). Thirdly, Tsai founded a journal to promote academic exchanges of the physiological science field among southwest China, domestic and overseas. *Proceedings of Chinese Physiological Chengtu Branch* started the publication work in June 1941, Tsai served as editor-in-chief, from then till June 1945, with 13 issues published in total. And this journal was the only physiological academic publication in the rear of the War of Resistance against Japanese Aggression. Finally, Tsai conducted blood research. Tsai, as the first author, participated in the academic writing of 11 papers, and guided students to publish more than 20 papers. Three themes were included in these papers: 1. Study on causes of increased erythrocytic fragility (Tsai et al., 1940, 1941, 1942; Puh et al., 1945). 2. Study on haemolytic and antihaemolytic (Lee and Tsai, 1942a, 1942b, 1943). 3. Study on Chinese hematology standards (Wu and Tsai, 1940) and the standardization of Chinese people's sensory measurements (Wu and Tsai, 1943). In 1948, Gregerson, President of the American Bureau for Medical Aid to China, led professionals to China to hold a research class to teaching blood related technology. Having had the laboratory led by Tsai as a base, they organized the members of the research class to determine the blood coefficients of dozens of young Chinese.

In 1943, accompanied by another six professors which included Xiaotong Fei (费孝通), Tsai went to America with the identity of Exchange Professor for one year. Tsai introduced nutritional problems of China, with respect to the economic situation, agriculture development, healthy habits and education problems, towards American people (Liu et al., 1946). Also, he studied vasoconstrictor substances with two scholars in the Physiology Department, Medicine College, Columbia University (Tsai et al., 1944), which promote the finding of 5-hydroxytryptamine (Zhang, 2013).

Tsai's earlier stage of academic work can be recognized from 1924 to 1948, whereby his research topic was foundation physiology. After 1948, his research turned to aviation physiology, navigation physiology with the need for national defense construction.

In 1948, Academia Sinica appraised and elected a total of 81 academicians, Tsai was one of 25 academicians in the biology group. And appointed as the Dean of the medical superintendent of Nanjing University in 1949, principal of The Fifth Military Medical University in 1952, as a leader, he built the concrete human hypobaric chamber, which promoted the development of altitude hypoxia experiments (Zhang, 1997). From 1952 to 1954, Tsai guided young and middle age science and technology personnel and graduates to conduct psychological research including the topic of sleep and pilots' nutritional situation, which composed the first Summary of preliminary report on aviation physiology research.

Between 1954 and 1968, Tsai, on basis of the nomination from Central Military Commission, became vice-president of the Academy of Military Medical Sciences of PLA and director of Military Performance and Physiology Institute which was renamed the Aerospace Medicine Institute in 1964. Between 1957 and 1966, Tsai leaded people to build numerous large scale pieces of equipment and formulated safeguard regulations and device requests suitable for the Chinese Military. At the same time, Tsai took the role of vice-president of Academy of Military Medical Sciences of PLA, and leaded research to conduct scientific research related with aviation physiology. Then they wrote 12 papers for the *Chinese People's Liberation Army Academy of Military Medical Sciences Proceedings*, for example, with the theme of tentative modelling and identification of flight fatigue, the human body's tolerance of high and low temperature. Tsai also took part in the International Physiological Science, Brussels Conference in 1956. From August 1956 to February 1957, as the deputy head of the Chinese Military Medical Delegation, Tsai conducted an integrated survey in the Soviet Union for around six months, then he participated in international conferences in Czechoslovakia and Poland. Tsai also attended the International Space Science Conferences in Warsaw, International Aviation Space Medicine Conference in Ireland, and he also visited a research laboratory in London. In 1966, China made the cosmomedicine plan drafted by Tsai, and this led to the establishment of the Cosmomedicine

Medical Engineering Institute, which contributed to the development of China's cosmomedicine career.

From 1966 to 1976, Tsai lost his medical career, but still insisted on writing *Fundamentals of Aviation and Space Medicine*, which was published in 1979. In 1978, Tsai reclaimed his title as vice-president at the Academy of Military Medical Sciences of PLA, where he guided research in the neurobiology laboratory and trained postgraduates. In 1981, Tsai retired from the Chinese Association for Physiological Sciences with the position of president of a council. On July 29, 1990, Tsai passed away.

Acknowledgements

The research of Yanyan Qian is supported by a post-graduate scholarship (PhD) of the China Scholarship Council (CSC).

Figures

Fig.1 The name of minor planet Cai Qiao stem from Chinese physiologist Tsai Chiao

Fig.2 Tsai Chiao (1897—1990)

Fig.3 Tsai (1st from the left) and the technicians were discovering the ventral mesencephalic tegmentum

Fig.4 In 1986, Tsai (right) and Zhijun Wang, a physiologist, who was elected as an academician in 1980 (Photo credit: Academician Library of Chinese Academy of Sciences)

Fig.5 In 1985, Tsai (front row, 2nd from left) and colleagues from the Academy of Military Medical Sciences (Photo credit: Academician Library of Chinese Academy of Sciences)

Fig.6 Group photo of teachers and students taken in the early day of the establishment of the Academy of Military Medical Sciences (Tsai is in the second row, 6th from the right) (Photo credit: Academician Library of Chinese Academy of Sciences)

Fig.7 In 1956, Tsai (front row, 1st from the left) took a group photo with the Soviet Union physiologists in the physiology department of the Kirov Academy of Military Medical Sciences in Leningrad, Soviet Union (Photo credit: Academician Library of Chinese Academy of Sciences)

Fig.8 In 1956, Tsai (1st from the right) visited the Soviet Union and took a photo in front of Leningrad Institute of Experimental Medicine (Photo credit: Academician Library of Chinese Academy of Sciences)

Fig.9 In September 1986, Tsai (front row, 4th from the left), as the judge of doctoral thesis defense, took a photo with the jury and the PhD graduates after the defense meeting (Photo credit: Academician Library of Chinese Academy of Sciences)

References

Chen TI, Tsai C (1948) The mechanism of haemostasis in peripheral vessels. J Physiol 107:280–288.

Evans CL, Tsai C, Young FG (1931a) The behaviour of liver glycogen in experimental animals: I. Methods: The effect of ether and amytal. J Physiol 73(1):67–80.

Evans CL, Tsai C, Young FG (1931b) The behaviour of liver glycogen in experimental animals: Ⅱ. Glycogen recovery after decapitation and decerebration. J Physiol 73(1):81–102.

Evans CL, Tsai C, Young FG (1931c) The action of adrenaline on glycogen distribution in the cat. J Physiol 73(1):103–114.

Hsü FY, Tsai C (1930) A note on the calcium content of the skeletal muscles after thyroparathyroidectomy and parathormone injection. Chin J Physiol 4:423.

Lee JS, Tsai C (1942a) Studies on the antihaemolytic properties of lecithin and cholesterol. Chin J Physiol 31:281–297.

Lee JS, Tsai C (1942b) The protective action of serum against natural haemolysin. Physiol Soc Chengtu Br 1:44–48.

Lee JS, Tsai C (1943) The development of a specific anticholesterol factor in the immune serum. Physiol Soc Chengtu Br 1:111–120.

Liu NC, Tsai C, Chu CK, et al (1946) Voice from Unoccupied China. Chicago: The University of Chicago Press.

Puh YC, Lee JS, Tsai C (1945) Blood platelet destruction as the main factor in increasing erythrocytic fragility during stasis. Physiol Soc Chengtu Br 11:149–155.

Qian YY, Chen W, Guo B (2020) Zing-Yang Kuo and behavior epigenesis based on animal experiments. Protein & Cell 11(6):387–390.

Tsai C (1924) A comparative study of retention curves for motor habits. Comp Psychol Monogr 2: 1–29.

Tsai C (1925a) The optic tracts and centers of the opossum, *Didelphis virginiana*. J Comp Neurol 39(2):173–216.

Tsai C (1925b) The descending tracts of the thalamus and midbrain of the opossum, *Didelphis virginiana*. J Comp Neurol 39(2):217–248.

Tsai C (1925c) The relative strength of sex and hunger motives in the albino rat. J Comp Psychol 5(5):407–415.

Tsai C (1930) The effect of thyroparathyroidectomy and parathormone administration on the gastric motility in dogs. Chin J Physiol 4:415–422.

Tsai C (1931) Action of narcotics on the conduction of nerve impulses from a single end-organ. J Physiol 73(4):382–404.

Tsai C (1933a) The total carbohydrate content of the liver tissue in the fasting rabbit: A method of estimation. Chin J Physiol 7:91–102.

Tsai C (1933b) Carbohydrate metabolism of the liver. Ⅰ. Glycogen and other carbohydrates in decapitate cats. Chin J Physiol 7:215–228.

Tsai C (1934) The question of the validity of using aqueous extracts for estimating glycogen and total carbohydrate of the liver. Chin J Physiol 7:343–352.

Tsai C (1935) An improved method of angiostomy. Chin J Physiol 9:355–362.

Tsai C (1937a) Studies on the combined glycogen of liver and muscle. Ⅰ. Nature. Chin J Physiol 11:87–92.

Tsai C (1937b) Studies on the combined glycogen of liver and muscle. Ⅱ. Variation under different physiological conditions. Chin J Physiol 11:93–102.

Tsai C, Chen CJ, Chiu KY (1941) The relation of increased erythrocytic fragility induced by stasis to lactic acid content of the blood. Physiol Soc Chengtu Br 1:19–43.

Tsai C, Chen CJ, Chiu KY (1942) Further observation on the various factors influencing the increase of erythrocytic fragility induced by stasis. Physiol Soc Chengtu Br 1:52–59.

Tsai C, Hsü FY (1929a) Studies on the pathogenesis of parathyroid tetany. Ⅰ. The effect of removal of large intestine. Chin J Physiol 3:183–196.

Tsai C, Hsü FY (1929b) Studies on the pathogenesis of parathyroid tetany. Ⅱ. The effect of ligation of the bile duct. Chin J Physiol 3:197–204.

Tsai C, Hsü FY (1929c) Studies on the pathogenesis of parathyroid tetany. Ⅲ. Influence of intestinal obstruction. Chin J Physiol 3:389–398.

Tsai C, Hsü FY (1929d) Studies on the pathogenesis of parathyroid tetany. Ⅳ. Influence of intestine putrefaction. Chin J Physiol 3:399–404.

Tsai C, Hsü FY (1930a) Plasma calcium and inorganic phosphorus following intravenous injection of parathyroid extraction: A study on the source of mobilized calcium. Chin J Physiol 4:265–272.

Tsai C, Hsü FY (1930b) The effect of intravenous injection of sodium oxalate and citrate on the concentration of plasma calcium and inorganic phosphorous. Chin J Physiol 4:273–288.

Tsai C, Lee JS, Wu CH (1940) The role of splenic action in altering erythrocytic fragility. Chin J Physiol 15:165–180.

Tsai C, McBride JJ, Zucker MB (1944) An ether-extractable substance from blood serum and buffy coat which contracts smooth muscle. Proc Soc Exp Biol Med 55:283–285.

Tsai C, Yi CL (1934a) Carbohydrate metabolism of the liver. Ⅱ. The sugar output. Chin J Physiol 8:245–272.

Tsai C, Yi CL (1934b) Carbohydrate metabolism of the liver. Ⅲ. The sugar intake during glucose absorption. Chin J Physiol 8:273–296.

Tsai C, Yi CL (1934c) Carbohydrate metabolism of the liver. Ⅳ. Sugar output in amytalized cats. Chin J Physiol 8:399–410.

Tsai C, Yi CL (1936a) Carbohydrate metabolism of the liver. Ⅴ. Sugar intake in the normal intact cat during glucose absorption. Chin J Physiol 10:87–104.

Tsai C, Yi CL (1936b) Carbohydrate metabolism of the liver. Ⅵ. Sugar output and intake in the daily digestive cycle. Chin J Physiol 10:105–118.

Wu CH, Tsai C (1940) Hematological standards for the Chinese. Chin J Physiol 15(3):289–299.

Wu CH, Tsai C (1943) Standardization of some sensory measurements for Chinese. Physiol Soc Chengtu Br 1:130.

Zhang H (2013) The Biography of Science and Education Elites in the Republic of China. Nanjing: Nanjing Press. (张衡. 2013. 民国科教精英百人传. 南京: 南京出版社.)

Zhang LF (1997) To commemorate the 100th anniversary of professor Tsai's birth. Chin J Aerosp Med 8(4): 196. (张立藩. 1997. 千秋风范照后人: 纪念蔡翘教授诞辰100周年. 中华航空航天医学杂志, 8(4):196.)

65. 冯德培：中国肌肉神经学事业开拓者

冯德培院士（图1）是我国著名生理学家和神经生物学家，也是我国近代生理学研究的开拓者之一。他的研究主要囊括神经–肌肉接头化学传递；快肌、慢肌的生理功能特性和脑内的长时程增强（LTP）功能。其中神经–肌肉传递机制是冯德培享誉国际的研究（Jin，2007）。对此，美国斯坦福大学的R. W. Tsien（钱永佑）高度赞许冯德培对中国神经科学的贡献："可以肯定地说，如果没有冯教授的组织、说服和强有力的个人榜样引领，中国神经科学的过去和未来将会非常不同。冯德培非凡的人生和职业生涯跨越了一个漫长的历史时期，无论是在世界还是在科学领域，其广度都不太可能再有人与他匹敌。"（Tsien，2007）冯德培的学生陈功认为，冯德培留下三份遗产："一，他的科学成就；二，在建设中国科学院上海生理研究所和促进我国生理学发展的道路上起领导者的作用；三，他在追求真理的道路上坚持不懈的精神。"（Chen，2007）

图1　冯德培院士（1907—1995）

冯德培的科学成就奠定了我国肌肉神经学研究的基础，其中神经–肌肉传递机制研究中的突触可塑性仍是当前的前沿领域。2018年中国人工智能大会（Chinese Congress on Artificial Intelligence，简称CCAI）上，蒲慕明（Poo Mu-Ming）院士的主题演讲重点提到突触可塑性问题。蒲慕明强调最值得注意的是，冯德培发现了强直后增强的现象——长时间的高频刺激导致突触传递效率的提高。这是第一次对活动依赖性突触可塑性的展示，这说明突触可塑性的现象最早是由中国科学家冯德培发现的（Feng，1941；Poo，2010），冯德培的名字也因此被牢牢刻进了神经科学的历史。直至1973年，Bliss和Lømo才在中枢神经系统中发现了类似的现象（Bliss and Lømo，1973）。时至

作者：谢伟[1]，郭本禹[1]，钱燕燕[2]

1　南京师范大学心理学院，南京 210097，中国

2　荷兰莱顿大学社会与行为科学学院，莱顿2333AK，荷兰

邮箱：psyqyy@163.com（钱燕燕）

2020年，冯德培于1940年在《中国生理学杂志》（英文版）上发表的与突触可塑性相关的文章仍被研究者引用（Ge et al., 2020）。这两个例子佐证了冯德培在突触可塑性研究方面的先驱性和前瞻性。

1907年，冯德培出生在浙江省临海县城一个富裕的家庭中。由于学业优异，中间跳级，11岁考入浙江省第六中学，1922年考入复旦大学文学院。1923年，受惠于刚刚留美归来的心理学教授郭任远的一堂“科普课”（Qian，1981；Feng，1986），冯德培于翌年转入心理学系。两年后，他又转而正式跟随蔡翘教授学习生理学。1926年，他从复旦大学毕业，获得了生物学学士学位（图2）。毕业后，蔡翘聘请冯德培为他的助教，任期为1年（1926—1927年）（The Editorial Department of *World Science*, 1985）。随后经蔡翘推荐，冯德培进入北京协和医学院（PUMC）生理系，在林可胜教授的实验室工作了两年，主要参与研究神经和体液对胃分泌的调控。

图2 1926年冯德培在复旦大学的毕业照

令人惊讶的是，早在1929年冯德培就做出了具有开创性的研究，他首次发现了胃的肠抑胃素分泌（enterogastrone）现象，并且首创“enterogastrone”这一术语，这也是他唯一的关于消化道系统生理学的研究工作（Feng，1929；Ji，2007）。除此之外，在北京协和医学院这两年中，冯德培广泛涉猎了生理学几乎所有领域，并且精通英语、德语、法语、西班牙语和意大利语5门外语（Feng，1986）。

1929年夏，冯德培顺利获得清华大学留美奖学金，同时由于先前对芝加哥大学Ralph Lillie的著作《原生质活动和神经活动》特别感兴趣，因而于1929年秋赴美跟随Ralph Lillie学习普通生理学。由于Ralph Lillie主要从事神经传导的金属线模型研究，而冯德培想研究的是真实的神经而非模型，因而很快决定跟随另一实验室的Ralph Gerard教授进行神经生物学研究，这也是他在这一领域的第一位导师。进入实验室后，冯德培主要进行的是神经窒息的机理研究，即一根窒息的神经通过浸泡在含有某些氧化性颜料（如亚甲基蓝）的无氧溶液中，而不是给它充氧，是否能恢复？在实验过程中，冯德培敏锐地发现，神经的结缔组织鞘是一种有效的扩散屏障，能阻碍亚甲基蓝到达神经纤维（Feng and Gerard，1930）。

1930年秋，获得芝加哥大学硕士学位之后，经北京协和医学院林可胜的推荐，冯德培转赴伦敦大学与A. V. Hill一起工作。期间，他主要围绕肌肉组织和神经组织的产热进行研究，在随后的两年半时间，他完成足够发表9篇文章的研究，其中独立撰写5篇。1932年是冯德培在留学期间十分重要的一年，此时他才25岁，也就是在这一年冯德培发现了静息肌肉被拉长时产热和氧耗量均明显增加的现象，即“牵张反应”（Feng,1932a）。后来，波兰的肌肉化学家J. K. Parnas以及整个国际生理学界称之为“冯氏效应”（Wei，1988）。随后冯德培在Hill的要求下对Lapicque的神经和肌肉在各种条件下的等时性研究理论进行验证，虽然最后冯德培发现Lapicque的理论是错的，但他觉得这并没有多大意义。因而他决定将研究转向乳酸在神经活动中的某种作用。1932年，在发现“冯氏效应”之后，冯德培又发现乳酸能使正常的神经实施长时间的持续功能，并将结论写成《乳酸在神经活动中的作用》一文并发表（Feng，1932b）。

1933年，冯德培在获得博士学位后，Hill又将他送去剑桥的E. D. Adrian实验室和牛津的C. S. Sherrington实验室（他们两人在1932年共同获得诺贝尔生理学或医学奖）各学习两个月以扩展研究经验，这使冯德培很快了解到30年代神经生物学的前沿问题，即神经肌肉接头信息传递究竟是电传递，还是电-化学传递（The Editorial Department of *World Science*，1985）？虽然在Adrian实验室仅待了两个月，但冯德培很快就帮Adrian解决了一个困扰他们许久的问题，即钾离子与神经损伤后恢复的关系。冯德培的实验结论经Adrian反复验证得到证实并被接受（Feng，1933）。1933年夏，在Hill的推荐下，冯德培又转而在Detlev Bronk 领导的美国费城新成立的约翰逊医学物理学基金会逗留了一年，期间他广泛地接触生物物理学方面的研究，并学习电子仪器的制作，为在中国建立实验室做准备。

抱着青年时的理想（图3），1934年夏，冯德培回到祖国并开始着手建立实验室。由于条件匮乏，实验室设在地下室。实验室初步建成后，冯德培立马着手神经肌肉相关的实验，随后冯德培得出一个重要的结论：直接刺激肌肉引起的肌肉收缩可以被一定频率的附加神经刺激所抑制。凭借自己敏锐的直觉，冯德培很快决定将研究主题转向神经-肌肉接头，随后提出“钙影响神经递质释放”这一诺贝尔奖级别的见解（Sa，2015）。基于前期的研究成果，1936年经Hill推荐，冯德培撰写了名为《神经放热》的综述性文章，发表在*Ergebnisse der Physiologie*杂志上，被学术界认为是神经能力学方面的权威性参考文献（Feng，1936）。

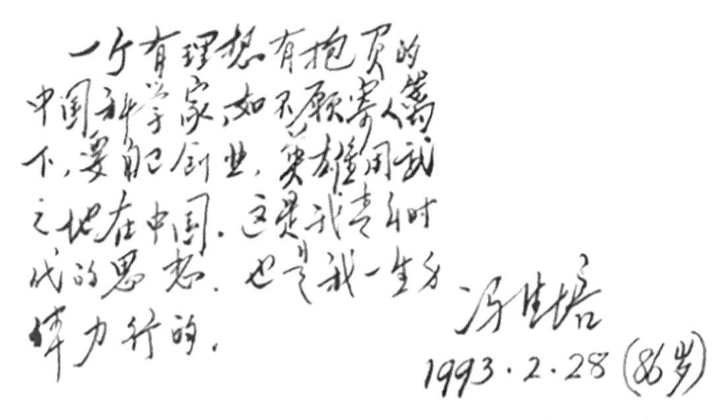
一个有理想有抱负的
中国科学家，如不愿寄人篱
下，要能创业，要有用武
之地在中国。这是我青年时
代的思想，也是我一生身
体力行的。
冯德培
1993·2·28（86岁）

图3　冯德培对人生目标的看法（图片来源：中国科学院院士文库）

从1936年到1941年，冯德培在《中国生理学杂志》（英文版）发表了至少26篇文章。后因抗日战争全面爆发，北京协和医学院被迫关闭，冯德培于1941年末离开北京协和医学院，研究被迫中断。在这26篇论文中（其中有些论文至今还在被人引用），冯德培主要研究的就是神经-肌肉接头的化学传递理论，该领域在当时还处于形成阶段，这些研究使他的实验室成为该领域一个国际瞩目的研究中心（图4和图5）。在他的回忆录《回顾，展望》中，他认为自己在1936年至1941年这段时间的研究主要有：① 高频间接刺激在神经肌肉接头产生抑制（现在称为接头抑制），与此相关，他们发现在神经末梢周围有一个局域性的收缩（Feng and Yang，1938；Feng，1988）。② 钙对神经肌肉接头有多种显著的作用，即钙使每个神经冲动在神经末梢处释放更大量或更浓缩的乙酰胆碱。③ 首次描述了终板电位的强直后易化或增强，这种效应可持续数分钟（Feng，1941）。④ 揭示了毒扁豆碱和箭毒的接头前效应及强直后增强效应（PTP），这是突触可塑性的开创性研究，同时也是后来许多神经药理学研究的起点（Feng，1988）。

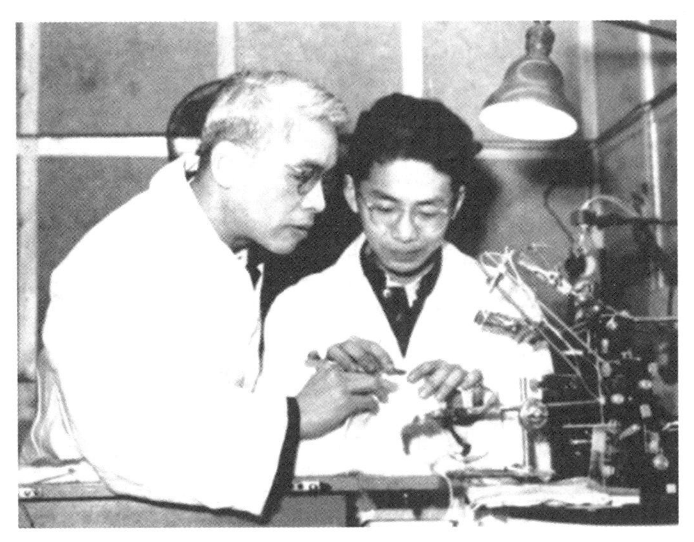

图4　冯德培（左）指导研究生进行兔脑解剖实验（图片来源：中国科学院院士文库）

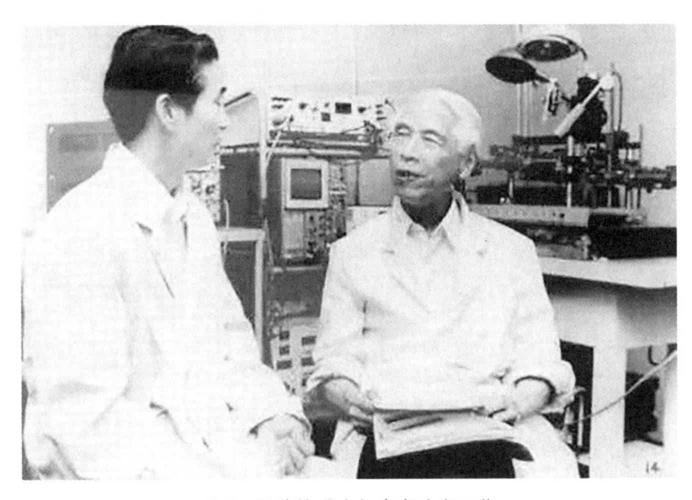

图5　冯德培（右）在实验室工作

1946年，冯德培访问美国，为新研究所购置设备和书籍，并了解科学进展。随后冯德培在洛克菲勒研究所与Lorente de Nó 共事了一年。期间两人由于蛙或牛蛙神经的结缔组织鞘是否是扩散屏障而发生争执，Lorente de Nó 认为这是完全不可能的（Lorente de Nó，1950），然而早在1930年冯德培就获得有关神经鞘作为有效扩散屏障的最初论断（Feng and Gerard，1930），于是他于1947年夏回到中国与助手刘育民博士重新研究，实验结果支持早期论断（Feng and Liu，1949a，1949b，1950）。1948年，冯德培当选中央研究院院士。1949年中华人民共和国成立，冯德培身兼数职。

20世纪40年代末至50年代初，冯德培在外周神经、中枢神经和肌肉方面的研究也取得了众多成就（Fan，2007）。1955年，他当选中国科学院学部委员（中国科学院院士）（图6）。冯德培创建和发展了中国科学院上海生理研究所，还和中国科学院海洋研究所合作。1956年，冯德培赴哥本哈根邀请张香桐回国共同开展神经生理学事业（图7）（Qian，1981；Zhang and Le，2011），张香桐后来组建中枢神经研究室。1958年，冯德培参加中国科学院代表团，赴苏联与苏联科学院商谈两院合作。从苏联回国后，冯德培于1959年成立青岛工作组，由他亲自领导并在青岛组建了我国第一个海洋动物生理实验室（Fan，2007，2011）。

图6 1992年，冯德培院士（前排左三）参加中国科学院第六届院士大会

图7 1990年，冯德培（右一）与张香桐（右二）参观上海植物园（图片来源：中国科学院院士文库）

1960年，冯德培开始考虑神经肌肉营养性关系的研究，根据以往经验，新项目必须建立在发现新现象的基础上。基于以往神经肌肉萎缩的老现象，冯德培带领荣辛未和毋望远用鸡背部的快、慢肌展开探索性实验，很快就发现鸡慢肌纤维去神经后肥大的新现象，这对阐明神经如何决定肌纤维类型的机制作出重要贡献（Feng et al.，1962）。1962年在布拉格召开的肌肉讨论会上冯德培报告了这一发现，如今学术界公认这一现象是冯德培首先发现的（The Editorial Department of *World Science*，1985）。1978年A. V. Hill去世前不久，他来信说冯德培（Hill称呼冯德培为T.P）是他这一生中最优秀的两个学生之一。

20世纪80年代以后，冯德培开始致力于开拓突触的细胞和分子生物学领域，特别是突触可塑性的研究（图8）。1985年9月，冯德培参加在上海举行的欧美同学会年会（图9）。1988年，冯德培回顾了他的整个科研生涯，并在《神经科学年刊》的邀请下撰写了传记文章《回顾，展望》（Feng，1988）。甚至在85岁高龄时，冯德培和他的学生发现了一个新现象，即突触后蛋白激酶C对长时程增强（LTP）的诱导和维持至关重要，并发表在*PNAS*上（Wang and Feng，1992；Feng，1995）。这是自1986年冯德培成为美国国家科学院外籍院士以来，在该杂志发表的第一篇也是最后一篇论文。1983年至1989年，他连续三届被选为国际生理科学联合会（IUPS）执行委员会成员。晚年由于疾病，他的气管已经被切开，他不能说话，但仍艰难地在纸上写下了学生最新研究成果发表时所使用标题的建议（图10）（Rong，2007）。1995年4月10日，冯德培病逝于上海，享年88岁。

图8　20世纪80年代，冯德培（前排左三）和Bernard Katz（前排左二）参加国际会议

图9　1985年，冯德培（右）与苏步青（左）在上海的欧美同学会年会上亲切交流

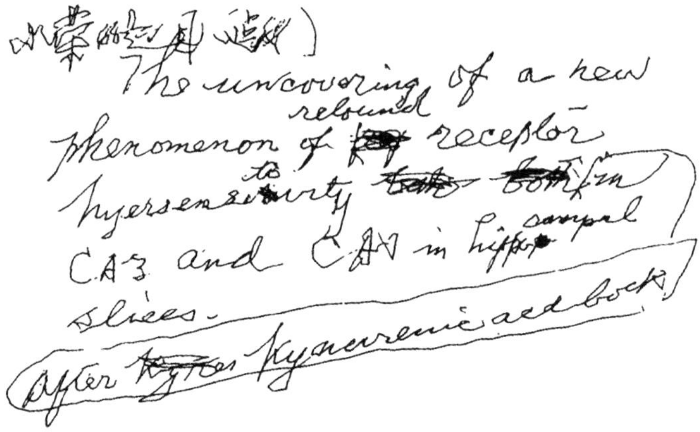

The uncovering of a new rebound phenomenon of receptor hypersensitivity to CA3 and CA1 in hippocampal slices.
After kynurenic acid block

图10　冯德培在病床上给学生拟的最后一个论文题目

65. Te-Pei Feng: A pioneer of neuromuscular physiology in China

Te-Pei Feng (冯德培, 1907—1995) was a famous Chinese neurophysiologist and also one of the pioneers of modern physiological researches in China. He mainly studied chemical transmission at the neuromuscular junction, the biological characteristic of high and low contraction speed muscles, and the long-term potentiation (LTP) function of the brain (Jin, 2007). Richard Wuyin Tsien (钱永佑) of Stanford University highly valued Feng's contributions to neurology in China, stating that, "It is safe to say that the past and the future of neuroscience in China would be very different had professor Feng not provided leadership, by dint of organization, persuasion and strong personal example. Moreover, Feng's remarkable life and career spanned a lengthy epoch of history, in both the world and scientific arenas, of scope unlikely to be matched again." (Tsien, 2007) Feng's student Gong Chen (陈功) said that Feng left three legacies: a. Feng's scientific achievements, especially in the field of physiology; b. His leading role in establishing the Shanghai Institute of Physiology and promoting the development of Chinese physiological sciences; c. His spirit of never bending on the way of seeking truth (Chen, 2007).

Feng's scientific achievements have laid the foundation for muscle neurology research in China, while the study of synaptic plasticity in nerve-muscle transmission mechanisms remains at the forefront of current research. At the Chinese Congress on Artificial Intelligence (CCAI) in 2018, Mu-Ming Poo (蒲慕明), an Academician of the Chinese Academy of Sciences, prominently mentioned the phenomenon of synaptic plasticity, which was first discovered by Feng in 1941 (Feng, 1941; Poo, 2010). Much later, in 1973, Timothy Vivian Pelham Bliss and Terje Lømo discovered a similar phenomenon in the central nervous system (Bliss and Lømo, 1973). Even as late as 2020, Feng's article published in *Chinese Physiology Journal* (English edition) in 1940 was still cited in a paper related to synaptic plasticity research (Ge et al., 2020). These two examples demonstrate the great enduring significance of Feng's pioneering work in the field of synaptic plasticity.

Wei Xie[1], Benyu Guo[1], Yanyan Qian[2]

1 School of Psychology, Nanjing Normal University, Nanjing 210097, China

2 Social and Behavioral Sciences Faculty, Leiden University, Leiden 2333AK, The Netherlands

Correspondence: psyqyy@163.com (Y.-Y. Qian)

Feng was born to a rich family in Linhai county of Zhejiang Province in 1907. At only 11 years of age, he demonstrated excellent scholastic achievements and skipped grade to the Sixth Middle School of Zhejiang Province. In 1922, Feng enrolled at the Department of Literature of Fudan University. The following year, Feng's scientific interest was piqued by a science course given by the psychologist Zing-Yang Kuo (郭任远), who had returned to China from the United States (Qian, 1981; Feng, 1986). Then, Feng transferred to the Department of Psychology from the Literature Department of Fudan. In his junior year of university, Feng became the formal student of Professor Chiao Tsai (蔡翘), studying physiology, after which he obtained a bachelor's in biology in 1926 from Fudan University. After that, Feng also has been the lecturer recruited by Chiao Tsai at Fudan University for one year (1926—1927) (The Editorial Department of *World Science*, 1985). In the autumn of 1927, Feng was recommended by Tsai to the laboratory of Robert Kho-Seng Lim (林可胜) at the Department of Physiology of Peking Union Medical College (PUMC). There, he worked under the guidance of Lim, who studied the nervous and humoral control of gastric secretion.

As early as 1929, Te-Pei Feng made a pioneering discovery on the physiology of the alimentary canal, revealing the enterogastrone hormones of the stomach, for which he coined the term (Feng et al., 1929; Ji, 2007). During his time at the PUMC, Feng read extensively about almost all fields of physiology. Additionally, Feng also mastered 5 languages, including English, German, French, Spanish and Italian (Feng, 1986).

In the summer of 1929, Feng obtained Tsinghua University Fellowship to study in the United States. Because of his interest in Ralph Stayner Lillie's book *Protoplasmic Action and Nervous Action*, he finally went to Chicago University and studied general physiology under biologist Ralph Stayner Lillie's guidance. At that time, Lillie was preoccupied with the iron-wire model of nerve conduction, but what Feng really wanted to explore was real nerves, and not the model. Therefore, he quickly made a choice to go to another laboratory, which was led by neurophysiologist and behavioral scientist Ralph Waldo Gerard, who finally became Feng's first tutor in neurophysiological research. Working with Gerard, Te-Pei Feng was mainly preoccupied with the mechanism of nerve damage due to asphyxiation. He investigated whether an asphyxiated nerve could be induced to recover by soaking it in an oxygen-free solution of certain oxidizing dyes like methylene blue, instead of giving it oxygen? Feng quickly discovered that the connective tissue sheath surrounding the nerve was an effective diffusion barrier, which could prevent the methylene blue from reaching the nerve fibers (Feng and Gerard, 1930).

After earning his master's degree from Chicago University in the autumn of 1930, Feng was recommended by Lim of PUMC to work with Archibald Vivian Hill at London University. There, Feng mainly worked on heat production, first in muscle and later in nerves. In the next two and a half years, Feng either conducted or participated in sufficient work for 9 papers, 5 of which he wrote up himself. In 1932 he discovered the increase of the resting heat production of muscle on passive stretch, which was called "stretch response" (Feng, 1932a). Later, muscle chemist Jakub Karol Parnas from Poland and the whole international physiology circle named it the "Feng

Effect" (Wei, 1988). After that, Hill requested Feng to look into Lapicque's controversial theory of isochronism of muscle and nerve under different conditions. Although he was convinced that Lapicque's theory is incorrect according to the experimental results, he felt what he had done was not meaningful. Therefore, Feng decided to shift his research direction to the role of lactic acid in nervous activities. The same year, he discovered another new phenomenon, showing that lactic acid enables nerves to perform normal continuous function for many hours. Later, this discovery was described in the paper "The role of lactic acid in nerve activity" (Feng, 1932b).

After Feng got his doctoral degree in Physiology from the University College of London in 1933, he was recommended by Archibald Vivian Hill to stay in the laboratories of Edgar Douglas Adrian at Cambridge University and Charles Scott Sherrington at Oxford University for two months each, so that Feng could expand his study experience (Adrian and Sherington shared the Nobel Prize in Physiology or Medicine in 1932). Feng quickly learned about the advanced issues in 1930s that was which the information delivered in the neuromuscular junction is electrical transfer or electro-chemical transfer in neurophysiology? (Editorial Department of *World Science*, 1985) Although Feng worked in the Laboratory of Adrian for merely two months, he solved a longstanding problem of Adrian that was related to the relationship between potassium and the recovery of injured nerves. Later, Feng's conclusion was verified repeatedly by Adrian and finally accepted (Feng, 1933). Before returning to China, Feng was recommended by Hill again to go back to the United States, where he spent one year working at the newly established Johnson Foundation for Medical Physics in Philadelphia, directed by Detlev Bronk. During that period, Feng devoted the majority of his time to learning how to assemble electronic instruments in preparation for establishing a laboratory in China.

With the idealism of the young adult period, Feng returned to China in the summer of 1934 and set about establishing his laboratory. Because of the difficult socioeconomic circumstances at the time, the laboratory was set up in the basement. In spite of the hardship, Feng immediately started his experiments on nerves and muscles. Afterwards, Feng came to an important conclusion that the contraction elicited by direct muscle stimulation could be inhibited by additional nerve stimulation at certain frequencies. Due to this bout of intuition, Feng switched his research topic to the neuromuscular junction, developing the Nobel Prize-level theory that calcium affects the release of neurotransmitters (Sa, 2015). Due to his previous achievements, and with recommendation of Hill, in 1936, Feng wrote a summary article "The heat production of nerves" , which was regarded by academic circles as an authoritative reference in the field of neural heat study and was published in the journal *Ergebnisse der Physiologie* (Feng, 1936).

From 1936 to 1941, Feng published at least 26 papers in the *Chinese Journal of Physiology*. However, because of full outbreak of the War of Resistance against Japanese Aggression, PUMC was forced to close and Feng had to leave in 1941, which interrupted his research. Among the crucial 26 papers, some of which are still cited today, is Feng's pioneering work in the theory of chemical transmission at the neuromuscular junction, which led his laboratory to become a research center highly regarded by international peers. In his autobiographical article “Looking back, looking forward” , Feng described his main areas of research between 1936 and 1941: a. Accompanying the inhibition produced at the neuromuscular junction by high-frequency indirect stimulation (also generally called junctional inhibition), they found a local contraction surrounding the nerve endings (Feng and Yang, 1938; Feng, 1988); b. Calcium was shown to have various striking effects on the neuromuscular junction, which means that calcium causes each individual nerve impulse to liberate a larger or more concentrated amount of ACh from the nerve terminus; c. The post-tetanic potentiation (PTP) or facilitation of the endplate potential, which can last many minutes, was described for the first time (Feng, 1941); d. A new pre-junctional aspect of the eserine and curare effects, in addition to the post-tetanic effects, was thus brought to light, which is one of the first concrete examples of synaptic plasticity and is also the starting point of neuropharmacology (Feng, 1988).

In 1946, Feng visited the United States to buy experimental equipment and books for a new institute, and at the same time acquire new information at the scientific frontier. Afterwards, Feng worked at the Rockefeller Institute with Lorente de Nó for one year. However, Feng quickly returned to China due to an argument regarding the question whether the connective tissue sheath of frog or bull-frog nerve could act as a diffusion barrier. Lorente de Nó considered this idea utterly impossible (Lorente, 1950), vigorously challenging Feng’s opinion and scientific conclusions. However, Feng and Gerard had made a preliminary conclusion that the connective tissue sheath of nerves could act as a diffusion barrier as early as 1930 (Feng and Gerard, 1930). Thus, in the summer of 1947, Feng returned to China and re-verified this conclusion with the help of his assistant, Dr. Yumin Liu (刘育民). Finally, the experimental results corroborated the original conclusion of Feng and Gerard (Feng and Liu, 1949a, 1949b; Feng and Liu, 1950). In 1948, Feng was selected as a member of Academia Sinica. Following the founding of the People’s Republic of China in 1949, Feng held multiple posts, and also made great strides in the study of peripheral nerves, central nerves and muscles (Fan, 2007). Feng was selected as a member of Academic Division of the Chinese Academy of Sciences in 1955. The following year, Feng went to Copenhagen and invited Hsiangtung Chang back to China to establish a neurophysiology research capacity with him (Qian, 1981; Zhang and Le, 2011). Chang came back to China in 1956 and set up the central nervous system research laboratory a few years later. In 1958, Feng joined the delegation of the Chinese Academy of Sciences to discuss the corporation with the Soviet Academy of Sciences. After returning from the USSR, Feng set up and directed the Qingdao working group, after which the first Chinese Marine Animal Physiology Laboratory was created in Qingdao in 1959 (Fan, 2007, 2011).

Feng studied nerve-muscle trophic relations in 1960. On the basis of his previous research experience, Feng considered that a new phenomenon should be discovered before launching a new project. Therefore, based on the known phenomenon of muscle atrophy following denervation, Feng led his assistants Xinwei Rong (荣辛未) and Wangyuan Wu (毋望远) to conduct experiments with high- and low-speed muscles, which have the most strikingly different contraction speeds among all muscles of chickens. Then, a surprising new phenomenon was discovered that the low contraction speed chicken muscle was strikingly hypertrophied. This discovery made an important contribution to elucidating the mechanism by which nerves determine the type of muscle fibers (Feng et al., 1962). Feng reported this new discovery in a muscle seminar held in Prague in 1962, and there is an academic consensus that this phenomenon was first discovered by him (The Editorial Department of *World Science*, 1985). In 1978, before Hill died, he wrote to Feng and told him that he is one of the two best students in his teaching career.

Since the 1980s, Feng mainly explored the field of cellular and molecular synaptic physiology, focusing on synaptic plasticity in central synapses. Feng devoted himself to the study of long-term potentiation (LTP) and made a significant contribution to the modern understanding of cellular mechanisms underlying LTP. In September 1985, Te-Pei Feng attended the annual meeting of the European and American Alumni Association in Shanghai. In 1988, Feng wrote a retrospective of his whole scientific career in his biographic article Looking Back, Looking Forward, in response to an invitation by the *Annual Review of Neuroscience* (Feng, 1988). Even at 85 years old, Feng discovered a new phenomenon with his students, reporting in *PNAS* that the postsynaptic protein kinase C is essential for the induction and maintenance of long-term potentiation (LTP) (Wang and Feng, 1992; Feng, 1995). This was his first and last paper in *PNAS* since Feng became a foreign member of the National Academy of Sciences, United States in 1986. From 1983 to 1989, Feng was selected as an executive committee member of the International Union of Physiological Sciences (IUPS) for three consecutive sessions. In the last phase of his life, he underwent a tracheotomy and could not speak, but still summoned the strength to write down the title for his students' latest research paper (Rong, 2007). On April 10 1995, aged 88, Feng passed away in Shanghai.

Figures

Fig.1 Te-Pei Feng (1907—1995)

Fig.2 Te-Pei Feng's graduation photo from Fudan University in 1926

Fig.3 Te-Pei Feng's view on life's purpose (Photo credit: Academician Library of Chinese Academy of Sciences)

Fig.4 Te-Pei Feng (left) instructed one graduate student to conduct anatomy experiments on the rabbit brain (Photo credit: Academician Library of Chinese Academy of Sciences)

Fig.5 Te-Pei Feng (right) still worked in the laboratory in old age

Fig.6 In 1992, Te-Pei Feng (3rd from the left, front row) joined the 6th general assembly of the Chinese Academy of Sciences

Fig.7 In 1990, Te-Pei Feng (1st from the right) and Hsiang-Tung Chang (2nd from the right) visited the Shanghai Botanical Garden (Photo credit: Academician Library of Chinese Academy of Sciences)

Fig.8 In the 1980s, Te-Pei Feng (3rd from the left, front row) and Bernard Katz (2nd from the left, front row) attended international conferences

Fig.9 In 1985, Te-Pei Feng (right) had a cordial exchange with Buqing Su (left) at the annual meeting of the European and American Alumni Association in Shanghai

Fig.10 On his hospital bed, Te-Pei Feng wrote down the last article title for a student

References

Bliss T, Lømo T (1973) Long-lasting potentiation of synaptic transmission in the dentate area of the anaesthetized rabbit following stimulation of the perforant path. J Physiol 232:331–356.

Chen G (2007) In memory of a great physiologist and my mentor Te-Pei Feng. Acta Physiol Sin 59(6):716.

Fan SF (2007) The scientific work of Mr. Te-Pei Feng from the late 1940s to 1960—commemorating the 100th anniversary of the birth of Mr. Te-Pei. Acta Physiol Sin 59(6):719–729.

Fan SF (2011) The history of Qingdao Laboratory of marine animal physiology. CSHST 32(1):99–104.

Feng TP (1932a) The effect of length on the resting metabolism of muscle. J Physiol 74:441–454.

Feng TP (1932b) The role of lactic acid in nerve activity. J Physiol 79:477–486.

Feng TP (1933) Reversible inexcitability of tactile endings in skin injury. J Physiol 79:103–108.

Feng TP (1936) The heat production of nerve. Ergebn Physiol 38:73–132.

Feng TP (1941) Studies on the neuromuscular junction. XXVI. The changes of the end-plate potential during and after prolonged stimulation. Chin J Physiol 16(3):341–372.

Feng TP (1986) Review and prospect for the past 60 years. Prog Physiol Sci 17(3):193–198.

Feng TP (1988) Looking back, looking forward. Ann Rev Neurosci 11:1–12.

Feng TP (1995) Academician Te-Pei Feng's speech at the commemoration of the 60th anniversary of the publication of Acta Physiologica Sinica in 1987. Acta Physiol Sin 6:524–526.

Feng TP, Gerard RW (1930) Mechanism of nerve asphyxiation: With a note on the nerve sheath as a diffusion barrier. Proc Soc Exp Biol Med 27:1073–1076.

Feng TP, Hou HC, Lim RKS (1929) On the mechanism of the inhibition of gastric secretion by fat. Chin J Physiol 3:371.

Feng TP, Jung HW, Wu WY (1962) The contrasting trophic changes of the anterior and posterior latissimus dorsi of the chick following denervation. Acta Physiol Sin 25:431–441.

Feng TP, Liu YM (1949a) The connective tissue sheath of the nerve as effective diffusion barrier. J Cell Comp Physiol 34:1–16.

Feng TP, Liu YM (1949b) The concentration-effect relationship in the depolarization of amphibian nerve by potassium and other agents. J Cell Comp Physiol 34:33–42.

Feng TP, Liu YM (1950) Further observations on the nerve sheath as a diffusion barrier. Chin J Physiol 17(3):207–218.

Feng TP, Yang YC (1938) Studies on the neuromuscular junction. A note on the local concentration of cholinesterase at the motor nerve ending. Chin J Physiol 13:141–144.

Ge D, Noakes PG, Lavidis NA (2020) What are neurotransmitter release sites and do they interact? Neuroscience 425:157–168.

Ji YH (2007) Gazing at the blue waves of the past sea of time—in memory of foundational contributions of Mr.

Feng Te-Pei and others who opened the precedent of "Enterogastrones" . Acta Physiol Sin 6:736–737.

Jin GZ (2007) Remembering several past things—commemorating the 100th anniversary of Mr. Feng Te-Pei's birth. Acta Physiol Sin 6:711–712.

Lorente de Nó (1950) The ineffectiveness of the connection sheath of nerve as a diffusion barrier. J Cell Comp Physiol 35:195–240.

Poo MM (2010) Neuroscience in China 2000—2009: Introduction. Sci China Ser Sci 53(3):301–303.

Qian WH (1981) In memory of Professor Te-Pei Feng. Chin J His Sci Technol 4:26–37.

Rong XW (2007) Accompanied by Mr. Feng Te-Pei in the study of neuromuscular junctions and synapses for 35 years. Acta Physiol Sin 59(6):733–735.

Sa S (2015) The fate between Te-Pei Feng and the Nobel Prize. Chin Sci Daily, 2015–10–3.

The Editorial Department of *World Science* (1985) Neurophysiologist Feng Te-Pei. World Science 1–2.

Tsien RW (2007) A remembrance of Professor Te-Pei Feng. Acta Physiol Sin 6:713–715.

Wang JH, Feng DP (1992) Postsynaptic protein kinase C essential to introduction and maintenance of long-term potentiation in the hippocampal in the Ca1 region. PNAS 89:2576–2580.

Wei NS (1988) A brief introduction of "Feng effect" . Bull Biol 8:44–45.

Zhang BY, Le K (2011) Chang Hsiang-Tung's study on dendritic function: Gold is shining everywhere. Protein & Cell 2(4):264–265.

66. 谢和平：中国感染病学的先驱

谢和平（Richard Ho-Ping Sia）（图1），福建厦门人，出生于1895年4月28日。1914年于武昌文华大学毕业后赴美国克利夫兰市西储大学医学院学习临床医学，1918年毕业后于克利夫兰城市医院完成1年临床轮转培训，1919年就职于北京协和医学院（后简称“协和”）内科，长期从事感染性疾病专业研究。1939年离开协和赴美国，先在夏威夷大学任细菌学教员和校医，1946年离开学校独自开业行医，并在檀香山的几家医院兼职（Sia，1971；张艳荣和李志平，2016）。谢和平的学术生涯基本上在协和完成，赴美后即主要开业行医。谢和平在协和内科工作长达20年，主要在感染性疾病领域。在当时感染性疾病占据社会主要病种的情况下，应该说对协和内科居功至伟。

图1　谢和平（1895—1970）

北京协和医学院由美国洛克菲勒基金会于1917年通过其全资注资的中华医学基金会创办，是我国最早设有八年制临床医学专业和护理本科教育的医学院校。谢和平于1919年入职协和内科任助理住院医师。而在1920—1921年北京协和医学院年报中，内科员工一共9人，谢和平即为其中之一且是唯一华人，和Otto Willner一起担任讲师职位，因此从某种意义上来说，谢和平是协和内科尤其是感染性疾病专业的创建者之一（图2）。其他7人分别为校长兼内科主任F. C. McLean教授，感染性疾病负责人O. H. Roberson副教授，神经精神科负责人A. H. Woods副教授，H. J. Smyly、C. W. Young、J. H. Korns和当时尚在美国的W. G. Lennox均为讲师。

1929—1930年，谢和平赴美进修，在纽约与O. T. Avery合作进行肺炎球菌的研究。回国后他即掌管感染性疾病部门，并一直作为传染病隔离病房的负责人，直至1939年

作者：刘旭东[1]，李圆梦[2]，李乃适[2,3]

1　中国医学科学院北京协和医学院，北京协和医院医学科学研究中心，疑难重症与罕见病国家重点实验室，北京100730，中国

2　中国医学科学院北京协和医学院，北京协和医院内分泌科，卫生部内分泌重点实验室，疑难重症与罕见病国家重点实验室，北京100730，中国

3　中国医学科学院北京协和医学院人文与社会科学学院，北京100730，中国

邮箱：LNS@medmail.com.cn（李乃适）

图2 谢和平在北京协和医学院传染病实验室

离开协和。那一时期的病案首页主治医师处署名几乎均为Dr. Sia，而谢和平医师查房也被大量病历所详细记载。1927年11月—1928年8月，谢和平还担任北京协和医学院病案委员会负责人。因此谢和平在协和内科尤其是感染性疾病领域承担了大量医疗工作；到1939年谢和平离开协和之时，他已经是整个内科中唯一工作达到20年的医师，因此他对协和内科的发展有着不小的贡献。

谢和平在刚入职协和不久即投入感染性疾病的研究中。谢和平于1921年就发表了2篇有关黑热病的论文（Sia，1921；Sia and Wu，1921），其中1篇是与吴宪合著。他们合作在《中华医学杂志》上发表了后来被称为“谢氏实验”的黑热病检测方法：将血浆滴入盛有去矿物质水的试管中，出现沉淀或絮凝为阳性反应；这种优球蛋白实验被认为可能是原发性巨球蛋白血症的最早筛查方法。

O. H. Roberson和谢和平在生物学实验室探讨后得出，我国的肺炎球菌以Ⅰ型和Ⅳ型最为多见。1923—1927年，谢和平和O. H. Roberson以“肺炎球菌的生长抑制”为题，设计了一系列试验，对该领域的科学问题进行逐一探索。其中，谢和平发现了明胶对悬浮液中肺炎球菌的保护作用，可使室温下悬浮液中肺炎球菌的存活时长由数小时延长至6～7天(Robertson et al., 1924)。为证实正常血清–白细胞混合物对肺炎球菌的生长抑制和杀菌作用，谢和平发明了一种能更相近地模仿动物体内条件的新技术，其中还包括一种可使血清–白细胞混合物保持恒定运动的新仪器，展现出强大的发明创造能力(Robertson and Sia, 1924a)。通过探索特异性抗肺炎球菌血清以及肺炎球菌可溶性物质对正常血清–白细胞混合物中肺炎球菌生长的影响，谢和平发现免疫血清及白细胞抑制和杀灭肺炎球菌的作用在本质上是定量的(Robertson and Sia, 1924b)，而肺炎球菌可溶性物质对促进肺炎球菌的生长有一定作用，且该作用对肺炎球菌的类型具有高度特异性 (Sia, 1926)。谢和平还进一步探索了血清–白细胞混合物抗肺炎球菌的机制，并发现某些哺乳动物对肺炎球菌感染的免疫差异主要是由血液中抗肺炎球菌调理素的浓度不同所造成(Robertson and Sia, 1927)。这一系列试验建立了若干新的试验方法，循序渐进地探索并分析了肺炎球菌生长抑制的相关问题，推进了肺炎研究的进展。

在这一系列研究之后，谢和平继续对肺炎领域进行深入探索。立足于既往动物试验的基础之上，谢和平进一步研究了获得性体液免疫与人类大叶性肺炎恢复机制之间的关系（Sia et al.，1928）。在纽约期间，他和两位同事再度发明了用于测定抗肺炎球菌血清中特异性可沉淀蛋白的新方法，且用该方法证明血清中特异性可沉淀蛋白与免疫保护在数量上存在定量关系（Heidelberger et al.，1930）。这一快速、简便、经济的新方法当时被推荐用于代替传统的小鼠保护试验。

1929—1930年，M. H. Dawson和谢和平进行了肺炎球菌的体外转化研究（Dawson and Sia，1931；Sia and Dawson，1931），这也是谢和平学术生涯中最为重要的贡献。该研究促使O. T. Avery对肺炎球菌的工作转向转化研究，最终在1944年做出了“DNA是遗传的化学基础”这一重要发现。谢和平在纽约的重要研究与他在协和的前期工作的联系，正如饶毅（2014）在博客中指出，至少有一方面有明确关系，即动物血清的选择（Dawson and Sia，1931）；因谢和平发现多种动物血清均有抗R物质（Sia，1926）。Dawson和谢和平在用马血清（Dawson and Sia，1931）进行深入研究后，否定了转化的物质为多糖的假说（Sia and Dawson，1931）。这是一系列后续试验中的重要一步，后来的试验证实脱氧核糖核酸是转化物质，以及DNA是遗传密码的载体。A. B. Hastings描述这项工作为“经典的却未被充分认识的试验，在此基础上的后续试验获得了三项诺贝尔奖”（鲍尔斯，2014）。尽管Avery与诺贝尔奖擦肩而过实在令人遗憾，但其团队的成就在世界医学史上的贡献是不可磨灭的。而谢和平和Dawson所完成的前期工作则构成了Avery团队贡献的研究基础，同样是不应被忽视的。

谢和平在协和工作长达20年，这在1942年以前的协和内科是独一无二的。而他在1927年被任命为病案委员会负责人，在1931年被聘为助理教授仅1年就晋升为副教授，1939年协和在他辞职前依旧做出续聘的决定，均说明他在协和是受到重视和认可的。他的主要研究工作均是围绕肺炎球菌进行，因此去洛克菲勒医学研究所时继续研究肺炎球菌也在意料之中，并得以与Dawson一起进行肺炎球菌体外转化研究。谢和平在著名国际期刊*Journal of Experimental Medicine*发表了9篇论文，在当时取得了较高的学术成就（饶毅，2014）。究其原因，一方面协和与洛克菲勒医学研究所的学术平台均属国际一流，为谢和平的科研打下了良好的基础；另一方面，*Journal of Experimental Medicine*创刊自约翰斯·霍普金斯大学医学院，首任主编正是William Welch；而后来期刊主办方即转至洛克菲勒医学研究所，时任所长Simon Flexner长年担任主编，而Welch和Flexner正是筹建协和过程中最负盛名的医学教育家，因此协和学者将重要稿件投稿至该期刊的倾向就很容易理解了。因此，谢和平是协和内科历史上的重要人物，进一步深入研究相关资料将对我国感染性疾病史产生非常重要的意义。

致谢

本文写作过程中得到了陈达维先生、蒋育红女士（北京协和医学院）、刘畅先生（美国得克萨斯大学奥斯汀分校）、翁琳倩女士（比利时鲁汶大学）的大力支持和帮助。

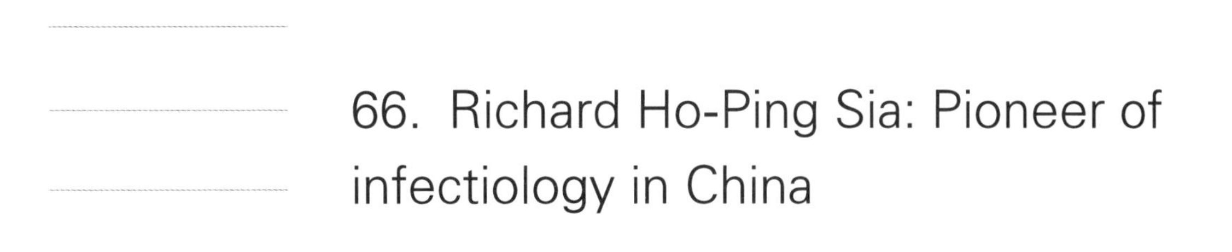

66. Richard Ho-Ping Sia: Pioneer of infectiology in China

Dr. Richard Ho-Ping Sia (谢和平) was born in Xiamen (Amoy) on April 28, 1895. After graduating from Boone University in Wuchang in 1914, he went to the Medical School of Western Reserve University in Cleveland, USA, to study clinical medicine, and received his MD degree in 1918. After completing 1-year clinical rotation training at Cleveland City Hospital, he returned to China in 1919 and joined Peking Union Medical College (PUMC) Hospital as an Assistant Resident. He was promoted to Assistant in 1920, and then worked as a physician treating infectious diseases in the Department of Medicine of PUMC Hospital. In 1939, he left PUMC for the University of Hawaii, where he served as a lecturer in bacteriology and school physician, and later practiced medicine in Honolulu until his retirement (Sia, 1971; Zhang and Li, 2016). Dr. Sia completed almost all of his academic career at PUMC. As he worked at PUMC Hospital for 20 years with a primary focus on infectious diseases, which were highly prevalent in China at that time. There is no doubt that he made significant contributions to the Department of Medicine of PUMC Hospital.

PUMC was founded by the Chinese Medical Board and was entirely supported by the Rockefeller Foundation of the United States. The foundation stone laying ceremony of new PUMC buildings was held in 1917. It was the first medical school in China with an eight-year medical education program and an advanced nursing program. In 1919, Dr. Sia returned to China as an assistant resident in medicine at PUMC Hospital. From 1920 to 1921, Dr. Richard Ho-Ping Sia was the only Chinese doctor among the nine internal medicine staff according to the annual report of PUMC Hospital. He held the position of Assistant together with Otto Willner. Therefore,

---

Xudong Liu[1], Yuanmeng Li[2], Naishi Li[2,3]

1 Medical Science Research Center, State Key Laboratory of Complex Severe and Rare Diseases, Peking Union Medical College Hospital, Chinese Academy of Medical Science and Peking Union Medical College, Beijing 100050, China

2 Department of Endocrinology, NHC Key Laboratory of Endocrinology, State Key Laboratory of Complex Severe and Rare Diseases, Peking Union Medical College Hospital, Chinese Academy of Medical Science and Peking Union Medical College, Beijing 100730, China

3 School of Humanity and Social Sciences, Chinese Academy of Medical Science and Peking Union Medical College, Beijing 100730, China

Correspondence: LNS@medmail.com.cn (N.-S. Li)

in a sense, Dr. Sia was one of the founders of the Department of Medicine, especially in the field of infectiology. The other seven were Professor Franklin C. McLean (President and Head of Medicine), Associate Professor Oswald H. Roberson (Head of Infectiology), Associate Professor Andrew H. Woods (Head of Neuropsychiatry), H. J. Smyly, Charles W. Young, John H. Korns and William G. Lennox, the latter four of whom were all Associates.

Dr Sia went to New York to collaborate with Oswald Avery from 1929–1930. After his return to PUMC, he was in charge of infectiology, and the attending physician of the infectious disease isolation ward until he left PUMC in 1939. His detailed comments on each patient in the ward were recorded in a large number of medical records. From November 1927 to August 1928, Dr. Sia also served as the head of the Medical Records Committee of PUMC. Therefore, Dr. Sia had undertaken a lot of clinical work in Internal Medicine, especially related to infectious diseases. By the time he left PUMC in 1939, Dr. Sia was the only physician in the entire Internal Medicine department who had worked for two decades. He had made remarkable contributions to the development of Internal Medicine at PUMC.

Dr. Sia devoted himself to the study of infectious diseases at the PUMC. He published two papers on kala-azar disease in 1921 (Sia, 1921; Sia and Wu, 1921), one of which was co-authored with Biochemist Hsien Wu (吴宪). In this collaborating work published in the *Chin. Med. J.*, he described a method for the detection of kala-azar, later known as "Sia's experiment" . Plasma was dropped into a test tube filled with demineralized water, and precipitation or flocculation indicated a positive reaction. This euglobulin test was considered to be the earliest screening method for primary macroglobulinemia.

After discussion with Roberson, Sia concluded that the most common types of pneumococci in China were type Ⅰ and Ⅳ. From 1923 to 1927, Drs Sia and Roberson designed a series of experiments for a study titled "inhibition of the growth of *Pneumococcus*" to systematically explore the scientific questions in this field. In this series of studies, they discovered the protective effect of gelatin, which extended the survival time of pneumococci in suspension at room temperature from a few hours to 6~7 days (Robertson et al., 1924). They also showed a strong ability to invent and create. To confirm the growth inhibition and bactericidal effect of a normal serum-leukocyte mixture on pneumococcus, they developed a new technique that most closely mimics the in vivo conditions in animals (Robertson and Sia, 1924a), which also included a method for keeping the serum-leukocyte mixture in constant motion. By exploring the effects of specific anti-pneumococcal sera and pneumococcal soluble substances on the growth of pneumococci in normal serum-leukocyte mixtures, Roberson and Sia found that the inhibitory and bactericidal effects of immune sera and leukocytes on pneumococci were quantitative in nature (Robertson and Sia, 1924b), whereas pneumococcal soluble substances can promote the growth of pneumococci to a certain extent (Sia, 1926), and it is highly specific to the type of pneumococcus. They also further explored the mechanism of the anti-pneumococcal effects of serum-leukocyte mixtures, and found that the difference in immunity of some mammals to pneumococcal infection was mainly due to the different concentrations of anti-pneumococcus opsonin in the blood (Robertson and Sia, 1927). Based on this series of experiments, they established several new experimental methods, examined problems related to the growth inhibition of pneumococci, and promoted the progress of pneumonia research.

After this series of studies, Dr. Sia continued to investigate bacterial pneumonia. Based on animal experiments, the relationship between acquired humoral immunity and the recovery mechanism of human lobar pneumonia was further studied (Sia et al., 1928). During his stay in New York, Dr. Sia and two colleagues developed a new method for measuring specific precipitable proteins in anti-pneumococcal serum to demonstrate a quantitative relationship between specific precipitable proteins in the serum and immune protection (Heidelberger et al., 1930). This fast, simple, and low-cost method was recommended to replace the traditional mouse protection test.

Between 1929 and 1930, Martin H. Dawson and Richard Ho-Ping Sia experimented with in vitro transformation of pneumococci (Dawson and Sia, 1931; Sia and Dawson, 1931), which is considered the most important contribution in Dr. Sia's academic life. The relationship between this research and pervious work at PUMC, as Rao (2014) pointed out, included at least the choice of animal serum (Dawson and Sia, 1931), because Sia found that the serum of various animals generally had anti-R properties (Sia, 1926). During their studies using horse serum, Sia and Dawson discovered that the converted substances were not polysaccharides (Sia and Dawson, 1931). This experiment was one step in the chain of studies that showed that deoxyribonucleic acid is the transforming substance, and that DNA is the carrier of the genetic code. The American Biochemist, A. B. Hastings, described this work as "the classic but insufficiently recognized studies on which subsequent research that earned three Nobel Prizes was based" (Bowers, 2014). Avery's loss of the Nobel Prize was regrettable, but his contributions were world-renowned, whereby the research done by Richard Ho-Ping Sia and Martin H. Dawson, which constituted the foundation of Avery's work, should not be neglected.

Dr. Sia worked at PUMC Hospital for 20 years, which was unique at the PUMC Department of Medicine prior to 1942. Many facts indicate that Dr. Sia was recognized by PUMC for his ability and achievements. In 1927, he was appointed as the head of the Medical Records Committee. In 1931, just one year after becoming assistant professor, he was promoted to associate professorship. Furthermore, in 1939, when Dr. Sia decided to leave, PUMC was still preparing to renew his employment for another 4 years. His main research work at PUMC focused on pneumococci, and he continued research in this field during his time as a visiting scholar at the Rockefeller Institute of Medicine, where he devoted himself to in vitro transformation of pneumococci. Dr. Sia published 9 papers in the *Journal of Experimental Medicine* (*J Exp Med*), which was a great academic achievement at the time (Rao, 2014). The academic platforms of PUMC and the Rockefeller Institute of Medicine were world-class, which laid a good foundation for his scientific research. Additionally, *J Exp Med* was founded by the Johns Hopkins University School of Medicine, with William Welch as the first editor-in-chief. When the sponsor of the journal was moved to the Rockefeller Institute of Medicine, the director of this academic establishment, Simon Flexner, served as the second editor-in-chief until 1935. William Welch and Simon Flexner were the most famous medical educators in North America at the time, and both were important medical scientists in the process of founding PUMC. This might explain why PUMC scholars like Dr. Sia preferred submitting to *J Exp Med*. As such, Dr. Sia is an important figure in the history of the Department of Medicine at PUMC. Further in-depth study of relevant data is of great significance to the history of infectiology in China.

Acknowledgments

We thank David Chen (grandson of Tsun-Chi Shen, US), Yuhong Jiang from Peking Union Medical College (Beijing, China), Chang Liu from University of Texas at Austin (Austin, US), and Linqian Weng from Katholieke Universiteit Leuven (Leuven, Belgium) for their assistance with shared opinions, data collection or language polishing.

Figures

Fig.1 Richard Ho-Ping Sia (1895—1970)

Fig.2 Dr. Sia in the infectious disease laboratory at PUMC

References

Bowers JZ ed. Jiang YH, Wu D, Zhang L (translation). 2014. Western Medicine in a Chinese Palace: Peking Union Medical College, 1917—1951. Beijing: Peking Union Medical College Press. (鲍尔斯. 蒋育红，吴东，张麟，译. 2014. 中国宫殿里的西方医学. 北京：中国协和医科大学出版社.)

Dawson MH, Sia RHP (1931) In vitro transformation of pneumococcal types: Ⅰ. A technique for inducing transformation of pneumococcal types in vitro. J Exp Med 54(5):681–699.

Heidelberger M, Sia RHP, Kendall FE (1930) Specific precipitation and mouse protection in type Ⅰ antipneumococcus sera. J Exp Med 52(4):477–483.

Rao Y (2014) The Peking Union Medical College and the advanced scientific fields in the world 80 years ago. (饶毅. 2014. 80年前的协和医学院与世界科学前沿.)

Robertson OH, Sia RHP (1924a) Studies on pneumococcus growth inhibition: Ⅱ. A method for demonstrating the growth-inhibitory and bactericidal action of normal serum-leucocyte mixtures. J Exp Med 39(2):219–244.

Robertson OH, Sia RHP (1924b) Studies on pneumococcus growth inhibition: Ⅲ. The influence of specific antipneumococcus serum on the growth-inhibitory and bactericidal action of normal serum-leucocyte mixtures. J Exp Med 40(4):467–485.

Robertson OH, Sia RHP (1927) Studies on pneumococcus growth inhibition: Ⅶ. The relation of opsonins to natural resistance against pneumococcus infection. J Exp Med 46(2):239–262.

Robertson OH, Sia RHP, Woo ST (1924) Studies on pneumococcus growth inhibition: Ⅰ. The protective action of gelatin for pneumococci in suspension. J Exp Med 39(2):199–218.

Sia CC (1971) Richard Ho-Ping Sia, M.D. 1895–1970. Hawaii Med J 30(1):59.

Sia RHP (1921) Rays "hemolytic" test in kala-azar. Chin Med J 35 (5): 397–399.

Sia RHP (1926) Studies on pneumococcus growth inhibition: Ⅵ. The specific effect of pneumococcus soluble substance on the growth of pneumococci in normal serum-leucocyte mixtures. J Exp Med 43(5): 633–645.

Sia RHP, Dawson MH (1931) In vitro transformation of pneumococcal types: Ⅱ. The nature of the factor responsible for the transformation of pneumococcal types. J Exp Med 54(5):701–710.

Sia RHP, Robertson OH, Woo ST (1928) A study of the mechanism of recovery from lobar pneumonia. J Exp Med 48(4):513–532.

Sia RHP, Wu H (1921) Serum globulin in kala-azar. Chin Med J 35(6): 527–532.

Zhang YR, Li ZP (2016)R.H.P. Sia: Being far away from the view of history. Med Phil 37(2):83–86. (张艳荣,李志平. 2016. 远离历史视野的谢和平(R.H.P.Sia). 医学与哲学, 37(2):83–86.)

67. 苏德隆：中国血防先驱

图1　苏德隆教授（1906—1985）

苏德隆（Teh-Long Su）是我国著名公共卫生学家、医学教育家和医学哲学思想家，我国流行病学奠基人之一（图1）。新中国首批一级教授和首批博士研究生导师，原上海第一医学院副院长，全国血吸虫病研究委员会副主任委员。他是国际流行病学学会荣誉会员，是44位获此荣誉的世界知名流行病学家中唯一的中国科学家。苏德隆是首位被世界卫生组织（WHO）提名莱昂·伯纳德基金奖的中国专家，该奖项享有盛誉，旨在表彰在社会医学领域作出杰出贡献的优秀科学家。苏德隆是世界上首位全面阐明钉螺分布规律和中国血吸虫病生态学的学者，并将其研究应用于我国消灭血吸虫病。他创立了生态学灭螺方法，筛选出多种灭螺药和预防血吸虫感染的方法，他发明的造价低廉的“防蚴衣”“防蚴笔”获全国医药卫生科学大会奖。他参与编写了我国第一部《流行病学》（全国统编教材），并三次担任主编，奠定了我国流行病学基础，创立了中华医学会流行病学学会并首任主任委员。苏德隆教授通过预防和控制血吸虫病、肝癌、副霍乱等重大流行病，为我国和世界的公共卫生事业作出了不可磨灭的巨大贡献，其中尤以消灭肆虐华夏大地2000多年的“瘟神”——血吸虫病的贡献最为突出，在公共卫生史上书写了浓墨重彩的一笔。

初识“瘟神”　情定公医

1906年，苏德隆出生于六朝古都南京的一个普通家庭，家境清贫，父母以小手工艺为生。苏德隆是家中存活的五个孩子中的长子，他从小聪慧过人，小学三年级即开始自学英文，品学兼优的他在教会中学毕业。而后因家贫失学2年，但他仍想尽办法自

作者：阙之玫，苏竹君
苏德隆家属，安亭路，上海
邮箱：mariana@126.com（阙之玫）

学。虽然长期接受的都是英文教育，但苏德隆的国文功底极好，他是一个内心秉持着“修齐治平”人生抱负的中国人。

1927年，苏德隆考入国立中央大学医预科，后转入国立上海医学院。大学期间，他加入中华自然科学社并参与创办了《科学世界》杂志，发表了众多科普文章，热情为民众普及医药知识，开展公共教育。1931年夏，千余民众因在上海郊区昆山县青阳港游泳而患急性血吸虫病，当时医学界对这种疾病所知甚少，难以诊断。苏德隆在学校做了一次读书报告，专门论述血吸虫病的危害。他认为这是种值得研究的疾病，这是他第一次关注这种对人民危害极大的传染病。

1935年，苏德隆以第一名的成绩毕业，取得医学博士学位，并荣获金质奖章及优等荣誉证书，留校任助教。他在上海颛桥乡开办了一个农村卫生所，开辟了国立上海医学院新的卫生教学基地并做流行病学调查，这段经历让他有了医治血吸虫病患者的丰富经验（图2）。

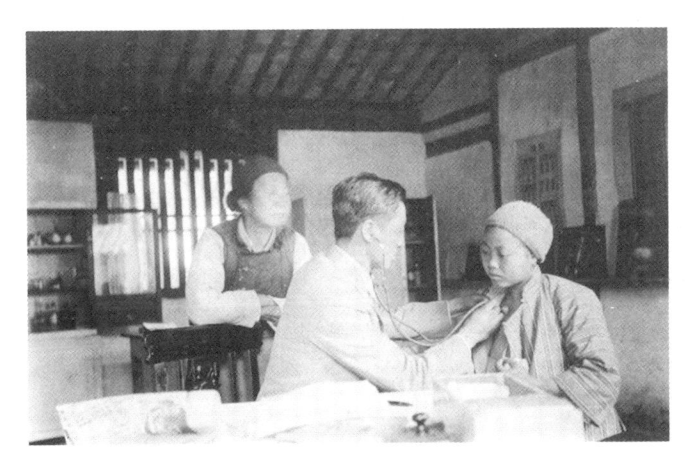

图2　1935年，苏德隆（中）在昆山县为农家义诊

去国怀乡　内外求索

1944年10月，苏德隆获美国洛克菲勒基金会资助，赴美国约翰斯·霍普金斯大学公共卫生学院攻读硕士学位。他的导师是大名鼎鼎的K. F. Maxcy教授和L. J. Reed博士。苏德隆在约翰斯·霍普金斯大学的学习经历帮助他形成流行病学的基础和方法论，建立了流行病学的数学建模概念，为今后的流行病学研究和公共卫生教育奠定了坚实的基础。1945年，苏德隆教授从约翰斯·霍普金斯大学毕业，获得公共卫生硕士学位。

1945年秋，苏德隆赴牛津大学继续博士学业。青霉素发明人之一、诺贝尔奖获得者H. W. Florey院士是他的导师。在牛津期间，苏德隆学习了化学、微生物学和病理学的实验技能，获得了坚实的实验医学基础，他还师从社会医学研究所所长J. A. Ryle教授学习医学统计学。1947年，苏德隆从污水中发现并分离出一种强力抗生素，命名为微球菌素（micrococcin）。这一发现属于世界首次，路透社将其作为科学新闻向全世界做了报道。同年，苏德隆获博士学位（图3），并同时入选英国皇家统计学会和微生物学会会员，这在当时的中国人中是极为罕见的。1947年9月，苏德隆代表中华医学会出席了在巴黎举行的世界医学会第一届会议。1948年6月24日，苏德隆出席了在日内瓦召开的联合国第一届世界卫生大会。

1948年底，苏德隆毅然谢绝了导师Florey教授的热情挽留，带着用省下来的生活费购置的科研仪器和试剂，回到灾难深重的祖国，受聘为母校国立上海医学院流行病学教授，担任公共卫生科主任并兼任微生物学科主任。他重建了国立上海医学院卫生科，在他主持的近40年间，将一个只有几个人的小科室发展壮大，成为具有世界影响力的著名公共卫生学院，创建了具有中国特色的公共卫生学科群，为国家培养了大量的医学、医学教育和公共卫生的专家、领军人物以及卫生行政官员，为公共卫生和医学的研究和发展作出了杰出的贡献。

图3 1947年，苏德隆获英国牛津大学博士学位

"一定要消灭血吸虫病"

血吸虫病在中国已有2000多年的流行历史，被视为肆虐华夏大地的"瘟神"。我国南方13个省（区、市）是我国重要的农业省份和经济命脉地区，在新中国成立初期，约有1亿人受危害，血吸虫病人达1300多万。整个20世纪40年代，我国学者发表关于血吸虫病的著作寥若晨星。苏德隆遍历第二次世界大战以后的西方文献，于1950年在《中华医学杂志》发表了长文综述《近年日本血吸虫研究之进展》，该论文对此后他个人及我国的血吸虫病研究起了重要的作用。国家卫生事业的需要以及党和人民的期望，使苏德隆毅然改变学术方向，踏上了消灭血吸虫病的漫漫征程。

血吸虫的唯一中间宿主是钉螺，消灭钉螺是消灭血吸虫病的重要措施，但当时尚无行之有效的灭螺方法和药物，前人的经验大多停留在假想和实验室阶段。苏德隆决心改变这种局面，他把广阔的农村看成一个"天然实验室"，他和他的学生在"试验田""教学基地"进行各种观察和实验，探寻既经济又有效的灭螺方法。根据苏德隆的建议，各地开展了大规模的卫生教育宣传运动，加强粪便管理，积极开展储粪灭虫卵运动，农村尽可能改用井水，将"阻断农民在生产、生活中接触疫水"列为预防血吸虫病的重点工作之一。

1953年，苏德隆的研究开启了我国钉螺生态学研究的发端。1954年，他用实验证明了人尿中的尿素分解产生的氨可以杀死血吸虫卵。这一不花钱的方法推动了江南水网地区的人粪无害化处理，成为阻断血吸虫病传播的一项有效措施。苏德隆通过实验室研究和实地考察，研究了钉螺对光照和温度变化的反应，判定钉螺最适应13℃的温度以及晨曦和黄昏时强度为3600Lux的光线。他还研究了钉螺在不同季节的变化规律，从而为钉螺在我国分布的地域特点提供了科学依据。当时人们只对河道水线以上进行灭螺，常年灭而不绝，花费了大量人力物力却收效甚微。苏德隆坚持用实验来证实事物的正确性，他推翻了教科书上钉螺有冬眠习性的说法，经过一年的连续实验，发现了水线上下钉螺分布的规律，这些发现对掌握钉螺生存条件和从根本上消灭钉螺具有重要意义。根据他的建议，江南水网地区彻底改变了河岸灭螺的策略。

与此同时，苏德隆开始研究并推广化学灭螺方法，从实验室的药物筛选到中型实验，再到现场实验，他都亲自参加，发现和创造了不同的灭螺途径，直至他意外去世前从未间断。他首创了“纸钱法”用于筛选灭螺药，这种多用途、高效且价廉简便的方法，博得了国内外同行的赞誉。在20多年的时间里，他筛选了多种灭螺药物，有效的如砷酸钙、亚砷酸钙、茶籽饼等，后来又先后找到杀螺有效、对鱼无毒的乙二胺和尿素。尿素用于鱼塘灭螺不仅对鱼安全无害，而且可以肥鱼。

从1950年起，苏德隆教授每年都到全国十大血吸虫病疫区之一的青浦县进行调研。1958年，青浦县划入上海，当时青浦有40%的人口感染了血吸虫病，病人和钉螺分布区域和整个日本的血吸虫病流行区差不多。苏教授提议在青浦县建立试验田作为研究重点，开展血吸虫病防治。他带领20多位青年教师和技术员，与当地血防干部一起，一个乡、一个乡地查灭钉螺，他的化学灭螺法取得了良好的成效（图4）。1963年，苏德隆所著的论文《钉螺的负二项分布》得以发表，这是世界上第一个全面阐述钉螺分布规律的研究成果。

图4 1965年，苏德隆（中）在青浦县朱家角血防现场指导灭螺

20世纪70年代早期，尽管身处逆境，苏德隆仍然坚持做实验，完成了氯硝柳胺防御血吸虫尾蚴的研究，并以此完成了造价低廉的“防蚴衣”“防蚴笔”两项发明。“防蚴衣”经泥浆摩擦、水冲数万次，露天悬挂数月，室内保存20余年，仍能防御血吸虫尾蚴侵入。“防蚴笔”仅几分钱一支，但效果显著，涂擦皮肤后经泥浆摩擦8小时仍有特效。该溶液可用作涂肤防护剂，还有预防稻田皮炎的作用，涂在皮肤上无不适感，也不会污染衣服，这项发明深受农民喜爱。1975年，他的发明通过鉴定被推广应用，并被带到非洲，防护效果受到援非水利建设者的欢迎。20世纪70年代后期，世界卫生组织、美国海军部及英国剑桥大学都纷纷来函索取样品，了解配方及转让技术。这两项发明在1978年荣获全国医药卫生科学大会奖及上海市重大科研成果奖。

1981年8月，苏德隆教授在爱丁堡举行的国际流行病学学会第九次科学大会上作了“中国血吸虫病的生态学研究”报告，我国防治结合、预防为主的血吸虫病防治科研成就，备受国际学术界重视和赞誉。苏德隆是世界上首位全面阐明中国血吸虫病生态学的学者，他的著作《中国血吸虫病生态学》为我国的血吸虫病防治事业指明了方向。他辩证地论述了建设三峡工程是否会引起血吸虫病的问题，得到国内外高度评价。1982年1月，苏德隆获世界卫生组织莱昂·伯纳德基金奖提名，他是首位被该奖提名的中国人。

1985年11月，上海市宣布消灭血吸虫病。为表彰苏德隆在血防战线上的巨大贡献，追授他“血防战线先进工作者”称号并记大功。2018年11月9日，中华预防医学会追授苏德隆教授“全国血防先驱”称号。

在中国血防纪念馆里，特设了“一门四代血防建功”展板，记录了以苏德隆教授为首的上海医学院师生70多年来为我国血防事业所作的突出贡献和他们的独特地位（图5）。在以苏德隆为代表的代代血防科研人员的努力下，我国的血吸虫病防治工作取得了巨大的成绩，成为全球“优等生”“好榜样”，众多科研成果在全球分享，我国的产品和标准应用到其他血吸虫病流行国家，为全球血吸虫病控制作出了重大贡献。

图5　中国血防纪念馆“一门四代 血防建功”展板

“祖国和人民的需要，就是我的研究方向。”苏德隆教授探索、追寻真理的故事至今仍被广泛传颂，他宽容以待的伟大人格让后人铭记。苏德隆教授倡导的“预防为主、改造环境消灭疾病”的理念深入人心，他的科学家精神也激励着一代代知识分子报国图强、奋发踔厉，为保护人类健康和建设中国特色世界一流医学院和世界顶尖大学接续努力！

67. Professor Delong Su: A pioneer in schistosomiasis eradication in China

Delong Su (Teh-Long Su, 1906—1985) was an internationally renowned public health expert, a medical educator and philosopher, and one of the major founders of classic epidemiology in China. He was among the first group of the national distinguished professors and doctoral supervisors appointed since the founding of the People's Republic of China. Dr. Delong Su served as vice-president of the Shanghai First Medical College and vice-chairman of the National Schistosomiasis Research Council. He was elected as an honorary member by the International Society of Epidemiology, the only Chinese epidemiologist among 44 world-renowned epidemiologists. He was the first Chinese expert nominated for Léon Bernard Foundation Prize, a very prestigious award for these with accomplished outstanding service in the field of social medicine by the World Health Organization (WHO) because he was the world's pioneer who fully elucidated the distribution of snails and ecology of schistosomiasis and applied his research in the elimination of schistosomiasis in China. Professor Su created an ecological method to eradicate snails and screened a series of measures on eliminating snails and preventing schistosomiasis infection. His economical innovation named "anti-cercaria coat" and "anti-cercaria pen" were awarded the National Science on Medical Science Prize. Dr. Delong Su edited the first national textbook *Epidemiology* (three times as chief editor), which laid the foundation for epidemiology research in China, and founded the Society of Epidemiology, Chinese Medical Association, serving as the first President of the Society of Epidemiology. Professor Delong Su made great contributions to public health in China and the world, such as preventing and controlling a variety of major epidemics, including schistosomiasis, a plague ravaging China for more than two thousand years, liver cancer, and cholera. Especially, his outstanding achievements in schistosomiasis control have made a significant and brilliant chapter in the history of public health in China.

Zhimei Que[1], Zhujun Su[1], Yuanyuan Meng[2]

1 Family of Prof. Delong Su, Anting Road, Shanghai

2 Beijing Institutes of Life Science, Chinese Academy of Sciences, Beijing 100101, China

Correspondence: mariana@126.com (Z.-M. Que)

Encountering the "pestilence" , falling in love with the public health

In 1906, Delong Su was born in a civilian family in Nanjing, the ancient capital of the six dynasties. His parents lived on small handicrafts and his family was very poor. Delong Su was the eldest of the five surviving children in the family. He was intelligent from an early age, and began to learn English in the third grade of primary school by himself. With excellent academic performance, Delong Su graduated from a Christian middle school with a scholarship. For the next two years, he dropped out of school because of poverty, but still tried his best to study by himself. Although Delong Su had received an English education for a long time, he had an excellent foundation in the Chinese language, and his heart was to become a leader who adhered to the core of four Chinese words "Xiu Qi Zhi Ping" , meaning self-cultivation, family harmony, national governance, and world peace.

In 1927, Delong Su was admitted to the pre-medical program of the National Central University in Nanjing and then admitted to the National Shanghai Medical College. When he was in college, he joined the Natural Science Society of China and participated in the founding of *Science World* magazine, published popular science articles, and enthusiastically popularized medical knowledge for public education. In the summer of 1931, over a thousand people suffered from acute schistosomiasis after swimming in Qingyang Port, Kunshan county, a suburb of Shanghai. At that time, the medical community knew little about the disease and it was difficult to diagnose. Delong Su presented a report at the school's journal club, elaborating the damage of schistosomiasis, and considering it as a disease worthwhile to study. This was the first time he paid attention to an infectious disease, which was harmful to human beings.

In 1935, Delong Su received his MD. degree and successfully graduated as a top student among his cohort. He received a Gold Medal and CUM LAUDE certificate because of his academic excellence. Dr. Su stayed in the same school as an Assistant Professor. He established a "Rural Health Center" in Zhuanqiao town, Shanghai where he started a new health education center of the National Shanghai Medical College, and conducted epidemiological investigations. This process gave him extensive experience in preventing and treating patients with schistosomiasis.

Studying overseas, exploring new knowledge

In October 1944, Dr. Delong Su was awarded a scholarship from the Rockefeller Foundation of the United States and continued his study as a Master's student at the School of Public Health, Johns Hopkins University in the United States. He was mentored by well-known Professors, Drs. Kenneth F. Maxcy and Lowell Jacob Reed. His study at Johns Hopkins University helped him learn epidemiological methods and establish the concept of mathematical modeling in epidemiology, which made a solid foundation for his further epidemiologic study and public health education. In 1945, Dr. Delong Su graduated from Johns Hopkins University with a

Master's degree in Public Health (MPH).

In the fall of 1945, Dr. Delong Su went to Oxford University to further pursue his doctoral studies. His doctoral advisor, Professor Howard Walter Florey was one of the inventors of penicillin and a Nobel Prize laureate. During his stay at Oxford University, Dr. Delong Su mastered basic experimental skills in chemistry, microbiology, and pathology, and gained a solid basis on experimental medicine. He also studied medical statistics from Professor John Alfred Ryle, the chair of the Institute of Social Medicine at the University of Oxford. In 1947, Dr. Delong Su first-ever discovered and isolated a powerful antibiotic from sewage in the world, and named it Micrococcin. Reuters reported this innovative discovery to the world as a piece of scientific news. In the same year, Dr. Delong Su received a DPhil degree and was selected as member of the Royal Statistical Society and Microbiological Society, which were extremely rare among Chinese at that time. Dr. Su participated in the First General Assembly of the World Medical Association in Paris in September 1947 representing the Chinese Medical Association. On June 24, 1948, he attended the First World Health Assembly of the United Nations convened in Geneva, Switzerland.

At the end of 1948, Dr. Delong Su declined the enthusiastic retention of his mentor, Professor Florey, and returned to China. He brought back scientific research instruments and reagents purchased by his personal savings. He was appointed as a professor of epidemiology at National Shanghai Medical College, his alma mater, and as the chair of the departments of public health and microbiology. For almost 40 years, Dr. Su rebuilt the department of public health of the National Shanghai Medical College from a very small department with only a few people into a well-known and influential school of public health in the world, created a group of public health disciplines with Chinese characteristics, and trained a large number of experts and leaders in medicine, medical education and public health, as well as health administrators for China. Dr. Delong Su had made outstanding contributions to the research and education for the development of public health and medicine.

"Schistosomiasis must be eliminated"

Schistosomiasis has been prevalent in China for at least 2000 years and was known as the "god of plague" that ravaged thirteen provinces (autonomous regions and cities) in southern China including important agricultural provinces and economic areas in China. In the early days after the People's Republic of China was established in 1949, approximately 100 million people were affected, and more than 13 million people suffered from schistosomiasis. Throughout the 1940s, almost no articles about schistosomiasis were published by Chinese scholars. Dr. Delong Su reviewed the western literature after World War Ⅱ and published a long review article, entitled "The progress of *Schistosoma japonicum* research in recent years" in *Chinese Medical Journal* in 1950. This article and the need to improve Chinese people's health played an important role in the process of changing his academic research direction on Chinese schistosomiasis research, prevention and control, and embarking on a long journey to eliminate schistosomiasis.

Snails were recognized as the "only intermediate host" of *Schistosoma*. Eradicating snails was a key preventive measure to control schistosomiasis at that time, while there were no proven measures and drugs to eliminate snails, because most of these measures stayed in the hypothetical and laboratory experimental stage. Dr. Delong Su was determined to change the situation by using the vast countryside fields as his nature laboratory. He worked with his students, conducted a variety of experiments, made observations in the experimental and education fields, and explored more economical and effective measures to eliminate snails. According to the suggestion of Dr. Su, people in most of the area in the village carried out extensive health education activities and campaigns of eliminating eggs of *Schistosoma* by storage-manure, reinforcing the management of feces, changing the consumption of river water to well water, and shutting off potential routes of infection with polluted water among villagers. These preventive measures had been considered the most important tasks in the prevention of schistosomiasis at that time.

In 1953, Dr. Delong Su initiated snail ecology research in China. In 1954, he proved in his experiments that ammonia produced from the decomposition of urinary urea in human urine can kill the eggs of *Schistosoma*. This cost-free method promoted the practice of feces bio-safety disposal, and became an effective way to block the transmission of schistosomiasis in the water-network region in the south of the Yangtze River. Dr. Delong Su observed the responses of snails in relation to changes of lights and temperatures through laboratory and on-the-spot fields. It was determined that the temperature of 13℃ and the light intensity of 3600 Lux at dawn and dusk were the most suitable for snails, making it easier to identify them. He also studied the changing rule of snails in different seasons, providing a scientific basis for regional characteristics of snail distribution areas in China. In the earlier days, people put a lot of manpower and resources into eliminating snails above the river waterline, but gained fruitlessly. Dr. Su insisted on an evidence-based prevention strategy by translating research results into preventive practice. He overturned the opinion in textbooks that snails had hibernating habits. After persistent research for one year, Dr. Su found the distribution regularity of snails above and under the river waterline, which was significant for mastering the living condition of snails and even eliminating snails. According to his suggestion, the water-network region in the south of the Yangtze River had completely changed the strategy for snail control and elimination.

At the same time, Dr. Delong Su started to test chemical methods for snail elimination. From chemical screening to laboratory experiments, and eventually on-site experiments, Dr. Su participated in all these steps in person. Dr. Delong Su would have never stopped his research experiments if he were still alive. He pioneered the "paper money method" to screen molluscicides, and this multi-purpose, efficient and inexpensive method won the praise from colleagues worldwide. In more than 20 years, he screened a variety of molluscicidal drugs, such as calcium arsenate, calcium arsenite, tea seed cake, and so on. Later, he successively discovered that ethylenediamine and urea were effective to snails and non-toxic to fish. More importantly, urea used for snail control in fish ponds was not only safe and harmless to fish, but also can fatten fish.

Since 1950, Delong Su had conducted research every year in Qingpu county, one of the top ten schistosomiasis endemic areas in China. In 1958, when Qingpu county was incorporated into Shanghai, 40 percent of the population were infected with schistosomiasis. The distribution area of patients and snails was similar to that of the endemic areas in Japan. Professor Delong Su proposed to establish an experimental field with a research focus in Qingpu county, aiming to carry out control against schistosomiasis. Leading more than twenty young teachers and technicians together with the local people and farmers, snails were eliminated in villages and towns one by one. In this process, he employed chemical methods that achieved great effects. In 1963, Dr. Delong Su's paper "Negative binomial distribution of snails" was published, which was the first research achievement in the world to comprehensively explain the distribution patterns of snails.

In the early 1970s, although facing adversity, Dr. Su still continued his research and completed a project of niclosamide against cercariae of schistosomes. According to that, Su invented the "anti-cercaria coat" and "anti-cercaria pen" . The efficacy of "anti-cercaria coat" remained unchanged after tens of thousands of times after mud grinding and water flushing. "Anti-cercaria coat" can still prevent the invasion of schistosome cercariae after it was hung in the open air for several months or preserved indoors for more than 20 years. The "anti-cercaria pen" owned a remarkable effect, although it only cost a few RMB cents. After rubbing the skin with mud for 8 hours, the "anti-cercaria pen" still had a protective effect. The solution can be used as a skin protective agent and can also prevent rice field dermatitis. It had no discomfort on the skin and did not pollute clothes. This invention was very popular with farmers. In 1975, his inventions passed technical appraisal and were promoted and brought to Africa. In the late 1970s, the World Health Organization, the Admiralty of the United States, and the University of Cambridge in the United Kingdom all sent letters to ask for samples, starving for the understanding of formulations and transferring the related technology. The two inventions were awarded the National Conference on Medical Science Prize and Shanghai Major Award for Scientific Research in 1978.

In August 1981, Professor Delong Su delivered a speech, entitled "Ecological study of schistosomiasis in China" , at the 9th Scientific Conference of the International Society of Epidemiology in Edinburgh. The scientific achievements of schistosomiasis control in China, which were based on prevention and control, had been praised by international academic experts. Dr. Delong Su was the first scholar who fully elucidated the ecology of schistosomiasis in China, and his publication, *Ecology of Schistosomiasis in China*, indicated the directions of schistosomiasis control. He dialectically analyzed the potential impact on schistosomiasis with the construction of the Three Gorges of the Yangtze River, which was enjoyed at a high degree of evaluation at home and abroad. In January 1982, Dr. Delong Su was nominated for Léon Bernard

Foundation Prize by WHO, who was the first Chinese to be nominated for this award.

In November 1985, Shanghai announced the elimination of schistosomiasis completely. In recognition of Dr. Delong Su's great contributions to schistosomiasis prevention and control, he was posthumously awarded the distinguished pioneer and researcher on schistosomiasis control and scored great meritorious service. On November 9, 2018, the Chinese Preventive Medicine Association posthumously awarded Professor Delong Su the title of "National Pioneer of Schistosomiasis Control" .

In China Schistosomiasis Prevention Museum, a special exhibition board, entitled "Achievements by four generations, Merits of schistosomiasis prevent" , had been set up to record the outstanding contributions and unique status made by the professors and students of Shanghai Medical College headed by Professor Delong Su for schistosomiasis prevention in China for over 70 years. Under the efforts made by generations of researchers in schistosomiasis control led by Dr. Delong Su, the schistosomiasis control program in China had gained great achievements and made China become a great example in the world. Most of the scientific achievements in schistosomiasis control in China had been shared in the world. Chinese prevention products and standards have been applied to other schistosomiasis endemic countries and made a great contribution to the global control of schistosomiasis.

"The needs of my country were my research goal" , Dr. Delong Su said. His lifetime exploring and pursuing truth in his research and practice was still widely recognized by the scientific community. His great personality of tolerance was remembered by his colleagues and future generations. His idea of giving a high priority to prevention and transforming the environment to eliminate diseases was deeply rooted in the hearts of public health researchers. His scientific spirit also inspired generations of intellectuals to work hard for people's health and to build world-class medical colleges with Chinese characteristics and the world's top universities!

Figures

Fig.1 Professor Delong Su (Teh-Long Su) (1906—1985)

Fig.2 Delong Su (middle) offered a free clinic for farmers of Kunshan in 1935

Fig.3 Delong Su received his DPhil. from the Oxford University in 1947

Fig.4 Delong Su (middle) instructed schistosomiasis control and snail elimination in Qingpu county in 1965

Fig.5 Exhibition board in China Schistosomiasis Prevention Museum

References

Su ZJ, Que ZM (2017) De Long Wang Zun, Qi Ji Ren Sheng. In: Yu SZ. Shang Yi Zhi Wei Bing—How the Scholars of Shanghai Medical College Contribute to the Preventive Medicine and Chinese Healthcare.

Shanghai: Fudan University Press. (苏竹君, 阙之玫. 2017. 德隆望尊的骐骥人生. 见: 俞顺章. 上医治未病——上医人送瘟神、降疫魔、讲卫生的历程(第七章). 上海: 复旦大学出版社.)

Su ZJ, Que ZM (2017) Ren Jie Qi Ji—A brief biography of Delong Su, the founder of preventive medicine in China. In: Shanghai Talented Scholars of Jiusan Society. Vol. 1. Beijing: Xueyuan Press. (苏竹君, 阙之玫. 2017. 人杰骐骥——我国预防医学奠基人苏德隆传略. 见: 九三学社上海先贤(第一辑). 北京: 学苑出版社.)

Xu ZY, Su ZJ (1995) Professor Delong Su and his academic contributions. In: Su DL. Selected Papers of Professor Delong Su. Tianjin: Tianjin Science and Technology Press. (徐志一, 苏竹君. 1995. 苏德隆教授及其建立的学派. 见: 苏德隆. 苏德隆教授论文选集. 天津: 天津科学技术出版社.)

68. 石赤不夺——张香桐对树突功能的研究给我们的启迪

在1900年，孟德尔去世后的第16年，他发现的遗传学规律才被大家普遍认同并接受；同样地，在1983年，经过了30多年时光的洗涤，Barbara McClintock的"跳动基因"学说才终于在科学圣殿中找到了自己的位置，个人独享了当年的诺贝尔生理学或医学奖。生命科学研究中，这些取得跨时代成就的杰出科学家给我们留下了深刻印象，他们的故事像神话一样广为传颂。而历史总在不断重演，鲜为人知的是，中国科学院院士、国际著名神经生理学家张香桐（图1）也有着类似的经历。

图1　张香桐教授（1907—2007）

张香桐是我国神经科学的奠基人，中国科学院上海脑研究所的创始人，也是国际上公认的树突生理功能研究的先驱者之一。他还是比利时皇家医学科学院外国名誉院士，曾任国际脑研究组织中央理事会理事以及世界卫生组织神经科学专家顾问委员会委员等。

1992年冬，张香桐突然接到一份来信，信中说国际神经网络学会即将授予他终身成就奖，希望他出席授奖仪式。张香桐知道，国际神经网络学会是一个总部设在华盛顿的学术团体，由国际上从事电脑研究的工程师组成。自己既不是工程师，又没有对电子计算机进行过研究。所以他认为该学会一定是闹了个误会。当他看到该学会颁发奖状上的颁奖词时，他才知道，这一切和他40年前从事的神经树突电位研究密切相关。颁奖词中特别指出："他关于大脑皮层神经元树突电位的研究形成了一个划时代的重要标志。这一卓越成就为我们将来发展使用微分方程和连续时间变数的神经网络计算机，而不再使用数字脉冲逻辑电子计算机奠定了基础。"

故事要从20世纪中期说起。当时，国际生理学界对轴突的电兴奋性和传导性有了初步的认识，但还很少有人真正认识到树突的功能和重要意义。1950年至1955年的这

译者：张保元[1]，郝宁[2]

1　中国科学院北京生命科学研究院，北京100101，中国

2　中国科学院生物物理研究所，北京100101，中国

邮箱：zhangby@biols.ac.cn（张保元）；haoning@ibp.ac.cn（郝宁）

段时间，张香桐几乎把主要的精力都倾注到了研究神经树突功能的工作上。在洛克菲勒医学研究所任副研究员期间，他利用电生理手段对大脑皮层锥体神经元顶树突的功能特点进行了研究，并相继发表了9篇这方面的研究论文，在国际生理学界产生了广泛影响。在1956年回国后，他依然以极大的热忱继续从事神经系统的研究工作（图2）。通过六年的艰苦努力，他创建了我国第一个神经细胞学实验室，我国许多杰出的神经生理学家都曾在他的实验室学习、工作，这其中包括吴建屏院士、陈宜张院士等知名学者。他被国际学术界公认为树突生理功能研究的先驱者之一，是“历史上第一个阐述了树突上突触连接的重要性的人”。可他从来也没有想到，自己的研究成果竟然有助于智能计算机的发展！

图2　张香桐教授在实验室

这完全是一次意外的获奖！然而张香桐并不认为这值得炫耀。在他看来，这只是他曲折迂回的脑科学研究路途中的一个小插曲。他没有声张，没有庆贺，甚至鲜有人知道他获奖。同时，这件事引起了张香桐无限的感慨和深思，他说：“基础理论研究的价值及社会意义，往往并不能立即被人们所认识与重视。必须等到社会文化及其他与之相关的分支学科都发展到一定程度时，它的光芒才会散发出来。”

这个看似意外的奖励，其实并不意外！回顾张香桐的一生，我们不难发现，他把毕生的心血都倾注到了脑神经生物学研究上，孜孜不倦，从未放弃。他曾以仙人掌自喻，在《仙人掌》一词中写道：“烈日严寒历尽，漫天风沙无情，行若无事，篱下自峥嵘。”这种在任何环境下都能屹立、成长的精神，使他在科研工作中常常能够以客观实验结果为依据，坚持自己的观点。他提出的树突有电兴奋性和能传导冲动的重要论断，尽管在40年来一直受到争议和反对，但最终被证实是完全正确的。自助者，天亦助之，张香桐身上体现的是一种高尚的科学操守：“石可破也，而不可夺坚；丹可磨也，而不可夺赤。”他的特殊经历也告诉我们：基本理论问题的解决需要坚持长期系统的研究，而不能仅仅追求短期时效。

“人，尤其是做学问的人，应该像乌贼那样，无论走到哪里，都要留下一丝墨迹。”这是张香桐的座右铭。他对于研究工作孜孜不倦地追求和奉献，留给我们的是丰富的学术遗产。当今生命科学的发展中，挑战与机遇并存，创新与责任并重。优秀的科学家必须立足于自己的研究领域，不断探索，矢志不渝，石赤不夺。真正的金子在哪儿都会发光，当系统性的科研成果大量涌现时，基础生命科学研究的光芒才会更加耀眼！

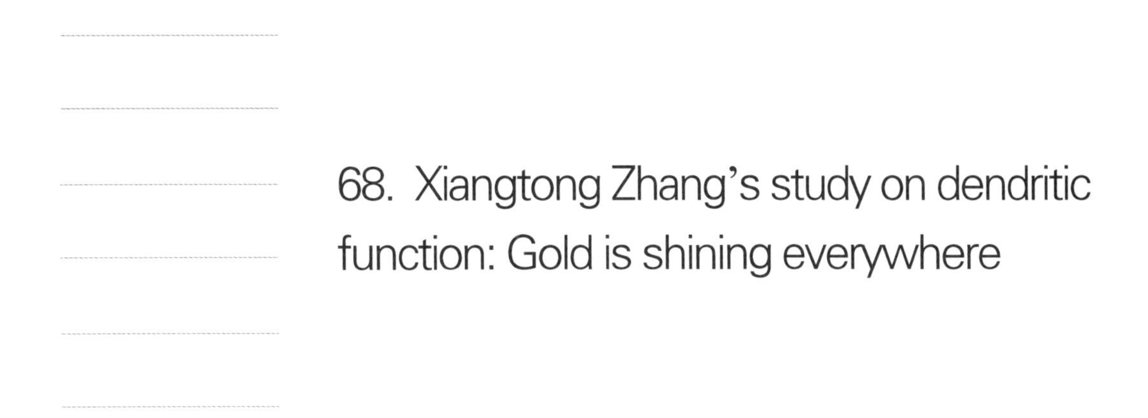

68. Xiangtong Zhang's study on dendritic function: Gold is shining everywhere

It was in 1900, 16 years after Gregor Johann Mendel's death, that Mendel's laws were widely accepted and the significance was well acknowledged by the scientific world. Similarly, it was in 1983, 30 years after Barbara McClintock's innovative discoveries, that she won an unshared Nobel Prize in Physiology or Medicine for the theory of jumping genes. Those touching stories have been talked a lot in scientific field, and those outstanding biologists have been memorized for their advanced theories, their persistence and their legacies. History always repeats itself. The neurophysiologist Xiangtong Zhang (Hsiang-tung Chang) also had a same experience, while he was rarely recognized by people.

Xiangtong Zhang, Academician of Chinese Academy of Sciences (CAS), was one of the founders for neuroscience research in China and for the establishment of Shanghai Brain Research Institute. His pioneer work on the physiological function of dendrites has been well-acknowledged by worldwide scientists. He was elected Foreign Honorary Academician of the Royal Academy of Medicine of Belgium, Member of the Central Committee of International Brain Research Organization, and Neurological Consultant of the World Health Organization.

In the winter of 1992, Xiangtong Zhang received a surprising letter, which informed that he would be honored with "Life Achievement Award" from the International Neural Network Society, and he was invited to the award ceremony. At first, he thought it must be a mistake by the Society because the Society was organized by engineers in computer studies, while he was neither an engineer nor a computer scientist. However, only after reading the letter did he realize that all the things were closely related to his study on nervous dendrites over 40 years ago. The certificate of commendation especially noted that he is the first to recognize the functional significance of dendrites in central nervous system, which was one of the important symbols for developing new generation computer, and his remarkable achievement laid the theoretic foundation for the application of neural network computer instead of electronic digital computer.

In the middle period of 20th century, many neurophysiologists had clearly realized that axon

---

Baoyuan Zhang, Le Kang
Beijing Institutes of Life Science, Chinese Academy of Sciences, Beijing 100101, China
Contact: lkang@ioz.ac.cn (L. Kang)

was one protoplasmic protrusion, carrying nerve impulses away from the cell body of neuron. But few realized that the dendrites were also able to conduct nerve impulses. From 1950 to 1955, Xiangtong Zhang devoted all his time to the study of dendrites. As an Associate Researcher at Rockefeller Institute for Medical Research, he investigated the functional characteristics of apical dendrites of pyramidal neurons in the cerebral cortex using an electrophysiological approach, and his nine publications on dendrites function had widespread influence in the neurophysiological field. After he returned to China in 1956, he continued his research on nervous system with great enthusiasm. With six years of arduous efforts, he built up the first laboratory to study nerve cells in China. Many outstanding Chinese neurophysiologists were trained in his laboratory, including Academician Jianping Wu, Academician Yizhang Chen, et al. For his brilliant contribution in neuroscience, he was regarded as one of the pioneers in the study of dendritic potentials and the first to purpose a fundamental distinction between axosomatic and axodendritic synapses. However, it was beyond his imagination that his research results were actually conducive to the development of intelligent computer.

Although the award was a big surprise, Xiangtong Zhang thought he did not have a lot to show off. In his mind, it was a slight incident on his tortuous path of brain research. He treated it as a quiet coming and a silent leaving, with no celebration, no pride, and even no one knowing it. He carefully summarized his experience, and pointed out that the social significance and the practical value of basic theory is often not recognized soon after its formation; instead, its light will radiate when social sciences and other related branch sciences reach the corresponding level.

Maybe some people would describe Xiangtong Zhang's prize as chance and serendipity. However, when we review his life, he really deserves this prestigious award. During the long years, from beginning to end, he focused his attention on the neurophysiological research to uncover the mysteries of nervous system. Even when encountering serious obstacles and enduring painful difficulties, he never gave up his research. His hypothesis that dendrites were able to conduct action potentials, in spite of controversies and objections in the past 40 years, was finally proven to be correct. God helps those who help themselves, and strong-willed men could change their fate in adversity. Just like what Xiangtong Zhang said: "Life is colorful; I could still learn a lot from it. My strong survival ability, much like the cactus, would hold on and thrive in any difficult circumstance whether it is the desert or the thorns." His particular experience emphasized again on the important inspiration: Science is a long-term and systematic research, and would not be accomplished in a short time; in no case should we pursue immediate successes or shortterm gains.

"Leave a trace wherever you go, like the cuttlefish" is Xiangtong Zhang's life-long motto. With his dedication to academic research, he has left us absolutely much more than a trace. His diligence and persistence will be engraved on our minds forever. Nowadays, with the development of life science, biologists not only have tremendous opportunities, but also face complex challenges. The secret of success in life is to be ready for the opportunity when it comes. Thus, it is important to be industrious, and it is more important to be patient and to be ready. After all, a rolling stone gathers no moss, and gold is shining everywhere. Only when great systematic work gather together, will life science be able to have a brilliant future.

Figures

References

Zhang W (2003) Xiangtong Zhang. Shanghai: Shanghai Science and Technology Education Press. (张维. 2003. 张香桐传. 上海: 上海科技教育出版社.)

Zhang XT (1995) The Road of Brain Research. Beijing: Scientific and Technological Literature Publishing House. (张香桐. 1995. 脑研究的崎岖道路. 北京: 科学技术文献出版社.)

69. 王振义：抗击白血病

王振义，中国工程院院士（图1），上海交通大学医学院终身教授，上海血液学研究所名誉所长。众所周知，他对医学最重要的贡献是开创了急性早幼粒细胞白血病（APL）的诱导分化疗法。

图1　王振义教授在办公室

1948年，王振义先生毕业于震旦大学医学院（今上海交通大学医学院），并获得医学博士学位，因成绩优异，留在广慈医院（瑞金医院前身）担任住院医师。1952年医院院系调整，由大内科趋向专业化，王振义先生根据指示带队专攻一个内科研究领域，他毅然投入血液病研究中，并开始了他的毕生追求。出乎意料的是，他这个决定背后的原因却是“非同寻常”的——他喜欢音乐，在他的脑海中，显微镜下的血液细胞就像音符一样简单而美丽。然而，当他开始在血液科的工作后，现实无情地打击了他——血液病既不简单也不美丽，它们在夺走生命时是那样无情。

在20世纪50年代，我国的白血病临床治疗方法相当有限，更不要说通过生物医学研究开发新治疗策略了。由于研究条件简陋，王振义先生开始通过阅读大量有关白血病特别是APL的书籍和研究文章进行自学和研究。随着学习和研究的深入，他不断问自己：治疗APL有没有传统细胞毒性化疗以外的办法？“坏”的癌细胞能否被“改造”进而“转化”为好的细胞，而不是被化疗直接杀死呢？到了20世纪70年代晚些时候，他发现他的观点与西方提出的诱导癌细胞分化的概念不谋而合。这引起了他的极大兴趣，并使他坚定信心使用HL-60细胞和新鲜的APL细胞筛选分化诱导剂。根据文献报道，有许多化合物能够诱导癌细胞成熟，包括丁酸、二甲基亚砜和13顺式视黄酸。幸运的是，13顺式视黄酸的异构体——全反式视黄酸（ATRA）在临床上被用于治疗皮肤病，他在上海就能得到。他“将错就错”将ATRA用于筛选实验，结果发现，ATRA能

译者：李柳
中国科学院北京生命科学研究院，北京 100101，中国
邮箱：liliu@ioz.ac.cn

够强烈诱发异常早幼粒细胞的终末分化，并且其诱导作用优于13顺式视黄酸和其他候选化合物。这些有意思的实验数据为使用ATRA治疗APL患者的临床试验提供了“助推器”。

时间来到1985年，一名不幸患上了APL的五岁小女孩被送至上海儿童医院，经过传统化疗，她的症状并没有缓解，病情迅速恶化，高烧不退，出血严重，家人已经绝望了。当医生提议ATRA试验时，她的父母为了给女儿尽到一切可能的努力，同意尝试ATRA治疗。口服ATRA三周后，患者的高烧退去，出血停止，症状明显好转。随后，她在症状缓解后，又接受了一年ATRA结合化疗的继续治疗。如今这个女孩已经年过而立，病情非常稳定，健康活泼。这是世界上第一例口服ATRA而成功治愈的APL病例，它极大地鼓舞了王振义先生和他的同事继续进行ATRA试验。到1988年，24例APL患者在上海接受了ATRA治疗，23人在治疗后完全缓解。1988年，这个重要的试验结果在国际学术期刊《血液》上发表，迄今这篇文章已被引2000余次，成为全球引证率最高和最具影响力的论文之一。ATRA对APL的治疗效果在欧洲和北美的其他医院和血液学中心也逐步得到证实。

在接下来的十年里，王振义先生和他的同事为优化ATRA治疗做出了艰苦的努力。在减少耐药性和病情复发方面，ATRA结合化疗被证明比单独使用ATRA效果更好。20世纪90年代中期，王振义先生的学生、原卫生部部长陈竺院士，证实了三氧化二砷（ATO）治疗耐药复发APL的功效。ATO是一种毒药，也是传统中药之一。后来，ATRA和ATO联合治疗APL的4年临床研究证明，联合治疗比单独使用ATRA或ATO更有效。联合应用ATRA和ATO治疗APL的方案被称为“上海方案”，也被确认为国际APL的治疗规范，使得APL这一致命疾病转化为高度可治愈的病症。

由于在抗击APL方面取得的巨大成就，王振义先生荣获许多奖项，包括1994年美国通用汽车癌症研究基金会的凯特林奖、1997年瑞士布鲁巴赫肿瘤研究奖、1998年法国西蒙台尔杜加科学奖、2012年全美癌症研究基金会第七届圣捷尔吉癌症研究进展大奖（与陈竺共享）以及2010年国家最高科学技术奖。

王振义先生无疑是我国最著名的血液学专家，他一生淡泊名利，追求宁静。尽管他获得了众多的奖项，但是在他的办公室里几乎找不到奖章和奖状，那里堆满了书籍和杂志。大部分奖金都被他分享给团队同事或者捐献给慈善机构。他也从未考虑过将ATRA治疗APL申请专利，只因那样会增加病人的经济负担。

对于王振义先生来说，病人的利益永远是第一位的。在1985年当他决定进行ATRA临床试验时，一些朋友和同事强烈建议他不要这样做，因为试验的风险太大了，如果临床试验期间有任何差错，他的名声将毁于一旦。但对他来说，挽救生命胜过他个人的得失。他抛开个人的整个职业生涯于不顾，坚持进行了这次试验，并取得了巨大的成功。因此，在某种程度上，正是他对病人的深切同情和关爱，使ATRA试验成为可能，并成就他成为一名伟大的医生。ATRA治疗的第一篇论文发表于1988年，之后王振义先生在全国范围内和国际上大力推广ATRA，使更多的患者能够早日接受有效治疗。20世纪90年代初，日本医生Ryuzo Ohno拟在日本开展ATRA的临床试验，但他很快意识到，通过任何官方途径从中国进口药物在当时是非常困难和费时的。他们在会议期间碰面，王振义先生听说了这件事，当即以个人礼物的名义慷慨地送给Ryuzo

Ohno一瓶ATRA药片，并在后来，在Ryuzo Ohno的朋友刚好访问上海时，将更多的ATRA作为礼物送给他，以便其做前瞻性试验。如果没有王振义先生的帮助，Ryuzo Ohno和他的团队就不可能那么早在日本进行ATRA试验。

王振义先生对于病人始终给予关心和爱护，他经常收到病人的来信和小礼物作为回报，其中有些人甚至在几十年前接受过他的治疗。“只有医生才能感受到这种感激和满足。”王振义先生如是说。

69. Dr. Zhen-Yi Wang: Fighting leukemia

Dr. Zhen-Yi Wang, Academician of the Chinese Academy of Engineering, a professor at the Medical School of Shanghai Jiao Tong University and the honorary director of the Shanghai Institute of Hematology, is best known for his contribution in developing a differentiation induction therapy of acute promyelocytic leukemia (APL).

Graduated with a MD degree from the Aurora University School of Medicine in Shanghai in 1948, Dr. Wang was recruited as a resident doctor by Rui-Jin Hospital. Upon his promotion in 1952, he was required to choose a field to major in. He chose hematology and it was the beginning of his life-long pursuit. The reason behind his decision was quite unusual—he loved music and in his mind, blood cells were just as simple and beautiful as musical notes when examined under a microscope. However, when he started working in the hematology department, reality finally hit him—blood diseases were neither simple nor beautiful; they showed no mercy when taking lives away.

Back in the 1950s, the clinical approaches for leukemia were quite limited in China, not to mention biomedical research for developing new strategies. Never discouraged by the poor research condition, Dr. Wang started a selftraining program by extensively reading books and research articles about leukemia, especially APL. The more he read, the more he asked himself: Was there any way to treat APL other than cytotoxic chemotherapy? Could the "bad" cancer cells be "educated" and "converted" to good cells, instead of being killed by chemo? Later in the 1970s, he found out that he happened to share the same view with Western researchers who raised the induced cancer differentiation concept. This further intrigued him and gave him firm confidence in the screen for differentiation inducers using HL-60 cells and fresh APL cells. A number of compounds were reported to be capable of inducing cancer cell maturation, including butyrate, dimethyl sulfoxide, and 13 cis-retinoic acid (13 cis-RA). Fortunate enough, an isomer of 13 cis-RA, all-trans retinoic acid (ATRA), was available in Shanghai as a skin

Ming Li[1], Guangbiao Zhou[2]

1 Beijing Institutes of Life Science, Chinese Academy of Sciences, Beijing 100101, China

2 Institute of Zoology, Chinese Academy of Sciences, Beijing 100101, China

Contact: gbzhou@cicams.ac.cn (G.-B. Zhou)

disease medicine. In Dr. Wang's screen, ATRA strongly induced terminal differentiation of abnormal promyelocytes, and its inducing effects were superior to 13 cis-RA and other candidate compounds. The intriguing data provided main impetus for the clinical trial of ATRA treatment on APL patients.

In 1985, a five-year-old girl who suffered from APL was admitted into Shanghai Children's Hospital. She had gone through chemotherapy but it did not alleviate her symptoms such as high fever and severe hemorrhage. Her parents practically gave up hope before the doctors told them about the ATRA trial. Wanting to make every possible effort for their daughter, the parents agreed to try ATRA. After orally taking ATRA for three weeks, her high fever was gone and the hemorrhage stopped. She was then treated with ATRA/chemo as a post-remission treatment for a year. The girl is now over 30-year-old and very stable and healthy. This was the very first case of ATRA treatment on an APL patient and the successful case encouraged Dr. Wang and his colleagues to continue the ATRA trial. By 1988, 24 APL patients had been treated with ATRA in Shanghai and 23 out of 24 went into remission after treatment. The significant results were reported in *Blood* in 1988, and the article has been cited for nearly 2000 times over the years and became one of the most cited papers on cancer research. The dramatic effects of ATRA on APL were also progressively confirmed by other hospitals and hematology centers in Europe and North America.

For the following decade or so, Dr. Wang and his colleagues made painstaking efforts to optimize the ATRA treatment. ATRA/chemo in combination was proven to provide better effects with less drug resistance and reduced relapse compared with ATRA alone. In the mid1990s, Dr. Zhu Chen, a former student of Dr. Wang and currently the Health Minister of China, proved the efficacy of Arsenic Trioxide (ATO), a well-known poison used as a traditional Chinese medicine, in treating replased APL patients. Later on, a synergistic combination of ATRA and ATO was tested in a four-year long clinical trial and proven to be more effective than ATRA or ATO alone. The ATRA/ATO combination, often referred to as "Shanghai Protocol" , has become the standard therapy for APL worldwide, and turned this deadly disease into a highly curable one.

For his achievements in fighting APL, Dr. Wang has received numerous prizes and awards, including the Kettering Prize from the General Motors Cancer Research Foundation USA in 1994, the Prize of Brupbacher from Switzerland in 1997, the Prize for Science from the Simmon Del Ducca Foundation of France in 1998, and the 7th annual Szent-Gyorgyi Prize from the National Foundation for Cancer Research of US in 2012 (shared with Dr. Chen). He also received the 2010 State Supreme Science and Technology Award of China.

Even though he is unquestionably the most famous hematologist in China, fame or fortune has never been Dr. Wang's aim in life. Despite all the prizes and award he has won, the medals and certificates of awards can hardly be found in Dr. Wang's office, which is packed with books and journals. Most of the award money was either shared with his colleagues or donated to charity. Dr. Wang has never considered the option of applying for a patent of ATRA treatment on APL, as it would only increase the financial burden of the patients.

The benefit of patients is always Dr. Wang's number one priority. When he decided to conduct the first clinical trial with ATRA in 1985, some friends and coworkers strongly advised him against it, as the trial would be too risky and if anything should go wrong during the trial, his reputation would be ruined. But for him, saving a life outweighed his personal gain and loss. Putting his whole career at stake, he conducted the trial which turned out to be a great success. So in a way, it was his deep sympathy and affections for patients that made the ATRA trial possible and made him a great doctor. After the first paper of ATRA therapy was published in 1988, Dr. Wang put great efforts into promoting ATRA nation-wide and internationally, so that more patients would receive the treatment sooner. When a Japanese doctor, Ryuzo Ohno, initiated a clinical trial of ATRA in Japan in the early 1990s, he realized that it was extremely difficult and time-consuming to import ATRA from China officially. Dr. Wang heard about this and gave Dr. Ohno a whole bottle of ATRA as a personal gift when they met at a meeting. He later sent more ATRA to Japan whenever possible, mostly through Dr. Ohno's friends who were visiting Shanghai. Without Dr. Wang's help, Dr. Ohno and his team could never have carried out the ATRA trial in Japan so early.

Love and care, that's what Dr. Wang gives to his patients without reservation. In return, he often receives letters and small gifts from patients, some of whom he treated decades ago. "Only a doctor can receive this kind of gratitude and satisfaction." said Dr. Wang.

Figures

Fig.1 Dr. Zhen-Yi Wang in his office

References

Wang Z (2008) Acute promyelocytic leukemia: From highly fatal to highly curable. Blood 111: 2505–2515.

70. 刘彤华：七厘米载玻片上献身临床病理学的人生

中国工程院院士，著名医学家、病理学家、医学教育家，北京协和医院病理科教授刘彤华（图1），因病医治无效，于2018年7月8日11时11分在北京协和医院逝世，享年89岁（北京协和医院，2018）。

刘彤华院士1929年11月13日出生于江苏无锡。1947—1953年就读于上海圣约翰大学医学院（方文钧等，2009）。20世纪50年代初，北京协和医学院病理系主任胡正详教授举办全国病理学"高级师资训练班"。按照当时的规定，所有的医学生只能报基础学科。"既然不能选择临床，那就选与临床离得最近的学科吧。"从小立志要做医生的刘彤华选择了介于基础与临床之间的病理学。1952年，刘彤华从上海来到北京，跟随胡正详做病理学研究（图2）。胡正详说过的一句话"研究科学的人要沉浸在科学里，里外渗透，不能分心"，让刘彤华铭记了一辈子，坚守了一辈子（段文利等，2016）。

图1　刘彤华教授（1929—2018）

图2　1952年全国病理学"高级师资训练班"师生，刘彤华院士（二排左三）和胡正详教授（一排左三）

译者：傅谭娉

中国医学科学院北京协和医院，北京100730，中国

邮箱：501776763@qq.com

1953年，刘彤华从训练班毕业。1953—1954年，刘彤华前往第六军医大学任病理系助教。1954—1956年，刘彤华担任第七军医大学病理系助教（方文钧等，2009）。1957年，北京协和医学院病理系并入中国医学科学院实验医学研究所（简称“实研所”）。刘彤华在1957—1969年任协和医学院病理系助教、实研所病理系助教及助理研究员。1969年，病理系随实研所迁往四川简阳，刘彤华因为是军人家属，不在迁离之列。她多年来一直负责北京协和医院（简称“协和”）的病理业务，便留在了北京协和医院（段文利等，2016）。

刘彤华克服艰难困苦创办了北京协和医院病理科。实研所病理系迁往简阳后，几乎所有的仪器设备、档案资料，包括尸检档案和尸检大标本全部都被带走了，只给刘彤华留下了几间空荡荡的屋子和两名技术员。“外检病理档案是协和病人的资料，是协和的财富，留在医院才能发挥更大的作用。”刘彤华据理力争，将27万份外检病理档案留在了协和。几年后，已搬到简阳的尸检档案也顺利地回到协和，这为以后的病理教学发挥了不可估量的作用（刘彤华，2010）。现在，1916年起的协和全部尸检档案，1917年起的全部外检病理档案，一共110多万份，全都完好无损地保存在协和的档案柜里（段文利等，2016）。

1978年，刘彤华担任北京协和医院病理科副主任；1985年任病理科主任（北京协和医院，2018）。在科室创建的早期，充满了艰辛与挑战。人手不够，刘彤华就亲自干起技术员的活。每天早晨7点，她准时第一个来到医院，把浸蜡的标本包埋成蜡块，便于技术员上班后切片制片，以节省时间。白天，刘彤华要处理大量的阅片及报告。遇上疑难病例，她晚上还要留下来反复查资料。刘彤华的勤奋努力已经成为协和病理科的传统（段文利等，2016）。

从事病理事业65年，经刘彤华之手阅过的片子、签发的报告达30万份之多。刘彤华签发的每一个病理报告都有明确的诊断，体现出干练、精准、坚定、果敢的“刘氏”风格。她对疑难病症的诊断率极高，尤其对淋巴结病理、消化道疾病病理、内分泌病理等病理诊断造诣精深，她的诊断被誉为“全国病理诊断的金标准”。20世纪90年代，一块辗转北京数家大医院均不能得到确诊的病理切片被送到刘彤华手中。会诊之后，经仔细阅片，她写下了“颈部淋巴结转移性鳞癌”（转移瘤）的诊断。可在接下来的全面体检中，临床医生始终找不到病人的原发肿瘤病灶。病人和临床医生都对转移瘤的诊断心存疑惑，再度请来刘彤华重新阅片。第二次阅片的结论没变，但刘彤华补进去“建议查口腔”五个字。最后，口腔科大夫在病人的牙龈处发现了一个很不起眼的原发肿瘤病灶（段文利等，2016）。

在科学研究上，她始终坚持基础和临床相结合的协和传统。20世纪80、90年代，她带领团队在胰腺癌及胰腺神经内分泌肿瘤分子生物学和分子遗传学研究方面取得一系列开创性成果。其中“胰头癌对胰内胆管环形壁内浸润”和“人胰腺癌细胞分子生物学及细胞生物学特性的研究”分别获1985年和1993年卫生部科技进步奖二等奖（Liu et al.，1983；Chen et al.，1993），“人胰腺癌细胞分子生物学及反义基因调控对其恶变表型的逆转”获1995年国家科学技术进步奖二等奖（Liu et al.，1995）。由于在胰腺癌病因学和基因治疗学领域的杰出贡献，1999年，刘彤华被增选为中国工程院院士。她也是中国工程院的第一位病理学院士（北京协和医院，2018）。

刘彤华始终站在病理学的制高点。20世纪90年代后期，刘彤华敏锐地看到分子生物学和细胞遗传学在病理学领域的前途和方向，随即多次派出年轻技师和医生学习分子遗传及分子生物学先进技术，组建了分子生物实验室，将传统的形态学病理与分子生物学紧密结合起来，将多项新技术应用到临床诊断中（图3）。步入21世纪，刘彤华看到了肿瘤生物靶向治疗的前景，提出了靶向治疗需要靶向诊断的观点（刘彤华，2008）。2003年，由刘彤华挂帅的国内第一个分子遗传病理实验室在协和建成，在肿瘤靶向治疗的基因检测和分析方面均居国内领先水平，吸引了多个国际知名药厂慕名前来，将其全球性肿瘤靶向药物的临床试验放在协和。协和关于靶向诊断的研究迅速在乳腺癌、胃癌、结直肠癌、肺癌等多种疾病领域铺开。这些尝试不仅拓展了病理学的发展方向，提升了病理医生在疾病诊治中的地位，更重要的是为发病率日趋增高的肿瘤患者带来了福音（方文钧等，2009）。

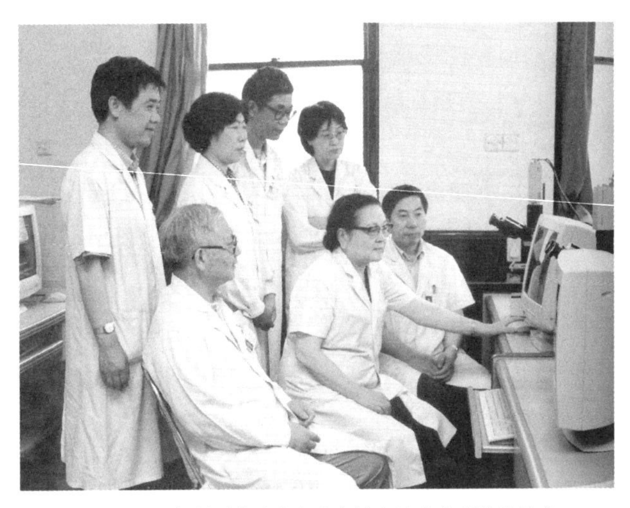

图3　2001年刘彤华院士和北京协和医院病理科的同事

刘彤华院士曾担任《中华病理学杂志》第三、四届常务编委和第八届名誉总编辑（霍临明等，2009），《诊断病理学》杂志名誉主编（刘彤华，2014），中华医学会病理学分会常务委员等。先后发表论文400余篇，主编专著4部。1994年主编的著作《诊断病理学》成为临床病理工作者的必备参考书，于1995年获评“全国优秀科技图书一等奖”（北京协和医院，2018）。为铭记刘彤华院士对我国病理学事业作出的卓越贡献，第四版《诊断病理学》特别更名为《刘彤华诊断病理学》（董琳和庞钧译，2018）。

刘彤华院士1995年被评为全国优秀教师、北京市优秀教师，1998年获得卫生部“有突出贡献专家”称号，2003年获“首都劳动奖章”。此外，她还获得了中国科学技术协会先进工作者、北京市三八红旗手标兵、北京市爱国立功标兵、“协和名医”、北京协和医院卓越贡献奖、杰出贡献专家等多项奖励（北京协和医院，2018）。

刘彤华院士一生与显微镜为伍、与放大镜做伴，为我国病理学事业发展和病理人才培养作出了巨大贡献，是后辈学习的典范和楷模（图4）（段文利等，2016）。

图4　2018年6月6日，刘彤华在家中阅读杂志，接受了北京协和医院口述历史项目的采访。这是她最后一张照片，一个月后她去世了

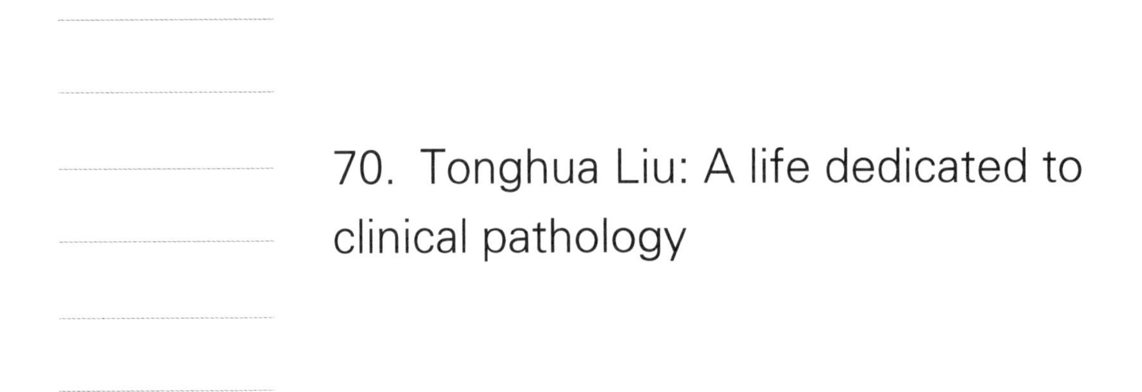

70. Tonghua Liu: A life dedicated to clinical pathology

Professor Tonghua Liu (刘彤华) was an academician of the Chinese Academy of Engineering, a renowned medical scientist, pathologist, medical educator and professor of the Pathology Department in Peking Union Medical College Hospital (PUMCH). She passed away from illness on July 8, 2018 at PUMCH. She was 89 (PUMCH, 2018).

Prof. Liu was born on November 13, 1929 in Wuxi, Jiangsu Province. She studied at the Medical School of Saint John's University in Shanghai from 1947 to 1953 (Fang et al., 2009).

In the early 1950s, Professor Zhengxiang Hu (胡正详), director of the Pathology Department of Peking Union Medical College (PUMC), was running a national Pathological Senior Teacher Training Program. Based on Liu's educational background, she could only pursue basic medicine sciences. Despite this, Liu still dreamed of becoming a doctor. As a result she made the second best choice and applied for Prof. Hu's program as a pathologist. This would allow her to remain close to both doctors and patients during clinical diagnosis. In 1952, Liu left Shanghai for Beijing to follow Prof. Hu for further studies. Prof. Hu's mentorship had a great influence on her. Prof. Hu's teachings always stuck with Liu: "A scientific researcher shall completely immerse himself in science with no distractions." (Duan et al., 2016)

In 1953, Liu graduated from Prof. Hu's program. She became a teaching assistant at the 6th Army Medical University and was there from 1953—1954. In the following two years she worked for the 7th Army Medical University as a teaching assistant. Today the 6th and the 7th Army Medical Universities are now known as the 3rd Army Medical University (Fang et al., 2009).

In 1957, the Pathology Department of PUMC was merged with the Experimental Medicine Institute at the Chinese Academy of Medical Sciences (CAMS). From 1957 to 1969 Liu worked as both a teaching assistant at the Pathology Department of PUMC and a teaching assistant and assistant researcher at the Pathology Department at the Experimental Medicine Institute of CAMS. In 1969, the Experimental Medicine Institute moved to Jianyang, Sichuan Province. As a spouse of an army man, Liu could not separate from her husband. Thus she decided to stay in

Lin Dong, Tanping Fu, Junyi Pang, Zhiyong Liang, Wenli Duan
Peking Union Medical College Hospital, Beijing 100730, China
Correspondence: 501776763@qq.com (T.-P. Fu)

Beijing to found a separate Pathology Department in PUMCH (Duan et al., 2016).

The Pathology Department of PUMCH started from scratch. Almost all equipment and archives, including all autopsy files and gross specimen, had been moved to Jianyang. All left was a few office rooms and two technicians. Liu insisted that the 270000 clinical pathological archives should be left to PUMCH, "Clinical pathological archives are critical to clinical studies. In order to make the best use of them, these files should be given to PUMCH." A few years later, the previous autopsy archives were also transported back from Jianyang to Beijing (Liu, 2010). These documents were invaluable to the development of clinical pathology in PUMCH, and pathology in China in general. Because of Prof. Liu's dedication, PUMCH now housed over 11 million pathological archives dating all the way back to 1916. The PUMCH pathological archive is a testament to Prof. Liu's professionalism (Duan et al., 2016).

Prof. Liu was appointed deputy director of the Pathology Department of PUMCH in 1978 and then director in 1985 (PUMCH, 2018). Prof. Liu overcame great challenges so that the Pathology Department of PUMCH could grow and thrive. She voluntarily took over part of the technicians' duties when the lab was severely understaffed during the initial years. She would come to the lab at 7 am every day to embed paraffin. This would allow the technicians to slice the paraffin and make section specimen at 8 am. After that, Liu would assess a great number of slices and generate reports. She always stayed late to review literature on difficult cases. Prof. Liu's diligence has become a tradition in the Pathology Department of PUMCH (Duan et al., 2016).

Prof. Liu signed her name on almost 1 million clinical pathology reports during her 65-year career. She always gave definitive diagnosis, embodying her capable and decisive temperament. Prof. Liu was famous for diagnosing rare and difficult cases, especially in the field of lymph nodes, digestive and endocrine diseases. Her diagnosis was held as the "golden standard" across the country. In the 1990s, Prof. Liu was consulted and made a diagnosis of cervical lymph node metastatic squamous cell carcinoma. After receiving Prof. Liu's report, doctors examined the patient carefully, seeking to identify the primary tumor lesion. However, they found nothing and doubted Prof. Liu's diagnosis. The doctors requested Prof. Liu review the case again. After re-reviewing the case, Prof. Liu's diagnosis remained unchanged but she added 4 words to her report, "examine patient's oral cavity" . At last, doctors were able to find an unconspicuous primary tumor lesion on the patient's gum. When it came to diagnosis, her suggestion had ascended from science to art (Duan et al., 2016).

Prof. Liu always believed in drawing on the PUMCH tradition of combining basic sciences with clinical research. In the 1980s and 1990s, her team achieved a series of innovative breakthroughs on molecular biology and molecular genetics of pancreatic carcinoma and pancreatic neuroendocrine tumours (PNET). Her team's study on "Pancreatic Dead Carcinoma's Ring Invasion into Intrapancreatic Bile Duct Wall" won the Second Prize of Science and Technology Progress Award of the National Health Commission in 1985 (Liu et al., 1983). Her team won another Second Prize of Science and Technology Progress Award of the National Health Commission in 1993 for their study of "Molecular Biology of Human Pancreatic Carcinoma Cell and Reversal of its Malignant Phenotype by Antisense Gene Regulation" (Chen et al., 1993). Another study on "Antisense Gene Regulation Reverse Malignant Phenotype of Human Pancreatic Carcinoma Cell" won the Second Prize of the National Science and Technology Progress Award in 1995 (Liu et al., 1995). In 1999, Prof. Liu was elected the academician of the Chinese Academy of Engineering for her outstanding achievements in pathogenesis and gene therapeutics of pancreatic carcinoma. She was the first pathologist to be elected as an academician of the Chinese Academy of Engineering (PUMCH, 2018).

Prof. Liu held a great vision for scientific development. In the late 1990s, Prof. Liu sensed the prospects of molecular biology. She sent young pathologists and technicians abroad to study in this field. By combing traditional pathomorphology with molecular biology, many new technologies were applied to clinical diagnosis. Entering the 21st century, Prof. Liu foresaw the future of biological targeted therapy. She put forward the notion that targeted therapy needed targeted diagnosis (Liu, 2008). In 2003 she established the first molecular genetics pathology lab in China. Many renowned international pharmaceutical companies sought for cooperation with her. Global clinical trials for new antineoplastic targeted drugs were undertaken in PUMCH. Prof. Liu led her team digging into the targeted diagnosis of breast cancer, gastric cancer, colorectal cancer, lung cancer and many other diseases. These efforts expanded the application of clinical pathology, improving pathologists' importance in the multidisplinary team, bringing hope to the increasing number of tumor patients in China (Fang et al., 2009).

Prof. Liu was an executive editor of the third and fourth editorial board of the *Chinese Journal of Pathology*, and the honorary editor-in-chief of its eighth editorial board (Huo et al., 2009). She was also the honorary editor-in-chief of *Chinese Journal of Diagnostic Pathology* (Liu, 2014) and a member of the standing committee of the Society of Pathology, Chinese Medical Association. She published more than 400 theses and was chief editor of four academic books. *Diagnostic Pathology* was issued under her general editorship in 1994. It won the National Excellent Science and Technology Book Award in 1995. This book has since become an essential reference for all Chinese clinical pathologists (PUMCH, 2018). In memorial of Prof. Liu's contribution, the fourth edition of the book was renamed to *Liu Tonghua Diagnostic Pathology* (Dong and Pang, 2018).

Prof. Liu also won many other titles and awards including the National Distinguished Teacher & Beijing Distinguished Teacher award in 1995, Expert of Outstanding Contribution by National Health Commission in 1998, Capital Labor Medal in 2003, Distinguished Worker from the China Association for Science and Technology, Beijing March 8 Red Flag Bearer, Beijing Patriotic and Contributing Model, Renowned Doctor of PUMCH, Distinguished Contribution Award and Distinguished Expert from PUMCH etc. (PUMCH, 2018)

Prof. Liu spent her whole life in the company of microscopes and magnifying lens, making significant contribution to China's pathology undertakings and education, setting an outstanding example for later generations (Duan et al., 2016).

Figures

Fig.1 Professor Tonghua Liu (1929—2018)

Fig.2 Faculty and students from PUMC Pathological Senior Teacher Training Program in 1952, Tonghua Liu is the 3rd from the left in the middle row, Prof. Zhengxiang Hu is the 3rd from the left in the front row

Fig.3 Tonghua Liu and her colleagues from Pathology Department of PUMCH in 2001

Fig.4 Reading magazine at her house on June 6, 2018, Tonghua Liu was interviewed by PUMCH's Oral History Program. This is her last picture. She passed away one month later

References

Chen J, Liu T, Wang Z, et al (1993) Expression of transforming growth factor and c-erbB-2 gene on human pancreatic carcinoma cell. Chin J Oncol 15(4):273–277.

Dong L, Pang JY (2018) Launching ceremony of Tonghua Liu Diagnostic Pathology took place in Chengdu. PUMCH Newspaper, 2018–11–12. (董琳, 庞钧译. 2018.《刘彤华诊断病理学》新书首发式在蓉举行. 北京协和医院院报, 2018–11–12.)

Duan WL, Liang ZY, Dong L, et al (2016) Tonghua Liu: A life on the 7 cm microslide. China Science Daily, 2016–6–20. (段文利, 梁智勇, 董琳, 等. 2016. 刘彤华:七厘米载玻片上的医学人生. 中国科学报, 2016–6–20.)

Fang WJ, et al (2009) As Is Our Pathology, So Is Our Medicine! Biography-in-photo of Tonghua Liu at Her 80th Birthday. Beijing: Peking Union Medical College Hospital (internal publications). (方文钧, 等. 2009. 彤映华章——刘彤华教授八十华诞纪念画册. 北京:北京协和医院内部刊物.)

Huo LM, Cai ZG, Chang XQ, et al (2009) Prof. Tonghua Liu's greatest contribution to Chinese Journal of Pathology is her diligence, devotion and sense of responsibility. China Medical News 2009–9–27. (霍临明, 蔡振国, 常秀青, 等. 2009. 用勤奋、敬业、责任心撑起学术天空——记为《中华病理学杂志》发展做出突出贡献的刘彤华院士. 中华医学信息导报, 2009–9–27.)

Liu TH (2008) Pathology in the era of personalized medicine. Chin J Pathol 37(4):217–218. (刘彤华. 2008. 个性化医学时代的病理学. 中华病理学杂志, 37(4):217–218.)

Liu TH (2010) 40th anniversary of the Pathology Department of PUMCH. Med J 1(1):1–4. (刘彤华. 2010. 北京协和医院病理科建科40周年. 协和医学杂志, 1(1):1–4.)

Liu TH (2014) Wish you further and greater progress: On the occasion of Chinese Journal of Diagnostic Pathology's 20th anniversary. J Diag Pathol 21(6):330. (刘彤华. 2014. 百尺竿头更进一步——写在《诊断病理学杂志》创刊20周年之际. 诊断病理学, 21(6):330.)

Liu TH, Chen J, Zeng X (1983) Morphologic study of pancreatic head and ampullary region basis of carcinoma with special reference to pathologic jaundice. Chin Med J 96(2):95–104.

Liu TH, Wang Z, Cui Q (1995) Significance of the detection of Ki-ras codon 12 mutation—the diagnosis of differential diagnosis of pancreatic carcinoma. Int J Surg Pathol 3(2):93–100.

PUMCH (2018) A biography of Tonghua Liu. PUMCH Newspaper, 2018–7–30. (北京协和医院. 2018. 刘彤华院士生平简介. 北京协和医院院报, 2018–7–30.)

71. 卢于道与中国神经科学的发展

卢于道（Yu-Tao Loo，1906—1985），我国著名神经心理学家，我国人类大脑皮层结构和功能研究的先驱之一（图1）（钱伟长，2013）。卢于道为神经解剖学和生理心理学领域的奠基作出了卓著的贡献。同时，他的理论研究拓展了这些领域的知识界限。此外，作为一个科学家，他撰写了大量科普文章和书籍来传播科学知识。他也是一个坚定的爱国主义者，对我国的建设抱有热忱。

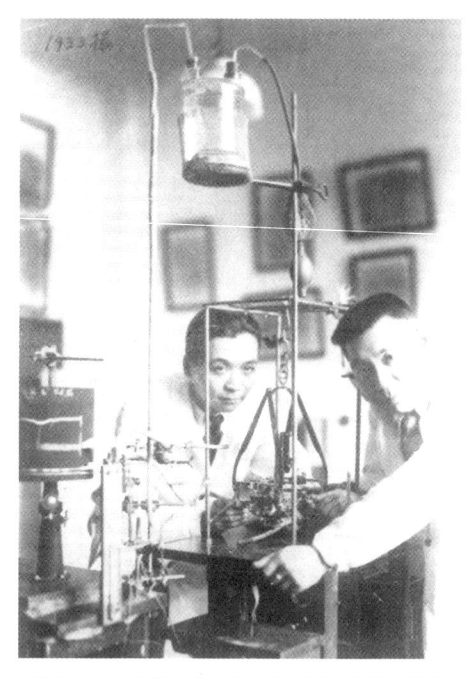

图1　卢于道博士（左）和他的同事朱鹤年博士（右）在一起（1933年）

1906年1月9日，卢于道出生于浙江鄞县。1921年，卢于道考入东南大学（即国立中央大学）心理学系，4年后进入该校生物系学习。1926年，获得理学学士学位之后，卢于道考取浙江省政府设立的官费留学基金，进入美国芝加哥大学医学院解剖学系。

幸运的是，卢于道师从世界著名比较神经解剖学家C. J. Herrick。1930年，在Herrick指导下，卢于道完成博士论文《北美负鼠的前脑》，以解剖模型展现负鼠前脑的内外结构。这篇论文的最后一段解释了微观解剖前脑内部结构的价值和重要性，即“对内部结构的检查将会促进这些研究的进一步开展。前脑的一般性特征和某些高度发展部位的缩小或衰退也会进一步帮助澄清前脑形态的一般规律”（Loo，1931）。后来的学者高度评价他的文章：“他的文章全面地描述了有袋类动物的整体解剖结构。”（Karlen and Krubitzer，2007）此外，科学引文索引数据库科学网（Web of Sciencc）显示，直至今天，卢于道的博士论文已经被引用了247次。基于卢于道对负鼠脑区的优秀研究结果，当代的科学家推进各自的研究，如丘脑核（Olokowicz

译者：钱燕燕

南京师范大学心理学院，南京210097，中国

邮箱：psyqyy@163.com

et al., 2008）和屏状核（Hinova-Palova et al., 2008；Mathur, 2014）。在Herrick推荐之下，卢于道的博士论文最终发表于《比较神经学杂志》。与这份文章相携而来的是，他获得了洛克菲勒基金会2万美元的资助。

1930年，卢于道回国后任国立中央大学医学院（上海第一医学院的前身）副教授。他的职责包括给大学生主讲实验解剖学，负责科研与教学工作。1931年至1939年，他担任中央研究院心理研究所的研究员，并主管神经解剖实验室。1937年，卢于道成为《中国心理学报》的主要撰稿人。在这份期刊上，他发表了《大脑皮层的机能：说明》，这篇文章是四卷《中国心理学报》中唯一一篇关于生理心理学的文章。

抗日战争时期，卢于道深入敌后，担任中国科学社代理总干事，从事社会活动。1941年春，卢于道迁往贵阳，任湘雅医学院（当时在贵阳）教授。1941—1942年，卢于道迁往重庆，担任中国科学社生物研究所副教授。1942年7月1日，卢于道成为复刊后的中国期刊《科学》的总主编，并发表了数篇科普文章。1942年春，卢于道在复旦大学担任领导角色，他连续任生物系主任、理学院院长、人体及动物生理教研组主任等职务。1954年，卢于道与孙宗彭开设人体及动物生理学专业。

1950年，卢于道参加全国自然科学工作者大会，并担任中国心理学会筹建委员会筹备处委员（Jing, 2001）。1949—1950年，中国科学院（2004）在全国范围内完成一份关于全国科学专家调查的综合报告，并将这些专家按照他们的学术贡献重要性进行排名。结果，卢于道在这次评选中获得了两个科学学科的提名，他在实验生物组排第50名，心理组排第23名。

新中国成立后，卢于道积极参加社会工作和科普工作。1947年，他被刚刚恢复的中国动物学会选为选举委员会委员。1947年7月，他成为中国解剖学会的第一任理事长。1950年，他成为中国心理学会15名负责人之一，并于5月与吴定良、刘咸等人共同参与创立中国人类学学会。

1954年8月，卢于道被选为上海市科学技术普及协会主席。1958年9月20日至23日，卢于道被选为上海市科学技术协会副主席，任期两届。1979年6月，卢于道和张香桐联名提出在中国科学院筹建脑研究机构。1980年即在上海成立了中国科学院脑研究所，标志着我国脑研究进入一个新阶段。

正如卢于道所言："自从回国后，曾做三方面的研究。第一方面是关于中国人脑；第二方面是继续Herrick学派的比较神经学研究；第三方面是神经细胞的显微化学研究。"（Loo, 1948）

首先，卢于道探讨人大脑皮层组织和结构的前沿问题。1929年，通过解剖胎儿大脑研究大脑皮层的发展，卢于道试图找到一种特殊的细胞——专门负责高级心理活动的不稳定型细胞，并想确定是否存在某些细胞比其他细胞保留有更强劲的胚胎潜能。卢于道把他的研究总结于论文《关于人类大脑皮层在个体发育形成时其各层机能的讨论》（Loo, 1929）。这篇文章促使他成为第一批探索中央神经系统结构和功能的中国科学家之一（Xu, 2000）。1931年，卢于道编写了我国第一部有关人体神经系统解剖学的教科书《神经解剖学》，系统介绍了人体中枢神经系统的结构与功能。

基于前期的人脑研究，卢于道将中国人的人脑与黑种人和白种人的人脑进行比较研究。1926年，学者J. L. Shellshear认为中国人脑更接近猿类（Shellshear, 1926）。这为

Shellshear于1934年在第一届国际人类学与民族学联合会的演讲中宣称中国人脑不如欧洲人脑埋下了伏笔。这件事使卢于道萌发了探索中西方人脑差异的念头（Loo，1934）。在美国俄亥俄州的凯斯西储大学，他观察并收集50个中国人脑，并与50个黑人人脑和50个白人人脑相比较。尽管卢于道的研究证实了Shellshear的实验数据，但是卢于道用一个细节性的事实"大脑的沟没有出现在胎儿大脑之中"来反对Shellshear的论断。此外，卢于道提出三个假设来驳斥Shellshear先前的论断，并解释了为何中国人人脑的月状沟比例高于其他人种：① 中国人的视觉区域发育更好；② 中国人的顶骨更发达；③ 中国人短头颅的纵轴更短（Loo，1933）。

其次，卢于道致力于为比较神经学作出卓著的贡献，对不同的哺乳动物进行了系统比较。从1932年至1937年，卢于道与他的同学朱鹤年一起使用霍斯利－克拉克（Horsley-Clarke）脑立体定位仪研究猫中脑的血管收缩反应，他们发现一个被称为"怒叫反应"的现象（Loo and Chu，1937a，1937b），这是怒叫反应的首次记录。更进一步，他们在猫的不同脑区嵌入电极进行慢性实验（Chu，1985）。1942年，卢于道在中国科学社生物研究所工作，根据解剖黄鼠狼、狸猫、豹及大熊猫等动物的脑的经验，著成《脑之进化》一书，获得国家自然科学奖二等奖（Hu，2014）。

1941年，他在国立中央大学心理研究所对不同动物（如蝙蝠、刺猬、鼹鼠、白鼠、露鼠、兔、穿山甲、猫和狗）的成年大脑的端脑部分进行大量的显微研究，以确定哺乳动物脑的进化（Loo，1941a）。卢于道指出，有一群在不同哺乳动物之中都有出现的细胞是退化的侧丝（Loo，1941d）。1947年，卢于道重新使用他于1941年采集到的哺乳动物的大脑，发现纹杏复合体（除了屏状核）会随着哺乳动物的进化而分化（Loo，1941b）。1957年，卢于道比较了猫、狸猫、熊和大熊猫的大脑，通过研究确定大熊猫属于熊类（Loo，1957a，1957b）。基于上述研究，卢于道认为直向演化理论不一定有效（卢于道，1948）。

再次，通过对大脑皮层内的神经细胞内胸腺核酸、核蛋白等的探索，卢于道推动了神经化学领域的发展。他发现细胞核内的胸腺核酸的分布会随细胞类型的差别而不同（Loo，1936a）。在另一个关于正常成年人神经细胞内胸腺核酸分布的研究之中，他发现胸腺核酸（这种核蛋白复合物的酸性部分）同时出现于某类特型细胞（如浦肯野细胞）的核内与核外（Loo，1936b）。

除了上述提及的研究成果，卢于道分析了神经系统视角下的人类体质。卢于道提出："人类体质是指生理结构和个体与健康和疾病倾向的相关关系。"（Loo，1941c）获得教育部的资助和马客谈、吴南轩、陈立（他们是中国科学社生物研究所动物科学系的同事）等人的帮助之后，卢于道检查了我国青少年不同生理或心理类型的生理或心理健康（卢于道，1945）。

最后，卢于道促进了科学普及工作的开展，并培育了下一代年轻学者。作为《科学画报》的创始人，卢于道影响了很多年轻人，激发他们对科学的兴趣。例如历史学家、古文字学家李学勤（李学勤和张耀南，2008）。张香桐自本科二年级起跟随卢于道学习神经解剖，并进入由卢于道建立的解剖实验室；赵翰芬亦受卢于道影响，成为神经组织学的技术员（张维，2003）。卢于道写了大量科普文章和部分图书章节，如《活的身体》《科学概论》。

卢于道是心理学家、生理解剖学家，也是爱国人士。对于他而言，心理学是通向科学研究道路的第一步。在某种程度上，神经解剖学是心理学的基石，这也是他学习神经解剖学的初衷。作为一个爱国人士，卢于道潜心学术二十载，推动了我国神经心理学的发展。

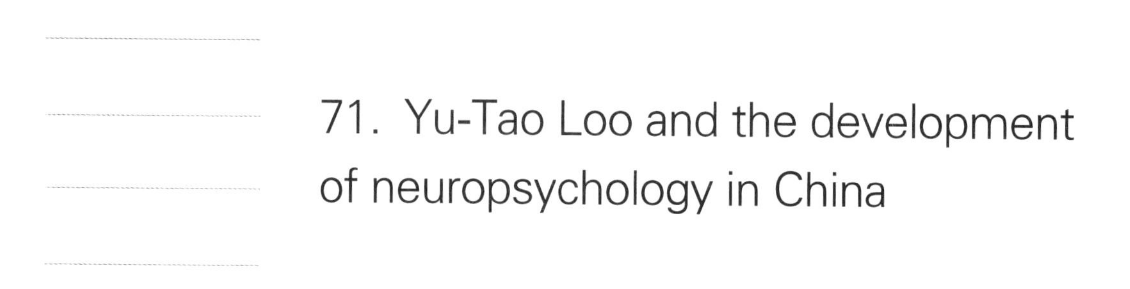

71. Yu-Tao Loo and the development of neuropsychology in China

Yu-Tao Loo (1906—1985) was one of the Chinese neuropsychologists who acted as a pioneer of structure and functions studies of human cerebral cortex (Qian, 2013) in China. Loo made great contributions in laying a solid foundation for neuroanatomy and physiological psychology in China. Meanwhile, his academic research broadened the horizons of this field. Besides, as a scientist, he wrote a series of popular-science articles and books to disseminate scientific knowledge. He was also a firm patriot and keen on construction in China.

Yu-Tao Loo was born on January 9, 1906 in Yinxian County, Zhejiang Province. In 1921, Loo got accepted into in the Psychology Department of Southeast University (i.e., National Central University), and transferred to Biology Department four years later. After receiving his Bachelor of Science in 1926, Loo won a scholarship sponsored by the Zhejiang Provincial Government to study abroad at the Department of Anatomy of University of Chicago.

Fortunately, Charles Judson Herrick, an illustrious anatomist who has made exceptional contribution to psychobiology, became Loo's supervisor. In 1930, under Herrick's guidance, Loo completed his doctoral dissertation The Forebrain of the opossum, *Didelphis virginiana*. In his study, Loo presented the gross and minute anatomy of the opossum endbrain. The final sentence in his doctoral dissertation revealed the real value and significance of the minute anatomy of the internal structure. That is, "It is hoped that this examination of its internal structure will facilitate the further prosecution of such studies. The generalized character of this brain and the great development of some parts which are reduced or rudimentary in higher mammals also aid in clarifying the general principles of forebrain morphology" (Loo, 1931). The scholars of later generations gave a wonderful comment that his article comprehensively described the anatomy of gross neuroanatomical structures in marsupials (Karlen and Krubitzer, 2007). Moreover, Web of Science shows that Loo's dissertation has been cited for 247 times so far. Even contemporary

Yanyan Qian[1], Wei Chen[2], Shengjun Wen[3]

1 School of Psychology, Nanjing Normal University, Nanjing 210097, China

2 Department of Psychology, Shaoxing University, Shaoxing 312000, China

3 Department of Cognitive Neurology, Hertie Institute for Clinical Brain Research, University of Tübingen, Tübingen 72076, Germany

Correspondence: anti-monist@163.com (W. Chen)

scientists are developing their scientific researches based on Loo's exceptional results on brain areas in *Didelphis*, such as thalamic nuclei and claustrum (Hinova-Palova et al., 2008; Mathur, 2014). Under Herrick's recommendation, Loo's doctoral thesis was published in the *Journal of Comparative Neurology*. Accompanying with this paper, he received a $20000 grant from the Rockefeller Brothers Fund.

In 1930, Loo was recruited as an adjunct professor by the Medical College of National Central University as soon as he returned to China. His responsibilities included giving lectures on experimental anatomy to the university students. Between 1931 and 1939, he worked as a researcher in Psychology Institute of Academia Sinica and took charge of the neuroanatomy laboratory. In 1937, Loo became the main copywriter of the *Chinese Journal of Psychology* where he published a paper The Functions of the cerebral cortex, an interpretation that was the only article about physiological psychology in four issues of this journal.

During the War of Resistance against Japanese Aggression, Loo was entrusted by Hong-jun Ren to hold the post as acting director-general of Chinese Science Society, developing social activities in the Rear Areas. In the spring of 1941, Loo moved to Guiyang where he worked as a professor at Xiangya School of Medicine. Between 1941 and 1942, Loo moved to Chongqing, and became an associate professor of Institute of Biology of Chinese Science Society. On July 1, 1942, Loo served as editor-in-chief for resume publication of Chinese periodical *Science* and published several scientific popular articles himself. Since the autumn of 1942, Loo took a leadership role in Fudan University. He successively held the post of director of Biology Department, assumed the officer of Dean of the School of Science and also served as head of the teaching and research section of Human and Animal Physiology Department. In 1954, Loo established the major of human body and animal physiology with Zongpeng Sun.

In 1950, Loo attended the National Congress of Natural Science Workers in China and became a committee member in preparatory office of the founding committee of the Chinese Psychological Society (Jing, 2001). The Chinese Academy of Sciences (2004) carried out a nationwide election to summarize the contributions of natural scientists from 1949 to 1950 and ranked the natural scientists according to the significance of their academic contributions. As a result, Loo was nominated in two scientific disciplines in this election. He ranked 50th in experimental biology group and 23rd in psychology group.

After the founding of the People's Republic of China, Loo took an active part in the work of Chinese societies and science popularization. In 1947, Loo was elected as the member of electoral commission of China Zoological Society that had just reestablished. In July 1947, Loo became the first chairman of board in Chinese Society for Anatomical Sciences. In 1950, Loo was one of the 15 principals to organize Chinese Psychological Society. And in May, Loo took part in the creation of Chinese Association of Anthropology with Tingliang Woo, Chungshee H. Liu and so on.

In August 1954, Loo was elected as president of Shanghai Science and Technology Popularization Society. Between November 20 and 23, 1958, Loo was elected as first vice-chairman of Shanghai Association for Science and Technology and stayed the course for two terms.

In June 1979, Loo and Hsiang-Tung Chang jointly presented proposals to set up an institution of brain research in Chinese Academy of Sciences. In 1980, Shanghai Brain Research Institute of Chinese Academy of Sciences was established in Shanghai. With their efforts, the brain research in China entered into a new stage.

As Loo said, "His research pattern can be categorized into three directions after returning to China. The first of these is on Chinese brains. The second aspect is on continuing to doing research on Herrick's comparative neurology. And the third aspect is on detecting nerve cells by means of microscopic chemistry." (Loo, 1948)

First, Loo made pioneering contributions in the study of tissue and structure of human brain. In 1929, through studying the development of the cerebral cortex by dissecting fetal brains, Loo wanted to find out a special kind of cells—labile type cells which are chiefly responsible for higher mental activity and knew if there are any cells remaining to have more embryonic potency than others. Loo summarized his research in a paper entitled On Formation of human cerebral cortex ontogenetic study with a discussion on the functions of different cortical layers (Loo, 1929). This paper has rendered him one of the first scientists engaged in structure and function in the central nerve system in China (Xu, 2000). In 1931, *Neuroanatomy*, compiled by Loo, became the first textbook on the anatomy of human nervous system in China that systemically introduced the structure and function of human central nervous system.

Building on previous human brain research, Loo compared the human brain of Chinese with that of Negro and White. In 1926, Shellshear Joseph Lexden, a scholar in Hong Kong, believed that Chinese brains were more anthropoid (Shellshear, 1926). This formed the basis of Shellshear's speech that Chinese brain is not as powerful as European at the First International Union of Anthropological and Ethnological Sciences held in London in 1934. This incidence solidified Loo's decision to explore the difference in the brain between Chinese and Europeans (Loo, 1934). During Loo's stay in the Case Western Reserve University, Cleveland, Ohio, United States, he observed and collected statistical data on fifty Chinese brains in comparison with fifty Negro and fifty White. Though Loo's research findings verified Shellshear's research data, Loo still rebutted Shellshear's claim with a supporting detail that sulcus does not appear in ontogentically fetal brains. Furthermore, Loo proposed three hypotheses to contradict Shellshear's previous conclusion that Chinese brains is inferior to Western brains and explained the higher percentage of sulcus lunatus in Chinese brains: ① visual area may be better developed in Chinese; ② parietal may be more advanced in Chinese; ③ the longitudinal axis of the brachycephalic brain is shorter in Chinese (Loo, 1933).

Second, Loo continued to make outstanding contribution to comparative neurology. Loo performed a systematic comparison in different mammals. From 1932 through 1937, Ho-Nien Chu, his fellow classmate collaborated with Loo to study the vasomotor responses of the midbrain with the Horsley-Clarke apparatus in the cat. They discovered a special phenomenon named groaning reaction (Loo and Chu, 1937a, 1937b). It was the first recorded observation of groaning response. Moreover, they imbedded electrode in cat's different brain regions to perform chronic experiment (Chu, 1985). In 1942, as an associate professor of Institute of Biology of Chinese Science Society, Loo completed a book entitled *The Evolution of the Brain* after cumulating experience in dissecting the brain of weasel, civet cat, leopard, panda and so on. This book won Loo the Second Class Prize of National Natural Science (Hu, 2014).

In 1941, while at the Institute of Psychology, Academia Sinica, Loo performed an extensive microscopical survey on the septum of adult brains in different mammals (e.g., bat, hedgehog, mole, mouse, rat, rabbit, pangolin, cat and dog) to expose the evolution of mammalian brains (Loo, 1941a). Loo identified a group of cells which were consistently present in various different mammal species as the vestigial paraphysis (Loo, 1941d). In 1947, Loo reused the brains of mammals that he collected in 1941, and discovered that the strio-amygdaloid complex (except the claustrum) underwent differentiation with the evolution of the mammals (Loo, 1941b). In 1957, Loo compared the brains of cats, leopard cats, bears and pandas, through these studies Loo found that pandas belong to bears (Loo, 1957a, 1957b). Based on above research, Loo proposed that the orthogenesis theory may not be valid (Loo, 1948).

Third, by focusing on thymus nucleic acid, and nuclear protein, etc. in brain cortex, Loo contributed in the field of neurochemistry. He found that the histochemical pattern of the distribution of the thymo-nucleic acid in the nucleus varied in different cell types (Loo, 1936a). In another study on the distribution of thymo-nucleic acid in the normal adult nerve cells, Loo showed that thymo-nucleic acid, the acid part of the nucleo-protein compound, presented both intranuclearly and extranuclearly in a specialized cell type (e.g., the Purkinje cell) (Loo, 1936b).

In addition to the contributions mentioned above, Loo analyzed human constitution from the perspective of nervous system. Loo (1941c) supposed that, "Human constitution is the physical make-up and their interrelationships of an individual with predisposition to health and disease" . With the financial sponsorship from the Ministry of Education and assistance from Ketan Ma, Nan-Hsuan M. Woo and Li Chen (fellows from Department of Animal Science, Institute of Biology, Chinese Society of Science), Loo examined physical and mental health of Chinese adolescents either in physiological or psychological type (Loo, 1945).

Finally, Loo contributed to the promotion of popularization of scientific knowledge, and fostered the next generation of young scholars. As one of the founding members of *Science Illustrated*, Loo influenced many young people to pay close attention to science, such as Xueqin Li, a well-known contemporary historian and expert in ancient writing (Li and Zhang, 2008). Loo also cultivated additional junior faculty members. For instance, Hsiang-Tung Chang learned neuroanatomy from Loo in his sophomore year and entered Anatomy laboratory established by Loo; Hanfen Zhao was influenced by Loo and became a technologist specialized in neurohistology (Zhang, 2003). Loo wrote a series of popular science articles and chapters, such as "The living body" , "The introduction to science" .

Loo is known as a psychologist, a neuroanatomist and a patriotic personage. For Loo, psychology was the first step on the road towards a career in scientific research. Neuroanatomy, to a certain degree, was a stepping stone to psychology, which made him study neuroanatomy. As a patriotic personage, Loo's rich accumulation had broken forth vastly for 20 years, which contributed tremendously to the development of neuropsychology in China.

Figures

Fig.1 Dr. Yu-Tao Loo (left) with his colleague Dr. Ho-Nien Chu (right) in 1933

References

A general report on national scientific expert by Chinese Academy of Sciences between 1949 and 1950 (2004) China Historical Materials of Science and Technology 25(3): 228–249. (中国科学院1949—1950年全国科学专家调查综合报告. 2004. 中国科技史料, 25(3): 228–249.)

Chu HN (1985) Observation and observation again, experiment and then experiment again, shall be courageous in envisioning and innovating. Prog Physiol Sci 16(4):289–291. (朱鹤年. 1992. 观察再观察, 实验再实验, 敢于设想, 敢于创新. 生理科学进展, 16(4):289–291.)

Hinova-Palova DV, Edelstein L, Paloff A, et al (2008) Neuronal nitric oxide synthase immunopositive neurons in cat claustrum—a light and electron microscopic study. J Mol Histol 39(4):447–457.

Jing QC (2001) A History of 80 Years on Chinese Psychological Society: 1921—2001. Beijing: People's Education Press. (荆其诚. 2001. 中国心理学会80年: 1921—2001. 北京: 人民教育出版社.)

Karlen SJ, Krubitzer L (2007) The functional and anatomical organization of marsupial neocortex: Evidence for parallel evolution across mammals. Prog Neurobiol 82(3):122–141.

Li XQ, Zhang YN (2008) A Talk on Chinese Civilization by Li Xueqin. Beijing: Orient Press. (李学勤, 张耀南. 2008. 李学勤讲中国文明. 北京: 东方出版社.)

Loo YT (1929) On formation of human cerebral cortex, an ontogenetic study with a discussion on the functions of different cortical layers. Anat Anz 68:305–324.

Loo YT (1930) The forebrain of the opossum, *Didelphis virginiana* Part I. Gross anatomy. J Comp Neurol 51(1):13–64.

Loo YT (1931) The forebrain of the opossum, *Didelphis virginiana* Part Ⅱ. Histology. J Comp Neurol 52(1):1–148.

Loo YT (1933) The fissural patterns of Chinese brains. Trans Sci Soc China 7:139–144.

Loo YT (1934) The cerebral cortex of a Chinese brain. Monog Natl Res Inst Psychol No. 6.

Loo YT (1936a) The thymo-nucleic acid in normal nerve cells. Chin J Zool 2:1–16.

Loo YT (1936b) A comparative study of the distribution of the thymo-nucleic acid in the cells of the cerebral cortex. Chin J Zool 2:187–194.

Loo YT (1937) Thymonucleic acid in Purkinje Cells. J Comp Neurol 67(3):423–431.

Loo YT (1941a) The mammalian endbrain I. The septum. Contrib Biol Lab Sci Soc China Zool Series 15(3):29–69.

Loo YT (1941b) The mammalian endbrain Ⅱ. The strio-amygdaloid complex. Contrib Biol Lab Sci Soc China Zool Series 16(1):1–25.

Loo YT (1941c) A neurological analysis of the constitution. Contrib Biol Lab Sci Soc China Zool Series 15(7):121–136.

Loo YT (1941d) The paraphysis in adult mammalian brains. Contrib Biol Lab Sci Soc China Zool Series 15(4):69–76.

Loo YT (1945) The physiological or psychological type of Chinese adolescents. Fudan J (2):259–280. (卢于道. 1945. 青年身心的类型问题. 复旦学报, (2): 259–280.)

Loo YT (1948) The anatomy in China in the past thirty years. Science 20:199–203. (卢于道. 1948. 三十年来国内的解剖学. 科学, 20: 199–203.)

Loo YT (1957a) External features of the brain of the giant panda, *Aeluropus melanoleucus*. Acta Anat Sin 2(3):221–231.

Loo YT (1957b) External features of the civet cat (*Viverra zibetha*) brain, and its characteristics in carnivora. Acta Anat Sin 2(4):301–306.

Loo YT, Chu HN (1937a) On the vasomotor centers in the forebrain and the midbrain. Chin J Physiol 11: 295–300.

Loo YT, Chu HN (1937b) The groaning response to mesencephalic stimulation. Chin J Physiol 11(3):301–304.

Mathur BN (2014) The claustrum in review. Front Syst Neurosci 8(48):1–11.

Qian WC (2013) Overview of Academic Achievement of Renowned Chinese Scientists in 20th Century. Biology, Part 2. Beijing: Science Press. (钱伟长. 2013. 20世纪中国知名科学家学术成就概览·生物学卷·第二分册. 北京: 科学出版社.)

Shellshear JL (1926) The occipital lobe in the brain of the Chinese with special reference to the sulcus lunatus. J Anat 61(1):1–13.

Xu K (2000) A memorial of the pioneer in neuroscience research in China—the centenary of Tao Lieh's birth. Chin J Neurosci 16(1):82–83.

Zhang W (2003) Xiangtong Zhang. Shanghai: Shanghai Science and Technology Education Press. (张维. 2003. 张香桐传. 上海: 上海科技教育出版社.)

72. 一位活了110岁的长寿智者——郑集：中国生物化学和营养学奠基人之一

你相信“食物相克”吗？你敢把螃蟹和柿子一起吃吗？其实，完全没有必要担心中毒。因为早在80年前就有人对此进行过实验验证，证明了所谓的“食物相克”是迷信。这个人就是我国著名的生物化学家、营养学家郑集，一位活了110岁的长寿智者。

郑集（图1）1900年出生于四川省南溪县。1928年毕业于国立中央大学，获学士学位；1931年在美国俄亥俄州立大学获得硕士学位；1934年在美国印第安纳大学获得博士学位。郑集于1934年回到中国，应秉志教授的邀请到中国科学社生物研究所负责创办生物化学研究室（Li and Kang，2010）。1936年，他被聘为国立中央大学医学院生化科教授兼主任（图2）。1945年，他在国立中央大学医学院创办生物化学研究所，这是中国教育史上第一个培养生物化学研究生的正式机构。1957年，郑集创办南京大学生物化学专业，两年后，正式担任南京大学生物系教授，并兼任生物化学教研室主任。郑集1934年被选为美国科学家荣誉学会会员，1956年被评为一级教授，1996年被选为美国国家科学促进会会员。

图1　郑集教授（1900—2010）

图2　20世纪30年代郑集教授在国立中央大学医学院工作

作者：张翮
中国科学技术大学，合肥230026，中国
邮箱：zhangheahu@163.com

郑集赴美留学期间，曾先后在芝加哥大学和耶鲁大学学习，有机会接触到当时著名的生物化学家，例如植物蛋白质化学和营养学家L. B. Mendel和H. B. Vickery，维生素B营养学家G. R. Cowgill，矿质营养学家A. H. Smith等。由于对蛋白质化学和营养学有着浓厚的兴趣，他选择到印第安纳大学R. T. Hartman博士的实验室从事大豆蛋白的提取和理化性质的研究，获得博士学位。之所以选择研究大豆蛋白，一个重要的原因是它是当时中国人膳食中蛋白质的主要来源。郑集的科学研究有一个很大的特点，就是理论联系实际，关注现实问题。1934年回国后，他密切结合当时的国情，在继续深入研究大豆蛋白营养效价的同时，开展了一系列的营养调查和食物分析研究，如“南京冬季膳食调查”（郑集等，1935）、“全米、全麦营养价值的幼龄大鼠生长试验”（Cheng and Tao，1935）等。这些研究工作针对改善国民营养状况面临的实际问题，收到了实效。尤其值得一提的是郑集的“食物相克”实验，以动物和人体实验的确实证据有力地驳斥了“食物相克”这个在民间流传甚久的谬论。1935年秋，南京民间传说香蕉和芋艿同食会因食物相克而导致中毒，影响颇大。郑集认为对于此类与民众日常生活关系密切的问题，应当通过科学实验明断是非，释疑解惑。他先是亲身实验证明香蕉和芋艿同食无毒，而后从我国古籍中搜集了历代流传的所谓“相克”的食物共计184对，再从中选出14对在日常生活中容易遇到的组合，包括螃蟹和柿子、花生和黄瓜等，用白鼠、猴子和狗进行动物实验。最后，还挑选出更为常见的7对进行人体实验，他自己和一名同事亲身参与了实验。在食用后24小时内详细观察实验动物和人的表情、行为、体温及粪便颜色与次数等，结果均正常，毫无中毒迹象（Cheng，1936）。

1974年，74岁高龄的郑集提出要开创我国衰老生化机制研究的新方向。他从调查研究入手，先是调查了100多位70岁以上的健康老人，分析他们健康长寿的原因，总结出10条健康长寿的经验。而后的十多年里，他与合作者共完成了十余篇论文，内容涉及衰老与红细胞及全脑、肝、胰、胸腺等器官中酶、核酸等成分变化的相关性，取得了许多有价值的成果。在一系列细胞和分子水平的生化研究基础上，郑集创造性地提出了衰老机制的“代谢失调学说”，为我国的衰老生物化学研究奠定了基础。他坚持理论与实践相结合，以自己抗衰老的杰出实践印证着理论。百岁高龄的他，身体健康，思维敏捷，仍经常去办公室工作（图3）。据说，常挂在他嘴边上的一句话是：“只要我明年健康允许，我要完成……”（萧信生和金以丰，2000）

图3　郑集教授百岁工作照

郑集一生著述颇丰，涉及专著、教材、科普作品等多种类型。1938年在四川成都华英书局出版的英文版《生物化学实验手册》(Cheng，1938)是我国第一本自编的生物化学参考书；1947年由重庆正中书局出版的《实用营养学》(郑集，1947)是我国最早正式出版的两本营养学专著之一；他主编的《普通生物化学》(第2版)(郑集，1985)教材获得全国普通高等学校优秀教材二等奖。临终前，他的系列科普书籍《最好的医生是养生》(Cheng，2010)由江苏教育出版社出版，为他长达110年的传奇人生做下最后一个注脚。

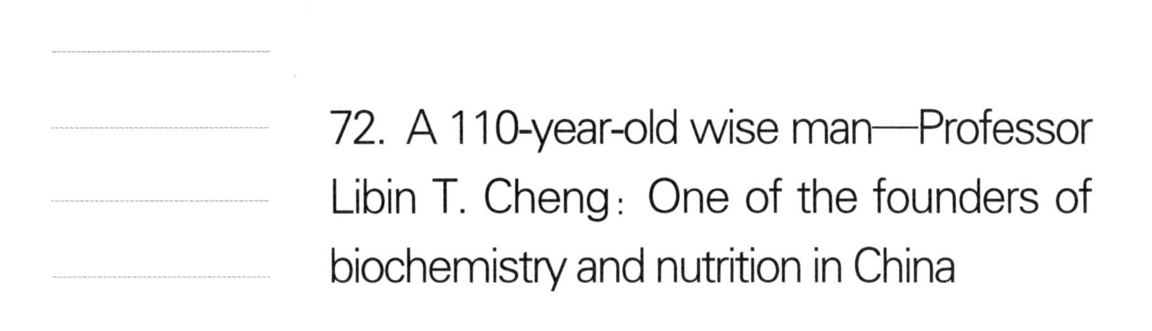

72. A 110-year-old wise man—Professor Libin T. Cheng: One of the founders of biochemistry and nutrition in China

Do you believe in so-called food-combination poisoning? Do you have the courage to eat crab and Chinese persimmon together? In fact, there is no need to worry about poisoning. As early as 80 years ago, it was proved that so-called food-combination poisoning is a superstition, through experiments by a man named Libin T. Cheng (1900—2010), who was one of the founders of biochemistry and nutrition in China and who also became a 110-year-old wise man.

Libin T. Cheng was born in Nanxi County, Sichuan Province, in 1900. He received his bachelor's degree from National Central University in 1928, master's degree from Ohio State University in 1931, and doctor's degree from Indiana University in 1934. In 1934, Cheng returned to China and took charge of establishing the Department of Physiological Chemistry in the Biological Laboratory of the Science Society of China, invited by Professor Bing Zhi (Li and Kang, 2010), a pioneer of modern biology in China. In 1936, Cheng was appointed as the professor and director of the Department of Biochemistry in the Central Medical School of National Central University. In 1945, he established a biochemistry research institute in the Central Medical School, which was the first formal organization to produce biochemistry postgraduates in the history of education in China. In 1957, Cheng established biochemistry as a major at Nanjing University. Two years later, he formally served as the professor of the Department of Biology and director of the biochemistry teaching and research office. In 1934, Libin T. Cheng was selected as a member of the Sigma Xi Society; in 1956, he was awarded as the First Grade Professor; in 1996, he was selected as a member of the American Association for the Advancement of Science.

During his postgraduate education in the United States, Cheng studied at the University of Chicago and Yale University successively and had the opportunity to contact with the famous biochemists at that time, such as plant protein chemistry experts and nutritionists L. B. Mendel and H. B. Vickery, Vitamin B nutritionist G. R. Cowgill, and mineral nutritionist A. H. Smith.

He Zhang[1,2]

1 Department for the History of Science and Scientific Archaeology, University of Science and Technology of China, Hefei 230026, China

2 School of Marxism, Bengbu Medical College, Bengbu 233030, China

Correspondence: zhangheahu@163.com

Because of a strong interest in protein chemistry and nutrition, he chose to conduct researches on the extraction and physicochemical properties of soybean protein in the laboratory of R. T. Hartman at Indiana University, from which he received a doctor's degree. One of his important reasons for choosing soybean protein as a research subject was that it was the main source of protein in Chinese people's diets at that time. Cheng's scientific research had the great feature of integrating theory with practice and focusing on practical problems. After returning to China in 1934, he continued to conduct in-depth research on the nutritive value of soybean protein in close combination with the national conditions. At the same time, he carried out a series of nutritional investigations and research on food analysis, such as "Survey on the winter diet in Nanjing" (Cheng et al., 1935) and "The nutritional value of whole wheat and whole rice in regard to the growth, hemoglobin and calcium and inorganic phosphorus of the serum and bone of the albino rat" (Cheng and Tao, 1935). Aimed at the practical problems faced in improving the national nutritional status, these studies had actual effects. It is particularly worth mentioning that Cheng's experiments on "food-combination poisoning" in animals and human beings provided substantial evidence that effectively disproved this long-rumored fallacy. In the summer of 1935, so-called poisoning by banana and yam was rumored among the folk of Nanjing and had a great influence at that time. Cheng believed that such problems, which closely related to the daily life of people, should be judged and explained through scientific experiments. At first, he ate banana and yam simultaneously in order to prove that they can be eaten together without causing poisoning. Then, he collected 184 pairs of so-called poisonous food combinations from the ancient Chinese books and selected 14 pairs from them, including crab and persimmon, peanut and cucumber, and so on, which were common in the daily life. Next, the food combinations were prepared and fed to albino rats, monkeys, or dogs for two days successively, according to the usual home method. He also selected the seven most common pairs for human experiments on himself and one of his colleagues. After intake of every food combination, the expression, behavior, body temperature, and color and frequency of the excreta of the animals and human beings were observed closely for 24 h. All of the results were normal and showed no noticeable symptoms of poisoning (Cheng, 1936).

In 1974, Cheng, 74 years old, created a new direction for research into the biochemical mechanisms of aging in China. Starting with investigation and research, he surveyed more than 100 healthy people over 70 years old, analyzed the reasons for their health and longevity, and summed up 10 health and longevity experiences. In the following 10 years, he completed more than 10 papers, together with his cooperators, with regard to the correlation of aging and enzyme, nucleic acid, and other compositional changes in the red blood cells, whole brain, liver, pancreas, thymus gland, and other organs, which yielded a lot of valuable results. Based on a series of biochemical studies at the cellular and molecular levels, Cheng creatively proposed a "metabolic imbalance theory" of the aging mechanism, which laid the foundation for research into aging chemistry in China. He insisted on the combination of theory and practice, and he verified the theory with his own outstanding practice of anti-aging. At the age of 100, he was healthy and quick-thinking, and he still went to work quite often at his office. He is quoted as always saying "As long as I am healthy in the next year, I want to complete this and that." (Xiao and Jin, 2000, p. 293)

Libin T. Cheng wrote numerous books, including monographs, textbooks, popular science books, and others. The English version of *A Laboratory Manual of Biochemistry*, published by Canadian Methodist Mission Press in 1938, is the first self-compiled biochemistry reference book in China (Cheng, 1938); *Applied Nutrition*, published by Chengchung Book Company in 1947, is one of the two earliest officially published nutrition monographs (Cheng, 1947); *General Biochemistry* (Version 2), under his general editorship, won the second prize for outstanding textbooks of the colleges and universities in China (Cheng, 1985). Just before his death, his series of popular science books *The Best Doctor is Health Preserving* was published by Jiangsu Education Publishing House, which created a last footnote for his legendary life (Cheng, 2010).

Figures

Fig.1 Professor Libin T. Cheng (1900—2010)

Fig.2 Libin T. Cheng at work in the Medical School, National Central University in the 1930s

Fig.3 Libin T. Cheng working at his office at the age of 100

References

Cheng LT (1936) Are the so-called poisonous food combinations really poisonous? Cont Biol Lab Sci Soc China, Zool Ser 11(9):307–316.

Cheng LT (1938) A Laboratory Manual of Biochemistry. Chengdu: Canadian Methodist Mission Press.

Cheng LT (1947) Applied Nutrition. Chongqing: Chengchung Book Company. (郑集. 1947. 实用营养学. 重庆: 正中书局.)

Cheng LT (1985) General Biochemistry (2nd edn). Beijing: Higher Education Press. (郑集. 1985. 普通生物化学(第2版). 北京: 高等教育出版社.)

Cheng LT (2010) The Best Doctor is Health Preserving. Nanjing: Jiangsu Education Publishing House. (郑集. 2010. 最好的医生是养生. 南京: 江苏教育出版社.)

Cheng LT, Tao H (1935) The nutritional value of whole wheat and whole rice in regard to the growth, hemoglobin and calcium and inorganic phosphorus of the serum and bone of the albino rat. Cont Biol Lab Sci Soc China, Zool Ser 11(3):97–107.

Cheng LT, Tao H, Zhu ZG (1935) Survey on the winter diet in Nanjing. Science 19(11):1753–1758. (郑集, 陶宏, 朱章庚. 1935. 南京冬季膳食调查. 科学, 19(11):1753–1758.)

Li M, Kang L (2010) Bing Zhi: Pioneer of modern biology in China. Protein & Cell 1(7):613–615.

Xiao XS, Jin YF (2000) Congratulation and best wishes to you, Professor Zheng Ji, on the occasion of your 100th birthday. Chin J Biochem Mol Biol 16(3):289–293. (萧信生, 金以丰. 2000. 祝贺生物化学家、营养学家郑集教授百岁华诞. 中国生物化学与分子生物学报, 16(3):289–293.)

73. 黎鳌：烧伤医学的先驱

2020年是黎鳌院士逝世21周年纪念。虽然他已经离我们远去，但是“黎鳌精神”一直激励我们继续勇攀科学高峰，为祖国和人民作出新的贡献。

图1 黎鳌院士（1917—1999）

黎鳌，我国烧伤医学的开拓者和奠基人，中国工程院院士（图1）。黎鳌院士原名黎升旭，1917年5月出生于湖南长沙，1941年毕业于国立上海医学院，曾任国立中正医学院副教授、第三军医大学（现为陆军军医大学）烧伤研究所所长、第三军医大学副校长等职务。黎鳌院士对烧伤吸入性损伤的发病机理和防治开展了多学科综合性研究，研究成果先后获得国际烧伤学术最高奖——伊文斯奖（1994年）、国家科学技术进步奖一等奖、军队科学技术进步奖一等奖等奖励。黎鳌院士不仅为我们留下了丰硕的科研成果，还有更为宝贵的精神财富——“黎鳌精神”：赤诚报国的爱国精神、艰苦奋斗的创业精神、勇闯一流的攻关精神、精心育才的人梯精神。

赤诚报国的爱国精神

黎鳌院士中学毕业时，因学习成绩优异曾被学校免试推荐到国立中央大学物理系或数学系就读。然而，当时的中国缺医少药、民不聊生的社会现实深深刺痛了他，加之父亲病情严重，怀揣济世救父愿望的他毅然弃“理”从“医”。1935年他以优异成绩考入国立上海医学院，从此开启了医学生涯。毕业后，他先后在国立中正医学院、第三军医大学等院校附属医院工作，救治了成千上万的伤病员。

20世纪50年代末，为救治因炼钢而出现的大量烧伤病人（李照洲等，1983），黎鳌院士临危受命，调整研究方向，集（第三军医大学）全校之力，从基础和临床应用

作者：廖亚玲，邹全明，顾江

陆军军医大学药学与检验医学系，重庆400038，中国

邮箱：jianggu2012@163.com（顾江）

角度，对烧伤的发病机制和救治措施开展深入研究，扭转了当时大面积烧伤治疗困难的被动局面，显著提高了治愈率，为保障人民健康作出了巨大的贡献。

艰苦奋斗的创业精神

1958年黎鳌院士筹建烧伤专科病房时，仅有6张床位、4位医生。如今，因为黎鳌院士的全身心投入和国家军队的重要支持，烧伤科已发展成国家重点学科，设立国家重点开放实验室。他通过严密观察，于1963年初步探索出一整套烧伤治疗方案（图2）；他首先提出“从整体出发，维护机体本身抗病能力”的观点；他提倡适当“简单清创”以减少对患者的干扰和刺激；他提倡尽早清除坏死组织、覆盖封闭创面，简化了烦琐的隔离措施；通过实测人体表面积，制定了适合我国人群烧伤面积计算的“新九分法”；通过大量的病例调查，总结出符合我国国情的休克期补液公式等。黎鳌院士的这些具有我国特色的烧伤治疗理念和方法，在国内得到同行认可且迅速推广，为我国烧伤医学的发展奠定了坚实基础（黎鳌，2001）。

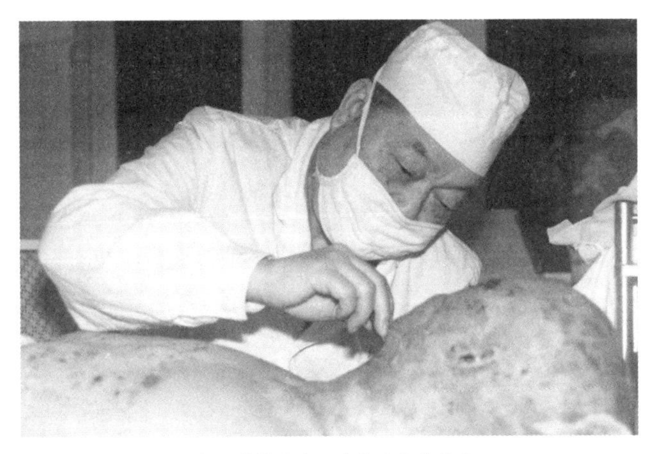

图2　黎鳌院士正在检查烧伤患者

勇闯一流的攻关精神

吸入性损伤病死率高，是当时国内外烧伤领域最为棘手的难题之一。其致病过程复杂，当时还缺乏有效的救治方案。为解决这一难题，黎鳌院士组织全校科研力量，包括16个科室近百名科研骨干进行课题攻关，取得了一系列重大科研突破，基本弄清了吸入性损伤的病理生理变化过程及机制，还发现了其疾病进展规律，总结出了一整套提高治愈率的早期治疗方案，撰写了《吸入性损伤》专著（黎鳌和杨宗城，1993），为广大临床医生和科研人员提供了翔实的第一手资料，获得了国家科学技术进步奖二等奖。

20世纪80年代，黎鳌院士与史济湘教授合作开始新的科研难题攻关：烧伤早期损害发病机理与创面愈合机理。经过研究发现并证实失控性炎性反应是严重休克、缺血缺氧和内脏器官损伤的主要原因，肯定了巨噬细胞在烧伤后失控性炎性反应中的启动作用；验证了血管内皮在烧伤后早期脏器损害发病中的核心作用；基本阐明了烧伤后早期肠道损伤和肠源性感染的发病机制（黄跃生等，2001）。这些研究发现大大缩短了我国与西方国家烧伤研究的差距，被授予国际烧伤学术最高奖——伊文斯奖，获得了国际同行的高度认可。

精心育才的人梯精神

黎鳌院士非常关心和注重青年人才的培养，主动放手让科研团队骨干挑起医疗、教学、科研的重担，为他们提供施展才华的舞台，甘为人梯，培养他们成为新的学科带头人和中坚力量（图3）。他培养的学生中，已有博士生导师5名，硕士生导师15名，博士生54名。1986年，他建立了全军青年烧伤医学优秀论文奖和黎鳌烧伤医学基金，用以支持和鼓励青年人才从事烧伤基础研究与临床工作。这些奖项的获奖者已经逐渐成为我国烧伤学科的新带头人，带领团队继续为我国烧伤医学的发展贡献力量。1996年黎鳌院士荣膺总后勤部“一代名师”称号，这是他精心育才最真实的写照。

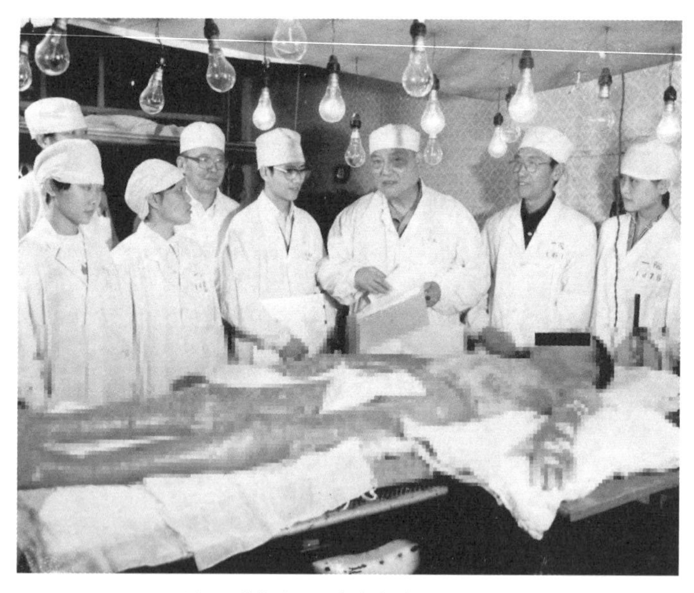

图3　黎鳌院士正在病房进行现场教学

“云山苍苍，江水泱泱，先生之风，山高水长。”黎鳌院士取得了辉煌的科学成就，开创了我国烧伤医学新的篇章，永远是我辈学习的楷模。“桃李不言，下自成蹊。”黎鳌院士奋斗一生所铸就的“黎鳌精神”必将继续发扬光大，指引一代又一代的青年学子，不畏艰难，攀登科学高峰，为新时代人民群众健康和中华民族伟大复兴事业作出更大的贡献。

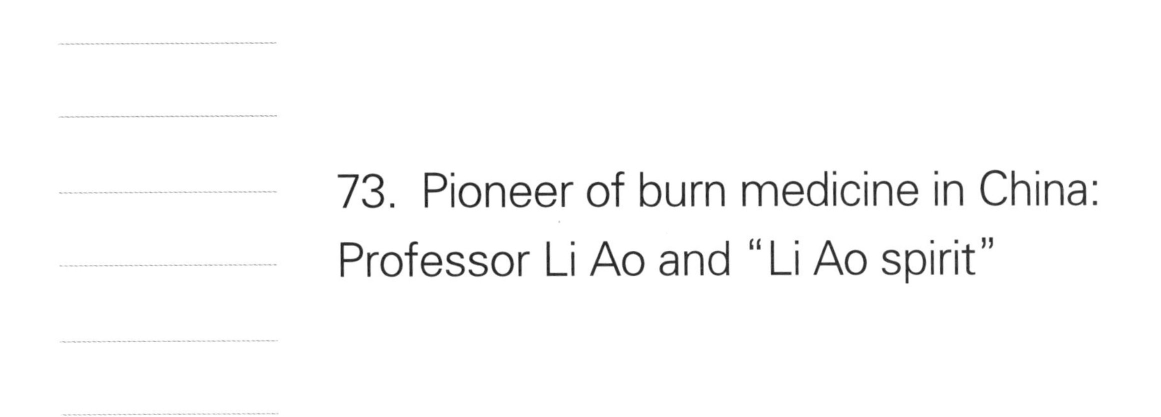

73. Pioneer of burn medicine in China: Professor Li Ao and "Li Ao spirit"

The 2019 marks twenty years since Professor Ao Li (黎鳌) passed away. Although he has left us, the "Li Ao Spirit" has lived on and continuously encourages us to climb higher in scientific research and make more contributions to our people and country.

Professor Li is a well-known Chinese expert in burn surgery and an Academician of the Chinese Academy of Engineering. Mr. Li, also named Shengxu Li (黎升旭), was born in Changsha, Hunan Province in May 1917. He graduated from the National Shanghai Medical College in 1941, and had a long and distinguished career in medicine. His various positions over the years included Associate Professor at the National Chung Cheng Medical College, Professor at the Third Military Medical University, Director of the Burn Research Institute of the Third Military Medical University, and the Vice President of the Third Military Medical University. As one of the pioneers and founders of burn medicine in China, professor Li led comprehensive multidisciplinary research on the pathogenesis and prevention of inhalation injury. For his great contribution to the field of burn research, he was not only the recipient of the Everett Idris Evans Memorial Lecture Award of the American Burn Association in 1994, but also won first prize in both the Chinese Science and Technology Advancement Award and the Military Scientific and Technological Progress Award among other accolades.

Li Ao Spirit: Selflessness, patriotism and devotion

After graduating from high school, Mr. Li was recommended to study in the Department of Physics or Mathematics at Nanjing National Central University. However, touched by the sufferings of the general public caused by a lack of medical knowledge and medicine and his father's illness, Mr. Li was determined to become a doctor. In 1935, he was given the opportunity to study at Shanghai Medical College, where he started his medical career. After graduating, he

---

Yaling Liao, Quanming Zou, Jiang Gu

National Engineering Research Center of Immunological Products, Department of Microbiology and Biochemical Pharmacy, College of Pharmacy, Army Medical University, Chongqing 400038, China

Correspondence: jianggu2012@163.com (J. Gu)

successively presided over general surgical work in the National Chung Cheng Medical College, the Third Military Medical University and other institutions, saving thousands of lives.

In the late 1950s, the number of burn patients increased drastically due to the country's efforts in increasing steel production, but there were no effective therapies available at that time (Li et al., 1983). In order to solve this urgent problem, Professor Li adjusted the direction of his research and began to investigate the mechanism of burns. He gathered the resources of the whole school (the Third Military Medical University) and achieved many advances in this field. Together with his team, Professor Li significantly improved the recovery rate of burn patients and made a great contribution to protecting people's health.

Li Ao Spirit: Hard-working entrepreneurship

In 1958, Professor Li began to build a burn ward, starting with only six beds and four doctors. From then on, he devoted all his time and energy to the research of burns. He stayed at the bedsides of burn patients day and night for first-hand observations. Moreover, in 1963, by collaborating with the other researchers in the Third Military Medical University, he successfully developed a set of novel and effective burn treatment programs. For example, he first put forward the idea that when transportation conditions are poor, patients with large-scale burns should not be transported over long distances due to the shock incurred. Instead, anti-shock treatment should be timely applied. In addition, he identified the type of bacterial pathogens and the of infections post burns after a large number of investigations. Moreover, he simplified the cumbersome isolation measures for the sake of earlier debridement. He also conducted measurements of the body surface area of Chinese people, and proposed China-specific formulations for the calculation of the burn area. On the basis of a large number of cases, he summarized the formula of fluid replacement for burn patients during the shock period (Li, 2001).

These measures greatly reversed the passive situation in treating patients with large-scale burns at that time. Professor Li's concepts and methods of burn treatment with Chinese characteristics were promptly accepted and promoted by his peers, laying a solid foundation for the development of burn medicine in China.

Li Ao Spirit: Embracing challenges

The high mortality rate of inhalation injury is a serious problem in burn treatment, and there was no systematic study focusing on this issue worldwide at that time for its complex pathogenesis. Despite his awareness of the difficulties, Professor Li chose inhalation injury as his main research direction. He organized nearly 100 researchers from 16 departments of the Third Military Medical University, and set up a burn collaboration group to focus on inhalation injury research. The research group achieved many important progressions.

For instance, this research project clarified the pathological morphology and pathophysiology of inhalation injury, revealed its pathogenesis and led to a set of early treatment measurements which reduced complications and improved the recovery rate of burn patients. These research progressions and treatment experiences were included in his monograph *Inhalation Injury* (《吸入性损伤》), which was published by People's Military Medical Press in 1993 (Li and Yang, 1993). The fruits of this research project won the 2nd Prize of Chinese Science and Technology Advancement Award.

In the 1980s, Professor Li found that severe shock, ischemia, hypoxia and subsequent uncontrolled inflammatory reactions are the main causes of internal organ damage in burn patients. To further investigate the mechanism, he and Professor Jixiang Shi (史济湘) launched the research project named "Study on the Pathogenesis of Early Burn Injury and Wound Healing Mechanism" . The project was sponsored by the National Natural Science Foundation. After five years of hard work, they discovered that macrophages act as an initiator in the uncontrolled inflammatory response after burns, verified that vascular endothelium played the central role in the pathogenesis of early organ damage, clarified the mechanism of the early intestinal damage and intestinal infections, and proposed corresponding treatment measures (Huang et al., 2001). These findings were highly recognized by international burn researchers and made Professor Li the recipient of the Everett Idris Evans Memorial Lecture Award of the American Burn Association in 1994.

Li Ao Spirit: Cultivating young talents

Professor Li attached great importance to the growth and cultivation of young scientists. He emphasized training in practice, encouraged young scientists to participate in clinical operation, classroom teaching and laboratory tasks. He provided a stage for the young generation to showcase their talents and helped them grow into principle academic leaders.

In 1986, Professor Li won the National Science and Technology Progress Award. He donated the bonus of the award to establish the Outstanding Paper Award in Burn Medicine for Young Army Scientists to encourage young scientists and technicians to engage in burn research. In 1996, Professor Li won the China Engineering Science and Technology Award and Military Technical Major Contribution Award. Again, he donated the bonus of both awards, a total of 150000 Chinese Yuan, to set up the Li Ao Burn Medicine Fund so as to reward young and middle-aged scientists who contributed to the development of burn medicine. Ten of the recipients of the Li Ao Burn Medicine Fund have become the leading scientists of burn medicine in China. In 1996, Professor Li was awarded the title of "Outstanding Mentor" by the General Logistics Department of Chinese People's Liberation Army.

Professor Li has devoted his whole life to the cause of burn medicine in China and made a lot of remarkable achievements. Although he left us twenty years ago, we are still always encouraged and inspired by the "Li Ao Spirit" of selflessness, diligence and entrepreneurship.

With the great efforts of generations of doctors and scientists, we believe burn medicine research in China will bring about more and more breakthroughs and play a leading role in the international arena.

Figures

Fig.1 Professor Ao Li (1917—1999)

Fig.2 Professor Ao Li was examing the change of burn patient for first-hand observations

Fig.3 Professor Ao Li was educating students beside the bed

References

Huang YS, Yang ZC, Xiao GX, et al (2001) Clinical analysis of measures for preventing early postburn damage in improving survival rate of burn patients. J Third Mil Med Univ 23(2):217–220. (黄跃生, 杨宗城, 肖光夏, 等. 2001. 烧伤早期损害的防治措施对提高烧伤存活率的作用. 第三军医大学学报, 23(2):217–220.)

Li A (2001) The past, present and future of burn treatment research in China. Chin J Burns 17(1):5–7. (黎鳌. 2001. 我国烧伤救治研究的过去, 现在和未来. 中华烧伤杂志, 17(1):5–7.)

Li A, Yang ZC (1993) Inhalation Injury. Beijing: People's Military Medical Press. (黎鳌, 杨宗城. 1993. 吸入性损伤. 北京: 人民军医出版社.)

Li ZZ, Wang JH, Zhu PF, et al (1983) Analysis of mortality rate of 3617 burned patients. J Third Mil Med Univ 5(3):182–190. (李照洲, 王甲汉, 朱佩芳, 等. 1983. 3617例烧伤病人病死率分析. 第三军医大学学报, 5(3):182–190.)

74. 朱锡侯：从生理学到心理学的传奇人生

为什么蜜蜂放弃睡觉、享用蜂蜜、爱情和美妙的悠闲时光？为什么它们要承受这么多的痛苦和烦恼，这个决定从何而来？因此，它们渴望的交配行为一定值得如此牺牲，它一定更美好，更快乐，并能做到一些它们做不到的事情。

——莫里斯·梅特林克，《蜜蜂的生活》，1901

朱锡侯（Si-Ho Tchou）（1914—2000），祖籍浙江绍兴，出生于吉林东北部，是中国生理学会、中国心理学会创始人之一，并担任该两会昆明地区负责人。1937年毕业于北平中法大学（图1），因学业优秀而被选送赴法留学，是中法庚子赔款的最后一批公派留学生之一。在求学法兰西的八年时间里，朱锡侯先是在法国里昂大学学习，获得生理学博士学位后，又到巴黎大学心理学院学习并获得心理学博士学位。1945年，欧亚交通刚刚恢复，获得了两个博士学位的朱锡侯，满怀一腔报国热情，几经曲折毅然回到了自己的祖国，受聘担任云南大学（简称“云大”）医学院的生理学教授和文法学院的心理学与美学教授。1956年之后，他在昆明医学院生理教研组任主任教授。1980年调到杭州大学（简称“杭大”）心理系任生理心理学教授，后又任中国心理学会生理心理学委员会委员。

图1 朱锡侯（1937年）

朱锡侯的事迹至今仍闪耀在华人世界的星空，成为后来学子的楷模。朱锡侯在留学的八年时间里取得了重要的学术成果。在里昂大学，朱锡侯先是在世界著名美学家、心理学家Étienne Souriau教授门下攻读美学、心理学（图2），继而转入医学院师从著名的生理学家Henri Cardot教授。后者是1913年诺贝尔生理学或医学奖获得者Charle Richet的高足（Stewart，2012）。

求学期间，朱锡侯经常利用假期跟随Cardot教授到位于地中海边的海洋生物观测

译者：陈巍

绍兴文理学院心理学系，绍兴312000，中国

邮箱：anti-monist@163.com

站进行专业实验，研究动物的神经细胞、神经纤维，花大量的时间记录神经电波，夜以继日地扑在小小的实验室里。他们当时关注的是海兔这种海洋无脊椎动物腺体的巨型神经元，每天都要出海捕捞海兔。朱锡侯的主要工作就是负责处理那些捕捞物，根据它们的年龄（重量）来分辨腺体和巨型神经元，对它们做出区分。朱锡侯还有一个研究课题，即不同重量的海兔的神经放电节律现象。1942年底，朱锡侯就以海兔神经细胞的生长规律及其节律性发电现象的相关研究成果获得了里昂大学理科博士学位。他的发现表明，海兔腺体和巨型神经元容易辨识和分类。另外，他还发现细胞大小只是细胞分化的一个粗略指征（Tauc，1966；Ingoglia and Sturman，1980；Ambron and Kremznes，1982；Abel and Kandel，1998）。

图2　朱锡侯在里昂

海兔能成为认知神经科学研究的重要模式生物，其实与其腹神经节含有巨型神经元有关。早在1942年，朱锡侯就研究了海兔巨型神经元，并和Arvanitaki一起发表了《海兔独立的中间神经元的增长定律》，对海兔神经系统以及个体不同时期神经元数量和相对大小的观察进行记录和定量（Arvanitaki and Tchou，1942）。在此基础上，法国科学家Tauc（1954年）和Arvanitaki（1955年）等最早开创了巨型神经元的微电极研究。得益于朱教授的方法，美国哥伦比亚大学教授E. R. Kandel所在的研究团队发现，习惯化和去习惯化并不由感受器或腮肌肉变化造成，而是由腮运动神经元（gill motor neurons）的突触电位所造成（Pinsker et al.，1970；Kupfermann et al.，1970；Carew and Kandel，1973；Castellucci and Kandel，1976；Kandel and Schwartz，1982；Voronezhskaya and Croll，2015）。Kandel研究了敏感化和习惯化两种相反作用的神经通路和细胞内分子机制。短期敏感化的记忆跟虹吸管的感知神经元和运动神经元之间的突触功效增强有关：感知神经元内的cAMP浓度水平增加，并导致钾离子通道关闭（Shapiro et al.，1980；Kandel et al.，1982；Abel and Kandel，1998）。因此，海兔的学习与记忆研究的重心转向运动神经元和中间神经元的分子机制。这种神经元细胞的可塑性与记忆现象或许存在着关联机制。时至今日，神经元可塑性尤其是神经突触的可塑性与记忆关系已经成为进展迅猛的记忆研究的一种重要假设。

科学家继续对海兔神经系统进行了较深入的研究，生理学和药理学研究成果数量均非常可观。但除了朱先生（Tchou，1942）和Arvanitaki的工作外，几乎没有哪个科学家对不同大小的海兔个体之间的神经系统相对大小或神经元的数量和大小进行定量比较研究（Hughes，1968）。神经元几乎是最难培养的细胞，所以神经生物学对特定研究方法的建立非常倚重。朱锡侯对海兔神经元的量化和放电节律研究，以及在研究过程中所建立的实验方法和标准，为后来海兔能够成为学习研究的黄金模式生物打下了坚实的基础。许多杰出神经科学家所做出的后续研究都多少受益于朱先生的开拓性工作。

例如，Kandel教授发现记忆现象与连接神经细胞的突触有关，并且揭示出其分子机制。这是一个具有划时代意义的发现，从而开启了一个全新的科学领域——认知分子生物学（cognitive molecular biology），Kandel以此获得了2000年诺贝尔生理学或医学奖的殊荣。而这种成功，显然与朱锡侯等人对海兔这种模式生物的探索和建立的基础工作不可分割（Coggeshall et al.，1966；Hughes，1968；Peretz and Lukowiak，1975；Shapiro et al.，1980）。Kandel没有选择小鼠、猴子或者人，而是睿智地选择了没有海马体的海兔，这是他成功的关键条件之一，这绝非偶然。朱锡侯在20世纪40年代对海兔腹神经节巨型神经细胞（包括R2）的研究，为海兔记忆研究从细胞水平深入到分子水平打下了基础。朱锡侯等人发现，R2神经元的直径可以增加一倍（Arvanitaki and Tchou，1942）。作为巨型胆碱能神经元的R2具有分子水平研究的极大优势，被用以研究细胞内代谢。在此基础上，包括Kandel在内的Coggeshall等（1966）和Hughes（1968）对这些巨型神经元簇进行了更详细的形态和功能研究。正是幼年海兔缩鳃反应较少的现象使得R2因为这种特异性变化而成为科学研究的对象（Ambron and Kremzner，1982）。腔壁脏神经节（parieto-visceral ganglion，PVG）的L7（控制着缩腮反应）和腹神经节的R2均具有随着个体生长而长大的特性，并对缩鳃反应有着不同表现（Peretz and Lukowiak，1975；Bailey and Chen，1988）。如果朱锡侯的工作能够继续下去，或许也可以在缩鳃反应的研究领域获得成功。

朱锡侯对生理学虽然痴迷，但他自己也说，他更大的兴趣在于实验心理学。因此，1943—1945年，朱锡侯进入巴黎大学心理学院（Sorbonne），先是师从Paul Fraisse教授和Rene Zazzo教授，开始了他在生理心理学和应用心理学方面的研究。随后师从国际上享有盛名的心理学家Henri Piéron教授。Piéron是巴黎大学心理学研究所的创始人，并在当时兼任所长，同时也是第十一届国际心理学大会主席。受到Piéron的深刻影响，朱锡侯坚定了心理学不能被还原为生理学的决心，走上了生理心理学研究的旅程，试图从人类心理活动来研究大脑内部各种神经细胞相关活动所产生的变化，以此揭开人类生命本源的奥秘。

1945年秋天，朱锡侯接受云南大学校长熊庆来的聘请。1945—1957年，在云大从教的那几年，朱教授正处于一生中最风华正茂、最能发挥作用和创造力的年纪。因此，虽然当时的各种教学环境和条件都相当艰苦，工作和生活亦有无数的困难与艰辛，朱锡侯一直都竭尽全力地奋斗在自己的工作岗位上（图3）。他怀着科学救国、教育救国的梦想，在困苦的条件下，开始了白手起家式的艰辛创业。没有教材，他就花费大量的时间和精力自己翻译编写；没有实验室，他就自己掏钱想办法弄来一些实验设备，可以说一切都是他亲力亲为从头开始的。正是这样，朱先生一手创建了云南大学医学院生理科，增设了科研仪器，建立技术室、慢性动物观察室等，并先后进行了“痛觉与皮肤反射”“大小肠与输尿管、膀胱吻合后适应机能”“针刺足三里对消化造血功能的影响”等研究，都达到较高水平。恰逢当时云大正在开展师培热潮，有了领导的支持与信任，朱锡侯在教学、实验和考试等方面所做的一系列改革都是卓有成效的，“短短几年中，云大医学院教师队伍中涌现了许多心系学校师培、勤奋学习和积极向上的好老师。例如朱锡侯教授，他利用外出学习机会主动为本院青年教师到先进院校进修牵线搭桥……”尽管艰辛，他的付出还是得到了应有的回报。他以那个时代的知识分

图3　1955年左右，朱锡侯和他的大女儿在云南大学

子独有的那份赤忱，为多灾多难的祖国和人民毫无怨言地奉献着自己的青春。

遗憾的是，朱锡侯在1957年无辜卷入政治漩涡，此后的20余年他的科学生涯完全中断。直至1980年，在遭遇了太多的人生磨难之后，回到杭大任教对于朱锡侯来说尤为珍贵（图4）。“说实话，我已经66岁了，我真的很希望在自己的有生之年，能够回归心理学，能够再为我国的教育事业尽一点绵薄之力……我只是渴望工作……”（朱锡侯，2011）当时心理学在我国已经被打入冷宫几十年，一直被当成资产阶级的伪科学看待。此时心理学重新起步，与国际上有很大差距，生理心理学课的开设以及有关的教材问题，虽已被逐步提到议事日程上来，但一时间国内尚无成形的教材可供参考。因此，朱教授寻找大量的国外教材、杂志，编译了一些生理心理学的基础教材，提炼出其中适合授课用的生理心理学部分，包括Allen Schneider和Barry Tarshis的《生理心理学概论》以及Charles Levinthal的《心理学的生理学基础》。1980年11月20日，在当时的南京师范学院举行了中国心理学会生理心理学专业委员会成立大会和第一次学术报告会。朱先生作为筹备委员会成员之一，译著的这两部生理心理学基础教材也被收录进了年会的论文集里。可以说，朱先生为开设这门生理心理学课前前后后做了大量的工作。生理心理学当时在我国是一个比较落后的心理学分支，研究工作与国际水平的差距极大，研究内容上也比较狭窄，这些都对朱教授开设这门课程造成了诸多不便，但是他都想方设法一一克服了。在生理心理学方面，朱锡侯无疑是我国20世纪80年代初期的开路先锋。

图4　朱锡侯在杭州大学

就其专业教育背景而言，朱锡侯是一位对西方哲学、美学、生理学与心理学都有专门研究与造诣的学者。他毕生孜孜不倦，为开拓心理学、生理学等专业领域，付

出了艰辛努力，一直致力于心理世界奥秘的探索。正如他在晚年总结的那样："有一条中心线索贯穿着我一生的学习生涯，那就是想认识人生的奥妙，想认识人的内心世界，人的心理活动，心理生活……我从哲学概论开始，一直到心理学、美学、神经生理学、生理心理学，最后走到生命科学的园地来，是一脉相通、顺理成章的。"（朱锡侯，2011）

致谢

本研究得到国家社会科学基金项目（16CZX015）的资助，对于朱锡侯先生的女儿朱新地女士授予照片的使用权谨致谢忱！

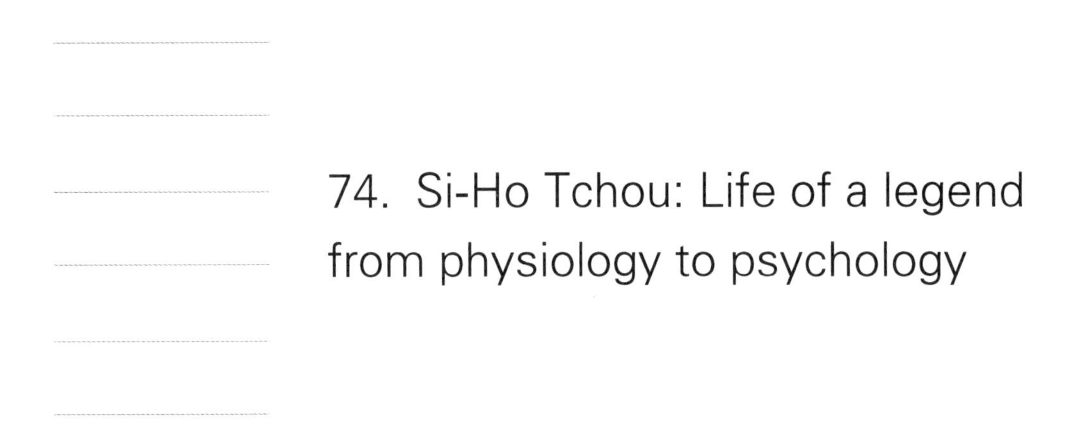

74. Si-Ho Tchou: Life of a legend from physiology to psychology

Why do the bees renounce sleep, the joys of honey, love, and divine leisure? Why do they get so much pain and effort, and where does such a decisiveness come from? So the sex they are dying for must deserve this sacrifice, it must be more beautiful, happier, and do something they can not do?

——Maurice Maeterlinck, *Das Leben der Bienen*, 1901

Dr. Si-Ho Tchou (朱锡侯, 1914—2000) was born in the northeast part of Jilin Province, although his ancestry can be traced to Shaoxing, Zhejiang Province. Dr. Tchou graduated from the L'Institut Franco-Chinois de Lyon (IFCL) in 1937 and was selected to continue his studies in France. Dr. Tchou spent eight years in France, where he first obtained his PhD in physiology from the University of Lyon and then received his PhD in psychology from the University of Paris. Dr. Tchou returned back to China in 1945 and was appointed as a professor in physiology at Yunnan University School of Medicine. Meantime, he was also appointed as a professor in psychology and aesthetics at Yunnan University, School of Humanities and Law. In 1980, he became the professor of physiology and psychology at the department of psychology in Hangzhou University. Then he served as a member of the physiological psychology committee in Chinese Psychological Society.

Dr. Tchou has made significant achievements during the eight years' study in France. At the University of Lyon, Dr. Tchou first studied aesthetics and psychology under the guidance of the famous aesthetician and psychologist Prof. Étienne Souriau and received master's degree in Arts. Then he transferred to Medical School to study under the supervision of a brilliant physiologist Prof. Henri Cardot. He is also the student of Prof. Charle Richet who was the Nobel Laureate in

Wei Chen[1,2], Xi Chen[3], Shengjun Wen[4]

1 Department of Psychology, Shaoxing University, Shaoxing 312000, China

2 Department of Psychology, The Education University of Hong Kong, Hongkong, China

3 College of Humanities, Shanghai University of Finance and Economics, Shanghai 200433, China

4 Department of Cognitive Neurology, Hertie Institute for Clinical Brain Research, University of Tübingen, Tübingen 72076, Germany

Correspondence: anti-monist@163.com (W. Chen)

Physiology or Medicine in 1913. Both of them are well known for their researches on the body's immune reactions to foreign substances (Stewart, 2012).

During the holidays, Dr. Tchou joined his mentor Prof. Cardot to the marine life station by the Mediterranean Sea, where he conducted researches on animal nerve cells and nerve fibers, spending days and nights recording nerve impulses in a tiny laboratory. At that time, their researches focused on the glandular giant neurons of marine invertebrates like *Aplysia*, and they had to go daily to the sea to capture *Aplysia*. Dr. Tchou was responsible to differentiate the glands and giant neurons of *Aplysia* based to their weight, and the other research project he conducted was to compare the neural discharge rhythms of *Aplysia*, based on their weight. At the end of 1942, Dr. Tchou was awarded a PhD degree in Science from the University of Lyon for his work on the growth pattern and rhythmic discharge phenomenon of neurons in *Aplysia*. His findings showed that the glands and giant neurons of *Aplysia* could be easily identified and sorted out. In addition, he found that the cell size was only a superficial feature of cell differentiation (Tauc, 1966; Ingoglia and Sturman, 1980; Ambron and Kremzner, 1982; Abel and Kandel, 1998).

As early as 1942, Dr. Tchou has studied the giant neurons of *Aplysia*, and together with Dr. Arvanitaki, published the paper "Laws of the relative individual growth of nerve cells in *Aplysia*" , where they recorded and quantified the number and relative size of individual neurons at different stages (Arvanitaki and Tchou, 1942). Based on this work, French scientists Dr. Tauc (1954) and Dr. Arvanitaki (1955) pioneered researches on single cell recording of giant neurons using microelectrodes. In addition, benefiting from the methods reported by Dr. Tchou, Prof. Kandel's team found that habituation and dishabituation was due to the synaptic potential of gill motor neurons (Pinsker et al., 1970; Kupfermann et al., 1970; Carew and Kandel, 1973; Castellucci and Kandel, 1976; Kandel and Schwartz, 1982; Voronezhskaya and Croll, 2015).

Scientists continued to conduct in-depth researches in *Aplysia* and made numerous significant research findings, however, as pointed out by Hughes (1968): "Apart from the work of Si-Ho Tchou (Tchou, 1942) and Arvanitaki few quantitative observations seem to have been recorded on the relative size of its nervous system or on the numbers and sizes of neurons in individuals of different sizes (p. 423)" . Thus, Dr. Arvanitaki and Dr. Si-Ho Tchou were the first ones who studied this feature in the abdominal ganglion of *Aplysia punctata* (Hughes and Tauc, 1961; Kandel and Tauc, 1965; Koester and Kandel, 1977; Adams and Benson, 1985; Bailey and Chen, 1988; Cleary et al., 1991). As neurons are very difficult to grow, neurobiologists heavily depended on the established research methods, and Dr. Tchou's work on the quantification of neuron cells and rhythmic discharge in *Aplysia*, as well as the experimental methods established during the researches, laid a solid foundation for the studies using *Aplysia* as a model organism. The following researches by many prominent neuroscientists has benefited from Dr. Tchou's pioneering work. Such as Prof. Eric R. Kandel at Columbia University, who found that memory was linked to the synapses that connect nerve cells and uncovered its molecular mechanisms in *Aplysia*. Prof. Eric Kandel was awarded the Nobel Prize in Physiology or Medicine in 2000. This was inseparable from the foundational work of Dr. Tchou and his colleagues on the investigation and establishment of *Aplysia* as a model organism (Coggeshall et al., 1966; Hughes, 1968; Peretz and Lukowiak, 1975; Shapiro et al., 1980).

Dr. Tchou had great interests in physiology, especially experimental psychology. Thus, from 1943 to 1945, Dr. Tchou enrolled in the University of Paris (Sorbonne), and began his researches on physiological psychology and applied psychology under the supervision of Profs. Paul Fraisse and Rene Zazzo, and later with the central figure of French psychology in the twentieth century, Prof. Henri Piéron, who was not only well known for his research in sensory physiology, but also the founder and director of the Institute of Psychology at the University of Paris. With the profound influence of Prof. Piéron, Dr. Tchou embarked on researches in physiological psychology and studied changes to various nerve cell activities in the brain resulting from human psychological activities in order to uncover the mystery origin of human life. In July 1945, Dr. Tchou decided to give up the plan to finish the PhD thesis in physiological psychology in Prof. Piéron's laboratory, rejected the invitation of Prof. Alfred Burloud to stay at Rennes University and returned back to China.

In the autumn of 1945, Dr. Tchou accepted the offer from Hiong King-Lai, the President of Yunnan University. From 1945 to 1957, Dr. Tchou worked hard at Yunnan University in spite of the poor teaching conditions and the difficulties he faced in his work and life. With the belief of saving the nation with science and education, he started his career under the harsh conditions. With the lack of teaching materials and laboratory equipments, Dr. Tchou spent a lot of time and energy translating and writing teaching materials and paid out of his own pocket for the laboratory equipments. Finally, Dr. Tchou established the department of physiology at Yunnan University School of Medicine, introduced scientific research equipments, and set up technical rooms and observation rooms. There, he conducted researches on "pain and skin reflex" , "functional adaptation after anastomosis of the intestines and the ureter and bladder" , and "the effect of acupuncture stimulation of Zusanli on digestive hematopoietic function" .

Unfortunately, Dr. Tchou was innocently involved in a political turmoil in 1957. After experiencing major hardships in life, Dr. Tchou went to Hangzhou University. "To be honest, I am already 66 years old. I really hope to be able to return to psychology in my remaining years and do my bit for education in China... I just want to work..." (朱锡侯, 2011, p. 197) At that time, psychology has been neglected in China for decades and was treated as a pseudoscience of the bourgeoisie. There was an enormous gap with the international community in this research area. Although the setting up of physiological psychology classes has been proposed, there were no references available domestically. Therefore, Dr. Tchou searched through a large number of foreign texts and magazines, excerpted the information that was suitable for the physiological psychology classes and translated some of the basic teaching materials such as *Introduction to Physiological Psychology* by Allen Schneider and Barry Tarshis; *Introduction to Physiological Psychology* by Charles Levinthal. On November 20, 1980, the inaugural assembly and symposium of the physiological psychology professional committee of Chinese Psychological Society was held in Nanjing Normal University, and as one of the organizing committee members, the two textbooks translated by Dr. Tchou were also included in the proceedings. Indeed, Dr. Tchou has paved the way for the development of physiological psychology and was certainly the pioneer of physiological psychology. While the teaching conditions were tough, it allowed him to develop a close relationship with his students. He often directed student essays, and discovered a constant relationship between time perception and the adjustment of knowledge and experience (朱琪, 1985).

Dr. Tchou made brilliant achievements in the fields of Western philosophy, aesthetics, physiology and psychology. With his in-depth researches in understanding the mysteries of human brain functions, he made outstanding contributions to the development of physiology and psychology in China. As he once said, "There is a common thread running through my lifetime's learning, that is, to know the mystery of life, to understand people's inner worlds, their psychological activities and psychological lives... It is the logical connection in my journey beginning from philosophy, to psychology, aesthetics, neurophysiology, physiological psychology, and finally, the life sciences." (朱锡侯, 2011)

Figures

References

朱琪. 1985. 幻嗅过程中知识和经验的调整作用与时间知觉恒常性. 心理科学, 6: 42–46.

朱锡侯（口述), 朱新地（整理）. 2011. 昨夜星辰昨夜风：八十自述. 北京：人民文学出版社.

Abel T, Kandel E (1998) Positive and negative regulatory mechanisms that mediate long-term memory storage. Brain Res Rev 26:360–378.

Adams WB, Benson JA (1985) The generation and modulation of endogenous rhythmicity in the *Aplysia*, bursting pacemaker neurone r15. Prog Biophys Mol Biol 46(1):1–49.

Ambron RT, Kremzner LT (1982) Post-translational modification of neuronal proteins: Evidence for transglutaminase activity in r2, the giant cholinergic neuron of *Aplysia*. PNAS 79(11):3442–3446.

Arvanitaki A, Tchou SH (1942) Les Lois de la croissance relative individuelle des cellules nerveuses chez l'Aplysie. Bull D'Histologieet de Tech Microsc 19:244–256.

Bailey CH, Chen M (1988) Long-term sensitization in *Aplysia* increases the number of presynaptic contacts onto the identified gill motor neuron L7. PNAS 85(23):9356–9359.

Carew TJ, Kandel ER (1973) Acquisition and retention of long-term habituation in *Aplysia*: Correlation of behavioral and cellular processes. Science 182(4117):1158–1160.

Castellucci V, Kandel ER (1976) An invertebrate system for the cellular study of habituation and sensitization. In: Tighe TJ, Leaton RN. Habituation: Perspectives from Child Development, Animal Behavior, and Neurophysiology. Hillsdale: Erlbaum.

Cleary LJ, Baxter DA, Nazif F, et al (1991) Neural mechanisms underlying sensitization of a defensive reflex in *Aplysia*. Biol Bull 180:252–261.

Coggeshall RE, Kandel ER, Kupfermann I, et al (1966) A morphological and functional study on a cluster of identifiable neurosecretory cells in the abdominal ganglion of *Aplysia Californica*. J Cell Biol 31(2): 363–368.

Hughes GM (1968) The left and right giant neurons (LGC and RGC) of *Aplysia*. In: Salánki J. Neurobiology of Invertebrates. Boston: Springer, 423–441.

Hughes GM, Tauc L (1961) The path of the giant cell axons in *Aplysia depilans*. Nature 191:404–405.

Ingoglia NA, Sturman JA (1980) Axonal transport of putrescine and the polyamines. Neurochem Res 5(8):913–914.

Kandel ER, Schwartz JH (1982) Molecular biology of learning: Modulation of transmitter release. Science 218(4571):433–443.

Kandel ER, Tauc L (1965) Heterosynaptic facilitation in neurones of the abdominal ganglion of *Aplysia depilans*. J Physiol 181:1–27.

Koester J, Kandel ER (1977) Further identification of neurons in the abdominal ganglion of *Aplysia* using behavioral criteria. Brain Res 121(1):1–20.

Kupfermann I, Castellucci V, Pinsker H, et al (1970) Neuronal correlates of habituation and dishabituation of gill-withdrawal reflex in *Aplysia*. Science 167(3926):1743–1745.

Peretz B, Lukowiak KD (1975) Age-dependent CNS control of the habituating gill withdrawal reflex and of correlated activity in identified neurons in *Aplysia*. J Comp Physiol 103(1):1–17.

Pinsker H, Kupfermann I, Castellucci V, et al (1970) Habituation and dishabituation of the gm-withdrawal reflex in *Aplysia*. Science 167(3926):1740–1742.

Shapiro E, Castellucci VF, Kandel ER (1980) Presynaptic membrane potential affects transmitter release in an identified neuron in *Aplysia* by modulating the Ca^{2+} and K^{+} currents. PNAS 77(1):629–633.

Stewart W (2012) Brain, Mind, and Medicine: Charles Richet and the Origins of Physiological Psychology. New Brunswick and London: Transaction Publishers.

Tauc L (1966) Physiology of the nervous system. In: Wilbur KM, Yonge CM. Physiology of Mollusca, vol 2. New York: Academic Press, 387–454.

Tchou SH (1942) Contribution à l'étude de la physiologie des cellules nerveuses chez l'aplysie. Lyon: Bosc Frères & Riou.

Voronezhskaya EE, Croll RP (2015) Mollusca: Gastropoda. In: Schmidt-Rhaesa A, Harzsch S, Purschke G. Structure and Evolution of Invertebrate Nervous Systems. Oxford and New York: Oxford University Press, 196–221.

75. 经利彬：中国著名生理学家

经利彬教授（Li-Pin King）（1895—1958），我国著名的生理学家，一生主要从事中药药理、实验生物学、现代医学等多方面研究，是我国最早用科学方法研究中药和较早从事现代医学科学研究的科研工作者，为我国近现代生理学的发展作出了卓越的贡献（罗尔纲，1993）。

经利彬教授于1895年出生于浙江，早年留学于法国里昂大学，获理学博士和医学博士双学位。毕业后，他在那里担任助教。经利彬教授归国后担任国立北平大学农学院教授，兼任生物系主任。1929年10月，国立北平研究院生物学研究所（后改名为北平研究院生理学研究所）成立，经利彬教授任所长。经教授还担任国立北平大学、北平女子文理学院、中山大学等多所学校的教授。不幸的是，1937年北平沦陷，经利彬教授随国立北平研究院前往云南昆明，担任云南大学生物学教授和中国医学研究所所长（图1）。1946年经利彬教授前往台湾。经教授于1958年在台湾逝世。

图1　经利彬教授1943年在云南

译者：龚伟
浙江师范大学，金华321004，中国
邮箱：yzugw@163.com

经教授开展了营养学、中医药学、实验生物学及现代医学的研究，并发表多篇论文和著述。据1948年国立北平研究院总办事处出版的《国立北平研究院出版品目录》统计显示，经利彬教授带领的生理学研究所发表的论文数量在十个研究所中位居第三，国外期刊发文数量排在第二位。所有这些出版物都极大地推动了我国生理学事业的发展（张琳，2014）。

经教授担忧国民的健康和营养问题，因此他专注于研究食物营养成分与中国人必需的营养元素之间的关系。例如，他研究了我国北方食物营养与人体血液中磷和钙质之间的关系，揭示平常食物对人体生长和健康的影响，并强调饮食的重要性。经教授还系统地开展中药的药理研究。他的研究阐明了部分中药的生理作用及其对人体血液、血压、器官等的影响（经利彬和石原皋，1934；经利彬，1935），开启了我国用科学研究方法研究中药药理的先河。他还将相关研究成果发表在国际杂志《生物学》上（张琳，2014），向外国人介绍我国传统中医的科学原理。经教授对实验生物学也颇有兴趣。他与动物学研究所的水产动物研究团队进行合作，开展水产生物实验生物学研究（北平研究院生理学研究所，1936），这极大地促进了我国实验生物学的发展。经教授还是我国最早从事现代医学研究的科学家之一。他研究并系统介绍了血清、红细胞、血块等知识。此外，他还将实验生物学与中药药理学相结合，研究了中药和麻醉剂对血液成分、血压等的影响，并将相关研究成果发表在最具影响力的期刊《科学》上。

经利彬教授回国后任国立北平研究院生理学研究所代理所长初期，该所研究人员仅有4人。因此，经教授开始培养和聘请科研人员，拥有了一支由17人组成的科学研究队伍，其中包括2名专任研究员，1名特约研究员，1名驻欧通讯员和几名助理人员（张琳，2014），为进一步的研究奠定了基础。经教授非常注重团队之间的合作，他大部分的研究成果是与研究团队成员一起进行的，如石原皋、侯玉清、赵燏黄，这极大地促进了年轻学者的发展。

经利彬教授一生从事生理学研究，著述众多，如《脊椎动物之脑量》《金鱼鳍及鳞之复生》《槐实之生理作用》等，对生理学有重要影响。经教授还与其他研究人员一起合著《滇南本草图谱》。这本书不仅是振兴中药事业的重要著作，也是植物研究的代表性著作。在《滇南本草图谱》第一辑中，经教授选取了《滇南本草》中的26种药物，绘制了植物的形状，列出植物的释名、基本特征、形态、相关研究、分布、药理和图版说明（张瑞贤和李国坤，2004）。由于其重要性，它已被多次重新出版（图2）。除学术著作外，经利彬教授独立出版了大学教材《普通生物学》，与魏春芝教授合作出版初中教科书《生理卫生学》，推动了我国生物学、生理学教育事业的发展。

经教授不仅是位出色的科研工作者，也是一位培养了许多年轻人才的人民教师。1925年9月，国立北平大学生物系成立，经教授担任讲师，为第一届3名学生张风瀛、石原皋和郝景盛讲授生理学及实习的课程。其中，石原皋后来加入经教授带领的生理学研究所的研究团队，跟随经教授研究中药药物的药理作用和北方食物的营养价值，也成为我国早期著名的医药科学家。此外，在云南期间，经教授为了支持学生学习，还将自己的藏书《王叔和脉经》赠予云南大学。

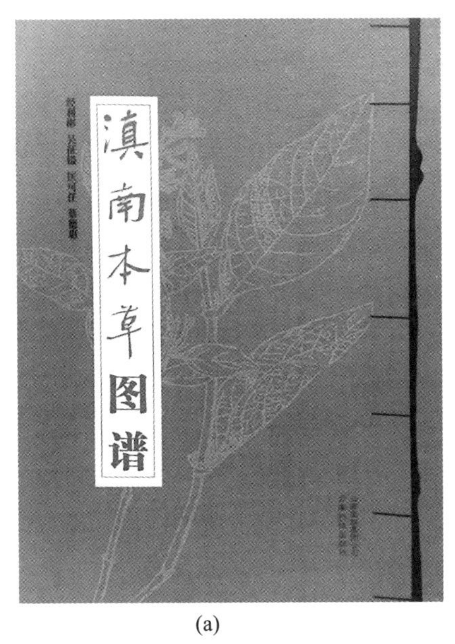

(a)

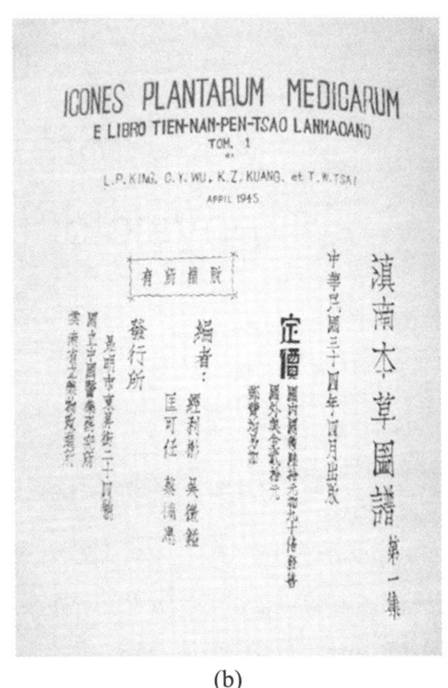

(b)

图2 《滇南本草图谱》(2007年拍摄)

经利彬教授还是中国动物学会的发起人之一。中国动物学会是由经利彬教授、童第周教授、朱洗教授等29人签名创立。1934年8月，中国动物学会在江西省成立，旨在促进我国动物学科的发展。经教授是该学会的第二届理事和第四届副会长。两年后，经教授还与罗宗洛、贝时璋和童第周创办了至今在国内外仍具有一定影响力的《中国实验生物学杂志》(后更名为《实验生物学报》，现为《分子细胞生物学报》)，推动了我国实验生物学事业的发展。

经利彬教授是我国伟大的生理学家，在学术研究、人才培养、期刊和学术团体创办等方面对我国生理学事业的发展作出了杰出贡献，尤其是在中药药理、实验生物学、现代医学等领域。他严谨求实的科研作风、合作研究的科研精神、成果共享的科研人格值得广大科研工作者学习。

75. Professor Li-Pin King: A famous physiologist in China

Professor Li-Pin King (经利彬, 1895—1958) is a famous physiologist in China. He devoted his entire life to the researches on Pharmacology of traditional Chinese medicine, Experimental Biology and Modern medicine. He made tremendous contributions to the development of Physiology in China. Prof. King is also the first scientist who applied the scientific methods to researches on traditional Chinese medicine and one of the earliest generations of Chinese researchers who studied modern medical science (Luo, 1993).

In 1895, Prof. King was born in Zhejiang, China. He studied abroad at the University of Lyon in France and graduated with degrees in both science and medicine. After graduation, he served as a teaching assistant there. Later on, he returned back to China and became a professor at the Agricultural College of Peiping University and also was appointed as the director of the Department of Biology. In October 1929, Institute of Biology, National Academy of Peiping (later renamed as Institute of Physiology, National Academy of Peiping) was established and Prof. King was appointed as the director. He was also a professor of Peiping University, Peiping Women's College of Arts and Sciences and Sun Yat-sen University. Unfortunately, Peiping was occupied by the invaders in 1937. Prof. King moved to Kunming, Yunnan together with the National Academy of Peiping and later served as professor in biology at Yunnan University and the director of the Chinese Medical Academy. In 1946, Prof. King moved to Taiwan. Prof. King passed away in Taiwan in 1958.

Prof. King carried out researches on Nutrition, Pharmacology of traditional Chinese medicine, Experimental biology and Modern medicine, and published many research articles and monographs. According to *The Catalogue of the Publications of the National Academy of Peiping* published by the General Office of National Academy of Peiping in 1948, the number of articles published by the Institute of Physiology led by Prof. King was in the third place among ten research institutes, and the number of articles published in foreign journals ranked in the second place. All those publications significantly promoted the development of Physiology in

Wei Gong, Fangfang Wang, Yike Ying
Zhejiang Normal University, Jinhua 321004, China
Correspondence: yzugw@163.com (W. Gong)

China (Zhang, 2014).

Prof. King concerned about the health and nutrition of Chinese, therefore he focused on the researches on the relationship between food nutrients and nutritious elements essential for Chinese. For example, he studied food nutrients in northern China and its effects on the levels of phosphorus, calcium in human blood, showing the effects of ordinary food on people's growth and health, and emphasizing the importance of diet. Prof. King also performed pharmacological researches on Chinese medicine systematically. His researches clarified the physiological effects of some Chinese herbs and their effects on blood components, blood pressure and organs in human (King and Shi, 1934; King, 1935) which opened the way of using scientific research methods in Pharmacology to study traditional Chinese medicine. He also published his research data in an international journal *Biology* (Zhang, 2014), introducing the scientific principles of traditional Chinese medicine to benefit the foreigners. Prof. King was also interested in Experimental Biology. He collaborated with the aquatic animal research team in the Institute of Zoology and conducted experimental biological researches about water organisms (Institute of Physiology, National Academy of Peiping, 1936), which significantly promoted the development of Experimental Biology in China. Prof. King was also one of the earliest scientists who engaged in the studies of modern medical research in China. He studied and introduced systematical knowledge of serum, red blood cells and blood clots. In addition, he combined experimental biology with pharmacology of traditional Chinese medicines, and studied the effects of Chinese medicines and anesthetics on blood components, blood pressure etc., and published his research data on *Science*, one of the most influential scientific journals.

At the time when Prof. King was appointed as the director of the Institute of Physiology, National Academy of Peiping, there were only 4 researchers in the institute. He thus began to recruit and cultivate researchers and had a scientific research team of 17 persons, including two full-time researchers, a special researcher, a correspondent in Europe and several assistants (Zhang, 2014), which laid great foundation for further researches. Prof. King paid great attention to the collaborations between team members. Most of his researches were performed together with his team members, such as Yuangao Shi (石原皋), Yuqing Hou (侯玉清) and Yuhuang Zhao (赵燏黄), which benefited greatly the development of young scholars.

Prof. King engaged in physiological researches all his life and compiled many books such as *The Brain Volume of Vertebrates*, *The Resurrection of Goldfish Fins* and *Scales*, *The Physiological Role of Sophora Fruit*, etc., which had great impacts on Physiology. Prof. King also completed the book *Icones Plantarum Medicarum* together with other researchers, which was not only an important monograph in revitalizing Chinese medicine industry, but also a representative book of plant researches. In the first chapter of *Icones Plantarum Medicarum*, Prof. King selected 26 kinds of drugs in *Materia Medica of South Yunnan*, and drew the shapes of original plants, wrote instructions about their names, original texts, forms, relative researches, distributions, pharmacology and illustrations (Zhang and Li, 2004). Because of its importance, it has been republished multiple times. In addition to academic researches, Prof. King also compiled *General Biology* independently, a university textbook, and collaborated with Prof. Chunzhi Wei (魏春芝) to compile *Physiological Hygiene*, a junior high school textbook, boosting biology and physiology education in China.

Prof. King was not only an outstanding researcher, but also an excellent educator who has cultivated many young talents. In September 1925, the Department of Biology of Peiping University was established and Prof. King was recruited as a lecturer to teach Physiology and Experimentation for the first three students of the Department: Fengying Zhang (张风瀛), Yuangao Shi (石原皋) and Jingsheng Hao (郝景盛). One of them, Yuangao Shi, joined the research team led by Prof. King to study the pharmacological effects of Chinese medicine and the food nutrition in northern China and later on became a famous medical scientist in China. In addition, to support the students' studies, Prof. King donated the book *Mai Jing* (Pulse Classic) written by Shuhe Wang (王叔和) to the Yunnan University during his stay in Yunnan.

Prof. King was also one of the founders of Chinese Zoological Society. Together with other 29 scientists including Profs. Ti-Chow Tung (童第周) and Tchou Si (朱洗), Prof. King signed to establish the Chinese Zoological Society. In August 1934, Chinese Zoological Society was officially established in Jiangxi Province, aiming to promote the development of Zoology in China. Prof. King was the second director and the fourth vice president of the society. Two years later, Profs. King, Tsung-Lê Loo (罗宗洛), Shitsan Pai (贝时璋) and Ti-Chow Tung set up *Chinese Journal of Experimental Biology* (later named as *Acta Biologiae Experimentalis Sinica*, now known as *Journal of Molecular Cell Biology*), which still has international influences and advances the development of Experimental Biology in China.

Prof. Li-Pin King is a great physiologist and made outstanding contributions to the development of physiology in China in terms of academic researches, cultivation of young talents, establishment of academic journals and research groups, especially in the areas of Chinese medicine pharmacology, experimental biology and modern medicine, etc. His rigorous research style, cooperative research spirit and collaborating research personality are worthy of appreciation and learning by scientific researchers.

Figures

Fig.1 Professor Li-Pin King in Yunnan, 1943

Fig.2 *Icones Plantarum Medicarum* (a photocopy in 2007)

References

Institute of Physiology, National Academy of Peiping (1936) Work Report of Institute of Physiology. 1:41. (北平研究院生理学研究所. 1936. 生理学研究所工作报告. 1:41.)

King LP (1935) The vomiting effect of *Pinellia*. J Inst Physiol Natl Acad Peiping 8:157. (经利彬. 1935. 半夏之制止呕吐作用. 国立北平研究院生理学研究所丛刊, 8:157.)

King LP, Shi Y (1934) Study on the physiological role of *Codonopsis*. J Chin Rep Inst Physiol Natl Acad Peiping 3:50. (经利彬, 石原皋. 1934. 党参的生理作用之研究. 国立北平研究院生理学研究所中文报告汇刊, 3:50.)

Luo EG (1993) Read "Gossip of Hu Shi" . Social Sci Front 6:175–181. (罗尔纲. 1993. 读《闲话胡适》. 社会科学战线, 6:175–181.)

Zhang L (2014) Research on National Academy of Peiping's biological sciences research and institutionalization (1929—1937). Taiyuan: Shanxi University. (张琳. 2014. 北平研究院生物学科学研究与体制化研究(1929—1937). 硕士学位论文. 太原: 山西大学.)

Zhang RX, Li GK (2004) Pioneer in Chinese medicine research in Beijing. J Jiangxi Univ Tradit Chin Med 3:23–28. (张瑞贤, 李国坤. 2004. 北京中医药科学研究的先驱. 江西中医药大学学院学报, 3:23–28.)

八、心理学

76. 汪敬熙与中国的生理心理学研究发展

汪敬熙先生（图1）作为中国和世界著名的心理学家，1948年当选中央研究院（1928年成立）院士，曾任中州大学、中山大学和北京大学教授，1953年后他在美国约翰斯·霍普金斯大学和威斯康星大学进行研究工作。汪敬熙先生是我国西方心理学研究的先驱，在我国生理心理学和神经科学等学科的创建中起到非常重要的作用。

图1　汪敬熙（1897—1968）

汪敬熙先生对心理学研究兴趣浓厚，但是他最初所学专业并非心理学，而是经济学。他1919年毕业于北京大学经济学系，作为青年学生在北京大学学习期间，他积极支持新文化运动，而且率先使用白话文，因此得到了我国著名文学家鲁迅先生的高度赞赏。他职业生涯的转折发生在1920年（图2），那时候他从北京大学选派到美国约翰斯·霍普金斯大学学习，师从精神病学家Adolf Meyer教授，从此开启了他在生理心理学和神经科学领域杰出而辉煌的职业历程。他在1923年从约翰斯·霍普金斯大学毕业并获得博士学位。

图2　汪敬熙（前排右一）（1920年）

译者：苏瑞凤

中国科学院北京生命科学研究院，北京100101，中国

邮箱：surf@biols.ac.cn

汪先生博士毕业以后于1924年回国，之后一直到1948年他相继在不同的大学和研究所担任领导职务，为我国的生理心理学教育及制度化作出了巨大贡献。他1924年在中州大学担任教育系主任。1927年他呕心沥血创建了中山大学的心理学系和心理学研究所，随后分别担任系主任和所长。在这里，他创建了我国第一个神经心理学实验室。1930年，北京大学心理学系邀请他加入并随后任命他为系主任，期间他建成了北京大学的第一个心理学实验室。1934年，汪先生担任中央研究院心理研究所所长，并在上海和南京创建了新的实验室。1948年，他当选中央研究院院士。

汪先生非常重视教育在心理学发展中的作用。1932年他在《独立评论》上发表的文章《提倡科学研究最应注意的一件事——人才的培养》（汪敬熙，1932）中提到，培养优秀的科研人才是保证科学发展的重要因素，甚至比提高科研的氛围本身更重要（杨翠华，1991）。

在这样的教育理念指导下，汪先生努力为学生提供最全面和最合适的学习条件。根据中山大学的历史资料记载，汪先生不但在设置的课程中包括广泛而具体的现代心理学知识，同时也非常重视心理学史和英语的教学。另外，他通过演讲来激发公众对心理学的兴趣。他的努力也收到了很好的效果，他培养出了多位非常杰出的心理学家和神经生物学家。中国科学院院士、世界著名神经生理学家张香桐教授，在回忆恩师汪敬熙先生时写道："为我指明了通向神经科学世界之路……在他的建议和鼓励下……汪先生给予的这种体验塑造了我以后科学生涯的思维方式。"（Squire，2001）

除了教学以外，汪先生取得众多令人瞩目的科研成果并得到了学界广泛的认可，其中部分重要的研究结果是他在我国当时非常简陋的实验环境下获得的。他是将电子设备引入国内应用于大脑功能研究的第一人；他是世界上最先研究皮肤电反射的几个人之一，在研究皮肤反射电位时他指出反射是由汗腺分泌引起而不是意识引起的；他通过创新性的方法来研究哺乳动物的行为，发现了雌性老鼠行动和发情周期之间的关系。此外，汪先生不仅是第一个描述光刺激视网膜后在上丘产生动作电位的科学家，也是第一位研究对瞳孔收缩和扩张做出反应的大脑皮层区域的科学家。

张香桐先生回忆恩师的文章在一定程度上对汪敬熙先生做了精确的总结评述："他是一位杰出的老师和优秀的科学家，他对科学研究和科学教育本质有着非凡的智慧和深入的理解。"毫不夸张地说，如果没有汪先生打下的研究基础，我国生理心理学和神经学研究领域的发展不会如此迅速。

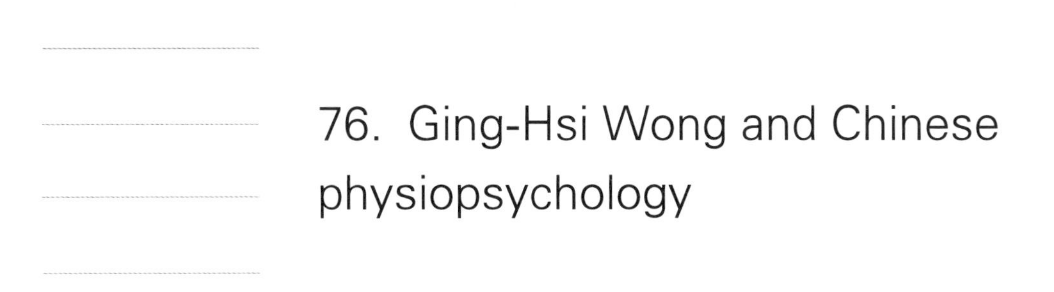

76. Ging-Hsi Wong and Chinese physiopsychology

Ging-Hsi Wong (1897—1968) was a Chinese psychologist, an academician of the Academia Sinica (founded in 1928), and a professor in several universities in China and the United States. As one of the pioneers of Western psychology in the country, he played an important role in establishing the fields of Physiopsychology and Neuroscience in China.

Wong's interest in Psychology, however, did not start from the beginning of his scholarship. In fact, he graduated from the Economics Department of Peking University in 1919. As a young student who passionately supported the New Culture Movement in China, Wong pioneered the use of modern vernacular Chinese and was regarded as "very promising" by renowned Chinese litterateur Lu Xun. The turning point of his career came in 1920, when he was sent by Peking University to Johns Hopkins University in the USA to study with the psychiatrist Adolf Meyer. Since then, he began his illustrious career in the fields of Physiopsychology and Neuroscience. He eventually received his Ph.D. from Johns Hopkins in 1923.

After his graduation, Wong spent most of his time in China from 1924 to 1948. By successively holding leadership roles in different universities and institutes, he made enormous contributions to the education and institutionalization of Physiopsychology in China. In 1924, Wong became the dean of the Education Department of Zhongzhou University in Henan Province. In 1927, he worked hard to establish the Psychology Department and the Psychology Institute of Sun Yat-sen University in Guangdong Province and was subsequently appointed as the director of both organizations. There, he also established the first laboratory of neurophysiology in China. The Psychology Department of Peking University invited Wong to be part of the faculty in 1930, and eventually appointed him to be the dean. During his tenure, he also established the first psychology laboratory in Peking University. In 1934, Wong became the director of the Institute of Psychology of the Academia Sinica, and in this capacity, he was able to establish new laboratories in Shanghai and Nanjing. In 1948, the Academia Sinica officially appointed Wong as an academician.

Shu Zheng
The Institute for the History of Natural Sciences, Chinese Academy of Sciences, Beijing 100190, China
Contact: zhengshu@ihns.ac.cn

Wong emphasized the importance of education in the development of psychology. In his article entitled "The most important consideration for promoting science—developing talents" (Wong, 1932), he regarded cultivating research talents as the most important element in ensuring the development of science, even much more important than developing the research environment itself (Yang, 1991).

With this spirit, Wong strived to offer his students the most comprehensive and appropriate learning experience. According to the historical materials of Sun Yat-sen University, Wong's course designs not only focused on the wide range of concrete modern psychological knowledge but also emphasized the history of psychology and English study. He also occasionally gave public speeches to promote public interest in psychology. His efforts paid off, and he had cultivated some of the most outstanding psychologists and neurologists in China. Hsiang-Tung Chang (1907—2007), a world-renowned Chinese neurophysiologist and an academician of the Chinese Academy of Sciences, recalled that Wong "pointed out the path leading to the world of neuroscience... At his suggestion and encouragement... this experience proved to be decisive in shaping my way of thinking in my later scientific career" (Squire, 2001).

Apart from teaching, Wong achieved widely recognized scientific accomplishments, some of which he obtained under rough experimental conditions during his stay in China. For example, he was the first to introduce an electronic instrument to the research of brain functions in China. He was among the first to study skin potential reflex and pointed out that the reflex was due to the secretion of sweat glands and not consciousness. He also discovered the relationship between the activity of a female rat and its oestrous cycle, using a revolutionary research method in the study of mammalian behavior. In addition, Wong was not just the first scientist to describe the action potentials in the superior colliculus produced by the light stimulation of the retina, he was also the first to study cortical regions responding to the contraction and dilation of the pupil.

Professor Hsiang-Tung Chang's description, to a certain extent, accurately portrays Professor Wong, "a good teacher and scientist with a remarkable acumen in research and deep understanding about the nature of science education" . Indeed, without the foundation he laid, the development of the fields of Physiopsychology and Neuroscience in China would have been considerably slower.

Figures

Fig.1 Ging-Hsi Wong (1897—1968)

Fig.2 Ging-Hsi Wong (front row, 1st from right) in 1920

References

Huang YX (1999) History of Sun Yat-sen University (1924—1949). Guangzhou: Sun Yat-sen University Press. (黄义祥. 1999. 中山大学史稿 (1924—1949). 广州: 中山大学出版社.)

Lu X (1935) The corpus of the Modern Literature of China—the second volume of fiction compiler. Shanghai: Shanghai Liangyou Press. (鲁迅. 1935.《中国新文学大系·小说二集》导言. 上海: 上海良友图书公司.)

Lu ZH (2001) The story of Wong Ging-Hsi. Neurosci Bulletin 17: 357. (鲁子惠. 2001. 汪敬熙先生传略. 中国神经科学杂志, 17: 357.)

Shepherd G (2009) Creating Modern Neuroscience: The Revolutionary 1950s. London: Oxford University Press.

Squire LR (2001) The History of Neuroscience in Autobiography. Vol 3. San Diego: Academic Press.

Vision DTW (1988) Vision: Structure and Function. Hartwell: World Scientific Publishing, 3–6.

Wong GH (1932) The most important consideration for promoting science—developing talents. Independent Critic 26: 10–14. (汪敬熙. 1932. 提倡科学研究最应注意的一件事——人才的培养. 独立评论, 26: 10–14.)

Yang CH (1991) Patronage of sciences: The China foundation for the promotion of education and culture. Special Issue of Academia Sinica Institute of Modern History 65. (杨翠华. 1991. 中基会对科学的赞助. 中央研究院现代史研究所特刊, 65.)

77. 郭任远与通过动物实验构建的行为渐成论

郭任远（Zing-Yang Kuo，1898—1970）（图1），字陶夫，享誉世界的中国心理学家。他是行为主义流派中激进派先驱之一，是行为主义心理学历史长河中最为激进的行为主义学家。他的研究焦点是动物心理学和比较心理学。动物实验是他偏爱的研究方法，并且他对多种动物都做过实验研究。郭任远的研究目的是探索行为形成的起源，1967年，他建构了行为渐成论以解释行为的成因。郭任远是唯一一个被《实验心理学一百年》以插画人物形式收录的中国心理学家，一向只刊载科学实验报告的《比较与生理心理学杂志》破例刊文纪念并赞誉："他以卓尔不群的姿态和勇于探索的精神为国际学术界留下了一笔丰厚的精神财富。"（Gottlieb，1972）

图1　郭任远（1955年）

1898年，郭任远生于广东省潮阳县铜盂村（今汕头市潮阳区铜盂村）的一个商贾家庭。1916年，郭任远进入复旦大学。两年后，郭任远留学美国旧金山湾区伯克利市的加利福尼亚大学伯克利分校。经历徘徊和犹豫后，他将心理学作为主修专业。同年，他认识了在伯克利分校任教的新行为主义旗手E. C. Tolman（1886—1959），亦是他的导师。1921年至1922年，郭任远在伯克利分校任助教。当他与校方在论文修改上意见相左之时，他拒绝按照校方意见修改自己的文章，遂放弃博士候选人的身份，并于论文答辩之前归国（郭任远，1940）。直至1936年，他才获得博士学位。

1923年，郭任远归国。1923年至1927年，他在复旦大学担任了一系列教职。1923年，任复旦大学教授；1924年任副校长，代理校长行使权力并承担责任，直至1926年，结束代理校长之务。在这期间，他于1924年开设心理学系；从族人处募款筹建"子彬医院"（图2）。《申报》称"该院的规模位居世界大学心理学院第三位"（胡寄南，1995），并且是当时我国唯一一所心理学院。

译者：钱燕燕
南京师范大学心理学院，南京210097，中国
邮箱：psyqyy@163.com

图2 子彬医院

次年，郭任远到国立中央大学任教，他创办了中央研究院心理研究所并担任所长。从1933年4月至1936年2月，他在浙江大学担任教职，教书育人。他担任浙江大学校长，在浙大创建了心理学系。1935年，郭任远当选中央研究院第一评议员。郭任远还参与心理学会的成立：1931年暑期，郭任远与其他8位心理学家（如郭一岑、艾伟、肖孝嵘）发起组织中华心理学会并于上海举办了筹备会议。

1935年12月10日，浙江爆发了由浙江大学学生领导的学生运动，这场运动并不被郭任远许可。郭任远与学生之间的冲突导致了郭任远的离职，离职消息宣布于1936年2月的行政院第257次例会上。这进而导致他的研究方法和研究前途的骤变。从那时起，郭任远成了一个居无定所的幽灵般的科学家。1936年至1946年，他在伯克利讲授生理心理学，在耶鲁大学的奥斯本动物实验室（1937—1938年）和华盛顿的卡耐基研究所（1938—1939年）进行实验，并前往各类大学进行巡回演讲（1941—1943年）。

1946年，郭任远任香港大学校董。在接下来的12年间，他忙于人类行为的研究，包括对中国人的民族性格的社会心理学分析（Gottlieb，1972）。1963年，郭任远留居美国教学来推广他的科学理念，包括在8月的华盛顿的国际心理学及动物学年会上演讲（Kuo，1967）。同年，郭任远回到中国香港，并于1970年8月14日去世。

图3 郭任远

郭任远的研究历程分三个阶段：理论建构、动物实验和理论成型（图3）（Gottlieb，1972）。在郭任远的理念之中，心理学必须是一个实证科学。他相信本能是对实证科学的偏离："无论本能支持者同意与否，我始终认为我们否定本能的最主要动机是将心理学从扶手椅猜想之中解救出来。我们力图将遗传心理学的障碍移除。"（Kuo，1922）于是，当郭任远还是一个大三学生时，他发表《取消心理学上的本能说》以求证明"本能是已完结的心理学"（Kuo，1922），并成为第一个反对本能概念和分类的学者。这篇文章获得了美国心理学学术界的广泛关注。随后，郭任远获得了其他反对本能概念的心理学家（如Allport，Ayers，Bernnard，Dunlap，Faris，Josey，Kantor等）的支持。相应地，赞成本能概念的世界著名心理学家（如William McDougall）为此写信回应郭任远，引发了著名的反本能运动。

郭任远颇费周章地实施了大量以动物行为反应为主题的实验。这些研究旨在观察和发现动物的发生和发展规律。他的系列研究可以分为三类。第一类关于猫对鼠反应的起源，这些实验探索了不同环境条件下猫对鼠的反应："我们的研究显示，猫可以去杀死鼠，可以去爱鼠，可以去恨鼠，可以去害怕鼠或是跟鼠玩耍。"（Kuo，1930）他指出，发生和发展要素在猫对鼠反应之中均有重要作用（Kuo，1938）。

第二类研究之中，他探索了影响动物搏斗的要素。郭任远选用各类动物，包括好斗的蟋蟀、暹罗斗鱼、日本灰鹌鹑和小鸡，在群体搏斗行为效应研究中，实验动物包括鱼（多个品种）、鸟（超过30个种类）、猫、鼠、兔和几内亚豚鼠，记录它们的行为并将这些行为分为23类。实验进一步揭示营养、生理、发展和环境要素在日本灰鹌鹑的搏斗行为模式之中有重要作用（Kuo，1960a，1960b，1960c，1960d），并且他描述了鱼类、鸟类和哺乳类动物的种间共存现象（Kuo，1960e，1960f，1960g）。

第三类对胚胎神经系统的行为和生理机能的研究，主要是通过一系列的鸡胚胎行为开展。郭任远以鸡胚胎行为发生学研究为主题系统地发表了数篇文章。他创设郭氏观察窗，这是他最为著名的实验，并且他提供了首个在不损伤胚胎前提下打开鸡蛋的方法。基于这些研究，郭任远成为伟大的科学家，不仅是作为心理学家，更是作为生物学家（Scheithauer et al.，2009）。Kiessling和Anderson高度赞许了他的工作："自亚里士多德之后，郭任远是第一个窥视到鸡蛋内部的人。"他的研究成果和研究方法被下一代学者所习得，例如，Gottlieb采用了郭任远的研究方法开始了他的一系列开创性实验，研究经验对幼鸭识别母鸭召唤叫声的影响（Logan and Johnston，2007）（图4）。

图4　郭任远和Gilbert Gottlieb在北卡罗来纳州（1963年）

经过多年对鸡胚胎行为发展、猫对鼠反应的发展以及不同动物的搏斗行为的研究，他在《行为发展之动力形成论》一书中提出了一个新理论，探讨了行为渐成论、行为梯度理论和行为潜势理论。郭任远将行为渐成定义为：每个动物都是它的发展历史和它在具体环境之中受具体刺激做出反应的成果。他用一个简单的公式来表达这个观点：

$$\underset{(1)}{(\mathrm{DH+ST})} + \underset{(2)}{(\mathrm{PE\ Set.+SS})} + \underset{(3)}{(\mathrm{BCF})} = \mathrm{Pre.Beh}^{①}.$$

① DH=发展历史，ST=特定训练（如搏斗、感觉区分等），PE Set.=物理环境设置，SS=特定刺激（如光亮、声音、食物、性、对象等），BCF=生物化学要素，Pre.Beh.=预测的行为。（1）与动物历史相关的要素，（2）与动物当前相关的要素，（3）与它们的生理机制相关。

除此之外，行为梯度理论呈现了关于行为发展次序的信息，行为潜势理论则阐述行为起源和行为发展次序的信息，如在发生学上的不同阶段某个特定行为模式出现的单个或多个原因（Kuo，1967）。

Logan和Johnston(2007)高度评价郭任远的工作："郭任远率先用实验阐述超越'先天和后天二元分化'的论述。"郭任远革命性的工作具化了如今我们称为"关系发展理论"(DSP)的远景(Overton and Lerner, 2012),这有益于发展科学之中的库恩范式转变(Greenberg, 2014)。总之,郭任远是心理学历史上有过璀璨光辉的中国心理学家,他的研究是可被深度发掘的资源宝库,可应用于多学科领域。

77. Zing-Yang Kuo and behavior epigenesis based on animal experiments

Zing-Yang Kuo (1898—1970), styled Taofu, was a world-renowned Chinese psychologist. He was one of the most extreme precursors of the behaviorism school and the most radical behaviorist in the history of behavioristic psychology. His main scope of research focused on animal and comparative psychology. Animal experimentation was his favorite research method, and he studied many different species of animals. The aim of Kuo's research was to determine the origin of the formation of behavior, and in 1967, he constructed a theory for this formation called behavior epigenesis. Kuo was the only Chinese psychologist profiled in *The First Century of Experimental Psychology*. In addition, *the Journal of Comparative and Physiological Psychology*, which usually only publishes scientific experiments, provided a few pages to discuss the life of Kuo and gave this high appraisal: "During his turbulent career he stood at the center of several important metatheoretical controversies and made unique investigative contributions to the study of behavior and the nervous system, particularly from the standpoint of developmental analysis." (Gottlieb, 1972)

In 1898, Kuo was born to a merchant family in Tongyu Village, Chaoyang County, Guangdong Province, China. He entered Fudan University in 1916 and left two years later to pursue advanced studies in the United States, where he enrolled in the University of California at Berkeley. After some vacillation, he settled on psychology as his major. In 1918, he met his supervisor, Edward Chace Tolman, a flagman in neo-behaviorism, who began at the University of California the same year. During the years of 1921 and 1922, he worked as an assistant at Berkeley, completed his studies and met the requirements for a Doctor of Philosophy in 1923. However, Kuo did not agree with the amendments put forward by school authorities on his doctoral thesis and instead persisted in his academic viewpoints. He thus gave up the chance to earn his Ph.D. and returned to China before his oral thesis defense (Kuo, 1940). He would not

Yanyan Qian[1], Wei Chen[2,3], Benyu Guo[1]

1 School of Psychology, Nanjing Normal University, Nanjing 210097, China

2 Department of Psychology, Shaoxing University, Shaoxing 312000, China

3 Department of Psychology, The Education University of Hong Kong, Hongkong, China

Correspondence: anti-monist@163.com (W. Chen)

earn a doctoral degree until 1936.

In 1923, Kuo returned to his homeland and took a series of positions at Fudan University between 1923 and 1927. He was a professor in 1923, acted as vice-president and exercised the power and duties of acting-president in 1924, and served as acting-president until 1926. During this time, he set up a department of psychology in 1924 and raised money from clansmen to establish Sub Bin hospital—the third largest Institute of Psychology in size according to *Shun Pao* (Hu, 1995) and the only Institute of Psychology in China at that time.

The following year, Kuo held a teaching post at Nanjing Central University, where he founded the Psychology Institute in Academia Sinica while acting as the head of this research institute. From April 1933 to February 1936, he held a position at Zhejiang University, where he imparted knowledge and educated people. He occupied the role of schoolmaster, launched the department of psychology, and was elected to the First Research Council of Academia Sinica in 1935. He also took part in the construction of a psychology association: Kuo and eight other psychologists (Kuo Itzen, Ai Joseph Wei, Hsiao Hsiao-Hung etc.) launched the Chinese Psychological Association and held a preliminary meeting in Shanghai during the summer session of 1931.

On December 10, 1935, a student movement erupted in Zhejiang, led by Zhejiang University students, that opposed by Kuo. This conflict caused Kuo's deposition which announced in 257th regular meeting of Executive Yuan in February, 1936. This further led to a change in his research methods and prospects. From then on, Kuo became a wandering scientist without a nation to call home. From 1936 to 1946, he gave lectures on psychobiology at Berkeley, performed research at the Osborn Zoological Laboratory of Yale University (1937—1938) and the Carnegie Institution of Washington (1938—1939), and went on a lecture circuit of several universities (1941—1943).

In 1946, Kuo became a member of the Board of Trustees of the University of Hong Kong. During the next twelve years, he occupied himself with studies of human behavior, including a social psychological analysis of the Chinese national character (Gottlieb, 1972). In 1963, Kuo stayed in America for months to popularize his scientific ideas, including making speeches at the meeting of the International Congress of Psychology and the International Congress of Zoology in August in Washington, DC (Kuo, 1967). That same year, Kuo returned to Hong Kong, where he passed away on August 14, 1970.

Kuo's research process can be divided into three stages: theoretical assumption, animal experimentation and theory formation (Gottlieb, 1972). In Kuo's opinion, psychology should be an empirical science. He believed that instincts led away from empirical science: "Whether all the opponents of instincts will agree or not, I believe our chief motive for denying instincts is to rescue psychology from armchair speculation. We mean to remove this stumbling-block from genetic psychology." (Kuo, 1922) Thus, he published Giving up instincts in psychology when he was a junior (3rd year) student to prove that "Instinct was a finished psychology" (Kuo, 1922) and became the first person to argue against the concept and classification of instinct. This paper garnered widespread academic attention in American psychology academy. Subsequently, Kuo received support from other psychologists (Allport, Ayers, Bernard, Dunlap, Faris, Josey, Kantor, etc.) who objected to the concept of instinct. In contrast, world-renowned psychologists such as William McDougall, who supported the concept of instinct, wrote articles in response to Kuo, spurring the famous anti-heredity movement.

Kuo made considerable efforts to carry out a large number of experiments on the animal behavioral response. These experiments sought to observe and determine the rules governing animal ontogeny and development. His series of experiments can be grouped into three categories. The first pertained to the genesis of the cat's response to the rat. These experiments showed the behavior of kittens toward rats and mice under different environmental conditions: "Our study has shown that kittens can be made to kill a rat, to love it, to hate it, to fear it or to play with it." (Kuo, 1930) He pointed out that both ontogeny and development play important roles in the formation of the cat's response to the rat (Kuo, 1938).

In the second vein of research, he studied the factors that determine fighting in animals. Kuo employed a large number of species, such as fighting crickets, Siamese fighting fish, Japanese grey quails, chickens, dogs, numerous species of fish, more than 30 species of birds, cats, rats, rabbits and guinea pigs, recording and classifying all fighting behavior patterns into 23 types. Experiments further revealed that nutritional, hormonal, developmental and environmental factors play an important role in the fighting behavior patterns of the Japanese grey quail (Kuo, 1960a, 1960b, 1960c, 1960d), and he described the phenomenon of inter-species coexistence in fish, birds and mammals (Kuo, 1960e, 1960f, 1960g).

Third, he studied the behavior and physiology of the embryonic nervous system, mainly through a series of behavioral studies on chick embryos, which he systematically presented in several articles on the subject of the ontogeny of embryonic behavior in Aves. He created the Kuo Observation Window, which is regarded as one of the most well-known experiments, and he provided the first record of a method for opening the egg and studying the embryo without inducing harm. Based on these studies, Kuo was recognized as a great scientist, not only by psychologists but also by biologists (Scheithauer et al., 2009). Kiessling and Anderson (2003) praised his work, saying "Kuo was the first to peer inside an egg since Aristotle did so" . Not only his research findings but also his experimental methods were taught to later generations of scholars, for example, "Gottlieb employed Kuo's methods to begin his pioneering series of experiments on the effects of experience on ducklings' recognition of their species' maternal assembly call" (Logan and Johnston, 2007).

Over many years of the study of the development of behavior in the avian embryo, the development of the behavior of the cat in relation to the rat, and particularly the fighting behaviors of a number of species, Kuo proposed new theoretical concepts in his book, *The Dynamic of Behavior Development*, which discussed behavior epigenesis, behavior gradients and behavioral potential. Kuo defined behavior epigenesis as the idea that every animal is a result of its developmental history and it will react to a specific stimulus in a specific environmental setting. He expressed this view in a simple formula:

$$\underset{(1)}{(DH+ST)} + \underset{(2)}{(PE\ Set.+SS)} + \underset{(3)}{(BCF)} = Pre.Beh^{①}.$$

Besides, the theory of behavior gradients present information about the developmental sequence of behavior, and the theory of behavioral potential provide information of the origin and developmental sequence of behavior, i.e., the cause or causes for the appearance of certain behavior patterns at different stages of ontogeny (Kuo, 1967).

Logan and Johnston (2007) spoke highly of Kuo's work: "Kuo was the first empirically supported statement of the necessity of transcending the separation between nature and nurture in order to understand behavior." Kuo's revolutionary work embodies what today we call a relational developmental systems (RDS) perspective (Overton and Lerner, 2012), and it contributed to the Kuhnian paradigm shift in developmental science (Greenberg, 2014). In a word, Zing-Yang Kuo was a special Chinese psychologist of great significance to the history of psychology and whose research represents a vast treasure of resources that could be developed and used in the future in many different fields.

Figures

Fig.1 Zing-Yang Kuo in 1955

Fig.2 Sub Bin hospital

Fig.3 Zing-Yang Kuo

Fig.4 Zing-Yang Kuo and Gilbert Gottlieb in North Carolina in 1963

References

Gottlieb G (1972) Zing-Yang Kuo: Radical scientific philosopher and innovative experimentalist (1898—1970). J Comp Physiol Psychol 80(1):1–10.

Greenberg G (2014) Emergence, self-organization, and developmental science: Introduction. Res Hum Dev 11:1–4.

Hu JN (1995) The Psychological Anthology of Hu Chi-Nan (Supplement edition). Shanghai: Xuelin Press. (胡寄南. 1995. 胡寄南心理学论文论选(增补本). 上海: 学林出版社.)

Kiessling AA, Anderson S (2013) Human Embryonic Stem Cells: An Introduction to the Science and Therapeutic Potential. Sudbury: Jones and Bartlett Publishers.

Kuo ZY (1922) How are our instincts acquired? Psychol Rev 29(5):344–365.

Kuo YZ (1930) The genesis of the cat's responses to the rat. J Comp Psychol 11(1):1–36.

① DH = developmental history, ST = special training (such as fighting, sensory discrimination, etc.), PE Set.=physical environment setting, SS = special stimulus (such as light, sound, food, sex, objects, etc.), BCF = biochemical factors, and Pre.Beh. = predictable behavior. Bracket (1) deals with the animal's past, (2) with its present, and (3) with its physiological mechanisms.

Kuo ZY (1938) Further study of the behavior of the cat toward the rat. Comp Psychol 25(1):1–8.

Kuo ZY(1940) Psychology and I. Digest (68–69): 1507–1508. (郭任远. 1940. 我和心理学. 文摘, (68–69): 1507–1508.)

Kuo ZY (1960a) Studies on the basic factors in animal fighting Ⅰ. General analysis of fighting behavior. J Genet Psychol 96(2):201–206.

Kuo ZY (1960b) Studies on the basic factors in animal fighting Ⅱ. Nutritional factors affecting fighting behavior in quails. J Genet Psychol 96(2):207–216.

Kuo ZY (1960c) Studies on the basic factors in animal fighting Ⅲ. Hormonal factors affecting fighting in quails. J Genet Psychol 96(2):217–223.

Kuo ZY (1960d) Studies on the basic factors in animal fighting Ⅳ. Developmental and environmental factors affecting fighting in quails. J Genet Psychol 96(2):225–239.

Kuo ZY (1960e) Studies on the basic factors in animal fighting Ⅴ. Inter-species coexistence in fish. J Genet Psychol 97(2):181–194.

Kuo ZY (1960f) Studies on the basic factors in animal fighting Ⅵ. Inter-species coexistence in birds. J Genet Psychol 97(2):195–209.

Kuo ZY (1960g) Studies on the basic factors in animal fighting Ⅶ. Inter-species coexistence in mammals. J Genet Psychol 97(2):211–225.

Kuo ZY (1967) The Dynamic of Behavior Development. New York: Random House.

Logan CA, Johnston TD (2007) Synthesis and separation in the history of "nature" and "nurture". Dev Psychobiol 49(8):758–769.

Overton WF, Lerner RM (2012) Relational developmental systems: A paradigm for developmental science in the postgenomic era. Behavioral and Brain Sciences 35(6):375–376.

Scheithauer H, Niebank K, Ittel A (2009) Developmental science: Integrating knowledge about dynamic processes in human development. Dynamic process methodology in the social and developmental sciences. In: Valsiner J. Thinking in Psychological Science: Ideas and Their Makers. New Brunswick: Transaction Publishers, 315–329.

78. 孙国华：中国生理心理学和儿童心理学的奠基者

孙国华（Kuo-Hua Sun）（图1），字晓孟，我国著名生理心理学家、儿童心理学家。

图1 孙国华（1902—1958）

1902年1月26日，孙国华出生于山东潍县。1914年至1923年就读于北京清华学校。1923年，进入美国俄亥俄州立大学教育学院学习，并于1928年和1929年分别获得心理学硕士和哲学博士学位。在此期间（1925—1926年），孙国华也在芝加哥大学生理学院进行了学习（图2）。1928年底回国，历任清华大学心理学系教授、系主任，北京大学、东北大学、北京师范大学心理学教授。1941—1946年因病暂离教学岗位，任国立编译馆编辑兼总务主任。中华人民共和国成立后，任清华大学心理学系教授、系主任；1952年，经院系调整，孙国华任北京大学校务委员、哲学系副主任，兼任中国科学院心理研究所研究员、学术委员、发生发展心理室领导，中国心理学会理事，《心理学报》常务编辑。1958年7月8日因心脏病发作在北京逝世（《心理学报》编辑部，2003）。①

孙国华的研究主要有两大领域：一是生理心理学，二是儿童心理学。在生理心理学领域，1926年，他在《美国生理学杂志》上发表《鸟类的瞳孔反射》。1931年，孙国华在《美国博物学家》发表《蚯蚓通过增加生殖环节来生长吗？》一文，文章表明蚯蚓长度增加的因素之一是新节的形成，然而此文也说明有两种新生蠕虫可能与成熟的蠕虫具有同样数量的生殖环节（Sun and Pratt，1931）。通过调查发现，在蠕虫的发展周期中有三个阶段：一是蠕虫破茧初始；二是未形成生殖带的蠕虫（从堆肥箱中获得）；三是形成生殖带的蠕虫——生殖带被认为是成熟的标志。文章研究结果表明：① 从蠕虫体节长度与环节数量的关系来说，不同的发育阶段之间存在着显著差异，只

作者：王伊萌[1]，钱燕燕[2]

1 南京师范大学心理学院，南京210097，中国

2 荷兰莱顿大学社会与行为科学学院，莱顿2333AK，荷兰

邮箱：m15651010370_1@163.com（王伊萌）；psyqyy@163.com（钱燕燕）

① 孙国华生平（1902.01.26—1958.07.08）及其文章《无头扁虫的食物反应》均记录于《中国大百科全书·心理学卷》。

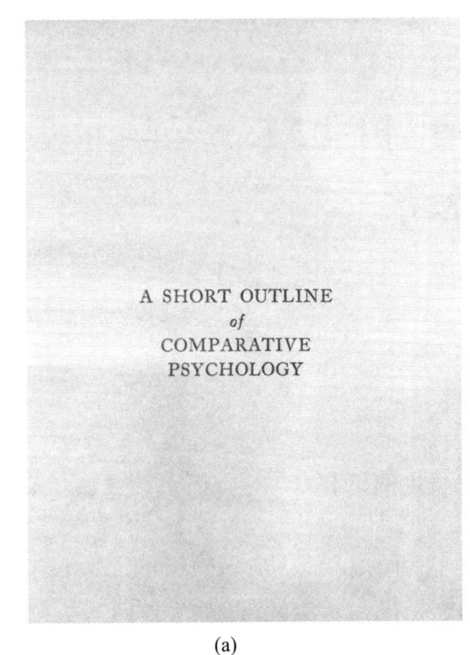

(a)

(b)

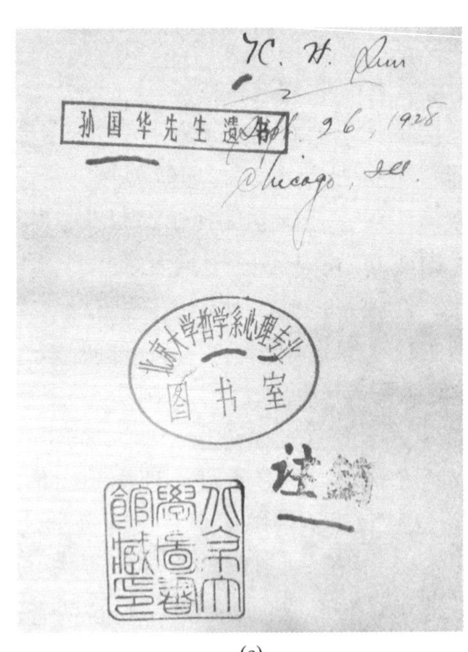

(c)

图2 孙国华教授在美国芝加哥读书时购买的书籍

有当幼虫达到一定的节数之后成熟，重新计算环节数量，才能说明蚯蚓是否能够通过增加环节数量来生长；② 具有生殖带的蚯蚓与不具有生殖带的蚯蚓不属于同一类型（Sun and Pratt，1931）。

1936年，孙国华、周先庚和陆志伟创建了第一个心理学双语期刊——《中国心理学报》，孙国华任编辑和出版人（忻志鹏，1991）。其后，孙国华陆续在《中国心理学报》上发表与他人合作的系列文章。例如，1936年，孙国华与敦福堂（Fwu-Tarng Dun）撰写《暗适应扁虫的向地性笔记》（Sun and Dun，1936），论证经过暗适应的扁虫

（*Planaria gonocephala*）在觅得食物之后未表现出对地心吸力的消极反应。此外，论文还表明，抵抗图形和地面亮度值反转时的混淆能力被用来测试对形状的感知。孙国华和他的合作者后来在一篇论文中对大鼠进行了研究，把它们与幼鼠进行了对比（Luh and Sun，1936）。1936年，孙国华与张民觉合作撰写《无头扁虫的食物反应》，发表于《中国心理学报》。研究发现，无头扁虫（*Planaria dorotocephala*）仍具有食物反应（Sun and Chang，1936）。

在儿童心理学领域，1927年冬季，孙国华参加学位答辩，硕士毕业论文为《感觉运动系统的种系发生》（List of Graduate Degrees and Dissertations in Education Educational Research Bulletin，1928）。1928年，孙国华发表《婴儿的视听觉的反应研究》（Sun，1928）。

1930年，孙国华与K. C. Pratt、A. K. Nelson合作，共同发表《初生儿的行为研究》（图3），解释了新生儿行为的一般发展规律：① 新生儿行为是普遍性的，给予新生儿任何部位刺激，都能引起其他所有部位的反应；② 被刺激部位的反应远远比其他部位反应强烈，且反应的强度与频率从刺激部位到周围部位逐渐降低。这是儿童心理学的长篇重要论文，收录在《俄亥俄州立大学研究：心理学贡献》丛书第10卷（Pratt et al.，1930）。1933年，《教育研究评论》高度评价了孙国华等人对婴儿从出生到青春期的心理发展进行的翔实而细致的观察（Frank and George，1933）。

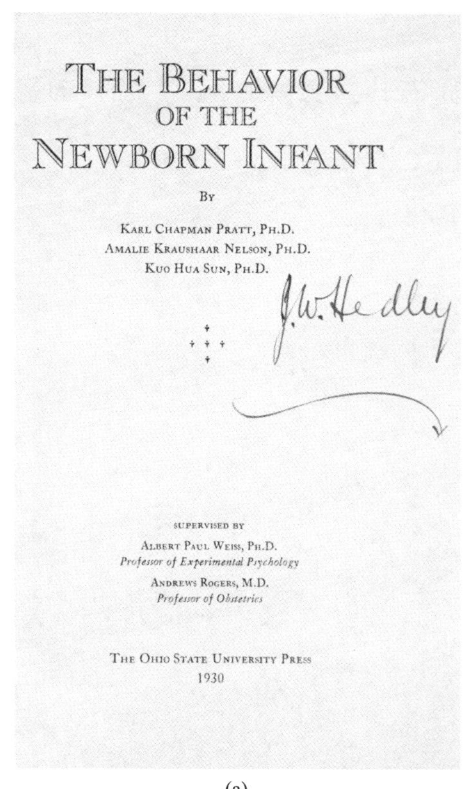
THE BEHAVIOR OF THE NEWBORN INFANT

BY

KARL CHAPMAN PRATT, PH.D.
AMALIE KRAUSHAAR NELSON, PH.D.
KUO HUA SUN, PH.D.

SUPERVISED BY
ALBERT PAUL WEISS, PH.D.
Professor of Experimental Psychology
ANDREWS ROGERS, M.D.
Professor of Obstetrics

THE OHIO STATE UNIVERSITY PRESS
1930

(a)

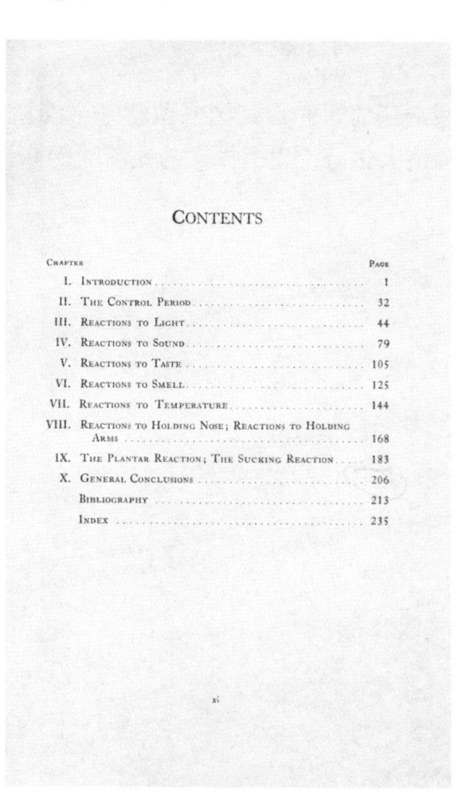
CONTENTS

xi

(b)

图3 孙国华博士撰写了《初生儿的行为研究》第三、四、五、六章，并对各仪器进行了描述和绘图，同时参与了资料准备和专著各部分的协调

孙国华也参与翻译了一系列的外文书籍，包括《心理学史》(Pillsbury，1929)、《现代猫：她的思想和行为——比较心理学的介绍》(Gates，1928)、《人类行为进程》(Sherman and Sherman，1929)、《动物心理学的介绍：大鼠的行为》(Munn，1933)以及《婴儿的感官和意志》(Preyer，1889)。

除了学术领域的工作，孙国华也参与学科建设的工作。1926年秋，孙国华与唐钺、周先庚等人创建清华大学心理学系，并与唐钺、周先庚、沈履以及陈立等人定期举行讲座(钱颖一和李强，2011)。清华大学心理学系是我国最早建立的大学心理学系之一，为清华大学理学院六大系之一。1952年，孙国华担任北京大学哲学系副主任，邵郊作为孙国华的助手也参与了实验室建设，并协助孙国华调整学科的划分，提供行政服务，举办讲座，开展高级神经系统活动、动物心理学和比较神经解剖学的实验(钱颖一和李强，2011；Wang et al.，2019)。一年后，孙国华和生理心理学家以及比较心理学家邵郊(Wang et al.，2019)与沈迺璋合作，建立了我国第一个动物条件反射实验室，后发展为生理心理实验室，该实验室在心理学成为国家重点学科过程中起到了关键作用(钱颖一和李强，2011；Wang et al.，2019)。1955年秋，孙国华教授在北京大学创建了儿童心理学实验室(后发展成为北京大学心理学系发展心理学实验室)，并促成了大量的领先研究，特别是在儿童语言领域。

致谢

钱燕燕的研究得到国家留学基金管理委员会研究生(博士)奖学金的资助。感谢南京师范大学博士生王振东与熊咪咪的修订。

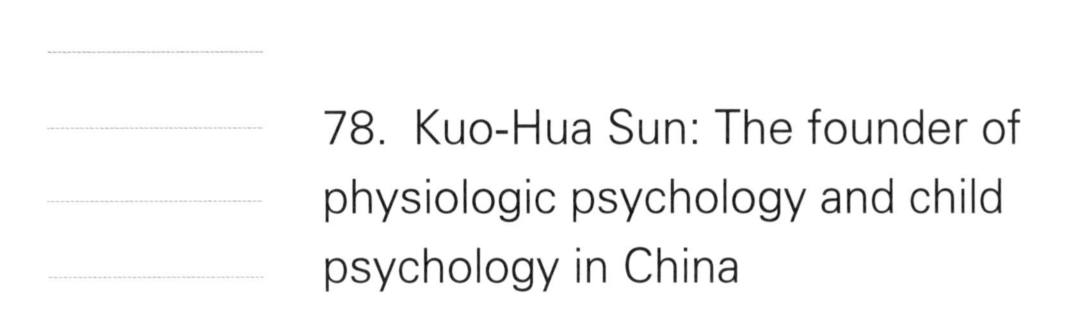

78. Kuo-Hua Sun: The founder of physiologic psychology and child psychology in China

Kuo-Hua Sun (孙国华, 1902—1958, courtesy name—Xiaomeng) was a Chinese psychologist mainly engaged in Physiological Psychology and Child Psychology.

Kuo-Hua Sun was born on January 26, 1902, in Weixian, Shandong Province. He studied in Tsinghua School from 1914 to 1923. At the same year, Sun attended the education school of Ohio State University, and then received a Bachelor of Psychology (1928) and a Doctor of Philosophy (1929) from Ohio State University. During this period, Sun studied in the Department of Physiology at Chicago University (1925—1926). At the end of 1928, Sun returned to China. He was appointed as the dean and professor of the Psychology Department at Tsinghua University, and a professor of Beijing Normal University, Northeastern University, and Peking University. From 1941 to 1946 he temporarily left the teaching profession because of his illness; during this time he served as an editor and general affairs director in the former National Compilation Library. In 1952, Sun served as the deputy head of the Department of Philosophy and was a member of the school council of Peking University. He was also an academic member of the Institute of Psychology of the Chinese Academy of Sciences (CAS), a leader of the Department of Occurrence and Development Psychology, a director of the Chinese Psychological Association, and a managing editor of the *Journal of Psychology*. On July 8, 1958, Sun died of heart attack at 56 years of age① (Editorial Board of *Acta Psychologica Sinica*, 2003).

Sun had two main research interests throughout his life: Physiological Psychology and Child Psychology. In the field of Physiological Psychology, he published the paper "The pupillary reflex in birds" in *American Journal of Physiology* in 1926. He later published "Do earthworms grow by adding segments?" in *American Naturalist* (Sun and Pratt, 1931), and it has indeed commonly been asserted that one of the factors contributing to the growth in length of earthworms

---

Yimeng Wang[1], Yanyan Qian[2]

1 School of Psychology, Nanjing Normal University, Nanjing 210097, China

2 Social and Behavioral Sciences Facility, Leiden University, Leiden 2333AK, The Netherlands

Correspondence: psyqyy@163.com (Y.-Y. Qian)

① One recording of Kuo-Hua Sun's lifetime, another statement is "Feeding behavior of headless *Planaria dorotocephala*". The 1902.01.26—1958.07.08 in *Encyclopedia of China*, Psychology Volume (《中国大百科全书 · 心理学卷》).

is the laying down of new segments. This paper, however, found that the number of segments of two newly emerged worms (probably *Helodrilus foetidus*) was essentially the same as that among mature specimens of the species (Sun and Pratt, 1931). Three stages of development were selected for investigation: (1) Worms just emerging from cocoons; (2) Worms having no clitella (collected from the compost heap); (3) Worms selected on the basis of clitella—these being considered indicative of sexual maturity. The results showed that: (1) As for the relationship between body length and segment number, there were significant differences among different stages of development, so whether earthworms can grow by increasing segments can be solved only when larvae with a certain number of segments can mature and recount; (2) Earthworms with zones (genital belts) do not belong to the same type as those without links (Sun and Pratt, 1931). Furthermore, according to Google Scholar, this article was cited 14 times, with the most recent citation recorded in 2015.

In 1936, as an editor and publisher, Sun cooperated with Siegen K. Chou (周先庚) and Chih-Wei Luh (陆志伟) founded the first bilingual journal of psychology—*The Chinese Journal of Psychology* in Beijing (Xin, 1991). Since then, Sun published dozens of articles in this journal, in cooperation with other researchers. Taking planaria research for example, Sun and Fwu-Tarng Dun (敦福堂) (1936) wrote the report "Note on Geotropism in the dark-adapted *Planaria gonocephala*", arguing that *Planaria gonocephala* did not show a negative response to geocentric attraction when it found food. Additionally, the paper showed that the ability to resist confusion at the reversal of brightness values of the figure and the ground is used as a test for the perception of form. Sun and colleagues later investigated this in rats with the paper "Brightness reversal in form discrimination of the rat as compared with a young child" (Luh and Sun, 1936). In 1936, Sun and Min-Chueh Chang (张民觉) co-authored "Feeding behavior of headless *Planaria dorotocephala*". The study showed that *Planaria dorotocephala* still had a food response (Sun and Chang, 1936).

In the field of Child Psychology, Sun was conferred the master's degree with his thesis "Phylogenesis of the sensorimotor system" in the winter of 1927 (List of Graduate Degrees and Dissertations in Education Educational Research Bulletin, 1928). In 1928, Sun was awarded his doctor's degree with his dissertation "A study of visual and auditory reactions in infants" (Sun, 1928).

In 1930, co-authoring with Karl Chapman Pratt and Amalie Kraushaar Nelson, Sun published "The behavior of the newborn infant" which explained the general developmental rules of the newborn infants' behavior: "(1) The infants behavior was generalized, and any stimulation of the body in any place would lead to almost all parts of the individuals' response; (2) The response of the stimulated parts was stronger than that of other parts, and the intensity and frequency of stimulation gradually decreased from the stimulated parts to the surrounding parts." This article, printed in a single book for *Ohio State University Studies: Contributions in Psychology*, Volume 10 (Pratt et al., 1930), was regarded as having great value in Child Psychology. To date, the Google Scholar search engine reports that the article has 169 citations.

In 1933, the *Review of Educational Research* highly praised Sun, Pratt, and Nelson's studies of psychological development of infants from birth to adolescence (Frank and George, 1933).

Sun was also involved in foreign academic resource translation, including *The History of Psychology* (Pillsbury, 1929), *The Modern Cat: Her Mind and Manners: An Introduction to Comparative Psychology* (Gates, 1928), *The Process of Human Behavior* (Sherman and Sherman, 1929), *An Introduction to Animal Psychology: The Behavior of the Rat* (Munn, 1933), and *The Sense and Will of Babies* (Die Seele des Kindes) (Preyer, 1889).

The construction of disciplines is another area of Sun's significant work. In the autumn of 1926, in collaboration with other academics, Sun founded the psychology department of Tsinghua University, where Sun gave lectures regularly, alongside Yueh Tang (唐钺), Siengen K. Chou, Lv Shen (沈履), and Li Chen (陈立) (Qian and Li, 2011, p. 286). It was one of the earliest psychology departments in Chinese universities and one of the six major departments in the School of Science of Tsinghua University. In 1952, led by Sun, the deputy director of the philosophy department at Peking University, Jiao Shao (邵郊) joined this institute as Sun's coadjutant and assisted Sun to adjust the division of disciplines, offer civil service, give lectures and conduct experiments on higher nervous system activity, animal psychology, and comparative neuroanatomy (Qian and Li, 2011; Wang et al., 2019). One year later, Sun, cooperating with physiological psychologists and comparative psychologists Jiao Shao (Wang et al., 2019) and Nai-Chang Shen (沈迺璋), established the first Animal Conditioned Reflex Laboratory in China which later developed into the Physiological and Psychological Laboratory, which played a key role in psychology for becoming a national key discipline (Qian and Li, 2011; Wang et al., 2019). In the autumn of 1955, Sun founded the Child Psychology Laboratory in Peking University, which grown into the Developmental Psychology Laboratory of the Psychology Department in Peking University. This laboratory prompted plentiful researches in a leading position, especially in the area of children's language.

Acknowledgements

The research of Yanyan Qian is supported by a post-graduate scholarship (Ph.D) of the China Scholarship Council (CSC). We thank Nanjing Normal University Ph.D candidate Zhendong Wang and Ph.D candidate Mimi Xiong for their language revision.

Figures

Fig.1 Kuo-Hua Sun (1902—1958)

Fig.2 Prof. Sun bought the books when he studied in Chicago, USA

Fig.3 Dr. Sun composed the Chapter Ⅲ, Ⅳ, Ⅴ and Ⅵ of *The Behavior of the Newborn Infant*, and made efforts to descriptions and drawings of all the apparatus, preparation of the data, and coordination of the different sections of the monograph

References

Editorial Board of *Acta Psychologica Sinica* (2003) A commemoration on 100th birthday of Teacher Kuo-Hua Sun. Acta Psychol Sin 1:77. (《心理学报》编辑部. 2003. 纪念孙国华先生诞辰100周年. 心理学报, 1:77.)

Frank N, George D (1933) Mental development from birth to puberty. Educ Res Rev 3(2):84–107.

Gates GS (1928) The modern cat, her mind and manners: An introduction to comparative psychology. Nature 124: 364–365.

Munn NL (1933) An Introduction to Animal Psychology: The Behavior of the Rat. Oxford: Houghton Mifflin.

Pillsbury WB (1929) The History of Psychology. New York: W W Norton & Co.

Pratt KC, Nelson AK, Sun KH (1930) The Behavior of the Newborn Infant. vol. 10. Ohio: Ohio State University Press, 206–212.

Preyer WT (1889) Die seele des kindes: mit materialien zur rezeptionsgeschichte. Berlin, Heidelberg: Springer.

Qian YY, Li Q (2011) Social Science of Old Tsinghua. Beijing: Tsinghua University Press, 286–287. (钱颖一, 李强. 2011. 老清华的社会科学. 北京: 清华大学出版社, 286–287.)

Sherman M, Sherman IC (1929) The Process of Human Behavior. New York: W W Norton & Co.

Sun KH (1928) A Study of Visual and Auditory Reactions in Infants. Ohio: Ohio State University Press.

Sun KH, Chang MC (1936) Feeding behavior of headless *Planaria dorotocephala*. Chin J Psychol 1(4):333.

Sun KH, Dun FT (1936) Notes on geotropism in the dark-adapted *Planaria gonocephala*. Chin J Psychol 1:1–9.

Sun KH, Luh CW (1936) Brightness reversal in form discrimination of the rat as compared with a young child. Chin J Psychol 1:101–114.

Sun KH, Pratt KC (1931) Do earthworms grow by adding segments? Am Nat 65(696):31–34.

Wang L, Qian Y, Su Y (2019) Jiao Shao: A forerunner of physiological psychology and comparative psychology in China. Protein & Cell 10(9):623–627.

Xin ZP (1991) Practical Clinical Psychological Medicine. Shanghai: Shanghai Medical University Press. (忻志鹏. 1991. 实用临床心理医学. 上海: 上海医科大学出版社.)

79. 丁瓒：中国现代医学心理学的倡导者

丁瓒（图1）是我国杰出的心理学家，现代医学心理学的开拓者，中国科学院创建时期领导人之一。但他的名字却不为人所熟知。

丁瓒出生于江苏省南通市，1931年考入国立中央大学（现东南大学）心理学系，1935年毕业。大学四年时间里，他认真系统地学习了国际心理学不同流派的各种学说，对与疾病相关的心理卫生和繁难精细的精神分析尤为用功。大三时，他与同学丁祖荫合译了美国科学家F. D. Brooks的《青年期心理学》，首次在我国提出了青春期心理学。

图1 丁瓒(1910—1968)

从国立中央大学毕业后，丁瓒在北京协和医学院进行了研究生阶段的学习，师从国际著名精神科专家R. S. Lyman教授。经过三年的刻苦学习，他因成绩优异，在北京协和医学院留校任教，讲授医学心理学，并延续了他的医学心理学研究工作。授课期间，协和医学院其他部门的同事也深受其影响，开始更多地重视疾病相关心理健康指导与医学心理学的理念。他与协和医院内科张孝骞、妇产科林巧稚、儿科诸福棠等医生联合开展了不同科室疾病的心理学诊断和治疗。不仅如此，他还经常去社区医院、学校以及工厂为精神疾病患者提供诊治和心理咨询服务。1936年，他在北平创办了中国精神卫生协会，对医学心理学应用于临床及社区服务的综合模式进行了探索。

1940年，受抗日战争影响，丁瓒不得不放弃了自己在协和医院的教研工作，从北平迁往重庆，再辗转至南京。即便在战火洗礼中，他仍坚持进行精神卫生和心理健康的科学普及工作，用通俗易懂的语言，通过演讲的方式向民众宣传民主科学思想和心理健康知识。1944年，他建立了我国第一个拥有咨询室和诊疗室的医学心理学实验室。根据他公开演讲的内容和诊疗实践案例，1945年和1947年分别整理出版的《心理卫生论丛》和《青年心理修养》，既是我国医学心理学和病理心理学早期重要著作，也是关

译者：张保元

中国科学院北京生命科学研究院，北京100101，中国

邮箱：zhangby@biols.ac.cn

于青年心理问题的最早的学术论著。

1947年，受世界卫生组织奖学金资助，丁瓒先后赴美国芝加哥大学心理学系和米歇尔·瑞茜医院进行访问深造。1948年底，回国之前，他还访问了英、法等国，参加了在英国伦敦举行的国际心理卫生大会。在美国与欧洲访问考察的两年中，他不仅在科研方面取得了重大进展，而且结识了很多在国外生活和工作的优秀中国科学家。

1949年回国后不久，由于他与身处国外的中国学者有着广泛的接触和联系，丁瓒参加了中国科学院（前身为中央研究院）的建院工作，与物理学家钱三强、化学家恽子强等人一起为召集海外精英学者回归祖国开展工作。当时正值新中国建设初期，筹建中国科学院的第一需要就是人才回归。回顾大多数爱国科学家从美国和欧洲回到祖国的原因，不难发现，他们在很大程度上是受到了丁瓒在1947年至1948年出国访问期间的影响和感染。当很多爱国科学家在不到半年的时间里纷纷回国，希望用自己所学重新建设祖国时，将这些科学家有序妥善地安排到不同的学科机构中进行工作和生活，无疑是一件极为耗费精力的事。为此，在半年多的时间里，丁瓒每天从早上6点工作到次日凌晨2点（依据丁瓒的司机回忆记录）。正是由于他与其他同事的不懈努力，中国科学院于1949年11月1日顺利成立，丁瓒担任党组副书记。

中国科学院成立初期，第一批组建的研究所共有17个，同时设立了3个研究所筹备处，心理所筹备处是其中之一。由于长期从事心理学研究，筹建中国科学院心理研究所（简称“心理所”）的想法是丁瓒提出的，但很多人（包括一些科学家）对这个提议持保留态度，甚至是反对。主要有两个方面原因：一方面，由于早先学科设置的偏斜和落后，科学界中不少人仅仅把心理学归属教育学或哲学之中，作为一个极不重要的可有可无的分支学科，认为心理学远不如物理学、化学、数学等古老学科重要，不能单独建立研究所。另一方面，经过数十年的战争和动乱，我国在恢复和平后的首要任务是改善民生，心理学与解决当时吃饭穿衣等国计民生的基本困难毫不相关，不能单独建立研究所。但丁瓒承受住了来自学界同行和朋友的巨大压力，高瞻远瞩、力排众议，最终于1951年12月成立了中国科学院心理研究所（薛攀皋，2006）。

从最初筹建开始，心理所由于历史和政治原因屡经沉浮。丁瓒的个人生活亦是如此，但无论条件多么艰苦，他都认真坚持自己的科学研究，至死不渝。1958年，由丁瓒提出的“人类健康是由社会、心理和生物医学因素共同决定的”模型学说，要比1977年由G. L. Engel提出的类似的生物心理社会学（BPS）模型领先19年。同年，丁瓒在北京大学组织大家对神经衰弱进行群体性预防和治疗，是我国首次开展的群体性心理疾病干预。此外，他还在我国组建第一个对心理疾病进行预防和治疗的社区门诊，这些工作在国际上也是很罕见的。不幸的是，他的研究成果由于英年早逝而没有及时发表，这同样也是很少有人知晓丁瓒这位优秀科学先驱的原因。但他对中国科学院和学界的贡献永远也不该被人遗忘！

致谢

感谢丁瓒先生的儿子丁宗一先生为本文提供素材。

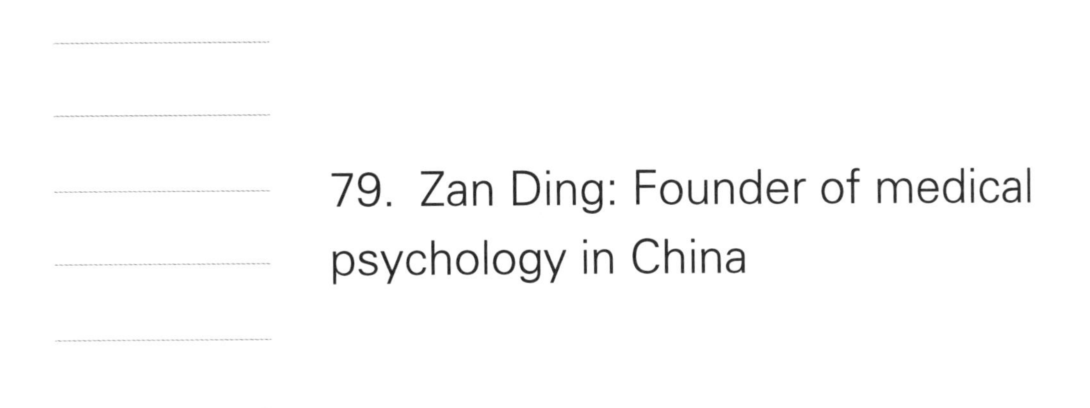

79. Zan Ding: Founder of medical psychology in China

Professor Zan Ding (1910—1968) is a distinguished psychologist, the founder of medical psychology in China, and one of the important founders of the Chinese Academy of Sciences in practice, even though his name is known to few of us.

Born in Nantong, Jiangsu Province, Zan Ding went to the National Central University (NCU) (now the Southeast University) in Nanjing to study psychology. During his four-year study at NCU, he learned a broad range of various schools of psychology and received rigorous training in the psychological labs, and was particularly interested in disease-related mental health. In his junior year, he and his friend Zuyin Ding translated "*The Psychology of Adolescence*" by Fowler D. Brooks. The translation was published and was the first introduction of adolescent psychology to Chinese readers.

After he graduated from NCU, Zan Ding continued his graduate study under Professor R. S. Lyman at the Department of Brain Science of Peking Union Medical College (PUMC). After he finished his three-year graduate training, he was offered a teaching position by the department. He accepted the offer and continued his medical psychological research in every possible way. While he was teaching the medical psychological course at PUMC, more and more colleagues from other departments of PUMC started to accept the concepts of disease-related mental health and medical psychology. Working together with Drs. Xiaoqian Zhang, Qiaozhi Lin and Futang Zhu, Zan Ding was able to carry out psychological diagnosis and treatment for patients from the departments of Internal Medicine, Obstetrics and Gynecology, and Pediatrics of the PUMC Hospital (PUMCH). Meanwhile, he also went to public clinics, schools and even factories, to provide counseling service and sometimes diagnosis and treatment of mental disorders. In 1936, he founded the Chinese Association for Mental Health in Beijing.

In 1940, Zan Ding had to give up his research work in Beijing due to the War of Resistance against Japanese Aggression and moved to Chongqing and later to Nanjing. Even during the war, he kept on teaching medical psychology and providing counseling services whenever possible. In

Ming Li
Beijing Institutes of Life Science, Chinese Academy of Sciences, Beijing 100101, China
Contact: liming@ioz.ac.cn

1944, he established the first laboratory of medical psychology in China, which also had affiliated counseling offices and clinics. He also published two books based on his public speeches and clinical cases: "*Essays on Mental Health*" in 1945 and "*Mental Cultivation of Youth*" in 1947. These are the earliest publications of medical psychology in China.

In 1947, Zan Ding went to the United States to continue his psychological research at the University of Chicago and Michael Reese Hospital as a visiting scholar, supported by the World Health Organization (WHO) Scholarship. Before he returned to China in late 1948, he also visited Europe to attend the International Conference of Mental Health. During these two years in the US and Europe, he not only made major progresses with his research projects, but also met and befriended a lot of Chinese scientists who lived and worked abroad.

Shortly after he returned to China, because of his extensive contacts with Chinese scholars, Zan Ding was appointed to participate in the establishment of the Chinese Academy of Sciences (formerly known as Academia Sinica), working with the famous physicist Sanqiang Qian(San-Tsiang Tsien), chemist Ziqiang Yun, et al. It was right around the time of the founding of the People's Republic of China and many patriotic scientists came back to China from the USA and Europe, some of whom were persuaded by Zan Ding during his tour in 1947 and 1948. Thousands of S&T elites were ready to work for their home country, but it was not an easy task to organize these scientists into different disciplines and different institutes. After half a year of hard work (which meant working from 6 AM to 2 AM next day, according to Ding's driver), the Chinese Academy of Sciences was established on November 1, 1949, and Zan Ding was later appointed as one of the leaders of the Academy.

When the Academy was founded, it had 17 institutes and preparatory offices of three more institutes. The Institute of Psychology was one of the three institutes to be established. The idea of establishing the Institute of Psychology was proposed by Zan Ding. Many people, including some scientists, were skeptical about the idea and even against it. Some felt that psychology was more in the category of philosophy and was not as important as classical disciplines such as mathematics and physics; others questioned the practicability of psychology. After decades of war and turbulence, the top priority of the new government was to improve people's livelihood. Would psychology give people food and clothing? The answer was No. But Zan Ding withstood the huge pressure from others, even from his friends, and the Institute of Psychology was finally established in December, 1951 (Xue, 2006).

From the very beginning, the Institute of Psychology went through a lot of ups and downs because of political reasons, so did Zan Ding himself. No matter how hard the external conditions were, he kept on with his research until he passed away in 1968. In 1958, Zan Ding proposed a model which stated that one's health was determined by social, psychological and biomedical factors combined. The model was quite ahead of the time and 19 years later a similar model, the Biopsychosocial (BPS) Model, was proposed by psychiatrist George L. Engel in 1977. In the same year of 1958, Zan Ding organized a mass prevention and treatment for neurasthenia at the Peking University, which was unprecedented in China and rarely seen internationally. He also established the first community clinic for prevention and treatment of mental disorders in China. Unfortunately, his research findings did not get published due to his early death. It was part of the reason why very few people know Zan Ding as a psychologist and the founder of medical psychology in China. His contribution to the establishment of the Chinese Academy of Sciences should not be forgotten, either.

Acknowledgements

The author is grateful to Dr. Zongyi Ding, Zan Ding's son, who provided most of the material for the article.

Figures

Fig.1 Zan Ding (1910—1968)

References

Li XT, Yue WH (2009) Medical Psychology. Beijing: People's Military Medical Press. (李心天, 岳文浩. 2009. 医学心理学. 北京: 人民军医出版社.)

Xue PG (2006) Autonomy and intervention: Psychology in China (1949—1976). Science and Culture Review 3: 111-121. (薛攀皋. 2006. 自主与干预: 心理学科在中国(1949—1976). 科学文化评论, 3: 111-121.)

80. 曹日昌：将辩证唯物主义应用于心理学研究的倡导者

图1　曹日昌先生（1911—1969）

“不要和我讲为什么，快点告诉我怎么做！”我们在实验室里经常能够听到学生这样急迫的诉求，可当他们按照师长手把手的指导完成实验后，取得的结果却往往不能令人满意；与此类似，很多同学阅读了大量文献，重复了许多实验，却对论文的立论假说和结论分析毫无头绪；被问到自己科研的前景和发展方向时，更是一脸茫然……

造成上述现象的原因有很多，但最重要的一条是：我们把科研停留在了被动学习阶段，很少积极主动地去自觉探索和尝试；学而不思则罔，思而不学则殆，思辨是科研工作中的一个非常重要的环节。我国著名心理学家曹日昌先生（图1），是最早将辩证唯物主义应用于心理学研究的倡导者之一，他在科研工作中运用的辩证观和方法论至今仍值得我们学习和借鉴（张侃，2006）。

曹日昌是新中国成立后心理学界的主要领导者和心理科学事业兴建、发展的主要奠基人之一，他主张心理学应当以马克思主义认识论为指导，研究依赖于社会实践的心理现象的普遍规律。他用事物运动发展中的三条普遍法则“矛盾统一律、质量互变律和否定之否定律”解释心理现象和心理学的发展，对我国的心理学发展起到极为重要的作用。

1943年，曹日昌运用唯物辩证法的观点对心理测验的若干重要问题进行剖析，发表了《心理测验的几个原则问题》，他以量与质、片段与整体、限定反应与自发表现、表象的测量与本质的理解等几对不同范畴的辩证关系，分析了在心理测验过程中存在的重重复次数而轻质量分析、重片段的反应而轻整体的特质、重限定反应而轻自发表现、重表象的测量而轻本质的理解等现象。这些辩证的分析为构建新的心理测验体系提供了坚实的理论基础，明确告诉我们科学研究一定要有正确的指导思想，必须重视

作者：张保元

中国科学院北京生命科学研究院，北京100101，中国

邮箱：zhangby@biols.ac.cn

对本质规律的理解和掌握。1965年，在他翻译的外文著作序言中，更是明确指出："从事心理学研究没有正确的哲学思想做理论指导，虽然有时也可以得到一些具体成果，但决不能得到关于心理学规律的本质的正确结论。""在已有的研究成果的基础上，运用辩证唯物主义观点和方法定会把心理学的实验研究提高到历史上未有的新水平。"

正是由于能够自觉地将辩证唯物主义的观点运用于研究实践，曹日昌在实验心理学领域，特别是学习、记忆研究方面取得了突出的成绩，对今天我国记忆心理学的发展仍有很大影响。

1945年至1948年，曹日昌在英国剑桥大学学习心理学，他的论文选题是"学习与记忆中的时间间隔"（赵莉如，2001）。期间，他从事了关于分配与集中学习的实验研究、图形记忆再认研究等，相关研究成果均发表在国外期刊上，而相关的实验设计和结果处理堪称我国早期有关学习和记忆实验研究的典范。20世纪60年代初，曹日昌在之前有关学习、记忆研究的基础上，研究了时间间隔对触觉的影响，实验的主试者和受试者只由他一人担任，实验设计令人耳目一新，具有独到之处。1962年，曹日昌指导了一个记忆研究专题组，一方面他指导有关人员以实验室方法，分别运用巴甫洛夫高级神经活动学说及信息理论，研究不同感觉通道（视觉、触觉、听觉）对识记的影响，以探讨识记的心理机制；另一方面，紧密结合教育实际，研究初中学生对四字短句、文言短文等的识记方法，并对文言文与白话文的识记过程做了比较分析，力图改进教学方法。这些工作虽然未能全部完成，但足以看出曹日昌对记忆心理学的研究与发展有他自己系统完整的设想，对记忆专题从不同方面各个角度都做了研究部署。他把记忆看成对输入信息的编码、存储和提取的过程，系统研究了瞬时记忆、短时记忆和长时记忆的不同机制，是我国最早运用信息加工理论研究记忆的开创者。不得不提的是，以这些长期积累研究、学习的成果为基础，由曹日昌主编的教材《普通心理学》，成为很多知名心理学家的启蒙教材，在我国广为流传，曾被重印了20余次！

除了在学术上的卓越成就，曹日昌对我国科学事业的发展同样贡献卓著。他历任中国科学院计划局副局长、中国科学院联络局副局长、中国科学院心理研究所所长和副所长等职务。新中国成立初期，他不遗余力地联络帮助了包括钱学森在内的多名在国外的科学家回国；负责筹建了中国科学院心理研究所和中国心理学会，为我国的心理学学科建设倾注了毕生精力。他将辩证唯物主义实践于心理学研究中，如他晚年所说："在心理学研究工作中要贯彻理论联系实际、为社会主义建设服务的方针。研究选题服从社会主义建设的需要，研究社会主义建设中需要解决的心理学问题。"这种治学的态度和思辨的思想，不仅是对他和他研究的心理学的自身要求，更是在启迪和要求我们每一位科研工作者在实践中用辩证的观点和方法去认真探索解决每一个科学问题。

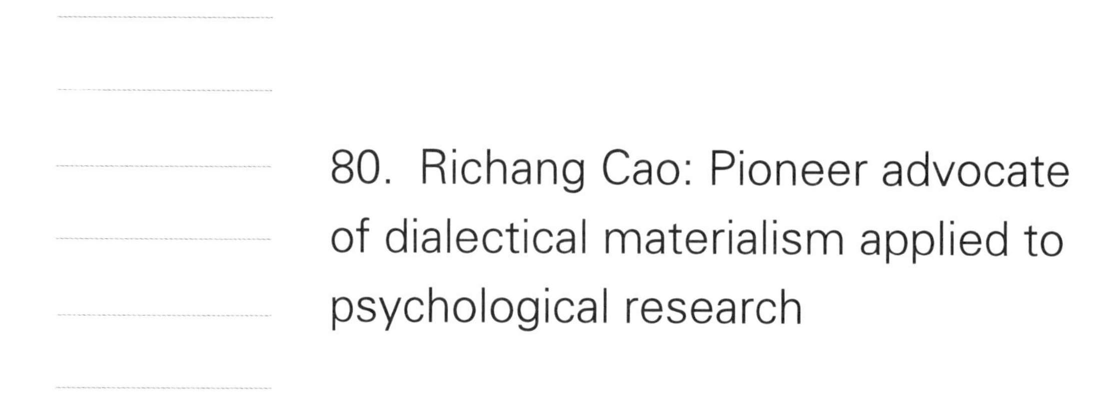

80. Richang Cao: Pioneer advocate of dialectical materialism applied to psychological research

"Don't tell me what for or for what, just tell me what I should do!" It is an urgent demand that we always heard in our labs. However, the results end up being far from satisfaction when postgraduate do experiments strictly by their adviser's guidance. There are many students who are completely indifferent to the theoretical hypothesis and concluding analysis of their research; even if they have read a lot of literature and repeated a lot of experiments in their daily routine. When we ask about theoretical foundations of their research, they are at a loss and can't answer at all.

Though there are many reasons for these disappointments, the most important one is because we are used to accepting or following tradition—what other people did—and rarely attempt to explore what other people have never done before. Confucius said: "Learning without thinking leads to confusion; thinking without learning ends in danger." Speculation is a vital part of doing research, like psychologist Richang Cao (Jih-Chang Tsao, 曹日昌), a pioneer advocate for applied dialectical materialism to psychological research, reviews his work and methodology which is still worth learning and using today (Zhang, 2006).

Prof. Cao is one of the main founders of the construction and development of psychological science after the founding of the People's Republic of China. He advocated that psychology should be used to study the universal law of psychological phenomena, depending on social practice, while guided by the epistemology of Marxism. He explained the development of psychological phenomena and psychology with three universal principles which are the law of unity of opposites, the law of qualitative and quantitative mutual change, and the law of negation of negation. These laws play an extremely important role in the development of psychology in China.

In 1943, Prof. Cao started to use material dialectics to analyze some important questions of psychological testing, and published Several questions of principle of psychological test. In his article, he analyzed the inadequacies of the existing psychological tests by a few common

Baoyuan Zhang

Beijing Institutes of Life Science, Chinese Academy of Sciences, Beijing 100101, China

Correspondence: zhangby@biols.ac.cn

categories of dialectical relationships including: quality and quantity, parts and whole, surface and essence, and so on. He clearly highlighted limitations of the existing tests, like they did only pay attention to the number of repetitions but didn't care about the quality of the analysis, only pay attention to individual parts but didn't care about the whole system, only pay attention to measurement of appearance but didn't care about the essence of things. Those dialectical analyses provided a solid theoretical basis for the construction of a new psychological test system. Clearly, we must have a correct guiding ideology, focused on understanding and mastering the law of nature in doing research. In 1965, Prof. Cao wrote in a preface of translating foreign books, "If we engaged in psychological research without the correct philosophical thought as theoretical guidance, although sometimes you may get some specific results, but we could not get the right conclusions about the nature of the laws of psychology" ; "On the basis of existing research results, the use of dialectical materialism will make the experimental study of psychology to a new level" .

Because of the conscious use of the dialectical materialism in research practice, Prof. Cao has made outstanding achievements in the field of experimental psychology, especially in learning and memory research. It still has a great impact on the development of memory psychology in current China.

From 1945 to 1948, Prof. Cao has been studying psychology at Cambridge University. His project focused on "The time between learning and memory" (Zhao, 2001). During this time, he published many results on the experimental study and graphics memory recognition research in foreign journals. These results are now used as a model of early experimental study of learning and memory in China. In the early 1960s, his work at Cambridge was used to study the effect of time interval on haptic. In his experiment, he was both the experimenter and subject. It was a very specific and unique design. In 1962, he guided a memory research group. On the one hand, he instructed the staff to study the effects of different sensory channels (visual, tactile and auditory) on memory and to explore the mental mechanism of memorization by Pavlov's theory of advanced neural activity and information theory. On the other hand, he also combined his research with practical education. He studied junior middle school students' memorizing methods of four-character phrases such as, classical Chinese essays. He did this while making a comparative analysis of the process of memorization of the classical and colloquial texts to improve teaching methods. Although his work was not able to be completed, it is important to know that Prof. Cao created his own complete system of ideas on memory in psychological research. He regards memory as the process of encoding, storing and extracting input information. He has also systematically explained the different mechanisms of instantaneous memory, short-term memory and long-term memory. There is no doubt that he is the pioneer of information processing theory in China. Based on these long-term accumulated research projects and learning achievements, the *General Psychology* course edited by Prof. Cao has become the enlightening material of many well-known psychologists. This book is widely circulated in China and has been reprinted more than 20 times!

In addition, Prof. Cao has contributed to the development of science and industry for our country. He served as the Planning Bureau, Liaison Office Deputy Director of Chinese Academy of Sciences (CAS), Deputy Director and Director of Institute of Psychology, CAS. In the early days of The People's Republic of China, he spared no effort to help drifters in a foreign country sent back to China, which included the famous scientist Hsue-Shen Tsien (Xuesen Qian, 钱学森). In the meanwhile, Prof. Cao is responsible for the preparation of the Institute of Psychology of CAS and the Chinese Psychological Society. He devotes all his life for the construction of psychology in our country. He put dialectic materialism in the research of psychology as he has mentioned in his later years, "In the work of psychological research, we should be upheld the guide line to carry out theory with practice, serve the socialist construction. Research topics should be satisfied to the needs of socialist construction, and the needs to solve the psychological problem in socialist construction." This study, which analyzes attitudes and philosophical thought, is not only the requirement for his research work, but also the inspiration for future researchers. It enlightens us to use dialectical methodology to seriously explore and solve scientific problems.

Figures

Fig.1 Prof. Richang Cao (Jih-Chang Tsao) (1911—1969)

References

Zhang K (2006) The admirable Prof. Richang Cao—Speeches at the publication symposium on Selected Works of Psychology by Prof. Richang Cao and Commemoration of the 95th Birthday of Prof. Richang Cao. Acta Psychol Sin 38(1):157-158. (张侃. 2006. 令人景仰的曹日昌先生——在纪念曹日昌先生诞辰 95 周年暨《曹日昌心理学文选》出版座谈会上的讲话. 心理学报, 38(1):157-158.)

Zhao LR (2001) The great contribution of Prof. Richang Cao to the development of Chinese psychology—in commemoration of the 90th Birthday of Prof. Richang Cao. Psychol Sci 24(3):323-325. (赵莉如. 2001. 曹日昌先生在中国心理学事业上的重大贡献——纪念曹日昌先生诞辰90周年. 心理科学, 24(3):323-325.)

81. 胡寄南：中国神经科学和认知科学的助力者

图1　胡寄南（1905—1989）

胡寄南（Chi-Nan Hu）（图1），我国著名动物心理学家、神经心理学家。1905年11月18日生于上海（当时直属江苏省），原籍安徽太平（今黄山市黄山区）。1987年被美国国际人际关系实验训练学会授予永久荣誉会员称号，1988年成为文科首批博士生导师之一。

胡寄南在神经科学领域的工作可以分为两个时间段。

第一个时间段是1922年至1946年，拓展和普及导师郭任远和K. S. Lashley的动物心理学的实验研究。

1922年春，胡寄南考上复旦大学心理学系，成为首届心理学专业学生。期间，他在被誉为“中国的华生”“激进的科学哲学家和革新的实验家”的比较心理学家和动物心理学家郭任远指导下接受了动物心理学研究的训练（Qian et al.，2018），参与“猫鼠同笼，天下大同”的实验研究，继续反对本能论，提倡“整体心理学”。1925年毕业，获学士学位，是当时我国仅有的四位心理学毕业生之一（胡寄南，1985）。毕业后，胡寄南积极投身我国的心理学事业，并作出重要贡献。其中，他和吴颂皋、黄维荣共同翻译了1928年由开明书店出版的郭任远的心理学论著，包括《我们的本能是怎样获得的》《一个心理学革命者的口供》《心理学的真正意义》《反对本能运动的经过和我最近的主张》。

1928年，胡寄南留学美国俄亥俄州立大学，并于1930年申请心理学硕士学位，次年毕业（胡寄南，1985；周文业，2015）（图2和图3）。1934年，他转入芝加哥大学，师从生理心理学家K. S. Lashley，研究大脑的记忆功能。1934年，发表博士毕业论文《大脑损伤对动物记忆力的影响》，刊载于1938年第18期《普通心理学》，作为优秀毕业论文，收入美国国会图书馆。该文章被Lashley教授的一篇重要论文所引用，指出大脑损伤的存在和范围对动物记忆力的影响，确立了针对记忆和遗忘因素，大脑损伤的

作者：王蕴瑾[1]，钱燕燕[2]

1　台州学院心理健康教育与咨询中心，台州318000，中国

2　荷兰莱顿大学社会与行为科学学院，莱顿2333AK，荷兰

邮箱：wyjpsy@163.com（王蕴瑾）；psyqyy@163.com（钱燕燕）

图2 胡寄南与妻子俞竹贞（1930年春）

图3 胡寄南全家福（1932年）（老斯，1998）

作用超过时间因素和插入活动因素。研究支持了Lashley的论点：大脑皮层在记忆方面具有广泛作用（Hu，1938）。

1936年，胡寄南于《教育杂志》第25卷第4期发表《Lashley教授的大脑均势说和实验》，普及Lashley的脑研究工作，因为胡寄南认为“Lashley的实验研究是最能抓住心理学重要问题核心的”（胡寄南，1985）。同年，胡寄南总结动物心理学的研究方法，在我国历史最为悠久、与北京大学共同誉为“中国近代文化的双子星”的商务印书馆《读书指导》第二辑上发表《动物心理学研究法》。这篇文章阐明了动物心理学研究的目的为：① 满足好奇心；② 与人类心理相比较；③ 提高实用性。方法为：① 轶事法；② 观察法；③ 实验法。注意事项为：① 问题的发生；② 历史的探讨；③ 解决的方法；④ 实验的进行（仪器的布置与材料的选择）；⑤ 结果与结论。

1938年，胡寄南于浙江大学担任教授，1945年于暨南大学担任教授、教务长（周文业，2015）。1946年，胡寄南应聘到中国生理和心理研究所，着手准备研究语言与大脑的关系（阎书昌，2015）。1948年，在复旦大学生物系担任教授和主任。1951年在中央研究院中国教育研究室（中国教育科学研究院前身）担任研究员。自1952年起，在华东师范大学担任教授（周文业，2015）。

第二个时间段从20世纪60年代开始，主要包括两部分工作。其一，翻译国外的脑研究著作和文献。例如，1965年，胡寄南翻译E. G. Boring撰写的《K. S. Lashley和大脑皮层的整合作用》发表于《心理科学通讯》，即现在的《心理科学》杂志；1981年，翻译J. W. Kalat《生物心理学》中的《学习的生物学基础》，发表于《自然杂志》第6卷第8期。其二，讲授生理学知识。例如，他先在复旦大学任教，1951年全国院系调整后，除1958年至1961年任北京中央教育科学研究所研究员和北京师范大学心理学教授，其余时间一直担任华东师范大学教授，是华东师范大学教育系心理学“五虎将”（萧孝嵘、张耀翔、左任侠、谢循初、胡寄南）之一。1952—1957年，在教育学专业下主讲人体解剖生理学（陈桂生，2018）。

20世纪60年代起，胡寄南从行为心理学转向认知心理学，思考人的神经性和心理性的融合。胡寄南认为，人的科学是一门社会生物科学，人的社会学和生物性矛盾的对立统一是人的心理变化发展的动力（胡寄南，1985）。心理活动的实质是信息的分析、综合和储存（胡寄南，1980）；而“信息是从物理过程到神经过程再到心理过程的转化物。反过来也是如此”（胡寄南，1982，1985）。因而，同一时期，胡寄南编译了精神分析领域涉及主观心理层面的文章以实现对心理活动的深度了解（沃尔曼和胡寄南，1961a，1961b；霍尔等，1961；奥尔塞和胡寄南，1963）。基于对人的神经性和心理性的深入了解，胡寄南在认知心理学的框架之下开展对抑郁症的研究。1993年至1995年，胡寄南与郑维廉、杨治良、蔡能合作的以抑郁症为主题的系列文章得以发表，文章主题包括：抑郁症与自尊心（郑维廉等，1993）、想象力（郑维廉等，1994b）、基本决策和再认功能（郑维廉等，1994a）、性别差异（郑维廉等，1995b）、自责心、想象力、谨慎度的相关研究（郑维廉等，1995a）。

1985年，80岁高龄的胡寄南教授远渡重洋，访问了波士顿大学的动物心理实验室，并在J. M. Harrison教授的陪同下与著名行为主义心理学家B. F. Skinner热切交流（老斯，1998）。1989年12月20日，胡寄南于上海病逝，享年84岁。令人钦佩的是，在逝世前，胡寄南教授笔耕不辍；翻译书籍，指导学生，一直奋战在心理学教育事业第一线（图4和图5）。1990年1月6日晚，举办了胡寄南先生的追悼会。苏渊雷教授致挽联为胡寄南先生作结：

人生如寄名利轻于鸿毛　两袖清风生时遗憾少
吾道其南事业重于泰山　光明磊落身后教诲长

图4　胡寄南在他的办公室（胡寄南，1985）

图5　胡寄南教授与他的同事Carl Ratner教授（照片所有人：王波）

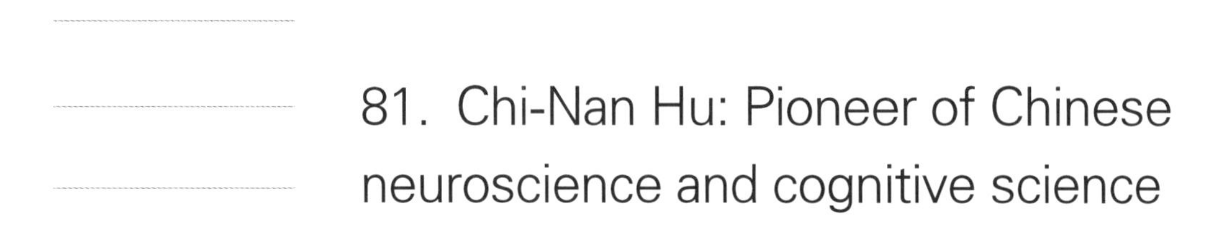

81. Chi-Nan Hu: Pioneer of Chinese neuroscience and cognitive science

Chi-Nan Hu (胡寄南, 1905—1989), the renowned Chinese animal psychologist and neuropsychologist, was born in Shanghai at November 18, 1905. He was awarded the lifetime honorary membership by the International Human Relations Training Institute (IHRT) in 1987 and became one of the first doctoral tutors in China in 1988.

Hu's research in neuroscience can be divided into two stages.

The first stage (1922—1946), Hu expanded and popularized the experimental research conducted by animal psychologists Zing-Yang Kuo (郭任远, 1898—1970) and Karl Spencer Lashley, both of whom were Hu's supervisors at different period.

In the spring of 1922, Hu studied psychology in Fudan University supervised by Dr. Zing-Yang Kuo, a world-famous comparative psychologist and animal psychologist known as Chinese Watson, one of the most extreme precursors of the behaviorism school and the most radical behaviorist in the history of behavioristic psychology (Qian et al., 2018). Hu participated in an experimental research "genesis of the cat's response toward the rat" and received the systematical training in the field of animal psychology research. Meanwhile, he inclined to the opposite point of the instinct theory and advocated holistic psychology. In 1925, Hu received his bachelor's degree. It was worth noted that he was one of the existing four psychology graduates in China at that time (Hu, 1985). From then on, he devoted himself to his scientific career. Together with Sung-Kao Wu (吴颂皋) and Wei-Jung Huang (黄维荣), he translated a series of books written by Zing-Yang Kuo and published the books by the Kaiming Press in 1928, such as "*How Are Our Instincts Acquired?*" "*The Net Result of the Anti-heredity Movement in Psychology*" .

Hu studied at Ohio State University in 1928, applied for the master's degree in psychology in 1930 and graduated the next year (Hu, 1985; Zhou, 2015). In 1934, Hu transferred to the University of Chicago, researching in the field of memory function of the brain supervised by Karl Spencer Lashley, a famous physiological psychologist. In 1934, Hu finished his doctoral thesis

---

Yunjin Wang[1], Yanyan Qian[2]

1 Center for Mental Health Education and Counseling, Taizhou University, Taizhou 318000, China

2 Social and Behavioral Sciences Facility, Leiden University, Leiden 2333AK, The Netherlands

Correspondence: psyqyy@163.com (Y.-Y. Qian)

on "The effects of brain injury upon retentiveness in the rat" . The thesis was published in the *Journal of General Psychology* (No.18, 1938) and selected as an excellent graduation thesis by the Library of Congress. This research, which pointed out cerebral lesions's great influence on the animal memory, "established the preponderant effectiveness of the factor of cerebral lesions in contributing to the function of retention or forgetting over those of time and interpolated activity. It was in favor of the mass action of the cerebral cortex in retention as proposed by Lashley" (Hu,1938).

In 1936, Hu published the brain equipotentiality and the experiments of professor Lashley (Vol. 25, No. 4) in *Education Journal* to popularize the brain research of Lashley. Hu postulated "the experimental research of Lashley could concerning the essence of the most important issues in psychology" (Hu, 1985). That year Hu also focused on the research methods of animal psychology, summarized and delivered a handbook, named "The methods of animal psychology research" , in which he elaborated on the goals, methods and important matters. Hu's main goals were firstly, to satisfy curiosity; secondly, to compare with the human psychology; thirdly, to improve practicality. Hu described three types of methods including anecdotal, observational and experimental research. Additional, he suggested there are five issues that require people to follow: firstly, raise of the problem; secondly, colloquy of history; thirdly, method of solving; fourthly, progress of the experiment (the arrangement of the instruments and the choice of materials); and finally, the results and conclusions. His thesis was included in "*Reading Guide Book* 1" published by Commercial Press, which is the oldest Press in China and jointly known as the "Gemini of China's Modern Culture" with Peking University.

In 1938, Hu became a professor at Zhejiang University, and joined Ji'nan University as dean in 1945 (Zhou, 2015). The next year, Hu worked at China Institute of Physiology and Psychology, where he started to investigate the relationship between language and brain (Yan, 2015; General situation of China Institute of Physiology and Psychology, 1946). In 1948, he worked as a professor and dean of biology department of Fudan University. In 1951, he served as a researcher at National Institute of Education Sciences. Since 1952, he occupied the position of professor at East China Normal University (Zhou, 2015).

Since 1960s, Hu launched the second stage of his career which consists two main parts. On one hand, he translated manifold foreign literature in the field of brain research. In 1965, Hu translated "K. S. Lashley and the integration of cerebral cortex" written by E. G. Boring and published the book in the *Psychological Science Newsletter* which was the predecessor of *Psychological Science*. In 1981, he translated "The biology foundation of learning" , selected from *Biopsychology*, written by James W. Kalat. The book was published on *Nature Journal* (Vol. 6, No. 8). On the other hand, he devoted himself to spreading the knowledge of physiology. He successively taught at Fudan University and East China Normal University, with a break between 1958 and 1961 when he served as a researcher in Beijing Central Institute of Educational Science and professor of psychology at Beijing Normal University. What's more, Hu was the member of "Five Valiant Tigers" , together with Hsiu-Rong Hsiao (萧孝嵘), Yao-

Hsiang Chang (张耀翔), Jen-Hsia Tso (左任侠), Hsun-Chu Hsieh (谢循初), those who were psychological professors in the department of education at East China Normal University. From 1952 to 1957, Hu persisted in giving lectures on human anatomy and physiology to students majoring in education (Chen, 2018).

Hu switched research focus from behavioral psychology to cognitive psychology ever since 1960s, considering the interaction of human neurology and psychology. Hu suggests that the psychology is science about human, based on the social science and physical science. The motivating force of advancement of human psychology is the unity of contradiction between human sociology and biology (Hu, 1985). The essence of psychological activity is analysis, synthesis and storage of information (Hu, 1980). However, "the essence of information is the conversion product from physical process to neural process and then to psychological process, vice versa." (Hu, 1982; Hu, 1985) Hu mainly focused on translating and editing articles in the field of psychoanalysis in order to achieve a deep understanding of psychological activity (Wallman and Hu, 1961a, 1961b; Hall et al., 1961; Orser and Hu, 1963). Based on the insightful and thoughtful view of human's neurology and psychology, he conducted the research about depressive disorder from a cognitive perspective. From 1993 to 1995, a series of articles on the subject of depression were published in collaboration with Chi-Nan Hu, Wei-Lian Zheng (郑维廉), Zhi-Liang Yang (杨治良) and Neng Cai (蔡能). These studies revealed the relationship among depression and self-respect (Zheng et al., 1993), basic decision-making and recognition function (Zheng et al., 1994a), imagination (Zheng et al., 1994b), self-blame, imagination and prudence (Zheng et al., 1995a), gender difference (Zheng et al., 1995b).

In 1985, at an age of 80, Hu visited the animal psychology laboratory in Boston University and met John Michael Harrison and Burrhus Frederic Skinner (Lao, 1998). Unfortunately, Hu passed away on 20 December, 1989 due to disease, at the age of 84. On the bright side he admirably lived the age of 84, he still insisted on conducting researches, translating literature and supervising his students. On the evening of 6 January, 1990, a memorial meeting was held, professor Yuanlei Su (苏渊雷) wrote an elegiac couplet for Hu:

Fame and gain are lighter than feather last for all his life (人生如寄名利轻于鸿毛).
Rectitude and sincerity throughout his lifetime without any regrets (两袖清风生时遗憾少).
Hu broaden career in South which is weightier than Mount Tai (吾道其南事业重于泰山).
Open and aboveboard, influence of education will exists forever (光明磊落身后教诲长).

Figures

Fig.1 Chi-Nan Hu (1905—1989)

Fig.2 In the spring of 1930, Chi-Nan Hu married with Chu-Chen Yu

Fig.3 In 1932, Chi-Nan Hu, Chu-Chen Yu and their daughter (Lao, 1998)

Fig.4 Chi-Nan Hu was doing research in his office (Hu, 1985)

Fig.5 Chi-Nan Hu and his colleague, Carl Ratner (Copyright of this photo is owned by Professor Bo Wang)

References

Chen GS (2018) The record event of department of education in East China Normal University at initial stage. J Sch Stud 15(1):109–112. (陈桂生 . 2018. 华东师范大学初期教育学系纪事(1951—1965). 基础教育, 15(1): 109–112.)

Hall, Lin DD, Wallman, et al (1961) The new-Freudian school's personality. Dig Foreign Soc Sci (11):26–32. (霍尔, 林德戴, 沃尔曼, 等. 1961. 新弗洛伊德主义学派的人格(个性)理论. 现代外国哲学社会科学文摘, (11): 26–32.)

Hu JN (1938) The effects of brain injury upon retentiveness in the rat. J Gen Psychol 18:267–304.

Hu JN (1980) The analyse, synthesize, storage of information——the essence of mental activity. J Psychol Sci (Annu Symp Pap) (1):61–65. (胡寄南 . 1980. 信息的分析、综合、储存——这就是心理活动的实质. 心理科学通讯(年会论文选), (1): 61–65.)

Hu JN (1982) Information is conversion product that from physical process to neural process and then arrive at psychological process. Nat J (5):344–345+353–400. (胡寄南 . 1982. 论信息是从物理过程到神经过程又到心理过程的转化物. 自然杂志, (5): 344–345+353–400.)

Hu JN (1985) A Collection of Selected Essays on Psychology(Supplement edition). Shanghai: Xuelin Press. (胡寄南 . 1985. 心理学论文选(增补本). 上海: 学林出版社.)

Lao S (1998) A Female General Born on the Other Side of the Ocean: Feipei Hu. Qingdao: Qingdao Publishing House. (老斯 . 1998. 出生在大洋彼岸的女将军: 胡斐佩传. 青岛: 青岛出版社.)

Orser M, Hu JN (1963) Allport: Pattern and growth in personality. Dig Foreign Soc Sci (12):33, 41. (奥尔塞, 胡寄南 . 1963. 阿尔泡特: 人格的型式与成长. 国外社会科学文摘, (12): 33, 41.)

Qian YY, Chen W, Guo BY (2020) Zing-Yang Kuo and behavior epigenesis based on animal experiments. Protein & Cell 11(6):387–390.

Wallman, Hu JN (1961a) Adler: Individual psychology. Dig Foreign Soc Sci (11):14–19. (沃尔曼, 胡寄南 . 1961a. 阿德勒: 个人心理学. 国外社会科学文摘, (11): 14–19.)

Wallman, Hu JN (1961b) Jung: Analytical psychology. Dig Foreign Soc Sci (11):20–25. (沃尔曼, 胡寄南 . 1961b. 荣格: 分析心理学. 国外社会科学文摘, (11): 20–25.)

Yan SC (2015) The History of Modern Chinese Psychology(1872—1949). Shanghai: Shanghai Education Publishing House. (阎书昌 . 2015. 中国近代心理学史(1872—1949). 上海: 上海教育出版社.)

Zheng WL, Hu JN, Yang ZL, et al (1993) The self-esteem control experiment on persons with a history of endogenous depression. J Psychol Sci (2):28–32, 67. (郑维廉, 胡寄南, 杨治良, 等. 1993. 内源性抑郁症史者的自尊心操纵实验研究. 心理科学, (2): 28–32, 67.)

Zheng WL, Hu JN, Yang ZL, et al (1994a) A study on some basic characteristics of policy decision and recognition ability of recovered depressions. Chin J Nerv Ment Dis (6):338–340. (郑维廉, 胡寄南, 杨治良, 等. 1994a. 有抑郁症史者的某些基本决策特征与再认功能的研究. 中国神经精神疾病杂志, (6): 338–340.)

Zheng WL, Hu JN, Yang ZL, et al (1994b) A study on the imaginative ability of persons with a history of endogenous depression. Chin J Nerv Ment Dis (3):157–159. (郑维廉, 胡寄南, 杨治良, 等. 1994b. 有抑郁症史者想象力的研究. 中国神经精神疾病杂志, (3): 157–159.)

Zheng WL, Hu JN, Yang ZL, et al (1995a) A study on correlation analysis characteristic of depressed patients' self-accusation, imagination and prudent. Chin Mental Health J 9(4):156–157. (郑维廉, 胡寄南, 杨治良, 等. 1995a. 抑郁症患者自责心、想象力、谨慎度特征的相关分析. 中国心理卫生杂志, 9(4): 156–157.)

Zheng WL, Hu JN, Yang ZL, et al (1995b) A study on sex differences of depression symptoms among different occupations and ages. Chin J Ergon (1):45–50+70–71. (郑维廉, 胡寄南, 杨治良, 等. 1995b. 不同职业、年龄段中抑郁症状的性别差异研究. 人类工效学, (1): 45–50+70–71.)

Zhou WY (2015) A Brief Introduction and Research on Modern Chinese Psychologists. Zhengzhou: Zhongzhou Ancient Books Publishing House. (周文业. 2015. 中国近代心理学家传略及研究. 郑州: 中州古籍出版社.)

82. 徐联仓：中国管理心理学的开创者

徐联仓先生出生于浙江海宁，是我国著名心理学家（图1），我国管理心理学的创始人，也是工业心理学和工程心理学的先驱。

1947年到1949年，徐联仓在南开大学哲学教育系[①]学习，后转入清华大学心理学系就读。1951年获得学士学位后，他开始了在中国科学院心理研究所（IPCAS）的学术生涯，此后在这里工作终生。20世纪50年代初期，徐联仓教授主要从事关于质量控制、安全生产和操作合理化的工业心理学研究。值得一提的是，他的研究成果在我国纺织和冶金行业得到了成功的推广（傅小兰，2015）。

为了继续深造，徐联仓于1958年去往苏联教育科学院的心理研究所[②]学习，并于1962年获得副博士学位（图2）[③]。在此期间，他创造性地将信息论引入心理学，并提出了一种生产线废品成因的分析方法。他的理论和方法被尝试性地应用于电视工厂的

图1　徐联仓教授（1927—2015）

图2　徐联仓先生在苏联教育科学院（1962年）

译者：李晶
南京师范大学，南京210097，中国
邮箱：lij@njnu.edu.cn

① 现为哲学院。

② 该所于1970年改名为苏联教育科学院普通心理学与教育心理学研究所，于1992年改为斯库奇纳心理所。

③ 相当于我国的博士学位。

生产中，提高了产品的质量。他的这项研究后来被写进由苏联教育科学院出版的教科书《劳动心理学》中。由于这些杰出的贡献，徐联仓被苏联教育科学院授予“乌申斯基奖”。

从苏联回国后，徐联仓先生继续研究信息论在心理学中的应用。他分析了刺激与反应的相容性问题，就控制设备在信息传递函数方面的差异、语言在人机交互过程的中介作用等问题进行了开创性的研究。随后，徐联仓先生将他的研究成果在《中国科学》上发表，引起了西方学者的广泛关注。20世纪60年代后期，徐先生参与了我国卫星“东方红1号”发射计划。作为该计划领导小组的成员，他和他的团队完成了中国第一个“动物上天”的联合实验。此外，他揭示了失重条件下实验动物的生理和心理变化（杨虹，1999）。

20世纪70年代，我国的心理科学遭到了严重的破坏，发展停滞。为了恢复中国科学院心理研究所和中国心理学会，徐先生和他的同事作了大量贡献。在极端困难的环境下，他仍然坚持完成了两项科研工作：激光的生物效应研究和通过观察动物的行为表现预报地震。1976年中国科学院心理研究所恢复后，他继续有关管理心理学的研究，并且对工人的工作和生活满意度进行了调查，被《纽约时报》评为“中国的盖洛普民意调查”。

1979年以来，徐联仓先生侧重以组织发展为主的行为研究。通过努力，他成功开发了可衡量具体管理措施有效性的测量工具，并在石油、煤炭、航空和铁路等行业系统进行验证，发现了经济绩效与人的管理之间的因果联系。此外，他和他的合作者还在我国开展了领导行为、管理决策、人才培养和风险意识等方面的研究，并将生产力、生产关系和文化相结合，提出了管理的三重性理论，引起了国内外学术界的广泛关注。徐联仓先生还与其他科学家（如Misumi Juuji、Hofstede和Federer）进行合作及开展跨文化研究，并致力于发现管理理论在不同国家的文化特异性和适用性。徐先生所有这些研究成果都为我国管理心理学学科建设作出了重要贡献，促进了我国管理科学的发展。

为了迎合经济体制改革的需要，促进行为科学的研究与应用，徐联仓先生和其他一些学者在20世纪80年代呼吁建立中国行为科学学会。同时，他重视管理研究，如人员培训和工作价值分析等。除了本土研究之外，他还关注其他国家的研究进展，例如，日本的PM法（领导的工作绩效和团体维系职能分析）和美国的汇编栅格方法。徐先生将这些方法引入我国，促进了我国工业心理学的发展。

徐联仓先生著作等身，其中包括《管理心理学》（1986年与卢盛忠合著）、《组织管理心理学》（1988年与凌文辁合著）、《领导行为研究》（1991年与杨林林合著）、《组织行为学》（1993年）、《管理心理学及其应用：服务于医院管理》（1993年）、《丛林中的管理心理学》（2007年）等十多本学术专著，并在国际和国内期刊上发表超过100篇论文（梁川，2009）。徐先生和他的学生时勘在领导行为和智能模拟训练方面的研究成果获得了三项科学技术进步奖：中国科学院科技进步奖二等奖、轻工业部[1]科技进步奖二

① 现已整合进工业和信息化部。

等奖和石油工业部①科技进步奖二等奖。为了表彰他在心理学方面的重要贡献，中国心理学会于1999年为徐联仓先生颁发了终身成就奖（图3）。

图3　中国心理学会主席陈永明为徐联仓先生（右）颁发终身成就奖（1999年）

除了研究工作外，徐联仓先生在许多国内外学术机构都担任过重要职务，包括：中国科学院心理研究所所长（1983—1987年），中国心理学会秘书长（1978—1985年），国际应用心理学会执行委员（1984—1998年），中国行为科学学会副会长（1985—1988年），中国社会心理学会副会长（1990—1993年），《心理学报》主编等（金会庆，1996）。

作为我国心理学界的典范，徐联仓先生努力探索，敢于创新，为心理学在我国的发展作出了杰出贡献，培养出了一批优秀的心理学博士和硕士。他因兢兢业业的工作态度和慷慨豁达的性格深受海内外学者的尊重。

---

① 现已整合进国家能源局。

82. Lian-Cang Xu: The founder of management psychology in China

Dr. Lian-Cang Xu (徐联仓, 1927—2015), born in Haining County, Zhejiang Province, China, is a well-known Chinese psychologist, the founder of management psychology in China, and also the pioneer of industrial psychology and engineering psychology.

From 1947 to 1949, Lian-Cang Xu studied in Department of Philosophical Education ① in Nankai University, and then transferred to Tsinghua University, majoring in psychology. After acquiring bachelor's degree in 1951, he began his academic career in the Institute of Psychology, Chinese Academy of Sciences (IPCAS), where he worked all his life. He was mainly engaged in industrial psychology studies about quality control, production safety, and operation rationalization in the early 1950s. What's more, these research finding were successfully popularized in Chinese textile and metallurgy industry (Fu, 2015).

Aiming to continue his education, Lian-Cang Xu went to Psychological Research Institute of the Union of Soviet Socialist Republics (USSR) Pedagogical Science Academy② in 1958, and received a Kandidat Nauk degree③ in 1962. During this period in USSR, he introduced information theory into psychological science creatively, and proposed a brand-new analytic method of scrap in production. His theory and method were applied in a television factory experimentally and that resulted in the improvement of product quality. This study was included in a textbook named "*Labor Psychology*" published by Pedagogical Science Academy later. Because of these outstanding performances, he was awarded with Ulsenski Prize by the academy.

After returning from USSR, Dr. Lian-Cang Xu continued to work on the application of information theory in psychology. Based on these work, he discovered the differences of information transmission function among various control equipment, and the results, which was published on *Science China*, received much attention from Western scholars. Besides that, he

---

Jing Li, Yongkang Lu

School of Psychology, Nanjing Normal University, Nanjing 210097, China

Correspondence: lij@njnu.edu.cn (J. Li)

① Now it is called Philosophy College.

② The institution was renamed General and Educational Psychology Research Institute of USSR Pedagogical Science Academy in 1970, and has been returned to its original name "L. G. Schukina's Psychological Institute" since 1992.

③ The Kandidat Nauk degree is equal to the doctor's degree.

focused on the stimulus-response compatibility and the mediating effect of language in human-computer interaction. In the late 1960s, Dr. Xu participated in the development project of Chinese first man-made satellite "Dongfanghong-I" . As a member of the leaders, he and his group members finished the first observation experiment of the animals under the weightlessness condition in China. Furthermore, he revealed the physiological and psychological changes of the experimental animals under such conditions (Yang, 1999).

The psychological science in China was damaged seriously in 1970s. To rehabilitate the IPCAS and Chinese Psychological Society (CPS) in a short time, Dr. Lian-Cang Xu tried his best to take a lot of measures with other colleagues. In such difficult environment, he still completed two projects: One was the biological effect of laser, specifically on vision, and the other was the earthquake forecasting through the observation of animals' reaction. As soon as IPCAS was formally reinstated in 1976, he went back to study management psychology and led a team to carry out a survey on employees' job and life satisfaction, which was appreciated as "Chinese Gallup Poll" by *New York Times*.

Since 1979, Dr. Lian-Cang Xu began to lay stress on the action research which centers on the organization development. A tool for measuring the effectiveness of concrete managing policies was developed through his efforts, and its validity was verified in oil, coal, aviation, and railway industries. According to the data obtained by this measuring tool, he discovered the causal relationship between economic performance and human resource management. Moreover, he and his partners conducted studies on leadership behavior, management decision-making, personnel training, and risk awareness, all of which were pioneering in China, and he also integrated the productivity, relations of production and culture to propose a "Tripility Theory" , triggering much attention from academia of China and other countries. Cooperating with other world-famous scientists, such as Misumi Juuji, Hofstede, and Federer, he did a series of cross-cultural researches and devoted to finding out whether Management Theory has cultural specificity and applicability in different countries. All these research findings made significant contribution to the discipline construction of management psychology and promoted the development of management science in China.

In order to cater for the needs of economic system reform and promote the research and application of behavioral science, Dr. Lian-Cang Xu and some other scholars appealed for the establishment of Chinese Behavioral Science Society in 1980s. Meanwhile, he paid more attention to the research of management, such as staff training, work values analysis, and other aspects. Apart from the indigenous study, he began to focus on the research progress in other countries, for example, the PM (Performance Maintain) method from Japan and the Repertory Grid method from U.S. The introduction of these methods to China enhanced the development of industrial psychology research effectively.

Dr. Lian-Cang Xu had abundant works through his life, including more than ten academic monographs, such as *Management Psychology* (1986) with Sheng-Zhong Lu (卢盛忠), *Organizational Management Psychology* (1988) with Wen-Quan Ling (凌文辁), *Research on*

Leadership Behavior (1991) with Lin-Lin Yang (杨林林), *Organizational Behavior* (1993), *Management Psychology and Its Application*: *Serve for the Management in Hospital* (1993), and *Management Psychology out of Jungle* (2007), and more than one hundred papers on international and Chinese journals (Liang, 2009). He and his student Kan Shi's (时勘) research achievements on leadership behavior and Intelligent Simulation Training were awarded three prizes for Science and Technology Progress (STP): one second class prize for STP of Chinese Academy of Sciences (CAS), one second class prize for STP of National Light Ministry of China①, and one second class prize for STP of National Petroleum Industry Ministry of China②. To honor his significant contribution in psychology, CPS conferred the Life Achievement Award in 1999 which is the top prize on him.

In addition to the research work, Dr. Lian-Cang Xu held many important positions in domestic and International academic institutions. His duties and titles included the director of IPCAS (1983—1987), secretary general of CPS (1978—1985), executive committee member of International Association of Applied Psychology (IAAP, 1984—1998), vice-president of Chinese Behavioral Science Society (1985—1988), vice-president of Chinese Association of Social Psychology (1990—1993), and the chief editor of *Acta Psychologica Sinica*, etc (Jin, 1996).

As a model in Chinese psychological circle, Dr. Lian-Cang Xu studied rigorously, explored diligently, and dared to innovate. For the development of Chinese psychology, he was strongly willing to be dedicated to his work, and trained a group of psychology doctorate and master for China. His conscientious working attitude and generous character were deeply respected all over the world.

Figures

Fig.1 Professor Lian-Cang Xu (1927—2015)

Fig.2 Mr. Lian-Cang Xu in USSR Pedagogical Science Academy in 1962

Fig.3 Mr. Lian-Cang Xu (right) was awarded Life Achievement by Yong-Ming Chen (陈永明), who was then the president of CPS, in 1999

References

Fu XL (2015) Special issue to commemorate Mr. Liancang Xu. Institute of Psychology, Chinese Academy of Sciences (Internal publications), Special issue No. 204. (傅小兰. 2015. 纪念徐联仓先生专刊. 中国科学院心理研究所(内部刊物), 第204期专刊.)

Jin HQ (1996) The resume of Mr. Liancang Xu. Chin J Ergon (3): 73–74. (金会庆. 1996. 徐联仓先生简历. 人类工效学, (3): 73–74.)

① It has been integrated into Ministry of Industry and Information Technology of China.

② It has been integrated into National Energy Administration of China.

Liang C (2009) Psychological Anthology of Liancang Xu. Beijing: People's Education Press. (梁川. 2009. 徐联仓心理学文选. 北京: 人民教育出版社.)

Yang H (1999) Plum blossom incense from the cold weather: The biography of Liancang Xu, an investigator of IPCAS. Science News (13): 19. (杨虹. 1999. 梅花香自苦寒来——记中国科学院心理研究所研究员徐联仓. 科学新闻周刊, (13): 19.)

83. 邵郊：中国生理心理学和比较心理学的先行者

图1　邵郊（1923—2017）

邵郊（图1），著名生理心理学家、比较心理学家，为我国生理心理学和比较心理学的学科发展、科学研究和人才培养作出了重要贡献（《心理学报》编辑部，2017）。他曾任第二届国务院学科评议组委员（教育学学科）、中国心理学会第三届和第五届常务理事、中国心理学会生理心理专业委员会主任（1984—1988年）、中国大百科全书出版社编辑咨询委员会成员、《中国大百科全书·心理学卷》编委、《心理学报》编委（1979—1991年）等职，荣任中国心理学会会士（2007年）。

1923年6月27日，邵郊出生于山东省济南市。1943年，考入清华大学心理学系。1948年毕业留校并担任助教。1952年，院系调整，邵郊从清华大学转入北京大学哲学系心理学专业，担任哲学系副主任孙国华的助手，协助安排学科调整，开展行政事务，从事高级神经活动学课程及实验、动物心理学和比较神经解剖学的教学工作（董耀会和《北大人》编辑部，1993；钱颖一和李强，2011）。

1953年，邵郊和沈迺璋、孙国华共同筹建全国第一个动物条件反射实验室（钱颖一和李强，2011；阎书昌和周广业，2013）。该实验室是生理心理实验室的前身，为心理学成为国家重点学科奠定了重要基础（《心理学报》编辑部，2017）。一方面，该实验室有助于理解并澄清巴甫洛夫实验的本质；另一方面，它为验证条件反射的生理机制提供了实验场所（张人骏和朱永新，1986；《北京大学哲学系史稿》编委会，2004）。在当时有限的实验室条件下，邵郊和沈迺璋、沈德灿共同撰写《在用电流刺激狗的乙状回时，电流的频率与刺激效果的关系》一文，并于1956年11月发表在《心理学报》第1卷第1期上。在邵郊的领导下，该实验室后来成为国家重点实验室（《心理学报》

译者：王礼军[1]，钱燕燕[2]

1　安徽师范大学教育科学学院，芜湖241000，中国

2　荷兰莱顿大学社会与行为科学学院，莱顿2333AK，荷兰

邮箱：nnuwlj8421@163.com（王礼军）；psyqyy@163.com（钱燕燕）

编辑部，2017）。

1978年，心理学系恢复后，邵郊一直工作于北京大学心理学系，任副教授、教授，直至退休。1975年，邵郊和孟昭兰用电击诱发情绪，探讨脉搏容积和呼吸是否反映心理紧张状态（邵郊和孟昭兰，1975）。1981年，邵郊曾到美国密歇根大学从事一年的研究工作，他与著名的生物心理学家E. S. Valenstein合作，对脑刺激产生的点燃效应的长期抑制进行了实验研究，其成果发表在美国《实验神经学》杂志上（Shao and Valenstein，1982）。1980年11月20日，邵郊参加了中国心理学会生理心理学分会成立大会暨第一次学术研讨会（图2）。1981年12月7日，在中国心理学会第三届理事会上，邵郊荣任理事（荆其诚和中国心理学会，2001）。1987年，他撰写的《生理心理学》由人民教育出版社出版，该书成为领域内的经典教材和“圣经”，至今仍再版（钱颖一和李强，2011）。2009年，邵郊成为中国心理学会评定的首届心理学家成员之一。2016年12月，获中国心理学会“优秀心理学家”称号。2017年9月18日，邵郊因病逝世。

图2 1980年11月20日，邵郊（二排右五）参加第一次全国生理心理学研讨会

在比较心理学领域，邵郊指导诸多学生从种系发生发展的视角探究低等动物向高等动物的进化，其研究对象包括草履虫、蚯蚓、白鼠、狗以及灵长类动物等，以期从心理学角度为进化论提供重要证据（钱颖一和李强，2011）。例如，邵郊、林国彬和陈劭夫在《文昌鱼的光动反应》一文中指出，文昌鱼表现出颜色偏好，这表明动物行为演化的焦点问题在于管状中枢神经系统是由何种自然选择的压力促成（邵郊等，1981）。另外，邵郊还关注动物学习以研究进化问题。如林国彬、万传文、邵郊和刘范以三只金丝猴（*Rhinopithecus roxellanae*）为对象，探究它们对实物、照片和画片识别的概括能力。研究发现，它们皆不能将从实物客体习得的辨别反应迁移到黑白照片和素描图画上（林国彬等，1982）。他们进一步指出，当实验对象由金丝猴转换为1岁半到2岁半的儿童时，儿童的概括能力与其语言发展水平同步（林国彬等，1982）。邵郊等人做出如下推测：动物智力的演化并非是一种系统发展的现象，而语言能力的缺乏是金丝猴不具备概括能力的重要原因之一（林国彬等，1981）。

1998年，邵郊和李量合作撰写英文文章《大鼠腹侧前额叶分区局限性损伤会阻碍逆向学习，但不影响视觉辨别学习》，发表在《生理与行为》杂志上。该研究结果表明，老鼠的内侧前额叶皮层对辨别学习并非是必要的，但内侧前额叶皮层、边缘前区和边缘下区的两个腹侧亚区在逆向学习中都发挥着关键作用（Li and Shao，1998）。这一结果与先前的研究发现不一致，即老鼠内侧前额叶皮层的大范围损伤会导致逆向学习障碍（Li and Shao，1998）。

在生理心理学领域，邵郊主要从事听源性癫痫发作的神经机制研究，其关于啮齿类动物听觉癫痫模型以及点燃效应的研究在国际上具有较大的影响（《心理学报》编辑部，2017）。邵郊等人的实验研究取得了一些重要结果：首先，大脑皮层下的电刺激不仅可以产生点燃效应，还可以产生抑制效应（薛祚纮和邵郊，1983）；其次，通过比较电刺激组和声音刺激组的老鼠发现，下丘是听觉癫痫的一个重要中枢（许小冬和邵郊，1986）；最后，通过对听源性癫痫易感老鼠的实验证实了下丘的作用（胡丹和邵郊，1986）。

此外，邵郊为心理学的学科建设和人才培养倾注了心血（图3），主要表现在两个方面：一方面，他与其他学者合作翻译了一些国外名著，一是A. R. Luria的《神经心理学原理》，二是D. A. Dewsbury和D. A. Rethlingshafer的《比较心理学——现代概观》，这两本译著皆由科学出版社出版。另一方面，他为心理学的发展培育了大量的新生力量，指导了许多青年学者（图4）。通过检索万方数据库发现，1994年至1998年，邵郊共指导了15位学生获得博士学位。例如，北京大学心理学系苏彦捷教授（图5）在邵郊的指导下开展猕猴研究，通过延迟比样任务探讨恒河猴（*Macaca mulatta*）和红面猴（*Macaca arctoides*）数目认知能力及其机制（苏彦捷，1992）。又如，邵郊曾指导张学明、陈炳卿开展关于可口可乐和鼠的系列研究，以检测可口可乐与鼠的听源性癫痫（张学明等，1985）和运动行为（张学明等，1988）之间的关系。

图3 1994年10月10日，邵郊（一排右五）参加中国科学院脑与行为心理研究中心成立大会

图4 邵郊（一排右二）及其学生

图5 邵郊（一排右三）参加苏彦捷（一排右二）的博士学位论文答辩会

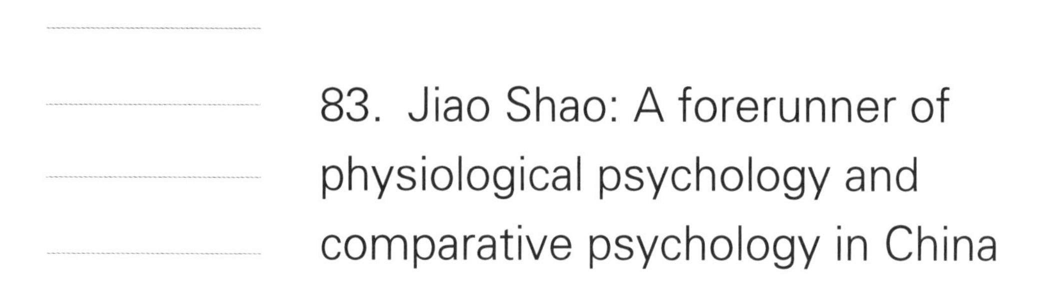

83. Jiao Shao: A forerunner of physiological psychology and comparative psychology in China

Jiao Shao (邵郊, 1923—2017), who was an expert in physiological psychology and comparative psychology in China, made important contributions to the development of physiological psychology and comparative psychology, scientific research and personnel training (Newsroom of *Acta Psychologica Sinica*, 2017). He was a member of the Second Discipline Review Group of the State Council (Pedagogy), the director of the third and fifth sessions of Chinese Psychological Society (CPS), the chairman of the Professional Committee on Physiological Psychology of the CPS (1984—1988), an editorial advisory board member of the *Encyclopedia of China* Publishing House (First Edition), an editorial board member of *Acta Psychologica Sinica* (1979—1991), and he was honored as a fellow of the CPS (2007).

On June 27, 1923, Shao was born in Jinan City, Shandong Province, where he lived until he entered the department of psychology at Tsinghua University in 1943. After graduating in 1948, he stayed at Tsinghua University to work as an assistant. Four years later, led by Kuo-Hua Sun (孙国华), the deputy director of the philosophy department at Peking University, Shao joined this institute as Sun's coadjutant. Shao assisted Sun by adjusting the division of disciplines; performing civil service; giving lectures; and completing experiments on higher nervous system activity, animal psychology, and comparative neuroanatomy (Dong and Newsroom of *Peking University People*, 1993; Qian and Li, 2011).

In the year of 1953, in collaboration with Nai-Chang Shen (沈迺璋) and Kuo-Hua Sun, Shao established the first national animal conditioned reflex laboratory (Qian and Li, 2011; Yan and Zhou, 2013), which, although it was later superseded by physiological psychology laboratory, was able to lay a strong foundation for psychology becoming a national key discipline (Newsroom of *Acta Psychologica Sinica*, 2017). This lab both helped to make clear the basic essence of Ivan Petrovich Pavlov's experimentation and provided a test site to verity the physiological mechanism

---

Lijun Wang[1], Yanyan Qian[2], Yanjie Su[3]

1 School of Educational Science, Anhui Normal University, Wuhu 241000, China

2 Social and Behavioral Sciences Facility, Leiden University, Leiden 2333AK, The Netherlands

3 School of Psychological and Cognitive Sciences and Beijing Key Laboratory of Behavior and Mental Health, Peking University, Beijing 100871, China

Correspondence: psyqyy@163.com (Y.-Y. Qian)

of conditioned reflexes (Zhang and Zhu, 1986; Editorial committee of *Historical Draft of Peking University Philosophy Department*, 2004). Under the special experimental conditions at that time, Shao published The relationship between the frequency of electrical current and the motor effect when the dog's sigmoid gyrus is under direct electrical stimulation in a joint work with Nai-Chang Shen and Te-Chan Shen (沈德灿) (Shen et al., 1956). Shao's lab later became a key state laboratory under the leadership of Shao (Newsroom of *Acta Psychologica Sinica*, 2017).

Shao worked at Peking University as an associate professor and then professor, starting in 1978 after the reconstruction of the department of psychology, until he retired. In 1975, Shao and Zhaolan Meng (孟昭兰) adopted the electric shock method to induce emotion and found the relationship between volume pulse, breathing and psychological stress (Shao and Meng, 1975). In 1981, Shao went to the University of Michigan to conduct a research program called Long-term Inhibition of Kindled Seizures by Brain-Stimulation for one year, where he cooperated with the world-renowned biopsychologist E. S. Valenstein. The results of this program were presented in *Experimental Neurology* (Shao and Valenstein, 1982). On November 20, 1980, Shao attended the founding congress of the Professional Committee on Physiological Psychology of CPS at its first meeting. On December 7, 1981, Shao was elected as the director of CPS at the third session of CPS (Jing and Chinese Psychological Society, 2001). In 1987, The *Physiological Psychology*, written by Shao, became a classic textbook and "*Holy Writ*" in the field of physiological psychology, and it is still in print today (Qian and Li, 2011). On October 10, 1994, Shao took part in the founding conference of the Center for Brain and Behavior Research in the Institute of Psychology, Chinese Academy of Sciences (IPCAS). Shao became one of the first members of psychologists assessed by CPS in 2009, and he was given the CPS Excellent Psychologist Award in 2016. Sadly, Shao passed away due to illness on September 18, 2017.

In the field of comparative psychology, Shao guided numerous students in analyzing the evolution of lower creatures to higher animals from the perspective of phylogeny. The research objects included paramecia, earthworms, white mice, dogs and primates. Shao and his students hoped to find some important pieces of evidence for the theory of evolution from a psychological perspective (Qian and Li, 2011). For instance, Shao and other researchers observed the color light preference of the amphioxus (Shao et al., 1981). Their research showed that the focus of animal evolution is natural selection shaped by the tubular central nervous system (Shao et al., 1981). In addition, Shao also focused on animal learning to study the issue of evolution. Shao, Lin (林国彬), Wan (万传文) and Liu (刘范) explored the capacity of three golden-haired monkeys (*Rhinopithecus roxellanae*) for generalization ability from three dimensional objects to two dimensional representative photographs and pictures. However, this research found that all the subjects showed no generalization ability (Lin et al., 1982). The researchers added children aged one-year-six-months to two-years-six-months as subjects, and found that they gained the generalization ability in parallel with their language development (Lin et al., 1981). According to Shao's formulation, the evolution of animal intelligence is not an orthogenesis phenomenon, and a lack of language is one of the important reasons for why golden-haired monkeys do not have the capacity of generalization.

In 1998, Shao collaborated with Li in writing Restricted lesions to ventral prefrontal subareas block reversal learning but not visual discrimination learning in rats, which was published in *Physiology and Behavior*. It is worth noting that this paper has been cited 59 times and 2018 is the newest citation year (George et al., 2018). The results from this study indicated that "The mPFC of rats is not essential for discrimination learning, but that each of the 2 ventral subareas of the mPFC, PL and IL plays a critical role in reversal learning" (Li and Shao, 1998), which is inconsistent with the previous finding that "Extensive damage to the medial prefrontal cortex (mPFC) of rats causes reversal learning deficits" (Li and Shao, 1998).

In the field of physiological psychology, Shao mainly explored the neural mechanism of audio-epileptic seizures. Shao's series of studies on rodents' auditory epilepsy modal and kindling effect have had considerable influences internationally (Newsroom of *Acta Psychologica Sinica*, 2017). These experiments yielded some important results: First, it was found that the electrical brain stimulation of subcortical could not only cause the kindling effect, but also an inhibitory effect (Xue and Shao, 1983). Second, it was demonstrated that inferior colliculus is an important main center of auditory seizures by comparing electrical stimulation and an acoustic primed group (Xu and Shao, 1986). Third, they proved the role played by the inferior colliculus through experiments performed on audiogenic seizure susceptible rats (Hu and Shao, 1986).

Shao devoted himself to the development of psychology, mainly focusing on disciplinary construction and personnel training. He translated some foreign classics with other scholars. One of these translations is *Principles of Neuropsychology*, written by A. R. Luria, and the other is *Comparative Psychology—A Modern Survey*, written by D. A. Dewsbury and D. A. Rethlingshafer, both of which were published by Science Press. Shao also offered guidance to many young scholars. We retrieved the Wanfang Data and found that Shao mentored 15 students to finish PhD dissertations. For example, under Shao's instruction, Yanjie Su (苏彦捷), who is now a professor at Peking University, analyzed *Macaca mulatta* and *Macaca arctoides's* achievements in a delayed match-to-sample task (Su, 1992). Shao also led Xueming Zhang (张学明) and Bingqing Chen (陈炳卿) in carrying out a series of experiments to detect the relationship between Coca-Cola and auditory epilepsy (Zhang et al., 1985), and between Coca-Cola and movability behavior (Zhang et al., 1988).

Figures

Fig.1 Jiao Shao (1923—2017)

Fig.2 On November 20, 1980, Jiao Shao (5th from right, in the second row) attended the national physiological psychology seminar at its inaugural meeting

Fig.3 On October 10, 1994, Jiao Shao (5th from right, in the first row) attended the founding conference of the Center for Brain and Behavior Research in IPCAS

Fig.4 Jiao Shao (2nd from right, in the first row) and his young students

Fig.5 Jiao Shao (3rd from right, in the first row) at the doctoral dissertation defense of Yanjie Su (2nd from right, in the first row)

References

Dong YH, Newsroom of *Peking University People* (1993) People in Peking University. Beijing: Huaxia Publishing House. (董耀会,《北大人》编辑部. 1993. 北大人. 北京: 华夏出版社.)

Editorial committee of *Historical Draft of Peking University Philosophy Department* (2004) Historical Draft of Peking University Philosophy Department. (《北京大学哲学系史稿》编委会. 2004. 北京大学哲学系史稿.)

George SA, Rodriguez-Santiago M, Riley J, et al (2018) D-cycloserine facilitates reversal in an animal model of post-traumatic stress disorder. Behav Brain Res 347:332–338.

Hu D, Shao J (1986) The inferior colliculus plays an important role in audiogenic seizures in rats. Acta Sci Nat Univ Peking (4): 58–65. (胡丹, 邵郊. 1986. 中脑下丘在听源癫痫中的作用. 北京大学学报(自然科学版), (4): 58–65.)

Jing QC, Chinese Psychological Society (2001) Chinese Psychological Society Has a History of 80 Years (1921—2001). Beijing: People's Education Press. (荆其诚, 中国心理学会. 2001. 中国心理学会80年(1921—2001). 北京: 人民教育出版社.)

Li L, Shao J (1998) Restricted lesions to ventral prefrontal subareas block reversal learning but not visual discrimination learning in rats. Physiol Behav 65(2):371–379.

Lin GB, Wan CW, Liu F, et al (1981) A comparative experimental study on the preliminary generalization of non-human primates (Ⅰ) Generalization of objects, photographs and pictures in golden haired monkeys (*Rhinopithecus roxellanae*). In: Digest anthology of the third member congress of Chinese Psychological Society and the 60th anniversary academic conference (the fourth national academic conference on psychology). (林国彬, 万传文, 刘范, 等. 1981. 灵长类初级抽象概括能力的比较实验研究(Ⅰ)金丝猴对实物、照片和画片的概括. 见: 中国心理学会第三次会员代表大会暨建会60周年学术会议(全国第四届心理学学术会议)文摘选集.)

Lin GB, Wan CW, Shao J, et al (1982) A comparative experimental study on the preliminary generalization of non-human primates (Ⅰ) Generalization of objects, photographs and pictures in golden haired monkeys (*Rhinopithecus roxellanae*). Acta Psychol Sin 14(2): 233–238. (林国彬, 万传文, 邵郊, 等. 1982. 灵长动物初级抽象概括能力的比较实验研究(Ⅰ)金丝猴对实物、照片和画片的概括. 心理学报, 14(2): 233–238.)

Newsroom of *Acta Psychologica Sinica* (2017) The obituary of Shao Jiao. The funeral committee Shao Jiao in College of Psychological and Cognitive Sciences, Peking University. (《心理学报》编辑部. 2017. 邵郊先生讣告. 北京大学心理与认知科学学院邵郊先生治丧委员会.)

Qian YY, Li Q (2011) The Social Science of Old Tsinghua. Beijing: Tsinghua University Press. (钱颖一, 李强. 2011. 老清华的社会科学. 北京: 清华大学出版社.)

Shao J, Lin GB, Chen SF (1981) The photo-kinetic reaction of amphioxus (*Branchiostoma belcheri*). Acta Psychol Sin 13(3): 358–363. (邵郊, 林国彬, 陈劭夫. 1981. 文昌鱼的光动反应. 心理学报, 13(3): 358–

363.)

Shao J, Meng ZL (1975) Does the reduction of volume pulse and the decrease of work level in manual operation reflect the psychological tension. In: The Papers from the first annual academic meeting of the Professional Committee on General Psychology and Experimental Psychology, Chinese Psychological Society. (邵郊, 孟昭兰. 1975. 手工操作中的脉搏容积的缩小和作业水平的下降是否反映心理的紧张状态？见：中国心理学会普通心理学与实验心理学专业委员会第一届学术年会论文.)

Shao J, Valenstein ES (1982) Long-term inhibition of kindled seizures by brain stimulation. Exp Neurol 76(2):376–391.

Shen NC, Shao J, Shen TC (1956) The relationship between the frequency of electrical current and the motor effect when the dog's sigmoid gyrus is under direct electrical stimulation. Acta Psychol Sin 1(1): 71–83. (沈迺璋, 邵郊, 沈德灿. 1956. 在用电流刺激狗的乙状回时,电流的频率与其刺激效果的关系. 心理学报, 1(1): 71–83.)

Su YJ (1992) Number cognition processing of *Macaca* in delayed match-to-sample task: A comparative study. Doctoral dissertation. Beijing: Peking University. (苏彦捷. 1992. 猕猴在延迟比样任务中数目认知过程的比较研究. 博士学位论文. 北京：北京大学.)

Xu XD, Shao J (1986) Comparison of audiogenic seizures induced by electrical stimulation of inferior colliculus with those by acoustic priming. Acta Psychol Sin 18(4): 403–408. (许小冬, 邵郊. 1986. 听觉诱发和电刺激中脑下丘诱发大鼠听因性癫痫发作的比较. 心理学报, 18(4): 403–408.)

Xue ZH, Shao J (1983) Inhibitory effect of the subcortical brain stimulation. Acta Psychol Sin 15(4): 474–479. (薛祚纮, 邵郊. 1983. 皮质下电刺激的抑制作用. 心理学报, 15(4): 474–479.)

Yan SC, Zhou GY (2013) Zhou Xiangeng's Collected Works. Vol. 1. Beijing: China Science and Technology Press. (阎书昌, 周广业. 2013. 周先庚文集. 卷1. 北京：中国科学技术出版社.)

Zhang RJ, Zhu YX (1986) A Dictionary of Psychological Figures. Tianjin: Tianjing Renmin Press. (张人骏, 朱永新. 1986. 心理学人物辞典. 天津：天津人民出版社.)

Zhang XM, Chen BQ, Shao J (1988) The Coca-Cola's influence on rat and mouse's physical activities. In: Collected papers of the second national conference of Chinese Medical Association on food sanitation. (张学明, 陈炳卿, 邵郊. 1988. 可口可乐对大鼠、小鼠运动行为的影响. 见：中华医学会第二届全国食品卫生学术会议科研论文集.)

Zhang XM, Shao J, Chen BQ (1985) Coca-Cola and rat's auditory epilepsy. In: Collected papers of the second national conference of Chinese Medical Association on food sanitation. (张学明, 邵郊, 陈炳卿. 1985. 可口可乐与大鼠听源性癫痫. 见：中华医学会第二届全国食品卫生学术会议科研论文集.)

84. 林仲贤：中国颜色心理学的奠基人

林仲贤教授（图1）是杰出的心理学家，是我国颜色视觉心理学的奠基人和领导者。他在彩色电视及中国人肤色测量方面做出了开创性研究。他是亚非心理学会（AAPA）执行委员会委员、国家标准化管理委员会委员、美国纽约科学院院士、美国科学促进会（AAAS）会士。

图1　林仲贤教授（1931—2011）

林仲贤于1931年12月3日出生于广州，并在那里度过了他的童年时光。1937年“七七事变”后，他随母亲回到老家恩平避难。1945年抗日战争胜利后，他回到广州完成高中学业，并于1951年考入清华大学攻读心理学，1952年转入北京大学心理学专业继续深入学习。大学期间修读的心理学、哲学和生理学等课程为他后来的研究工作打下了坚实的基础。1955年，他以优异成绩从北京大学毕业，并被分配到中国科学院心理研究所（简称“心理所”）工作。在那里，他历任研究助理（1962年）、副研究员（1979年）和研究员（1986年）。

早年，他在心理所曹日昌所长指导下开展航空心理学的研究。他在《心理学报》和《航空知识》等杂志上发表了十多篇文章。同时，他参与了与军事部门的合作项目，并发表了三篇综述。他阐述了飞行员着陆时高度判断的机制（林仲贤，1961），并坚信飞行着陆过程中高度判断的困难是导致飞行事故的主要原因之一。他提出着陆距离判断是一种复杂的条件反射活动，是动力定型的统一综合。林仲贤和荆其诚系统总结了飞行中高空高度飞行、低空高度飞行、海上飞行和黑夜飞行的视觉问题以及飞行中的眼睛运动问题（林仲贤和荆其诚，1962）。此外，为了选拔飞行员，他建议，应当使用实验心理学的方法来预测飞行所要求的认知活动和动作反应，使用个案分析来评估学员的情绪、意志和个性等心理特征（荆其诚和林仲贤，1962）。他与曹日昌、封根泉合

译者：李海峰
福建师范大学心理学院，福州350117，中国
邮箱：haifeng.li@fjnu.edu.cn

作的预防和克服飞行错觉的实验研究更是获得了1965年国防部科学技术进步奖三等奖和中国科学院重大科技成果奖。

图2 肤色测量（傅小兰，2012）

1972年，林仲贤及其同事倡导恢复并重建心理所。林仲贤立即投身于颜色心理学及相关标准化研究（图2）。颜色心理学是研究色彩产生、接受及其应用规律的学科。他发表了一系列代表性的著作（如《视觉及测色应用》《颜色视觉心理学》）和代表性的文章（如《中国人肤色色度测定》《中国学前儿童颜色命名及知觉辨认》）。他主持或参与制定的国家标准有《反射式彩条测试图》（1987年）、《反射式肤色测试图》（1987年）、《透射式彩条测试图》（1987年）、《透射式肤色测试图》（1987年）、《彩色电视接收白场色温及其宽容度》（1989年）、《灯光信号颜色》（1987年）、《同色异谱的目视评价方法》（1995年）以及《中国颜色体系》（1995年）等。这些开创性研究成果为心理学、光学、电影电视等众多专业领域的发展与生产提供了重要理论基础和标准支撑。1976年9月毛泽东主席逝世后，林仲贤因这些卓越成就被特聘为专家参与主席遗体保存工作，负责主席的遗容肤色测量工作。

除了航空心理学和颜色心理学，林仲贤还开创了我国高原作业心理学研究。他证明了在大约海拔4000米的高度，低氧会对驾驶员的感觉机能和大脑功能产生影响（林仲贤等，1981）。他发现，深度知觉误差会随着海拔的升高而增加（林仲贤，1986）。不论是在平原地区还是在高原地区，高的工作负荷都能降低驾驶员的深度知觉机能。他还总结和分析了人类在高原地区的生理和心理状况，包括低氧对感知觉的影响，低氧对复杂心理活动的影响，以及低氧对中枢神经系统的影响（林仲贤，1988）。这些成果为高原地区发展作出了实质性贡献。

林仲贤为中国心理学会的恢复和发展倾注了自己的一生。他先后担任中国心理学会副秘书长（1984—1989年）和秘书长（1989—1993年），并因出色的领导、组织协调能力于1993年10月当选第六届中国心理学会理事长。任期内，他基于我国大多数省份的心理学发展状况做出重要的规划及布局，使得中国心理学会得到长足的进步和发展。

此外，林先生积极推动我国心理学走向国际。他多次出席国际心理学大会，先后前往十多个国家进行学术交流，介绍中国心理学发展现状和成就。由于上述杰出贡献，2004年，林先生获得中国心理学会终身成就奖，并当选中国心理学会会士。林先生的事迹被收入《中国科学院科学家名录》《亚太地区国际名人录》《中国当代教育名人传略》《当代中国科学家与发明家大辞典》《英国剑桥传记中心（IBC）国际名人录》和《美国传记中心（ABI）国际名人录》等。林先生的杰出贡献将永载我国心理学史册。

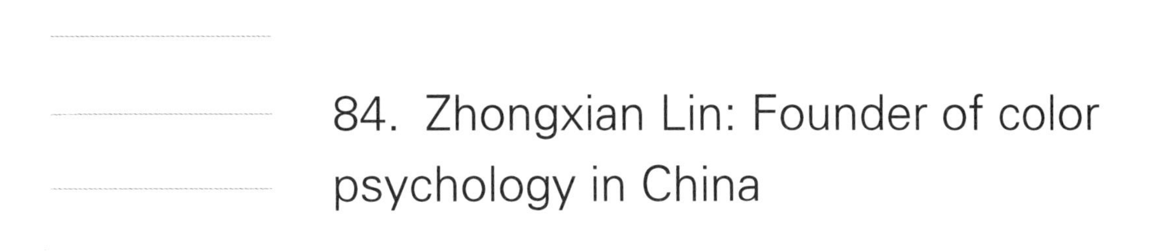

84. Zhongxian Lin: Founder of color psychology in China

Professor Zhongxian Lin (林仲贤, 1931—2011) is the founder and the leader of color psychology in China. He made pioneering researches on the determination of the color vision and the Chinese skin color. He is a member of executive committee of the AFRO-ASIAN Psychological Association (AAPA), the National Color Standardization Committee, the New York Academy of Sciences, and the American Association for the Advancement of Science (AAAS).

Born on December 3, 1931 in Guangzhou, Professor Lin spent his childhood there. After the full outbreak of the War of Resistance against Japanese Aggression in 1937, he took refuge with his mother in their hometown, Enping county. After the victory of the War of Resistance against Japanese Aggression in 1945, he came back and finished his senior high school in Guangzhou. In 1951, he went to Tsinghua University to study psychology, and continued his study in Peking University the next year. During that time, he took a broad range of courses, including psychology, philosophy and physiology, which laid a solid foundation for his future researches. In 1955, he graduated from Peking University with honors, and was assigned to work at the Institute of Psychology, Chinese Academy of Sciences (IPCAS). There, he was promoted as professor in 1986.

At the beginning of his research career at IPCAS, Professor Lin studied aviation psychology, which was supervised by Professor Richang Cao (曹日昌). He published more than 10 articles in several journals in this area, such as *Acta Psychologica Sinica* and *Aviation Knowledge*. Meanwhile, he participated in the key research projects collaborated with military forces, and published 3 review articles. He elaborated the mechanism of altitude estimation during aircraft landing (Lin, 1961), and stated that one of the main reasons for aircraft accidents was attributed to the difficulty for the pilot to estimate the altitude of the aircraft when landing. He proposed that estimation of landing distance is a complex conditioned reflex behavior, and a unified comprehensive dynamic stereotype. Professors Zhongxian Lin and Qicheng Jing (荆其诚)

Haifeng Li[1], Xiaomei Li[1], Buxin Han[2]

1 Department of Psychology, Fujian Normal University, Fuzhou 350117, China

2 Key Lab of Mental Health, Institute of Psychology, Chinese Academy of Sciences, Beijing 100101, China

Correspondence: haifeng.li@fjnu.edu.cn (H.-F. Li)

systematically summarized vision problems under different flight conditions, such as high speed, low speed, oversea and night flights, as well as the eye movement problems during the flights (Lin and Jing, 1962). In addition, he suggested that methods of experimental psychology should be utilized to predict the cognitive activities and motor responses when selecting pilots and case analysis should be used to evaluate the candidates' emotion, will, and personality (Jing and Lin, 1962). Working together with Professors Richang Cao and Genquan Feng (封根泉), Professor Lin's research on The experimental study of preventing and overcoming flight illusion (1965) won the Prize of the Science and Technology Achievement Award of the Ministry of Defense, and the Significant Achievement Prize of the Chinese Academy of Sciences.

On the initiative of Professor Lin and his colleagues, IPCAS was reestablished in 1972. At that time, Professor Lin started his research on color psychology which studies the production, acceptance and application of color, and its standardization. He published a series of representative books, such as *Vision and Color Measurement Application*, and *Psychology of Color Vision*, and representative articles, such as "Chinese skin color measurement" , and "Color naming and perceptual discrimination of Chinese preschool children" . He hosted and participated in the determination of several national standards, including *Reflection Color Bar Test Card* (1987), *Reflection Flesh Tone Test Card* (1987), *Transparency Color Bar Test Card* (1987), *Transparency Flesh Tone Test Card* (1987), *Color TV Receiver-Color Temperature of White Point-Reference White and Its Tolerance* (1989), *Colors of Light Signals* (1987), *Method for Visual Evaluation of Metamerism* (1995) and *The Chinese Color System* (1995). These pioneering achievements provided significant theoretical basis and national standards for the development of several professional areas, such as psychology, photology, film, and television. For his outstanding achievements on color psychology, after Chairman Mao's death in September, 1976, Professor Lin was invited to measure his skin color for the body reservation.

Besides aviation psychology and color psychology, Professor Lin also opened up a new avenue for the psychological researches on working on the plateau. He proved that drivers' sensory and brain functions can be affected by the low-oxygen at the altitude around 4000 meters (Lin et al., 1981). He also found that depth perception error increased with the increase of altitude (Lin, 1986). Whether in the plains or on the plateau, high work load would result in decreased depth perception function of the drivers. He also summarized and analyzed the physical and mental status of the human beings on the plateau, including the influences of low-oxygen on sensory perception, complex mental activity, and central nervous system (Lin, 1988). These studies made substantial contributions to the development of scientific research in the plateau areas.

As a leading scientist for the researches on the Chinese psychology, Professor Lin devoted almost all his life to the recovery and development of the Chinese Psychological Society (CPS). He served as the deputy secretary general (1984—1989) and the secretary general (1989—1993) for the CPS, and was elected as the sixth President of the CPS in October, 1993 based on his outstanding contributions to the field and his excellent capabilities on leadership, organization, and coordination. He investigated the developmental status of psychology in the majority of provinces in China, and made the significant layout for the development of Chinese psychology, which drove the development of the CPS considerably.

In addition, Professor Lin actively promoted the Chinese psychology to the world. He attended the International Congress of Psychology several times, and carried out academic exchanges with more than 10 countries, and introduced the current situations and achievements of the Chinese psychology to the world. Due to his prominent contributions, he was awarded the Lifetime Achievement Award of the CPS, and was elected as the Fellow of the CPS in 2004. He was included in *The Scientists Directory of the Chinese Academy of Sciences*, *Who's Who of the Asian Pacific Rim, The Brief Biographies of Chinese Contemporary Education Celebrities, Dictionary of Contemporary Chinese Scientists and Inventors*, *Who's Who of the International Biographical Centre of Cambridge*, *Who's Who of the American Biographical Centre*. His outstanding contributions have been recorded in the history of the Chinese psychology.

Figures

Fig.1 Professor Zhongxian Lin (1931—2011)

Fig.2 Skin color measurement (Fu, 2012)

References

Fu XL (2012) Special issue to commemorate Mr. Zhongxian Lin. Institute of Psychology, Chinese Academy of Sciences (Internal publications), Special issue No. 186. (傅小兰. 2012. 纪念林仲贤先生专刊. 中国科学院心理研究所(内部刊物), 第186期专刊.)

Jing QC, Lin ZX (1962) The question of psychological prediction on flight ability. Acta Psychol Sin 6(3): 248–261. (荆其诚, 林仲贤. 1962. 关于飞行能力的心理学预测问题. 心理学报, 6(3): 248–261.)

Lin ZX (1961) The mechanism of height judgement in flight landing. Acta Psychol Sin 5(4): 259–266. (林仲贤. 1961. 飞行着陆高度判断的机制问题. 心理学报, 5(4): 259–266.)

Lin ZX (1986) The influence of low-oxygen and work load on depth perception. J Psychol (1): 24–27. (林仲贤. 1986. 低氧及劳动负荷对深度觉的影响. 心理学杂志, (1): 24–27.)

Lin ZX (1988) Physical and mental changes in high altitude hypoxic conditions. J Psychol (1): 30–38. (林仲贤. 1988. 高空低氧条件下的生理心理变化. 心理学杂志, (1): 30–38.)

Lin ZX, Jing QC (1962) Some vision problems in flight. Psychol Res Trends (2): 13–17. (林仲贤, 荆其诚. 1962. 飞行中的几个视觉问题. 心理学研究动态, (2): 13–17.)

Lin ZX, Ma MC, Hong SQ, et al (1981) The influence of low-oxygen on brain function in plateau area. Acta Psychol Sin 13(2): 192–200. (林仲贤, 马谋超, 洪生勤, 等. 1981. 高原低氧对脑功能的影响. 心理学报, 13(2): 192–200.)

九、生命科学领域重要成就与重要事件

85. 创造合成结晶牛胰岛素

40年前，我国取得了一项生命科学的伟大成就——结晶牛胰岛素（图1）的全合成，这让中国科学家感到无比兴奋和自豪。胰岛素是胰腺β细胞分泌的一种激素。在胰岛素用于临床以前，糖尿病是一种经常导致死亡的可怕疾病。1868年，Paul Langerhans，一名在柏林的医学生，发现了胰腺中存在的细胞簇（Langerhans，1868）。这些细胞簇后来被称为"朗格汉斯小岛"。这些细胞中的一些最终被证明可产生胰岛素。"胰岛素"（insulin）这一术语起源于"Insel"，这是"小岛"的德语单词（Sakula，1988）。F. G. Banting，一名年轻的加拿大医生，利用J. J. R. Macleod教授提供的实验设施，在Macleod教授的学生C. H. Best的辅助下，于1921年在多伦多大学首先从胰管手术结扎的狗胰腺中提取了胰岛素。生物化学家J. B. Collip帮助纯化了提取物。因为这一工作，Banting和Macleod分享了1923年诺贝尔生理学或医学奖（Nobelprize.org，2014）。

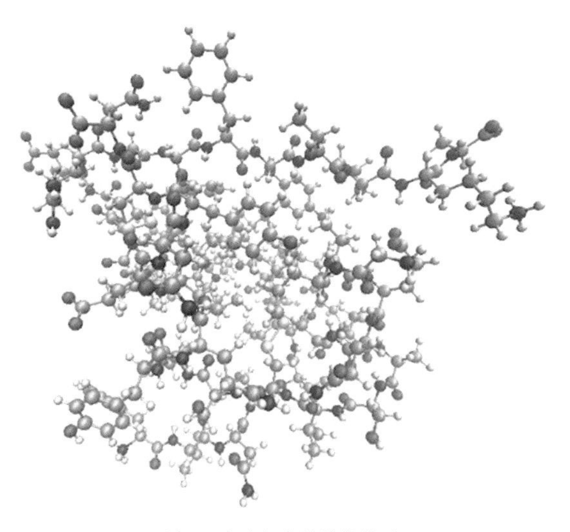

图1　牛胰岛素的结构模型

作者：孙业平
中国科学院微生物研究所病原微生物与免疫学重点实验室，北京100101，中国
邮箱：sunyeping@im.ac.cn

英国分子生物学家Frederick Sanger经过10年的努力测定了胰岛素的一级结构。它由两条链组成：A链和B链；A链含有21个氨基酸，而B链有30个氨基酸；两条链通过两个二硫键相连接，A链有一个链内二硫键（图2）（Sanger，1959）。这是人类历史上测定的第一个蛋白质结构，Sanger因此获得了1958年诺贝尔化学奖。

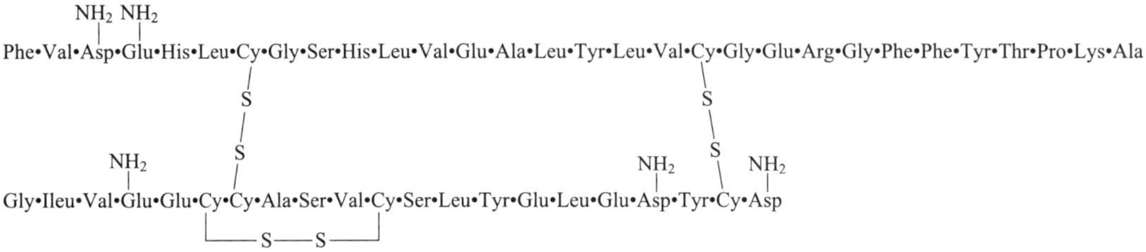

NH_2 NH_2
Phe•Val•Asp•Glu•His•Leu•Cy•Gly•Ser•His•Leu•Val•Glu•Ala•Leu•Tyr•Leu•Val•Cy•Gly•Glu•Arg•Gly•Phe•Phe•Tyr•Thr•Pro•Lys•Ala
S S
S S
NH_2 NH_2 NH_2
Gly•Ileu•Val•Glu•Glu•Cy•Cy•Ala•Ser•Val•Cy•Ser•Leu•Tyr•Glu•Leu•Glu•Asp•Tyr•Cy•Asp
S—S

图2 胰岛素的一级结构

1958年，中国科学院上海生物化学研究所和北京大学提出“我国应该人工合成胰岛素”并得到了政府的支持。这一项目在1959年启动，但是当时缺少足够的设备、氨基酸原料和必要的试剂。因此，合成这样一个大的复合物是一个艰巨的任务。当时采取的策略是人海战术，八个研究机构的几百名人员参与了这个项目。他们夜以继日地制备氨基酸和其他试剂、纯化溶剂并合成短肽。

1963年，美国匹兹堡的P. G. Katsoyannis（Katsoyannis et al.，1963）和德国亚琛的Helmut Zahn（Zahn and Schade，1963）报道合成了具有弱活性的胰岛素。这些报道对我国的研究者造成了巨大的压力，也给了他们继续完成这一项目并首先获得具有完全活性胰岛素的动力。为了完成这一任务，选出了三个团队在上海协同工作：北京大学的邢其毅教授带领其团队前往上海与中国科学院上海有机化学研究所的汪猷教授团队合作，负责A链的合成；中国科学院上海生物化学研究所的钮经义教授团队负责B链的合成；中国科学院上海生物化学研究所的邹承鲁教授团队负责A链和B链的拆分和重组。

1964年8月，钮经义教授成功合成了B链并将其与天然胰岛素的A链组装成半合成牛胰岛素（Niu et al.，1964）。1965年5月，北京大学与中国科学院上海有机化学研究所联合团队完成了胰岛素A链的合成（Wang et al.，1966）。与此同时，邹承鲁教授团队大大增加了A链和B链重组的得率（Institute of Biochemistry，Chinese Academy of Sciences，1966）。最后，在1965年9月17日，全合成的牛胰岛素终于问世了。这是世界上第一个具有全部生物活性、免疫原性和化学性质的结晶形式（图3）的全合成的牛胰岛素。

这一工作发表在《中国科学》上（Kung et al.，1965），引起了国际上的极大关注。*Science*杂志在1966年7月报道了这一成就（McElheny，1966）。著名的肽化学发言人H. N. Rydon教授在*New Scientist*杂志上述评了为什么中国科学家的竞争者没能获得具有完全活性的胰岛素，并称中国科学家的这项成就“具有重大的意义，它将刺激并鼓励合成更大、更典型的蛋白质”（Rydon，1966）。

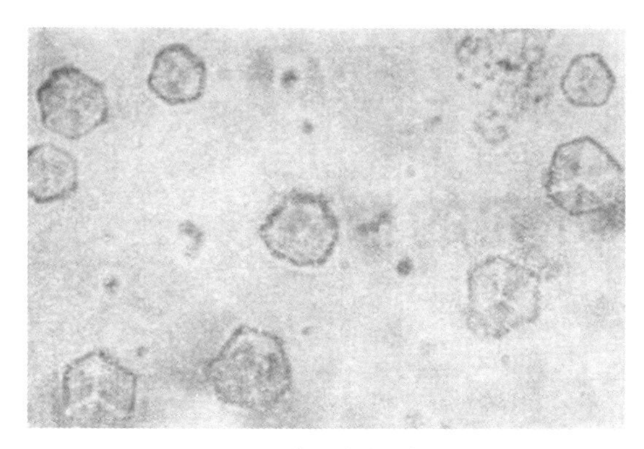

图3 结晶牛胰岛素

在仅六年的时间里合成了完整的具有全部活性的胰岛素是令人难以置信的成就，因为预计需要的时间要久得多。为什么这样一个了不起的成就首先在中国这个科学研究基础相对较弱的发展中国家而不是像美国和德国这样的发达国家实现？除了科研管理部门及时的决策和策划，最重要的决定因素可能是当时科学家值得赞颂的精神。这些精神可以概括为“胰岛素精神”，它包括：① 无私奉献。参加项目的所有人员都全身心投入项目中，而不考虑自身的利益。② 诚实。在超过200步的合成过程中，每一个中间体都必须严格地鉴定，这些鉴定过程稍有问题就会导致整个项目的失败。③ 密切合作。从三个不同的研究单位组织起来的三个团队具有明确的责任分配并为项目的共同目标协同工作，因而取得了高效率。④ 敢于迎接挑战。合成一个由51个氨基酸组成的蛋白质是一项极其困难的任务。因此，是敢于迎接挑战的精神帮助中国科学家赢得了世界的尊重。今天，胰岛素精神在建设科研体系和管理科研项目中仍然具有重大的价值。最重要的是，它已经成为每一个中国研究者信心和力量的源泉。

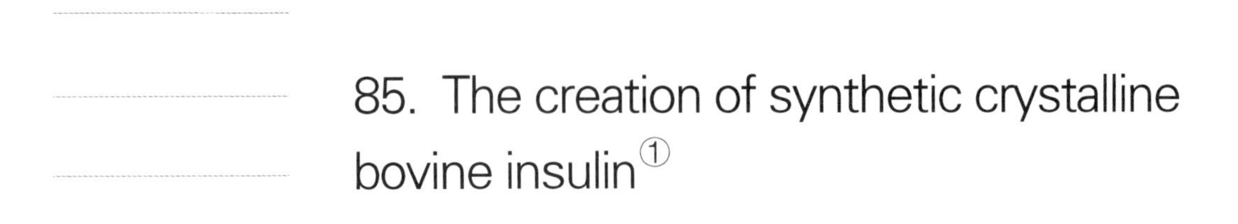

85. The creation of synthetic crystalline bovine insulin[1]

Fifty years ago, a great achievement in life science occurred in China—the complete synthesis of crystalline bovine insulin—which gave Chinese scientists a sense of great elation and pride. Insulin is a hormone secreted by β cells in pancreas. Before the clinical application of insulin, diabetes was a feared disease that commonly led to death. Insulin has been studied since 1868 when Paul Langerhans, a medical student in Berlin found clusters of cells in the pancreas (Langerhans, 1868). These were later called "Islets of Langerhans". Some of these cells were eventually shown to produce insulin. The term "insulin" origins from "Insel", the German word for "islet" or "small island" (Sakula, 1988). Frederick Grant Banting, a young Canadian physician first extracted insulin from the pancreas of a dog whose pancreatic duct had been surgically ligated at University of Toronto in 1921, with the experimental facilities provided by Prof. John James Rickard Macleod and the assistance of one of Macleod's students, Charles H. Best. Biochemist James Bertram Collip helped purify the extract. For this work Banting and Macleod shared the 1923 Nobel Prize in Physiology or Medicine (Nobelprize.org, 2014).

British molecular biologist Frederick Sanger determined the primary structure of insulin through 10 years of research: It comprises of two chains, chain A and chain B; chain A contains 21 amino acid residues while chain B consists of 30 residues; the two chains are linked by two disulfide bonds and there is an intra-chain disulfide bond in chain A (Sanger, 1959). This is the first protein structure determined in human history, which Sanger was awarded the 1958 Nobel Prize in Chemistry.

Driven by the "great leap forward" campaign in 1958, Shanghai Institute of Biochemistry, Chinese Academy of Sciences and Peking University proposed that China should artificially synthesize insulin and obtained the support of the government. The project started in 1959, however, at that time, there was a lack of adequate equipment, the raw materials of amino acids

---

Yeping Sun

CAS Key Laboratory of Pathogenic Microbiology and Immunology, Institute of Microbiology, Chinese Academy of Sciences, Beijing 100101, China

Correspondence: sunyeping@im.ac.cn

[1] This paper was kindly revised by Joel Haywood (CAS Key Laboratory of Pathogenic Microbiology and Immunology, Institute of Microbiology, Chinese Academy of Sciences, Beijing 100101, China).

and other necessary reagents. Consequently, synthesis of such a large compound represented a formidable task. The strategy adopted was to involve as many capable scientists as possible with eventually several hundreds of participants from eight different institutes participating in the project. People worked day and night preparing amino acids and other reagents, purifying solvents and synthesizing small peptides.

In 1963, Panayotis G. Katsoyannis (Katsoyannis et al., 1963) at Pittsburgh in U.S.A and Helmut Zahn (Zahn and Schade, 1963) at Aachen in Germany reported that they had synthesized insulin with weak activity. These reports placed a great deal of pressure on Chinese researchers and gave them further motivation to complete their project and be first to synthesize insulin with full activity. In order to accomplish this task three teams were chosen to work together synergistically in Shanghai: Prof. Qiyi Xing of Peking University was to lead his group to Shanghai and combine with Prof. You Wang's group in Shanghai Institute of Organic Chemistry, Chinese Academy of Sciences where they were to be responsible for the synthesis of chain A; Prof. Jingyi Niu of Shanghai Institute of Biochemistry, Chinese Academy of Sciences and his team would complete the synthesis of chain B; and Prof. Chenglu Zou's team in Shanghai Institute of Biochemistry, Chinese Academy of Sciences would be responsible for the split and recombination of chain A and B.

In August of 1964, Prof. Jingyi Niu successfully synthesized chain B and assembled it with chain A of natural insulin into semi-synthesized bovine insulin (Niu et al., 1964). In May of 1965, the combined team of Peking University and Shanghai Institute of Organic Chemistry finished the synthesis of Chain A of insulin (Wang et al., 1966). At the same time, Prof. Chenglu Zou's team greatly increased the yield of the recombination of chain A and B (Institute of Biochemistry, Chinese Academy of Sciences, 1966). And finally, the fully synthetic bovine insulin came into production on September 17, 1965. This is the first totally synthetic insulin in crystallized form with full biological activity, immunogenicity and chemical property in the world.

This work was published in *Scientia Sinica* (Kung et al., 1965) which aroused a great deal of international interest. *Science* magazine reported this achievement in July, 1966 (McElheny, 1966). Prof. Henry Norman Rydon, a celebrated statesman in peptide chemistry, reviewed in *New Scientist* magazine the reasons why the competitors of the Chinese scientists had failed to obtain insulin with full activity and commented that the achievement of the Chinese scientists was "a truly seminal piece of work which would stimulate and encourage work directed towards the synthesis of larger, more typical, proteins" (Rydon, 1966).

The synthesis of a complete and active insulin in only six years was a fantastic achievement given that it was predicted to take much longer. Why was such a brilliant and awesome achievement first accomplished in China, a developing country where the basis of scientific research was relatively weak, rather than in developed countries such as America and Germany? Besides the timely decision-making and strategic planning of the scientific administrative department in the government, the most important determinants might be the laudable mentality of Chinese scientists at that time, which can be summed up into "insulin spirit" , which includes four aspects: (1) Selfless dedication. All people involved in the project devoted all themselves into the demands of the project without considering their own interest; (2) Honesty. Every intermediate in more than 200 steps of the synthetic procedure had to be rigorously identified, so even a slight problem on these identification procedures might lead to total failure of the whole project; (3) Close cooperation. The three teams organized from three different institutes had clear assignment of their responsibilities and worked synergistically for the common aim of the project, so that high efficiency was achieved; (4) The spirit of welcoming challenges. To synthesize a protein consisting of 51 amino acids was a formidable task, so it was the spirit of welcoming a challenge that helped the Chinese scientists gain the respect of the world. Today, the "insulin spirit" is still of great value in constructing our research systems and managing research projects. Most importantly, it has become the source of confidence and strength of every Chinese researcher.

Figures

References

Institute of Biochemistry, Chinese Academy of Sciences (1966) Resynthesis of insulin from its A and B chains. Chin Sci Bull 11:246–255.

Katsoyannis PG, Fukuda K, Tometsko A (1963) Insulin peptides 9. Synthesis of A-chain of insulin and its combination with natural B-chain to generate insulin activity. J Am Chem Soc 85:2863–2865.

Kung YT, Du YC, Huang WT, et al (1965) Total synthesis of crystalline bovine insulin. Sci Sin 14:1710–1716.

Langerhans P (1868) Ueber die Nerven der menschlichen Haut. Archiv für Pathologische Anatomie und Physiologie und für Klinische Medicin 44:325.

McElheny VK (1966) Total synthesis of insulin in China. Science 153:281–283.

Niu CI, Chen CC, Wang KZ, et al (1964) Synthesis of peptide fragments of B-chain of insulin Ⅸ. Synthesis of B-chain of insulin Ⅸ. Its reconstitution with natural A-chain to regenerate insulin activity. Acta Biochemica et Biaphysica Sinica 4:726–728.

Nobelprize.org (2014) The discovery of insulin. Nobel Media AB Web. 7 Sep 2015.

Rydon HN (1966) First synthesis of a protein. New Scientist: 766–767.

Sakula A (1988) Paul Langerhans (1847—1888): A centenary tribute. J R Soc Med 81:414–415.

Sanger F (1959) Chemistry of insulin. Science 129:1340–1344.

Wang Y, Hsu JZ, Chang WC, et al (1966) Synthesis of A-chain of bovine insulin and partial synthesis of crystalline bovine insulin from synthetic A- and natural B-chains. Acta Chim Sin 32:276–283.

Zahn H, Schade F (1963) Chemische Modifizierung Von Insulin, Seidenfibroin, Sehnenkollagen Und Wollkeratin Mit Nitrophenylestern. Angewandte Chemie-International Edition 75:377.

86. 王德宝：中国核酸研究的奠基人

图1　王德宝在美国时的照片

王德宝院士于1918年5月7日出生于江苏泰兴，18岁考入国立中央大学农学院农业化学系，并于1940年毕业留校任助教。1943年夏，王德宝转入生物医学院生物化学科，师从郑集教授（Zhang，2020）。此时，怀揣着为国家发展作贡献的梦想，他渴望出国学习其他国家的先进技术。

1947年初，王德宝来到了美国南部的路易斯安那州立大学学习（图1），在制糖专业攻读一年多以后，于1948年秋转入华盛顿大学生物化学系学习，师从当时获得诺贝尔生理学或医学奖的C. Cori夫妇。1949年夏，他随导师转学至凯斯西储大学，并于1951年秋获得博士学位。同年，王德宝赴美国约翰斯·霍普金斯大学的麦克勒姆·普拉特研究所从事博士后研究。王德宝在美国学习期间，首创了从烟酰胺腺嘌呤二核苷酸（NAD）直接合成烟酰胺腺嘌呤二核苷酸磷酸（NADP）的制备方法，30年来为世界各大药厂采用，改变了NADP价格居高不下的形势。此外，他还发现了核苷水解酶、胞嘧啶脱氨酶、尿嘧啶氧化酶等几种对生物化学产生深远影响的新酶。

1955年，王德宝回到祖国后，到中国科学院生理生化研究所（简称“生化所”）工作，并建立了我国第一个核酸研究组。1961年，该组被扩建为我国第一个核酸研究室，王德宝担任室主任（图2）。更重要的是，他专注于转移RNA（tRNA）的结构和功能，在我国开始了核酸的研究。

在从事理论研究的同时，王德宝也十分重视生产问题。王德宝指导他的学生开展自溶法生产4种5′-核糖核苷酸的研究，并与味精厂合作生产特鲜味精，这项工作获得了1978年国家重大科技成果奖。由此，王德宝成为我国生产核苷酸类助鲜剂的创始人。王德宝在生物化学事业中作出的巨大贡献有目共睹，这使他收获了众多赞誉，并于1980年当选中国科学院学部委员（院士）。

作者：马中良，王显仪，李艳利

上海大学生命科学学院，上海200444，中国

邮箱：liyanli@shu.edu.cn（李艳利）

图2 王德宝（中）与他的团队在上海核酸研究室讨论实验方案

王德宝率领课题组展开了对tRNA结构与功能的研究，成为我国开展核酸生化研究工作的第一人。经过13年的不懈努力，课题组终于在1981年11月中旬完成了酵母丙氨酸转移核糖核酸（$tRNA^{Ala}$）的人工全合成，这是世界上首次人工全合成的与天然tRNA结构和组成完全相同并具有全部生物活力的tRNA（Wang et al.，1983a，1983b；刘新垣，2002）。由于RNA中核糖的第二个碳含有羟基，因此必须保护该基团以减少不需要的反应。为了解决这个难题，他们选择分别合成3′-tRNA和5′-tRNA半分子，最后将这两个半分子连接成一个完整的酵母$tRNA^{Ala}$。

他们提出，这一人工合成的酵母$tRNA^{Ala}$具有与天然酵母$tRNA^{Ala}$相同的化学成分和结构，这使它们具有了生物活性（Wang et al.，1983a）。而且他们发现，酵母丙氨酰tRNA的反密码子环不参与来自大鼠肝脏的丙氨酰tRNA合成酶和酵母丙氨酰tRNA之间的相互作用（Jin，1987）。此项工作获中国科学院重大科技成果奖一等奖、国家自然科学奖一等奖和陈嘉庚生命科学奖，使我国人工合成生物大分子化合物的研究水平保持世界领先地位。

1983年，王德宝以"酵母$tRNA^{Ala}$的人工全合成"为题向第十届国际tRNA学术研讨会作报告，获得国内外一致重视和赞扬。王德宝团队对这些人工合成模拟物的研究揭示了生物复制、翻译和代谢途径之间的联系，为探索生命进化的过程提供了宝贵的工具。酵母$tRNA^{Ala}$全合成工作完成以后，王德宝课题组继续在酵母$tRNA^{Ala}$的结构和功能方面进行研究，阐明了各部分结构的生物学功能，对tRNA发生及转运氨基酸的机制进行了进一步研究（陈莉等，2000）。

此外，王德宝十分重视培养我国的核酸研究与教学人才，众多研究生、年轻的大学毕业生和进修生在他的指导下快速地成长起来，并成为我国核酸研究与教学工作的骨干力量。1961年，生化所举办了第一次全国性的生化训练班，他花了不少力气，编写了我国第一本核酸讲义《核酸——结构、功能与合成》，在我国生物化学教学和研究工作中产生了深远的影响。

2002年11月1日，王德宝院士逝世，享年84岁，国内外学者深切哀悼。中国科学院院士、中国科学院生物化学与细胞生物学研究所所长李伯良给予高度评价："先生无论是做学问还是做人，都是学界楷模，后辈榜样。"分子生物学国家重点实验室主任金由辛教授缅怀道："师恩难忘，楷模永继。"目前上海大学生命科学学院非编码RNA与癌症实验室正在研究tRNA衍生片段（tRF）在癌症中的作用，这是金教授的想法，也是王德宝院士tRNA科学探索的新延续。近年来已在tRF方面取得了进展，包括一项授权专利和多篇SCI论文。

王德宝院士深爱祖国，毕生致力于科学研究，对我国生命科学的发展产生了重大影响。在王院士及其团队的指导下，tRNA的结构和功能得到了很好的诠释，这极大地推动了非编码RNA的研究。最重要的是，王德宝院士给我们留下了大量宝贵的精神财富，值得我们后辈学习和传承。

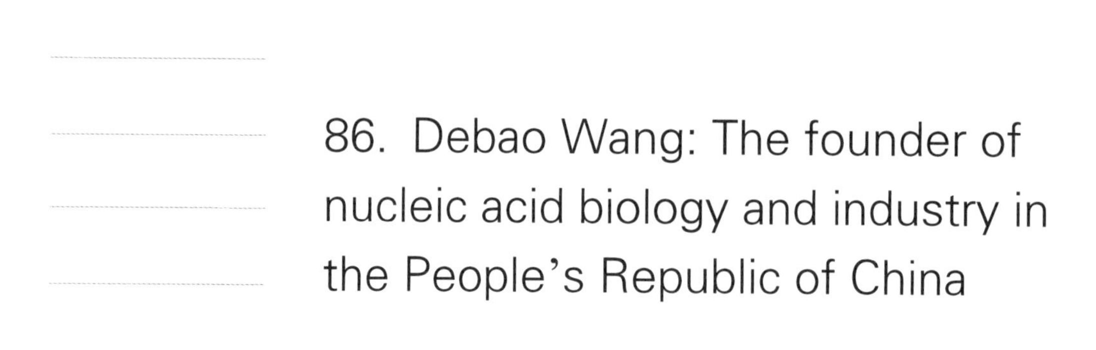

86. Debao Wang: The founder of nucleic acid biology and industry in the People's Republic of China

Dr. Debao Wang (王德宝) was born on May 7, 1918, in Taixing, Jiangsu Province, China. At the age of 18, he was admitted to the Agricultural Chemistry Department of the National Central University, and graduated in 1940. In the summer of 1943, Dr. Wang worked in the Department of Biochemistry, School of Biological Medicine, under the supervision of Prof. Libin T. Cheng (Zhang, 2020). At that time, he had a dream to contribute to the development of his motherland, which inspired him to study aboard.

In early 1947, he left to study at Louisiana State University, USA. In the autumn of 1948, he studied at the Department of Biochemistry, University of Washington, under the supervision Prof. C. Cori, who won the Nobel Prize in Physiology or Medicine. In the summer of 1949, he moved to the Case Western Reserve University with Dr. Coci, where he received his doctorate in the fall of 1951. In the same year, Dr. Wang started his postdoctoral research at Johns Hopkins University. During his study in the United States, Dr. Wang pioneered the synthesis of nicotinamide adenine dinucleotide phosphate (NADP) from NAD, which was later used by major pharmaceutical companies in the world for 30 years. His industrial methods greatly reduced the price of NADP.

In addition, he also discovered nucleoside hydrolase, cytosine deaminase, uracil oxidase and several other new enzymes, which had a profound impact on biochemistry.

In 1955, he decided to return and devote his life to scientific research in his motherland, China. He worked at the Institute of Physiology and Biochemistry of Chinese Academy of Sciences (CAS), and established the first nucleic acid research group in China. In 1961, Dr. Wang served as the director of a nucleic acid research laboratory. What's more, Dr. Wang focused on the structure and function of transfer RNA (tRNA), starting nucleic acid research in China.

Dr. Wang also used his knowledge to produce some things people needed. He and his students modified the production procedures of 4 kinds of 5′-ribonucleotides using the autolysis method, and obtained the flavor enhancer monosodium glutamate in cooperation with

Zhongliang Ma, Xianyi Wang, Yanli Li
Lab for Noncoding RNA & Cancer, School of Life Sciences, Shanghai University, Shanghai 200444, China
Correspondence: liyanli@shu.edu.cn (Y.-L. Li)

an MSG factory. Their work won the 1978 National Science and Technology Achievement Award and made him the founder of the production of flavor enhancers in China. Due to his great contributions to biochemistry, he was elected an academician of the Chinese Academy of Sciences in 1980.

In addition, Dr. Wang investigated the synthesis of yeast $tRNA^{Ala}$. After 13 years of unremitting efforts, his team finally completed the artificial total synthesis of yeast $tRNA^{Ala}$ in mid-November 1981. This was the first artificial synthetic tRNA in the world, and they put a lot of efforts into it due to the complexity of RNA structures (Wang et al., 1983a, 1983b; Liu, 2002). Because the second carbon of ribose in RNA contains a hydroxyl group, it is necessary to protect the group to reduce unwanted reactions. In order to solve this difficult problem, they chose to synthetize 3′-half and 5′-half tRNA, respectively, and finally connected these two halves to a complete yeast $tRNA^{Ala}$.

They proposed that the synthetic yeast alanine tRNA has the same chemical composition (containing 9 modified nucleotides of 7 different species) and structure as the natural yeast alanine tRNA, making it biologically active (Wang et al., 1983a). A comparison of the charging activities of yeast alanyl tRNA and reconstituted natural yeast tRNA revealed that the anticodon loop of yeast alanyl tRNA was not involved in the interaction between alanyl-tRNA synthetase from rat liver and yeast alanyl-tRNA (Jin, 1987). This work has won the first prize of science and technology achievements of the Chinese Academy of Sciences, the first prize of the National Natural Science Foundation, and the Tan Kah Kee Award in Life Sciences. In 1983, he made a report to the 10th International tRNA Symposium on the topic of "Artificial Total Synthesis of Yeast $tRNA^{Ala}$" which won unanimous attention and praise at home and abroad. The study of these analogues revealed the link between biological replication, translation, and metabolic pathways, providing a valuable tool for exploring the evolutionary process of life (Chen et al., 2000).

After the completion of yeast alanine tRNA synthesis, Dr. Wang led his team to unmask the structure and function of yeast alanine tRNA. Not only did they clarify the biological functions of yeast alanine tRNA components, but also further studied the mechanism of tRNA generation as well as its function in transporting amino acids.

In addition, Dr. Wang made great contributions to nucleic acid research and teaching talents in China. Many students, including young university undergraduates and post-graduates, have grown up quickly under his guidance and have become the backbone of nucleic acid research and teaching in China. In 1961, the Institute of Biochemistry held the first national biochemical training course. He made great efforts to compile the first nucleic acid lecture in China, "*Nucleic Acids—Structure, Function, and Synthesis*" , which had a profound impact on biochemistry education and research in China.

Dr. Wang passed away at the age of 84 on November 1, 2002. He was deeply mourned by scholars at home and abroad. Professor Boliang Li, director of the Institute of Biochemistry and Cell Biology, an academician of Chinese Academy of Sciences, highly praised him with the words: "Whether it is learning or being a human being, Dr. Wang is an academic model and an example for the younger generations." Professor Youxin Jin, the director of State Key Laboratory for Molecular Biology, remembered him with the words: "The kindness of teacher Wang is hard to forget, and his qualities will be inherited forever." Our lab is studying the role of tRNA-derived fragments (tRF) in cancer, which was Professor Jin's idea, but an update of an original tRNA research idea of Dr. Wang. In recent years, we made progress on tRF, including a patent and a number of published papers.

Dr. Wang loved his country deeply and dedicated his life to scientific research, which had a significant effect on the development of life sciences in China. Under the guidance of Dr. Wang and his team, the structure and function of tRNAs were explained and the research on noncoding RNAs was greatly promoted. He led us into a new field of RNA research. Above all, Dr. Wang left us a great deal of valuable spiritual wealth that is worthy of being learned and inherited by our younger generations forever.

Figures

References

Chen L, Jin YX, Wang DB (2000) Application prospect of tRNA. Chem Life 20(3):103–105. (陈莉，金由辛，王德宝. 2000. tRNA的应用前景. 生命的化学，20(3): 103–105.)

Jin Y, Qiu M, Li W, et al (1987) Effect of the anticodon loop size of yeast alanyl tRNA on its biological activity. Anal Biochem 161(2):453–459.

Liu XY(2002) Tribute to my tutor, Dr. Wang. Chin J Biochem Mol Biol 18(6): 3–4. (刘新垣. 2002. 悼念我的导师王德宝院士. 中国生物化学与分子生物学报，18(6): 3–4.)

Wang D, Qiu M, Liang Z, et al (1983b) Synthesis of the 3′-half molecule of yeast alanine tRNA. Sci Sin B 26(5):482–494.

Wang D, Zheng K, Qiu M, et al (1983a) Total synthesis of yeast alanine transfer ribonucleic acid. Sci Sin B 26(5):464–481.

Zhang H (2020) A 110-year-old wise man: Professor Libin T. Cheng, one of the founders of biochemistry and nutrition in China. Protein & Cell 11(2):79–81.

87. 国际“人类基因组计划”（HGP）和中国的贡献

我不希望在那一天，当我把HGP作为人类登月一样的故事在课堂上说给学生听的时候，一个学生会问：“老师，那时的中国为什么没有参与？”假设我的回答只能是这样的：“孩子，你们不知道啊，那时我们的国家是如何贫穷，我们的人民是如何愚昧，我们的决策者是如何短视，我们的科学家又是如何令人失望，我们古老的民族就这样又失去了一次历史机遇……”我将会抱憾终身。

——一位中学教师谈中国参与HGP（1999年）

人类基因组计划（Human Genome Project，HGP）是人类历史上对生命奥秘最大规模的全面探索。HGP与“原子弹计划”“登月计划”并称为20世纪最为重要的“三大计划”（Lambright，2002），被誉为生命科学史上继发现DNA双螺旋结构这个“第一次革命”之后的“第二次革命”（Sharp，2014）。

HGP的启动和国际化

有关国际HGP的意义、可行性以及相关的伦理和社会问题的讨论，开始于20世纪80年代。经过长达6年的全球性讨论，美国首先在1990年10月启动HGP。

1996年2月28日，由英国Wellcome基金会的Micheal Morgan提议并主持，“人类基因组测序第一次国际战略会议”在百慕大召开，史称“百慕大会议”。英国不仅动员法国、德国和日本的代表参会，还促成了历史性的“百慕大原则”，因而使HGP国际化。随后在1999年的“人类基因组测序第五次国际战略会议”上（图1），经过认真答辩和慎重讨论，中国作为“最后一个贡献者”被正式接纳，作为唯一的发展中国家加入了“国际人类基因组测序协作组”，HGP成为人类历史上第一个由发展中国家和发达国家共同参与、平等合作的大规模国际合作计划。

译者：王晓玲
深圳华大生命科学研究院，深圳518083，中国
邮箱：wangxiaoling@genomics.cn

图1 1999年8月31日，英国欣克斯顿，人类基因组测序第五次国际战略会议

中国的贡献

中国的具体分工是3号染色体短臂端粒一侧约30 cM（厘摩，遗传图距单位）区域的测序、组装和注释任务，约占整个人类基因组测序和注释工作的1%，又称“1%项目”，或称“HGP的中国卷”。同时，也因所有用于测序的细菌人工染色体（BAC）克隆都标以“Beijing”字样而被称为人类基因组的“北京区域”。

中国参与HGP和美国等其他国家刚刚启动HGP时一样，争议颇多。现在仍能从当时的出版物中读到“（人类基因组测序）由国际上大实验室之间瓜分和大公司抢占，我们不必去参与竞争”以及“人类基因组是伪科学……这些人硬是把苹果排列在地上的顺序说成是它们在树上的顺序”之类的不同意见。

中国承担的具体任务意味着从1999年10月至2000年3月的6个月里必须完成50万个成功的Sanger测序反应。“1%项目”由国家科技部的863项目、国家自然科学基金委员会和中国科学院共同资助。中国人类基因组协作组（Chinese Human Genome Consortium，CHGC）主要成员有国家人类基因组北方研究中心（北京）、国家人类基因组南方研究中心（上海）、中国科学院遗传研究所人类基因组研究中心暨北京华大基因研究中心（BGI，简称华大基因），以及包括西安交通大学和东南大学等在内的15家合作单位。华大基因负责从种子克隆库中通过原位杂交和图谱等方法筛选“延伸克隆”，经验证后提供给CHGC各成员单位测序，再运用诸多改进的生物信息学工具对所有成员单位递交的“下机序列”的每一碱基进行准确度（质量）评估和拼接，并按照国际协作组的要求在24小时之内完成递交。更重要也更艰难的是对序列进行反复组装并以“限制图”进行定位，再进行文库筛选，或以“反向PCR”等技术进行克隆内和克隆间的“补洞”。所有的组装序列再以“限制图”进行验证。

从技术角度说，“1%项目”几乎是从零开始。为了“祖国的荣誉”和证明“别人能做到的我们也一定能做到”，所有参与者都在最短的时间内通过严格训练，基本掌握了所需的全部技能。他们开创了一条既经济又高效的“具有中国特色”的测序“生产线”，为克服经费不足的重重困难，实验人员用设备包装箱搭建起临时实验台（图2）。我们还能找到很多记录了实验人员如何夜以继日艰苦奋斗的老照片。以当时的测序技术，整个项目至少消耗了1500万个“吸枪头”就是一个很好的例子（图3）。

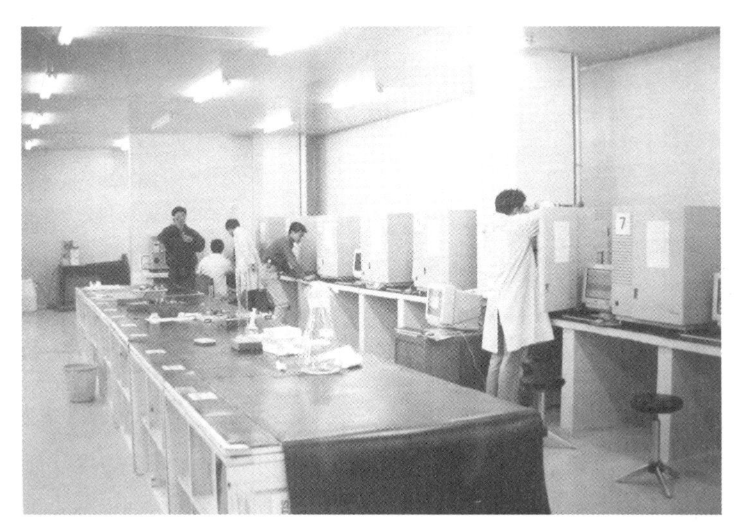

图2 “1%项目”的实验室

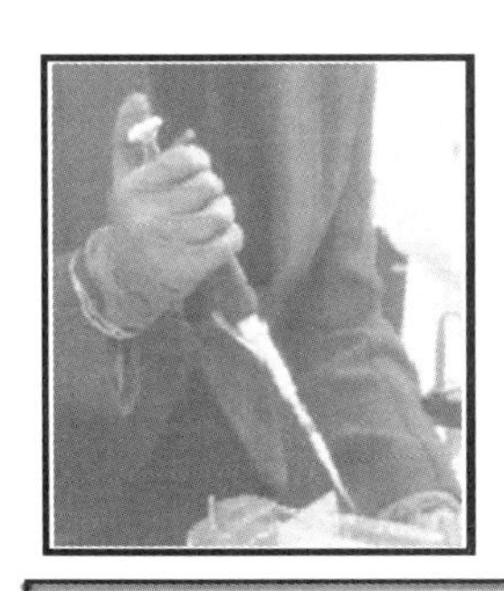

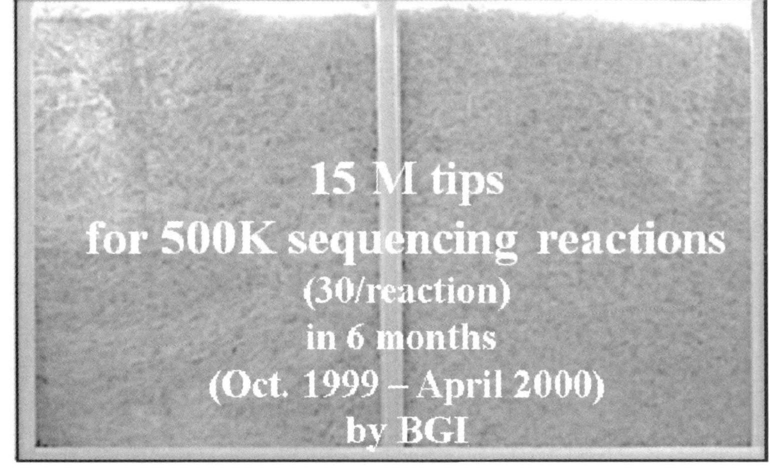

图3 “1%项目”用过的吸枪头样本小样

经过CHGC全体成员的共同努力，2000年6月26日庆祝“人类基因组序列草图”发表之时，中国一共为人类基因组序列的草图递交了64 Mb（百万碱基对）的原始序列数据。2004年10月“人类基因组序列精细图谱”在《自然》杂志发表之时（International Human Genome Sequencing Consortium，2004），中国递交了38 Mb的完成图序列数据。“北京区域”至今仍是人类基因组“参照序列”中最为准确的基因组区域之一（图4）。由CHGC参与完成的人类基因组3号染色体基因组测序和分析论文于2006年4月发表在《自然》杂志上（Muzny et al.，2006）。

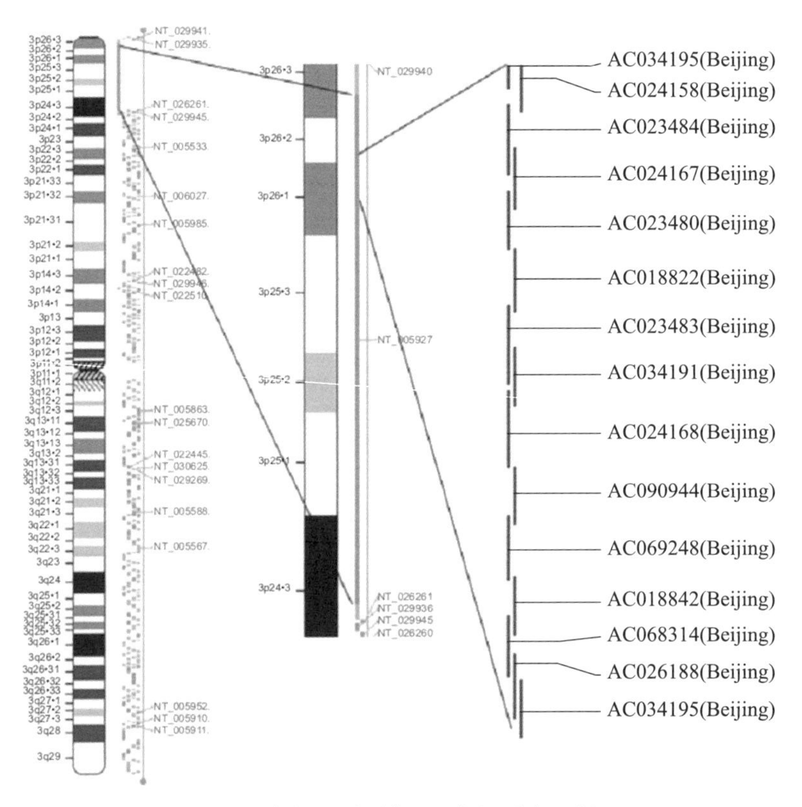

图4　人类基因组序列精细图谱的“北京区域”

2003年4月14日，HGP的6个成员国领导人联合签署了“人类基因组计划宣言”（图5），标志着HGP最终落下帷幕。时任美国总统、英国首相以及其他参与国领导人都对中国科学家所作的努力和贡献表示充分的认同和衷心的感谢。

中国的参与不仅提高了HGP的国际代表性，使HGP成为人类自然科学史上第一个由发达国家和发展中国家一起参与的国际合作科研计划，也是中国生命科学和生物产业走向国际科技大舞台的又一重要转折点。

除了科学贡献外，中国科学家还积极参加“人类基因组属于全人类”的有关讨论，并促成了2000年5月7日联合国教科文组织（UNESCO）关于“支持人类基因组数据免费分享”的声明。诺贝尔奖获得者、英国HGP负责人John Sulston曾赞扬说：“我要特别感谢中国的同事，因为他们不仅对国际合作的HGP，而且对保证人类基因组属于全人类作出了重要贡献。”

中国参与HGP这一历史事件已被铸刻在北京中华世纪坛的青铜甬道之上。

440 *Apr. 14 / Administration of George W. Bush, 2003*

Joint Proclamation by the Heads of Government of Six Countries Regarding the Completion of the Human Genome Sequence

April 14, 2003

We, the Heads of Government of the United States of America, the United Kingdom, Japan, France, Germany, and China, are proud to announce that scientists from our six countries have completed the essential sequence of three billion base pairs of DNA of the human genome, the molecular instruction book of human life.

Remarkable advances in genetic science and technology have been made in the five decades since the landmark discovery of the double-helix structure of DNA in April 1953. Now, in the very month and year of the 50th anniversary of that important discovery by Watson and Crick, the International Human Genome Sequencing Consortium has completed decoding all the chapters of the instruction book of human life. This information is now freely available to the world without constraints via public databases on the World Wide Web.

This genetic sequence provides us with the fundamental platform for understanding ourselves, from which revolutionary progress will be made in biomedical sciences and in the health and welfare of humankind. Thus, we take today an important step toward establishing a healthier future for all the peoples of the globe, for whom the human genome serves as a common inheritance.

We congratulate all the people who participated in this project on their creativity and dedication. Their outstanding work will be noted in the history of science and technology, and as well in the history of humankind, as a landmark achievement.

We encourage the world to celebrate the scientific achievement of completing the Human Genome Project, and we exhort the scientific and medical communities to rededicate themselves to the utilization of these new discoveries to reduce human suffering.

His Excellency
Jacques Chirac,
President of the French Republic.

The Honorable
George Bush,
President of the United States of America.

The Right Honorable
Tony Blair, M.P.,
Prime Minister of the United Kingdom.

His Excellency
Gerhard Schroeder,
Chancellor of the Federal Republic of Germany.

His Excellency
Junichiro Koizumi,
Prime Minister of Japan.

His Excellency
WEN Jiabao,
Premier of the State Council of the People's Republic of China.

NOTE: An original was not available for verification of the content of this joint proclamation.

图5 人类基因组计划宣言

中国参与HGP的意义

近20年前落下帷幕的HGP对生命科学和生物产业的影响随处可见，并会随着时间的推移越发显现。

HGP培育了一个新学科——基因组学，它把生命科学几乎所有学科都“组化”和“组学化”了，例如蛋白质组学、外饰基因组学、癌症基因组学、表型组学等。基因组学以“生命是序列的”和“生命是数字的”的理念为基石和支柱，扩展到所有生物表型的数字化，为生命科学的新阶段奠定了生物信息学基础。

HGP催生了一门新技术——测序，这一将生命“序列化”和“数字化”的过程，彻底改变了21世纪的生物学和医学。

最重要的是，HGP倡导了一种新的文化——合作的文化，“共需、共有、共为、共享”的“HGP精神”正是由中国科学家提出，并得到国际协作组的广泛支持和一致认可。HGP为其他影响人类进步的科学发展计划建立了国际合作的范例。以此为例，随后的“国际HapMap计划”“国际千人基因组计划”“国际癌症基因组计划”“全球生物基因组计划”以及其他国际合作项目陆续启动。

HGP与很多新兴技术，例如诱导多能干细胞(iPS)/干细胞和动物克隆、基因组编辑和合成、大数据、人工智能以及其他“组学”等，正在影响生命科学和生物技术的今天和未来，为建设美丽的生命世界和人类更好的未来带来新的机遇和信心。

87. The international Human Genome Project (HGP) and China's contribution

I can't imagine something like this: One day (the future which is now) a pupil asked such a question in class when I taught about the HGP, "My teacher, why didn't China do something for the HGP?" I would deeply regret for the rest of all my life if my response could only be the following: "My kids, how can you know, ah, that time, how poor our country was, how illiterate our people were, how short-sighted our decision-makers were, and how disappointing our scientists were, ... we just saw our great nation thus losing another historic opportunity without doing anything, ..." No, it will never come true.

—— A Chinese middle school teacher (1999)

The HGP is the first and greatest endeavor so far to understand ourselves and all other types of life through the vast international collaboration. It is widely acknowledged as one of the three most important projects in natural sciences of the 20th Century (Lambright, 2002). It is also generally accepted by the scientific community as "the 2nd revolution" in life sciences, following "the 1st revolution" discovering DNA double helix (Sharp, 2014).

The initiation and globalization of the international HGP

The worldwide discussion about the significance and feasibility, as well as the relevant ethical and social issues, of the HGP began at the beginning of the 1980s. The first official meeting about the HGP was held in 1984 in Utah, USA. The USA became the first country to initiate the HGP in October, 1990.

The HGP was joined by the UK, France, Germany, and Japan in 1996 at the 1st "International Strategic Meeting on Human Genome Sequencing" (1st ISMHGS, also called "Bermuda Meeting"), in Bermuda, proposed and organized by Dr. Michael Morgan, Director of Wellcome Trust, UK. The HGP was further globalized by China's participation at the 5th

Xiaoling Wang, Zhi Xia, Chao Chen, Huanming Yang
BGI-China, Shenzhen 518083, China
Correspondence: yanghuanming@genomics.cn (H.-M. Yang)

Strategic Meeting in Hinxton, UK, after a cautious discussion and serious defense, and became the "latest contributor" of the HGP and the only developing country in the International Human Genome Sequencing Consortium.

China's contribution to the HGP

China was committed to sequencing, assembly, and analysis of a region of approximately 30 cM (centimorgan, a unit of "genetic unit" measuring the size of a genome) on the tip of the short arm of chromosome 3, which was then estimated to account for about 1% of the entire human genome, thus called the "1% Project" , or the "Chinese Chapter of the Human Genome Sequence" , or the "Beijing Region" of the human genome because all of the sequenced overlapping BAC (Bacterial Artificial Chromosome) clones are labeled "Beijing" .

The debates on the HGP and China's participation were no less than those in the US and other countries since the very beginning. We can now still read from the publications at that time, such as "(the HGP) has been 'cut as a watermelon' by big labs and monopolized by a few big companies, we should not be involved in the competition" , as well as "the human genome sequencing is a pseudoscience, ... those people arrange the apples on ground and try to convince us it is the sequence of apples on the tree" and so on.

The commitment to the HGP means that China had to complete 500000 successful Sanger sequencing reactions within six months between October, 1999 and March, 2000. The "1% Project" was officially sponsored by the 863 Project of the Chinese Ministry of Science and Technology (MOST), and followed by China Natural Science Foundation (NSFC) and Chinese Academy of Sciences (CAS). The Chinese Human Genome Consortium (CHGC) was composed of 15 teams from the Northern Center of National Human Genome Center (Beijing), the Southern Center of National Human Genome Center (Shanghai), the Human Genomics Center of the Institute of Genetics, Chinese Academy of Sciences, and BGI, as well as Xi'an Jiaotong University, Southeast China University, and other institutions. BGI was responsible for providing all BAC clones after careful selection of the "seed clones" by means of *in situ* hybridization and other mapping methods, remapping and confirmation of all "extended clones" for all other participating teams, as well as the assembly and filling in both "inter-clone gaps" and "inner-clone gaps" through bioinformatic tools and other techniques.

Technically, the "1% Project" began almost from "Ground Zero" . Bearing the "for glary of the motherland" and "proving ourselves that we can do others are able to do" in minds, all the participants have a good mastery of all required skills after intensive and strict training in a short time. They developed a pipeline "with Chinese characteristics" both efficient and economic. They even made the working benches with packing boxes, to overcome the difficulties of insufficient funding without any intention to show off. We can also see many pictures showing how they were working day and night those days. Just take it as an example, more than 15 million of pipetting tips of various size have been used up for the project by means of the technologies that time.

Finally, through the efforts of all teams, the CHGC submitted 64 Mb (a million of base pairs) raw data for the human genome "draft sequence" before the joint celebration of the "Human Genome Draft Sequence" on June 26, 2000, and 38 Mb data of "finished sequence" without even a single "gap" in the whole region for the paper of the human genome fine sequence published in *Nature* in October, 2004 (International Human Genome Sequencing Consortium, 2004). The fine sequence of the human chromosome 3 was finally published in *Nature* in April, 2006 (Muzny et al., 2006).

The international HGP was officially closed with the "Joint Proclamation by the Heads of Government of Six Countries Regarding the Completion of the Human Genome Sequence" on April 14, 2003. The former US president and UK prime minister, as well as other state heads, have fully acknowledged the contribution by the Chinese scientists on different occasions.

China's participation not only improves the international representation of the HGP and made it the first vast internationally collaborative project joined by both developed and developing countries in history, but also marks an important starting point for Chinese scientists to play their important roles on the international science stage as they are doing now.

In addition to the substantial scientific contribution, Chinese scientists made irreplaceable contribution to the international supports by making the UNESCO statement on May 7, 2000 published supporting the free-sharing principle of human genome sequences, as well as many other efforts. Sir John Sulston, a Nobel Prize Laureate and the British HGP leader, said, "I especially salute the Chinese colleagues, who have contributed so much to the international genome effort... and affirmed its common ownership by all humankind."

The contribution by China to the HGP is carved on the bronze road of the Chinese Millennium Monument, as one of the important events in Chinese history.

The impacts of the HGP and China's participation

The HGP came to an end about one decade ago, its impacts on life sciences and bioeconomy can be seen everywhere, and will surely be felt with time going on.

The HGP has cultivated a new field, GENOMICS, which has made almost all branches of the life sciences "omicized" , e.g., proteomics, epigenomics, canceromics, phenomics, and numerous others. Genomics, with "life is in sequences" and "life is digital" as its belief and pillars, is extended to make all phenotypes digitalized, laying foundation for bioinformatics in a new phase of life sciences.

The HGP has fueled the development of a technology, SEQUENCING, which has made life "sequencized" and "digitalized" , thus revolutionized biology and medicine for the 21st Century.

More importantly, the HGP has nurtured a CULTURE of collaboration under the HGP Spirit: "Owned by All, Done by All, and Shared by All" , as proposed by the Chinese scientists and endorsed by the HGP community. The HGP has established a brilliant example of global collaboration for the sake of humanity and scientific advancement, and followed by the International HapMap Project, the International 1000 Genome Project, the International Cancer Genome Project, the Global BioGenome Project, and dozen of other internationally collaborative projects going on now.

The HGP, together with other newly emerging technologies, such as iPS/stem cell and animal cloning, genome editing and writing (synthetic genomics), big data, deep-learning and AI (Artifical Intelligence), as well as other-omics, has "changed biology and biotech forever" and has brought us a new opportunity for building a more beautiful life world and our confidence of a better future of mankind.

Figures

Fig.1 The International Human Genome Sequencing Consortium at the 5th Strategic Meeting, Aug. 31, 1999, Hinxton, UK

Fig.2 A lab for the "1% Project"

Fig.3 A small sampling of used tips for the "1% Project"

Fig.4 The "Beijing Region" of the fine map of the human genome sequence

Fig.5 Joint Proclamation by the Heads of Government of Six Countries Regarding the Completion of the Human Genome Project

References

Administration of George W. Bush (2003) Weekly compilation of presidential documents.

International Human Genome Sequencing Consortium (2004) Finishing the euchromatic sequence of the human genome. Nature 431(21):931–945.

Lambright W (2002) Managing "big science" : A case study of the Human Genome Project. CEPA, Syracuse.

Muzny DM, Scherer SR, Wang J, et al (2006) The DNA sequence, annotation and analysis of human chromosome 3. Nature 440(7088):1194–1198.

Sharp P (2014) 2014 AAAS annual meeting.

88. 20世纪40年代初在昆明进行的青霉素试验生产

1928年，Alexander Fleming划时代地发现了青霉素。大约10年后，由Howard Florey爵士、Ernst Chain教授和他们的助手等组成的“牛津小组”在青霉素的生产尤其是提取、纯化和鉴定等方面进行了详细研究并取得了突破性的进展，他们的研究结果发表在1941年8月16日的《柳叶刀》杂志。

为了躲避第二次世界大战期间德国空军的轰炸，Howard Florey爵士前往美国寻求合作以大规模生产青霉素。美国政府意识到青霉素会在战争中发挥重要作用，于是美国医学研究委员会组织美国的研究所（例如北方地区研究所）和制药公司（例如默克、施贵宝、辉瑞）开展大规模工业化生产青霉素的研究。1941—1944年，在战时生产委员会的支持下，青霉素产业化生产获得成功，从二战后期开始成为军队和家庭日常临床用药。

我国的科学家在看到“牛津小组”关于青霉素的研究进展后，对这个神奇的药物充满了兴趣并决定尝试自己生产。作为国家级的生物制品研制单位，中央防疫处承担起了研制青霉素的任务，并取得了可喜的进展。

中央防疫处是我国主要的防止传染病流行的研究机构，1919年在北京成立并于1935年迁往南京。1937年抗日战争全面爆发后，中央防疫处被迫于1938年迁往长沙，我国著名的微生物学家汤飞凡先生临危受命担任处长，随后又被迫迁往昆明。在汤飞凡先生的领导下，中央防疫处在战争时期艰苦的条件下重新建立了实验室并邀请了多名科研和技术人员加入，形成了初具规模的微生物学和免疫学研究小组。

在昆明，汤飞凡先生组织研究小组每周研读和学习国外发表的科研论文。在1941年秋的一次文献学习时，他们读到了“牛津小组”关于青霉素的研究论文，并对青霉素产生了极大的兴趣，决心开始自己的青霉素研制工作。

经典的青霉素研制步骤主要分成两部分。第一部分是要选取合适菌株。从1941年到1945年，他们共获得了30个本地菌株和10个外国菌株，其中包括2株当时已经应用于美国大规模工业化生产青霉素的菌株NRRL 832和NRRL 1229-B21。经过对比，由我国科学家自己分离的本地菌株第22号产生青霉素的效价最高，随后的实验也采用此菌

译者：程浩

中国科学院北京生命科学研究院，北京100101，中国

邮箱：chengh@biols.ac.cn

株继续进行（图1）。他们在自己研制青霉素的同时，也一直关注国外的研究进展并与留美的青霉素研究学者童村先生保持密切联系。

图1　中央防疫处生产青霉素

第二部分就是进行抗生素的提取纯化，由汤飞凡先生和朱既明先生、黄有为先生、樊庆笙先生等科学家完成了最后关键的步骤。中央防疫处采用的主要方法和规模不但参考了“牛津小组”的经验，还加入了他们自己的创新设计。例如，他们采用萨氏培养基进行菌种培养，然后将菌种的孢子置于有花生油和含锌离子的玉米浆培养基，其中将花生油加入玉米浆培养基是他们的首创，而采用含锌离子的玉米浆培养基则是借鉴了美国大规模产业化生产青霉素的方法。

不仅如此，因为战时的物质条件太艰苦，常用的仪器和设备极为缺乏，在各种困难面前，我国的实验人员因地制宜地进行了提取和纯化方法的简化和改进。例如，常用的浓缩步骤所需的氧化铝吸附柱当时难以获得，中央防疫处的研究人员就简化了浓缩步骤，采用两次乙酸戊酯加两次乙醚的方法提取青霉素，最开始甚至连乙酸戊酯也没有，只能用乙酸乙酯来代替。在青霉素生产不同阶段的效价检测时，由于自制的牛津小杯工艺精度不能满足要求，无法采用杯碟法，实验人员只好采用自己改良的孔碟法进行测定。如果没有冷冻干燥机，辛苦纯化的青霉素无法从溶液中提取，就无法制成产品。黄有为先生自力更生设计了一个简易的低温冷冻干燥装置，并由当地工厂制造，很幸运的是这台简易的设备正好可以完成最后的干燥纯化工作。

1944年9月5日，第一批共五瓶青霉素粗制品终于制作完成，每瓶含青霉素5000牛津单位，总量约为140毫克。他们送了2瓶分别到英国牛津大学（由Norman Heatley进行检测）和美国威斯康星大学进行检验，英美双方都给予肯定和鼓励。第二批青霉素制剂被送到昆明的惠滇医院和昆华医院用于临床试验。1944—1945年，中央防疫处仍然继续改进青霉素的生产方法并制备小量的青霉素制剂。

正是由于具有上述的研究工作基础，美国医药助华会在1945年向中央防疫处捐赠了一套发酵生产设备，并于抗日战争胜利后运至北京安装。这套小型设备在1947年生产了新一批的青霉素。这里也成为我国抗生素工业先驱和专家的摇篮。

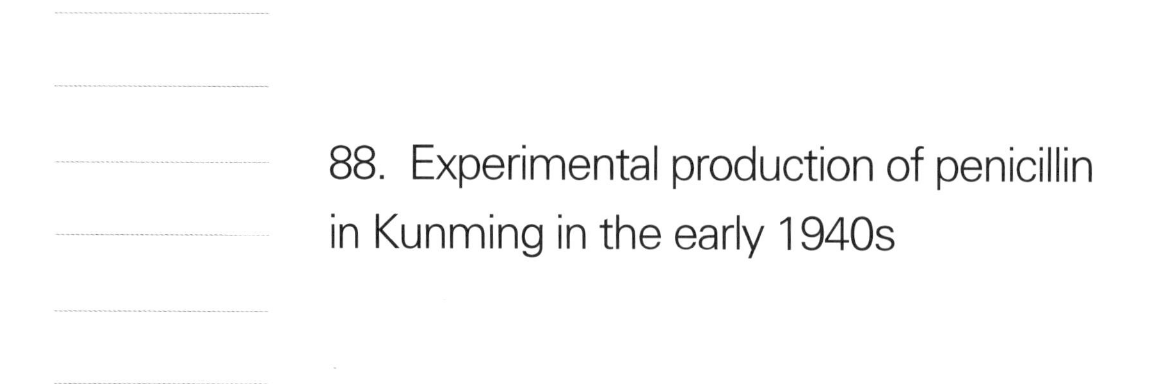

88. Experimental production of penicillin in Kunming in the early 1940s

In 1928, Sir Alexander Fleming made the profound discovery of penicillin. After approximately 10 years, the Oxford team, including Sir Howard Florey, Professor Ernst Chain, and their assistants, made progress in the methods of penicillin production, especially in extraction, purification, and assay. They then published their research in detail in *The Lancet* on August 16, 1941.

Threatened by the Luftwaffe bombings, Florey came to the United States to seek opportunities for the mass production of penicillin. Aware of the possible massive usage of penicillin during the war, the Committee of Medical Research of the U.S. government organized research institutes (e.g., North Region Research Laboratory) and pharmaceutical corporations (e.g., Merck, Squibb, and Pfizer) to study and produce penicillin in an industrial scale. From 1941 to 1944, and supported by the War Production Board, the industrialization was completed, and penicillin became a common antibiotic applied to both military and everyday life during the latter half of the Second World War.

After learning about the achievement of the Oxford team, Chinese scientists became interested in this magical drug and attempted to produce it themselves. The major research institute designated for this purpose is the National Epidemic Prevention Bureau (NEPB), which generated successful results.

As the chief institute for the prevention of infectious diseases and the production of biological products in China, the NEPB was established in Peking in 1919 and was moved to Nanking in 1935. After full outbreak of the War of Resistance against Japanese Aggression in 1937, the Bureau evacuated to 1938. In Changsha, renowned Chinese microbiologist Tang Feifan was appointed the director of the Bureau. Tang transferred the Bureau to Kunming in 1939. Led by Tang Feifan, the NEPB built a new laboratory and recruited new staff members, including scientists and technicians. Thus, a small group was formed for microbiological and immunological research even in the rear area of the war.

Dingding Xu

The Institute for the History of Natural Sciences, Chinese Academy of Sciences, Beijing 100190, China

Contact: xudd@ihns.ac.cn

In Kunming, Tang suggested to set up weekly seminars on thesis presentation from foreign academic journals. In one of these seminars in the autumn of 1941, they read about the Oxford team's work, which motivated them to manufacture penicillin by themselves.

According to history, the entire experimental production of penicillin comprised two steps. The first step involved finding and selecting the appropriate Penicillium strain. From the winter of 1941 to 1945, they compared 30 local strains and more than 10 foreign strains, including the NRRL 832 and NRRL 1229-B21. These strains were used in mass production in the United States at the time. The Chinese scientists found that the local No. 22 was the best strain and used it in their subsequent experiments. At the same time, they continued to focus on foreign thesis and maintained communication with Dr. Tung Tsun, who was studying penicillin in the United States at the time.

The second step involved the extraction and purification of penicillin, which was completed mainly by Chu Chi-Ming and Wong You-Wei from the NEPB, Fan Ching-Seng from the Chinese Blood Bank, and Tang Feifan himself. As for the major technical flow and the size of experiment, the work of the NEPB team was similar to that of the Oxford team, but the NEPB researchers adopted some developed technologies. For example, to grow the mould, they used Sabouraud medium to cultivate spore and inoculated it with peanut oil to the corn steep liquor medium containing traces of Zinc ion. As an effective improvement, they invented the oil inoculation method, whereas the corn steep liquor and the addition of Zinc ion were based on the mass production procedure in the United States.

However, due to the extreme lack of the necessary apparatus and equipment, they encountered obstacles, which compelled them to adjust or simplify the methods based on local conditions, especially in production. For instance, given the shortage of adsorption column of alumina used for the concentration of penicillin, the NEPB researchers simplified the extraction method. They used two times amyl acetate and two times ether to extract penicillin, and adopted ethyl acetate instead of amyl acetate due to the unavailability of such material at the beginning. They modified the agar cup method of qualitative assay to replace the cylinder plate method used for quality test due to the inadequate precision of self-made cylinder when they attempted to imitate the cylinder cup method to assay titration of penicillin at different stages. Without a freeze dryer, the production process had to be suspended at the end of purification despite the achievement of the extraction solution. Thus, Wong You-Wei designed a simple instrument and produced it in a factory. Finally and fortunately, it fulfilled the requirements.

The first batch of five ampoules of penicillin was produced on September 5, 1944. Each bottle contained 5000 Oxford units, which the total weight of penicillin was approximately 140 milligrams. They sent one sample to Oxford University (for examination by Norman Heatley) and another to Wisconsin University. Both institutions sent their congratulations and encouragement. The second batch of production in clinical tests generated good results in Huidian Hospital and Kunhua Hospital in Kunming. From 1944 to 1945, the NEPB continued to produce small amounts of penicillin with some improvements. Given their performance, a small suit of equipment from the American Bureau for Medical Aids to China was sent to them in the spring of 1945, which was installed in Peking after the victory of the War of Resistance against Japanese Aggression. This pilot plant, which probably produced the first batch of penicillin in 1947, became the cradle of pioneers and experts of the antibiotic industry for the People's Republic of China.

Figures

Fig.1 Manufacture of penicillin at the NEPB

References

Chu CM, Wong YW, Fan CC, et al (1945) Experimental production of penicillin in China. Chinese Med J 64: 89–101.

Ligon BL (2004) Penicillin: Its discovery and early development. Semin Pediatr Infect Dis 15: 52–57.

Richards AN (1964) Production of penicillin in the United States (1941—1946). Nature 201: 441–445.

Tang FF (1949) The retrospect and prospect of self-production of penicillin in China. Sci World 18: 3–5. (汤飞凡. 1949. 吾国自制青霉素的回顾与前瞻. 科学世界, 18: 3–5.)

89. 青蒿素背后的故事

2011年9月，屠呦呦教授（中国中医科学院，北京）荣获2011年度拉斯克–狄贝基临床医学研究奖，以表彰她在青蒿素的发现及应用于疟疾治疗方面所作出的杰出贡献（Zhang，2011）。

疟疾是一种通过蚊虫传播的寄生虫病，根据世界卫生组织发布的《2010年世界疟疾报告》，2009年，全球共有2.25亿疟疾病例，78.1万人死亡。20世纪50—60年代，由于抗药性的产生，疟疾的治疗面临更严峻的挑战，研制新的抗疟疾药物已是刻不容缓。1967年5月23日，我国政府启动了旨在对抗疟疾的“523项目”。这个项目历经十多年，全国有60多个研究机构、500余名不同学科领域的研究人员共同参与。每位科学家都对青蒿素的发现默默作出贡献。有人负责青蒿素的分离纯化，有人负责测定青蒿素的分子结构，还有人负责合成青蒿素的衍生物，而另外一些人负责将青蒿素及其衍生物应用于临床试验。这是科技部有史以来规模最大的团队项目之一，而梁丽和李鹏飞（已故）则非常荣幸地参与其中。

在1975年，中国中医科学院中药研究所的屠呦呦及其同事与梁丽和李鹏飞取得了联系。那时，屠呦呦及其同事已经分离和结晶了青蒿素，并已推测了其化学成分，也试图在中国科学院上海有机化学研究所专家的帮助下，通过质谱技术来探明青蒿素的分子结构。然而，在对分子结构和确切构象一无所知的前提下，青蒿素的进一步功能研究举步维艰。因此，中药研究所向曾利用X射线晶体学方法成功测定胰岛素三维结构的中国科学院生物物理研究所求助。生物物理研究所两位年轻的科学家——梁丽（图1）和李鹏飞承担了测定青蒿素结构的重任。

利用生物物理研究所的四圆X射线衍射仪，实验人员测得了一组青蒿素晶体的衍射强度数据。下一步将通过这些衍射数据进一步解析其分子结构。李鹏飞和梁丽都想到了直接法——一种基于概率关系从衍射强度数据中获取相位数据的数学方法。这种方法对于青蒿素这类小分子的结构测定是非常有效的。它最初是由H. A. Hauptman和Jerome Karle提出的，他们也因此在1985年获得了诺贝尔化学奖。

译者：郝宁

中国科学院生物物理研究所，北京100101，中国

邮箱：haoning@ibp.ac.cn

图1　梁丽博士1981年在纽约州立大学水牛城分校Philip Coppern实验室做访问学者。背景是一台四圆X射线衍射仪

到1975年为止，直接法已经在西方被应用于X射线晶体学超过10年，但还没有被引入中国。在没有任何实际经验且缺乏技术支持的情况下，梁丽和李鹏飞不得不一切从零开始，在参考直接法相关文献的基础上，独立编写算法和程序。非常幸运的是，虽然机时很紧张，但是他们还是得到了使用北京计算中心计算机的授权。他们的上机时间一般被安排在午夜。每个夜晚，他们都乘最后一班公交车前往计算中心，第二天清晨，再乘最早一班公交车回到生物物理研究所。不同于现在广泛使用的电脑，当时使用的TQ-16计算机是我国早期的一种初级计算机，他们不得不使用穿孔纸带将程序传输到计算机上，巨大的工作量对于现在已经习惯了21世纪电脑技术的我们来说是难以想象的。繁重的工作加上肾病旧疾，李鹏飞终因过度劳累、不堪重负，于1976年初春被送进了医院。梁丽独自继续调试程序。在计算中心度过了又一个漫长的黑夜后，梁丽最终完成了直接法程序的纠错和试算工作。第二天清晨，她迫不及待地奔向医院，想告诉李鹏飞这个令人兴奋的消息，却被告知，李鹏飞刚刚在数小时前过世了！李鹏飞未能亲眼见证他和梁丽设计的程序成功解析青蒿素分子结构就英年早逝，令人唏嘘。而这也是我国第一个用直接法解析的三维晶体结构！

最初，青蒿素试用结构如同其他测定的晶体结构，*R*-因子（晶体学中评估结构模型与X射线衍射数据符合程度的数值）仍是20%左右，与国际上可接受的标准相差甚远。因此，梁丽参阅文献，将最小二乘法应用到该计算程序中，以修正结构模型。她忘我地沉浸在工作中，以至于蹒跚学步的儿子都知道“最小二乘”这个名词——因为妈妈在睡梦中都会念叨。1977年，她最终完成了青蒿素结构（图2）的精确修正，成功地将*R*-因子降低到了5%（中国科学院生物物理研究所青蒿素协作组，1979），突破了我国晶体结构测定的最高精度，达到当时的国际水平。青蒿素的晶体结构及其化学绝对构型为世界卫生组织认可青蒿素打下了基础，并获得了1979年国家发明奖二等奖。

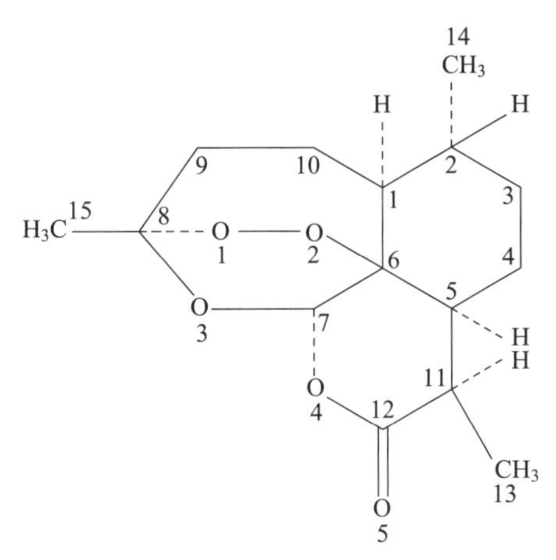

图2　青蒿素分子结构（中国科学院生物物理研究所青蒿素协作组，1979）

文章一开始就提到，梁丽和李鹏飞以及他们所做的工作，只是庞大的“523项目”中的一小部分，整个项目包含了500多位科学家废寝忘食的不懈努力和付出。这部分地解释了为什么这样一个伟大的发现能够在当时的艰苦条件下诞生。另一个原因则应该归于政府的大力支持。首先，“523项目”是由周恩来总理亲自负责监督的；同时，虽然20世纪60、70年代的中国，物资非常匮乏，但政府依然为研究机构提供了足够的资金，以资助它们购买先进设备，例如生物物理研究所购买的四圆X射线衍射仪。而我国在过去几十年中对科研经费的支持迅速增长，我们现在已经拥有比以往任何时候数量都多的科研工作者。鉴往知来，我们完全有理由展望并期许，在不久的将来，我国将会产生更多的重大科学突破！

致谢

感谢梁丽博士通过电话及E-mail为本文提供背景故事。

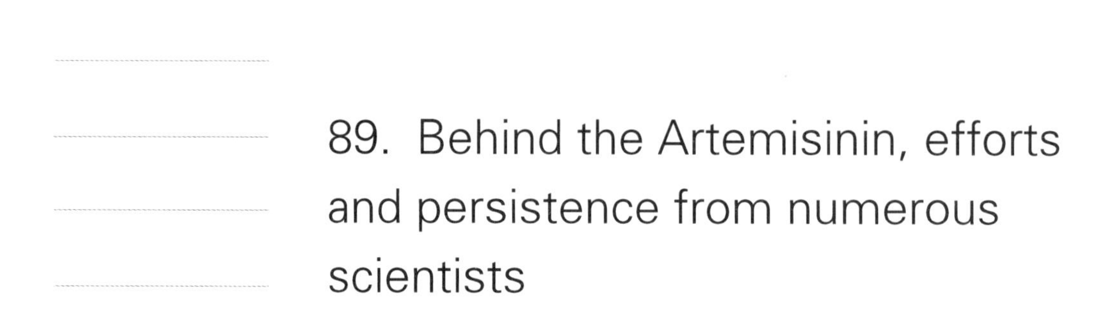

89. Behind the Artemisinin, efforts and persistence from numerous scientists

Professor Youyou Tu (China Academy of Chinese Medical Sciences, Beijing) was awarded the 2011 Lasker-DeBakey Clinical Medical Research Award for her contribution toward the discovery of Artemisinin, also known as Qinghaosu (Chinese: 青蒿素) and its utility for treating malaria (Zhang, 2011) in September 2011.

Malaria, a mosquito-borne parasitic disease, caused an estimated 225 million cases of infection and over 781000 deaths in 2009 according to the 2010 World Health Organization (WHO) *Malaria Report*. Back in 1950s and 1960s, the treatment of malaria became more challenging due to drug resistance, demanding urgent need for new antimalaria medicines. On May 23, 1967, the Chinese government launched a national project, Project 523, to fight against malaria. The project lasted over ten years and involved more than five hundred scientists of various disciplines from around sixty institutions throughout the country. Each of the scientists contributed to the discovery of Artemisinin in their own way, some isolating Artemisinin, some determining the molecular structure of Artemisinin, some synthesizing the derivatives of Artemisinin, while the others applying Artemisinin and its derivatives in clinical trials. It was one of the biggest team projects in China's S&T history ever, and Li Liang and Pengfei Li (deceased) were proud to be part of the team.

It was in 1975 when Li Liang and Pengfei Li were first contacted by Youyou Tu and her colleagues from the Institute of Chinese Materia Medica (ICMM). By then, Tu and her colleagues had isolated and crystallized Artemisinin, and had deduced the chemical composition and tried to elucidate its structure by Mass Spectrometry with aid from specialists of Shanghai Institute of Organic Chemistry (SIOC), Chinese Academy of Sciences (CAS). However, without knowledge on its molecular structure and absolute configuration, it was difficult to advance the functional study of Artemisinin. Therefore, ICMM asked for help from the Institute of Biophysics (IBP), CAS, an institute experienced in X-ray crystallographic determination of 3-dimensional Insulin structure. Two young scientists at IBP, Li Liang and Pengfei Li, took on the mission of determining the

Ming Li
Beijing Institutes of Life Science, Chinese Academy of Sciences, Beijing 100101, China
Contact: liming@ioz.ac.cn

structure of Artemisinin.

A set of diffraction intensities of the Artemisinin crystal had been collected from a 4-circle X-ray Diffractometer at IBP. The next step would be solving the molecular structure with the diffraction intensities, and Li and Liang both thought of the Direct Methods, a mathematical approach from probability relationships to glean phase data from the diffraction intensities. It is a perfect approach for structural determination of small molecules like Artemisinin. This method was originally developed in the 1950s by Herbert A. Hauptman and Jerome Karle who were awarded the Nobel Prize for Chemistry in 1985 for their work.

By 1975, the Direct Methods had been applied in X-ray crystallography in the Western countries for over a decade, but had not yet been brought into China. With no prior experience and little technical support, Liang and Li had to start from point zero, i.e., writing their own algorithms and computer programs based on the Direct Methods published in literature. Fortunately enough, they had access to the computers in the Beijing Computer Center despite of their busy schedule. Their computing times were routinely scheduled in the middle of the night. Every night, they would take the last bus to the Computer Center and then take the earliest bus back to IBP the next morning. Unlike the computers available nowadays, they had to use the punched paper tapes to feed their programs into the computer TQ-16, an early Chinese Minicomputer Mode. The workload was unimaginable to those of us who are accustomed to computer technology in the 21st century. With the heavy workload and his pre-existing kidney disease, Li finally succumbed to overfatigue and was hospitalized in the early spring of 1976. Liang continued working on debugging the programs. After another long night at the Computer Center, Liang finally finished debugging the Direct Methods program. The next morning, she rushed over to tell Li the good news, but was told that Li had just passed away hours before. It was regrettable that Li did not get to see the Artemisinin molecular model determined with his and Liang's programs. This was the very first crystal structure determined by the Direct Methods in China.

The *R*-factor (a measure of the agreement between the structural model and the experimental X-ray diffraction intensities in crystallography) of Artemisinin structure at the time was around 20%, far from the acceptable international standard for small molecule structures. Thus Liang introduced and applied the advanced Least-Square approach to refine the structural model by her programs. She was so immerged in her work that even her toddler son knew the term "Least-Square" because "Mommy mumbled it in her sleep" . Finally, in 1977, she completed the structural determination of Artemisinin with the *R*-factor down to 5% (Qinghaosu Coordinating Research Group at the Institute of Biophysics, 1979), which was the first crystal structure with such a low *R*-factor in China. The established crystal structure and chemical absolute configuration, which laid foundation for the recognition of Artemisinin by the WHO, won the National Technology Invention Award (second place) in 1979.

As mentioned at the beginning of this article, Liang and Li and their work were only a small part of Project 523, which included more than five hundred scientists with their diligent contribution. That partly explained why such an important discovery could be made during the difficult time. Another reason was the full support by the government. The Project 523 Office was under direct supervision of the late Prime Minister Enlai Zhou; and even though China was short of every kind of resources in 1960s and 1970s, the research institutes were provided with sufficient funding to purchase advanced equipments such as the 4-circle X-ray Diffractometer at IBP. Back to the present, China's R&D investment over the past decade has been tremendously increased, and China now has more scientific researchers than ever before. Taking all of these into account, we can expect more and more critical scientific breakthroughs in China in the near future.

Acknowledgements

The author is grateful to Dr. Li Liang who provided the majority of the story over phone interviews and E-mail communications.

Figures

Fig.1 Dr. Li Liang in Dr. Philip Coppern's Lab when she was a visiting scholar at the State University of New York at Buffalo in 1981. A 4-circle X-ray Diffractometer is shown in the background

Fig.2 Molecular structure of Artemisinin (Qinghaosu Coordinating Research Group at the Institute of Biophysics CAS, 1979)

References

Qinghaosu Coordinating Research Group at the Institute of Biophysics CAS (1979) Crystal structure and absolute configuration of Qinghaosu. Scientia Sinica 11: 1114–1128. (中国科学院生物物理研究所青蒿素协作组. 1979. 青蒿素的晶体结构及其绝对构型. 中国科学, 11: 1114–1128.)

Zhang X (2011) Youyou Tu honored by the Lasker award. Protein & Cell 2: 773.

90. 20世纪30年代中国描述生物学和实验生物学之间的一场难忘的辩论

20世纪初，随着新的实验方法的出现，遗传学、发育生物学和代谢组学等新兴学科的飞速发展，推动了生物学研究取得了巨大进步。与此同时，20世纪之前还没有出现在我国的现代生物学研究，在20世纪20年代到1937年开始进入繁荣时期。

在众多海外留学归来的年轻科学家的努力下，我国不少大学建立了生物系，最先建立的是东南大学生物系。1922年，中国科学社在南京建立了一个重要的生物学研究所——中国科学社生物学实验室。我国生物学的迅速发展引起了社会各界的关注，得到了诸多基金会的大力资助，例如中华教育文化基金会（简称中华基金会），这是一个1924年成立的机构，负责管理庚子赔款。到20世纪30年代，我国建立了四个生物学研究机构，主要从事分类学和形态学研究。随着在大学里不断扩增的标本收藏，分类学成为当时我国生物学研究最先进的学科。

当分类学在我国取得重大进展时，一些受过国外教育尤其是从事实验生物学研究的生物学家，开始分析我国生物学研究与西方国家之间的差异。其中最大胆直言的是汪敬熙博士，这位在美国留学的心理学家，最初受中山大学聘请回国，后来在北京大学任心理学教授。

从1932年开始，汪敬熙先生在《独立评论》上发表了一系列文章。这个期刊倡导的是言论自由，并为学者提供一个开放平台进行辩论。汪先生首先指出，西方国家生物学研究的发展趋势是用实验生物学（遗传学、生理学等）取代描述性和传统的生物学（分类学、形态学、动植物学等）。尽管我国生物学的发展才刚刚开始，但重要的是它应该朝着正确的方向发展（即实验生物学）。此外，他写道："与主要描述和命名有机体物种的分类学相比，基于实验的因果解释在理论和实践方面都被证明更有价值。"他还解释了为什么我国的实验生物学家无法像分类学家那样获得更多的经费支持——实验生物学的结果（数字、公式和图表）对于外行人来说似乎不如分类学（主要基于动物和植物标本）更吸引人，因此其价值被大众低估了。不过，他补充道："因为许多有才华的海外华人科学家回国，我国的实验生物学发展有着良好的开端，迫切需要经费支持来推动进一步的发展。"

译者：曲静
中国科学院北京生命科学研究院，北京100101，中国
邮箱：quj@biols.ac.cn

不出所料，一石激起千层浪，汪先生的文章引起了我国生物学家的广泛讨论。中国植物学家胡先骕博士，中国科学社生物学实验室联合创始人之一，是第一位反驳这一观点的学者。他的第一个论点是，尽管实验生物学是国际研究的主导模式，但在任何国家，分类学从未落后。像大多数西方国家一样，中国的科学发展不应该跳跃式地发展。尤其对于拥有很高物种多样性的中国来说，可信的生物学研究取决于准确的物种鉴定和可靠的科学命名。他还对为什么分类学得到了更多的经费支持表达了自己的观点：科学学科的发展主要依赖于科学家投入多少努力，社会团体倾向于支持更成功的学科是合理的。事实上，考虑到我国庞大的生物资源，生物分类学得到的经费仍然不足以支持研究本土的生物物种。

这场辩论从1932年到1936年一共持续了四年。不少生物学家参与了辩论，包括汪振儒和张作人。一些基金会出资人和外国顾问也参与了辩论。美国昆虫学家J. G. Needham认为，中国大学使用的大多数英文教科书里对植物学和动物学的描述都是基于北美和欧洲的自然环境，而不是中国自己的，而本地物种的鉴定和调查将为未来的研究提供一个坚实的基础。与此同时，还有其他声音指出，中华基金会在选择学科进行资助时过于单一，资助范围应从分类学扩展到所有生物学分支学科。

这次持久的辩论难分输赢。毕竟，辩论一直都是对事不对人。而且，辩论双方都有共同的目的——试图促进我国生物学研究的发展。不幸的是，那时的中国正处于非常不稳定的时期，对于踌躇满志的科学家来说，科研经费的支持远远不够。一个学科相对于其他学科占主导地位也并不是任何人的错，正是由于第二次世界大战前缺乏对科学政策的协调，才会导致这种不平衡。

您可能会问，辩论后发生了什么事？汪敬熙先生从来没有得到中华基金会的资助，他最终离开中国，自20世纪40年代起一直留在美国。他的同行，如遗传学家谈家桢先生和植物生理学家罗宗洛先生，尽管为了生存历经艰辛，却坚持留在了中国，并最终在他们自己的领域取得了重大成就。胡先骕先生也开始招募像遗传学家陈桢先生这样的非分类学家加入国立中央大学（现东南大学）的生物系。然而，这次辩论对中国生物学发展史的影响不应只看短期效应，而应该从长远来看。第二次世界大战后，许多从事不同领域研究的海外华人科学家回国，通过他们的培训，最终有更多的年轻科学家成为各个领域的专家。中华人民共和国成立后，我国大部分生物学部门的课程设计更加均衡合理，我国在遗传学、生物化学、生理学和生态学等实验生物学领域也取得了一定进展。特别是20世纪70年代以来，我国的生物学研究开始飞速发展，各学科取得了令人瞩目的成就。我们相信，像汪敬熙先生和胡先骕先生这样的前辈们，无论他们在辩论中站在哪一方，都会赞同和欣赏当今我国生物学的发展。

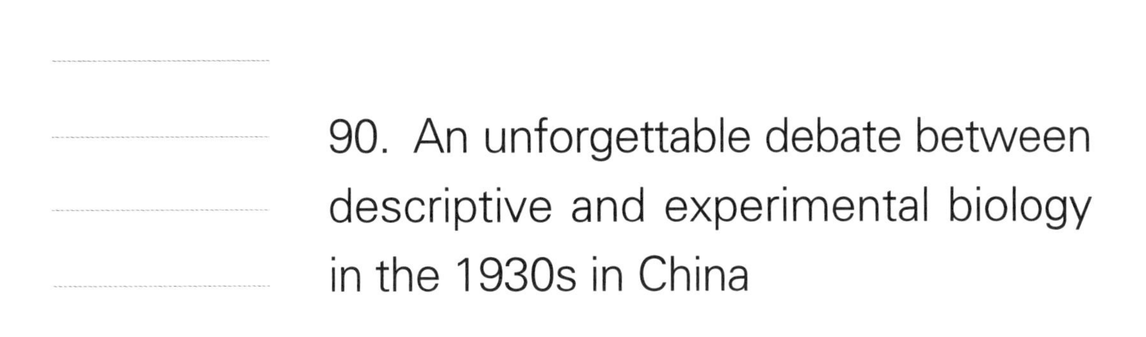

90. An unforgettable debate between descriptive and experimental biology in the 1930s in China

At the early 20th century, biological research was advancing with big steps, as new disciplines like genetics, development and metabolism had achieved tremendous success thanks to new experimental approaches. Meanwhile, China's biological research, which did not exist in many senses before the 20th century, was also going through a prosperous period between the 1920s and 1937.

With the efforts of numerous overseas-trained young scientists, quite a few Chinese universities set up biology departments, the first one being the Department of Biology at the Southeast University, Nanjing. In 1922, the Science Society of China established a major biological research institute in Nanjing—the Biological Laboratory of the Science Society of China (BLSSC). The rapid development of Chinese biology attracted attention from all sectors of the community, thus receiving major financial support from foundations such as the China Foundation for the Promotion of Education and Culture (the China Foundation in short), an organization set up in 1924 to manage the Boxer Indemnity Fund. By the 1930s, there have been four biological institutes in China, mainly working on taxonomy and morphology. Together with the ever-expanding specimen collections in universities, taxonomy was reportedly the most advanced discipline of Chinese biological research then.

When taxonomy was making major progress in China, some foreign-educated biologists, especially the ones working on experimental biology, started to sense the differences between Chinese biological research and that of Western countries. The most outspoken person was Dr. Jingxi Wang, an American-trained psychologist who was first recruited back by Sun Yat-sen University, Guangzhou, and later worked at Peking University as a professor of psychology.

Starting in 1932, Jingxi Wang published a series of articles in *Independent Review*, a journal advocating freedom of speech and providing an open platform for scholars to debate. Wang first pointed out that the trend of development of biological research in the Western

---

Ming Li[1], Zonggang Hu[2], Le Kang[1]

1 Beijing Institutes of Life Science, Chinese Academy of Sciences, Beijing 100101, China

2 Lushan Botanical Garden, Jiangxi Province and Chinese Academy of Sciences, Jiujiang 332900, China

Contact: lkang@ioz.ac.cn (L. Kang)

countries was replacing descriptive/classical biology (taxonomy, morphology, fauna and flora, etc.) with experimental biology (genetics, physiology, etc.). Even though Chinese biology was merely at the starting point, it was important that it should be steered toward the right direction (i.e. experimental biology). Furthermore, he wrote, experiment-based causal explanations were proven more valuable in both theoretical and practical aspects, compared with taxonomy, which mainly described and named organism species. He also explained why experimental biologists could not get as much funding as taxonomists in China—the results of experimental biology (numbers, formulas, and figures) seemed less interesting to laymen than those of taxonomy (mainly based on animal and plant specimens), thus were underappreciated by the community. However, he added, experimental biology in China had a good start as many talented overseas Chinese scientists in the area had been recruited back to China, and it was in urgent need of financial support to be pushed forward.

Unsurprisingly, Wang's articles caused controversial discussion among Chinese biologists. Dr. Xiansu Hu, a Chinese botanist who was the co-founder of the BLSSC, was the first one to refute. Hu's first argument was that even though the experimental biology was the dominant mode of research internationally, taxonomy was never falling behind in any countries. Instead of jumping with leaps, scientific development in China should take its course, like most Western countries. Especially for China, a country home to species diversity, credible biological research depended on accurate species identification and reliable scientific nominations. He also made comments about why taxonomy was getting more financial support: The development of a scientific discipline was mainly dependents on how much effort the scientists put into it, and it was reasonable that the community tended to support more successful disciplines. In fact, considering the vast biological resources in China, the funding for taxonomy was hardly sufficient to investigate local biological species.

The debate lasted four years, from 1932 to 1936. Quite a few biologists got involved in the debate, including Zhenru Wang and Zuoren Zhang. It also caused a debate among the benefactors and foreign advisors. The American entomologist J. G. Needham argued that most English-language botany and zoology textbooks used in Chinese universities were based on the natural environments of North America and Europe, not on that of China's own and that local species identification and investigation would provide a solid foundation for any further research. Meanwhile, there were other voices pointing out that the China Foundation was too singleminded in the way it chose one discipline over the others to support and that the scope of its beneficiary should expand from taxonomists to all biologists.

It was hard to say who won and who lost in the long debate. After all, the debate was never directed towards anyone involved personally; instead, both sides shared the same purpose—trying to promote the development of Chinese biological research. Unfortunately China was at a very unstable stage politically then and limited funding for scientific research was far from enough for ambitious scientists. It was not anyone's fault that one discipline was dominant over the others. It was the lack of coordination by the government's science policy that caused the imbalance among them before the Second World War.

You may ask what happened after the debate. Well, Jingxi Wang never got the funding from the China Foundation and he eventually left China to permanently stay in the United States since 1940s. His peers such as geneticist Jiazhen Tan and plant physiologist Zongluo Luo stayed in China despite the hard struggle they had to go through to survive, and finally both made major achievements in their own areas. And Xiansu Hu also started to recruit non-taxonomists like geneticist Zhen Chen to join his department at National Central University, Nanjing (formerly known as the Southeast University). However, the impact of the debate on the history of Chinese biology should be examined not in the short term but in the long run. After the Second World War, a large number of overseas Chinese scientists of many different disciplines returned to China and through their training, more young scientists became experts in various areas eventually. After the founding of the People's Republic of China, the curriculums of most biology departments in China were designed in a more balanced way; and Chinese experimental biology made some progress in genetics, biochemistry, physiology, ecology, etc. Especially since the 1970s, Chinese biological research started to advance at full speed and have made remarkable achievements in all disciplines. We believe that our predecessors like Jingxi Wang and Xiansu Hu would approve and appreciate today's Chinese biology, no matter which side they were on during the debate.

References

Hu ZG (2005) Unforgettable Xiansu Hu. Wuhan: Changjiang Literature and Art Publishing House. (胡宗刚. 2005. 不该遗忘的胡先骕. 武汉: 长江文艺出版社.)

Qian YQ, Wang YH (2004) Chinese Academic Canon in the 20th Century: Biology. Fuzhou: Fujian Education Press. (钱迎倩, 王亚辉. 2004. 20世纪中国学术大典: 生物学. 福州: 福建教育出版社.)

Schineider L (2005) Biology and Revolution in Twentieth-century China. Maryland: Rowman & Littlefield Publisher.

91. 1956年青岛遗传学座谈会：中国生物学的重要转折点

李森科主义，尽管现在大多数人已不知道了，但它曾经是20世纪中叶的热门词。许多年轻的中国人甚至从来没有听说过李森科或者李森科主义，更不知道李森科主义对苏联、东欧和中国生物学发展史产生的影响。

简单来说，李森科是苏联的一个农艺师，他坚持生物进化中的获得性遗传观念。从20世纪30年代开始，他得到了苏联领导人的政治支持，并逐渐被提拔为全苏列宁农业科学院院长。利用手中的政治影响力和权力，李森科不断打击学术上的反对者以及反对观点（例如孟德尔遗传学），使他的学说成为苏联生物遗传学的主流。接着发生的事情，更多的是政治运动而非仅仅是科学探讨。苏联在1948年全面禁止讲授孟德尔遗传学，3000多名遗传学家失去了在大学和科研机构中的本职工作。许多苏联科学家因反对李森科而被投入监狱，相当多的人受到严重迫害，其中包括国际著名的生物学家Nikolai Vavilov。

1949年中华人民共和国成立后，李森科主义开始影响到中国的生物学发展。作为一个刚从战火中重生的国家，新中国必须从头来过，亟须工业和科技的发展，因此掀起了"全盘学苏"的热潮，而苏联也慷慨相助。数以千计的苏联专家和顾问被派到中国，带来了苏联主流的知识和技术，其中也包括李森科的理论。数以百计的苏联生物学家，其中大部分是李森科的追随者，在中国的大学和研究院所开设了入门课程，他们的讲座讲义被翻译成中文教科书，在国内广为流传。在一段时间里，人们在同样的生物教科书中，可能会同时发现李森科的理论和孟德尔遗传学的知识。

然而，李森科生物学和孟德尔遗传学在中国生物学课程中的共存并没有持续很长时间。因为大多数中国主流生物学家在英、美等西方国家受过教育，有些人甚至曾在摩尔根的实验室工作过，所以他们不能完全同意李森科的理论，一些人公开发表了他们的意见和看法。不幸的是，李森科主义在中国的影响不仅是科学的，而且也涉及政治。

1952年，政府媒体尖锐批评了孟德尔遗传学及其支持者，随后，孟德尔遗传学的知识内容从所有的中国生物教科书中被删除了。很多生物学家只能保留他们不同的观

译者：李柳
中国科学院北京生命科学研究院，北京100101，中国
邮箱：liliu@ioz.ac.cn

点，通过使用自己重新编写的教科书来应对这种情况，并悄悄地继续他们的遗传学研究。但有些人却相当不幸，例如植物遗传学家鲍文奎，当时，他对小黑麦多倍体育种的研究已经到了关键期并取得了阶段性突破进展，但由于上级领导下令制止，他的试验田被铲除，实验也被迫停止。幸运的是，与苏联不同，李森科主义并没有发展成一个在中国普遍存在的、严重的政治运动。

1956年，苏联的政治形势发生了重大变化，李森科被迫辞去全苏列宁农业科学院院长的职务。与此同时，中国领导人认真研究中国的科技水平和实力，开始反思科技发展是否有必要全面依赖苏联的经验，并提出了“百花齐放，百家争鸣”这一发展艺术和科技的根本方针，鼓励社会各界对国家政策发表意见，科学研究的不同学派可以自由争论。正是在这样的情况下，中国科学院会同教育部请示上级中央，请求重新考虑遗传学问题。1956年8月10—25日，遗传学座谈会在山东青岛召开。青岛，中国东部沿海的美丽城市，因这次会议变得更加知名。

这次会议历时16天，邀请了54位中国生物学家，其中不仅有遗传学家（包括李森科学派和孟德尔学派），还有来自其他研究领域的生物学家（胚胎生物学家、生物化学家、细胞生物学家、生态学家等）。来自政府各部门的70多名代表也出席了会议，但只是旁听。我国著名生物学家、时任中国科学院生物地学部副主任的童第周先生致开幕词。他说明了会议的目的和基本原则，并号召大家消除隔阂，大胆地畅所欲言。这是七年来，科学家第一次在没有任何政治干扰的情况下自由讨论现代遗传学，就共同关注的问题，如“遗传的物质基础”“遗传与环境的关系”“遗传与个体发育”“遗传与系统发育”和“关于教学上存在的问题”等，展开热烈的讨论。孟德尔派学者提供了每个生物学领域有关遗传学的详细实验数据，以及来自西方国家有关遗传学的最新报告；李森科派学者更多的是进行反驳或只是围绕主题重复他们的立场。无论如何，这次会议是一次公开的研讨，而非一场辩论，目的不是证明理论的孰对孰错。会议结束时，所有参会的生物学家都就未来的学术自由和相互理解达成了共识。

后来，会议相关材料被广泛印制和分发宣传，使得更多的人了解到会议情况。大多数中学和高等教育教科书又开始教授孟德尔遗传学和李森科理论知识。虽然中国的遗传学花了一段时间才回到正轨，但是1956年青岛遗传学座谈会在中国生物学发展的曲折道路上绝对是一个转折点。这次会议不仅使中国的遗传学走上了正确的道路，也为中国的科技发展开创了一个新时代。自从这次会议之后，中国的科技政策制定开始遵循一条基本规则——让科学家（而不是政府）决定重要的科技问题，坚决不将学术争论变为政治运动。因此，新一代中国科学家在享受中国学术自由氛围的同时，不应忘记这次“非同寻常”的会议。

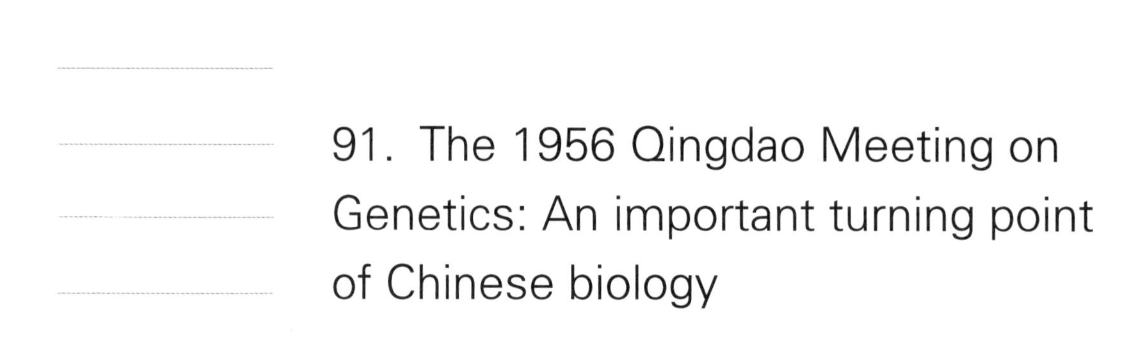

91. The 1956 Qingdao Meeting on Genetics: An important turning point of Chinese biology

Lysenkoism once was a popular word in the middle of the 20th century, but is now long forgotten by most people. Many young Chinese have never heard of T. D. Lysenko or Lysenkoism, neither do they know the impact Lysenkoism had on the history of biology in the Soviet Union, in Eastern Europe, and in China.

To put a long story short, Trofim Denisovich Lysenko was a Soviet agronomist who believed in inheritance of environmentally acquired traits. Starting in the 1930s, he received political support from the Soviet leaders and was gradually promoted to the top leader of the Academy of Agricultural Sciences of the Soviet Union. With the political influence and power in his hand, Lysenko began to promote his theory throughout the nation and eliminate any "harmful" ideas (i.e. Mendelian genetics) among Soviet biologists. The following events were more of a political movement than a scientific discussion, as Mendelian genetics was officially banned in the Soviet Union in 1948 and over 3000 scientists were expulsed from their positions. Many Soviet scientists were thrown into prison for expressing opinions dissent from Lysenko's, and quite a lot suffered persecution severely, including the internationally renowned biologist Nikolai Vavilov.

The Lysenkoism started to affect Chinese biology after the founding of the People's Republic of China in 1949. As a wartorn nation, China had to re-build literally everything from scratch, especially industry, science and technology. "Learning from the Soviet Union" was a major policy for Chinese industry and S&T, and the Soviet Union was generous to help. Thousands of Soviet experts and consultants were sent to China, bringing with them knowledge and technologies that were mainstream in the Soviet Union, including Lysenko's theories. Hundreds of Soviet scientists in biology, most of whom were Lysenko's followers, gave introductory courses in Chinese universities and institutes, and their lectures were translated into Chinese textbooks and widely spread in China. For a while, one may find both Lysenko's theories and Mendelian genetics in the same biology textbooks in China.

However, the co-existence of Lysenkoist biology and Mendelian genetics in Chinese biology

Ming Li, Le Kang

Beijing Institutes of Life Science, Chinese Academy of Sciences, Beijing 100101, China

Contact: lkang@ioz.ac.cn(L. Kang)

curriculums did not last long. As most senior Chinese biologists were westerntrained returnees and some of them even had worked in the Morgan laboratory, it was understandable that they could not completely agree with Lysenko's theories and some openly expressed their opinions about it. Unfortunately the influence of Lysenkoism in China was not only scientific but also political. In 1952, the government media sharply criticized the Mendelian genetics and its supporters, and the Mendelian genetics was forcedly taken out of all Chinese biology textbooks afterwards. Most biologists managed to cope with the circumstances—they taught the newly-edited textbooks and kept their different opinions to themselves, and quietly continued their genetic research. But some were not that lucky, including plant geneticist Wenkui Bao. His research of rice polyploid had been making steady progress and was reaching a critical point when the authorities ordered him to stop his work and destroyed all of his experimental plots. Unlike the Soviet Union, Lysenkoism did not develop into a severe political movement in general in China.

In 1956, the political atmosphere in the Soviet Union was going through abrupt changes and Lysenko was dismissed from his position as the leader of the Academy of Agricultural Sciences of the Soviet Union. Meanwhile, Chinese leaders were re-evaluating China's S&T power and started to consider whether the full dependency on Soviet expertise was necessary for the S&T development. The government decided to start a campaign called "Hundred Flowers Blossom, and Hundred Schools Contend" , encouraging a variety of views on national policy issues, including S&T issues. Seizing the moment, leaders of Chinese Academy of Sciences and the Ministry of Education wrote to the higher authorities and asked them to reconsider the issue of genetics, which eventually led to the Qingdao Meeting on Genetics on August 10—25, 1956. Qingdao, a beautiful city in the east coast of China, became more well-known due to the important meeting.

The 16-day long meeting invited 54 Chinese biologists, including not only geneticists (both Lysenko's followers and Mendelian geneticists), but also biologists from all other disciplines (embryo biologists, biochemists, cell biologists, ecologists, and so on). Over 70 representatives from various governmental departments also attended the meeting, but they were only to observe. At the opening ceremony, Chair Dizhou Tong, a leader and biologist of the Academic Division of Biology in the Chinese Academy of Sciences, delivered an opening speech about the goals and the ground rules of the meeting and asked all the participating scientists to give over their inhibition and speak out freely and boldly. For the first time in seven years, scientists were able to discuss modern genetics without any political distortion. Almost all the discussions during the meeting were technical and they were organized into the following sessions: The material basis of inheritance, the relationship between genetic mutation and environment, inheritance and individual development, inheritance and phylogeny, education of genetics and S&T policy. The Mendelian geneticists presented detailed experimental data from each field of biology related to genetics, as well as the latest reports on genetics from western countries. Most often, their Lysenkoist antagonists had to take a defensive posture or merely repeat their position on the main topic. Anyhow, the meeting was an open forum, not a debate, and the purpose was not to prove one theory was right and the other was wrong. At the end of the meeting, all the participating biologists reached an agreement for future academic freedom and mutual understanding.

Later on, the meeting materials were printed and widely distributed to educate a broader public. And most Chinese textbooks for secondary and higher education started to teach both Mendelian genetics and Lysenko's theories. Although it took some time later for Chinese genetics to go back to the right track, the 1956 Qingdao Meeting on Genetics was definitely a turning point along the twisted path of Chinese biology development. Not only did the meeting bring Chinese genetics to the right path, it also started a new era for the overall Chinese S&T development. Since the meeting, a basic rule was set for Chinese S&T policy—let the scientists (not the government) decide on important S&T issues and never turn academic controversy into political movements. For this reason, the meeting should not be forgotten by the new-generation Chinese scientists while they enjoy the academic freedom in China.

References

Li PS, et al (1985) Hundred Schools Contend—The Only Correct Way for the Development of Science: Memorandum of Qingdao Meeting on Genetics, August, 1956. Beijing: The Commercial Press. (李佩珊, 等. 1985. 百家争鸣——发展科学的必由之路: 1956年8月青岛遗传学座谈会纪实. 北京: 商务印书馆.)

Li PS, Xu LY (1985) History of Science and Technology in the 20th Century. Beijing: Science Press. (李佩珊, 许良英. 1985. 20世纪科学技术简史. 北京: 科学出版社.)

十、中外学术交流

92. 祁天锡对中国近代生物学的贡献

图1　祁天锡（1876—1937）（摄于1933年）

美国生物学家祁天锡在中国生活了几十年（图1）。他建立了中国最早的大学生物学系，并把生物学的研究生教育带到中国来，为中国近代生物学的发展作出了杰出贡献。

1876年4月20日，祁天锡出生于美国南卡罗来纳州的Union小镇。1892年考入伍福德学院，并在那里获得硕士学位。在哥伦比亚学院和其他学校短暂工作后，祁天锡在1901年来到中国，并被聘为东吴大学的自然科学教授。他把自己的大部分时间都投入了中国的生物学教育事业中，中间回国照顾生病的妻子也只花了几个月时间。1921年他的妻子去世后，他又回到中国，先是被聘为斯宾塞透镜公司的代表，随后被洛克菲勒基金会所属中华医学董事会聘请。1932年，祁天锡离开中国返美，去世前一直担任伦敦学院生物学教授（James，1938；王志稼，1940）。

祁天锡是我国第一个来自海外且具有大学教学水平的科学教师。作为一名生物学家，祁天锡在生物学教育、动植物调查、生物学研究的建制化等方面推动了中国生物学的发展。

祁天锡从1901年开始在东吴大学担任自然科学教授，多年来教授生物学、物理学、化学和其他学科。1912年，开始担任东吴大学生物系系主任，这是中国最早建立的生物系。他非常强调科学研究，并建立了当时中国最先进的生物实验室。他带领学生开展野外实习，并提醒学生要注意生物农业。他为大学生编写了《植物学教科书》，并在这本书中介绍了不少中国本土的植物。施季言和胡经甫是东吴大学生物系培养的中国最早的生物学硕士。1919年毕业后，施季言留校担任校监，1952年东吴大学在台北复校后，他出任校长。胡经甫在1922年获得康奈尔大学博士学位后回国，是中国昆虫分类学的奠基人（Li，2012）。祁天锡的许多学生都成长为中国大学生物系的负责人，如

作者：付雷
浙江师范大学，金华321004，中国
邮箱：ful527@163.com

胡经甫在燕京大学，陈纳逊在金陵大学，王志稼在上海大学，陈子英在厦门大学，朱元鼎在圣约翰大学，等等。这些著名的生物学家极大地推动了中国近代生物学的发展。直到祁天锡离开后，东吴大学生物系一直保持在生物学上的卓越地位，并陆续培养了一批生物学家和教育家，如高尚荫、刘建康、谈家桢等。

祁天锡非常重视科学普及和基础教育。他曾在东吴大学附属中学任教多年，并面向大众和中学生撰写了多部著作。在他出版于1904年的《昆虫学举隅》一书中，引用了不少中国的典故，如“螳螂捕蝉，黄雀在后”等（祁天锡，1904）。他给中学生编写了一本英文生物学教科书《生物学初桄》（图2），1913年由商务印书馆出版。他在这本教科书中呈现了大量的照片和插图，其中有一些是他在东吴大学和圣约翰大学拍摄的。为了有助于中国学生学习，他还邀请吴继杲给这本书添加了中文术语（祁天锡，1913）。他还编写了《格致读本》和《益智读本》等读物，向中国大众介绍本土的动植物资源。除此之外，祁天锡还给商务印书馆的杂志撰写了自然与科学类的科普文章。他的这些作品广受欢迎，并被不少学校作为教科书使用。

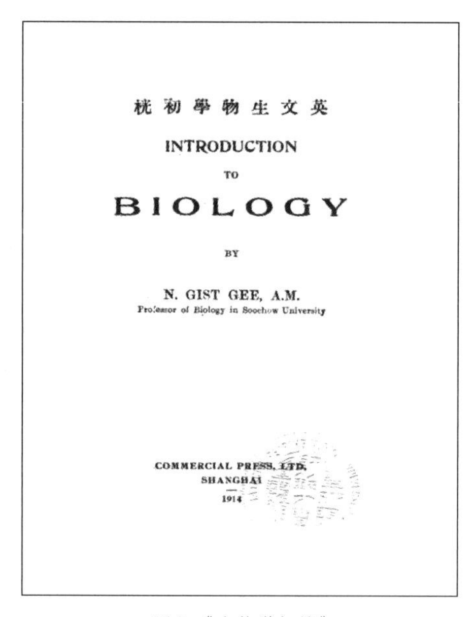
英文生物學初桄

INTRODUCTION

TO

BIOLOGY

BY

N. GIST GEE, A.M.

Professor of Biology in Soochow University

COMMERCIAL PRESS, LTD.

SHANGHAI

1914

图2 《生物学初桄》

从在大学工作开始，祁天锡就对动植物分类感兴趣。他在业余时间调查动植物种属状况，用中国本土的生物学资源教授学生，推动中国生物分类学的发展。从1913年开始在*The National Review*、1919—1921年在《科学》杂志，他陆续发表《江苏植物名录》。他在《科学》杂志上的文章由钱崇澍翻译，后结集由中国科学社出版。此外，《江苏植物名录》的一部分被上述植物学教科书收录并作为附录。

祁天锡对水生生物有浓厚兴趣，一直希望组建一个淡水生物研究所。他与外国的生物学家和博物馆交换了大量的淡水海绵标本，并在这一研究领域发表了十余篇文章。通过他的研究工作，中国的生物资源逐渐被世界所了解，祁天锡自己也成为淡水海绵的研究权威。他的学生胡经甫等继承了他的工作。

祁天锡对鸟类也非常感兴趣，与慕维德（L. I. Moffett）等人合著了《长江下游的鸟类索引》《中国鸟类》等。他的工作为后来的研究建立了典范。他还发表了有关中国两栖动物、爬行动物、哺乳动物等的文章，其中关于两栖动物的一些文章是与燕京大学的博爱理（A. M. Boring）共同发表的（Zheng，2015）。

祁天锡认为，中国生物学研究机构的建立至关重要。他加入了中国科学社，与中国的科学家保持联系，并与葛利普（A. G. Grabau）、秉志等科学家在1925年发起成立了北京博物学会。他们发行杂志，发表研究论文。同时，他们也热衷于科学普及。作为学会的会长和秘书，祁天锡致力于加强国内外科学家之间的联系，他把这个学会建成了博物学研究者重要的研究和交流平台。

当祁天锡还在为洛克菲勒基金会工作的时候，他就为推动北平静生生物调查所的建立不遗余力。这个调查所是秉志和胡先骕在1928年建立的，祁天锡是委员会中唯一的外籍成员。

祁天锡是伟大的科学家，对中国的生物学研究和教育作出了杰出贡献。他和他的工作在西方生物学的传入过程中起到了重要作用。

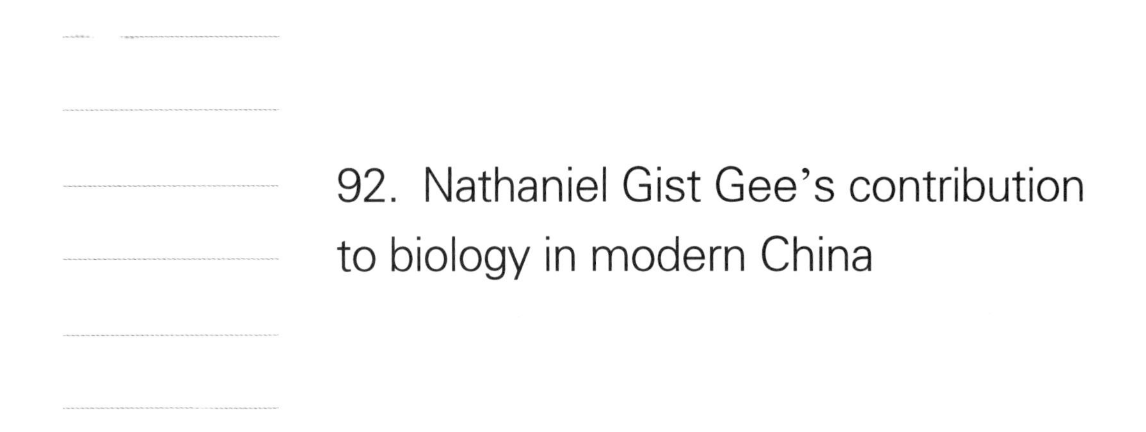

92. Nathaniel Gist Gee's contribution to biology in modern China

Nathaniel Gist Gee (祁天锡, 1876—1937) was an American biologist who lived in China for several decades. He greatly contributed to the development of biology in modern China, including the establishment of the first Biology Department in China and introducing the biology graduate education to China.

N. Gist Gee was born in Union, South Carolina, USA on April 20, 1876. He attended Wofford College in 1892 where he received his Master's Degree. In 1901, after working at Columbia College and other schools, he came to China and was appointed Professor of Natural Sciences in Soochow University. He devoted most of his time to the biology education of China, spending only a few months leave in his native country taking care of his sick wife. He returned to China after his wife's death in 1921. First serving as representative of the Spencer Lens Company, he was then hired by the China Medical Board of the Rockefeller Foundation. He left China in 1932, and worked as Biology Professor in London College until his death of illness (James, 1938; Wang, 1940).

N. Gist Gee was the first science teacher from abroad who reached the level of university instructor in China. As a biologist, N. Gist Gee promoted the development of biology in modern China, including biology education, investigation of plants and animals, and the institutionalization of biology research.

N. Gist Gee served as a professor in natural sciences at Soochow University from 1901, where he taught biology, physics, chemistry and other subjects for years. Then he became the first Director of Biology Department of this university in 1912, which was the first independent biology department in China. He put a special emphasis on scientific research and founded the most advanced biology laboratory in China at that time. He took students on field trips and led them to pay attention to biological agriculture. He wrote *A Textbook of Botany* for students, and introduced much knowledge about native plants. Shih Ji-Yan (施季言) and Wu Chen-Fu Francis (胡经甫) were the first Chinese biology masters cultured in his department and graduated in

Lei Fu

Zhejiang Normal University, Jinhua 321004, China

Correspondence: ful527@163.com

1919. After graduation, Shih Ji-Yan remained working as school superintendent in Soochow University. He was later appointed President of Soochow University, which was reestablished in 1952, in Taibei. Wu Chen-Fu Francis received his Ph.D. Degree in 1922 from Cornell University and then returned to China. Wu was the founder of modern entomotaxonomy in China (Li, 2012). Many of N. Gist Gee's students became leaders of biology departments at other universities of China, such as Wu Chen-Fu Francis at Yenching University, Nelson Chen (陈纳逊) at the University of Nanking, Wang Chu-Chia (王志稼) at Shanghai University, Chen Tze-Ying (陈子英) at Xiamen University, Chu Yuan-Ting (朱元鼎) at St. John's University, and so on. These renowned Chinese scientists greatly promoted the development of biology in modern China. The department maintained its excellent education on biology after N. Gist Gee's leaving the university, and several distinguished biologists and educators were cultivated there, including Harry Zan-Yi Gaw (高尚荫), Liu Chien-Kang (刘建康), and Tan Chia-Chen (谈家桢), et al.

N. Gist Gee also stressed the importance of the popularity of biology and elementary education. He was a biology teacher in the middle school attached to Soochow University for a few years and wrote books for public and middle school students. In 1904, he published *Life Histories of A Few Common Insects*, in which he quoted some Chinese idioms such as Mantis stalks the cicada oriole in the post (Gee, 1904). He prepared a biology textbook in English for middle school students named *Introduction to Biology*, published by Commercial Press in 1913. He presented many photos and pictures in this book, some of which were taken at Soochow University and St. John's University in Shanghai. In order to help Chinese students, he invited K. K. Woo (吴继杲) to add Chinese technical terms to this textbook (Gee, 1913). He compiled *Science Reader*, *The Useful Knowledge Reader* and other books to introduce the Chinese public the knowledge about native Chinese plants and animals. Furthermore, Gist Gee wrote articles about nature and science for students in magazines published by Commercial Press. All of these publications were very popular and adopted by several schools.

Since the beginning of his college career, Gist Gee was interested in plant and animal taxonomy. During his spare time, he traveled to investigate and collect plant and animal specimens so that he could instruct his students with local biological resources, thus advancing the development of biology taxonomy in China. He published Catalogue of plants in Jiangsu Province in *The National Review* from 1913 and in *Science* from 1919—1921. Gist Gee's publications in *Science* were translated by Chien Shung-Shu (钱崇澍) and published by the Science Society of China. Additionally, most of the book was appended to *A Textbook of Botany* cited above.

N. Gist Gee was keen to aquatic organisms and hoped to found an institute to research fresh-water organisms. He exchanged a lot of fresh-water sponge specimens with foreign biologists and museums and also published dozens of articles in this research field. Through his work and articles, Chinese biological resources gradually became known by the world, and N. Gist Gee became a well-known authority on fresh-water sponges. His work was succeeded by Wu Chen-Fu Francis and his other students.

N. Gist Gee was interested in birds as well, and he compiled *A Key to the Birds of Lower Yangtse Valley* and *A Tentative List of Chinese Birds* with Lacy I. Moffett and others. His work set examples for latter research. He also published several articles on Chinese amphibians, reptiles, mammals and other animals, with a few of the articles on amphibians accomplished with A. M. Boring from Yenching University (Zheng, 2015).

N. Gist Gee regarded it very important to the institutionalization of biology research in modern China. He joined the Science Society of China to connect with Chinese scientists, and also launched the Peking Society of Natural History with A. G. Grabau (葛利普), Ping Chi (秉志) and other scientists in 1925. They distributed a magazine and published many research articles. At the same time, they were active for the popularity of science. As secretary and president of the society, Gist Gee worked hard to communicate with scientists at home and abroad. He made the society an important research and communication platform for naturalists.

When he worked for the Rockefeller Foundation, N. Gist Gee made great efforts to promote the establishment of Fan Memorial Institute of Biology. The institute was set up in 1928 by Ping Chi and Hu Hsen-Hsu (胡先骕), and N. Gist Gee was the only foreign member of the board.

N. Gist Gee was a great scientist and made outstanding contributions to the development of biology education and research of China. He and his work were important in the transmission of modern biology to China.

Figures

Fig.1 N. Gist Gee (1876—1937) in 1933

Fig.2 *Introduction to Biology*

References

Gee NG (1904) Life Histories of A Few Common Insects. Educational Association of China. (祁天锡. 1904. 昆虫学举隅. 益智书会.)

Gee NG (1913) Introduction to Biology. Beijing: Commercial Press. (祁天锡. 1913. 生物学初桄. 北京: 商务印书馆.)

James GN (1938) Nathaniel Gist Gee. Science 88: 250.

Li M (2012) Detour and return—Chenfu Wu's interdisciplinary research journey. Protein & Cell 3:481–482.

Wang C (1940) A brief history of Dr. Nathaniel Gist Gee. Science 24:69–70. (王志稼. 1940. 祁天锡博士事略. 科学, 24:69–70.)

Zheng S (2015) Alice M. Boring: A pioneer in the study of Chinese amphibians and reptiles. Protein & Cell 6:625–627.

93. 博爱理：中国两栖和爬行动物研究的先驱

博爱理先生（Alice Middleton Boring）是美国生物学家和爬行动物学家，但是她的主要学术研究生涯却是在中国度过的（图1和图2）。她不仅投身于中国生物学的研究和教育，也是中国两栖和爬行动物研究的先行者。

图1　博爱理（1883—1955）

图2　博爱理（左）1939年在燕京大学

博爱理先生1910年获得美国布林茅尔学院的博士学位，学习期间她也得到了著名生物学家T. H. Morgan和遗传学家N. M. Stevens的指导。在美国开展了几年的教学和科研工作后，她来到中国并开启了在中国的研究生涯（罗桂环，2014）。1918年到1920年，她在由洛克菲勒基金会资助的北京协和医学院做生物学助教工作。博爱理先生1923年起开始担任燕京大学生物系主任，直到1941年该大学被日本侵略者关闭。她于1946年重返燕京大学工作，直至1950年离开中国回到美国。

博爱理先生最初学习和研究的领域是遗传学和细胞生物学，但是她来到中国后的学术兴趣就转向了两栖和爬行动物分类学和地理分布区系研究。中国爬行动物区系的多样性让她对研究工作充满了热情，她与生物学家N. G. Gee、C. H. Pope和胡经甫以及她的学生一起开展研究工作并发表了多项科研成果。她曾到中国许多地方进行标本采集，包括江西、浙江、安徽等。她通过努力工作取得大量重要研究成果，燕京大学也

译者：苏瑞凤

中国科学院北京生命科学研究院，北京100101，中国

邮箱：surf@biols.ac.cn

发展成中国两个两栖爬行动物研究中心之一（Adler and Zhao，1993）。她是北京自然历史学会的发起人之一，1929年到1950年学会发表了她21篇科研论文和1部手册（Ogilvie et al.，1999）。1945年，她编纂的另一重要书籍《中国两栖动物——活体与化石》正式出版。此外，她还推动了中国和美国博物馆和科学家交流并共享标本与数据。她的努力进一步拓宽了世界对中国两栖类和爬行类动物的了解和认识。

除了研究外，教学是博爱理先生在中国的另一项主要工作。学生口中的博爱理先生既严格又温暖，她对自己学生的实验操作和学术研究提出了非常高的要求，但是在严厉治学的同时，她也热情关怀着学生的日常生活。她坚持在自己家里与每一个生物系的学生每年至少共进一次午餐，并亲自教授学生西餐礼仪，以便他们做好赴国外学习的充足准备（Ogilvie et al.，1999）。她的学生在中国乃至世界的医学和生物学领域都作出了杰出贡献和瞩目的成就。其中，中国爬行动物学的奠基人刘承钊先生（1900—1976）就是一位代表。他曾经在四川省发现了一个蟾蜍新种并将其命名为博氏峨眉髭蟾（*Vibrissaphora boringii*）（图3），以表达他对博爱理先生的无限感激（燕京研究院，2001）。

图3 博氏峨眉髭蟾

值得一提的是，博爱理先生工作和生活的时代，中国社会环境动荡，物资匮乏，她遇到了很多挑战，但她依然放弃了在美国大学的职位和安定的生活，选择来到这个充满战争威胁的东方国度。第二次世界大战期间，博爱理先生一直坚持她在燕京大学的研究和教学工作。1943年，美国和日本的战争迫使她离开中国入住临时的安置点，战乱结束后她又立即返回中国，她在给家人的信件中写道："我属于这里。"除了她与中国同事和学生之间的友情，她的愿望是用超越物质匮乏的科学知识帮助中国人民，以此充实自己。这些信念也正是支持她一直留在中国工作的原因。

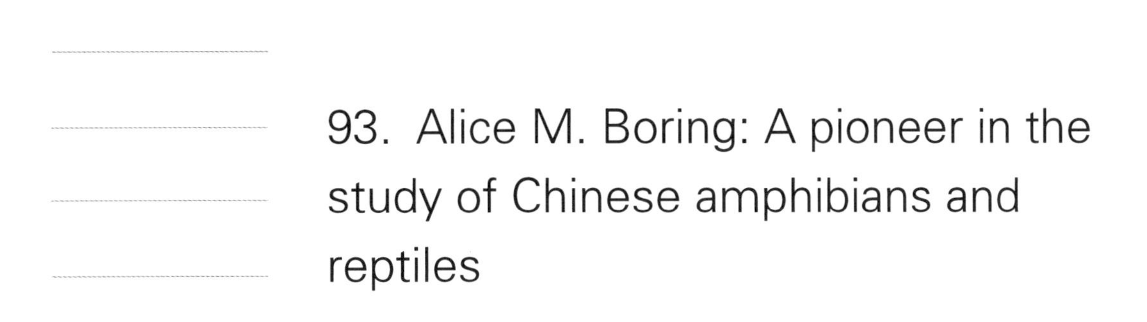

93. Alice M. Boring: A pioneer in the study of Chinese amphibians and reptiles

Alice Middleton Boring (博爱理, 1883—1955) was an American biologist and herpetologist who spent much of her academic life in China. She devoted herself to improving biology research and education in China, and is a pioneer in the study of Chinese amphibians and reptiles.

Boring received her Ph.D. in 1910 from Bryn Mawr College, where she studied under biologists including Thomas Hunt Morgan and Nettie M. Stevens. After some years of teaching and researching biology in the United States, she started her career in China (Luo, 2014). From 1918 to 1920, she worked as Assistant in Biology at Peking Union Medical College, which was funded by Rockefeller Foundation. In 1923, she was appointed the head of biology department at Yenching University, where she worked until Japanese closed the university in 1941. She returned to Yenching University in 1946, and stayed there till she went back to America in 1950.

Although originally trained and worked in the field of genetics and cytology, Boring's academic interest shifted to taxonomy and distribution of reptiles and amphibians during her stay in China. The variety of Chinese herpetofauna offered her an exciting research field, where she worked and published with biologists Nathaniel Gist Gee, Clifford H. Pope and Hu jingfu (胡经甫), as well as her students. In order to collect specimens, Boring had expeditions in different parts of China, such as Jiangxi, Zhejiang, Anhui, etc. Boring's hard work yielded substantial results. Her program in Yenching University became one of two major centers in China for the study of amphibians and reptiles (Adler and Zhao, 1993). She was one of the charter members of Peking Natural History Society, and had twenty-one papers and one handbook published by the Society between 1929 and 1950 (Ogilvie et al., 1999, p. 110). In 1945, she published another important Bibliography: *Chinese Amphibians: Living and Fossil Forms*. Boring also provided American museums and scientists with specimens and data in China. Her efforts further broadened the global understanding of the Chinese amphibians and reptiles.

Apart from research, teaching was always Boring's primary work in China. Her students described her as both strict and warm. She asked for very high standards when it came to

Shu Zheng
The Institute for the History of Natural Sciences, Chinese Academy of Sciences, Beijing 100190, China
Correspondence: zhengshu@ihns.ac.cn

laboratory operation and academic study. Behind her tough facade, she cared about her students warm-heartedly. She insisted to have dinner with every biology student at her house at least once a year. In order to prepare her students to study abroad, she would personally teach them western table manners (Ogilvie et al., 1999, p. 125). Boring's students made outstanding contributions to medicine and biology in China and abroad. One of them is Liu Chengzhao (刘承钊, 1900—1976), who is considered to be a founder of Chinese herpetology. Liu once discovered a new species of toad in Sichuan Province in China, and named it *Vibrissaphora boringii*, to express his enormous gratitude to Boring (Yenching Institute, 2001)

It's noteworthy that Boring's life and career in China had always been challenged by the unstable social environment and inconvenient material condition. After giving up her secure life as an Associate Professor in an American university, she arrived in an oriental country full of civil strife. During the Second World War, Boring insisted to continue her research and teaching in Yenching, in spite of the Japanese aggression. She had to stay in an internment camp before she was forced to leave China due to the war between America and Japan in 1943. She returned to China as soon as the chaos was over. There were reasons why she felt so "belong here" as she wrote to her family. Apart from her friendship with colleagues and students in China, her willing to fulfill herself by helping Chinese people with scientific knowledge transcended material hardship.

Figures

Fig.1 Alice M. Boring (1883—1955)

Fig.2 Alice M. Boring (left) at Yenching University, 1939

Fig.3 *Vibrissaphora boringii*

References

Adler K, Zhao EM (1993) Herpetology of China. Oxford: Society for the Study of Amphibians and Reptiles.

Luo GH (2014) The Development of Modern Biology in China. Beijing: China Science and Technology Press, 78-80. (罗桂环. 2014. 中国近代生物学的发展. 北京:中国科学技术出版社, 78-80.)

Ogilvie MB, Clifford J, Choquette A (1999) Dame Full of Vim and Vigour: Biography of Alice Middleton Boring (1883—1955), an American Biologist in China. London: Harwood Academic.

Yenching Institute (2001) Biographies of People in Yenching. Vol. 1. Beijing: Beijing University Press, 26-27. (燕京研究院. 2001. 燕京大学人物志. 第一辑. 北京:北京大学出版社, 26-27.)

94. 康奈尔大学与中国近代生物学的渊源

中国近代生物学的发展可以追溯到1910年初。第一批中国青年被派往国外接受高等教育。鉴于这批青年主要获得庚子赔款和其他美国慈善基金会的支持（例如洛克菲勒基金会的中国医学委员会和中国基金会），因此绝大多数青年在哥伦比亚大学和康奈尔大学等美国著名大学接受了科学教育。

秉志是康奈尔大学最著名的中国校友之一，1913年获得康奈尔大学学士学位。1918年，在J. G. Needham教授的指导下获得康奈尔大学博士学位。1915年，他在美国伊萨卡联合创立了中国第一个非政府科学团体“中国科学社”。他回到中国后，于1921年创办我国国立大学第一个生物系——东南大学生物系。1922年建立我国第一个生物学研究机构——中国科学社生物研究所。1934年创立中国动物学会。此外，他还是我国第一本科学杂志《科学》的联合创办人。

从1910年到1930年，康奈尔大学培养了大量中国留学生。他们后来成为中国生物学的中坚力量。这些校友包括昆虫学家邹树文、胡经甫和刘崇乐。胡经甫出版的《中国昆虫名录》，记录了昆虫25目、392科、4968属、20069种，是当时有关中国昆虫最全面的出版物。中国科学院动物研究所第一任所长、中国动物遗传学奠基人之一的陈桢，也曾在康奈尔大学农业系学习。另外，专门研究两栖动物和爬行动物的动物学家刘承钊，任中国科学院微生物研究所第一任所长的真菌学家戴芳澜和副所长的真菌学家邓叔群，都曾在康奈尔大学接受培养。还有一些担任高级政府官员的校友，例如钱天鹤，他在1940—1947年担任农林部常务次长；还有杨显东，他在中华人民共和国成立后被任命为农业部副部长。

除了动物学家和真菌学家之外，康奈尔大学还培养了中国的第一批植物遗传学家，包括沈宗瀚、金善宝、李先闻、李竞雄和冯泽芳。他们都是从事食品和经济植物农业应用的植物遗传学家，例如，小麦育种专家沈宗瀚和金善宝，棉花育种专家冯泽芳。20世纪30—40年代，更多的学生被派往康奈尔大学学习农业科学，其中大部分学生来自金陵大学（图1）。

译者：张菁菁

中国科学院北京生命科学研究院，北京 100101，中国

邮箱：zhangjj@biols.ac.cn

金陵大学成立于1910年，是一所美国教会学校。通过“作物改良合作计划”（1925—1930），金陵大学与康奈尔大学建立了密切联系。根据该计划，康奈尔大学植物育种系的高级教员来到中国，担任金陵大学的教师、管理者、顾问和研究人员。同时，中国学生被派往康奈尔大学学习和进修。截至1946年，60多名学生被派往康奈尔大学植物育种系学习。康奈尔大学的H. H. Love教授（图2）不仅在金陵大学工作，同时还担任国民政府实业部的农业和作物改良顾问。他还是政府资助的研究机构“中央农业实验所”的第一任总技师。他根据在中国的经历出版了《康奈尔大学和金陵大学的故事：康奈尔大学的首个农业国际技术合作计划》一书。

毫无疑问，康奈尔大学对中国近代生物学的贡献是巨大的。它不仅培养了大量科学家和教育工作者的骨干，而且在战争动荡时期对中国农业产生了积极影响。

图1　金陵大学北楼（建于1919年）

图2　H. H. Love教授（右）

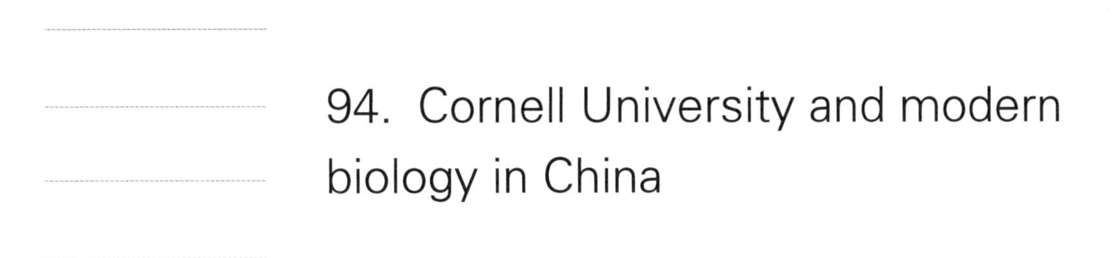

94. Cornell University and modern biology in China

The development of modern biology in China can be traced back to the early 1910s, when the first group of young Chinese was sent abroad for higher education. They were mainly supported by the Boxer Indemnity Fund and other American philanthropic funds, such as the Rockefeller Foundation's China Medical Board and the China Foundation. Therefore, most of these young Chinese received their scientific training in certain American universities like the Columbia University and Cornell University.

Chi Ping (秉志) is one of the most famous Chinese alumni of Cornell University. Obtaining his B.S. degree in Cornell in 1913 and his Ph.D. degree in 1918 under the guidance of his advisor Prof. J. G. Needham, Chi Ping returned to China to establish the first biology department of the national university in China, the Department of Biology at the Southeast University, in 1921. He also founded the first biological research institute in China, the Biological Laboratory of the Science Society of China, in 1922 and later the China Zoological Society (founded in 1934). He even co-founded the first non-governmental scientific society of China, the Science Society of China, in 1915 in Ithaca, US, while he was studying at Cornell. Moreover, he was also the co-founder of "*KeXue*" ("科学" meaning "Science" in Chinese), the first Chinese scientific journal.

From the 1910s to 1930s Cornell University trained many Chinese students, who turned out to be the backbone of Chinese biology later. These alumni include entomologists Shuwen Zou (邹树文), Chen-Fu F. Wu (胡经甫) and Chung-Lo Liu (刘崇乐). Chen-Fu F. Wu published *Catalogus Insectiorum Sinensium* which recorded a total of 25 orders, 392 families, 4968 genera, 20069 species of insects and was the most complete publication of Chinese insects of his time. Shisan C. Chen (陈桢), the first director of the Institute of Zoology of the Chinese Academy of Sciences (CAS) and one of the founders of animal genetics in China, also studied at the Department of Agriculture of Cornell University. In addition, zoologist Cheng-Chao Liu (刘承钊), who specialized in amphibian and reptant animals, and mycologists Fan-Lan Tai (戴芳澜) and

Ming Li

Beijing Institutes of Life Science, Chinese Academy of Sciences, Beijing 100101, China

Contact: liming@ioz.ac.cn

S. C. Teng (邓叔群), who were respectively the first director and deputy director of the Institute of Microbiology of the CAS, all received their trainings at Cornell University. Among the Cornell alumni, there were also high-ranking government officials, such as T. H. Chien (钱天鹤), who was the vice minister of the Ministry of Agriculture and Forestry of the Nationalist government from 1940—1947, and Xiandong Yang (杨显东), who was appointed the vice minister of the Ministry of Agriculture after the founding of the People's Republic of China.

Besides the zoologists and mycologists, Cornell University also trained the first plant geneticists of China, including Zonghan Shen (沈宗瀚), Shanbao Jin (金善宝), Hsien-Wen Li (李先闻), C. H. Li (李竞雄) and Che-Fang Feng (冯泽芳). They were all plant geneticists toward agricultural application of food and industrial plants. For example, Zonghan Shen and Shanbao Jin were specialized in wheat breading, and Che-Fang Feng was an expert in cotton breeding. In the 1930s and 1940s, even more students were sent to Cornell to study agricultural science, most of whom were from the Nanking University.

Nanking University, founded in 1910, was an American missionary school that developed a close bond with Cornell University through a program called "Plant Improvement and Genetics Program" during 1925—1930. According to the program, senior Cornell faculty from the Plant Breeding Department came to China working as teachers, organizers, advisors and researchers at the Nanking University. Meanwhile, Chinese students were also sent to Cornell. By 1946, over 60 students were sent to study at the Plant Breeding Department of Cornell University. Among the Cornell faculty was Prof. Harry H. Love, who not only worked at Nanking University, but served as an advisor on agriculture and crop improvement for the Ministry of Industry of the Nationalist government. Prof. Love was also the first chief technician of the Central Laboratory of Agriculture in China, a government sponsored research institute. Later on, he published a book based on his experience in China: *The Cornell-Nanking Story: The First International Technical Cooperation Program in Agriculture by Cornell University.*

It is unarguable to say that the contribution to modern biology in China from Cornell University was substantial. It not only trained the core group of scientists and educators, but also positively affected the general agricultural application in China, despite the turbulent times of war.

Figures

Fig.1 The north building of Jinling University (built in 1919)

Fig.2 Prof. H. H. Love (right)

95. 哈佛大学阿诺德树木园与中国的密切联系

图1 陈焕镛（1890—1971）

哈佛大学阿诺德树木园位于美国波士顿的牙买加平原和罗斯林德尔区，因其丰富的北美和东亚林木种类和植物学研究而世界闻名。在近150年的历史中（1872年建立），它与中国植物学和中国植物学家有着千丝万缕的联系。

19世纪晚期，阿诺德树木园首任园长C. S. Sargent对中国东部的植物群和树木种类非常感兴趣，并开始从中国收集标本。其中，英国园艺学家E. H. Wilson被Sargent派遣到湖北和四川采集和引种植物，结果满载而归，阿诺德树木园中中国的植物种类数量大增，因此研究中国植物的分类、生态和地理分布也成为该树木园的特色之一。这也是不少中国的年轻学者在1910—1930年到阿诺德树木园学习植物学的原因，这些学者包括中国现代植物学研究的奠基人陈焕镛、钱崇澍、钟心煊和胡先骕等。

陈焕镛先生（图1）在阿诺德树木园学习后于1919年回国，在美国奖学金的支持下，他来到前人没有研究过的海南岛进行植被情况的考察和植物标本的采集。9个月的艰苦卓绝的工作，陈先生取得了丰硕的成果，收集了数千份的植物标本并出版了一本海南岛的植物学专集。1921年他转到位于南京的东南大学任教，1922年他出版了第一本由中国学者撰写的英文植物学专著《中国经济树木》。陈焕镛随后到中山大学任教授，于1929年创立了中国最重要的植物学研究机构——农林植物研究所（中国科学院华南植物园的前身），并任所长。

另一位著名的中国植物学家胡先骕先生，也曾在哈佛大学阿诺德树木园学习并获得博士学位。回国后，胡先生仍与阿诺德树木园的工作人员保持密切的联系，其中就包括他的老师J. G. Jack和好友E. D. Merrill。1928年，胡先生协助中国著名生物学家秉志先生创办了当时我国最大的生物学研究机构——北平静生生物调查所（中国科学院

译者：程浩

中国科学院北京生命科学研究院，北京100101，中国

邮箱：chengh@biols.ac.cn

动物研究所和植物研究所的前身），他任植物部主任。胡先生在哈佛大学攻读博士期间，就萌生了效仿阿诺德树木园的做法，创建自己国家植物园的愿望。直到1934年，胡先生终于实现了自己的夙愿，在江西创建了中国第一座森林植物园——庐山植物园。北平静生生物调查所和庐山植物园都与阿诺德树木园有着密切的联系，包括相互交换植物标本和种子，它们的规模也得以迅速发展和扩大。

胡先骕先生最著名的成果之一是在20世纪40年代和郑万钧先生一起鉴定并命名植物界的活化石——水杉。但很少有人知道，水杉标本采集的经费正是来自胡先生在阿诺德树木园的好友Merrill教授的资助。胡先生后来专门邮寄了一袋水杉种子给阿诺德树木园，水杉也在Merrill教授的帮助下推广到了全世界，引起全世界植物学界的轰动。

哈佛大学阿诺德树木园对中国植物学家的影响远超过了简单的合作交流和互换标本。英国园艺学家Wilson把他在湖北和四川收集植物标本的经历编纂成书出版，这本书在中国植物学家中广为流传，很多中国学者以这本书作为科研圣经，继续在这些省份进行植被考察和植物标本采集。即便到了今天，已经过去了100多年，还有植物学家按照Wilson的考察路线重新进行区域生态环境变化的研究。

总之，尽管距离遥远还经历了艰难的战争时期，阿诺德树木园与中国的植物学家一直保持着友好和密切的联系，并且在中国植物学人才的培养、植物学研究的示范和激励中国建立植物园方面都产生了深远的影响。

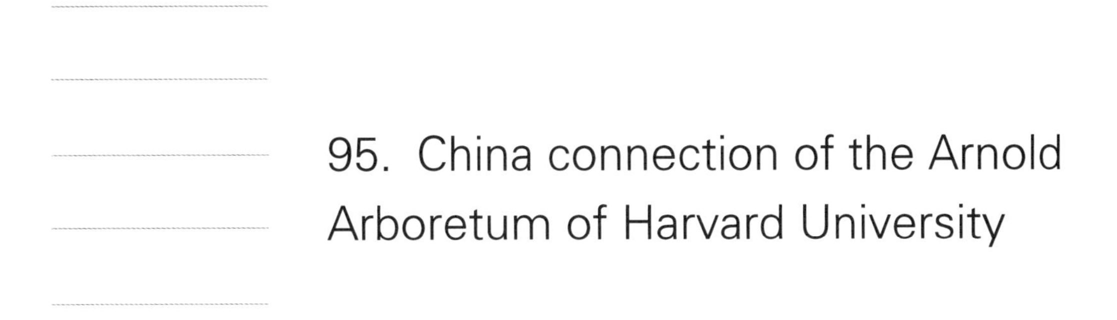

95. China connection of the Arnold Arboretum of Harvard University

The Arnold Arboretum of Harvard University, located in the Jamaica Plain and Roslindale sections of Boston, is famous for its collection and scientific study of ligneous species in North America and Eastern Asia. In the 150-year history of the Arboretum (established in 1872), it has built a strong connection with Chinese botany and Chinese botanists.

In the late 19th century, the first director of the Arnold Arboretum, Prof. Charles S. Sargent became interested in the flora and woody plants of eastern China and started to collect plant species from China. The collection expanded rapidly after Sargent hired British horticulturist Ernest H. Wilson to explore and collect specimens from Hubei and Sichuan provinces in China, and the Arboretum soon became famous for its specialty in Chinese plant taxonomy, ecology and biogeography. This is also why quite a few young Chinese students chose to study botany in the Arboretum in the 1910s and 1920s, including Woon-Young Chun, Sung-Shu Chien, Hsin-Hsüan Chung and Hsien-Hsu Hu, who later became the founders of Chinese botany.

After he finished his studies in the Arnold Arboretum and returned home in 1919, Woon-Young Chun immediately started to collect plant specimens in Hainan Island in South China, supported by the Shelden Fellowship from the US. His 9-month exploration in Hainan was quite fruitful with thousands of specimens collected and one book published. In 1921, Chun accepted an offer from the National Southeastern University in Nanjing and started teaching there. He also published an English book *Chinese Economic Trees* in 1922, the first botanical book in English by a Chinese author. He later joined the Sun Yat-sen University in Guangzhou and founded the Institute of Agriculture and Forestry in 1929, which became one of the most important botany research institutions in China.

Hsien-Hsu Hu is another famous Chinese botanist who received his doctor degree in the Arnold Arboretum. Even after he graduated and returned to China, he still maintained a close friendship with the Arboretum and its faculty, including John G. Jack and Elmer D. Merrill. In

---

Guihuan Luo[1], Ang Li[1], Ming Li[2]

1 The Institute for the History of Natural Sciences, Chinese Academy of Sciences, Beijing 100190, China

2 Beijing Institutes of Life Science, Chinese Academy of Sciences, Beijing 100101, China

Correspondence: guihuanl@ihns.ac.cn(G.-H. Luo)

1928, he co-founded the Fan Memorial Institute of Biology with Chi Bing and was appointed the director of the Botany Department, which later became a part of the Institute of Botany of Chinese Academy of Sciences. Inspired by the Arnold Arboretum, Hu had always hoped to build an arboretum in China. His dream came true in 1934 when he co-founded the Lushan Arboretum in Jiangxi Province, the first botanical garden in China. Both the Fan Memorial Institute of Biology and the Lushan Arboretum have expanded their collections by exchanging specimens and seeds with the Arnold Arboretum.

Hsien-Hsu Hu is also well known for the identification of a living fossil, *Metasequoia glyptostroboides* (the dawn redwood, or Shui-Shan in Chinese), with Wan-Chun Cheng in the 1940s. Few people know that the field work where the *Metasequoia* specimens were collected was sponsored by Hu's friend at the Arnold Arboretum, Prof. Elmer D. Merrill. Hu also sent Merrill a package of *Metasequoia* seeds, which Merrill helped to distribute to the arboretums throughout the world.

The influence of the Arnold Arboretum on Chinese botanists was far more beyond collaboration and exchanges. The book written by Ernest H. Wilson based on his exploration in Hubei and Sichuan provinces was widely spread among Chinese botanists, some of whom even followed Wilson's steps and kept on exploring the plants in those areas. Even today, botanists would re-visit Wilson's exploration sites and study the ecological changes in the areas over a hundred years.

All in all, despite the long distance and the war years, there has always been a strong connection between the Arnold Arboretum and Chinese botany/botanists, through education, collaboration and inspiration.

Figures

Fig.1 Woon-Young Chun (1890—1971)

96. 由崎岖开端展开的一段成功国际合作

"科学没有国界"，在传染性疾病研究领域中，没有比这更准确的说法了。众所周知，传染性疾病不仅可以通过被感染的人类或其他动物传播，还可以通过商业运输等途径蔓延。在全球化的时代，我们不可能只在一个国家与传染性疾病作斗争，而不去关心世界其他地方发生类似的事情。新发再发传染性疾病（例如2003年暴发的非典型病原体肺炎，即SARS）给整个亚洲社会带来了恐慌。为了抗击这些疾病，中国科学院（CAS）和东京大学（UT）于2005年启动了中日在新发再发传染性疾病方面的合作，该合作是日本文部科学省（MEXT）所资助"新发再发传染性疾病研究基地建设项目"的一部分。

中日两国在科技合作方面有着悠久的历史，但是早期的合作非常简单，主要涉及派遣中国的学生和年轻科学家到日本接受科学培训。后来，随着中国在科技领域取得巨大进步，中日科技合作变得更加频繁和多样。但是在生命科学领域，大部分合作仅在两国科学家之间进行，并且规模相对较小。毫无疑问，中国科学院和东京大学在新发再发传染性疾病方面的合作使中日在生命科学领域的合作迈上了一个新的台阶。

在日本文部科学省基金的支持下，东京大学医科学研究所（IMSUT）在中国科学院微生物研究所（IM）和中国科学院生物物理研究所（IBP）建立了两个联合实验室（图1），其中一个实验室侧重分子免疫学和分子微生物学，另一个侧重结构病毒学和免疫学。

在过去五年，日方派出五名日本科学家在中国的两个联合实验室开展研究。与此同时，一些中国科学家一直在日本的联合实验室工作。基于这两个联合实验室，双方合作已经延伸到我国其他医疗院所，例如北京地坛医院和中国疾病预防控制中心。2006年，日本文部科学省和中华人民共和国科学技术部（MOST）共同承认这两个联合实验室为中日政府之间的科技合作项目。这两个联合实验室有效推动和促进了中日在微生物学、免疫学和流行病学领域的研究合作。在联合实验室成功运行五年之后（2010年），中国科学院和东京大学共同承诺，未来五年将继续开展合作（图2），而且合作将扩大到更广泛的生命科学领域（图3）。

译者：张菁菁
中国科学院北京生命科学研究院，北京100101，中国
邮箱：zhangjj@biols.ac.cn

图1 中国科学院与日本东京大学联合实验室成立仪式

图2 中国科学院副院长李家洋（前排右）和东京大学副校长田中明彦（前排左）签署新一轮合作协议

图3 中国科学院北京生命科学研究院（BIOLS，CAS）院长康乐（前排左）和东京大学医科学研究所所长清木元治（前排右）签署生命科学领域的合作协议

这个合作之所以非常成功，作为启动该项目的关键人物之一，东京大学医科学研究所岩本爱吉（Aikichi Iwamoto）教授功不可没，他于2010年获得中国科学院国际科技合作奖（图4）。2011年，他获得“中国政府友谊奖”，并被提名为“中华人民共和国国际科学技术合作奖”候选人，这是中国授予外国科学家的最高级别奖项。

但是，这样一个成功的合作开端并不顺利，因为2005年的国际形势可能不是中国科学院与东京大学开展大型合作的最佳时机。尽管如此，中国科学院仍决定推进合作，并邀请了东京大学校长小宫山宏（Hiroshi Komiyama）于2005年4月访华，参加中日新发再发传染性疾病合作项目的启动仪式。在访问北京期间，小宫山宏校长还被授予中国科学院研究生院（中国科学院大学前身）名誉教授（图5）。

图4　岩本爱吉教授（前排左）获得中国科学院院长路甬祥（前排右）颁发的2010年中国科学院国际科技合作奖

图5　中国科学院院长白春礼（右六）授予小宫山宏（右五）名誉教授证书

仪式结束后，小宫山宏校长说："在我来这里之前，很多人建议我取消或者推迟我的访问。但我坚信中日科学家之间的合作与交流已经持续了数十年，我们之间的长期友谊不会受到一些短期事件的影响。"时间证明了一切。六年之后，中国科学院和东京大学的学术友谊比以往更加深厚。中国科学院和东京大学的合作将不断促进中日科学家共同面对科学挑战，从长远来看也必然有助于中日两国的友谊进一步加强。

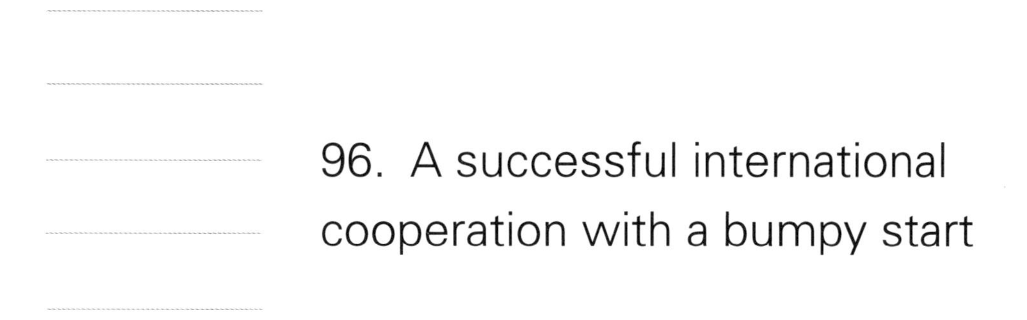

96. A successful international cooperation with a bumpy start

Science has no boundaries. The saying could not be more accurate in the case of infectious disease research. It is publicly known that infectious diseases can travel, not only through infected people or other animals, but also by commercial shipments. In the era of globalization, it is not possible to fight infectious diseases in one nation without worrying about what is happening elsewhere in the world. Emerging and re-emerging infectious diseases such as the outbreak of SARS in 2003 are likely to panic the Asian society at any time. To fight these diseases, the Chinese Academy of Sciences (CAS) and the University of Tokyo (UT) launched China-Japan Collaboration on Emerging and Re-emerging Infectious Diseases in 2005, as part of the Program of Founding Research Centers for Emerging and Re-emerging Infectious Diseases supported by Japan Ministry of Education, Science, Sports and Technology (MEXT).

China and Japan has a long history of collaborating in science and technology, yet the earlier collaboration was quite simple, and mostly involved sending Chinese students and young scientists to Japan for scientific training. Later on, along with the tremendous progress China made in S&T, the China-Japan scientific cooperation has become more mutual and versatile. However, in the field of life science, most of the collaborations are between individual Chinese and Japanese scientists and relatively "small-scaled" . Doubtless to say, the CAS-UT collaboration on emerging and re-emerging infectious diseases has taken the China-Japan collaboration in life science to a whole new level.

Supported by a contract fund from Japan's MEXT, the Institute of Medical Science of UT (IMSUT) has established two joint laboratories in the Institute of Microbiology (IM) and the Institute of Biophysics (IBP) of CAS, with one lab focusing on molecular immunology and molecular microbiology and the other focusing on structural virology and immunology.

In the past five years, the Japanese side has sent five Japanese scientists to do research at the joint labs in China. At the same time, several Chinese scientists have been working at the labs in Japan. Based on the two joint laboratories, the collaboration has been extended to

Ming Li
Beijing Institutes of Life Science, Chinese Academy of Sciences, Beijing 100101, China
Contact: liming@ioz.ac.cn

some Chinese hospitals such as Beijing Di Tan Hospital, as well as China CDC. The two joint laboratories, recognized in 2006 by both Japan's MEXT and Ministry of Science and Technology of the People's Republic of China(MOST) as Sino-Japan Intergovernmental S&T Collaboration Projects, have effectively promoted and facilitated research collaboration in the fields of microbiology, immunology and epidemiology between the two sides. After five years of successful run of the joint labs, in 2010, the CAS and the UT made a joint commitment to collaborate for a further five years and the collaboration has since been expanded to a broader area of life science.

The cooperation turns out to be such a success that in 2010, one of the key persons who initiated the project, Professor Aikichi Iwamoto from IMSUT, won the CAS Award for International Cooperation in Science and Technology. Later, he was granted the 2011 Friendship Award of China, and recently nominated as a candidate for the 2011 China International Science and Technology Cooperation Award, China's highest and most esteemed honor for foreign scientists.

However, such a successful collaboration did not necessarily have a smooth start. Under the international circumstances in 2005, it might not be the perfect time for CAS to start a big cooperation with UT. Nevertheless, the CAS still decided to push ahead on the collaboration and invited UT President Hiroshi Komiyama to China for the launching ceremony of the China-Japan Collaboration on Emerging and Re-emerging Infectious Diseases in April 2005. Meantime, President Komiyama was also bestowed the title of honorary professor during his visit to Beijing. After the ceremony, President Komiyama said, "Before I came here, a lot of people advised me to cancel or postpone my visit. But I firmly believed that the cooperation and exchanges between Chinese and Japanese scientists have been going on for decades, and that the long-term friendship between us won't be affected by some short-term events." Well, time tells everything. After six years, the academic friendship between CAS and UT is stronger than ever. And the cooperation between CAS and UT will continuously bring Chinese and Japanese scientists together to face scientific challenges, and help strengthen the Sino-Japan friendship in the long run.

Figures

Fig.1 Opening Ceremony of the Chinese Academy of Sciences and the University of Tyoko Joint Laboratories

Fig.2 CAS Vice President Jiayang Li (front right) and UT Vice President Akihiko Tanaka (front left) signing a new agreement to continue the collaboration

Fig.3 Beijing Institutes of Life Science (BIOLS, CAS) President Le Kang (front left) and IMSUT Dean Motoharu Seiki (front right) signing an agreement to collaborate in various areas of life science

Fig.4 Prof. Aikichi Iwamoto (front left) was granted the 2010 CAS Award for International Cooperation in Science and Technology by CAS President Yongxiang Lu (front right)

Fig.5 CAS President Chunli Bai (6th from right) conferred the certificate of honorary professorship to Prof. Hiroshi Komiyama (5th from right)

97. 詹姆斯·沃森对中国生命科学的热情与关爱

詹姆斯·沃森（James Waston）是当今生命科学领域中活跃的泰斗级人物之一。1953年，他与弗朗西斯·克里克（Francis Crick）一起发现了脱氧核糖核酸（DNA）的双螺旋结构（图1和图2），开启了现代分子遗传学的大门，也因此获得了1962年的诺贝尔奖，被人们尊称为“DNA之父”。1968年，他进入冷泉港实验室，经过40年的努力，将实验室从一个濒临破产的小机构变成了世界上最具影响力的综合性科研教育中心。他是“人类基因组计划”的主要发起者和推动者，也是第二位公布个人基因组测序数据的人。在赢得众多的尊敬与荣誉的同时，他对中国生命科学的发展也给予了极大的关注。在这里，笔者将向大家介绍他与中国保持的长期友谊，以及他对中国生命科学事业的热心帮助。

图1　沃森和克里克在DNA模型前

图2　沃森和克里克在剑桥（照片来自沃森档案馆冷泉港实验室）

沃森与中国的联系开始于20世纪50年代，他对一位在剑桥大学遇见的中国生物化学家印象深刻（由于时隔久远，我们很难查证到这位生物化学家的确切名字，但根据沃森的描述，这位学者可能是王应睐——中国人工合成牛胰岛素的发起人和主要贡

译者：张保元
中国科学院北京生命科学研究院，北京100101，中国
邮箱：zhangby@biols.ac.cn

献者之一）（图3）。20世纪80年代初，沃森首次访问中国，到访的城市有：北京—合肥—上海—桂林—香港。这些城市中，他最难忘的一站是上海。在中国科学院上海生物化学研究所，他重逢了剑桥大学的同窗好友，并在好友极为简陋的家中进行了令人难忘的小聚，他们花了整晚的时间，在一个小方椅上绘制了各种分子结构图。每次沃森与笔者提到此次到访中国，他都会满怀欢喜，眼中总会闪烁出难以掩饰的兴奋，他对中国生命科学的未来充满了信心。

沃森从首次访华的行程中获取了当时中国生命科学发展状况的第一手资料。他认为提高中国科研水平最好和最快的方法是让中国学者前往发达国家学习，给他们提供正式的培训机会。回到美国后不久，沃森便立即开始实施邀请中国学者到冷泉港实验室进行学习、培训或参加会议的计划，并亲自为此落实具体费用。1982年1月29日，沃森直接写信给时任美国驻华大使恒安石，告知他将邀请45名学者从上海到美国的想法。不久后，更多中国学者也开始前往美国冷泉港进行学习深造，有些人有幸直接接受沃森的指导，中国工程院院士李载平先生就是其中之一。

为了及时给大家提供交流学习的机会，在冷泉港实验室工作职位有空缺时沃森会直接写信邀请中国学者（参见1983年1月31日沃森的信），当他发现有天赋的中国学者时又会主动留意冷泉港是否有合适的位置给他们（参见1983年9月16日沃森的信）。此外，他还经常把冷泉港出版社的最新图书寄到复旦大学和中国科学院上海生物化学研究所（参见1982年3月26日给王珩先生和1982年11月12日给谈家桢先生的信）（图4，图5，图6和图7）。

2006年，应时任中国科学院北京基因组研究所所长杨焕明院士的邀请，沃森第二次来到中国进行长期访问。此次他参观了北京、上海、杭州和西安等地的科研机构。当得知他在剑桥大学结交的第一位中国挚友已逝去，他再也找不到几十年前那次重聚的简陋小屋，沃森颇为伤感。然而，中国在科学技术领域的飞速进步让他感到非常惊喜，尤其是上海交通大学贺林院士开展的神经退行性疾病的遗传学研究和中国科学院北京基因组研究所杨焕明院士进行的基因组学研究都给他留下了深刻印象。

图3　沃森与王应睐先生（照片来自沃森档案馆冷泉港实验室）

图4　沃森（左三）和谈家桢先生（左四）2006年在上海

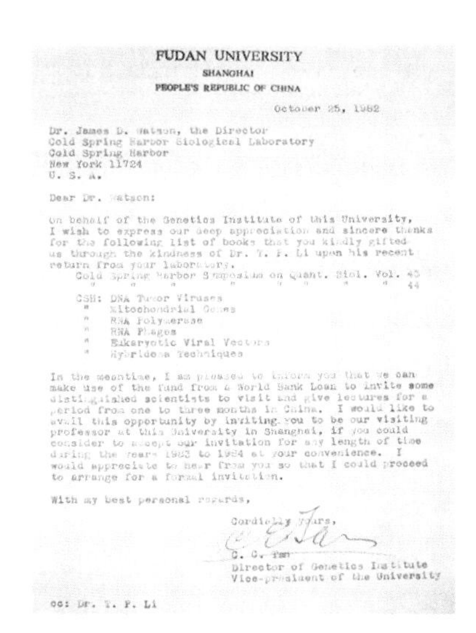

FUDAN UNIVERSITY
SHANGHAI
PEOPLE'S REPUBLIC OF CHINA

October 25, 1982

Dr. James D. Watson, the Director
Cold Spring Harbor Biological Laboratory
Cold Spring Harbor
New York 11724
U. S. A.

Dear Dr. Watson:

On behalf of the Genetics Institute of this University, I wish to express our deep appreciation and sincere thanks for the following list of books that you kindly gifted us through the kindness of Dr. Y. F. Li upon his recent return from your laboratory.

Cold Spring Harbor Symposium on Quant. Biol. Vol. 45
" " " " " " " 44

CSH: DNA Tumor Viruses
" Mitochondrial Genes
" RNA Polymerase
" RNA Phages
" Eukaryotic Viral Vectors
" Hybridoma Techniques

In the meantime, I am pleased to inform you that we can make use of the fund from a World Bank Loan to invite some distinguished scientists to visit and give lectures for a period from one to three months in China. I would like to avail this opportunity by inviting you to be our visiting professor at this University in Shanghai, if you could consider to accept our invitation for any length of time during the years 1983 to 1984 at your convenience. I would appreciate to hear from you so that I could proceed to arrange for a formal invitation.

With my best personal regards,

Cordially yours,

C. C. Tan
Director of Genetics Institute
Vice-president of the University

cc: Dr. Y. F. Li

图5 谈家桢先生寄来的一封信，感谢沃森从冷泉港实验室寄来的书（照片来自沃森档案馆冷泉港实验室）

图6 沃森收到了一位中国科学家赠送的DNA模型（照片来自沃森档案馆冷泉港实验室）

图7 1981年，沃森一家与中国科学家在一起（照片来自私人收藏，Liz Watson）

像许多其他杰出科学家一样，沃森对全世界的生命科学研究的热点十分关注。近年来，清华大学罗永章教授在抗癌药物研发方面取得的重大突破给沃森留下了深刻的印象。他亲自邀请罗永章赴冷泉港班伯里会议中心（素有生命科学“戴维营”之称）参加冷泉港班伯里高峰会议。为此，他直接写信给美国驻华大使馆，请他们尽快协助罗永章取得赴美签证。冷泉港会议结束后，他又跟罗永章约定将在第二年10月再次访问罗的实验室。

2006年6月26日，当沃森听到克林顿总统在白宫向包括中国在内的六国科学家致谢，感谢他们为第一张人类基因组框架图的公布作出重要贡献时，他感到由衷的高兴，欣然提笔致信国家主席江泽民、国务院总理朱镕基、科学技术部部长朱丽兰、中国科学院院长路甬祥等领导，向中国科学家表示祝贺。随着中国国力的崛起，沃森认为中国的科学研究必将占据同其在全球相匹配的地位。他在与时任国务委员的陈至立会谈时曾说：“中国是一个创造了辉煌文明的伟大国家。在过去短短的30年时间，我亲眼目睹了她从无到有令人惊奇的复兴过程。回顾过去，展望未来，我相信你们必将在世界科学版图上占有更加重要的地位！”

在过去的一个世纪里，冷泉港会议和培训课程在推动生命科学的发展方面起到了非常重要的作用。在与英国剑桥大学建立合作后，冷泉港一直希望在亚洲寻找一个合作伙伴。在亚太地区的几个国家中，沃森特别希望能与中国合作。2006年，他亲自勘察了苏州的地理位置、文化氛围及行政效率并做了唯一的正式报告。赴华前，沃森为此事做了很认真的准备，基本不用电脑和手机的他，破例使用了幻灯片文件制作正式报告。在报告中，他回顾了早年进入科学领域的缘由，详细介绍了冷泉港的历史文化，回顾了生命科学及生物科技发展的过程。他还欣然接受了时任苏州工业园区管理委员会主任马明龙的盛情邀请，成为苏州工业园区首席科学顾问。

他对苏州的重视及礼遇，同样体现在了一个月之后苏州代表团访美期间，他特别邀请代表团所有团员去他家做客（图8），如同接待老朋友一般，不仅亲自在大门口迎接客人，还为此举行了酒会及午宴，让中国代表团参观他家的书房和卧室，讲解了他收藏的私人照片、书籍、奖章等。据悉，这是他在家招待过的最大的海外代表团。饭后，他又亲自驾车早早地在冷泉港另一校区等待苏州代表团继续参观。在场的每个人都因沃森的热情好客而深受感动。

图8 沃森在他冷泉港的家中接待了来自苏州的代表团（照片来自私人收藏，Maoyen Chi）

在大家的共同努力下，冷泉港实验室决定在苏州设立其在亚洲的第一个也是唯一的分支机构——冷泉港亚洲。这在冷泉港百年历史上尚属首次，而沃森为这一合作开了绿灯。沃森对冷泉港亚洲的关心不只在决策层面，在操作层面也费尽心思。合作早期在美国时，他曾数次请笔者去他家中讨论，包括冷泉港亚洲科学顾问委员会候选人的提名、今后操作的重心、各分支领域的平衡问题、如何吸引公众注意力等。不仅如此，沃森对冷泉港亚洲所用建筑的风格、功能甚至装饰布置都非常关心。苏州建筑设计人员去美商讨时，他召集他们去办公室亲自讲述他希望看到的建筑效果，叮嘱室内该如何设计，怎样做到科学与建筑美观相融合等（图9和图10）。

图9 2006年，沃森在苏州调查了冷泉港亚洲的位置（照片来自私人收藏，Maoyen Chi）

图10 沃森在冷泉港的家中与苏州工业园区的马明龙先生交谈（照片来自私人收藏，Maoyen Chi）

2010年4月6日，沃森82岁生日当天，冷泉港亚洲发布了以沃森、克里克两人命名的大会，即“詹姆斯·沃森癌症研讨会”和“弗朗西斯·克里克神经科学研讨会”。2009年圣诞期间，沃森曾两度邀笔者到他家中商讨本次会议开幕式的种种细节，如美方学术界致开幕词的人选、议题的设立、如何请艺术家做快速素描以便将来可进行永久艺术性收藏、是否请亚裔音乐家同台助兴以活跃学术会议的严肃气氛等。他的每一句话都饱含了对发展亚洲生命科学、促进世界科学家交流、建立良好全球学术氛围的期望。尽管已82岁高龄，但他仍携妻子来中国主持“詹姆斯·沃森癌症研讨会”开幕式，见证冷泉港在中国乃至亚太地区的正式运行。

本文的最后，笔者想和大家分享一些沃森率真的个性。沃森有一个绰号叫“直率吉姆”，真实地概括了他的个性，率性、认真、做事单刀直入。笔者去他的办公室讨论事务，往往是坐下便说，说完便走。通电话也是一样干干脆脆。不了解他的人会觉得他很不耐烦，甚至很傲慢，但实际上，他的效率非常高，从不敷衍了事。最近，他写了一本书——《如何避免无趣之人》，该书主旨便是如何专注于做自己该做的，避免虚耗时间和精力于无味之事或人。

沃森的率性也给他造成了很多麻烦。在一次重要的国际会议上，沃森毫不客气地指责了一位官员，因为他所在的部门拒绝了冷泉港中一位知名研究员的经费申请。事后沃森很后悔，便邀请笔者与这位官员一起喝酒，边喝边说着自己的担忧及烦恼。他说如果因为此事使得冷泉港断了财源他会很内疚。此时，他对冷泉港的感情跃然脸上，闪烁在眼里，如同对待自己孩子一般。正是由于沃森的这份率真和耿直，冷泉港实验室成为当前生命科学研究的伊甸园。

沃森酷爱科学。他说他最大的乐趣是沿着班格敦街（冷泉港的主路）一路漫步，沿途询问每个实验室的最新科研结果。尽管年事已高，沃森仍然经常去听学术报告，并总是坐在前排，如果报告无趣他会悄然退出，甚至睡觉。沃森即使年过八旬，好奇心依然强烈，可谓童心未泯。在笔者和沃森所有的交往中印象最深的是这样一个镜头：他双手搓着双腿，眼睛闪烁着，专注地看着我说："科学真有趣！"完全像一个小男孩在得到他梦寐以求的一堆玩具的刹那，跃跃欲试。

沃森对科学的热爱还体现在当他听到任何有趣的报告时，他常会幽默地或着急地发问。当问题引起公众共鸣时更会扭转上身得意地朝着背后的听众笑着，眼神发出他招牌的亮光。有一次中国台湾的江安世先生来冷泉港作报告，沃森不停地打断江教授并提出问题。江先生不得不笑着说："沃森，请您耐心一点，等几秒钟，您就会得到所有答案。"报告结束后，沃森意犹未尽，又请江先生去他家边喝香槟边继续讨论问题。

在率性好奇之外，沃森又极具亲和力（图11和图12）。有一则流传很广的故事。20世纪70年代，冷泉港边上栗欧德港有位名叫罗伯逊的富翁。罗伯逊夫人过世后，留下大笔地产，即班伯里庄园，当时价值数千万美元。罗伯逊先生低调地物色着能接受这笔地产的合适单位以使其有更好的用途。远在加利福尼亚州的沃森得知此事后，连忙飞回冷泉港，再开车去栗欧德港。在进入罗伯逊家大门时发生了经典的一幕：沃森停了下来，他把头发弄乱，卷起裤腿，松开鞋带，然后才进门。他调皮地把自己打扮成一个大众心目中的科学家形象，不修边幅，不谙世故。最后该庄园成了冷泉港的班伯里会议中心，不仅如此，罗伯逊先生为此又追加了百万美元的运行经费。

图11　在沃森冷泉港的家中，他与一位在冷泉港实验室工作的中国科学家共进圣诞晚餐

（图中的这位女士已经回到中国科学院神经科学研究所，成为一名研究员）

沃森也是个十分幽默的人。他说要使中国百姓关注基因组研究的最好办法是高调测试姚明的基因组。年轻时沃森其貌不扬，不容易吸引女性。后来他曾自嘲地说："我发现了蕴含基因的DNA结构，但这一发现对我传播自己的基因没有太多帮助。"在规划冷泉港新园区时，有人提议为沃森立一座雕像，他幽默地拒绝道："我不同意！通常死了的人才立雕像，而我还有好多年要活。最不济也不要用我现在的模样，将来非要做的话记着用我年轻时的样子！"

这就是詹姆斯·沃森（图13），一位投身科学的大师，也是中国最好的朋友。他对冷泉港亚洲的贡献和他与中国的长期友谊，将是中国生命科学发展的宝贵礼物。

图12　沃森夫妇在聚会上（冷泉港实验室覃宏涛博士摄）

图13　沃森生活照（照片来自私人收藏，Maoyen Chi）

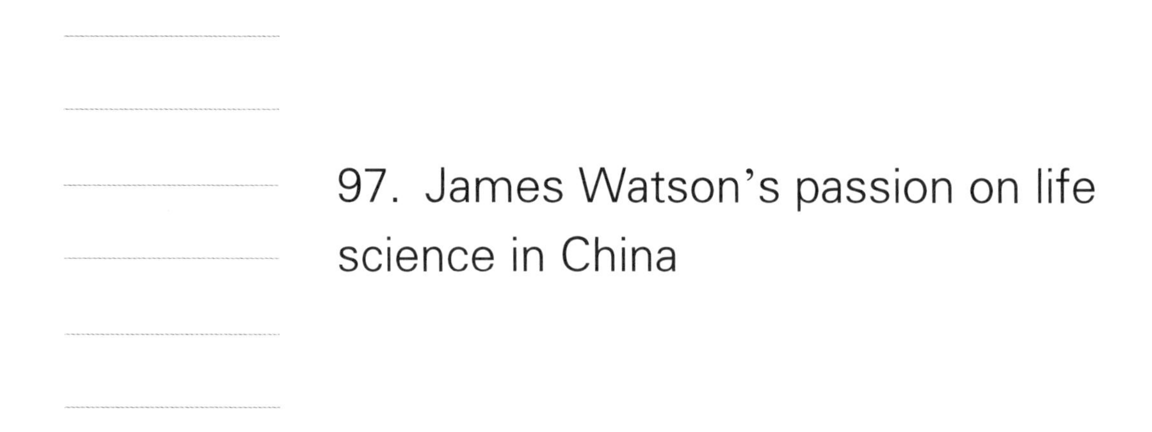

97. James Watson's passion on life science in China

James Watson is one of the few active giants in the current life science field. Together with Francis Crick, he opened the door to molecular genetics by presenting the double helix structure of DNA, and won the 1962 Nobel Prize in Physiology or Medicine. Since 1968, he devoted himself for nearly 40 years to developing Cold Spring Harbor Laboratory, which has become an unparalleled comprehensive research, education and conference center in the world. He initiated the Human Genome Project and was the second person to publish his fully sequenced genome. In addition to the numerous honors and titles, he also paid a lot of attention to the life science development in China. Here, I would like to tell about his lifetime friendship with China and his passion for life science in China.

Jim's connection with China started in the 1950s, when he met a Chinese biochemist at University of Cambridge (It is hard to know the precise name of this biochemist due to the long time, but based on the description of Jim, he may be Mr. Yinglai Wang, a founder of Bovine Insulin Synthesis Project in China). During his first visit to China in the early 1980s, Jim traveled to several cities, including Beijing, Hefei, Shanghai, Guilin and Hong Kong; however, the most unforgettable stop for him was Shanghai, where he met his old friend from Cambridge again at Shanghai Institute of Biochemistry, Chinese Academy of Sciences. Although the home of his friend was very tiny and simple, they were both so happy about their reunion that they spent the whole evening drawing molecular structures on a small table. Every time Jim mentioned this visit to me, he was so full of pleasure that his unconsealable excitement kept lightening up his eyes.

From this trip, Jim gained firsthand knowledge of the scientific research situation in China. He believed that the best and quickest way to improve the quality of research in China is to invite Chinese scholars to developed countries and provide formal training opportunities. Soon after returning to the U.S., Jim planned to invite Chinese scholars to Cold Spring Harbor Laboratory for education, training and/or conferences, and tried to look for related financial supports. On January 29, 1982, Jim wrote to the U.S. Ambassador of China, Arthur Hummel, and informed him

Maoyen Chi

Cold Spring Harbor Asia, Suzhou 215123, China

Correspondence: chi@cshl.edu

that he would be inviting 45 scholars from Shanghai to the U.S. Soon after, more Chinese scholars began to appear at the Cold Spring Harbor campus and some were directly supervised by Jim. One of them is Mr. Zaiping Li, current Academician of the Chinese Academy of Engineering.

To promote their communication, Jim contacted Chinese scholars directly if Cold Spring Harbor had a vacancy (refer to the letter on January 31, 1983), and searched out appropriate positions in Cold Spring Harbor if he found a distinguished Chinese scholar (refer to the letter on September 16, 1983). In addition, Jim often mailed newly published books from Cold Spring Harbor Press to Fudan University and Shanghai Institute of Biochemistry, Chinese Academy of Sciences (refer to the letters on March 26, 1982 to Mr. Heng Wang, and on November 12, 1982 to Mr. Jiazhen Tan).

On 2006, upon the invitation of Mr. Huanming Yang, President of Beijing Institute of Genomics, Jim took another long visit to China. This tour included a number of research institutes in Beijing, Shanghai, Hangzhou and Xi'an. Jim was very sad that his first Chinese friend from Cambridge had passed away and he was no longer able to find the tiny and simple room where they reunited decades ago. However, Jim was astonished by the rapid development of science and technology in China, especially the neurodegenerative disease research led by Mr. Lin He at Shanghai Jiao Tong University, and the genomics research led by Mr. Huanming Yang at Beijing Institute of Genomics.

Like many other outstanding scientists, Jim pays close attention to the life science research worldwide. He was deeply impressed by a recent breakthrough in the field of anticancer therapy made by Mr. Yongzhang Luo of Tsinghua University, and Jim sent a personal invitation to Mr. Luo to attend the Cold Spring Harbor Laboratory's Banbury Conference. Jim also wrote a letter directly to the U.S. Ambassador of China to facilitate the visa application. After the meeting in Cold Spring Harbor, Jim insisted on visiting the Luo Laboratory by himself in the following October.

On June 26, 2006, when Jim heard that President Bill Clinton acknowledged the six-nation (including China) scientists for their contribution to the Human Genome Project, he was so excited that he immediate wrote congratulation letters to the several Chinese leaders in political and scientific fields. Jim believes that with the current rapid economic development, China will soon find a matched position in scientific research. In a talk to Ms. Zhili Chen, ex-State Councilor of China, Jim said, "China is a nation that created glorious civilization. In the past 30 years, I witnessed the shocking renaissance of China. Looking to the past and future, I believe China will play more significant roles in scientific world."

In the past century, the conferences and training courses at Cold Spring Harbor have played significant promoting and leading roles in life science research. After the establishment of collaboration with Cambridge in the UK, Cold Spring Harbor sought out an Asian partner. Among the several countries in Asian–Pacific, Jim strongly recommended China; in 2006, Jim investigated the conditions of Suzhou by himself and gave the only official presentation. Actually, Jim seldom uses a notebook or cell phone; while he did prepare his presentation slides at Suzhou that outlined his academic experience, the development of biological sciences and biotechnology, and the history and culture of Cold Spring Harbor. He also accepted the invitation to be the Chief Scientific Consultant for Suzhou Industrial Campus.

One month after his visit to Suzhou, when the delegation from Suzhou Industrial Park (SIP) visited the United States, Jim invited all the members to his home at Bullybung. He treated everyone like old friends: He welcomed the delegation at the front door by himself; he prepared a big dinner and reception party; he invited the guests to visit his whole house, including his bedroom; he also showed his private collection of photographs, books and medals, etc. As far as I know, this is the biggest foreign group at Jim's house. After the party, he drove to the other campus of Cold Spring harbor to accompany the delegation for the rest of the trip. Everyone was deeply touched by Jim's kindness and hospitality.

With everyone's effort, Cold Spring Harbor decided to establish the first and only Asian branch at Suzhou: Cold Spring Harbor–Asia. This is an adventure in the history of Cold Spring Harbor, and it is Jim who made it happen. Furthermore, Jim cares about every detail in this project. In the United States, he often talked to me about Cold Spring Harbor–Asia, everything, from the scientific committee member nominee to the future directions of this Asian branch, the balance of different areas in life science, the approach to bring public attention, the design of architecture, the style of decoration, etc. Jim was excited upon every version of architectural draft. When designers visited the United States, Jim always invited them to his office to discuss the progress and provide his opinions.

On April 6, 2010, the 82nd birthday of Jim, Cold Spring Harbor–Asia will launch with two conferences: The James Watson Cancer Symposium and the following Francis Crick Neuroscience Symposium. During the short Christmas vacation of 2009, Jim invited me twice to his house to discuss the details about these conferences, including whom to select for the speakers of the opening ceremony, which topic to set up for each session, how to invite artists to make sketches for a permanent collection, whether to invite Asian musicians to make an active environment, etc. Every word showed the close attention that Jim paid to the development of life science in Asia, to the communication of worldwide scientists and to the establishment of beneficial academic culture. In spite of being 82 years old, Jim brought his wife to China and host the opening ceremony for the James Watson Cancer Symposium. This witnessed the connection of Jim and Cold Spring Harbor with life science in China.

Finally, I would like to tell some stories about Jim's distinct personality. Jim has a nickname "Honest Jim" , which precisely reflects his straightforwardness, diligence and genuineness. When I went to his office, I always directly hit the points and quickly left. Phone calls are the same direct and short. Some people felt he is boring or even arrogant, but actually, he is highly efficient but not perfunctory. Recently, he wrote a book called "*How to Avoid Boring People*" , which just tells how to be focused, without wasting time and energy on meaningless matters.

This sometimes brings trouble to Jim. Once in an important international conference, Jim ruthlessly criticized an officer because his department rejected a grant application from a renowned principle investigator in Cold Spring Harbor. Later, Jim was quite regretful and invited me to have a drink with the officer. He worried that if this cut the financial support of Cold Spring Harbor, he would have to put more efforts to collect private donations to fill the deficit. At that moment, his affection to Cold Spring Harbor was just written on his face. It is this affection that makes Cold Spring Harbor Laboratory an Eden for current life science research.

Jim is highly enthusiastic about science. He said his biggest joy is to walk down Bungtown Street (the main street in Cold Spring Harbor) and ask around about the updated progress of each laboratory. In spite of his old age, Jim still often goes to seminars and always sits in front, while he would leave or even sleep through the boring ones. He is always curious, like a child, and the curiosity never fades, however old as he is. The most impressive memory in my experience with Jim is a moment where he rubbed his legs with both hands, with eyes flaring and staring at me, proclaiming "Science is so FUN!" That is just like the moment when a boy gets his dream toy and is ready to give it a try.

Jim loves science so much that he often rushes to questions when hearing anything interesting. If the audiences have a positive response, he will turn around and show a proud smile on his face and the signature flare in his eyes. Once, during a seminar given by Mr. Ann-Shyn Chiang from Taiwan, China, Jim kept interrupting the presentation and asking questions. Mr. Chiang had to say, "Jim, please be a little patient, just wait for a few seconds and you will get all the answers." Jim definitely did not get all the answers during the seminar, and he finally invited Mr. Chiang to his house for champagne to keep discussing.

Jim is full of a sense of humor. He said that the best way to bring Chinese people's attention to genomics is to sequence Mr. Ming Yao's genome. Jim also jokes that he was not very attractive to girls at a young age. Later, he made the mockery, "I discovered the structure of DNA but it did not help me much in passing down my DNA." Someone suggested to have a sculpture of Jim made when designing the new campus for Cold Spring Harbor. Jim smiled, "No. A sculpture is usually for dead persons, but I still have a few years left. At least, it cannot be of my current appearance; if there must be one, please use my young face."

This is James Waston, a dedicated scientist and a best friend to China. His contribution to Cold Spring Harbor–Asia and his longtime connection to China will be a precious gift to the life sciences of China.

(The original article was written in Chinese and translated by Xiaoxue Zhang and Chris Vavricka)

Figures

Fig.1 Watson and Crick in front of a DNA model

Fig.2 Watson and Crick at Cambridge (photo from Cold Spring Harbor Laboratory, Watson Archive)

Fig.3 Watson and Mr. Yinglai Wang (photo from Cold Spring Harbor Laboratory, Watson Archive)

Fig.4 Watson (3rd from left) and Mr. Jiazhen Tan (4th from left) in Shanghai in 2006

Fig.5 A letter from Mr. Jiazhen Tan for thanking Watson for the books he sent from Cold Spring Harbor Laboratory (photo from Cold Spring Harbor Laboratory, Watson Archive)

Fig.6 Watson received a DNA model from anonymous Chinese scientist (photo from Cold Spring Harbor Laboratory, Watson Archive)

Fig.7 Watson family with Chinese scientists in 1981 (photo from Liz Watson, private collection)

Fig.8 Watson received the Delegation from Suzhou at his Cold Spring Harbor home (photo from Maoyen Chi, private collection)

Fig.9 Watson surveyed the possible site for Cold Spring Harbor–Asia in Suzhou in 2006 (photo from Maoyen Chi, private collection)

Fig.10 Watson talked with Mr. Minglong Ma of SIP in his home at Cold Spring Harbor (photo from Maoyen Chi, private collection)

Fig.11 Christmas dinner with a Chinese scientist working in Cold Spring Harbor Laboratory at Watson's Cold Spring Harbor home (The lady in this picture has returned to the Institute of Neuroscience, CAS in China to become a PI)

Fig.12 The Watsons in a party (photo from Dr. Hongtao Qin, Cold Spring Harbor Laboratory)

Fig.13 A casual profile of Watson (photo from Maoyen Chi, private collection)

98. Michael Rossmann的中国结：我与Rossmann的初次相识

“如果这故事不是Dorothy Hodgkin亲口告诉我们，没有什么人会相信的。”普渡大学的Michael Rossmann以肯定的口气对我说。这是1980年，我第一次拜访他在West Lafayette的实验室，我们正在一个很舒适的餐厅用早餐，一边享用着他最喜欢的苹果酱，一边看着园里的野兔在灌木下穿行。他是在说关于诺贝尔奖获得者Dorothy Hodgkin在一次国际会议上宣布了中国科学家在20世纪70年代关于胰岛素晶体结构的研究成果。这项重要的成果是在中国还深陷动乱时取得的，而且当时也只有少数几个国家的科学家（包括Michael Rossmann）解出了屈指可数的几个蛋白质的结构。

我是1979年秋天第一次赴美，中国那时刚刚打开国门，中国科学院就立即决定派访问学者出国。我当时在中国科学院生物物理研究所（简称生物物理所）工作，可能是最早派去美国的学者之一。那时中国科学家与外面的世界断绝联系已经有十多年了，也没有人知道如何与外国科学家重建联系。我的想法很简单，从胰岛素课题中，我已经了解了一点蛋白质结构，所以我想了解更多关于核酸的知识，我也知道美国有两个在结构病毒学（RNA病毒）研究中领先的实验室：普渡大学的Michael Rossmann实验室和哈佛大学的Steve Harrison实验室。麻省理工学院的Alexander Rich也在研究tRNA的结构。由于当时微妙的政治环境，我不知道是否应该直接与西方世界接触，所以我决定把申请交给中国科学院，过了半年多，也没有回复。但机会终于来了，当威斯康星大学校长率团访问中国时，他表示欢迎中国访问学者去他的大学继续发展职业生涯。在这个大学的教授名单中，我发现了Sundaralingam教授的实验室正在研究tRNA的结构。在政府的资助下，我来到了麦迪逊这个美丽的中西部小城。很快，Sunda（实验室的人都这么叫他）因为看到我在他实验室努力工作而不计报酬觉得过意不去。有一天，他告诉我他决定付我工资。我告诉他，他给我所有的钱都必须上交中国科学院，这就变成美国国家健康研究院给中国科学院发钱了。于是他想出了一个绝好的折中办法，用他的钱资助我去美国主要的结构生物学实验室访问。30年前，美国还没有那么多一流的结构生物学实验室，而Michael Rossmann的实验室是我的第一选择。

在我两天的访问中，Michael和我有过多次长谈，在他办公室、家里、中餐馆，以

译者：何士刚

中国科学院生物物理研究所，北京100101，中国

邮箱：shiganghe@moon.ibp.ac.cn

及在树林中散步时（散步是Michael最大的爱好）。他告诉我他早年在英国剑桥大学的经历，甚至在餐巾纸上画了蛋白质晶体学诞生地——著名的卡文迪许实验室的布局。而我则告诉了他中国的青年科学家如何在完全与外界隔绝的情况下，做出了胰岛素结构。当他听到许多引人入胜的故事后，由衷地期望能够访问北京，亲眼看看北京的实验室。一天晚上，在参观了他妻子在地下室的陶瓷工坊后，我们坐下喝茶，他突然问："你能不能给你北京的实验室打个电话，安排我去访问？"这个现在看来轻而易举就能满足的要求，在当时却让我进退维谷。20世纪80年代初，中美关系刚刚恢复正常，在美国可能只有数百名来自中国的学生和学者。但我看着他真诚而焦急的表情，实在难以拒绝他试一下的要求，他抓住了这个几乎转瞬即逝的机会，指着茶几上的电话，以坚定的口气说："家槐，就用它打吧。"

这是我终生难忘的经历，当我把听筒放在耳边，听到了来自大洋彼岸的声音。1990年前，生物物理所没有自己的大楼，北京胰岛素实验室的主要部分位于中关村微生物楼的一层，只在走廊里有一个公用电话。我的一个同事接的电话，我开始对着电话喊："我是王家槐，从美国打来的电话，老梁在吗？"我是问梁栋材博士，胰岛素组的组长。梁栋材刚好出去了，课题组的另一个领导接的电话，我向他解释了我在哪里和为什么打电话，特别是告诉他Michael非常希望访问北京。他毫不犹豫地说："我们非常欢迎Michael来访问。"家槐从美国打电话来了！消息像头条新闻似地传开了。我妻子是大学化学老师，她经常在生物物理所阅览室备课，因为这比我们邻近的蜗居更安静、更宽敞。她听到消息后冲到走廊，抓住了这个机会，开始滔滔不绝地谈起许多关于我们淘气的儿子和可爱的女儿的事情，Michael看见我说了这么久，有点不安了。他指着电话说："家槐，这一定很贵的。"我快速让我妻子停了下来。说实话，我从来没有告诉Michael到底是谁让这个电话打了这么久。

Michael得知通过这个简单的私人电话，事情进展得如此顺利后，兴奋异常。第二天早上，他几乎不能控制自己，见人便说："我要去中国了。"听起来似乎与阿波罗登月的成就也相差无几了。在最初的兴奋消退后,他意识到一个非常现实的问题。在20世纪80年代的中国，没有几个人会讲英语，Michael当然是一点儿中文都不会。他问我，从机场到研究所，他是否会迷路。这确实是个问题，我觉得责任重大。忽然我有了主意。我要了张纸，在一面写上"我是普渡大学的Michael Rossmann教授，我是来访问生物物理所的"，在另一面用英文写了一遍，我告诉Michael必要时可以举着这张纸。

一年后，当我们在意大利西西里的一个小村中举行的一次会议上重逢时（图1），Michael讲了中国之行许多故事，这确实是一次探险！他的飞机提前到达了北京机场，栋材还在去机场接他的路上。Michael出了海关后发现没有人来接他，感到非常紧张。而那时的北京总共可能也就几百辆出租车，他倒是还找到了一辆，毫无疑问，司机不会说英语。Michael给司机看了我给他的纸。幸运的是，司机知道生物物理所的位置。当他下车时，许多人围住了他。因为当时，尽管是中国科学院，外国客人也十分罕见。Michael又开始紧张了，不知道能做什么，只是举着那张纸，不停地喃喃道："我是Michael Rossmann，我是来访问生物物理所的。""他是我们的客人，我们在等他呢！"我在胰岛素组的一个朋友在人群中叫了起来。Michael接下来的访问就非常愉快了。他极为享受与中国同事之间的科学讨论，结交朋友和游览美丽的北京。自从那次以后，

图1　王家槐（左）与Michael Rossmann（右）在西西里

胰岛素组中的几个成员去了Michael的实验室做访问学者。Michael也被邀请成为实验室的顾问。以后他又成功地培训了许多来自中国的研究生。20年后，我去West Lafayette参加Michael的75岁寿辰，非常高兴地看到他以前的中国学生中的许多人在他们自己的领域已经卓有成就。

“家槐，你为什么最初没有选择来我的实验室呢？”在我1980年访问时的一次晚餐上，Michael盯着我，充满疑惑地问。“那可能是你的错啊！”我开始逗他，“我给你发过申请，但你没有回复。”“什么？你发过吗？”他告诉我从来没有收到过我的申请。Steve Harrison和Alexander Rich以后也给了我同样的答案。显然，我把申请信寄错了地方。1981年，我接受了Steve的邀请，参加了激动人心的关于阻抑蛋白如何识别一小段DNA的结构生物学研究，所以一直没有机会到Michael实验室工作。回首往事，我总是为没有机会和Michael一起工作而遗憾。

不料20年以后，我竟然在细胞间黏附分子1（ICAM-1）的结构上成了Michael的竞争对手。Michael对这个蛋白感兴趣主要是因为它是引起感冒的病毒——鼻病毒主要的受体。他试图理解病毒是如何与受体相互作用的。而我在Dana-Farber的主要研究聚焦在细胞黏附分子的结构。我在医学院的主要合作者Timothy Springer博士第一个描述了ICAM-1的功能。ICAM-1在感染或受伤时，从血液中招募白细胞，以及在形成使T细胞激活的免疫突触中起了重要作用。我们希望从免疫学的角度来理解ICAM-1的结构。充满巧合的是，分别在双方实验室做结构研究的博士后竟是来自西班牙的朋友。最终，我们两个实验室几乎同时解出了结构，并同意在*PNAS*上同时发表论文。

1980年后，我多次拜访了Michael，包括一次应邀作报告介绍ICAM-1的结构。当Michael向听众介绍我时，他说：“家槐是我的第一个中国朋友。”显然，他实验室的很多成员，尤其是中国学生和博士后已经知道了我初次访问的故事。他接着说：“朋友之间也是可以竞争的，就像家槐和我一样，但是我们是以很绅士的方式竞争的。”我被他的话深深感动。无疑，我们都是严肃的科学家。“我们都热爱科学”，正像Michael有一次对我强调的（图2）。

图2 Michael Rossmann在第十届生物大分子结晶国际会议上发言（2004年）

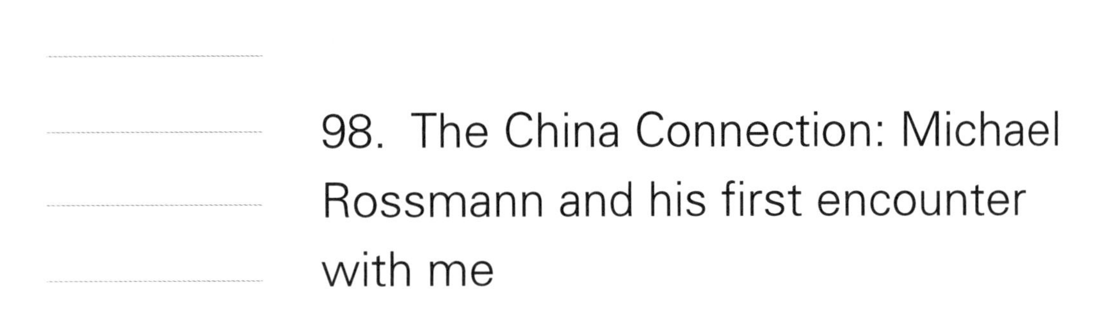

98. The China Connection: Michael Rossmann and his first encounter with me

"We would have never belived it, if it were not told by Dorothy Hodgkin." Michael Rossmann of Purdue University said to me. He was referring to what the late Nobel laureate Dorothy Hodgkin announced in an international meeting about the achievements of Chinese scientists in determining the crystal structure of insulin in early 70's. This significant contribution occurred at the time when China was still deeply entangled in the chaos, while even in the international community there were less than a dozen protein structures solved by scientists (Michael included) from a handful of countries. Michael and I sat in his cozy dinning room. While enjoying a breakfast with his favorite applesauce, observing a wild rabbit sneaking underneath the bush in the garden, we had a long chat. That was during my first visit to his lab in West Lafayette in 1980.

I first came to the United States in the autumn of 1979, when China had just begun to open its doors. The Chinese Academy of Sciences (CAS) immediately decided to send visiting scholars abroad. I was working in the Institute of Biophysics, CAS, and was among the early group of people preparing to go to the United States. Since Chinese scientists had no contacts with the outside world for more than a decade, no one even knew how to re-establish a connection with foreign scientists. My thought was simple. From the insulin project, I knew at least a little about protein structure, so ideally I would like to learn something about the nucleic acid. To my knowledge, in the U.S. there are Michael at Purdue and Steve Harrison at Harvard—the two pioneers in the field of structural virology—working on the RNA virus. Alexander Rich of MIT was also working on the structure of tRNA. Given the delicate political situation at the time, I was not sure whether it was appropriate to communicate directly with the Western world. I decided to write my application to CAS. Half a year passed, no one replied. Then the opportunity arrived, when the President of University of Wisconsin led a large delegation to China. He expressed his enthusiasm to welcome Chinese visiting scholars to his University for pursuing career advancement. In his faculty roster, I discovered that Professor Sundaralingam's lab was also

Jiahuai Wang

Dana-Farber Cancer Institute, Harvard Medical School, Boston MA02115, USA

Correspondence: jwang@red.dfci.harvard.edu

working on the tRNA structure. Sponsored by the Chinese government, I arrived at Madison—a beautiful midwestern town. Soon after, Sunda, as people normally called him, felt pretty upset by seeing me working hard at his lab without paying me a penny. One day he informed me that he decided to give me some stipend. I told him that any money he pays me would have to be handed back to the CAS, which would result in a situation of the NIH paying the CAS. Eventually we reached a nice arrangement: To use his fund for my extensive academic travel to other major structural biology labs in the U.S. Thirty years ago, even in the U.S., there were not many first class structural biology labs. I was keen to see their efforts in advancing this important field in modern biology. Michael's lab was the first one I picked for a visit.

During my two-day visit, Michael and I had several long chats, in his office, at his house, while dining in a local Chinese restaurant, and strolling in the wood (strolling is one of Michael's favorite hobbies). He described to me his early years at Cambridge, U.K. He even made a napkin sketch of the lab setting of the prestigious Cavendish laboratory, the birthplace of protein crystallography. I elaborated to him the absolutely isolated working conditions of the young Chinese group that did the insulin structure. Having heard all these amazing stories, Michael eagerly expressed his desire of paying a Beijing visit. He really wanted to see the lab through his own eyes. One night, after a visit to his wife Audrey's pottery workshop in the basement, we sat down for a cup of tea. He suddenly lit up with an exciting idea. "Can you make a phone call to your lab in Beijing to arrange a visit for me?" Michael asked. What seems matter-of-fact now was a big question mark for that era. Remember, this was at a time when the Sino–US relationship had only a couple of year history. In the U.S. in 1980, there were probably no more than a few hundred Chinese scholars and students from China. I frankly was clueless as to how, who, or if it was possible to make such a call at all. I looked into Michael's sincere and anxious face, and could not refuse to give it a try. Michael quickly seized my instant consent. He pointed to the telephone set at the coffee table, and said resolutely: "Just try this one, now, Jia-huai!"

It was an unforgettable experience, when I put my ear against the receiver and heard the ringtone from the other side of the world! Until the 1990's, the Institute of Biophysics never had her own building. The main part of the Beijing Insulin lab was located on the first floor in the old Microbiology Building in CAS's Zhongguancun campus. There was only one communal telephone set in the entire corridor. One of my colleagues answered the phone. I began to shout: "I am Jiahuai calling from USA. Is Lao Liang there?" I was referring to Dr. Dongcai Liang, the head of the Insulin group. Dongcai happened to have stepped out. Another senior lab member took over the phone. I explained to him where and why I was making the call. In particular, I told him that Michael was very eager to pay a visit. He replied without hesitation: "Michael would certainly be very warmly welcomed!" Jiahuai called from America! News rippled through like a front-page headline. My wife was a college chemistry teacher. She liked to prepare her lectures in the Biophysics Institute's library, a much quieter and more spacious space than my nearby tiny apartment. Learning the news, she dashed into sight and seized the opportunity. She chatted on with updates about our trouble-making son and the lovely little daughter. Standing by and watching me on the phone for so long, Michael started to get a bit upset. He pointed to the phone set: "Jiahuai, this must be terribly expensive···" I had to quickly stop my wife. Frankly, I have never told Michael who actually extended the phone call longer than it should have been.

Michael was so excited to learn that things went so smoothly through a simple private phone call. By next morning he could hardly contain himself: "I am going to visit China!" It probably sounded like just a slightly smaller feat than the Apollo trip. As the initial exhilaration subsided, he soon realized he's faced with a very practical issue. In the China of 1980, there are hardly any English-speaking Chinese to be found. Michael, of course, knew zero about Chinese. He asked me what if he gets lost from the airport to the Institute? This was a real problem indeed. I felt a serious responsibility. Then an idea popped into my head. I asked Michael to hand me a large piece of paper. On one side, I wrote in English: "I am Professor Michael Rossmann from Purdue University of the United States. I come to visit Institute of Biophysics." On the other side, it was a Chinese translation. I told Michael to hold the paper high whenever necessary.

It turned out to be a real adventure! Michael recounted the story to me one year later when we met again at a meeting in a small village in Sicily. His airplane arrived at Beijing ahead of schedule. Dongcai was still on his way to the airport. Michael passed through custom only to find no one from the Institute to greet him. He became very nervous, and decided to look for a taxi. There were probably only a few hundred taxis altogether in the entire city of Beijing at the beginning of 80's. Michael managed to find one. Needless to say, the driver did not speak any English. Michael showed the driver the paper's Chinese side. Fortunately the taxi driver knew where to go. When Michael got off the taxi, a huge crowd immediately gathered around him. After all, even at CAS, foreign visitors were a rare sight at that time. Michael started to get nervous again. He did not know what to do except for holding up the paper, and murmured: "I am Michael Rossmann. I am here to visit the Institute of Biophysics." "He is our guest. We have been expecting him!" One of my friends in the Insulin group shouted in Chinese. The rest of Michael's visit was unbelievably pleasant. Michael immensely enjoyed discussing science, making friends and seeing the beautiful Beijing. Since then several lab members in the Insulin group arranged to go to Michael's lab as visiting scholars. Michael was invited to serve as a consultant to the lab. Michael later started to successfully train many graduate students from China. More than two decades later, I went to West Lafayette again for Michael's 75th-year birthday celebration. It was a great pleasure to meet many of his former Chinese students, who had become well established in their own right!

"Jiahuai, why you did not choose to join my lab in the first place?" Michael stared at me with a puzzling expression at the dinner table during my 1980-visit. "That was probably your fault." I began to tease Michael. "I sent you an application, but you never replied." "What, did you?" Michael said he never received any applications from me. Steve Harrison and Alex Rich later gave me the same answer. Apparently, I sent my application letters to the wrong places. Since I had already got an offer from Steve to join an extremely exciting project in 1981 to carry out structural studies on how a repressor protein specifically recognizes a small piece of

DNA, the operator site. In retrospect, I always felt so regret to miss the opportunity to work with Michael.

I actually ended up competing with Michael 20 years later on the structure of ICAM-1 (intercellular adhesion molecule-1). Michael became interested in this protein largely because ICAM-1 is the receptor of the major group of rhinovirus, a virus that causes the common cold. Michael solved the structure of rhinovirus. He was eager to understand how the virus interacts with the receptor. My major focus at Dana-Farber has been the structural investigation into cell adhesion molecules. One of my major collaborators within our medical campus, Dr. Timothy Springer had first characterized ICAM-1. ICAM-1 plays an important role in leukocyte recruitment from blood stream to the site of infection or injury, and in the formation of immunological synapses for T cell activations. We wanted to see the structure of ICAM-1 more from the immunological perspective. Coincidentally, the postdoctoral fellows who worked on the structure on both sides also happened to be friends from Spain! Eventually we both solved the structure about the same time and agreed to publish in *PNAS* back-to-back in the same issue.

Since 1980, I have visited Michael quite a few times, including one visit at Michael's invitation to give a talk on the structure of ICAM-1. When Michael introduced me to the audience, he said: "Jiahuai is my first Chinese friend." Apparently many of his lab members, particularly those Chinese students and postdoctors already knew the story about my first visit. Michael continued: "Friends may also compete, like what Jiahuai and I did. But we did it in a very gentleman style!" I was pretty touched. After all, we are serious scientists. "We all love science" , as Michael once emphasized to me.

Figures

Fig.1 Jiahuai Wang (left) and Michael Rossmann (right) in Sicily

Fig.2 Michael Rossmann on the 10th International Conference on the Crystallization of Biological Macromolecules (ICCBM) in Beijing (2004)

99. 美国生物化学家在中国的机遇与体会

有些事情很有趣，一些看起来好像由偶然因素而导致的事件，回头来看，却似乎像是一个完美的计划。在北京的那些年，我被中国科学院微生物研究所特聘为国际青年科学家。有很多人，特别是中国的研究生曾问我，为什么一个土生土长的美国人，会不远万里来到中国做博士后研究。我认为，探索新事物的强烈好奇心是成为一名成功科学家的基本素质，我就是带着这种强烈的好奇心来到中国，开展研究工作的。2005年本科毕业的时候，我决定继续留在弗吉尼亚理工大学攻读生物化学博士学位。有很多同事提醒我，最好不要留在同一所大学继续博士学习，但在那个时候，我并没有太在意别人的建议。然而，在中国完成了两年博士后的研究学习以后，我才感受到当时那些同事的提议是正确的。虽然我的科学研究并没有因为继续留在同一所大学而止步，我也非常感谢在攻读博士研究生学位过程中获得的经验和知识。但是，我也深刻地体会到，毕业以后来到中国进行博士后的研究学习是一个明智的选择。崭新的研究环境与工作经验使我成为一名更优秀的青年科学家（图1）。

图1　在中国科学院微生物研究所的一个蛋白质晶体储藏区，Christopher拿着一些流感神经氨酸酶的晶体筛选板

译者：张萱
邮箱：zhangxuan0811@gmail.com

或许最初我来中国的原因是出于对传统中医和中国功夫的热爱。中国是世界上历史最悠久的文明古国之一，已有超过数千年的历史，并且在很多方面仍然处于世界的前列。虽然在过去的一二百年，中国面临过许多的挑战，但是今天的中国有了巨大的成长与发展的潜力。有些人还把中国比作100年前处于宏伟发展和进步时期的美国。用拿破仑的话来说："中国是一头沉睡的狮子，当这头睡狮醒来时，世界都会为之发抖。"有很多外国专家被聘请到中国，也有许多来自世界顶尖科研机构的中国专家回归祖国进行科学研究。

在完成博士学业以前，我就一直对中国文化有着浓厚的兴趣。事实上，我也曾很认真地考虑过放弃在美国攻读研究生的机会，来中国潜心学习中医或者做外教。虽然后来我选择了在弗吉尼亚理工大学基因组研究中心的Rich Helm博士和Malcolm Potts博士的实验室以及从事植物信号研究的Glenda Gillaspie博士的实验室做轮转，但最终我还是决定，与从事蛋白质生物化学领域研究的李建永教授一起完成学位论文研究。李教授是一位来自中国北京的优秀科学家，我选择跟随他做研究是出于我们对芳香族氨基酸代谢酶有共同的研究兴趣（Vavricka et al.，2010a，2010b，2011a）。当然，将来可以在中国找到合适的工作职位或从事交换项目也是我当时考虑的因素之一。李教授和同实验室的高级研究员韩谦博士都曾在中国农业大学（简称"农大"）从事兽医学研究，是他们把我介绍给他们在农大的同事，也就是我现在的导师高福研究员（图2）。

图2　在中国科学院微生物研究所门前，Christopher（左四）和高福实验室的成员在一起

高福研究员的实验室主要从事蛋白质结构与功能的相关研究，我决定博士毕业以后到他的实验室做博士后。那个时候的我对未来一无所知，所以我竭尽所能地以积极的态度和开放的心态工作。出生以后，我一直都在美国生活，在来中国之前的9年里，我一直生活在弗吉尼亚州布莱克斯堡的小镇上。而北京却是一个与之前完全不同的生活环境，我必须努力成长才能适应新的环境。每天的日常生活和简单的语言交流对我而言都是非常艰难的任务，为了弄清楚新实验室的工作流程，我也花费了不少精力。学习新的东西，保持开放的心态，适应新的环境，虽然有时候会感到很无助，但是我

也在这个过程中成长了许多。我认为，对一个研究领域的专家而言，失去了创新的态度与开拓精神，就不能产生新的想法，这是非常危险的。我的这个理念源于“禅者的初心”思想，一个公元6世纪发展于中国的生活哲学与工作哲学思想。当接触一个新的科研课题，或者与同事讨论研究思路与方向的时候，我总是尽量保持这种开放而灵活的思维方式，尽可能地多学习，保持一颗“禅者的初心”、一颗“科学家的初心”。中国有许多顶尖的科学家都曾花费大量的时间，留学于世界各地的研究机构。我想，保持这样不变的初心就是我与他们这些优秀科学家的共同之处。

尽管从美国生物化学实验室来到中国生物化学实验室的适应过程是非常艰难的，甚至有些时候让人沮丧，但在那些日子里我的论文产出量非常高。我一直积极参与关于流感病毒神经氨酸酶的功能及抑制剂的研究，该蛋白是目前针对季节性流感与大流行性流感的主要药物靶点，在加入高福研究员实验室的流感病毒神经氨酸酶研究组以后，我在许多高水平科学期刊上都发表过科研论文（Li et al.，2010；Liu et al.，2010；Vavricka et al.，2011b，2011c；Wang et al.，2011）。我很荣幸地出席了“第24届保加利亚索菲亚抗病毒研究国际会议”“2011年美国新奥尔良晶体学学会会议”以及“中日联合病毒学会议”等研讨会，并有幸在研讨会上做学术报告，介绍自己的科研工作。我非常荣幸能与来自中国、日本、美国、加拿大、澳大利亚等世界各地的顶尖科学家一起合作（图3）。

图3　中国科学院路甬祥院长（右一）、高福研究员（右二）和Christopher（左一）在国际科学院委员会新发传染病监测研讨会

我在中国实验室学到的最重要的事情就是努力工作与团结协作。高福研究员实验室的所有工作都是以团队为导向的，这是大多数研究课题取得成功的关键因素。以蛋白质结构的解析为例，无论从其中的哪一步（例如蛋白质表达、纯化、功能研究、结晶、数据收集、数据处理）开始一直到准备研究论文，几乎不可能靠自己一个人就能成功完成其中所有的重要步骤。仅就蛋白质纯化、功能分析及晶体筛选这些实验而言，由于很多蛋白质不稳定且易在较短时间内变性，实验者在大多数情况下必须争分夺秒。因此，团队合作对于提高工作效率和实验成功率尤为重要。在科学研究中，几乎所有的发现都是在前人工作的基础之上。而最大的发现，总是基于许多人、许多年的共同

努力。从广义的层面来说，世界各地的科学家一起进行合作，对于新的科学突破和改善人类生活质量至关重要。

在此，我衷心地希望美国和中国继续共同推动、支持科学家之间的交流与合作，从而进一步促使世界科学事业蓬勃发展。

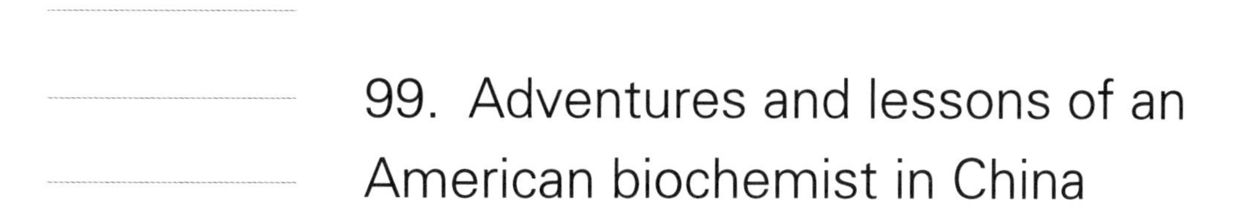

99. Adventures and lessons of an American biochemist in China

It's very interesting how events often lead up to each other seemingly by chance, but appear almost perfectly planned out when looking back. For the past two years, I have worked as an International Young Scientist Research Fellow at the Institute of Microbiology, Chinese Academy of Sciences in Beijing, China. Many people ask me, especially Chinese graduate students, why an American would choose to come to China as a postdoctoral researcher. I think that a strong curiosity to explore something new should be a fundamental quality of any successful scientist. In 2005, when I decided to stay at Virginia Tech (Virginia Polytechnic Institute and State University) to pursue my PhD studies in biochemistry, I was actually warned by many of my colleagues about continuing my research in the same university as my undergraduate studies. At that time, I did not take this very seriously; however, after spending the last two years working as a postdoc in China, I think I can now really appreciate this advice. Although I would never have ended up where I am now if I did not stay at Virginia Tech and I am truly grateful for the experience and knowledge I gained from my graduate studies in biochemistry, I really think that the mind-opening experience of moving to China has helped me become a much better scientist.

Perhaps it was my fascination with Traditional Chinese Medicine and Chinese Gongfu that originally sparked my interest in coming here. China is one of the oldest surviving civilizations and cultures in the world with a continuous history of over thousands years and for a majority of world history has been at the forefront of technology and society. Although China has faced many challenges over the past 100~200 years, today China has enormous potential for growth and development and some people even compare China to America as it was 100 years ago during a time of great development and advancement. Even Napolean Bonaparte is quoted as saying, "Let China sleep, for when China wakes, the world will shake." In terms of scientific research output, China is projected to overtake the United States in total number of scientific publications by 2013. Additionally, today many foreign experts are being recruited to China and many skilled

Christopher J. Vavricka

CAS Key Laboratory of Pathogenic Microbiology and Immunology, Institute of Microbiology, Chinese Academy of Sciences, Beijing 100101, China

Correspondence: chris@im.ac.cn, cjv@vt.edu

Chinese scientists from top research institutes across the world are returning home to carry out their research in Chinese institutes.

I had been interested in coming to China for many years before completing my graduate studies. Actually, I even seriously considered formally studying Traditional Chinese Medicine or taking an English teaching job in China instead of pursuing graduate studies in biochemistry. Although I chose my graduate lab rotations in the Virginia Tech Center for Genomics with Dr. Rich Helm and Dr. Malcolm Potts, and with Dr. Glenda Gillaspie's plant signaling research lab, at the end of my second rotation I decided to do my dissertation research in the field of protein biochemistry with Professor Jianyong Li, a scientist originally from Beijing, China. My choice to work with Professor Li was based primarily on our common interest in enzymes involved in aromatic amino acid metabolism (Vavricka et al., 2010a, 2010b, 2011a); however the possibility to find an exchange program or future position in China was also a consideration for me. Professor Li and another Senior Research Scientist in our lab, Dr. Qian Han, both of whom had worked as veterinary research scientists at China Agricultural University in Beijing, introduced me to my current advisor, Professor George Fu Gao, who is a colleague of theirs from China Agricultural University.

Professor Gao's research lab also has a focus on protein structure and function and after completing my PhD dissertation I decided to come to his lab as a postdoc. I really had no idea what to expect so I did my best to arrive with a positive attitude and an open mind. After spending my entire life in the US and the past 9 years in the small town of Blacksburg, Virginia, living and working in Beijing really presented a completely different environment to adapt and grow in. Even everyday life and simple language became a difficult task and figuring out how things worked in my new lab also took a lot of effort to adjust to. I had to learn so many new things that I was forced to keep an open mind and adapt or become helpless. As scientists who become experts in our fields of research, there is a real danger of losing the creative attitude necessary to be open to novel ideas that can push the boundaries of the established understanding. When approaching a new scientific problem or debating some research ideas with colleagues or even reviewers, I always try to stay open minded and flexible to learn as much as I can. Many top Chinese scientists spend time studying abroad in research institutes all over the world, and perhaps this is an experience that we have in common.

Although coming from an American biochemistry lab and adjusting to a Chinese lab can be very frustrating in many aspects, the last two years have been extremely productive. I have been heavily involved in studies that provide insight into the function and inhibition of the influenza neuraminidase, which is currently the major drug target against seasonal and pandemic flu. Through my participation in the influenza neuraminidase research group in Professor George Fu Gao's lab, I have been published in many quality science journals (Li et al., 2010; Liu et al., 2010; Vavricka et al., 2011b, 2011c;Wang et al., 2011). I have also been given the pleasure of presenting this work at the 24th International Conference on Antiviral Research in Sofia Bulgaria, the 2011 American Crystallographic Association Meeting in New Orleans as well as a few China–Japan joint virology meetings. Furthermore, I have had the pleasure of collaborating with many top scientists from China, Japan, United States, Canada and Australia.

The most important things I have learned working in a large Chinese lab are the values of hard work, collaboration and team work. Almost all of the work going on in Professor Gao's lab is very team orientated, which is a key factor in the success of most of the projects. For protein structure experiments, as an example, whether it's starting from protein expression systems, purification, functional studies, crystallization, data collection and processing, and all the way to manuscript preparation, it is virtually impossible for a single individual to successfully complete all the necessary steps on their own. Even for the protein purification, functional analysis, and crystal screening process alone, in most cases this must be done against the clock, as many proteins are unstable and denature rather quickly. Therefore the ability to work in groups is extremely valuable for increasing the efficiency and success of many experiments. In science, practically all discoveries are based upon some knowledge that was discovered previously. Furthermore, the biggest discoveries are almost always based upon the efforts of many individuals over many years of time. On the large scale, it is extremely important that scientists all over the world can collaborate to provide new scientific breakthroughs that can improve the quality of human life. I sincerely hope that both the United States and China will continue to promote and support collaboration and exchange between our scientists to foster the advancement of science for people everywhere.

Acknowledgements

For the last 2 years Christopher Vavricka has been solely supported by the Chinese Academy of Sciences Fellowship for Young International Scientists (2010Y2SB12 and 2009Y2BS2) and National Natural Science Foundation of China Research Fund for International Young Scientists (31150110147 and 31050110126).

Figures

Fig.1 Christopher holding some crystal screening plates for influenza neuraminidase in one of the protein crystal storage areas at the Institute of Microbiology, Chinese Academy of Sciences

Fig.2 Christopher (4th from left) with some lab members of the George Fu Gao's lab in front of the Institute of Microbiology, Chinese Academy of Sciences

Fig.3 A toast from Christopher (1st from left) with Yongxiang Lu (1st from right), President of Chinese Academy of Sciences, and George Fu Gao (2nd from right), at the banquet for the IAC Workshop on Backyard Surveillance of Emerging Infectious Diseases in Beijing, China

References

Li Q, Qi J, Zhang W, et al (2010) The 2009 pandemic H1N1 neuraminidase N1 lacks the 150-cavity in its active site. Nat Struct Mol Biol 17: 1266–1268.

Liu J, Wu P, Gao F, et al (2010) Novel immunodominant peptide presentation strategy: A featured HLAA* 2402-restricted cytotoxic T-lymphocyte epitope stabilized by intrachain hydrogen bonds from severe acute respiratory syndrome coronavirus nucleocapsid protein. J Virol 84: 11849–11857.

Vavricka CJ, Christensen BM, Li J (2010a) Melanization in living organisms: A perspective of species evolution. Protein & Cell 1: 830–841.

Vavricka CJ, Han Q, Huang Y, et al (2011a) From L-dopa to dihydroxyphenylacetaldehyde: A toxic biochemical pathway plays a vital physiological function in insects. PLoS One 6: e16124.

Vavricka CJ, Liu Y, Li Q, et al (2011b) Special features of the 2009 pandemic swine-origin influenza A H1N1 hemagglutinin and neuraminidase. Chin Sci Bull 56: 1747–1752.

Vavricka CJ, Qi J, Wu Y, et al (2011c) Structural and functional analysis of laninamivir and its octanoate prodrug reveals group specific mechanisms for influenza NA inhibition. PLoS Pathog 7(10): e1002249.

Vavricka CJ, Ray KW, Christensen BM, et al (2010b) Purification and N-glycosylation analysis of melanoma antigen dopachrome tautomerase. Protein J 29: 204–212.

Wang M, Qi J, Liu Y, et al (2011) Influenza a virus N5 neuraminidase has an extended 150-cavity. J Virol 85: 8431–8435.

100. 在中国的机遇与挑战

35岁的时候，我面临着科研生涯的转折点。同世界上很多国家一样，荷兰当前的学术体制很难让科研工作者一直以博士后的身份从事研究（Powell，2015）。所以，一个科研工作者如果没有在科研机构拿到永久职位，就只能选择离开学术圈，去药企或其他地方另谋发展。虽然我没能在前者上取得成功，但我也不愿意因此选择后者而离开学术圈。所以我决定另辟蹊径去中国。

我觉得中国是从事科学研究的好地方。世界上只有少数几个国家在增加而不是缩减科研经费的投入，中国是其中之一（图1）（Van Noorden，2014）。我有幸联系到中国科学院微生物研究所的高福教授，他建议我申请国际青年科学家奖学金，到他的结构病毒学实验室工作。这个奖学金只是中国科学院众多奖学金项目之一，旨在吸引外国科学家到它下属的各个研究所工作。

图1　2014年中-俄地区国际流感研讨会在中国科学院微生物研究所举行

奖学金项目申请获得成功后，我办理了相关手续，和妻子（她也找到了一份工作）在2012年7月来到了北京。在此之前，我与一位来自美国的博士后Christopher Vavricka

译者：程浩

中国科学院北京生命科学研究院，北京100101，中国

邮箱：chengh@biols.ac.cn

有过邮件交流，他从2009年起就在高福教授实验室工作（Vavricka，2011）。虽然他提供的信息有助于我初步了解中国，但来到北京后我还是感受到了很大的文化冲击。幸运的是，中国的同事给予我很多帮助。几周以后，我就开始了自己的第一个实验，研究蝙蝠体内新发现的流感病毒蛋白（Tefsen et al.，2014；Wu et al.，2014）。在中国科学院微生物研究所工作期间，我亲身感受到了实验室里学生和工作人员的刻苦与奉献精神。我也了解到了原汁原味的中国文化，品尝了美味可口的中国美食。

现在回想起来，我确实来到了最好的实验室。2013年的春天，在中国发现了感染人的H7N9流感病毒。流感病毒正是高福教授实验室的一个主要研究方向。除了要负责这方面的研究，当时高福教授还担任中国疾病预防控制中心（简称疾控中心）的副主任，指导流感疫情的防控。那段时间非常忙碌，我的很多同事夜以继日地工作，解析不同H7N9临床分离株表面的血凝素和神经氨酸酶的结构（Shi et al.，2013；Wu et al.，2013），以溯源病毒（Liu et al.，2013）。

幸运的是，那次H7N9的疫情相对较小，很快就得到了控制。这得益于中国政府对疫情信息的充分共享和采取的各种有效措施，2002—2003年暴发的SARS疫情让政府在这方面积累了很多经验。当时，高福教授在《科学》杂志上发表了一篇文章，呼吁关闭活禽市场以阻止新的H7N9疫情发生（Gao，2014）。为此，荷兰一位知名荣誉退休教授在当地的报纸专栏上特地提到了他（Borst，2014）。这是我作为一个在中国实验室工作的荷兰人的一个意外收获。

在此期间，其他病毒也在威胁全球。SARS相关的冠状病毒MERS就是一例，它在中东地区引发了几例感染者的死亡。这也给高福教授实验室里面专注于解决问题的科学家提出了新的挑战。实验室里的逯光文和其他几个同事完成了一项杰出的工作，他们解析了MERS病毒表面刺突蛋白与受体CD26的复合物结构（Lu et al.，2013）。

在2014年3月，另外一场更具毁灭性的疫情暴发。埃博拉病毒肆虐利比里亚、塞拉利昂和几内亚，夺去了很多生命（Alexander et al.，2015）。这些国家急需国际援助。中国向受灾国家派去了医疗小组。2014年，由高福教授领导的中国疾病预防控制中心小组远赴塞拉利昂，为当地提供医疗援助。他们在受灾地区待了两个月（Gao and Feng，2014）。我的另外一位同事刘军（图2），首先去了几内亚。几个月以后，他也去了塞拉利昂援助控制疫情（Tong et al.，2015）。

图2 刘军（右三）和疾控中心团队的其他成员前往非洲帮助抗击2014年的埃博拉疫情

在抗击埃博拉的动荡时期，我即将离开北京去西交利物浦大学任副教授。这所大学位于苏州，靠近上海。这所年轻的大学是在2006年由中国C9联盟高校之一的西安交通大学与英国的利物浦大学共同创立的，旨在融合两所学校的优势，建立一所世界一流的研究型大学。它给我提供机会建立自己的实验室，似乎一切都为我准备好了（图3）。

图3 西交利物浦大学实验室和我的两名学生

在苏州待了一年半以后，我更加确信当初的选择。我喜欢向大学里求知若渴的学生传授知识，虽然英文授课有时会给部分学生造成困难。西交利物浦大学是中国政府开展的一项实验，旨在探究有别于传统的新的教育模式。其中一个主要目的就是激发学生的创造性思维。

建立自己的实验室是一项很有激情的工作，因为我完全可以按照自己的兴趣开展研究。我非常享受这个过程。我现在主要研究分枝杆菌的细胞被膜合成机制。目前感染耐多药结核分枝杆菌的病人在日益增多，希望我的研究能够帮助这些人对抗感染（WHO，2014）。我在中国已经工作和生活了三年半，我想说的是这里非常适合我，我对未来充满期待。

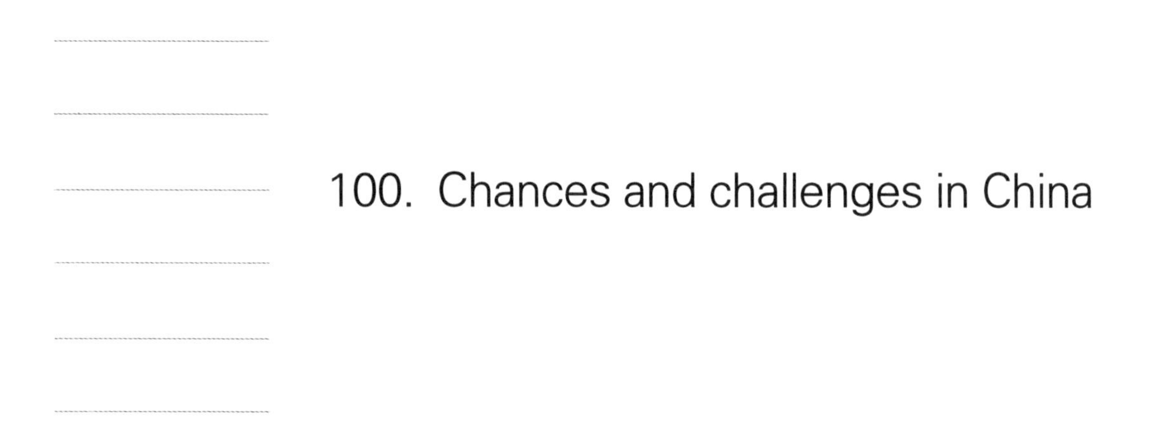

100. Chances and challenges in China

So there I was, 35 years old and at the crossroads of my scientific career. The current academic system in the Netherlands, like in many other places around the world (Powell, 2015), is not designed to have scientists staying in a postdoc position forever. So, you are either deemed good enough to become a permanent staff member, or you have to leave academia at some point and try your luck in a pharmaceutical company or elsewhere. I had not succeeded in the former, but I also did not want to do the latter, so I took a less threaded path and headed to China.

China seemed a good place to take my chances, as it is one of the few countries where funding for R&D is growing instead of shrinking (Van Noorden, 2014). I was lucky to get in touch with Professor George F. Gao in Beijing, who advised me to apply for a Fellowship for International Young Scientists and come work in his Structural Virology lab. This type of fellowship is one of the many funded by the Chinese Academy of Sciences with the aim to enhance the influx of foreign scientists into its institutes.

After having received the positive outcome of my application and arranging the necessary paperwork to be able to work in China, my wife (who had found herself a job as well) and I left for Beijing in July 2012. Before arriving at the Institute of Microbiology (IMCAS), I had been exchanging emails with Christopher Vavricka, an American postdoc who was working in the same lab since 2009 (Vavricka, 2011). Although the information provided by him helped me to shape my expectations, the culture shock was still considerable. Fortunately, my Chinese colleagues were also very helpful and after a few weeks, I could already conduct my first experiments on one of the internal proteins of an influenza-like virus discovered recently in bats (Tefsen et al., 2014; Wu et al., 2014). During my period at the IMCAS, I witnessed the hard work and dedication of the students and staff in the lab. I also got a real taste of Chinese culture, as many events in Chinese life are accompanied by very delicious food.

In retrospect, I could not have come to a better laboratory, as the first cases of human-infecting H7N9 influenza strains made their appearance in China in the spring of 2013. Not only

Boris Tefsen

Department of Biological Sciences, Xi'an Jiaotong–Liverpool University, Suzhou 215028, China

Correspondence: boris.tefsen@xjtlu.edu.cn

is influenza one of the main research topics in the lab of Gao, but he also serves as the Deputy Director-General of the Chinese Center for Disease Control and Prevention and therefore has the responsibility to supervise the battles against epidemics. During this hectic period, many of my colleagues were working around the clock to solve the structures of the influenza surface molecules hemagglutinin and neuraminidase identified from the different clinical H7N9 isolates (Shi et al., 2013; Wu et al., 2013) and to track down the origins of the virus (Liu et al., 2013).

Luckily, the H7N9 epidemic was relatively small and quickly under control, helped by adequate sharing of information and measures taken by the Chinese government, which had learned from the SARS epidemic that affected the country from 2002 to 2003. For me as a Dutchman working in a Chinese lab, an interesting by-product of a letter in *Science* about closing down live poultry markets to prevent new infections by H7N9 (Gao, 2014), was that Gao was mentioned in a Dutch national newspaper in a column by a prominent emeritus professor (Borst, 2014).

In the meantime, other viral threats emerged around the globe. The SARS-related corona virus MERS was causing several fatalities in the Middle East and it posed new challenges for the eager scientists in the Gao lab to solve. One extraordinary result was the elucidation of the complex structure of viral surface spike protein together with its CD26 receptor by Guangwen Lu and several other colleagues (Lu et al., 2013).

In March 2014, another and far more destructive epidemic started. The Ebola virus struck Liberia, Sierra Leone and Guinea with great force and claimed many lives (Alexander et al., 2015). International aid was (and is) desperately needed in these countries and China offered help by sending medical teams to the stricken countries. The CDC-team going to Sierra Leone in 2014 was led by Gao for two months (Gao and Feng, 2014). One of my other colleagues, Jun Liu, first went to Guinea and a few months later he went to help in Sierra Leone as well (Tong et al., 2015).

Amidst all this global turmoil, I was about to leave Beijing to become an Associate Professor at the Xi'an Jiaotong-Liverpool University (XJTLU) in Suzhou, an ancient city close to Shanghai. This young university was founded in 2006 in a joint effort by Xi'an Jiaotong University, one of the C9 League universities in China, and the University of Liverpool and it aims to blend the best of both education systems into its own and be a world class research-led university. It offered me the chance to build up my own research lab and it seemed like everything had fallen into place for me.

After one and a half year in Suzhou, I can only confirm my initial feelings. I enjoy teaching the eager students at my university, although the language—the curriculum is taught in English—poses sometimes difficulties to some of them during assessments. XJTLU is one of the experiments by the Chinese government to test education styles that differ from the classic Chinese education system and one of the main goals is to improve creative thinking amongst its student population.

Setting up my own laboratory has also been very stimulating and I thoroughly enjoy the freedom of finally being able to completely follow my own research interests. My research focuses on the synthesis of the mycobacterial cell envelope and will hopefully contribute to combating the growing number of cases of patients that are infected with multi-drug resistant *Mycobacterium tuberculosis* (WHO, 2014). Three and a half years after going to China, I can only conclude that it worked out great for me and I am excited to see what the future will bring.

Figures

Fig.1 A joint meeting between China and Russia on influenza research held at the Institute of Microbiology, Chinese Academy of Sciences in 2014

Fig.2 Jun Liu (3rd from right) and other members of the CDC-teams that went to Africa to help combat the Ebola outbreak in 2014

Fig.3 Posing in my lab at XJTLU together with two students, Xinzhu Fei (middle) and Jing Zhang (right)

References

Alexander KA, Sanderson CE, Marathe M, et al (2015) What factors might have led to the emergence of Ebola in West Africa? PLoS Negl Trop Dis 9: e0003652.

Borst P (2014) Li en ik. NRC Handelsblad.

Gao GF (2014) Influenza and the live poultry trade. Science 344:235.

Gao GF, Feng Y (2014) On the ground in Sierra Leone. Science 346:666.

Liu D, Shi W, Shi Y, et al (2013) Origin and diversity of novel avian influenza A H7N9 viruses causing human infection: Phylogenetic, structural, and coalescent analyses. Lancet 381:1926–1932.

Lu G, Hu Y, Wang Q, et al (2013) Molecular basis of binding between novel human coronavirus MERS-CoV and its receptor CD26. Nature 500:227–231.

Powell K (2015) The future of the postdoc. Nature 520:144–147.

Shi Y, Zhang W, Wang F, et al (2013) Structures and receptor binding of hemagglutinins from human-infecting H7N9 influenza viruses. Science 342:243–247.

Tefsen B, Lu G, Zhu Y, et al (2014) The N-terminal domain of PA from bat-derived influenza-like virus H17N10 has endonuclease activity. J Virol 88:1935–1941.

Tong YG, Shi WF, Liu D, et al (2015) Genetic diversity and evolutionary dynamics of Ebola virus in Sierra Leone. Nature 524:93–96.

Van Noorden R (2014) China tops Europe in R&D intensity. Nature 505:144–145.

Vavricka CJ (2011) Adventures and lessons of an American biochemist in China. Protein & Cell 2:601–603.

WHO (2014) Global Tuberculosis Report. Geneva: World Health Organization.

Wu Y, Bi Y, Vavricka CJ, et al (2013) Characterization of two distinct neuraminidases from avian-origin human-infecting H7N9 influenza viruses. Cell Res 23:1347–1355.

Wu Y, Tefsen B, Shi Y, et al (2014) Bat-derived influenza-like viruses H17N10 and H18N11. Trends Microbiol 22: 183–191.